ALIGNED WITH
PRECISION EXAMS
by youscience

Seventh Edition

Agriscience
Fundamentals and Applications

L. DeVere Burton
Renee Peugh

 Cengage

Australia • Brazil • Canada • Mexico • Singapore • United Kingdom • United States

Agriscience Fundamentals and Applications, Seventh Edition

L. DeVere Burton and Renee Peugh

SVP, Product: Erin Joyner

VP, Product: Thais Alencar

Portfolio Product Director: Jason Fremder

Portfolio Product Manager: Nicole Robinson

Product Assistant: Eduarda David

Learning Designer: Jennifer Starr

Content Manager: Kristen Mellitt

Digital Project Manager: Scott Diggins

Director, Product Marketing: Jeremy Walts

Product Marketing Manager: Antonette Adams

Content Acquisition Analyst: Erin McCullough

Production Service: Lumina Datamatics Ltd.

Designer: Felicia Bennett

Cover Image Source: Billion Photos/Shutterstock.com

Last three editions, as applicable: © 2015, 2010, 2007

© 2024 Cengage Learning, Inc. ALL RIGHTS RESERVED.

No part of this work covered by the copyright herein may be reproduced or distributed in any form or by any means, except as permitted by U.S. copyright law, without the prior written permission of the copyright owner.

Unless otherwise noted, all content is © Cengage, Inc.

The names of all products mentioned herein are used for identification purposes only and may be trademarks or registered trademarks of their respective owners. Cengage Learning disclaims any affiliation, association, connection with, sponsorship, or endorsement by such owners.

> For product information and technology assistance, contact us at
> **Cengage Customer & Sales Support, 1-800-354-9706
> or support.cengage.com.**
>
> For permission to use material from this text or product, submit all requests online at **www.copyright.com**.

Library of Congress Control Number: 2022924070

ISBN: 978-0-357-87557-5

Cengage
200 Pier 4 Boulevard
Boston, MA 02210
USA

Cengage is a leading provider of customized learning solutions. Our employees reside in nearly 40 different countries and serve digital learners in 165 countries around the world. Find your local representative at: **www.cengage.com.**

To learn more about Cengage platforms and services, register or access your online learning solution, or purchase materials for your course, visit **www.cengage.com.**

Printed in the United States of America
Print Number: 01   Print Year: 2023

Table of Contents

Preface ix
Chapter Organization ix
About the Authors xiv

Unit 1 — Agriscience: The Future of Food 3

Chapter 1   The Science of Living Organisms   4
Agriscience Defined 7
Agriscience Around Us 11
Agriscience and Other Sciences 14
A Place for You in Agriscience 16

Chapter 2   Agriscience and Living Conditions 20
A Broad Range of Living Conditions 21
The Homes We Live In 22
Key Environmental Factors 25
Other Environmental Factors 28
Our Shared Living Environment 31
Agriscience in Our Growing World 32
Impact of Agriscience 35
Improving Life Through Agriscience Research 41
Breakthroughs in Agriscience 42
Agriscience and the Future 45

Chapter 3   Biotechnology 48
Historic Applications of Biotechnology 49
Improving Plant and Animal Performance 49
Improving Plants and Animals 54
Safety in Biotechnology 58
Ethics in Biotechnology 58

Unit 2 — Agriscience as a Career Choice 63

Chapter 4   Career Options in Agriscience 64
Preparing for an Agriscience Career 75

Chapter 5   Supervised Agricultural Experience 82
The Total Agriscience Program 83
SAE for All 84
Foundational SAE 85
Immersion SAE 87
Types of Immersion SAEs 88
Plan an SAE 95
Propose an SAE 95
Conduct an SAE 96

Chapter 6   Leadership Development in Agriscience 100
Leadership Defined 101
Leadership Development Opportunities 103
Public Speaking 111
Parliamentary Procedure 115
Conducting Meetings 117

Unit 3 — Natural Resources Management 123

Chapter 7 Maintaining Air Quality 124
Air Quality 125

Chapter 8 Water and Soil Conservation 140
The Nature of Water and Soil 142
Relationships of Land and Water 146
Conservation and Quality Improvement of Water 148
Land Erosion and Soil Conservation 151

Chapter 9 Soils and Hydroponics Management 160
Plant-Growing Media 161
Origin and Composition of Soils 164
Land Capability Classification System 168
Physical, Chemical, and Biological Characteristics 171
Amendments to Plant-Growing Media 175
Soil Chemistry 178
Hydroponics 184

Chapter 10 Forest Management 194
Forest Regions of North America 196
Relationships Between Forests and Other National Resources 200
Important Types and Species of Trees in the United States 200
Tree Growth and Physiology 207
Properties of Wood 209
Woodlot Management 211
Seasoning Lumber 216

Chapter 11 Wildlife Management 220
Characteristics of Wildlife 221
Wildlife Relationships 222
Relationships Between Humans and Wildlife 224
Classifications of Wildlife Management 226
Approved Practices in Wildlife Management 228
The Future of Wildlife in the United States 236

Chapter 12 Aquaculture 240
The Aquatic Environment 241

Unit 4 — Integrated Pest Management 257

Chapter 13 Biological, Cultural, and Chemical Control of Pests 258
Types of Pests 260
Integrated Pest Management 268
Pest-Control Strategies 271

Chapter 14 Safe Use of Pesticides 276
History of Pesticide Use 278
Herbicides 279
Insecticides 282
Fungicides 283
Pesticide Labels 284
Risk Assessment and Management 290
Special Gear 291
Pesticide Storage 293
Health and Environmental Concerns 293

Unit 5 — The Quest for Better Plants 301

Chapter 15 Plant Structures and Taxonomy 302
The Plant 303
Roots 304
Stems 308
Leaves 310
Flowers 315
Plant Taxonomy 317

Chapter 16 Plant Physiology 322
Photosynthesis 324
Respiration 325
Transpiration 326
Soil 327
Air 328
Water 328
Plant Nutrition 329
Food Storage 335

Chapter 17 Plant Reproduction 338
Sexual Propagation 339
Asexual Propagation 344
Cuttings 344
Layering 349
Division 350
Grafting 351
Tissue Culture 353

Unit 6 — Crop Science 363

Chapter 18 Home Gardening 364
Analyzing a Family's Gardening Needs 365
The Garden Plan 366
Locating the Garden 366
Preparing the Soil 366
Common Garden Crops and Varieties 369
Cultural Practices for Gardens 372
Harvest and Storage of Garden Produce 375
Cold Frames, Hotbeds, and Greenhouses for Home Production 376

Chapter 19 Vegetable Production 382
Vegetable Production 383
Identifying Vegetable Crops 384
Planning a Vegetable-Production Enterprise 387
Preparing a Site for Planting 390
Planting Vegetable Crops 392
Cultural Practices 396
Harvesting, Marketing, and Storing Vegetables 399

Chapter 20 Fruit and Nut Production 406
Career Opportunities in Fruit and Nut Production 407
Identification of Fruits and Nuts 407
Planning Fruit and Nut Enterprises 410
Soil and Site Preparation 412
Planting Orchards or Small-Fruit and Nut Gardens 413
Cultural Practices in Fruit and Nut Production 414
Harvesting and Storage 416

Chapter 21 Grain, Oil, and Specialty Field-Crop Production 422
Major Field Crops in the United States 423
Growing Field Crops 437

Chapter 22 Forage and Pasture Management 448
Forage and Pasture Crops 449
Growing Forages 456

Unit 7 — Ornamental Use of Plants 469

Chapter 23 Indoor Plants 470
Popular and Common Indoor Flowering Plants 471
Popular and Common Indoor Foliage Plants 474
Selecting Plants for Indoor Use 477
Uses of Indoor Plants 479
Growing Indoor Plants 480
Interior Landscaping or Plantscaping 486

Chapter 24 Turfgrass Use and Management 492
The Turfgrass Industry 493
Turfgrass Growth and Development 494
The Turfgrass Plant 495
Turfgrass Varieties 498
Turfgrass Cultural Practices 503
Turfgrass Establishment 508

Chapter 25 Trees and Shrubs 516
Trees and Shrubs for Landscapes 517
Forest Resources 518
Plant Selection 519
Use 519
Plant Names 523
Obtaining Trees and Shrubs 524
Planting Trees and Shrubs 526
Mulching 528
Staking and Guying 528
Fertilizing 528
Pruning 530
Insects and Diseases 531

Unit 8 — Animal Sciences 535

Chapter 26 Animal Anatomy, Physiology and Nutrition 536
Nutrition in Human and Animal Health 537
Animal Anatomy and Physiology 538
Major Classes of Nutrients 545
Sources of Nutrients 548
Symptoms of Nutrient Deficiencies 549
Feed Additives 549
Composition of Feeds 549
Classification of Feed Materials 551

Chapter 27 Animal Health 554
Animal Health Indicators 555
Signs of Poor Health 555
Healthful Environments for Animals 557
Animal Diseases and Parasites 559
Treatment and Prevention 562
Veterinary Services 567

Chapter 28 Genetics, Breeding, and Reproduction 570
Role of Breeding and Selection in Animal Improvement 572
Principles of Genetics 572
Reproductive Systems of Animals 578
Systems of Breeding 580
Methods of Breeding 582
Selection of Animals 583

Chapter 29 Small Animal Care and Management 588
Poultry 589
Rabbits 595
Types and Uses 597
Honeybees and Apiculture 599
Other Small Animals 603
Pet Care and Management 604

Chapter 30 Dairy and Livestock Management 610
Dairy Cattle 611
Beef Cattle 616
Swine 622
Sheep 626
Goats 629
Importance of Livestock 631

Chapter 31 Horse Management 634

Horses 635
Anatomy 637
Types and Breeds 638
Riding Horses 642

Safety Rules 645
Approved Practices 645
Purchasing a Horse 648

Unit 9 The Future of Food Is Now 653

Chapter 32 The Food Industry 654

Economic Scope of the Food Industry 655
Quality Assurance 657
Food Commodities 660

Workplace Safety 664
Operations in the Food Industry 667
The Food Industry of the Future 674

Chapter 33 Food Science 676

Nutritional Needs 677
Food Customs of Major
 World Populations 682
Methods of Processing, Preserving, and
 Storing Foods 683
Food Preparation Techniques 687

Food Products From Crops 688
Food Products From Animals 689
Grades and Market Classes of Animals 698
By-products—How Waste Products
 Are Used 701
New Food Products on the Horizon 702

Unit 10 Planning with Purpose 707

Chapter 34 Agribusiness Planning 708

Common Business Structures 709
Agribusiness Management 710
Characteristics of Decisions
 in Agribusiness 712

Fundamental Principles of Economics 714
Agribusiness Finance 716

Chapter 35 Marketing in Agriscience 728

Marketing Starts with a Strategy 729
Marketing Plan 733
Promotion 736
Commodity Pricing 738
Pricing Strategies 740
Direct Marketing 741

Intermediate Marketing 743
Wholesale Marketing 743
Retail Marketing 744
Vertical Integration 744
Futures Markets 745
Export Marketing 746

Chapter 36 Entrepreneurship in Agriscience 750

The Entrepreneur 751
Entrepreneurship 751
Selecting a Product or Service 755
Organization and Management 755

Small-Business Financial Records 757
Forms of Employment 759
Contributions of Small Businesses 760

Appendix A: Reference Tables 764
Appendix B: Supervised Agricultural Experience 768
Appendix C: Ensuring Safety in SAE Lab Work 776
Appendix D: Building Birdhouses for SAE 778
Appendix E: Developing a Personal Budget 782
References 784
Glossary / Glosario 786
Index 828

Preface

Welcome to the agriscience world of the future! *Agriscience: Fundamentals and Applications, Seventh Edition* is dedicated to today's high-school agriscience students, for you are the agriculturists, scientists, and innovators of tomorrow. Your generation will be called on to feed the world as the human population nearly doubles to 10 billion people. To do this, you must learn more than any other generation has ever learned and discover more ways to increase food production than any other generation has discovered. You must accomplish this using marginal land because many of our fertile farms have been swallowed up to build cities and towns. *Agriscience: Fundamentals and Applications, Seventh Edition* is the modern agriscience textbook that will introduce a new generation to agricultural careers. Your generation will also lead the industry that the people of the United States depend on to feed and clothe themselves and to export surplus agricultural products to other regions of the world.

In keeping with this new generation, the seventh edition of *Agriscience: Fundamentals and Applications* has been revised to reflect how new technologies, trends, and scientific discoveries are shaping the agriculture of today and tomorrow. The focus on science has been strengthened by updating statistics, integrating new examples of agricultural applications of science and technology, and adding new lab exercises. The result is an engaging and highly practical resource for introductory-level agriscience classes in the ninth and tenth grades, one that addresses the most basic levels of agriscience using language and examples that are matched to the needs of beginning agriscience students. *Agriscience: Fundamentals and Applications, Seventh Edition* is part of a series of contemporary agricultural resources offered by Cengage.

Chapter Organization

This edition of *Agriscience: Fundamentals and Applications* is organized into 10 units and 36 chapters. Each unit introduces the subjects that will be covered in the individual chapters. The text and rich variety of graphics for each unit have been revised. Each chapter opens with a stated objective and a list of competencies to be developed. Important terms are listed at the beginning of each unit and highlighted in the text. Terms and definitions are also included in both English and Spanish in the Glossary/Glosario to support learners. Each chapter includes features on science, careers, and agriculture and concludes with an in-depth chapter review, including student activities and assessment. Select chapter reviews include connections to SAE, FFA, and accompanying agriscience videos.

Features of This Edition

Chapter Objective
An overarching goal for learning is clearly stated at the beginning of each chapter. A list of competencies to be developed breaks down the objective into manageable steps.

Key Terms
Each chapter highlights essential terms that students will learn to define in context.

Feature Articles
High-interest features explore STEM topics in-depth and cover ongoing breakthroughs in agriscience.

Preface **xi**

Career Exploration
Agri-Profiles provide career descriptions, data on salary ranges and outlook based on the U.S. Bureau of Labor Statistics, and career education and training resources.

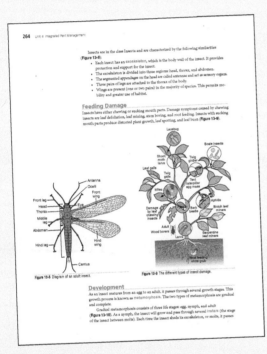

From the Field
Hundreds of updated photos as well as clear, detailed graphics provide students with relevant information at a glance.

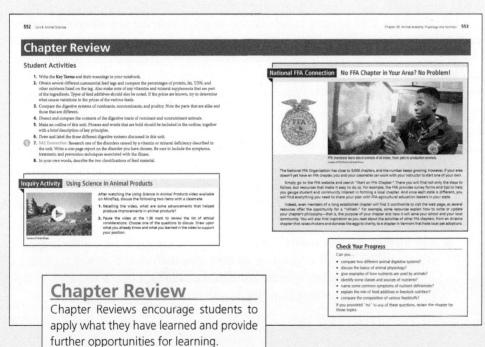

Chapter Review
Chapter Reviews encourage students to apply what they have learned and provide further opportunities for learning.

New to This Edition

The seventh edition of *Agriscience: Fundamentals and Applications* features many new exciting enhancements:

Current Information from the Industry includes the latest statistics from the USDA and reports of scientific advancements across all agricultural areas.

Expanded Information on the National FFA describes the various opportunities and programs designed to further student success in the agricultural education curriculum and beyond.

National FFA Connection at the end of select chapters highlights the programs related to participation in the National FFA, including Career Development Events, Proficiency Awards, and other FFA-related activities.

The Supervised Agricultural Experience is highlighted through an all-new chapter with special features and resources, two comprehensive new appendices, and an SAE Connection feature that identifies activities relevant to the SAE experience.

Inquiry Activity at the end of select chapters guides students to view informative short videos available on MindTap that build on the chapter topic, and then complete a brief activity to evaluate their understanding. Select chapters also include progress checks.

Enhanced Art and Design allow students a practical way of connecting concepts with real-world applications through hundreds of new photos while also engaging students in the content through a visually appealing and modern design.

Extensive Teaching/Learning Package

A complete supplemental package is provided to support *Agriscience: Fundamentals and Applications, Seventh Edition*. It is intended to assist teachers as they plan their teaching strategies while also providing students with additional opportunities to apply what they have learned in the chapters. This supplement package includes the following resources:

Lab Manual

The lab manual has been updated to correlate to the chapters in the new edition. This comprehensive lab manual reinforces key concepts presented in the chapters. It is recommended that students complete each lab to confirm understanding of essential science content. Great care has been taken to provide instructors with low-cost, strongly science-focused labs to help meet the science-based curriculum needs of this agriscience course in secondary schools.

Companion Site

The accompanying Companion Site includes online access to many helpful tools for teaching and learning, including:

- **Teacher Resource Guide** – Includes teaching tips, additional activities, assessments, and projects, standard correlations, and answers to end-of-chapter questions for each chapter.

- **Lab Manual Teacher Guide** – Provides safety information for conducting exercises in the lab as well as guidance on delivering individual exercises, including a list of required materials, specific instructions, additional online resources to support learning, and answers to questions with each lab exercise.

- **PowerPoint® Presentations** – Outline each chapter and include photos and illustrations to reinforce key points
- **Testbanks** – Provided in multiple formats compatible with a wide variety of LMS platforms, these question banks are drawn from the online testing powered by Cognero and allows flexibility in delivering online assessment.
- **Correlation Guides** – Correlates content from the seventh edition to several industry standards, including AFNR, NGSS, and Precision Exams.

Agriscience: Fundamentals and Applications, Seventh Edition combines current top notch content with information aligned to YouScience Precision Exam for *Agricultural Science I*, part of the *Agriculture, Food and Natural Resources* Career Cluster. The *Agriculture, Food and Natural Resources* pathway connects industry with skills taught in the classroom to help students successfully transition from high school to college and/or career.

Testing Powered by Cognero a flexible online system, provides chapter-by chapter quizzes, and enables teachers to:

- Author, edit, and manage test bank content from multiple sources
- Create multiple test versions in an instant
- Deliver tests from teacher/school-specific learning management system (LMS) or classrooms

MindTap for Agriscience: Fundamentals and Applications, Seventh Edition

The MindTap for *Agriscience: Fundamentals and Applications, Seventh Edition* features an integrated course offering a complete digital experience for the student and teacher. This MindTap is highly customizable and combines assignments, online lab exercises, videos, interactivities, and quizzing along with the enhanced ebook to enable students to directly analyze and apply what they are learning and allow teachers to measure skills and outcomes with ease.

- **A Guide:** Relevant interactivities combined with prescribed readings, featured multimedia, and quizzing to evaluate progress, will guide students from basic knowledge and comprehension to analysis and application.
- **Personalized Teaching:** Teachers are able to control course content – hiding, rearranging existing content, or adding and creating own content to meet the needs of their specific program.
- **Promote Better Outcomes:** Through relevant and engaging content, assignments and activities, students are able to build the confidence they need to ultimately lead them to success. Likewise, teachers are able to view analytics and reports that provide a snapshot of class progress, time in course, engagement and completion rates.

About the Authors

L. DeVere Burton

L. DeVere Burton is a former Director of Research and State Supervisor of Agricultural Science and Technology with the Idaho State Division of Professional-Technical Education, Dean of Instruction at the College of Southern Idaho, and President of the National Association of Supervisors of Agriculture Education (NASAE). He has participated in national curriculum-related task forces for the National Agricultural Education Council and served as an adult consultant to the National FFA Nominating Committee. Dr. Burton earned a BS in Agricultural Education from Utah State University, an MS in Animal Science from Brigham Young University, and a PhD in Agricultural Education from Iowa State University. He has authored four textbooks and edited an anthology, including *Agriscience & Technology, Fish and Wildlife: Principles of Zoology and Ecology, Introduction to Forestry Science, Agriscience Fundamentals and Applications*, and *Environmental Science Fundamentals and Applications*.

As an agricultural educator, Dr. Burton taught at several schools and a leading land grant university, supervised agricultural programs, and played a leading role in attaining graduation and college entrance credit in science for selected agriscience courses.

Raised on a farm in western Wyoming that bordered on forest lands, the author later gained invaluable experience testing milk for butterfat content; caring for livestock on a ranch; working in maintenance and the warehouse in a feed mill; managing a dairy farm; assisting in animal research; managing a purebred sheep and row crop farm; and working in the forestry, food processing, metal fabrication, and concrete construction industries. Dr. Burton has also traveled to Europe and Asia to learn about international food systems and production practices.

Renee Peugh

Growing up on a sheep and crop farm, Renee Peugh became interested in the science that drives agriculture. As a youth, she participated in 4-H and FFA and received a top place at the national level for her agriscience supervised agricultural experience (SAE). She earned a BS in Biology, with an emphasis on plant and animal sciences, from Boise State University. She holds a teaching certificate with endorsements in agriculture, natural science, health, and biology. Over the years, she has worked as a science consultant on textbook revisions and authored the *Agriscience Fundamentals and Applications* lab manual. Ms. Peugh is pursuing a master's degree in instructional design at the University of Idaho.

Ms. Peugh also teaches agriscience in an urban multi-teacher program. As an FFA advisor, she has coached FFA competitors for National FFA conventions and evaluated FFA proficiency and national chapters awards at the state level. Ms. Peugh has served on several agricultural education committees, including as chair, addressing such issues as alternative agriculture education certification and professional development. Her greatest career satisfaction comes from working with students in the classroom, visiting students' SAE projects, and coaching FFA, CDE, and LDE individuals and teams.

Acknowledgments

The authors and publisher wish to express their appreciation to the many individuals, FFA associations, and organizations that have supplied photographs and information necessary for the creation of this edition. A very special thank you goes to all the folks at the National FFA organization and the USDA photo libraries, who provided many of the excellent photographs found in these chapters. Because of their efforts, this content provides a visually engaging and authentic learning experience.

The author and publisher also gratefully acknowledge the unique expertise provided by the agricultural educators who contributed to the development of this edition:

Annette Applegate
Forest Park High School
Ferdinand, IN

Jennifer Crane
Eastmont Jr. High School
East Wenatchee, WA

Stacy DeVore
Cheney High School
Cheney, KS

Jessica Devries
Unadilla Valley Central School
New Berlin, NY

Neil Fellenbaum
Penn Manor High School
Millersville, PA

Chelsie Fugate
Payette High School
Payette, ID

Ethan Hunziker
New Madrid County Central
New Madrid, MO

Brandon Jacobitz
Adams Central High School
Hastings, NE

Stephanie Jolliff
Ridgemont High School
Mt. Victory, OH

Mike Jones
Adams Central High School
Hastings, NE

Leah Lucero
East Valley High School
Spokane Valley, WA

Rachel Payne
Clover High School
Clover, SC

Joel Rudderow
Penns Grove High School
Penns Grove, NJ

Paul Perry
Madison Central School
Madison, NY

Kris Spath
Waverly High School
Waverly, NE

Shannon Thome
Wellington High School
Wellington, OH

Lee Wright
Tygarts Valley Middle High School
Mill Creek, WV

Unit 1

Agriscience: The Future of Food

Science and technology are modern marvels that have opened the doors to new areas of research, dramatically extending the potential of agriscience. Space station research, new frontiers to investigate, and our never-ending quest for knowledge have exploded into many new and exciting careers.

You could become one of the people growing plants or animals in a space station high above the Earth. Or you might become an engineer who designs the animal- or plant-growing module of a space station, or a molecular geneticist or plant breeder designing new plants to grow well in low gravity, or a food scientist developing packaging for space-grown produce. Closer to home, you might discover ways to prevent plant or animal diseases. Perhaps you will become a researcher who discovers a better way to preserve food or a safe way to sanitize fresh fruits and vegetables. You may have personal attributes and skills that will propel you to become a teacher of agriscience, giving you an opportunity to have a positive influence on the lives of many students.

One career area in ever-expanding demand is plant science. As you will learn, plants are "green machines" that capture, package, and store energy from the sun through photosynthesis. They supply food and fiber for animals and humans to help sustain life. However, human knowledge and energy are also required to help plants function in the overall "green machine" that constitutes our food, fiber, and natural resources system. As students of the twenty-first century, you will become the agricultural professionals of the twenty-first century. You will become the agricultural producers, processors, marketers, and scientists who discover new ways to feed the citizens of the United States and the world. This you will accomplish by conducting basic research and applying it to the agricultural food system.

Whether you choose a career in plant or animal science, sales and marketing, mechanics, or processing, it is certain to be rewarding. By studying agriscience, you are opening the door to exciting educational programs and careers that contribute to better living conditions for people everywhere. What role will you play in the challenging task of producing the food and fiber that will be required by future generations?

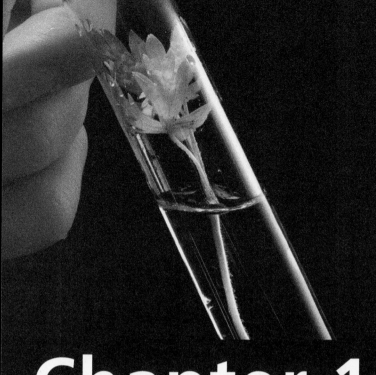

Chapter 1

The Science of Living Organisms

Objective
To recognize the major sciences contributing to the development, existence, and improvement of living things.

Competencies to Be Developed
After studying this chapter, you should be able to:
- define agriscience.
- discover agriscience in the world around us.
- relate agriscience to agriculture, agribusiness, and renewable natural resources.
- name the major sciences that support agriscience.
- describe basic and applied sciences that relate to agriscience.

Key Terms

agriscience
agriculture
agribusiness
renewable natural resources
technology
high technology
aquaculture
agricultural engineering
animal science technology
crop science
soil science
integrated pest management
water resources
environment
biology
chemistry
biochemistry
entomology
agronomy
horticulture
ornamentals
animal sciences
mathematics
statistics
geography
sociology
psychology
agricultural economics
agricultural education

Life in the United States and throughout the world is changing every moment of our lives. The space we occupy, as well as the people we work and play with, may be constant for a brief time. However, these are quick to change with time and circumstances. The things we need to know and the resources we have available to use are constantly shifting as the world turns.

Humans have the gift of intelligence—the ability to learn and to know (**Figure 1-1**). This permits us to compete successfully with the millions of other living organisms that share Earth's environments with us (**Figure 1-2**). In ages past, humans have not always fared well in this competition. Wild animals had the advantages of speed, strength, numbers, hunting skills, and superior senses over humans. These superior senses of sight, smell, hearing, heat sensing, and reproduction all helped certain animals, plants, and microbes to exercise control over humans to meet their own needs.

The cave of the cave dweller, lake of the lake dweller, and cliff of the cliff dweller indicate early human reliance on natural surroundings for basic needs (food, clothing, and shelter) (**Figure 1-3**). Those early homes gave humans some protection from animals and unfavorable environmental conditions. However, they were still exposed to diseases, the pangs of hunger, the stings of cold, and the oppressions of heat.

The world of agriscience has changed the comfort, convenience, and safety of people today. According to the USDA/Economic Research Service, Americans spent only 10.3 percent of our wages to feed ourselves in 2021 (**Figure 1-4**). Despite fluctuations in the percentage of income that is spent for food, the percentage of annual income spent for food in the United States has tended to decrease. People in many nations spend more than half of their incomes on food. We are fortunate that our scientists have discovered new ways to produce greater amounts of food and fiber (such as cotton) from each acre of agricultural land. They have done this by finding ways to stimulate growth and production of animals and plants and to reduce losses from diseases, insects, parasites, and storage. We have learned to preserve our food from one production cycle until the next without excessive waste; however, spoilage of stored food remains high among agriscience research priorities. The agriscience, agribusiness, and renewable natural resources of the nation provide materials for clothing, housing, and industry at an equally attractive price.

Figure 1-1 Humans have the gift of intelligence—the ability to learn and to know.

iStock.com/FatCamera

Figure 1-2 The gift of intelligence has permitted humans to compete with and benefit from animals, even though most animals are superior to humans in other ways.

Photo by Scott Bauer. USDA/ARS K7102-12.

Figure 1-3 Early humans had to rely on features in their natural environment to shield them from danger and the elements.

Figure 1-4 Americans spend 10.3 percent of disposable income on food.

BarbaraGoreckaPhotography/Shutterstock.com

Hot Topics in Agriscience — World Food Crisis

Unemployment and high food prices drive those who are most affected to seek food from charitable organizations.
James Steidl/Shutterstock.com

Serious food issues surface regularly as the world supply of rice, wheat, and corn declines to dangerously low levels, or wars and civil disruption interfere with food distribution. The usual result is substantial worldwide increases in the purchase price for all grains. This crisis continues in various degrees to this day, driving prices beyond expectations. In the United States and other nations, the cost of bread and other grain products increase as food processors adjust the prices of their products to compensate for the high cost of raw materials and transportation. The cost of grain and energy has also affected the price of meats, eggs, milk, and other foods, driving prices upward.

Political turmoil across the world due to economic recessions and in the form of revolutions has become a serious deterrent to affordable food prices in other ways. The price of oil fluctuates due to political manipulations, and is often associated with military conflicts in oil-producing nations. This raises production costs for most food items due to the cost of fuel.

For the poorer nations of the world and people living on fixed incomes or in poverty here at home, obtaining enough food to meet the needs of individuals and families becomes difficult. What should be done to overcome and resolve a world food crisis?

Hot Topics in Agriscience — Agricultural Research: Feeding a Hungry World

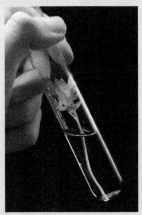

Environment refers to all the conditions, circumstances, and influences surrounding and affecting an organism or group of organisms.
© Garsya/Shutterstock.com.

The world's population has surpassed 7 billion people, and it is projected that it will reach 8 billion in 2025. During the same period, the amount of land and fresh water per person continues to decrease. Food production must become much more efficient if the people of the world are to have enough food to eat. During the past 50 years, food production has increased at a rate that is greater than the increases in the domestic population; however, food shortages and famine still exist in the world. Agricultural production is driven by a worldwide market.

Agricultural research has resulted in greater productivity of food, plants, and animals, and new technologies have made it possible for farmers to perform their work with greater efficiency. The key to an adequate food supply for the growing human population is agricultural research. New agricultural technologies that lead to the development of more efficient plants and animals and more efficient agricultural machinery will be needed. In addition, we will need to discover new food sources and maintain a healthy environment as the population approaches 10 billion people.

Agriscience Defined

Agriscience is a relatively new term that you may not find in your dictionary. **Agriscience** is the application of scientific principles and new technologies to agriculture. **Agriculture** is defined as the activities involved with the production of plants and animals and related supplies, services, mechanics, products, processing, and marketing (**Figure 1-5**). Actually, modern agriculture covers so many activities that a simple definition is not possible. Therefore, the U.S. Department of Education has used the phrase *agriculture/agribusiness and renewable natural resources* to refer to the broad range of activities in agriculture.

What Is Agriculture?

Animal Production · Marketing · Products · Services · Agriculture · Related Supplies · Mechanics · Crop Production · Processing

Figure 1-5 Agriculture consists of all the steps involved in producing a plant or animal and getting the plant or animal products to the people who consume them.

Agriculture generally has some tie-in or tieback to animals or plants. However, production agriculture, or farming and ranching, accounts for only 10.9 percent of the total jobs in agriculture (**Figure 1-6**) and 1.3 percent of the U. S. labor force. The other 89.1 percent of the jobs in agriculture are nonfarm and nonranch jobs, such as sales of farm equipment and supplies, plant and animal research, processing of agricultural products (**Figures 1-7** and **1-8**), agricultural education, and maintaining the health of plants and animals. **Agribusiness** refers to commercial firms that have developed in support of agriculture (**Figure 1-9**).

Figure 1-6 Farming and ranching account for approximately 10.9 percent of the agricultural jobs in the United States.

Nathan Allred/Adobe Stock Photos

Figure 1-7 Agricultural education teachers and agricultural extension educators are among those whose careers are related to agriculture.

Courtesy of DeVere Burton.

Figure 1-8 Food products must be sampled and tested to assure food safety and consistent quality.

SatawatK/Shutterstock.com

Figure 1-9 Agribusinesses are important to the people and the stability of most communities.

© iStock.com/David Sucsy

Hot Topics in Agriscience | New Border and Agriculture Labor Crisis

The United States border control and the needs for the services of agricultural workers are converging as foreign workers find it harder to obtain work permits and fewer citizens enter the agricultural work force.
©BearFotos/Shutterstock.com.

The number of American-born workers has been declining in the agriculture sector for quite some time. The result has been that many jobs, especially in farming and food processing, are now filled by foreign workers. This is true with seasonal jobs during the harvest season, but it is becoming more common in non-seasonal jobs such as food processing. The United States Department of Labor Statistics has noted that many agricultural employers actively seek workers from outside the country to get perishable crops and animal products harvested and processed before they spoil.

While it is important to secure our borders against terrorists and criminals, it is also important to be able to bring enough willing and documented workers across our international borders to sustain our domestic needs in farms, fields, and factories.

Renewable natural resources are the resources provided by nature that can replace or renew themselves. Examples of such resources are solar energy, wind energy, hydro-electric energy, wildlife, trees, and fish (**Figure 1-10**). Some emerging occupations in renewable natural resources are development and maintenance of wind towers and solar farms.

Technology is defined as the application of science to solve a problem. The application of science to an industrial use is called *industrial technology*. *Agriscience* was coined to describe the application of high technology to agriculture. **High technology** refers to the use of electronics and state-of-the-art equipment to perform tasks and control machinery and processes (**Figure 1-11**). It plays an important role in the industry of agriculture.

Figure 1-10 As mature trees are harvested, sunlight on the forest floor stimulates the growth of seeds and seedlings, providing a renewable source of wood for the future.
Courtesy of DeVere Burton.

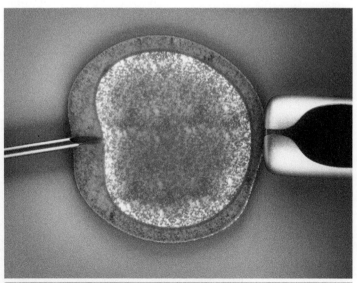

Figure 1-11 High technology plays an increasingly important role in agriculture. For example, it is possible to create identical cloned animals by dividing the cell mass of a growing embryo.
© digitalbalance/Shutterstock.com.

Agri-Profile Career Area: Agriscientist

Students experience the wonder of living things.
Courtesy of USDA/ARS K5304-17.

Science plays an essential role in the lives of plants and animals and the people around them. These living bodies include plants ranging in size from microscopic bacteria to the huge redwood and giant sequoia trees. They include animals from the one-celled amoeba to elephants and whales.

Increasingly, most notably during the 2020 Covid pandemic, science has clarified the nature of some viruses and permitted humans to observe the submicroscopic world in which they exist. The electron microscope, radioactive tracers, computers, electronics, robotics, artificial intelligence, nanotechnology, and biotechnology are just a few of the developments that have revolutionized the world of living things. We call this the world of science. Agriscience is a part of this world. Through agriscience, humans can control their destinies better than at any time in known history.

Agriscience spans many of the major industries of the world today. Some examples are food production, processing, transportation, sales, distribution, recreation, environmental management, and professional services. Study of and experiences in a wide array of basic and applied sciences are appropriate preparations for careers in agriscience.

Agriscience includes many endeavors. Some of these are aquaculture, agricultural engineering, animal science technology, crop science, soil science, biotechnology, integrated pest management, organic foods, water resources, and environment. **Aquaculture** means the growing and management of living things in water, such as fish or oysters. **Agricultural engineering** consists of the application of mechanical and other engineering principles in agricultural uses. **Animal science technology** refers to the use of modern principles and practices for animal growth, production, and management (**Figure 1-12**). **Crop science** refers to the use of scientific principles in growing and managing crops. **Soil science** refers to the study of the properties and management of soil to grow plants. Biotechnology refers to the management of the genetic characteristics transmitted from one generation to another and its application to our needs. It may be defined as the use of cells or components of cells to produce products and processes (**Figure 1-13**). You will explore biotechnology in more depth in Chapter 3.

Figure 1-12 Veterinarians use animal sciences to help keep our pets and production animals healthy.
iStock.com/Peeter Viisimaa

Figure 1-13 Genetic engineering and other forms of biotechnology have become some of the most important priorities in research today.
Courtesy of USDA/ARS K-5011-19.

The phrase **integrated pest management** refers to combining two or more different control methods to control insects, diseases, rodents, and other pests. *Organic food* is a term used for foods that have been grown without the use of chemical pesticides. **Water resources** cover all aspects of water conservation and management. Finally, **environment** refers to all the conditions, circumstances, and influences surrounding and affecting an organism or group of organisms (**Figure 1-14**). This generally means air, water, and soil, but it may also include such things as temperature, presence of pollutants, intensity of light, and other influences.

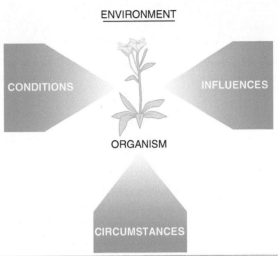

Figure 1-14 The term *environment* refers to all the conditions, circumstances, and influences surrounding and affecting an organism or group of organisms.

Agriscience Around Us

Agriscience and technologies have helped humans change their living conditions from dependence on hand labor to a highly mechanized society. In the process, food and fiber production has become much more efficient. This has allowed more people to pursue new careers because they are no longer required to spend time finding or producing food for themselves and their families. Unsurprisingly then, fewer than 2 percent of the people in the United States are farmers. On average, each farmer produces enough food for approximately 167 people. The large surplus of food that is produced in the United States is shipped to many other countries in the world.

Whether you live in a rural or urban area, you are surrounded by the world of agriscience. Plants use water and nutrients from the soil and release water and oxygen into the air. Animals provide companionship as pets and assistance with work. Both plants and animals are sources of food. Many microscopic plants and animals are silent garbage disposals (**Figure 1-15**). They assist in the process of decay of the unused plant and animal residue around us. This process returns nutrients to the soil and has many other benefits to our environment and our well-being.

Agriscience encompasses the wildlife of our cities and rural areas, and the fish and other life in streams, ponds, lakes, and oceans. Plants are used extensively to decorate homes, businesses, shopping malls, buildings, and grounds. When the use of one crop is lessened, another takes its place. This occurs even where the land resource changes from farm use to suburban and urban uses.

Figure 1-15A The process of composting uses bacteria in moist and aerated conditions to break down plant residue.

imging/Shutterstock.com

Figure 1-15B The material that remains when the composting process is complete is used to provide nutrients to crops and gardens.

Courtesy of DeVere Burton.

Corn and soybeans are extensively grown and considered to be high-value crops in the United States. Yet, in some states, including Texas and Virginia, turf grass is the number-one agricultural crop. Turf is a grass that is used for decorative as well as soil-holding purposes. This change has occurred as more land is used for roads, housing, businesses, institutions, recreation, and other nonfarm uses (**Figure 1-16**).

Agriculture and the agriscience activities that support it extend far beyond the borders of the United States. Many nations throughout the world depend on agriscience to improve

Figure 1-16 Turf has become an important crop, especially in areas near large cities where it is used to establish new lawns.

Courtesy of USDA/ARS CS-0311.

Figure 1-17 Flowers are often imported to the United States during seasons when local florists are unable to produce them at competitive prices.

coronado/Shutterstock.com

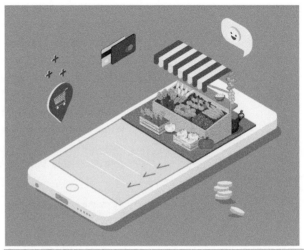

Figure 1-18 Marketing in agriscience has become a big business.

© Yana_Shadu/Shutterstock.com.

the production of their crop and livestock industries. Agriculture is a global industry, and although the United States exports many of its agricultural products, it also imports many agricultural products from other parts of the world. Many of the flowers used by florists in the United States come from South America, Colombia, and other foreign countries (**Figure 1-17**). Bulbs come from Holland, and meat products are imported from Brazil, Australia, and Argentina. China and Japan import timber from the United States.

Many of the logs that are shipped from Pacific Northwest ports never arrive in Japan. Japanese offshore processing ships anchor off the coast where timber, exported by the United States and Canada, is processed into lumber products. Once processed, these wood products are imported back to the United States.

Markets are affected by many things. For instance, a tariff imposed on China for wheat may reduce its price in other countries. A decrease in the price of sow bellies or an unexpected change in the price of grain futures at the Chicago futures market can affect business and investment around the world (**Figure 1-18**).

The great water-control projects on the Colorado River have permitted the transformation of the American Southwest from a desert to irrigated lands (**Figure 1-19**). This is now an area of intensive crop production that has stimulated national population shifts. Water management has transformed the great dust bowl of the American West into the "breadbasket" of the world.

Figure 1-19 The great Southwestern desert has been transformed into highly productive land, using irrigation water from the huge dams on the Colorado River and its tributaries.

©Tim Roberts Photography/Shutterstock.com.

Agriscience enterprises extend beyond farming to such fields as journalism and communications. Digital and print agricultural publications, such as magazines, journals, and newspapers provide information to farmers, helping make farm production more efficient. Radio and television programs—and a growing number of podcasts—provide similar services to agriculture. They provide a communications link among such people as agricultural specialists, agricultural extension educators, wildlife biologists, and others to communicate the latest information to farmers and other managers of natural resources. While such subjects as plants, animals, wildlife, market reports, gardening, and lawn care are popular "Saturday morning" topics, news of the latest agricultural topics and trends is now accessible 24 hours a day.

Agriscience and Other Sciences

Agriscience is really the application of many sciences. Colleges of agriculture and life sciences perform dual roles of conducting research and teaching students in these sciences. **Biology** is one of the three basic sciences. It derives from two Greek words: *bios*, meaning "life," and *logy*, meaning "to study." It is the science that studies all living things (organisms) and the environment in which these organisms live. Biology emphasizes the structures, functions, and behaviors of all living organisms. It focuses on the traits that organisms have in common as well as their differences. Developing an understanding of biology is important to you, the agriscience student. New biological discoveries can affect many areas of your life. Examples include the choice to plant a new variety of flower in the front yard, changes in the way food is processed that result in a fresher product with a long shelf life, and new medical treatments that can help keep you healthy.

Chemistry is another basic science. It is the branch of science that studies the nature and characteristics of elements or simple substances. Chemists study the changes substances undergo when they react with other substances. These changes are responsible for compounds that have been used by humans for thousands of years, whereas other compounds, like artificial sweeteners, are relatively new discoveries. Chemistry currently benefits us in many ways; for example, it is responsible for the creation of new medicines, textiles, fuels, and fertilizers.

Biochemistry is the last of the three basic sciences. It is a combination of biology and chemistry. Recall the definition of *bio*, which means "life"; when added to the word *chemistry*, it is easy to see that the science of biochemistry is the study of the chemical activities or processes of living organisms. These chemical activities take place in the cells and molecules of living organisms. This science is responsible for explaining things such as brain function, how genetic traits are passed from one generation to another, and how cells communicate and work together inside an organism. Applied science uses the basic sciences in practical ways. For instance, **entomology** is the science of insects, the most abundant species on the planet. Insects account for more than 3 million human deaths per year, they transmit diseases, and they are our principal competitors for food. Insects, however, are required for pollination by half the plants on Earth to produce seeds. It is important to find ways to help control problem insects safely without causing secondary problems such as halting pollination of plants.

There are many other applied sciences. Knowledge of biology, chemistry, and biochemistry is important in entomology and to the other applied sciences listed below.

Agronomy is the science of soil management and crops. Its focus is on the growth, management, and improvement of field crops such as wheat and corn. The goal is to increase food production and quality while maintaining a healthy environment. **Horticulture** is the science and art involved in the cultivation, propagation, processing, and marketing of flowers, turf, vegetables, fruits, nuts, and ornamental plants. **Ornamentals** are plants grown for their appearance or beauty. Examples are flowers, shrubs, trees, and grasses. Horticulture is a unique science because it also incorporates the art of plant design.

Science Connection: The Agriscience Project

The scientific method is an excellent and widely used method for systematic inquiry and documentation of new findings. The agriscience student is encouraged to learn and use the scientific method for classroom, laboratory, and field studies. The following procedures will guide you in your quest for new knowledge in agriscience.

Step 1. Identify the Problem

Decide precisely and specifically what it is that you wish to find out, for example, "How much nitrogen fertilizer is needed to grow healthy corn plants?" Be careful to limit your topic to a single researchable objective. Your teacher can suggest other topics that could be researched.

Step 2. Review the Literature

Reviewing the literature simply means reading up on and becoming well informed about the topic. See what is already known about it. Reputable websites like those of the USDA, magazines, newspapers, reference books, encyclopedias, science journals, computer information systems, television, podcasts, cooperative extension meetings, remote conferences, and personal interviews may all be sources of appropriate information. Be sure to seek information on appropriate ways to conduct research on the type of problem you have chosen.

The written report completes the research project and enables others to benefit from new information and knowledge.
Photo by Scott Bauer. USDA/ARS K8329-2.

Step 3. Form a Hypothesis

After learning as much as you can about the topic, develop a hypothesis or statement to be proven or disproved, which will solve the problem. For example, "Trout grown in 60° F water will grow faster than trout grown in 75° F water."

Step 4. Prepare a Project Proposal

Prepare a proposal outlining how you think the project should be done. Include the timelines, facilities, and equipment required as well as anticipated costs and a description of how you will do the project.

Step 5. Design the Experiment

Considering the information gathered in Step 2, develop a plan for carrying out the project so you can test the hypothesis. This is the most critical step in your research project. If this is not done correctly, you may invest considerable time, work, and expense and end up with incorrect or invalid conclusions. The method or procedure should be carefully thought out and discussed with your teacher or other research authorities. This is to ensure that your design will actually measure what you are testing.

Step 6. Collect the Data

In this phase, you conduct the experiment. Here you test and/or observe what takes place and record what you measure or observe.

Step 7. Draw Conclusions

Summarize the results. Make all appropriate calculations. Determine if the information allows you to accept or reject the hypothesis or if the information is inconclusive.

Step 8. Prepare a Written Report

The written report provides you, your teacher, and other interested parties with a permanent record of your research. From this, you can report to your peers, get course credit, and possibly apply for awards. For scientists, the written report becomes a permanent document that is kept by the research institution and becomes the basis for articles in research publications for the world to see. The results become part of the "literature" on the topic.

The **animal sciences** are applied sciences that involve growth, care, and management of domestic livestock. They include veterinary medicine, animal nutrition, animal reproduction, and animal production and care. Animal scientists work to discover scientific principles related to animals. Scientific principles are then applied to animal management plans to improve animal health and production.

Mathematics is a discipline that deals with quantity (numbers), space, structure, and their relationships. It is sometimes referred to as the science of numbers. Mathematics is integral to agriscience because it gives us an ability to apply numbers, measure, and calculate. Each of these functions is used extensively in engineering new products or evaluating business practices. Examples include calculating productivity, measuring reliability of products, and predicting economic cycles, among others.

Statistics is a branch of mathematics that is used to collect large amounts of data, organize it in number form, and analyze it. Statistics are used to predict trends for entire populations by studying small representative samples of the population. The amount of corn produced in a harvest season is predicted before the corn harvest is complete, based on measurements of yields from representative samples gathered from a few fields. Predicted yields make it possible to establish a fair price early in the harvest season. Statistics are also used to measure the reliability of agricultural research methods and results.

Geography is a social science related to the interactions of people and the different features of the earth. Examples of ways that geography applies to agriculture include the issues surrounding national parks where large predators are protected and the agricultural areas adjacent to the park boundaries where domestic livestock is at risk. Social sciences study human societies and their relationships. **Sociology** focuses on human issues and problems. Rural sociology relates to agriculture because it deals with problems in farm communities. For example, how can rural communities deal with problems associated with drought or wildfires? **Psychology** deals with how we think about things. It becomes especially important in rural areas when times of natural or economic disaster add to the existing stressors of farming. Consider how the 2020 coronavirus pandemic, and the related financial challenges, impacted the well-being of farm families.

Economics is the study of how societies use available resources to meet the needs of people. **Agricultural economics** relates to factors that affect the management of agricultural resources, including farms and agribusinesses, to meet the needs of the human population. Farm policy and international trade are important components of agricultural economics. **Agricultural education** is one of the most unique programs available to students. It offers organized instruction, supervised agricultural experience (SAE), FFA, and extension education activities. Agricultural communications, journalism, and community development are also components of agricultural education. These and other disciplines are part of the dynamic study known as agriscience.

A Place for You in Agriscience

What about career opportunities in agriscience? The nation's agricultural colleges report steady demand for graduates in agricultural disciplines. A U.S. Department of Agriculture (USDA) study group forecasted a national shortage of more than 4,000 agricultural and life sciences graduates per year. Employers are offering higher salaries and more job variety to agriscience college graduates than ever before. "Future development of our complex global food system requires the brightest minds from a wide range of backgrounds, cultures and disciplines working together to solve the challenges before us," said Parag Chitnis, acting director of USDA's National Institute of Food and Agriculture.

Career opportunities continue to be strong in the agricultural sector. Consequently, high school agricultural, horticultural, or other agriscience program participants gain opportunities to obtain good jobs and rewarding careers as they advance in experience and education (**Figure 1-20**).

Figure 1-20 Plenty of good jobs are available for agricultural graduates, whether at the technical degree, Bachelor of Science degree, or graduate degree level of education.
© Alexander Raths/Shutterstock.com.

One of the solutions to the shortage of agriscience graduates has emerged as the traditional roles of women have changed. Today, university majors in animal and dairy sciences, and agricultural education, communication, and leadership, are dominated by women. In addition, the United States Department of Agriculture reported in 2019 that 51 percent of the U.S. farm operators were women (**Figure 1-21**).

These opportunities are described in later chapters in this text. By studying agriscience, you open the door to exciting educational programs and careers that contribute to professional satisfaction and prosperity.

Figure 1-21 Women in agriculture can be found in careers at every level of the industry. Dr. Chavonda Jacobs-Young (center), administrator of the Agricultural Research Service of the USDA, meets with agriscientists and engineers in Washington, DC.
Archistoric/Alamy Stock Photo

Chapter Review

Student Activities

1. Write the **Key Terms** and their meanings in your notebook.
2. List examples of animals that have superior senses than do humans. Indicate the sense(s) along with the animals.
3. Write a paragraph or two on (1) cave, (2) lake, and (3) cliff dwellers. Explain how the types and locations of their homes provided protection from (1) animals and (2) unfavorable weather. Access to an online encyclopedia would be a good resource for this activity.
4. Ask your teacher to assign you to a small discussion group to talk about the responses to Activity 3.
5. Place a map of your school community on a bulletin board. Insert a colored map pin in every location of a farm, ranch, or agribusiness in your school community.
6. Talk to your County Extension Agent or another agricultural leader regarding the importance and role of agriscience, agribusiness, and renewable natural resources in your county.
7. Select one of the sciences mentioned in this chapter. Prepare a written report on the meaning and nature of that science. Report to the class.
8. **SAE Connection:** Identify several emerging technologies with agricultural applications and list them in the left-hand column of a spreadsheet titled "Analysis of Emerging Agricultural Technologies." Add columns titled "Benefits" and "Limitations." Record each applicable observation you can find in the appropriate cell. Use the data to form and record an opinion on the relative value of each technology.

National FFA Connection — What Do You Know About the FFA?

Courtesy of National FFA; FFA #148

Extracurricular opportunities to explore agriscience are invaluable, informative, and fun!
© Roman Samborskyi/Shutterstock.com

As part of this program, you will be exploring the opportunities available to you through the National FFA Organization (FFA), an intracurricular youth organization for students enrolled in agriscience programs. With your classmates, discuss what you already know about the FFA. For example, have you ever participated in any FFA agriscience programs? If so, describe the program and share what the experience was like. Next, work together to find out more, such as by visiting the FFA website and finding out about FFA events in your region or state.

Inquiry Activity: Agricultural Research

Courtesy of USDA/ARS K-5011-19.

After watching the *Agricultural Research* video available on MindTap, answer the following questions. Feel free to watch the video again to locate the answers.

1. How did agricultural research address a food-related problem that affected many countries in the 1950s? What was the outcome?
2. What is mastitis? How much of an impact has agricultural research had on the occurrence of mastitis?
3. Why did scientists engineer a parasite that affects chickens?
4. Consider how agricultural research balances benefits and risks. According to the video's conclusion, what does the research aim (1) to increase and (2) to decrease?

Check Your Progress

Can you . . .

- define agriscience?
- give examples of agriscience in the world around us?
- relate agriscience to agriculture, agribusiness, and renewable natural resources?
- name the major sciences that support agriscience?
- describe basic and applied sciences that relate to agriscience?

If you answered "no" to any of these questions, review the chapter for those topics.

Chapter 2
Agriscience and Living Conditions

Objective
To determine important elements of a desirable environment and explore efforts made to improve the environment.

Competencies to Be Developed
After studying this chapter, you should be able to:
- describe the conditions of desirable living spaces.
- discuss the influence of climate on the environment.
- compare the influences of humans, animals, and plants on the environment.
- examine the problems of an inadequate environment.
- identify significant world population trends.
- identify significant historical developments in agriscience.
- state practices used to increase productivity in agriscience.
- identify important research achievements in agriscience.
- describe future research priorities in agriscience.

Key Terms
sewage system
polluted
condominium
townhouse
famine
contaminate
parasite
reaper
combine
moldboard plow
cotton gin
corn picker
milking machine
legume
tofu
Katahdin
BelRus
Russet
Green Revolution
feedstuff
selective breeding
genetic engineering
monoclonal antibody
hybrid
laser

Living conditions in the world vary extensively. In almost all countries, there are some very desirable places to live and work (**Figure 2-1**), yet even in highly developed countries, there are areas of poverty. How do you explain differences in living conditions from one location to another? Why do living conditions vary from one community to another, from one neighborhood to another, or from one house to another? The wealth and preferences of individuals explain some of the differences. Yet, the environment or area around us influences the quality of life. It also has much to do with the way we feel about ourselves and others. In addition, the way we feel about ourselves is often expressed in the way we treat and care for our environment.

Figure 2-1 People of all cultures appreciate pleasant surroundings for their homes and work.
© benchart/Shutterstock.com.

A Broad Range of Living Conditions

The population of the world surpassed 7 billion people in 2011 (**Figure 2-2**). It is predicted that the population will increase to 8 billion in 2025. Across the world, several babies were given the distinction of being the 7 billionth human being living

Figure 2-2 In 2025, the population of the world is expected to exceed 8 billion people.
© Dmitry Nikolaev/Shutterstock.com.

on planet Earth (**Figure 2-3**). What will the home and community be like where "Baby 8 Billion" is born? Will there be adequate food? Will that child be warm but not too warm, or perhaps cool but not too cold? Will the child be kept free from serious illness? Will pandemics such as Covid-19 continue to afflict people across the world? Will their family have a house or good living space they call home? Will they have clothing to permit them to live and work outside the home in relative comfort? What will be the quality of life of others who live near the 8 billionth human being? Will that child survive, and will they go on to live a happy and productive life? Positive answers to these questions would indicate a good environment for a person. These same questions should be asked for the rest of humankind.

Figure 2-3 What will life be like on planet Earth when Baby 8 Billion is born?
Courtesy of NASA.

The Homes We Live In

Homes of the Most Impoverished

Homes of the poorest people range in size and quality from nothing to a piece of cardboard or a scrap of wood on the ground. Many survive freezing winters with only a tattered blanket on the warm sidewalk grates of our modern cities. For others, housing may take the form of a grass hut or a shack made of wood, cardboard, plastic, or scraps of sheet metal. These people often depend on the outdoors to provide water and washing areas and as a receptacle for human waste. Large families often share such homes with pets, poultry, or other livestock (**Figure 2-4**).

In cities, the poor frequently live in old buildings that are in unsafe condition and with plumbing that does not work. Drugs, crime, poisonous lead paint, polluted drinking water, and disease are typical hazards for these people. The steamy streets and sidewalks provide little relief from oppressive summer heat.

Homes of the Disadvantaged

People with modest sources of income may have homes that are simple but provide basic protection from the elements. Such homes may be of wood, stone, masonry blocks, concrete, sheet metal, brick, or other fairly permanent materials. The presence of windows and doors may provide protection from the elements and some privacy.

Figure 2-4 The poor of the world live in substandard housing, and it is not uncommon for homes to be shared with pets, livestock, or poultry.

Courtesy Elmer Cooper.

These people frequently have access to water that is safe to drink, but bathrooms may not exist or toilets may not have safe sewage disposal systems. A **sewage system** receives and treats human waste (**Figure 2-5**). To be regarded as safe, a sewage system must decompose human waste and release by-products that are free from harmful chemicals

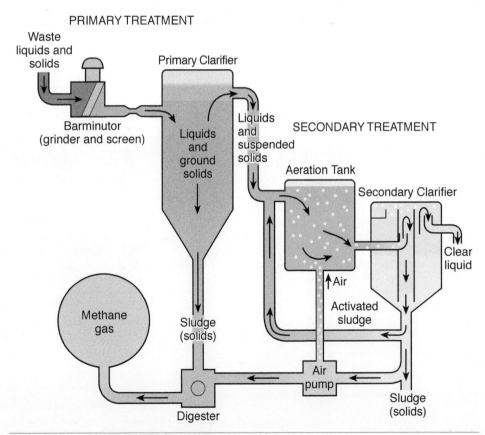

Figure 2-5 Safe sewage disposal is essential to good health.

and disease-causing organisms (**Figure 2-6**). In most countries of the world, people rely on rainwater containers, streams, or rivers to supply their drinking water, bathe the family, wash the clothes, and carry away the human waste.

People with low or modest incomes also may live in housing with bathrooms and running water. However, maintenance of the systems may be poor, and the users' lack of knowledge may cause conditions that are hazardous to health.

In poor neighborhoods, as well as in developing countries, the lower classes are fortunate if there is a source of safe water (water that is free of harmful chemicals and disease-causing organisms) at the village center (**Figure 2-7**). Modest and simple running-water facilities are

Figure 2-6 This modern sewage treatment plant captures energy from sewage sludge by using the sludge to produce methane gas. The gas is used as fuel for engines that drive generators, producing electricity.

Courtesy of DeVere Burton.

Figure 2-7 For much of the world's population, even a single community source of safe water is a luxury.

© iStock.com/Peeter Viisimaa.

generally the first evidence of community development in many such areas. Even a single faucet with unpolluted water is a major step forward for many communities. Communities that do not have a source of safe water must use whatever water they have available to them. Too often, water is **polluted** or unsafe to drink because it contains waste materials, chemicals, or unhealthful organisms.

Homes of the Middle and Upper Classes

The middle- and upper-class people of the world can afford and enjoy housing that is clean, safe, and convenient. Such living spaces are often found as single houses, in both rural and urban areas. In towns, villages, and urban areas, homes may also be in the form of townhouses, condominiums, or apartment buildings. A **condominium** is a building with many individually owned living areas or units. All living space of a single unit is generally on one floor. A **townhouse** is one of a row of houses connected by common side walls. Each unit is usually two or three stories high, giving the occupants more variety of living space.

Key Environmental Factors

Food Supply

Even today, much of the world still goes to bed hungry. Only a few countries have sufficient food for their people, and even some prosperous nations have problems with distribution. Not everyone has food of sufficient amount and quality for proper nutrition. Today, major **famines** (widespread starvation) are still a fact of life and death. Even in the modern world, serious famines continue to occur in various countries.

United Nations scientist John Tanner concluded that, in theory, the world could feed itself; but in practice it could not. It is estimated that nearly a billion people are not getting enough food for an active working life. Although some countries enjoy an adequate food supply from their own production and imports, most nations have many individuals who do not receive proper nutrition.

Family

Family life may well be the dominant force that shapes the environment for most individuals. The family has control of the household activities and sets the priorities of its members. For some, the family chooses the neighborhood and community where they live. A wise choice, however, is based on having the knowledge of better opportunities and the necessary resources to move to a better living environment. For most of the world's population, the communities where individuals are born are the communities where they are raised and spend their lives.

Neighborhood and Community

The neighborhood and community have substantial influence on the environment in which we live. Some communities have tree-lined country roads with attractive fields, pastures, or woodlands to provide variety in the landscape. Other communities may have the advantage of attractive homes, businesses, or community centers (**Figure 2-8**). Urban areas may boast high-rise buildings for work and residence. These provide beautiful vistas of city lights or harbor scenes of commerce and recreation.

Neighborhoods and villages are parts of larger communities. These communities are influenced greatly by the families who live in the immediate area. If families work together toward common goals, they can shape the character, education, religious activities, social outlets, employment opportunities, and other broad aspects of their environment.

Figure 2-8 Good communities provide pleasant, attractive environments for living and working.
© Robert Keneschke/Shutterstock.com.

Hot Topics in Agriscience

Preserving the Environment Using Intensive Farming Practices

Responsible use of intensive farming practices may be our best option to preserve forests and other natural environments.
Courtesy of DeVere Burton.

When the first colonists arrived in America from Europe, abundant land resources were available for food production. Trees were removed to make way for the farms. Midway through the twenty-first century, as the world's population approaches 10 billion people, the forests of the world will again become endangered unless we can continue to increase the food production of current farmland. We will need to increase the efficiencies of our existing farmlands and production methods to produce an adequate food supply for a growing world population. If we fail to increase the food production of our land, forests will probably be converted to farms because more land will be required to produce the additional food that will be needed.

Responsible use of intensive farming practices is likely to play a big role in preserving our forest lands. Applications of agricultural chemicals to crops contribute to high production of food by controlling weeds and insects. The application of fertilizers to farmland is also a proven method for sustaining high levels of production. However, good judgment must always be exercised in the application of fertilizers and chemicals to ensure that they are used safely and that they do not pollute the environment. It is an interesting paradox to consider that farm fertilizers and pesticides may be our best hope for preserving the forests and other natural environments in the world. As many foreign governments can attest, preservation of the environment becomes a low priority to people who are starving.

Climate and Topography

Climate and topography are also important factors affecting our environment. Climate is the average yearly temperature and precipitation for a region. Unlike topography, which is the physical shape of Earth, the climate of a given area is shaped by many factors. The movement of heat by wind and ocean currents, the amount of heat absorbed from the sun, latitude, and the amount of precipitation received are all factors that influence climate. The climate affects what kinds of crops can be raised. The tropical areas of the world produce crops such as pineapples and bananas, whereas the mild climates found in the U.S. mainland are better adapted for crops such as corn and wheat (**Figure 2-9**). Average annual temperatures are very high near the equator. Yet people living near ocean waters, even in tropical areas, enjoy a moderate climate with cool breezes most of the time. Inland, the inhabitants are likely to experience hot, humid weather with high rates of rainfall. The high rainfall, in turn, stimulates heavy plant growth, resulting in jungle conditions. Similarly, sea-level elevations may create balmy 80° F temperatures, whereas a short trip to the top of a nearby mountain may reveal snow on its peak (**Figure 2-10**).

Northern areas, such as Alaska, may border on the Arctic Circle and have long, frigid winters. Yet those same latitudes enjoy summers suitable for short-season crops (**Figure 2-11**). People inhabit most areas of Earth, so the climate and topography where they find themselves create environmental conditions that influence their quality of life.

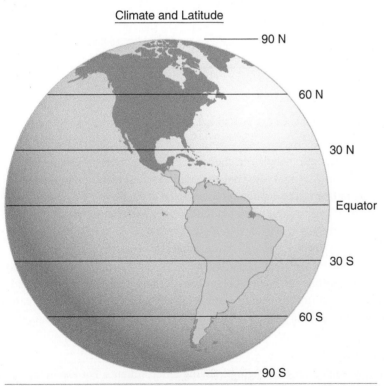

Figure 2-9 The distance from the equator is measured by degrees, with zero being the equator and 90 degrees North being the north pole. Distance from the equator affects the climate of a region.

Figure 2-10 Topography is an important factor that affects the temperature of the environment.

© Russell Illig/Stockbyte/Getty Images.

Figure 2-11 The climate near Anchorage, Alaska, is influenced by a short growing season; however, during the growing season, the number of hours of daylight per day is high, contributing to excellent growth conditions for some food crops.
© ARENA Creative/Shutterstock.com.

Other Environmental Factors

Humans are the only living creatures who have the ability to make choices that affect their living environments. For example, people can choose to protect or even to clean up the environments around them. They may also choose to damage their living environments. Living in an environment that is free of pollution is a choice of the people who live there. It may be influenced by politics or other causes, but until people exercise the choice to reduce pollution in their neighborhoods, it is likely to remain.

Scientific processes have been developed to remove waste products, poisons, and disease organisms from food, water, and the air (**Figure 2-12**). However, the body is limited in its capacity to remove poisons and harmful organisms. To remain healthy, humans and animals must limit their exposure to disease organisms and poisons. One important reason to protect living environments is to keep pollutants from entering food and water supplies.

Figure 2-12 Processes have been developed that will remove contaminants such as heavy metals and harmful organisms from water.
Don Farrall/Stockbyte/Getty Images

Humans and Animals

Some human activities can be very damaging to living environments. For example, exhaust fumes from cars are known to cause acid precipitation. Leaky fuel tanks pollute the soil, which in turn pollutes the drinking water supply. Improper use of chemicals (lawns, gardens, farms, and factories) pollutes rivers, streams, and lakes. Poor soil management results in erosion of the soil and contamination of streams, lakes, and reservoirs with silt.

A major problem for humans and animals is to avoid **contaminating** (adding material that will change the purity or usefulness of a substance) food and water with secretions from their own bodies. Urine and feces are liquid and solid body wastes. They are serious contaminants of food and water. Diseases are often spread by body contact, by eating impure food and water, or by breathing contaminated air.

There are serious animal diseases that spread from animal to animal by contact with body wastes. If animals have plenty of living space, this generally does not cause serious problems. But, as with humans, when animals are concentrated, health hazards increase. Fortunately, most diseases are spread among only a given species of animal and not from one species to another. For instance, most diseases of dogs do not spread to cats. Similarly, most diseases of animals do not infect humans. Brucellosis is an example of an animal disease that may be transferred from animals to humans, creating serious health problems. It kills unborn animals and humans by spreading the disease through the milk and birth fluids of infected individuals.

However, some animal disorders can cause human sickness. **Parasites** are organisms that live on or inside other organisms called hosts with no benefit to those hosts. The parasite is an unwelcome guest because it always causes some kind of harm to its host by feeding on it.

Insects

Insects impact heavily on our environment. Some cause damage to our living environment and others help to improve it. The cockroach is an unwelcome guest in many households of the world (**Figure 2-13**). Some cockroaches feed on human waste and then on the food of humans. They transmit disease from waste material to food and water. In poor housing conditions, they can move from household to household. In doing so, they often leave disease organisms and illnesses in their wake.

Insects cause damage to the environment in other ways. They damage and kill trees and other plants. A population of harmful insects can expand quickly, and damage to trees can be extensive unless steps are taken to control them. For example, pine beetles are capable of killing entire populations of pine trees (**Figure 2-14**). Imagine the damage to the environment of a community when many of the trees in parks, yards, and streets are of a single variety that is susceptible to a highly destructive insect pest. When a tree or other plant is not affected by the presence of a harmful insect, it is described as having

Figure 2-13 Cockroaches are household pests that pollute living environments and contaminate foods and beverages. The cockroach feeding station, pictured here, lures these insects inside and poisons them.

Photo by Scott Bauer. USDA/ARS K7233-6.

Figure 2-14 Pine beetles have destroyed a forest of pine trees that were first weakened by drought and then infested with beetles.

Courtesy of DeVere Burton.

immunity. Many plants in the environment are immune to particular insects, but they are often vulnerable to other insect species.

Not all insects are destructive to the environment. Some insects, such as bees and other pollinators, are useful to living environments. Without these insects to carry pollen from one flower to another, many plants could not reproduce. Some insects, such as the lady bird beetle, prey on harmful insects. These insects play important roles in keeping populations of harmful insects in check. Insects do have profound effects on the living environments that surround us.

Hot Topics in Agriscience: Pesticide Reduction and Biotechnology

The Colorado potato beetle can be controlled in genetically modified potatoes without the use of pesticides.
iStock.com/BasieB

A significant and well-documented outcome derived from the adoption of genetically engineered plants is the declining use of pesticides. This should be no surprise considering that many genetic modifications have introduced pest resistance into some of our most widely used food plants.

One such plant is the potato. This crop is highly vulnerable to the Colorado potato beetle, and an entire field can be destroyed by this pest in a very short period of time. The adult and immature forms of this beetle eat the stems and foliage and destroy the ability of the plant to engage in photosynthesis. The traditional treatment for these pests has been to apply pesticides to the crop. Potatoes that have been genetically modified to resist the beetles are able to produce a chemical that does not affect humans, but which poisons the beetles as they eat the foliage.

The end result is that pesticides are no longer needed to control this pest in potato varieties that carry the resistant gene. According to the Food and Agriculture Organization of the United Nations:

"Thus far, in those countries where transgenic crops have been grown, there have been no verifiable reports of them causing any significant health or environmental harm."

Chemicals

Chemicals can have both helpful and harmful effects on living environments (**Figure 2-15**). Oil spills and industrial chemical discharges have caused serious problems in oceans, lakes, rivers, and streams (**Figure 2-16**). Improperly used chemical pesticides continue to threaten wildlife, fish, shellfish, beneficial insects, microscopic organisms, plants, animals, and humans. In the mid-1960s, American biologist Rachel Carson shocked the world with her book *Silent Spring*. This was one of the first books to provide convincing evidence of environmental damage due to pesticides.

In 1972, DDT (dichloro-diphenyl-trichloroethane) was banned in the United States because of its damaging effects on the environment. This insecticide had been used to control mosquitoes, which carried the dreaded malaria organism. DDT was also a very effective chemical used against flies, and it enjoyed widespread use in homes, on farms and ranches, and wherever flies were a problem. Yet, because it was determined that DDT was responsible for interfering with the reproduction of birds by weakening their eggshells, it was discontinued; safer substitutes have been found.

Careful management and control of the different chemicals we use is absolutely essential. It requires the utmost care to avoid unacceptable damage to our environment.

Figure 2-15 Chemicals are needed in our modern society, but they may threaten the health of animals and people if misused or abused.

Photo by Jeff Vanuga, USDA Natural Resources Conservation Service.

Figure 2-16 Spills of petroleum or chemicals in water environments cause serious water pollution and damage populations of wild animals and plants.

© Gl0ck/Shutterstock.com.

Our Shared Living Environment

All of the members of the plant and animal kingdoms, including humans, must share the living environments that are available on Earth. Some living environments are more friendly to the inhabitants than others, and plant and animal life tends to be concentrated in the warm, temperate regions of the world. Other environments, such as the Arctic and Antarctic regions, do not support the variety of plants and animals that are found in other places. The same is true of the high-altitude, mountainous areas. One thing is certain: The living environments of Earth will never grow any larger. In fact, the habitable regions may decline for many species of plants and animals as humans pave Earth and create cities. Only a few species may be able to adapt to the less favorable environments as they are crowded from their natural ranges.

As we ponder the life of "Baby 8 Billion" (see earlier), we must wonder if we are doing our part to preserve and enhance the environment. Plants, animals, insects, soil, water, and air must be kept in reasonable balance, or all will suffer. Plants are generally considered to improve living environments, but excessive plant growth can infringe on the space for humans and animals. Humans and animals can damage a plant species until it is unable to adequately reproduce itself. Too many animals in a shared environment can compete excessively with humans for food, water, and space. Some species of insects are regarded as harmful by people because they feed on desirable crops or afflict humans or livestock. However, many species of insects are beneficial to plants, animals, or humans.

Humans and animals tend to consume or remove plants, which hold soil in place thus preventing erosion from wind and water (**Figure 2-17**). For instance, during the 1960s, most of the forests of China were cut and not replanted. Rapid and alarming soil erosion followed. The government then placed a high priority on reforestation and reversed the trend. Soil is needed to hold nutrients until plants require them. Similarly, we need the soil to filter and store clean water for plant growth and human and animal consumption. Plants take water from the soil and release water and oxygen to the air, which benefits humans and animals.

Figure 2-17 Soils that are located on steep slopes are vulnerable to severe erosion problems. Note the use of perennial grass in the eroded area between crops. The permanent plant cover binds the soil particles with roots, reducing the tendency to erode.

Courtesy of DeVere Burton.

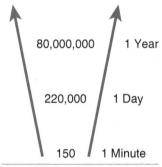

Figure 2-18 Earth's population is expected to reach 8 billion by 2025.

The number of human beings in the world is growing at the rate of 150 every minute; 220,000 a day; 80 million a year (**Figure 2-18**). At the current rate of growth, Earth's population will reach 8 billion by 2025. Can Earth sustain such population growth? Will humans find enough to eat? Will we learn to protect our environment, or will we destroy the system that supports life itself? Will we survive the competition of such population growth but sacrifice our quality of life? Might we, in fact, improve our quality of life by using our intelligence to stabilize and improve our environment?

Humans are the only living creatures who can choose to improve the living environments for themselves and other living organisms. The motivation to improve the environment is seldom present when people are hungry, however. As food production becomes more efficient, less of the total land area is required for the production of food. This allows some land areas and the living environments in the region to be preserved in their natural condition. For example, the national system of parks and monuments includes large areas where living environments support native plants and animals (**Figure 2-19**). Vast regions have been designated as "wilderness," and the type of human activity in these areas is strictly controlled to favor wild creatures. Humans also intervene in damaged environments to clean and restore them. Favorable economic conditions, in general, increase our ability and motivation to restore natural living environments and to preserve others.

Agriscience in Our Growing World

The keys to a prosperous future, indeed the bottom line for survival of the world's population, can be found in agriscience. Agriscience is the science of food production, processing, and distribution. It is the system that supplies fiber for building materials,

Figure 2-19 National parks and areas that are set aside as wilderness are intended to preserve the environment and the plants and animals that live there.
NPS photo by Jim Peaco.

rope, silk, wool, and cotton. It supplies medicines that are derived from animals, plants, and micro-organisms. It provides the grasses and ornamental trees and shrubs that beautify our landscapes, protect the soil, filter out dust and sound, and supply oxygen to the air.

Agri-Profile   Career Area: Environmental Management

Environmental management requires skills in observation, analysis, and interpretation.
© William Perugini/Shutterstock.com.

Management of the environment requires the attention of consumers as well as professionals. However, specialists in air and water quality, soils, wildlife, fire control, automotive emissions, and factory emissions all help maintain a clean environment against tremendous population pressures in many localities. Helicopter, airplane, and satellite crews gather important data for scientific analysis to help monitor the quality of our environment.

Individuals in environmental careers may work indoors or outdoors; in urban or rural settings; or in boats, planes, factories, laboratories, or parks. Careers range from laborer to professional. Environmental concerns are high on global agendas today as nations attempt to reduce global hunger and pollution and manage climate change.

According to a 2022 Feeding the Economy study by the American Farm Bureau, approximately 11 percent of food and agricultural jobs are in farming-related activities, while 89 percent of food and agricultural jobs are downstream in the supply chain.

It is one factor that permits the United States and other developed countries of the world to enjoy high standards of living. It is an element in a system that developing countries are using in their efforts to feed and clothe their bulging populations. People look to agriscience for the necessary technology to compete on a par with other nations in the twenty-first century. We must look to agriscience to maintain and improve our quality of life.

The United States is a major world supplier of food. It is also a major supplier of fiber for clothing and of trees for lumber, posts, pilings, paper, and wood products. The use of ornamental plants and acreage devoted to recreation has never been greater in the history of our country.

Changing Population Patterns

The United Nations has reported that more children than ever before are surviving to adulthood. It also indicates that on average, adults tend to live longer (**Figure 2-20**). Together, these trends mean more population growth and more pressure on the environment. Advancements in medical science and services have made good health and longer lives a reality, but only for those who can afford good nutrition and modern health services. Similarly, through agriscience, we have made substantial gains in providing food, fiber, and shelter for the world.

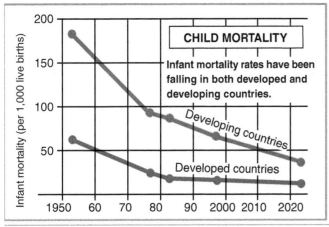
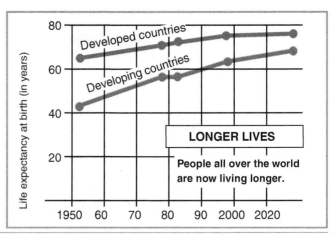

Figure 2-20 Worldwide child mortality rates and life expectancy estimates.
Based on data from the United Nations Fund for Population Activities.

In the past, individuals younger than 25 years constituted the world's largest population group. Children provided their labor in making the family living. Children and young adults were engaged in a nation's labor force and provided the manpower for armies. In most countries, the young respected their elders and provided for the needs of the elderly within the family.

The age profiles of people in developed countries are quite different from those of developing countries. Honduras has the traditional population pattern, with its largest number of citizens younger than 5 years. The number per age group then decreases to the smallest number, which occurs in the age group older than 80 years. When the Honduras population groups are displayed by sex in a bar graph, the graph takes the shape of a pyramid (**Figure 2-21**). Canada's pattern is slightly different. Its greatest population group is around the 20-year mark. Its graph reminds you of a Christmas tree, with its narrow bottom and cone appearance. The population of Sierra Leone illustrates the reduction of the 20- to 30-year-old population due to civil war. Sweden, a country known for its excellent health services and high survival rate, has age brackets that are about equal. The population graph for that country resembles a column.

China has about one-fifth of the world's population, yet it has been reasonably successful at feeding its population by keeping about 70 percent of its work force on farms. In contrast, less than 2 percent of the work force in the United States is necessary to operate the nation's farms.

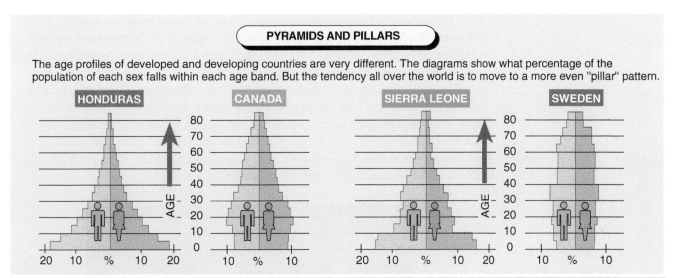

Figure 2-21 Age profiles and population patterns for developing and developed countries.
Based on data from the United Nations Fund for Population Activities.

Between 1980 and 2015, China implemented a policy whereby each couple was limited to one child. They called it the 4–2–1 policy. This means that extended families consisted of four grandparents, two parents, and one child. What would be the outcome if such a policy were strictly enforced for several generations? You would expect the pyramidal shape of China's population graph to change to the shape of a Christmas tree, and, in time, to an upside-down pyramid! What would be the implications of feeding a nation with a population of mostly elderly people? In 2015, the government replaced the policy with a two-child limit, and in 2021, the two-child limit was replaced with a three-child limit.

Impact of Agriscience

History records little progress in agriculture for thousands of years. Then, starting in the early 1800s, the use of iron spurred inventions that revolutionized agriculture in the United States, British Isles, and northern Europe (**Figure 2-22**). However, for most of the world, progress has been much slower. In some nations, government leaders are slow to implement agriscience because the nation initially would experience massive unemployment as machines displaced human labor.

Early American Agriculture

The first European immigrants to America were subsistence farmers whose most important work was to provide food and shelter for their families. They used the same farming methods as their parents had done in the lands from which they came. The National FFA parent organization was organized in Kansas City, Missouri in 1928. Considering that farm families in that era often included adult children who shared the workload, we often overlook the fact that the production of each farmer would feed approximately four people. Science, technology, and free enterprise have increased that output to 167 people in the first quarter of the twenty-first century.

Progress Through Agricultural Engineering

Mechanization through inventive engineering was an important factor in the United States' agricultural development. The change from 90 percent to less than 2 percent of = workers

Figure 2-22 The inventions of the 1800s brought revolutionary changes in agriculture in the United States and Europe.
Courtesy of USDA 01di1473.

being farmers evolved over a 200-year period. Machines helped make this possible. The old saying that "necessity is the mother of invention" suggests the relationship between an inventor's problem and the use of previously acquired skills to solve that problem. The solution is frequently a new device, machine, or process.

One of the most significant technologies to increase the efficiency of farm production was the generation and distribution of electricity to rural farming areas. Many of the labor-intensive jobs that were performed by hand 75 years ago are now performed by machines that are powered by electricity. Some examples include grain augers, milking machines, water pumps, fans, conveyor belts, power tools, and many other machines. The electric motor revolutionized the world (**Figure 2-23**).

Figure 2-23 Electric power has replaced human labor for many agricultural tasks.
Courtesy of DeVere Burton.

American Inventors

The United States is home to the inventors of many of the world's most important agricultural machines. In 1834, Cyrus McCormick invented the reaper, a machine to cut small grain (**Figure 2-24A**). Later, a threshing device was added to the reaper, and the new machine was called a combine. The reaper cut and bundled the grain in the field. Today, grain is harvested with a machine called a combine, which cuts and threshes in a single operation (**Figure 2-24B**). One modern combine operator can cut and thresh as much grain in one day as 100 individuals could cut and bundle in the 1830s.

Thomas Jefferson's invention, in 1814, of an iron plow to replace the wooden plow (which had been in use since 1794) was of great significance. In 1837, a blacksmith named John Deere experienced the frustration of prairie soil sticking to the cast-iron plows of the time. It became apparent that Jefferson's invention would not work in the rich prairie soils of the Midwest. Through numerous attempts at shaping and polishing a piece of steel cut from a saw blade, the steel moldboard plow evolved. That plow permitted plowing of the rich, deep prairie soils for agricultural production and launched the beginning of the John Deere Company.

In 1793, Eli Whitney invented the cotton gin. The cotton gin separated the cotton seeds from cotton fiber. This paved the way for an expanded cotton and textile industry. In 1850, Edmund W. Quincy invented the mechanical corn picker, which removed ears of corn from the stalks. During the same era, Joseph Glidden developed barbed wire, with

Figure 2-24 Cyrus McCormick's reaper (A) led to the development of the modern grain combine (B).

Courtesy Elmer Cooper. © Ivan Soto Cobos/Shutterstock.com

sharp points to discourage livestock from touching fences. This effective fencing permitted establishment of ranches with definite boundaries.

In 1878, Anna Baldwin invented a milking machine to replace hand milking (**Figure 2-25**). Today, robotic systems represent one of the newest developments for milking cows. Sensitive robotic arms, guided by electronic sensors, attach the milking apparatus to the udder of the cow without human intervention (**Figure 2-26**). This invention has the potential to greatly reduce the amount of labor required to produce milk products.

In 1904, Benjamin Holt invented the tractor, which became the source of power for belt-driven machines as well as for pulling.

Figure 2-25 The invention of the mechanical milking machine has greatly reduced the amount of human labor required to care for dairy animals.

Photo by Keith Weller. USDA ARS K8082-1.

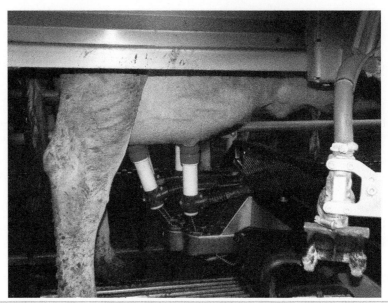

Figure 2-26 Modern robotics systems now make it possible for robotic arms to wash each cow in preparation for milking, attach milking equipment to the cow, and release her when the job is finished.
Courtesy of DeVere Burton.

Formation of Machinery Companies

Many of the early inventors worked alone or with one or two partners. They were agricultural mechanics, and as such, involved in agriscience. By the early 1900s, such inventors or other enterprising people had formed companies to produce agricultural machinery or process agricultural products. This helped make invention a continuing process. Successive inventions were used to improve earlier inventions and to develop new equipment and supplies to meet the needs of a changing agricultural industry.

The development of mechanical cotton pickers and corn harvesters greatly expanded the output per farm. Significant expansion of U.S. agriculture also resulted from the development of irrigation technology. Since the end of World War II, the mechanization of U.S. agriculture has moved at a breathtaking pace (**Figure 2-27**).

Figure 2-27 Since World War II, U.S. agriscience has progressed at a breathtaking pace.
© chantal de bruijne/Shutterstock.com.

Mechanizing Developing Countries

In the developing countries of the world, many engineers, teachers, and technicians have sought simple, tough, reliable machines to improve agriculture. In such countries, the United States' highly developed, complex, computerized, and expensive machinery does not work for long. Most countries do not have people trained for the variety of agriculture mechanics jobs that are needed to support U.S. agriculture.

A machine with rubber tires is useless if a tire is damaged and repair services are not available. Similarly, failure of an electronic device may cause expensive machines to become junk in the hands of an unskilled person in a country without appropriate repair facilities. This is the case in most developing countries in Central and South America, Eastern Europe, Asia, and Africa. For the developing nations of the world, other aspects of agriscience must become the vehicles for advancing agricultural productivity.

Bio-Tech Connection: Agriscience and Biotechnology

Biotechnology provides valuable mechanisms to improve life.
Photo by Scott Bauer. USDA/ARS K 5011-19.

Agriscience is heavily impacted by biotechnology, which addresses the continuation of life. In ornamental horticulture, one uses a myriad of plants to beautify the interiors of homes, businesses, and institutions. Such plants also consume carbon dioxide gas and supply oxygen for humans, animals, insects, and other living organisms. In outdoor settings, ornamental trees, shrubs, and turfgrasses beautify our yards, streets, parks, and other public areas. Our highways rely on plants to screen off oncoming traffic, provide living hedges, absorb sound, prevent soil erosion, and create a stimulating environment to keep motorists alert. Fruits, vegetables, grains, and forage crops of gardens, ranches, and farms provide the backbone of the world's food supply. Trees provide wood, paper, and other fiber products. The productive capability of plants has been greatly improved through the efforts of scientists, technicians, and growers.

Similarly, many animal species have been modified over the centuries by humans through domestication, selection, breeding, and care. Currently, biotechnology is improving the productivity of plants and animals and providing new foods and medicines to enrich our lives. Genetic engineering enables humans to modify and utilize microorganisms in our fight against harmful insects and other pests. Today, Earth is providing food, shelter, habitat, health care, and other essentials to more people than at any time in history. However, there is much malnutrition and starvation in the world. The causes tend to be rooted in deficiencies in government, national infrastructure, poverty, and lack of education. The technology of food production is believed to be capable of meeting the world food needs as modern technology is applied to agricultural production opportunities throughout the world.

Improving Plant and Animal Performance

Humans have improved on nature's support of plant and animal growth since they discovered that the loosening of soil and planting of seeds could result in new and better plants. Even before that discovery, they aided plant growth by keeping animals away from them until fruit or other plant parts edible to humans were harvested.

The human touch has permitted plants and animals to increase production and performance to the point where fewer people are needed to produce the food supply for the United States and other developed countries. Surplus food is exported to many other nations.

One of the remarkable occurrences of the twentieth century was the mechanization of agriculture. The twenty-first century is likely to be remembered for refining precision farming methods. The many technologies that were developed for the agricultural industry have contributed to larger farms. Many people have been displaced from their family farms because they were slow to adopt the new technologies and farming practices that were needed to make their farms more efficient. Many of these people have learned trades other than farming and have become productive citizens in other industries. Without the farming revolution of the last 70 years, our citizens would not be free to pursue other occupations. The U.S. space program is possible partly because our scientists do not have to produce their own food. The efficiency of U.S. farms has contributed to the freedom of our citizens to engage in many new and exciting occupations. These include the development of computers and other technologies that have resulted in the current "information age."

Application of Precision Technologies

Precision agriculture, also known as site-specific farming, combines the Global Positioning System (GPS) and Geographic Information Systems (GIS). This allows accurate information to be collected for each sector of a field. Such a high-tech system is capable of identifying the exact field location of weeds, insects, drought damage, soil problems, etc. These systems are also capable of recording crop yields for each sector of a field. When crop yields are known, precision application systems are used to place exact amounts of chemicals and fertilizers in the precise locations where they are needed to support maximum production. Site-specific farming is friendly to the environment in that residual amounts of fertilizers and chemicals can be controlled. When applications of fertilizers are based on yields, contamination of ground water and surface environments can be significantly reduced or eliminated.

Hot Topics in Agriscience | Precision Agriculture

Precision farming depends on precise information. This system uses electronic sensors associated with satellites, drones, aircraft, stationary positions, and other emerging technologies to detect exact information about each sector of a field. Such information may include soil factors such as soil type, fertility, moisture, and surface temperature. Plant health conditions such as disease outbreaks, competition from weeds, nutrient deficiencies, and chemical damage are all critical elements that affect productivity in crops. The status of each of these and other factors can be measured precisely in each sector of a field for any given moment in time. For example, such precision in measurement along with exact applications of plant nutrients, pesticides, water, and other inputs can be matched with the needs of the plants in each individual sector of a field.

Precision farming requires expensive machinery. It must be equipped with technologies that are capable of selectively applying exact amounts of crop inputs (such as fertilizer) according to the exact requirements of the plants that are growing in each particular sector of a field. Such precision has the advantage of avoiding chemical waste or pollution to soil and ground water due to over-application material. It also assures that applications of plant nutrients and other materials are adequate to promote healthy plant growth and production.

Improving Life Through Agriscience Research

Unlocking the Secrets of the Soybean

Americans have long appreciated the extensive research on the peanut done by the American scientist George Washington Carver. Carver is credited with finding more than 300 uses for the peanut. These include food for humans, feed for livestock, cooking fats and oils, cosmetics, wallboard, plastics, paints, and explosives.

Less known are the secrets of the soybean. The Chinese have known for centuries that the soybean is a versatile plant with many uses. Calling it the "yellow jewel," the Chinese are said to have grown the soybean 3,000 years ago. The strong flavor of the soybean itself is not appealing, but the bean is a legume and is nutritious. A legume is a plant that hosts nitrogen-fixing bacteria. These bacteria convert nitrogen from the air to a form that can be used by plants. They also convert atmospheric nitrogen to excellent sources of protein for humans and animals.

A Chinese scholar is believed to have first made tofu from soybeans in 164 bc. Tofu is a popular Chinese food made by boiling and crushing soybeans, coagulating the resulting soy milk, and pressing the curds into desired shapes. Today, tofu is a major food in the diet of China's huge population. Tofu contributes to a reasonably healthful diet. It can be fermented; marinated; smoked; steamed; deep-fried; sliced; shredded; made into candy; or shaped into loaves, cakes, or noodles.

Soybean oil is the world's most plentiful vegetable oil. It is first extracted from the soybean, and the material that is left is processed into a protein-rich livestock feed known as soybean meal. The components of the soybean are used for hundreds of items. These range from food products to lubricants, paper, chalk, paint, printing ink, and plastics (**Figure 2-28**).

Figure 2-28 The soybean is the world's most important source of vegetable oil, and it provides the basic materials for hundreds of products.
Courtesy of the American Soybean Association.

Baked Potatoes

Many improvements in our way of life can be traced to agriscience research. For instance, the U.S. Department of Agriculture developed many pest-resistant varieties of potatoes. A case in point is the work with the Katahdin, a popular potato variety of the 1930s. From the Katahdin, scientists developed the BelRus, a superior baking variety bred to grow well in the Northeast. In a similar manner, the Russet potato grown in the volcanic soils of Idaho and the Northwest has been improved through research. Selection of parent stock has increased its resistance to diseases and insects, resulting in greater yields.

Turkey for the Small Family

In your grandparents' time, Thanksgiving was probably observed by having all the relatives visit to consume the typical 30-lb. turkey. As families became smaller and more scattered, the need for such large birds decreased; but even people with small families liked turkey. The 30-lb. bird was too much, so the problem was to develop a breed of turkey that weighed 8 to 12 lb. at maturity (**Figure 2-29**). A solution was the Beltsville Small White turkey, named after the Beltsville Agricultural Research Center in Maryland, where the breed was developed. Further research and development have yielded meat animals with high yields of lean meat and less fat.

Figure 2-29 The Beltsville Small White turkey was developed to meet the needs of small families.
Courtesy of USDA/ARS.

The Green Revolution

During the 1950s, starvation was rampant in many countries of the world. A major question was: "Could the world's agriculture sustain the new population growth?" The solution was partly in the development of new, higher-yielding, disease- and insect-resistant varieties of small grains for developing countries. The result was the Green Revolution, a process whereby many countries became self-sufficient in food production in the 1960s by using improved plant varieties and proven management practices.

Cultivated Blueberries

Wild blueberries were enjoyed in early times when people had time to pick the tiny berries growing in the wild. But labor costs became too high to harvest such berries for sale. The solution was development of high-quality, large-fruited blueberry varieties from the wild. This started today's new and valuable cultivated-blueberry industry.

Nutritional Values

Until recently, animal and human nutrition were based on poor methods of feed and food analysis. The problem was the following: How can one recommend what to feed or what to eat if the content of food for humans and crops for livestock cannot be accurately determined? The solution was to develop detergent chemical methods for determining the nutritional value of **feedstuff** (any edible material used in the diets of animals). The procedures are now used widely throughout the world in both human and animal nutrition.

Biological Attractants

The use of chemical pesticides usually provides short-term solutions to many insect-control problems. However, it has become apparent that chemicals have some disadvantages and that additional means of control must be found. A partial solution was to discover chemicals that insects produce and give off to attract their mates (**Figure 2-30**). These chemicals are now produced in the laboratory. Laboratory production of these chemicals has permitted mass trapping of insects to survey insect populations for integrated pest-management programs.

Figure 2-30 Attractants are valuable nonpolluting chemicals that lure insects to traps, bait, or a system of control through sterility.
© iStock.com/GaryAlvis

Breakthroughs in Agriscience

Genetically Engineered Tomato

Calgene, an agricultural biotechnology firm in Davis, California, developed a bio-engineered tomato that resists rotting. The new tomato was developed by turning off the gene that causes the tomato to soften and rot. The new tomato lasts longer on the shelf at the grocery store, retains its flavor longer, and has proven superior in taste tests.

Natural Rubber Production

Scientists at the USDA Western Regional Research Center at Albany, California, have modified a scrubby bush called "guayule" by genetically engineering methods to produce up to 1,000 kg of rubber per hectare from the plant (an increase from 200 kg/hectare from native plants). The plant looks like sagebrush. This new technology makes it possible to produce a domestic supply of natural rubber in the United States.

Hot Topics in Agriscience: Genetically Modified Foods

Humans have been modifying the genes of plants and animals for centuries using a technique called **selective breeding**. Plants or animals that exhibit desired genetic characteristics are selected as the parents of the next generation. For example, if a scientist wanted a wheat plant that was resistant to a disease, they would plant a field of wheat and expose it to the disease. Then the researchers would find the plants that survived the disease and use them as parent stock for the next generation. This process is then repeated until a variety that has a high resistance to the targeted disease is found. This process is difficult and can take years.

Genetic engineering is a technique that allows scientists to physically put specific genes into the cells of a plant or animal without the rigorous process of selective breeding. As a result, genetic engineering is much faster than selective breeding. This technique has revolutionized agriculture. One common genetically modified crop is herbicide-resistant wheat. A gene that is resistant to the deadly effects of plant-killing herbicides is positioned into the DNA of the wheat. As a result, a farmer can spray an entire field with an herbicide, killing only the weeds and not the valuable crops. Genetic modification is a powerful tool that has resulted in more efficient production of a number of agricultural products.

This new advancement does not come without controversy. The United Kingdom, as well as other countries, placed bans on genetically modified crops. Activist groups, together with media, lobbied the governments of these countries to halt the sale of genetically modified products because of perceptions that the products are not safe for people to use. The controversy over genetically modified food is not isolated to Europe. In the United States, special interest groups continue to lobby the Food and Drug Administration (FDA) to require food labeling for all foods containing bioengineered ingredients. Legislation relating to labeling of genetically engineered food products has been introduced and approved in some of the states.

Bioengineered Designer Foods

By altering the genetic structure of food products, scientists have created new foods such as crispy vegetables, sweeter carrots, leaner meats, high-protein milk, longer-lasting melons, and healthier cooking oils.

Monoclonal Antibodies in Goats' Milk

Monoclonal antibodies are natural substances in blood that fight diseases and infections. Transgenic goats have been developed by inserting a gene into the goats' DNA, causing them to produce and secrete up to 4 g of the monoclonal antibody in each liter of milk. This level of antibody production is 10 to 100 times greater than traditional methods of production from cell cultures. This early work with transgenic goats produced an anticancer antibody.

Bio-diesel from Animal Fat

Excess animal fat (tallow) that is trimmed from the carcasses of meat animals is a low-value by-product of the meat-processing industry. A process has been developed that converts tallow to bio-diesel, a product much like the diesel fuel extracted from crude oil (**Figure 2-31**). The fat is heated to a liquid, followed by a purification process. The purified fat product is then mixed with methyl alcohol and a chemical catalyst. The bio-diesel that is produced from this process has approximately the same heating value and power potential as traditional diesel fuel, and it will burn in an ordinary diesel engine.

Figure 2-31 Forms of bio-diesel made from animal fat or plant oils show promise as a fuel for diesel engines.
Courtesy of DeVere Burton.

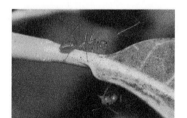

Figure 2-32 A new synthetic material has been identified for control of fire ants by increasing the ratio of drones to worker ants.
© injun/Shutterstock.com.

Figure 2-33 New varieties of exotic flowers known as impatiens have been developed.
© Michael Shake/Shutterstock.com.

Mastitis Reduced

The mastitis organism has always been a serious problem for dairy farmers. Mastitis is an infection of the milk-secreting glands of cattle, goats, and other milk-producing animals. The resulting loss of milk production adds millions of dollars yearly to the cost of milk in the United States. Recent research efforts resulted in the development of abraded plastic loops for insertion into cow udders. The procedure resulted in a reduction in clinical mastitis of 75 percent. The reduction in infections resulted in increased milk production, averaging nearly 4 lb. of milk per cow per day.

Fire Ant Control

Fire ants infest more than 367 million acres in 15 states in the United States (**Figure 2-32**). Their presence in the warmer climates of the world is a constant threat to the well-being of humans and livestock. A new synthetic control for fire ants increases the ratio of nonproductive drone ants to worker ants. This ratio change gradually weakens the colony and causes it to die.

Exotic Flowers

Horticulturists, gardeners, and hobbyists will be delighted with the new varieties of impatiens (a popular, easy-to-grow, summer-flowering plant) (**Figure 2-33**). Plant explorers introduced exotic new germ plasm, and plant breeders developed a new technique called ovuleculture to develop hybrids and new kinds of impatiens. A hybrid is the offspring of a plant or animal derived from the crossing of two different species or varieties.

Satellites and Nitrogen Gas Lasers

Nutrient deficiencies and other abnormalities in growing crops are not easy to detect from the ground. A deficiency occurs when a nutrient is not available in the amounts that are needed for optimum growth. New technology now permits the monitoring from satellites of deficiencies of iron, nitrogen, potassium, and other nutrients using nitrogen gas **lasers** (devices used to determine wavelengths given off by the plants). These wavelengths indicate the levels of various nutrients in plants. Crop damage due to insects, animals, chemical over-spray and other causes can also be detected using satellite imaging (**Figure 2-34**).

Figure 2-34 Infrared photos taken from satellites help diagnose problems such as ant infestations in fields and pastures.
Courtesy of USDA/ARS K3957-11.

Sugar Beet

The development of new varieties is a technique that has been used in agriscience for many decades to improve plant performance (**Figure 2-35**). A recent breakthrough has provided a sugar beet hybrid with a high ratio of taproot weight to leaf weight. The hybrid is known to yield higher sugar content per acre than previous varieties.

Another breakthrough in sugar beets is the release of a genetically engineered variety that causes the plant to be tolerant to the herbicide, Roundup. This makes it possible to control the weeds in a field without hand labor. Resistance to the new variety continues in the legal system, however, as advocacy groups continue to oppose the variety from being produced in the United States.

Rice Hybrids

On the other side of the world, Chinese agronomists developed hybrid rice. Hybrid rice is capable of yielding up to 40 percent more rice per acre than traditional varieties. The combination of hybrid semidwarf rice varieties, improved irrigation, and chemical fertilizers has increased world rice production from 240 million metric tons in the 1960s to 497.8 million metric milled tons in 2020.

A study of USDA publications reveals great numbers of improved varieties, new products, and superior processes that have been discovered or developed through agriscience research.

Figure 2-35 Sugar beet hybrids have improved sugar yields.
© luis c. jimenez del rio/Shutterstock.com

Agriscience and the Future

In 2021, the average American farmer was capable of producing enough food and fiber for approximately 167 people. Agriscience will become even more important in the next 100 years. As the world's population increases, it will require a highly sophisticated agriscience industry to provide the food, clothing, building materials, ornamental plants, recreation areas, and openspace needs for the world's billions. Americans will have to work more in the international arena as more countries become highly competitive in agriscience and as trade barriers are removed. Research and development will continue to play a dominant role as Americans lead the way in agriscience expansion in the future.

The USDA has developed the following mission statement to guide the agency: We provide leadership on food, agriculture, natural resources, and related issues based on sound public policy, the best available science, and efficient management.

The twenty-first century brings a new set of challenges to United States and world agriculture. The international business economy is a dominant factor in marketing agricultural products. A major share of U.S. agricultural production is now consumed in foreign countries, and this trend is expected to increase the volume of agricultural commodities that are sold outside U.S. borders. Canada and Mexico have become two of our biggest markets. They have also become successful competitors with the United States for a world market share in some agricultural commodity markets.

The future of U.S. agriculture will require that farmers become even more efficient in the production of food and fiber crops. Animal agriculture will depend on scientific improvements in production methods and in the genetic superiority of food animals to improve the quality of animal products. The efficiency with which they are produced must also improve for animal products to continue to demand a strong market share of the food supply in a world crowded with humans.

Chapter Review

Student Activities

1. Write the **Key Terms** and their meanings in your notebook.
2. Develop a bulletin board that illustrates the components of our environment.
3. **SAE Connection:** Collect newspaper articles that describe environmental problems in your community.
4. Prepare a two- or three-page paper describing a good environment in which to live. Include factors such as home, community, air, water, cleanliness, wildlife, plants, and animals.
5. Ask your teacher to invite a public health official to your class to discuss health problems in the community and how they could be reduced by improving the environment.
6. Draw a chart that illustrates some relationships among plants, animals, trees, soil, water, air, and people.
7. Look up three prominent U.S. inventors and describe the events that led to the inventions that made them famous.
8. Make a model of one of the machines that strongly influenced agriscience development.
9. Make a collage depicting some important discoveries, inventions, and developments in agriscience.
10. In groups of three or four students, develop an idea for a new piece of farm equipment that has potential to improve the harvest or production of a locally grown crop, farming technique, or ranching practice.

National FFA Connection — The History of the FFA

Courtesy of National FFA; FFA #148

The organization now known as the National FFA has been associated with advances in agriculture since its founding.
© AlexanderZam/Shutterstock.com.

The Future Farmers of America (FFA) was founded in 1928 to prepare farm boys for careers in production agriculture. Later, the name changed to The National FFA Organization, expanding the vision to prepare young men and women for careers in agriculture and natural resources. Today, the FFA has more than 800,000 student members in nearly 9,000 chapters. Visit the FFA website to view an illustrated timeline of its history. Select two milestones you consider important and share your reasons as you discuss them with your class.

Inquiry Activity: Advances in Farm Technology

Courtesy of DeVere Burton.

Watch the *Advances in Farm Technology* video available on MindTap. Then, use a T-chart like the one shown below to take notes as you watch again.

Advances in Farm Technology	
Category	**Example(s):**
Farm Machinery	

- In column 1, list the four main categories of advances, or improvements, in technology. For example, the first category is *Farm Machinery*.
- In column 2, list examples of advances for each category.
- Then, add a few more details about one or two of the examples.
- **Tip:** Create a print or digital portfolio where you can store copies of your written work. This will give you a source of research topics for future assignments.

Check Your Progress

Can you…

- describe the conditions of desirable living spaces?
- discuss the influence of climate on the environment?
- compare the influences of humans, animals, and plants on the environment?
- examine the problems of an inadequate environment?
- identify significant world population trends?
- identify significant historical developments in agriscience?
- state practices used to increase productivity in agriscience?
- identify important research achievements in agriscience?
- describe future research priorities in agriscience?

If you answered "no" to any of these questions, review the chapter for those topics.

By studying today, you are planting the seeds for tomorrow's success!

Chapter 3
Biotechnology

Objective
To examine elements of biotechnology

Competencies to Be Developed
After studying this chapter, you should be able to:
- define biotechnology, DNA, and other related terms.
- compare methods of plant and animal improvement.
- discuss historic applications of biotechnology.
- explain the concept of genetic engineering.
- describe applications of biotechnology in agriscience.
- state some safety concerns and safeguards in biotechnology.

Key Terms

biotechnology
improvement by selection
selective breeding
genetics
heredity
genes
generation

progeny
deoxyribonucleic acid (DNA)
nucleic acid
base
chromosome
gene
gene splicing

recombinant DNA technology
gene mapping
clone
genetic engineering
bovine somatotropin (BST)
porcine somatotropin (PST)

What is **biotechnology**? The Greek root word *bio* means "life;" biotechnology is the application of living organisms to technology, such as the use of microorganisms, animal cells, plant cells, or components of cells to produce products or carry out processes. In agriculture, this ranges from the use of traditional breeding methods to the genetic engineering of pest-resistant crops. A division of biology, biotechnology is where science and engineering meet.

The impact of biotechnology on agriscience has been significant. With its ever-expanding range of tools, it has delivered benefits to farmers and consumers through unprecedented advancements in plant and animal improvement, pest control, environmental preservation, and life enhancement. However, there are also risks that this new power over life processes could lead to unmanageable consequences in careless, uninformed, or criminal hands. Therefore, governments, scientists, agencies, corporations, and individuals have moved cautiously to pursue new benefits through biotechnology in tandem with safety considerations.

Historic Applications of Biotechnology

Figure 3-1 Many foods owe their texture and taste to microorganisms, such as yeast and bacteria.
Courtesy of USDA/ARS #K36072-2.

Despite its futuristic connotations, biotechnology actually has a long history. Living organisms have been used for centuries to alter and improve the quality and types of food consumed by humans and animals. Examples include the use of yeast to make bread rise, bacteria to ferment sauerkraut, bacteria to produce dozens of types of cheeses and other dairy products, and microorganisms to transform fruit and grains into alcoholic beverages (**Figure 3-1**). Similarly, green grasses and grains have been stored in airtight spaces and containers, such as silos, where bacteria convert sugars and starches into acids. The acids provide a desirable taste and protect the feed from spoilage by other microorganisms. The converted feed is called *silage* (**Figure 3-2**).

Improving Plant and Animal Performance

Improvement by Selection

History documents the domestication of the dog, horse, sheep, goat, ox, and other animals thousands of years ago. Improvement by selection soon followed. **Improvement by selection** means picking the best plants or animals for producing the next generation

Figure 3-2 Silage often consists of grains and/or green plant material preserved by the action of bacteria in an airtight environment.
© Wallentine/Shutterstock.com. © Frontpage/Shutterstock.com.

(**Figure 3-3**). As people bought, sold, bartered, and traded, they were able to get animals that had desirable characteristics, such as speed, gentleness, strength, color, size, and milk production. By mating animals with characteristics that humans preferred, the offspring of those animals would tend to exhibit the characteristics of the parents and further intensify the desired characteristics. Whether by accident or through a very basic understanding of heredity, the owner was practicing **selective breeding**, or the selection of parents to get desirable characteristics in the offspring.

The chariot armies of the Egyptians and Romans, the might of the Chinese emperors, the speed of the invading barbarians into northern Europe, the strength of mounts carrying armored knights into battle, and the proud Bedouins of the desert all provide convincing testimony to early successes at breeding horses for specific purposes.

Figure 3-3 Improvement by selection means picking the best plant or animal specimens for breeding purposes.
Courtesy of National FFA; FFA #162.

Improvement by Genetics

An Austrian monk named Gregor Johann Mendel is credited with discovering the effect of genetics on plant characteristics. **Genetics** is the science of heredity. **Heredity** is the transmission of characteristics from an organism to its offspring through genes in reproductive cells. **Genes** are components of cells that determine the individual characteristics of living things. Mendel experimented with garden peas. He observed that there was definitely a pattern in the way different characteristics were passed down from one generation to another. **Generation** refers to the offspring, or **progeny**, of common parents.

In 1866, Mendel published a scientific article reporting the results of his experiments. He had discovered that certain characteristics occurred in pairs, for example, short and tall in pea plants. Furthermore, he observed that one of those characteristics seemed to be dominant over the other. If tall was the dominant characteristic, then tall plants crossed with tall or short plants produced mostly tall plants, but some plants would still be short. It was observed that the short characteristic could be hidden in tall plants in the form of a recessive gene. Such recessive genes could not express themselves in the form of a short plant unless both genes in the plant cells were the recessive genes for shortness. He also observed that short plants crossed with short plants always had short plants as offspring.

Figure 3-4 Mendel's extensive experimentation, observation, and recordkeeping provided the foundation for the modern science of genetics.

Courtesy of National FFA; FFA #199.

This happened because there were no tall characteristics in either parent to dominate the characteristic of the offspring. Mendel's work provides an excellent example of the power of the written word. His discoveries and conclusions would have been lost if they had not been recorded. The usefulness of his discoveries was not recognized until long after his death. In 1900, other scientists reviewed his writings and built upon the observations and conclusions he had reported. Today, biologists credit his work as being the foundation for the scientific study of heredity (**Figure 3-4**). Principles of heredity apply to animals as well as plants. More information on the principles of heredity can be found in subsequent chapters of this text.

DNA—Genetic Code of Life

Of the estimated 300,000 kinds of plants and more than 1 million kinds of animals in the world, all are different in some ways. Conversely, plants and animals have certain similar characteristics that lend themselves to classification and permit prediction of characteristics of offspring by viewing the parents—that is, the individual fertilized cell, called the embryo, contains coded information that determines what that cell and its successive cells will become. The coded material in a cell is called DNA. DNA is the acronym, or abbreviation, for **deoxyribonucleic acid**.

It is believed that a universal chemical language or code unites all living things. It was observed in the early 1800s that all living organisms are composed of cells, and that cells of microscopic organisms, as well as larger plants and animals, are basically the same. In 1867, Friedrich Miescher observed that the nuclei of all cells contain a slightly acidic substance. He named the substance **nucleic acid**. Later, the name was expanded to deoxyribonucleic acid, or DNA. DNA in all living cells is similar in structure, function, and composition and is the transmitter of hereditary information. A gene is a small section of DNA that is responsible for a specific trait. Chromosomes are rod-like structures made of DNA and other substances that hold genes.

Agri-Profile Career Area: Genetic Engineering

Using a DNA probe, an animal physiologist examines film showing the gene patterns of various animals.

Courtesy of USDA/ARS #K-1968-13.

Genetic engineering cuts across many fields of endeavor. Procedures for genetic modification of organisms have been developing for more than a decade. Biologists, microbiologists, plant breeders, and animal physiologists are some examples of specialists who might use genetic engineering in their work. The work settings include the field, laboratory, classroom, and commercial operations.

People who work in genetic engineering usually have advanced degrees, and they are highly specialized in a narrow area of research, such as cellular biology. Other career opportunities include applied research in areas such as weevil control in small grains, nutrient requirements of small grains, or reproductive problems in dairy cattle. Others may be crop, animal, or pest control technicians who help manage the plants or animals that are the subjects of research. Still others may work in laboratories and devote most of their lives to analysis and observation.

Because genetic engineering is a relatively new field and the applications are so numerous, opportunities are expanding as the field develops.

Science Connection | Trait Predictability

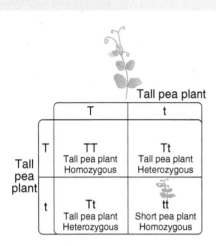

A Punnett square shows all possible outcomes for a genetic cross between two parent organisms.

A *trait* is a characteristic of an organism. Mendel discovered that parents pass their own traits to their offspring. In this type of reproduction, each parent contributes one half of the genes. The result of parental combinations can be shown and predicted in a Punnett square. The Punnett square represents the results of some of Mendel's experiments. An *allele* is one of the forms in which gene pairs can occur. In this example, there are three possible alleles. (1) "TT" represents a pea plant that is tall. Homozygous means that an organism has the same alleles for a given trait. (2) "tt" represents a homozygous pea plant that appears short. (3) "Tt" represents a heterozygous pea plant that is tall. The gene for "tall" is expressed because "T" is dominant over "t." Because of this dominance, both "TT" and "Tt" will result in a tall plant, and "tt" will result in a short plant. Heterozygous means that an organism has different alleles for a given trait. Because the alleles for this trait are heterozygous, or different, this organism is called a hybrid. The Punnett square cross is done one box at a time; by taking the parental alleles to the left and top of each box, the offspring allele combinations are found inside each of the four boxes.

The Structure of DNA

DNA occurs in pairs of strands intertwined with each other and connected by chemicals called **bases**. The pairs of DNA strands may be likened to the two sides of a wire ladder. The bases may be likened to the rungs of that wire ladder. The different bases are composed of only four chemicals: (1) **adenine**, (2) **guanine**, (3) **cytosine**, and (4) **thymine**. The first letters of each name of the bases—A, G, C, and T—have become known as the genetic alphabet of the language of life (**Figure 3-5**).

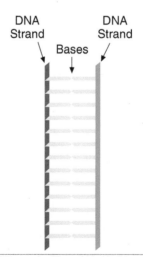

Figure 3-5 The components of DNA may be likened to a ladder with very close rungs.

If one end of the wire ladder is held while the other end is twisted, the resulting shape would be called a double helix. This is the shape of DNA strands in a cell. Two strands of DNA and the bases between the strands compose a specific gene (**Figure 3-6**).

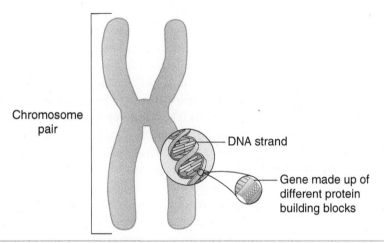

Figure 3-6 A **chromosome** is a structure containing the genetic information of a cell. A **gene** is a small segment of DNA located at a specific site on a chromosome that influences or controls a hereditary trait. There are thousands of genes on each strand of DNA.

The order or sequence of the bases between the DNA strands is the code by which a gene controls a specific trait. Therefore, each rung with its accompanying side pieces of DNA constitutes a gene containing the genetic code to a single trait (**Figure 3-7**). The genetic material in the cells of a given microbe, plant, animal, or human can be isolated and observed. The trait or traits that a given gene controls can be identified, and the combination of genes that influence a single trait can be determined.

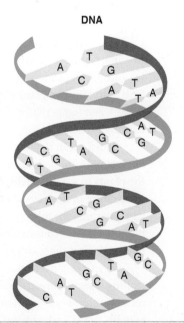

Figure 3-7 DNA is a double-stranded helix. The two strands are connected by the chemical based A, C, G, and T. A pairs with T; G pairs with C.

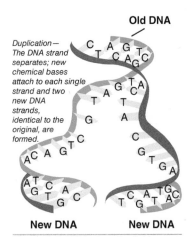

Figure 3-8 DNA strands divide, and bases attach themselves to the new strands to form identical genes for new cells.

Some examples of individual traits are hair color, tendency for baldness in humans, height of plants at maturity, and tendency of females to have multiple births in contrast with single births.

As a cell divides, the DNA strands separate from each other and create duplicate strands to go to each of two new cells. Therefore, the genetic codes are duplicated and passed on from old cells to new cells as growth occurs and individuals reproduce (**Figure 3-8**).

Scientists can now identify an individual gene carrying specific genetic information and replace it with a gene containing a different genetic code. By doing so, a given characteristic or performance can be altered in the organism. For instance, plants that are susceptible to being eaten by certain insects may be altered so they will have resistance to that insect. The process of removing particular DNA segments and inserting new genes into a DNA sequence is called gene splicing, or recombinant DNA technology. The process of identifying and recording the location of specific genes on chromosomes is called gene mapping.

Scientists have a working knowledge of how genetic information is stored in a cell, duplicated, and passed on from cell to cell as cells divide and new organisms are formed. Furthermore, the process of transmitting genetic codes from parents to offspring and from parent to clone is common scientific knowledge. A clone is an exact duplicate of something. A major breakthrough was made in the early 1980s as scientists developed the process of genetic engineering. Genetic engineering is the movement of genetic information in the form of genes from one cell to another.

Improving Plants and Animals

Scientists have learned to modify plants, animals, and microbes for specific purposes by manipulating the genetic content of cells. This procedure permits more choices for the researcher and more rapid observation of results. With the manipulation of cellular material, the scientist can now alter the characteristics of microorganisms, as well as those of larger plants and animals. This new capability has some amazing implications for human efforts to improve the quality of life.

Plant Applications

In 1988, California scientists made the first outdoor tests of a product called ice-minus. Ice-minus is a product containing bacteria that have been genetically altered to retard frost formation on plant leaves. Synthetic chemicals are now available to protect fruit crops when temperatures fall 4 to 6 degrees below what would normally interfere with the fruiting process.

Genetic material in plants themselves can also be manipulated to modify certain traits. For example, the Colorado potato beetle is a highly destructive insect that can destroy all the plants in a potato field. A gene has been identified that causes potato plants to produce a substance in the leaves that is toxic to potato beetles, but which does not affect humans who consume the tuberous roots. New potato varieties are created by inserting this gene into the DNA of commercial potato varieties. The new potato varieties are toxic to the beetles that eat the leaves of the modified plants.

Genetic engineering and other forms of biotechnology hold great promise in controlling diseases, insects, weeds, and other pests. The plants that we nurture for food, fiber, recreation, and preservation can benefit from biotechnology as well as humans. Potential benefits to the environment include less frequent use of chemical pesticides and increased use of biological controls (**Figure 3-9**).

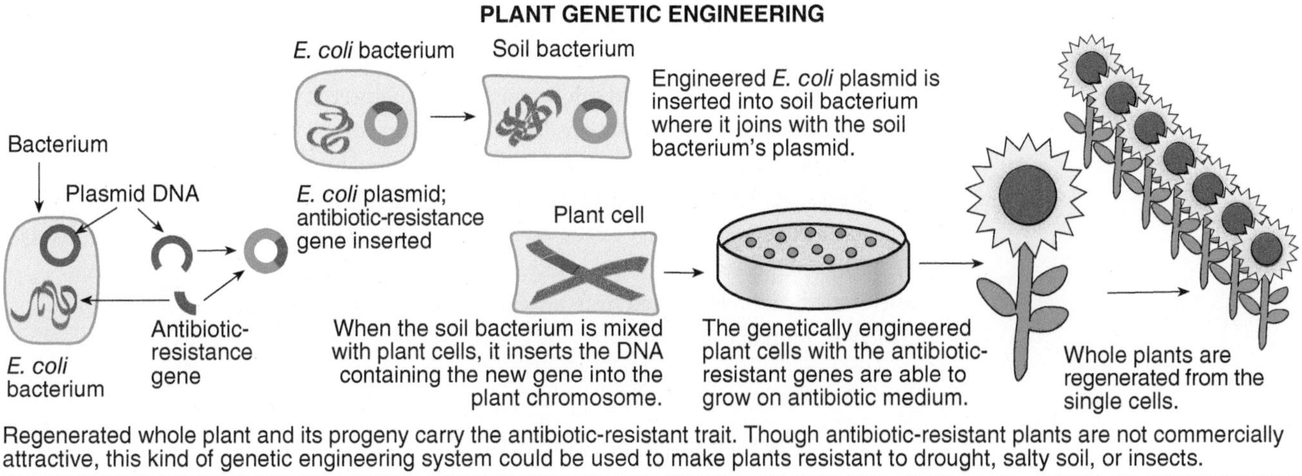

PLANT GENETIC ENGINEERING

Regenerated whole plant and its progeny carry the antibiotic-resistant trait. Though antibiotic-resistant plants are not commercially attractive, this kind of genetic engineering system could be used to make plants resistant to drought, salty soil, or insects.

Figure 3-9 Plant genetic engineering is widely used in the battle against diseases, insects, weeds, and other pests.

Animal Applications

In animal science, the hormone **bovine somatotropin (BST)** has long been known for its stimulation of increased milk production in cows. However, it was not available for commercial use until bacteria were altered to produce the hormone at a reasonable cost. Another example of hormone production by means of genetically altered bacteria is the use of an animal hormone called **porcine somatotropin (PST)**, which increases meat production in swine.

Each time humans or animals are exposed to a disease, some individuals do not become infected. Sometimes an entire population is found to be resistant to a disease that is highly contagious to other populations of the same species. In some instances, disease resistance is due to a single gene that has mutated or changed. It is now possible to identify the location of a resistant gene on a chromosome and to isolate it. This new genetic material can be transferred successfully to the chromosomes of an organism that is susceptible to the disease. Such a genetically altered individual is capable of passing the disease resistance to its offspring.

Hot Topics in Agriscience — Transgenic Animals: A New Kind of Farming

Transgenic animals are those that have been modified using genetic engineering methods to express genes not naturally found in the animal. Because the new gene is inserted into the chromosomes of the transgenic animal, it may be passed on to its offspring. Insulin and growth hormone are some of the first products for human use to be produced in this way. Most of the medical products produced by transgenic animals consist of proteins that are separated from milk or blood. These products include Human Protein C, an anticlotting protein that dissolves blood clots in humans. Other products include hemoglobin, a blood substitute produced in transgenic pigs, and Factors VIII and XI, which cause clotting in human blood to stop severe bleeding in people who have disorders such as hemophilia. All of these drugs are tested to ensure they are safe, and they offer hope for medical breakthroughs of even greater magnitude. The transgenic animal becomes a "living drug factory" that produces human proteins in milk or blood.

Solving Problems with Microbes

Microscopic plants and animals lend themselves well to genetic engineering. These microbes reproduce quickly and can be genetically engineered to produce products needed by other plants, animals, and humans. For example, one of the first commercial products made by genetic engineering was insulin, a chemical used by people with diabetes to control their blood sugar levels. Previously, insulin was available only from animal pancreas tissue. It was in short supply and very expensive. However, a bacterium called *Escherichia coli* was genetically engineered to produce insulin for human use (**Figure 3-10**). This important diabetes medication is isolated from a solution containing the engineered bacteria and purified for human use.

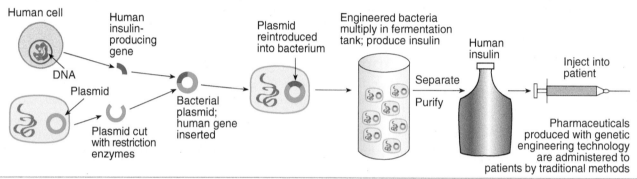

Figure 3-10 As a result of genetic engineering of bacteria, insulin is now readily available and relatively inexpensive.

Waste Management

Environmental pollution and the elimination of waste products from home, business, industry, utility, government, military, and other sources has become a major problem throughout the world. Landfills are becoming full, and pollutants from garbage are causing problems, such as leakage into the groundwater. Old dump sites are creating new problems, waste is piling up, and sewage and chemical disposal is a constant problem. Large floating masses of garbage pollute the world's oceans. Today, over 94 percent of solid waste in U.S. towns and cities is recycled or composted compared to 6 percent in 1960. However, no matter what we do in the United States, waste management is an international problem that must be solved by the nations of the world.

Biotechnology can help solve such waste disposal problems. Genetically altered bacteria are used to feed on oil slicks and spills, transforming this serious pollutant into less harmful products. Similarly, bacteria have been developed that are capable of decomposing or deactivating dioxin, polychlorinated biphenyl (PCB), insecticides, herbicides, and other chemicals in our rivers, lakes, and streams. Bacteria are capable of converting solid waste from humans and livestock into methane fuel that is used to generate electricity or to heat buildings.

Although great progress has been made in pollution reduction in some areas, pollution is still one of the world's greatest problems. However, biotechnology has brought about some spectacular breakthroughs in the use or decomposition of waste materials (**Figure 3-11**).

Figure 3-11 Biotechnology is used to modify bacteria, making it possible for them to break down crude oil. These bacteria have become important in reducing the harmful effects on the environment when oil or chemical spills occur.

Courtesy of USDA #CS 336.

Bio-Tech Connection: The Cloning of a Mule

One of the most remarkable scientific accomplishments of the twentieth century was the successful cloning of "Dolly" the sheep. Dolly was cloned in the laboratory from a single cell obtained from the udder of another sheep. A clone is an exact genetic copy of another living organism, so Dolly was genetically identical to the sheep from which the cell was obtained. Now scientists at the University of Idaho have cloned a mule named Idaho Gem. He is the first clone from the horse family. He also is a clone of a hybrid animal that is incapable of reproduction. This particular mule is cloned from the tissue of a fetus of the same parents as a champion racing mule named Taz. The DNA from his mule heritage was fused with an egg cell from a mare of the horse species, and a mare carried him through pregnancy and nurtured him as a newborn foal. In addition, two more identical clones were born later. Idaho Gem and one of his identical brothers, Idaho Star, raced successfully on the professional mule racing circuit.

What is the significance of this and other cloning events? It is now possible to produce a new generation of animals, with each animal identical to the most productive animal in the herd. Imagine having a whole herd of cows just like your best cow. Confusing, perhaps? Even their spots would be the same. You might have to give each animal a permanent microchip (or other identification) just to identify it.

Idaho Gem is the first mule ever cloned. His DNA came from a male champion racing mule, and his surrogate mother is a mare.

© Getty Images/Handout/Getty Images News/Getty Images.

Safety in Biotechnology

To sustain benefits and mitigate risks, federal and state governments monitor biotechnology research and development very closely. Much anxiety has been expressed about the perceived dangers of genetically modified organisms. Therefore, appropriate policies, procedures, and laws have been developed as biotechnology has evolved. Many of these regulations have been developed by the Environmental Protection Agency (EPA). Research priorities and initiatives require discussion and interaction by scientists, government agencies, and other authorities. Products are tested in laboratories, greenhouses, and other enclosures before being approved for testing outdoors and in other less controlled environments. Even then, outdoor tests are first conducted on a small scale in remote places under careful observation. Under these conditions, the efficiency, safety, control, and environmental impact of new organisms are determined. If the new organism poses an unmanageable threat, it can be destroyed.

Customer resistance to new food products developed through biotechnology has been demonstrated since the first of these foods arrived at supermarkets. For example, some customers demanded that milk from cows treated with the biotech product BST be labeled to distinguish it from other milk. Some believed that it should not be marketed at all. Despite assurances from the Food and Drug Administration (FDA) that no differences exist in comparisons of BST milk and milk from untreated cows, some customer resistance to BST persists. A few milk processors initially joined the resistance movement because they were afraid that their products might be boycotted by consumers. In some instances, these processors refused milk shipments from dairy farms that treated their cows with BST. However, biotechnology is rapidly becoming an important part of our daily lives. Many of its potential benefits have already been realized, and most people believe that we have only scratched the surface. With proper safeguards, we can look confidently to a bright future in this emerging field. Many other applications of biotechnology are cited throughout this text.

Both in the United States and in other nations. Some nations that refused bioengineered products are no longer doing so. However, there remains an active coalition of citizens and organizations that is actively attempting to place regulatory controls on genetically modified food products.

Ethics in Biotechnology

Ethics is a system of universal principles that define actions as right or wrong. Encompassed within that general definition is a wide range of norms, from documented legal and professional codes of ethics to the ideals, values, and standards of groups and individuals. While reflection on values and practices in agriculture has a history as ancient as agricultural itself, agricultural ethics as a field of formal study did not emerge until the late twentieth century.

Today, scientists, academics, policymakers, and others around the world recognize the critical role of systematic thinking about values and norms associated with the food system. A widely goal of agricultural ethics is to develop clear, comprehensive, and universal standards for determining right and wrong actions and policies. Among the ethical issues under discussion are several that apply directly to agricultural biotechnology, including but not limited to food safety, environmental impact and sustainability, global food security, and concerns about the implications of cloning and genetic engineering for human life.

In short, the ability to manipulate the genetics of living organisms raises important ethical questions about how the technology should be used.

It seems appropriate that a discussion of ethics should be part of the biotechnology revolution that is occurring. Such a discussion would help scientists and consumers decide how ethical issues related to biotechnology should be handled. At the very least, we can expect new laws to be passed and courtroom decisions to be rendered on the basis of ethics in biotechnology.

Science Connection: "Fingerprinting" Organisms

DNA analysis makes "fingerprinting," or individual identification, possible in all organisms.
Courtesy of USDA/ARS #K-4767-1.

USDA microbiologists at the National Animal Disease Center in Ames, Iowa, have been able to crack the mysteries of some of the worst known cases of food poisoning in North America. DNA matching, or "fingerprinting," is used to link (1) the persons who became ill, (2) the contaminated food that caused the poisoning, (3) the place where the food had been contaminated, and (4) the materials from which the food poisoning organisms had originated and spread. Food poisoning is a life-threatening condition resulting from eating food containing toxic material produced by bacteria under unsanitary conditions. *Salmonella, Campylobacter, Staphylococcus aureus, Clostridium perfringens, Vibrio parahaemolyticus, Listeria monocytogenes, Bacillus cereus,* and enteropathogenic *E. coli* are the most common bacteria to infect food.

Food-borne illnesses, though rare, have been known to be deadly. According to the U.S. Centers for Disease Control and Prevention, up to 48 million people get sick each year from contaminated food. Of these, approximately 228,000 are hospitalized, and around 3,000 die from pathogen-caused food-borne illnesses. Not all of these instances are traced back to farms and processors; a significant number occur because food is not properly refrigerated or prepared at home. Leaving food out on the counter overnight or failing to cook it adequately gives toxin-forming organisms a head start.

Chapter Review

Student Activities

1. Write the **Key Terms** and their meanings in your notebook.
2. Read an article in an encyclopedia or other reference on the process of genetic engineering.
3. Report your findings from Activity 2 to the class.
4. Make a collage depicting some important discoveries, inventions, and developments in biotechnology.
5. Form a discussion group to explore the benefits and hazards of biotechnology.
6. Arrange for a resource person to speak on the ethical and moral issues surrounding developments in biotechnology.
7. Organize a class debate on the ethical and moral issues regarding research and the use of new discoveries in biotechnology.
8. **SAE Connection:** Explore whether the career of an agricultural geneticist is right for you. Identify SAE programs of interest by exploring genetic engineering in agriscience using online resources.

National FFA Connection — Agricultural Issues Leadership Forum

Courtesy of National FFA; FFA #148

Leadership development events foster skilled-decision making.
© Gustavo Frazao/Shutterstock.com.

A critical element in career preparation is to study controversial issues with open minds, learning to understand the pros and cons of proposed actions. The Agricultural Issues Forum Leadership Development Event (LDE) challenges students to seek out and understand fundamental issues and consult with industry professionals to discover and evaluate possible solutions. Participants are organized into teams as they communicate their understandings of conflicting viewpoints through questions and portfolios to a panel of judges.

With your classmates, plan a debate. Work as a team to identify topics in this chapter that are likely to prompt opposing viewpoints such as genetically engineered animals. Write a resolution that begins: *Team A agrees that... Team B disagrees that...* Next, form two teams. Research the issue and share your findings as a team. Your instructor can advise you on how to cite industry experts. Then, prepare and practice your arguments. Invite the rest of the class to moderate the debate.

Inquiry Activity: Genetics in Agriscience

Courtesy of USDA.

Replay the *Agriscience Research* video available on MindTap.

- Then, apply what you have learned by listing 3-6 plants that are familiar to you. For example, your list might include food plants, native herbs, nonnative grasses, cover crops, and so on.

Next, answer these questions:

- If you could engineer the DNA of one of these plants to make it pest-resistant, which plant would you choose? Why?
- Which plant would you not want to make pest-resistant? Why?
- How has learning about genetic engineering affected your ideas about genetically modifying food from plants and animals?

Check Your Progress

Can you...

- define biotechnology, DNA, and other related terms?
- compare methods of plant and animal improvement?
- discuss historic applications of biotechnology?
- explain the concept of genetic engineering?
- describe applications of biotechnology in agriscience?
- state some safety concerns and safeguards in biotechnology?

If you answered "no" to any of these questions, review the chapter for those topics.

Tip: Reviewing the chapter headings and subheadings can help you recall information.

Unit 2
Agriscience as a Career Choice

On farms and beyond, career opportunities in agriculture cover an astonishingly wide range. Today's agricultural industry has strong demand for educated workers with specialized and general training in science, engineering, business and financial management, production, renewable natural resources, communications, and other knowledge fields. The U.S. Department of Agriculture (USDA) reports show annual job growth related to agricultural sectors— 22 million in 2019—continues to outpace the thousands of students who graduate with degrees in agriculture each year.

There is no lack of challenging and rewarding careers.

Futurists predict that over the next decades, the most important discoveries in genetic engineering will be made in agriscience. In agriscience, you could help develop plants and animals that grow better and more efficiently, even in adverse situations. As a nutritionist or food scientist who studies the links among food, diet, and health, you might work to ensure that new convenience foods are nutritious and healthy; or you might design a new way of preserving, processing, or packaging food. Agriscience educators teach in high schools, colleges, and universities, and may also work for land-grant universities as extension educators. The objective of a career in agriscience education is to teach new and proven scientific practices to students and farm families.

The business of agriculture is another key area of career opportunity. As a financial manager, you might work for a bank or a credit agency as an agricultural loan officer or for a company that sells supplies to producers and growers. As credit manager, loan officer, financial analyst, or marketing specialist, you could utilize your knowledge of business and finance, as well as production, processing, or distribution.

Farmers, ranchers, timber producers, and other growers are the foundation of the food and fiber system. To join their ranks is to fulfill an essential role, one that is critically important to the very future of our planet. This also holds true for career seekers who love the outdoors and who work as foresters, range managers, game managers, fish and wildlife managers, park rangers, and crop consultant. Like other aspects of our world that have been impacted by science and technology, the food, fiber, and renewable natural resources system is changing rapidly. Because it contributes to, and is strongly influenced by, trade and consumer lifestyles, the system must adjust continuously. To do so, it depends on qualified professionals.

In short, with appropriate training and experience, you may discover that any of these careers could be yours.

Chapter 4

Career Options in Agriscience

Objective
To survey the variety of career opportunities in agriscience, observe how they are classified, and consider how you can prepare for careers in agriscience.

Competencies to Be Developed
After studying this chapter, you should be able to:
- define agriscience and its major divisions.
- describe the opportunities for careers in agriscience.
- compare the scope of job opportunities in farm and off-farm agriscience jobs.
- list activities in middle school, high school, technical college/university to help prepare for agriscience careers.
- identify resource people for obtaining career assistance in agriscience.

Key Terms
production agriculture
agricultural processing, products, and distribution
horticulture
forestry
agricultural supplies and services
agricultural mechanics
profession
agriscience professions

Chapter 4 Career Options in Agriscience

> **How Much Is One Trillion Dollars?**
>
> You can count $1 trillion ($1,000,000,000,000) by using the following procedure:
>
> - One dollar bill every second
> - Sixty bills per minute
> - Thirty-six hundred bills per hour
> - Eighty-six thousand per day
> - Thirty-one million five hundred thirty-six thousand per year
> - And continue counting for thirty-one thousand seven hundred and ten years!

Figure 4-1 The agricultural industry in the United States has assets in land and equipment of approximately $2.6 trillion ($2,600,000,000,000).

Life is sustainable without many of our modern conveniences, but not without food. An adequate supply of suitable food and other products of the soil, air, and water is essential to life. This includes food for nourishment, fiber for clothing, and trees for lumber. Less obvious are alcohols for fuel and solvents, oils for home and industry, and oxygen for life itself. The industry that provides these vital basic commodities is agriculture. American agriculture is the world's largest commercial industry, with assets of $2.6 trillion in 2021 (**Figure 4-1**).

Agriscience includes all jobs that are related in some way to plants, animals, and renewable natural resources. Such jobs occur indoors and outdoors and include such sectors as banking and finance; radio, television, and satellite communications; engineering, technology, and design; construction and maintenance; research and education; and environmental protection.

Plenty of Opportunities

Approximately 22.2 million people are employed in agriscience careers. About 400,000 people are needed each year to fill positions in this field. Of those vacancies, only 100,000 are currently being filled by people trained in agriscience. That means there are many opportunities for you. You can use what you learn today in your current job, on your farm, or in agriscience classes to transition directly into full-time employment. However, if you choose to pursue a college degree in agriscience, many additional career opportunities will be open to you.

About 20 percent of careers in agriscience require a four-year college or university degree. These careers are in the fields of education, marketing, communications, production, social services, finance, management, science, engineering, and many others (**Figure 4-2**). Additional career opportunities are available to students who graduate with technical college certificates and degrees. Among these opportunities are careers as mechanics, sales representatives, field representatives, agriscience laboratory technicians, insurance adjustors, and many others. Many of these careers offer financial opportunities that are equivalent to those available to university graduates.

Careers That Help Others

Helping others is one of the bonuses of a career in agriscience. Many of the jobs in processing, marketing, production, natural resources, mechanics, banking, education, writing, and other areas are people-oriented jobs. This means you have the extra benefit of being a product or process specialist and receive the special appreciation of others.

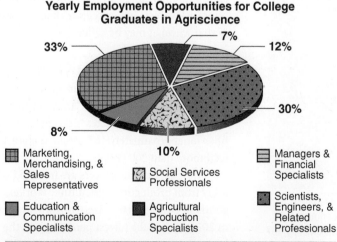

Figure 4-2 A college education opens additional doors in agriscience.

Careers That Satisfy

You might ask, "How can an agriscience career benefit me? What's in it for me?" Of course, there's the money. Salaries in agriscience vary tremendously from job to job. Generally, the better-qualified individuals will earn more. Agriscience industries employ one-fifth of all workers in the United States. And there are job openings for skilled individuals at various levels of expertise. You can be hired at the entry level as an agricultural mechanic right out of high school, or you can work for an advanced degree and become an agricultural engineer. Ultimately, you can do what you decide is best for you. Before you invest in a technical college or university education, it is a good idea to assess your talents and interests. What kind of work do you think you will enjoy doing? What special skills and talents do you have? How can your interests, skills, and talents be matched to a career in which your skills and talents can be developed and expressed? These are important questions to ask yourself before you enter into advanced education or technical education programs.

Some valuable resources are available to help you explore your career options. One of these may be as close as the counseling center in your high school. Many high school counselors are trained to use computer programs to help identify careers that match your interests and talents. If your school does not have a computer-assisted program such as Career Information System (CIS), you may be able to identify some career choices using the Internet.

Hot Topics in Agriscience: Agriscience Careers

- Food Technician/Scientist
- Environmental Technician
- Computer Technician
- Animal Technician/Scientist
- Plant Technician/Scientist
- Global Positioning System Technician
- Biotechnology Engineer
- Veterinarian/Technician
- Farm/Ranch Managers
- Urban Forester
- Soil Technician/Scientist
- Genetic Engineer

Figure 4-3 The average U.S. farmer produces enough food and fiber for approximately 167 people.
Courtesy of USDA/ARS.

The Wheel of Options

Agriscience is like a wheel with a large hub. The hub of that wheel is production agriculture, farming and ranching. The rest of the wheel consists of the nonfarm and nonranch careers in agriscience. Because so many opportunities for rewarding careers exist in that wheel, it may be called a wheel of options.

Production Agriculture

Production agriculture is farming and ranching. It involves the growing and marketing of field crops and livestock. Careers in this area account for 10.9 percent of all jobs in agriscience. Some estimates indicate the average U.S. farmer produces enough food and fiber for approximately 167 people (**Figure 4-3**). Large farm operators produce enough to feed more than 200 people.

Most agriscience careers are involved with goods and services that flow toward or away from production agriculture. Workers in those careers permit U.S. farmers to supply goods so efficiently that U.S. consumers spent only 10.3 percent of their income on food in 2021. This is the lowest percentage in the world. Out of six workers in agriscience, five have jobs that are not on farms. The nonfarm agriscience jobs may be in rural, suburban, or urban settings. The agriscience wheel of options contains a hub with one-sixth of the

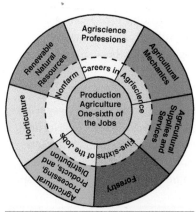

Figure 4-4 A wheel of options illustrates the wide range of agricultural careers.

agriscience workers in production on farms and ranches. The rest of the wheel contains the five-sixths of the agriscience jobs in nonproduction-type careers that are off the farm (**Figure 4-4**).

Agricultural Processing, Products, and Distribution

Agricultural processing, products, and **distribution** (**Figure 4-5**) are those parts of the industry that transport, grade, process, package, and market commodities from production sources. Pick any item of food, clothing, or other commodity, and then trace it back to its source. Except for metals and stone, most objects can be traced back to a farm, ranch, forest, greenhouse, body of water, or other agricultural production facility. If you consider a deluxe hamburger, you can trace the beef, mayonnaise, tomato, lettuce, pickle, catsup, mustard, relish, bun, and sesame seeds back to farms where they were produced (**Figure 4-6**). The same is true of many ingredients in soda, coffee, hot chocolate, or any other beverage you choose.

Check the label in your coat. Is it made of cotton, leather, vinyl, rubber, or wool? Each of these materials can be traced to a farm, ranch, or plantation. Your search may take you to a Maryland farm, a California ranch, a Colombian rubber plantation, or a Utah mink farm.

People with careers and jobs in agricultural processing, products, and distribution make consumer goods available to the public that were nonexistent fifty years ago. From transporting to selling, processing to merchandising, inspection,

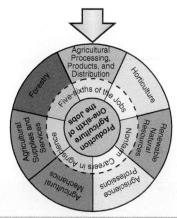

Figure 4-5 There are numerous opportunities in agricultural processing, products, and distribution.

Ag Establishment Inspector
Butcher
Cattle Buyer
Christmas Tree Grader
Cotton Grader
Farm Stand Operator
Federal Grain Inspector
Food & Drug Inspector
Food Processing Supervisor
Fruit & Vegetable Grader
Food Distributor
Fruit Press Operator
Flower Grader
Grain Broker
Grain Buyer

Grain Elevator Operator
Hog Buyer
Livestock Commission Agent
Livestock Yard Supervisor
Meat Inspector
Meat cutter
Milk Plant Supervisor
Produce Buyer
Produce Commission Agent
Quality Control Supervisor
Tobacco Buyer
Weights & Measures Official
Winery Supervisor
Wood Buyer

Figure 4-6 The components of a deluxe cheeseburger may have come from several states, but each ingredient came from a farm.
Courtesy of National FFA; FFA #18.

and research—the commodity moves from its source to consumption (**Figure 4-7**). The USDA reports that the producer's share of the food dollar is approximately 14.3 percent. The remainder is for handling, processing, and distribution.

Horticulture

Horticulture (**Figure 4-8**) includes producing, processing, and marketing fruits, vegetables, and ornamental plants such as turfgrass, flowers, shrubs, and trees (**Figure 4-9**). Horticultural production is a farming enterprise, but because it is generally done on small plots, the production of horticultural crops is classified with horticulture rather than farming. Horticultural commodities are high-labor and high-income commodities. The landscape

---

Hot Topics in Agriscience — A Career in Food Science

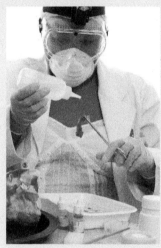

Food science has become a popular college major and is one of the "hottest" careers in the agricultural industry.
Leah-Anne Thompson/Shutterstock.com.

Food science has become one of the hottest career fields in agriculture. It combines the career fields of science and agriculture with the development of new food products. It also extends into the processing, packaging, distribution, and marketing of food products. This career requires a high degree of creativity in designing new products. Preparation for this career requires a college education with a strong emphasis in chemistry, bacteriology, and the biological sciences. Salaries in this career field rank high in comparison with most other agricultural careers, and college graduates in this field are in high demand.

Examples of jobs in food science include:

- Food microbiologist, making sure food is safe to consume
- Food engineer, designing systems for the safe processing, storage, packaging, and transportation of food
- Food chemist, analyzing methods of processing, canning, cooking, and packaging food
- Food technician, using knowledge of raw materials and food processing to assist product developers

---

Figure 4-7 Raw farm products usually require processing before they become available to the consumer.
© Paula Cobleigh/Shutterstock.com. © Richard Thornton/Shutterstock.com.

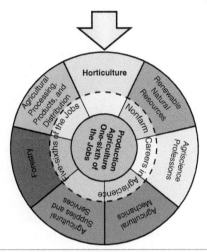

Figure 4-8 Careers in horticulture include:

Floral Designer
Floral Shop Operator
Florist
Golf Course Superintendent
Greenhouse Manager
Greenskeeper
Horticulturist
Hydroponics Grower
Landscape Architect
Landscaper
Nursery Operator
Plant Breeder
Turf Farmer
Turf Manager

Figure 4-9 Ornamental plants are used to create beautiful landscapes and pleasant outdoor environments.
© Paul Vinten/Shutterstock.com.

Figure 4-10 Turf farming is the production of grass that is harvested, roots and all, and transplanted to provide "instant lawns."
Courtesy of DeVere Burton.

designer, golf course superintendent, greenhouse supplier, greenhouse manager, flower wholesaler, floral market analyst, florist, strawberry grower, vegetable retailer, and turf farmer (**Figure 4-10**) are all horticulturists. Recent statistics indicate that approximately 83,000 people are employed by the floral industry in the United States each year.

Forestry

Forestry (**Figure 4-11**) is the industry that grows, manages, and harvests trees for lumber, poles, posts, panels, pulpwood, and many other commodities. Americans have huge appetites for wood products.

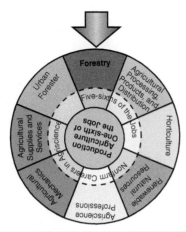

Figure 4-11 Careers in forestry include:

Forester	Log Grader	Nursery Operator	Tree Surgeon
Forest Ranger	Logging Operations	Park Ranger	Urban Forester
Heavy Equipment Operator	Inspector	Plant Breeder	
	Lumber Mill Operator	Timber Manager	

Figure 4-12 The forest industry provides many opportunities for outdoor work.

Courtesy of National FFA; FFA #187.

Careers in forestry range from growing tree seedlings to marketing wood products. Many jobs in forestry are outdoors and require the use of large machines to cut trees, retrieve logs, and load trucks (**Figure 4-12**). Other jobs are service oriented, such as the state or district forester whose job is to give advice and administer governmental programs. Many find enjoyable careers in forestry research, teaching, wood technology, and marketing.

Renewable Natural Resources

Renewable natural resources (**Figures 4-13** and **4-14**) involve the management of wetlands, rangelands, water, fish, and wildlife. All of these career choices draw people who appreciate and value natural and scientific knowledge of plants and animals. This area of agriscience

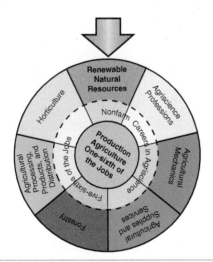

Figure 4-13 Careers in renewable natural resources include

Animal Behaviorist	Forest Fire Fighter/Warden	Resource Manager
Animal Ecologist	Forest Ranger	Soil Conservationist
Animal Taxonomist	Game Farm Supervisor	Trapper
Environmental Conservation Officer	Game Warden	Water Resources Manager
Environmentalist	Ground Water Geologist	Wildlife Manager
Fire Warden	Park Ranger	
	Range Conservationist	

Figure 4-14 Effective management of natural resources begins with education about resource management, and care. Hunter education is one approach to teaching respect for nature's resources.

LawrenceSawyer/E+/Getty Images.

is attractive to those who enjoy working in parks, on game preserves, or with landowners to preserve and enhance natural habitat, plants, and wildlife. Water quality and soil conservation are state and regional concerns of high priority. New career opportunities in natural resource management are resulting from renewed efforts to save our natural environments, soils, oceans, lakes, wetlands, rivers, and bays.

Agricultural Supplies and Services

Agricultural supplies and services (**Figures 4-15** and **4-16**) are businesses that sell supplies or provide services for people in the agricultural industry. Examples of supplies are seed, feed, fertilizer, lawn equipment, farm machinery, hardware, pesticides, and building supplies. These businesses are operated by owners, managers, mill operators, truck drivers, sales personnel, bookkeepers, field representatives, clerks, and others.

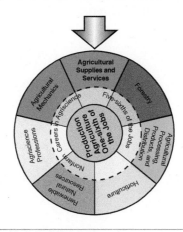

Figure 4-15 Careers in agricultural supplies and services include:

Aerial Crop Duster	Artificial Breeding	Dairy Management	Field Inspector	Harvest Contractor	Poultry Hatchery
Ag Aviator	Technician	Specialist	Field Sales Repre-	Horse Trainer	Manager
Ag Chemical Dealer	Artificial Inseminator	Dog Groomer	sentative, Agricul-	Insect & Disease	Poultry Inseminator
Ag Equipment Dealer	Biostatistician	Farm Appraiser	tural Equipment	Inspector	Sales Manager
Animal Groomer	Chemical Applicator	Farm Auctioneer	Field Sales	Kennel Operator	Salesperson
Animal Health	Chemical Distributor	Farrier	Representative,	Lab Technician	Service Technician
Products	Computer Analyst	Feed Mill Operator	Animal Health	Meteorological	Sheep Shearer
Distributor	Computer Operator	Feed Ration	Products	Analyst	
Animal Inspector	Computer	Developer &	Field Sales	Pest Control	
Animal Keeper	Programmer	Analyst	Representative,	Technician	
Animal Trainer	Computer	Fertilizer Plant	Crop Chemicals,	Pet Shop Operator	
Artificial Breeding	Salesperson	Supervisor	Machinery	Poultry Field Service	
Distributor	Custom Operator	Fiber Technologist	Harness Maker	Technician	

Figure 4-16 Agricultural supplies and services provide the vital materials and services to keep a trillion dollar industry moving.

Courtesy of National FFA.

People in these jobs provide the supplies for the agricultural industry. However, many who work in agriscience seldom handle the commodities themselves. Instead, they provide a service. For example, they may provide legal assistance, write agricultural publications, or recommend financial strategies. Agricultural consultants who advise producers on crops, livestock, pest control, or soil fertility are working in service occupations. Such jobs are ideal for those who are more people-oriented than commodity-oriented.

Agricultural Mechanics

Are you fascinated by tools and equipment? Are you challenged by something that does not work? Are you creative and like to build things? If so, a career in agricultural mechanics may be for you. **Agricultural mechanics** (**Figures 4-17** and **4-18**) includes the design, operation, maintenance, service, selling, and use of power units, machinery, equipment, structures, and utilities in agriculture.

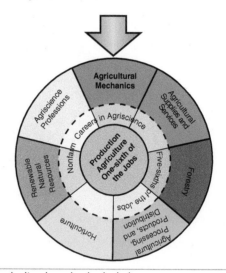

Figure 4-17 Careers in agricultural mechanics include:

Ag Construction Engineer
Ag Electrician
Ag Equipment Designer
Ag Plumber
Ag Safety Engineer
Diesel Mechanic
Equipment Operator

Hydraulic Engineer
Irrigation Engineer
Land Surveyor
Machinist
Parts Manager
Research Engineer
Safety Inspector
Soil Engineer
Welder

Figure 4-18 Careers in agricultural mechanics are varied and challenging.
© Dmitry Kalinovsky/Shutterstock.com.

Agricultural mechanics includes the use of hand and power tools, woodworking, metalworking, welding, electricity, plumbing, engine and machinery mechanics, hydraulics, terracing, drainage, painting, and construction. Choose your level—indoors or outdoors. Choose your role—employee, employer, or professional.

Agriscience Professions

The word **profession** means an occupation requiring specialized education, especially in law, medicine, teaching, or the ministry. **Agriscience professions** (**Figures 4-19** and **4-20**) are those professional careers that deal with knowledge and understanding of agriscience. They cut across all sectors of agriculture.

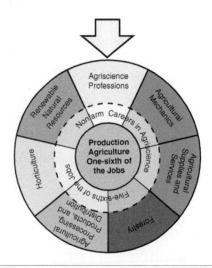

Figure 4-19 Agriscience professions include

Ag Accountant	Avian Veterinarian	Ichthyologist	Plant Nutritionist
Ag Advertising Executive	Bacteriologist	Information Director	Plant Pathologist
	Biochemist	International Specialist	Plant Taxonomist
Ag Association Executive	Bioengineer	Invertebrate Zoologist	Pomologist
	Biophysicist	Land Bank Branch Manager	Poultry Scientist
Ag Consultant	Botanist		Public Relations Manager
Ag Corporation Executive	Computer Specialist	Limnologist	
	Credit Analyst	Magazine Writer	Publicist
Ag Economist	Dairy Nutrition Specialist	Mammalogist	Publisher
Ag Educator	Dendrologist	Marine Biologist	Reproductive Physiologist
Ag Extension Educator	Electronic Editor	Marketing Analyst	
Ag Extension Specialist	Embryologist	Media Buyer	Rural Sociologist
Ag Journalist	Environmental Educator	Microbiologist	Satellite Technician
Ag Lawyer	Entomologist	Mycologist	Scientific Artist
Ag Loan Officer	Equine Dentist	Nematologist	Scientific Writer
Ag Market Analyst	4-H Youth Assistant	Organic Chemist	Silviculturist
Ag Mechanics Teacher	Farm Appraiser	Ornithologist	Software Reviewer
Ag News Director	Farm Broadcaster	Ova Transplant Specialist	Soil Scientist
Agriculture Attaché	Farm Investment Manager		Vertebrate Zoologist
Agronomist		Paleobiologist	Veterinarian
Animal Cytologist	Food Chemist	Parasitologist	Veterinary Pathologist
Animal Geneticist	Foreign Affairs Official	Pharmaceutical Chemist	Virologist
Animal Nutritionist	Graphic Designer	Photographer	Viticulturist
Animal Physiologist	Herpetologist	Plant Cytologist	Vocational Agriculture Instructor/FFA Advisor
Animal Scientist	Horticulture Instructor	Plant Ecologist	
Agriculturist	Hydrologist	Plant Geneticist	

Consider the agriscience teacher and the Cooperative Extension educator. Both must have a bachelor's or master's degree to be qualified. They may teach or consult about subjects in several or all divisions of agriscience.

Consider the veterinarian, agricultural attorney, research scientist, geneticist, or engineer—all of these professions require advanced degrees and high levels of education and skills. If you can meet the standard, then a career in an agriscience profession may be for you.

Figure 4-20 Professional workers are more in demand than ever before as agriscience has embraced the information age.
Courtesy of USDA/ARS K-3401-10.

Agri-Profile Soil Science Career

Soil science degrees focus on how plant life is supported by soil, and the effects of soils on water resources. College graduates enter such professional careers as soil scientists, soil surveyors, ecologists, environmental consultants, conservationists, and university extension specialists. Advancement in these careers often requires master and PhD degrees. Employment opportunities are trending higher with growth in this sector at near 7 percent per year. Median compensation ranges near $61,000.00 per year for scientists and conservationists. This career involves a significant amount of outdoor field work along with laboratory analysis of soil samples. It also involves one-on-one interactions with farmers and ranchers and group instruction as part of the university extension system.

Technology and Computing in Agriscience

The use of computers is extensive in agriscience. This means there are many opportunities to combine computer skills with agriscience settings. Among the agricultural uses of computers are machinery management, farm financial records, livestock management, crop management, commodity marketing, farm/ranch inventory management, agricultural business management, taxes, irrigation management, and precision farming. In addition, satellites provide data to airplanes, tractors, combines, drones, and other mechanized systems equipped with computer and global positioning systems and other high-technology devices that are highly sophisticated. These systems make it possible to operate the machines more efficiently and to incorporate precision farming practices. Computer skills are essential in nearly every aspect of an agriscience career. Every student should be certain that they become proficient in the use of computers.

Agri-Profile Career Area: Agriscience Technician or Professional?

Whether you prefer to be a technician or a professional, agriscience offers a broad array of career possibilities.

RubberBall Productions/Brand X Pictures/Getty Images.

Recent assessment of career opportunities in agriscience by the USDA reveals a bright outlook for college graduates. Other studies indicate a need for workers trained in agriscience for technical-level jobs.

The agriscience technician may be broadly trained in plant and animal sciences and employable in many fields. For the sharp individual with good work habits and a broad background in plant and animal sciences, the choices are extensive. Add agriscience mechanics skills to the package and the individual has access to dozens of career options.

Education for agriscience careers should begin at the high school level or earlier. If one plans to work at the technician level, an early start is especially helpful. The technician is expected to have firsthand experience and detailed knowledge of procedures and techniques. Such knowledge comes with experience in shops, laboratories, farms, greenhouses, fisheries, and on-the-job training situations as well as in the classroom. Technicians generally have specialized training beyond the high school level, whereas professionals are required to obtain degrees at the bachelor's, master's, or doctorate levels.

Preparing for an Agriscience Career

Career education is an important part of public education today. Since the early 1970s, the career education movement has spread from occupational education programs to the general curriculum. Many school systems now emphasize career education from kindergarten through adulthood.

As you contemplate a career in agriscience, it is important to consider how to meet the requirements to get started in that career. Some young people have an early start on careers in agriscience. They may have grown up on a farm or ranch. Their parents may have worked in one of the other areas of agriscience such as horticulture, resource management, business, or teaching. Perhaps they have worked for friends or neighbors who had agriscience businesses. Following are some suggestions for preparing for a career in agriscience.

Agriscience Career Portfolio

An agriscience career portfolio is a collection of your best work on agriscience projects and other career-related materials that you have developed. The portfolio will be used to sell your skills to a prospective employer. Only your best work should be included because a mediocre portfolio is an indicator of mediocre or average job skills.

Agri-Profile: What Skills Will You Bring to Your Career?

First published several years ago, the SCANS report continues to prove its value as a tool for assessing skills. Developed for the Department of Labor by the Secretary's Commission on Achieving Necessary Skills (SCANS), the report reflects the input of business owners, unions, workers, supervisors, and public employers. Together they developed the following five competencies that students need to achieve in order to be prepared for productive careers:

- **Resources: Identifies, organizes, plans, and allocates resources**
    A. Time—selects goal-relevant activities, ranks them, allocates time, and prepares and follows schedules
    B. Money—uses or prepares budgets, makes forecasts, keeps records, and makes adjustments to meet objectives
    C. Material and Facilities—acquires, stores, allocates, and uses materials or space efficiently
    D. Human Resources—assesses skills and distributes work accordingly, evaluates performance, and provides feedback

- **Interpersonal: Works with others**
    A. Participates as a Member of a Team—contributes to group effort
    B. Teaches Others New Skills
    C. Serves Clients/Customers—works to satisfy customers' expectations
    D. Exercises Leadership—communicates ideas to justify position, persuades and convinces others, responsibly brings new ideas forward to improve existing procedures and policies
    E. Negotiates—works toward agreements involving exchange of resources, resolves divergent interests
    F. Works with Diversity—works well with people from diverse backgrounds

- **Information: Acquires and uses information**
    A. Acquires and Evaluates Information
    B. Organizes and Maintains Information
    C. Interprets and Communicates Information
    D. Uses Computers to Process Information

- **Systems: Understands complex interrelationships**
    A. Understands Systems—knows how social, organizational, and technological systems work and operates effectively with them
    B. Monitors and Corrects Performance—distinguishes trends, predicts impacts on system operations, diagnoses deviations in systems' performance, and corrects malfunctions
    C. Improves or Designs Systems—suggests modifications to existing systems and develops new or alternative systems to improve performance

- **Technology: Works with a variety of technologies**
    A. Selects Technology—chooses procedures, tools, or equipment, including computers and related technologies
    B. Applies Technology to Tasks—understands overall intent and proper procedures for setup and operation of equipment
    C. Maintains and Troubleshoots Equipment—prevents, identifies, or solves problems with equipment, including computers and other technologies

Source: The Secretary's Commission on Achieving Necessary Skills. *What Work Requires of Schools, A SCANS Report for America 2000*. U.S. Department of Labor, pg. x. Retrieved from http://wdr.doleta.gov/SCANS/whatwork/whatwork.pdf

Some of the materials that you may want to include in your print or digital portfolio include the following:

- Resumé or vita
- FFA Agriscience Scholarship application
- Articles and papers written by you (published and unpublished)
- Photographs and written reports of projects that you have completed
- Documentation of participation in community and public service activities
- Personal letters and citations for services you have performed for other people
- Letters of recommendation
- Personal and career goals
- Action plan for accomplishing your goals
- Newspaper and website articles, video clips, and sound bites of media coverage of your projects

Career Plan While in Middle School

- Plan and conduct science projects with plants, animals, soil, water, energy, ecology, conservation, and wildlife.
- Research and report on the projects listed earlier that interest you most.
- Join 4-H or Scouts and choose agricultural projects and merit badges.
- Volunteer to work on lawn, garden, greenhouse, farm, or conservation projects.
- Enroll in agriscience or other career education programs.

Career Plan While in High School

- Enroll in agriscience classes, including plant science, animal science, agricultural mechanics, agribusiness, and farm management.
- Enroll in college-preparatory and/or dual-enrollment courses in English, Introduction to Business and STEM—science, technology, engineering, and math.
- Join the FFA organization and participate in leadership, citizenship, and agriscience activities.
- Develop a broad, supervised, occupational agriscience experience program.
- Acquire hands-on, skill-development experiences.
- Conduct an agriscience research project, and enter it in the FFA Agriscience Scholarship and Agriscience Fair programs.

Career Plan After High School

- Obtain an agricultural job and plan ways to get additional training while on the job.
- Enter a community college and take courses that will transfer to the college of agriculture or life sciences at your state university.
- Enter a two-year program in technical agriculture.
- Enter a college of agriculture or life sciences, and obtain a bachelor's degree (BS), a master's degree (MS), and/or a doctorate (PhD).

You can obtain information on careers, schools, and colleges from many sources. Seek out information from those who have experience in the career field you are considering. Listed below are the kinds of professionals who can be helpful to you as you choose a college major and a career:

High School Agriscience Teachers

- Agricultural mechanics
- General agriscience
- Animal and plant sciences
- Horticulture
- Biotechnology

High School Counselors

- Career brochures, bulletins
- Career information system
- Aptitude tests
- College/university catalogs

Cooperative Extension Service

- Listed in your phone book or found on the Internet under county or city government
- 4-H career bulletins

State Department of Education

- Specialist in agriscience, agribusiness, and renewable natural resources: scholarship, internships
- FFA State Executive Secretary: FFA career materials

Community Colleges, Technical Colleges, and Other Postsecondary Institutions

- Occupational Dean, Community College: program information, scholarships
- Director or Dean, Institute or Technical School: career/program information Typical programs include agriscience business management, animal/crop production and management, ornamental horticulture, water resources, forestry, and wildlife resources
- Dean, College of Agriculture or Life Sciences Typical programs include agricultural education, agricultural and resource economics, agronomy (crops and soils), animal sciences, food science, forestry, horticulture, natural resources management, and poultry science
- Agricultural Education Coordinator, Department of Agricultural and Extension Education, Land-Grant University: agricultural education career information

As new technologies and job opportunities emerge, so will the need for well-trained and educated new people. Agriscience is a diverse field with job opportunities available at all levels. Pick your area of interest, determine the level at which you wish to operate, obtain the appropriate education for the job, and follow through with a rewarding career!

Agri-Profile: Agriscience Teacher

Students of agriculture learn best by doing, and teachers seek ways to teach relevant skills.
© Goodluz/Shutterstock.com

Every student who has enrolled in an agricultural program has benefited from the experience of *learning by doing*. It is a curious thing that one of the critical worker shortages in the agricultural professions exists among teachers of agriscience programs. Too few students are enrolling as majors in university agricultural education programs, and too few graduates are entering the agricultural education profession after they graduate with their degrees. There is also high demand for experienced agriscience teachers in related agricultural careers.

As a result, some high school agriculture programs have been closed due to a shortage of qualified teachers. Top students would do well to consider teaching agriscience as their first career choice. Such students should express their interest to their teacher or a teacher educator at the university. This career requires a strong background in agriculture and related sciences and a four-year degree in agricultural education. Most agriscience educators are endorsed by their state to teach agriculture and science classes.

Pursuing a career as an agriculture teacher can give you the opportunity to:

- teach by doing, not just explaining.
- share your personal interest in agriculture.
- explore hot topics, like genetics, satellite mapping, robotics, biofuels, alternative energy, and more.
- travel in your state, your nation, and even internationally.
- investigate new technologies being developed by agribusinesses.

According to the National Association for Agricultural Educators, "Ag teachers never have the same day twice. One day they might be in a classroom or laboratory, the next visiting students in the field, preparing teams for a FFA Career Development Event, or leading a community service activity with their FFA Chapter."

Chapter Review

Student Activities

1. Write the **Key Terms** and their meanings in your notebook.
2. Calculate the amount of time needed to count the dollar value of agricultural assets in the United States, as suggested in Figure 4-1.
3. Develop a bulletin board that illustrates the "wheel of options," with its listing of the broad categories of jobs or divisions in agriscience.
4. Develop a collage that illustrates the many careers in agriscience.
5. Write the names of the divisions of agriscience, such as "Production Agriculture," "Agricultural Processing, Products, and Distribution," "Horticulture," and so on, and list five careers under each that may interest you.
6. **SAE Connection:** Choose a career from the lists you developed for Activity 5 and conduct online research about it. Then, write a one-page description for that career. Include the following sections in your description:
   a. Career title
   b. Education/training required to enter the career and advance in the field
   c. Working conditions
   d. Advantages/benefits
   e. Disadvantages of the career
   f. Salaries of beginning and advanced workers in the field
   g. Aspects of the career that you like
7. Using the career you researched for Activity 6, or another job or career area, list the things you should do during and after high school to prepare for a career in that area.
8. Find the name and address of an appropriate official of a school or college and request information from that person about educational opportunities for you in their institution. Be sure to specify your career interests.
9. Develop a list of agriscience careers in which computers and other high technology are used.
10. At fast food restaurants, a popular order is a cheeseburger, fries, and a soda. Make a list of possible agriscience jobs that were involved in making this order a reality. For example, a wheat farmer was responsible for growing the wheat that is in the bun.
11. Make an outline of the chapter. Phrases and words that are bold or in color should be included along with a brief description of each.

Inquiry Activity — Careers in Agriscience

Courtesy of USDA/ARS K-3401-10.

Watch the *Careers in Agriscience* video available on MindTap. Then, create a 3-column chart like the one shown below. Allow plenty of space for each row. Use the chart to take notes as you watch again.

Type of Agricultural Job	Example(s)	My Notes
1: Production		
2:		
3:		
4:		
5:		
6:		
7:		
8:		
9:		

- In column 1, list the nine types of agricultural jobs. For example, the first category has been listed for you: *Production*.
- In column 2, list one or more examples of agricultural jobs for each category.
- Next, highlight or put checkmarks by the jobs that most interest you.
- Then, use column 3 to briefly jot down why each job interests you.
- Finally, exchange lists with a classmate. Compare and discuss your choices. Here are some questions to help you get started: What do your selected jobs or job types have in common? Which job or job type interests you the most? Why? What level of education might it require? What more would you like to know about that job or job type?

Check Your Progress

Can you . . .

- define agriscience and its major divisions?
- describe the opportunities for careers in agriscience?
- identify job opportunities in farm and off-farm agriscience jobs?
- name activities in school, college, or university that can help you prepare for an agriscience career?
- identify resource people for obtaining career assistance in agriscience?

If you answered "no" to any of these questions, review the chapter for those topics.

Chapter 5
Supervised Agricultural Experience

Objective
To learn the rationale for and plan a supervised agricultural experience.

Competencies to Be Developed
After studying this chapter, you should be able to:
- define Supervised Agricultural Experience (SAE) and related program terms.
- determine the place and purposes of SAEs in agriscience education programs.
- explain the benefits and characteristics of an effective SAE program.
- identify the responsibilities of those involved in an SAE.
- determine the types of SAE programs.
- maintain, organize, and use records.
- plan an Immersion SAE.

Key Terms
Supervised Agricultural Experience (SAE)
career exploration
Foundational SAE
agricultural literacy
Career Plan
Immersion SAE
project
Placement/Internship SAE
SAE Agreement
Ownership/Entrepreneurship SAE
production agriculture enterprise
agribusiness
Research SAE
School-Based Enterprise SAE
Service-Learning SAE
recordkeeping

Three components form the total program of agricultural education: classroom and laboratory instruction, **Supervised Agricultural Experience (SAE)**, and the National FFA Organization (FFA). An SAE is a student-led, instructor-supervised, work-based learning experience. It is based upon Agriculture, Food, or Natural Resources (AFNR) Technical Standards and Career Ready Practices.

The Total Agriscience Program

By integrating the three components of the total agriscience program, students receive the best experience agricultural education has to offer. For teachers, this program ensures the most efficient use of their time and resources (**Figure 5-1**).

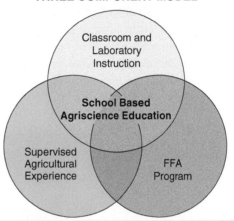

Figure 5-1 A comprehensive agriscience program provides students opportunities to learn through classroom and laboratory instruction, SAEs, and the FFA.

Figure 5-2 The best instruction includes hands-on activities.

© Blaj Gabriel/Shutterstock.com.

Classroom Instruction and an Effective SAE Program

The Foundational SAE is an integral part of classroom instruction. Students learn how to plan and implement an effective Immersion SAE that aligns to their career interests. Their agricultural education teacher is their primary supervisor, but the students are responsible for making their own decisions throughout their SAE. In class, the teacher provides necessary instruction and arranges for small group and individual instruction (**Figure 5-2**). Students have access to the classroom and laboratory space and resources to solve problems encountered in the SAE.

The National FFA Organization

Membership in the National FFA Organization (FFA), an intracurricular youth organization for students enrolled in agriscience programs, complements the SAE program (**Figure 5-3**). The FFA provides many instructional resources and activities to help students make the most of their SAEs. It also hosts competitive career development events (CDEs) to develop students' skills and formally honors achievements in SAEs. Further details on FFA resources, events, and awards are provided throughout this textbook.

Figure 5-3 NATIONAL FFA MOTTO

SAE for All

FFA partnered with the National Council for Agricultural Education (NCAE) on a renewal of the SAE program, and in 2011, *SAE for All* was introduced. The *SAE for All* model brings options to students who may not have the ability or resources to take on a project outside of the classroom. It is inclusive and provides *all* students the opportunity to develop work-readiness skills through individual SAE projects and school-based learning.

After students enroll in an agricultural education program, they launch into the Foundational SAE portion of the *SAE for All* model. They work through the components and transition to one or more Immersion SAEs (**Figure 5-4**). The Foundational SAE focuses on career exploration, while the Immersion SAE focuses on career preparation. The Immersion SAE is designed for students to develop and apply the knowledge and skills they learned in an agricultural education classroom out in the real world. The results of an SAE experience are measurable outcomes based on AFNR standards, which are set in advance and align to a student's Career Plan. The SAE provides the following benefits to students: real experience, real employability skills, and real earnings while learning. Upon completion of the Immersion SAE, students are college- and career-ready and have real-world skills and hands-on experience (**Figure 5-5**).

Figure 5-4 Agriculture teachers and students hone specific skills prior to participating in competitive events.
Courtesy of USDA/ARS #K-3405-11

Figure 5-5 SAE is based on real agricultural work opportunities.
© Kurhan/Shutterstock.com

Foundational SAE

Getting Started

In a **Foundational SAE**, students explore a variety of AFNR subjects and careers. This SAE is conducted by all students in an agricultural education program. There are three levels of the Foundational SAE: Awareness, Intermediate, and Advanced. Awareness is for beginners or students in grades 6–9 who are enrolled in agriculture education and have not decided on a specific career path. Intermediate is intended for students in grades 9–11 who have completed Awareness and have identified a career area of interest. Advanced is for students in 11–12 grades who have completed both Awareness and Intermediate and have plans for a career in a specific area.

Components

Within each level, there are five components of a Foundational SAE (**Figure 5-6**):
- Career Exploration and Planning
- Employability Skills for College and Career Readiness
- Personal Financial Management and Planning
- Workplace Safety
- Agricultural Literacy

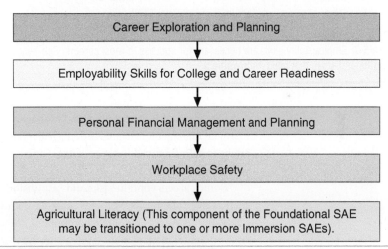

Figure 5-6 The five components of a Foundational SAE.

During the Career Exploration and Planning component, students research and explore the following AFNR career pathways:
- Agribusiness Systems
- Animal Systems
- Biotechnology Systems
- Career Ready Practices
- Environmental Systems
- Food Products and Processing Systems
- Natural Resources Systems
- Plant Systems
- Power, Structural, and Technical Systems

They also complete Career Interest Inventories and identify a career goal. For example, students can access a Career Interest Inventory on the AgExplorer website. They can then use the website's Career Finder feature or the school's Career Planning system to complete one.

The Employability Skills component prepares students with the skills they need to succeed in post-secondary education and their career. These skills include responsibility, communication, innovation, critical thinking, and collaboration. Students complete self-assessments of their employability skills to track their growth and progress (**Figure 5-7**).

EMPLOYABILITY SKILL	IN ACTION	USUALLY	SOMETIMES	RARELY	NEVER
Collaboration	I work well with others and contribute to the team.				
Communication	I present my thoughts clearly both in writing and while speaking.				
Critical Thinking	I am able to solve problems using a decision-making process.				
Innovation	I find new ways to solve problems, and/or I come up with new ideas.				
Responsibility	I am responsible for my actions including arriving on time and completing tasks in a timely manner.				

Figure 5-7 Employability skills are necessary for success in most careers. Self-evaluation of these skills helps an individual identify strengths and areas that need improvement.

The focus of Personal Financial Management and Planning is to create a personal financial management plan. Students learn budgeting, saving, and how to responsibly use credit to gain financial independence. For more information about developing a budget, see **Appendix E**.

Some careers contain tasks and responsibilities where employees encounter hazards, such as toxic chemicals. Students must have knowledge of and experience with workplace safety. In the Workplace Safety component, they learn the importance of health, safety, and environmental managements systems in the workplace. For more information on safety in the workplace, see **Appendix C**.

Agricultural literacy is knowledge about different areas of the agricultural industry. In this component, students research and analyze how issues, trends, technologies, and public policies impact AFNR systems. They also learn about how agriculture impacts society, the environment, and the economy.

Career Plan

The final product of a Foundational SAE for all levels is a Career Plan—a documentation of completed work from activities and the authentic experience (Day in the Life) from each component. It can be used as a planning tool as well as a record of learning and accomplishment. A Career Plan should align to any assigned requirements. The outcomes and documentation for the activities and authentic experience tasks vary between the three levels, but may include the following: Career Interest Inventory, assessment of employability skills, personal budget, résumé, or academic plan.

Using a Career Planning system compile the items from the Foundational SAE to create a Career Plan. It is important to note that this final product is not truly final. A Career Plan is designed to be modified as the student develops their skills and discovers new interests over the course of the agricultural education program and Immersion SAE.

Agri-Profile — **Career Area: Connecting Personal Interests to a Career Path**

Connecting career possibilities to a student's personal interests and skills is part of an SAE. For example, a love of animals might lead to a career in animal science.
© T.Den_Team/Shutterstock.com

For many people, the ideal career is one that builds on their personal interests as much as on their skills. Agriscience teachers and SAEs provide the guidance and resources to help students achieve that ideal. Here are some real-life examples of students' experiences with SAEs.

- A student of animal science visits AgExplorer, an online career resource of the National FFA Organization. She then clicks on Career Finder to identify careers that are most likely to connect with her love of animals. Later, she completes several Foundational SAE activities, including shadowing a veterinarian.

- Through an Ownership SAE, one student turns the chore of mowing the family lawn into the beginnings of a business. His SAE includes creating and acting on a detailed business plan. He also decides to take college classes in landscape architecture.

- A desire to help relatives with their dairy farm prompts one student to begin her first SAE. Drawing on her math and science skills, she conducts an Analytical Research SAE to find out which milk replacer would be most beneficial for the livestock. She decides to pursue a degree in Dairy Science.

- A student who grew up riding horses on her family's cattle ranch develops an SAE Service Learning Plan. This prepares the student to help provide therapeutic horseback riding to disabled youth in her community.

These are just a few of the many exciting possibilities that students have discovered through their combined participation in agriscience classes and SAEs. For more information, and to read detailed accounts of SAEs in students' own words, search online "SAE for All." To find out how a career in agriscience might align with your personal interests, search "AgExplorer" and then launch Career Finder.

Immersion SAE

Next Steps

An **Immersion SAE** builds upon the Agricultural Literacy component of the Foundational SAE. Students select a career within an AFNR career pathway based on their interests. They should use this SAE to explore, develop, and advance their career. They also deepen their critical thinking, communication, and management and leadership skills. With the help of their agricultural education teacher, students develop a plan so they can begin their Immersion SAE. Additionally, students identify the project or enterprise they will include in their SAE. A **project** describes a series of activities related to a single objective or enterprise, such as raising rabbits, fabricating a trailer, or improving wildlife habitats. A list of project examples for each type of Immersion SAE can be found in **Appendix B**.

Responsibilities of Trusted Adults

Parents or guardians, school guidance counselors, and employers or experienced mentors, in addition to an agriscience advisor, provide advice and support as the student conducts their SAE (**Figure 5-8**). Parents or guardians review and support the student's Career Plan and help them plan and implement their Immersion SAE. The school guidance counselor guides the student on exploring careers. An advisor evaluates SAE projects, which may be a graded component in an agriculture course.

Figure 5-8 A good SAE plan draws on the ideas of the student, as well as the teacher, employer, and other trusted adults who act as mentors.
Courtesy of National FFA.

Types of Immersion SAEs

Students' SAEs, while unique to them, should be based on their career interests within the AFNR Career Pathways and be agriculturally related. Based on their Career Plans and interests, students may choose from a variety of SAE projects. Students also choose which type of SAE to conduct. There are five categories of Immersion SAEs (**Figure 5-9**).

TYPES OF IMMERSION SAEs
Ownership/Entrepreneurship
Placement/Internship
Research
School-based Enterprise
Service Learning

Figure 5-9 Types of Immersion SAEs

Placement/Internship

In a **Placement/Internship SAE**, students work for an AFNR-related business, enterprise, or agency. Students may be placed at agencies such as Cooperative Extension Program, Farm Service Agency (FSA), Forest Service (FS), wildlife and environmental agencies, and agriscience laboratories. Many agricultural businesses sponsor internships within their organizations. Students may work as a paid employee or a volunteer.

Figure 5-10 The employer acts as a mentor and supervisor to a student who participates in a Placement SAE.

© Goodluz/Shutterstock.com.

In a Placement SAE, the emphasis is on gaining experience, learning job-related skills, and becoming proficient in a chosen agriculturally related career. The employer assigns tasks to the student or employee. These tasks must be essential to the operation of the business. The employer regularly evaluates the student's development (**Figure 5-10**). Students must document hours worked, income received, tasks completed, and knowledge and skills attained. A Placement SAE may involve working for a nature center, maintaining school grounds, or assisting Christmas tree farmers.

Students are encouraged to transition a Placement SAE into an Internship SAE. An Internship SAE is more involved than a Placement SAE and focuses on meeting the specific training needs of the student. The student has an SAE Training Plan, which is an agreement that defines the scope of the experience, knowledge and skills to be developed, responsibilities and roles, and safety issues to address. This plan is created with the help of the agricultural education instructor and employer. The student must also document their experience by writing a summary and reflections. The employer evaluates the student regularly and provides feedback to assist in the student's improvement.

An **SAE Agreement** must be completed for all Placement/Internship SAEs. This document defines the scope of the experience, responsibilities, and roles, and identifies any safety issues to address. The agriculture education teacher and employer help the student create it. The SAE Agreement helps finalize the plans for placement on a job or internship (**Figure 5-11**). Such agreements need the signature of the student, parents or guardians, employer, and teacher. Once all parties agree regarding the student's placement experiences, the chances for success will be enhanced.

Ownership/Entrepreneurship

An **Ownership/Entrepreneurship SAE** refers to supervised activities conducted by students as owners or entrepreneurs of their own for-profit business (**Figure 5-12**). Emphasis is placed on developing business administration skills and working in a profitable, professional manner.

In an Ownership/Entrepreneurship SAE, students create, own, and operate a business or enterprise in production agriculture or agribusiness. A **production agriculture enterprise** is a business that deals with the growing and marketing of plants and livestock. Production agriculture enterprises include ones that raise and sell animals, such as dairy, beef, or poultry, and ones that grow and sell plants, such as corn, hay, turf, or poinsettias. Students are encouraged to work in a production agriculture enterprise as it provides opportunities to learn about agriculture through hands-on experience (**Figure 5-13**).

Students may also choose an enterprise within agribusiness. **Agribusiness** is an industry that includes dealing with the operations of a farm; the distribution of farm equipment and supplies; and the processing, storage, and distribution of farm commodities. Agribusinesses include ones that offer products, such as buying and reselling of feed, seed, or fertilizer, and building and selling agricultural equipment, and ones that provide services, such as landscaping or equine training.

In an Ownership SAE, a student creates, owns, and operates an individual business that provides goods or service. They make operational and risk management decisions and provide the labor resources for the operation of the SAE. They must also maintain financial records and perform analysis on the productivity and profitability of the enterprise at the completion of the business cycle. In addition, the student must identify and account for all resources used in the business.

SAE AGREEMENT

A. This agreement is in effect for the _____ school year.
  1. **Description of SAE**
     Describe the placement/internship.
     _____
     _____

  2. **Roles and Responsibilities of Employee/Intern**
     List tasks to be performed.
     _____
     _____

  3. **Resources and Materials**
     List items required for the SAE. Note who will provide them.
     _____
     _____

  4. **Profit/Loss Responsibility**
     If the experience may involve potential for profit and loss, state who will receive the profit or incur the liability.
     _____
     _____

B. **SAE Risk Assessment Results and Safety Plan**
   Complete the *Safety in Agriculture for Youth SAE Risk Assessment* with your parent or guardian, teacher, and employer. Report findings and action items.
   _____
   _____

C. **Other Requirements**
   List additional requirements specific to school district or employer.
   _____
   _____

Signatures

I, _____, as the **employee/intern**, will uphold my roles, responsibilities, and duties outlined in this SAE Agreement.
Print Name:_____

I, _____, as the **employer/provider**, will uphold my roles, responsibilities, and duties outlined in this SAE Agreement.
Print Name:_____

I, _____, as a **parent/guardian** am aware and approve of this work-based learning experience.
Print Name:_____

I, _____, as the **agricultural education teacher** of this employee/intern, will supervise and help the student gain employability skills and abilities through this experience.
Print Name:_____

Figure 5-11 An SAE Agreement includes a description of the placement/internship, roles and responsibilities, and necessary resources. It addresses profit, liability for loss, and workplace safety. Signatures from the employee/intern, employer/provider, parent/guardian, and teacher are required.

EXAMPLES OF ENTREPRENEURSHIP

A production enterprise is a crop, livestock, or agribusiness venture. The student may own or be employed on a production enterprise.

Types of Crop Enterprises
- Corn
- Soybeans
- Small Grains
- Greenhouse
- Nursery
- Hay
- Vegetable
- Fruit
- Forestry
- Christmas Tree

Production Agriculture: Animals
- Commercial Cow Calf
- Registered Breeding Stock
- Market Beef
- Dairy
- Feeder Pig
- Market Swine
- Sheep
- Dairy or Market Goat
- Poultry
- Horse
- Rabbit

Agribusinesses
- Lawn Service
- Custom Farm Work
- Trapping and Pelt Sales
- Hunting Guide Service
- Tree Service
- Farm and Garden Supply Service
- Artificial Insemination Business
- Animal Care and Boarding
- Winery
- Fishing and Crabbing for Sales

Figure 5-12 Agriscience students have a variety of production agriculture enterprises and agribusinesses from which to choose.

A student's Ownership SAE may transition into an Entrepreneurship SAE, which requires creating and annually updating an SAE Business Plan. This plan provides information on continued growth and expansion of the operation.

Figure 5-13 Agriscience students are encouraged to own or work with an agricultural production enterprise to learn about agriculture through hands-on experience.
© iStock.com/BrandyTaylor

An Ownership/Entrepreneurship SAE is similar to a Placement/Internship SAE in that all students are working in a business or enterprise that sells goods or provides services. In a Placement/Internship SAE, the student is the employee and performs tasks assigned by their employer. These positions are not always paid. In an Ownership/Entrepreneurship SAE, the student is the owner of the business or enterprise and makes a profit.

Hot Topics in Agriscience — Immersion SAE—Inside the Dairy Business

The Dairy Replacement Heifer Project is a program that students can participate in across the United States. It is designed to increase the knowledge and interest of young people in the dairy industry. The project begins when the participant purchases a quality heifer calf from a program-approved seller. The participant raises and cares for the animal. They are guided as they perform the typical procedures a dairy or replacement heifer farmer would perform.

Specific care must be provided, vaccinations must be given, and a magnet needs to be placed in the heifer to prevent hardware disease. Quality feed must be provided to ensure proper weight and good health. The participant is required to keep detailed records of the animal. In some instances, the heifer is bred to an approved sire, and the project culminates in the sale of the pregnant heifer when she is presented for show and sale.

The Dairy Replacement Heifer Project is a challenging SAE that allows each participant to gain an understanding of dairy heifer husbandry and management. Contact the agriculture education teacher at school or the local extension office for more information about a program in the area.

Research: Experimental, Analysis, or Invention

In a **Research SAE**, students are involved in the investigation of materials, processes, and information, which they use to establish new knowledge or to validate previous research. The research topic must be within the AFNR industry. Research may be conducted in school laboratories, at home, on the job, or wherever suitable facilities may be found. There are three types of Research SAEs; analytical, experimental, and invention.

In an Analytical Research SAE, students begin with a question and seek to find why or how something occurs. They collect data using qualitative and quantitative methodologies. Students analyze data, facts, and other information to determine the answer.

In an Experimental Research SAE, students plan and conduct a major agricultural experiment using the scientific process—a method for answering scientific questions. Students control variables and manipulate others to observe the outcome. Students must define a hypothesis and determine an experimental design. Then they collect data and draw conclusions from that data. Students gain hands-on experience in verifying, demonstrating scientific principles, and discovering new knowledge.

A mentor can help students design an accurate experiment or testing procedure. The mentor can be a high school teacher, a college professor, or a science professional in a business or industry. In many cases, the mentor helps find a laboratory for the student to work in and equipment that can be used to conduct experimental testing.

In an Invention Research SAE, students create a product or service that meets an identified need. Students design, engineer, and create a prototype. At this point, it is suggested that students perform a patent check to ensure their invention does not conflict with a previously registered invention. Before introducing the new product or service to potential customers, tests must be performed to verify its usefulness. Corrections and adjustments should be made before introducing the new product or service to a focus group that will evaluate the invention and provide feedback.

All three types of Research SAEs can be tailored to a wide range of situations and goals, such as:

- to meet an individual need, such as researching ways to keep livestock eating when temperatures exceed 90°F for long periods of time.
- to function as one part of a larger community improvement activity, such as partner with a university extension office to research insects' impact on native plants.
- to contribute to state-level research efforts, such as stream monitoring, weather watch, forest fire watch, crop scouting, insect or weed monitoring, and crop reporting.

For all Research SAEs, students must have an approved SAE Research Plan. They also must follow the scientific process or accepted best practices for conducting research and conduct peer reviews with their supervisor and other professionals at multiple stages. At the conclusion of the Research SAE, students deliver a summary presentation to a local committee organized by their teacher.

The ultimate goal for any FFA member with an AFNR-related Research SAE is to enter it in a chapter, state, or National FFA Agriscience Fair or apply for a Proficiency Award. As a special incentive to develop research skills, the FFA offers SAE grants and distributes awards to outstanding agriscience students annually.

School-Based Enterprise

A **School-Based Enterprise SAE** is similar to an Entrepreneurship SAE, but this operation is based at the school or FFA chapter and may involve a group of students working cooperatively. These student-led enterprises provide goods and services and use the facilities, equipment, and other resources of the school campus, FFA chapter, or the agriculture department. The enterprise may be owned by the school or chapter but is managed and operated by students, or it may be structured as a partnership or cooperative between students. A written cooperative or agreement defines how duties, responsibilities, profits, and costs are divided among the students. Students operating a school-based enterprise must create and update an SAE Business Plan annually.

Science Connection: Research Assistant Opportunities

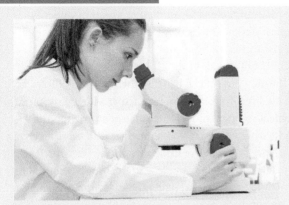

In USDA-ARS internships, students can gain hands-on, real-world experience in the research field.
luchschenF/Adobe Stock Photos

USDA agencies, such as the Agriculture Research Service (ARS), provide students with federal internship and employment opportunities to develop work experience in the field of their interest. In ARS internships, students can develop their research skills alongside mentor scientists and their research teams. The Pathways Internship Program provides paid work experiences for current degree-seeking high school students. Internships may be seasonal or year-round. According to the USDA website, "From the classroom to the workplace, [we] support engagement, recruitment, retention, and agricultural workforce development."

The ARS has numerous research laboratories throughout the United States, including the Arthropod-Borne Animal Diseases Research Laboratory in Laramie, Wyoming, the Children's Nutrition Research Center in Houston, Texas, the Plant Genetics Resources Unit in Geneva, New York, and many more. Research opportunities in these laboratories can be found in the following fields: agronomy, animal science, biological science, chemistry, computer science, engineering, food technology, hydrology, plant pathology, soil science and conservation, and veterinary medicine. In past internships, students helped conduct research on topics such as the following:

- Soybean plants, to track down the genes responsible for disease resistance and nitrogen fixation
- Plant response to soil moisture and drought, to improve agricultural production with limited water supplies
- Biological control of pest species via predation
- Insect pests and diseases of livestock and poultry

Participants in these programs do work of scientific significance, make original contributions to science, and gain hands-on, real-world experience in the research field. For information on research assistant internship opportunities for students, visit the USDA-ARS website.

It is important to understand that permission needs to be granted before students create a school-based enterprise. Students must seek approval from the FFA advisor, school principal, and other district leadership as needed. They also must operate the business under the supervision of a board of directors.

School-Based Enterprise SAEs are beneficial to students who would not usually be able to conduct an Ownership/Entrepreneurship SAE. Students learn skills required to create a product or marketable service in the school agriscience classroom or laboratory. They learn customer service, responsibility, cooperation, business management, and leadership.

An example of a School-Based Enterprise SAE is a student-operated flower shop. In the classroom, students learn floral design from their teacher or FFA advisor. Under the supervision of their advisor and a board of directors, students take online orders. After school, students work in the flower shop filling online orders. They use the chapter's wholesale account to order flowers and supplies. After the flowers are sold, the chapter is reimbursed for the cost of all used materials, and the students divide the income proportionately to the hours each individual worked. This SAE is beneficial to the community as it fills an undermet need and provides students with paid experience in a career field they enjoy.

Service Learning

The FFA motto is "Learning to Do, Doing to Learn, Learning to Live, Living to Serve." A **Service-Learning SAE** embodies these ideals. In a Service-Learning SAE, students solve or improve community and agricultural issues. A student or group of students plans, conducts, and evaluates a project designed to provide a service to a local school, community organization, religious institution, or nonprofit organization. Their FFA chapter, however, cannot be the recipient of this service.

There are three ways a service-learning project can benefit others: direct, indirect, and advocacy (**Figure 5-14**). Students have a direct impact when interacting with the recipient(s) of service efforts or improving a physical component within an area of an environment. SAEs with an indirect impact are those where students do not interact with those receiving the service. Lastly, an advocacy experience is one where students bring awareness regarding an issue to others who take actions to address it. For example, a student starts a community canned food drive to benefit the local food bank. The student promotes the drive, and members of the community donate the food.

Type of Service-Learning Impact	Examples
Direct	• Gain permission from a city agency to create a garden space for asylum seekers and help families prepare the plot to grow their customary foods. • Partner with the community parks department and a local community service club to repair broken playground structures, benches, and picnic tables. • Learn about reclamation and teach a lesson to an outdoor recreation group. • Work alongside class members to clean and repair unhealthy riparian areas of a pond, stream, or riverbed. • Learn how to perform simple bicycle repairs. Partner with local recycling centers to donate broken bicycles. Volunteer to teach bike repair skills to middle school children in an after-school program.
Indirect	• Start vegetable seeds in containers and donate them to a local refugee center to be distributed to families in need. • Learn about local native plants. Clean up a vacant lot and plant low-maintenance native vegetation. • Learn how to make simple wooden toys. Organize volunteers to paint them. Donate the toys to a local children's hospital or similar organization.
Advocacy	• Design and promote a garden sharing program where people with extra garden space can connect with families who need a place to grow a garden. • Create an adopt-a-grandparent program where volunteers can help elderly community members with yard work or minor home repairs. • Start a social media campaign highlighting a local charity. Make regular posts promoting its cause. Include a link to the organization's donation website in each post. • Raise funds to finance a community service project.

Figure 5-14 A Service-Learning SAE must benefit an organization, group, or individuals.

The student or group of students must develop an SAE Service Learning Plan, operate the project under supervision, provide reports to the supervisor, submit a summary report of the impact of the project, and write a reflection paper.

Service learning should not be confused with community service. Community service opportunities are short, preplanned volunteer activities that are typically completed in a day or less. Unlike most community service volunteer work, Service-Learning SAEs include an education component where students learn a new skill while volunteering for a nonprofit entity.

Plan an SAE

Planning is an important part of the SAE program, and it is important to follow each step before moving onto the next one.

Steps in Planning an SAE

The following are the steps of planning an SAE:

1. Identify your career interests in agriculture. Your SAE should be something about which you can get excited. You will be more successful in your career if you enjoy what you do.
2. Review the job responsibilities of career-interest areas you may have. You might find that you like the idea of a particular project or career, but that you don't like the specific task involved.
3. Identify SAE programs of interest by interviewing friends who have an SAE or by viewing suggestions found in various resources.
4. Develop a timeline for your SAE program. Include the different experiential learning activities and projects as well as a completion date. Electronic portfolios, recordkeeping systems, or even a simple calendar system can be used for this step.
5. Building on step 4, develop a long-range plan for your SAE program. Projects and activities occur over a short period of time. An SAE program, however, occurs over a longer frame and grows in scope and diversity.
6. Develop the first-year plan, starting now. This is your annual plan. It will keep you on track and help you make decisions about your SAE. The annual plan includes a calendar, description of projects, budget, and supplementary skills, but these items may vary depending on the category of your SAE.
7. Replan on a regular basis. Part of good planning is reviewing your activities in light of your plan and then adjusting your plan accordingly.

Make It Meaningful

To ensure a meaningful SAE, students should address the following elements while planning and conducting their project:

- Identify possible SAE choices.
- Choose activities for their own SAEs.
- Make appropriate arrangements/agreements with parents or guardians, teachers, and employers.
- Conduct teacher/parent or guardian follow-ups and evaluations at regular intervals.

Propose an SAE

Following the planning stage, students should write a statement justifying their plans with a proposal. This proposal should explain why the SAE was selected. It should also include the following: the SAE category; the purpose of the SAE; the goals for the project; dates the SAE will begin and end; who will supervise or provide assistance, and the form this will take; and any pre-arranged agreements, such as the SAE Agreement. Students should provide an estimated time investment, an estimate of the cost of the SAE, and the source of the funding. Lastly, students should list the knowledge and skills they expect to learn while completing the SAE.

Conduct an SAE

The first step to conducting an SAE is to just get started after all that planning and decision-making. During the SAE experience, students must document it and hone their recordkeeping skills as well as be evaluated.

Document an SAE

Documenting the SAE is important because it gives students an opportunity to see the plan come to life. It also creates a permanent record that they can use to learn and improve their project.

Some helpful tools for documenting are a calendar, journal, and portfolio. These tools can be paper or electronic. A calendar is helpful for planning as well as documenting hours invested, skills learned, and tasks performed. Students can also use a journal or portfolio. A journal is a written record of experiences and observations. A journal could be as simple as a notebook. A portfolio, usually housed online, stores all the student's documentation compiled over a period of time. The portfolio can also be used for recordkeeping, too.

Agricultural Experience Tracker (AET) is a personalized online FFA Record Book System. Students use AET to create records of their educational experience and develop their digital portfolio. The site's features also help students manage their time and financial resources, apply for FFA awards, and connect with colleges and companies that interest them. AET's availability as an app makes it especially convenient to maintain up-to-date records.

However, it should be noted that some programs may require that students keep paper records. Each state or individual agriscience program chooses the format that works best for their situation.

Keep Detailed Records

Recordkeeping is a useful and highly transferrable skill. Students can use this skill within AFNT careers and beyond. Foundational SAEs introduce students to an organized recordkeeping system. Students hone their skills in recordkeeping while conducting their SAE. For every project, students need to capture the following for all activities or items: the date, the name or description, the hours spent, and expenses or income that resulted. Be sure to add up the hours invested in all activities.

It is imperative to track finances and enter financial transactions as soon as possible. It is a good idea to keep receipts, paycheck stubs, loan payment records, education, and capital item investments in the records.

Being proficient at keeping accurate records enables students to do the following both in their SAE and in their career:
- Determine if money was made or lost.
- Ensure fair compensation with each transaction.
- Make management decisions.
- Document net worth for loans.
- Prepare tax returns.
- Plan for future events.
- Budget for the following year.
- Evaluate the strengths and weaknesses of different parts of the business.
- Help document activities for FFA recognitions and degree purposes.

Evaluate an SAE

Throughout the SAE, the teacher or advisor visits the student. They may need to travel to off-campus experience locations for an official SAE visit to monitor progress and provide helpful advice for improving the project. On-site visits may include meeting with the student and a parent or guardian, mentor supervisor, or boss to collaborate and support their student efforts. The advisor visit is a time for the student to showcase their SAE and proudly share the knowledge and skills they have learned through the SAE.

The evaluation looks different for different types of SAEs. See **Appendix B** for the evaluation components for each category of the Immersion SAE.

Chapter Review

Student Activities

1. Write the **Key Terms** and their meanings in your notebook.
2. Describe the relationships between classroom instruction, SAEs, and FFA.
3. Create a multimedia presentation or a bulletin board showing the relationships of the components of a total agriscience program presented in Figure 5-1.
4. Complete a Career Interest Inventory online or through your school's Career Planning System. Determine whether your interests are more in agribusiness, horticulture, production agriculture, or another area of agriscience.
5. Complete an Employability Skills in Action self-assessment (Figure 5-7). Save your results and reevaluate yourself periodically.
6. Interview a friend, family member, FFA member, or fellow student who has completed an Immersion SAE. Ask what type of Immersion SAE they completed, what led them to start their SAE, what they liked most about it, and how it affected their plans for the future. Choose a format for sharing your interview with your class, such as an audio or multimedia recording, or as a print-out for display in the classroom.
7. **SAE Connection:** Work out an SAE Agreement with an employer (Figure 5-11).
8. Study Figure 5-12, and write five ideas for Ownership/Entrepreneurship SAE projects in production agriculture and/or agribusiness to discuss with your teacher.
9. Brainstorm three Service-Learning SAE projects for each area of impact (Figure 5-14).
10. Talk with your teacher and parents or guardians. Develop an annual plan for your SAE including the projects and activities you plan to complete and skills you plan to develop during the current year.
11. Review the job responsibilities of the career areas in which you are interested using an online resource such as AgCareers or a similar website. Determine if you would enjoy performing the tasks that the jobs in these career areas require.
12. Explore the AET website, making notes on features that are useful for recordkeeping. Then answer this question: How would the AET website help a student keep records for an SAE?
13. If you have not already done so, review **Appendix C: Ensuring Safety in SAE Lab Work.**

Check Your Progress

Can you . . .

- define Supervised Agricultural Experience (SAE) and related program terms?
- explain some of the benefits and characteristics of an effective SAE program?
- identify some of the responsibilities of those involved in an SAE?
- compare two or more types of SAE programs?

If you answered "no" to any of these questions, review the chapter for those topics.

National FFA Connection — The National FFA Agriscience Fair

Courtesy of National FFA; FFA #148

The National FFA Agriscience Fair is a great way to take your Research SAE skills to the next level.
U.S. Department of Agriculture (USDA)

Students interested in Research SAEs often also participate in science fairs. The National FFA Agriscience Fair is a competition that is available to FFA members enrolled in middle school and high school agriscience programs. Competition begins at the local level and progresses to state- and national-level events. Specific rules and instructions are available on the National FFA Organization website in the form of a handbook containing the required documents and forms. A qualified experimental research project must be teacher supervised and include the following elements:

- Identify a current agricultural/scientific question or issue.
- Define objectives for research that are focused and specific.
- Implement a step-by-step process to collect and analyze relevant data.
- Commit sufficient time to complete the work.

The project elements in a complete entry for a local, state, or national agriscience fair include the following:

- **Logbook**—accurate, detailed notes of observations, procedures, and results
- **Written report**—a paper including a title page, abstract, introduction, literature review, materials and methods, results, discussion and conclusion, references, and acknowledgments
- **Display**—an exhibit including the specific identification information along with information and objects relevant to the study
- **Interview**—finalists are interviewed by the judges to explain their project

Work collaboratively with your classmates to find out more about the above rules and elements. For example, you might each volunteer to research one item in more detail and then present your findings to the class. You might also find out more about a winning project to analyze how the entrant or entrants met the requirements.

Chapter 6
Leadership Development in Agriscience

Objective
To develop effective leadership skills

Competencies to Be Developed
After studying this chapter, you should be able to:
- define leader and leadership.
- explain why effective leadership is needed in agriscience.
- list some characteristics of good leaders.
- describe the opportunities for leadership development in FFA.
- demonstrate positive leadership skills.

Key Terms

leadership	listen	order of business
plan	Cooperative Extension System	gavel
citizenship	4-H	motion
integrity	Girl Scout	main motion
knowledge	Boy Scout	amend
courage	extemporaneous speaking	refer
tact	parliamentary procedure	lay on the table
enthusiasm	business meeting	point of order
selflessness	presiding officer	adjourn
loyalty	minutes	

What is leadership in agriscience? Agriscience has been described as a broad and diverse field. It is not just horticulture or supplies and services; not just professions or products, processing, and distribution; not just mechanics or forestry; and not just renewable natural resources or production. Agriscience is all of these. Then, what is leadership in agriscience?

Leadership Defined

Leadership may be defined as the capacity or ability to solve problems and to set a direction. To lead is to show the way by going in advance or guiding the actions or opinions of others. To demonstrate such leadership skill in agriscience, you must have knowledge of technical information and people. You must know how to organize and manage activities. At the same time, a leader does not "go it alone." Most jobs are too big for one person. As individuals, we can do only part of what needs to be done. Therefore, we need to work in collaboration with others.

A leader depends on the knowledge and skills of others to achieve a common goal. In sports for instance, a quarterback on a football team uses leadership skills to coordinate the team players to achieve a touchdown. Similarly, the wise batter in baseball hits the ball in a place that not only gets the batter on first base but also permits other runners to advance around the bases. A properly placed hit supports the common goal of achieving runs. A single base hit, where everyone advances and no one gets out, may be the best for the team. Conversely, a line drive that doesn't quite make the fence may result in a third out, with no chance for other players to score.

Why Leadership in Agriscience?

Agriculture is a highly organized industry. It involves people and complex processes. Leadership skills are necessary whenever people are assembled. Those who teach in agriscience are part of a team of teachers, principals, supervisors, community advisory groups, and others. Those in agribusiness are typically part of teams consisting of the manager, office staff, sales representatives, field personnel, and board of directors (**Figures 6-1** and **6-2**). Those on farms may be the owner, manager, spouse, children, hired help, or neighbors who assist at times. The manager of a farm or business must plan (think through, determine procedure, assemble materials, and train staff to do a job). Once a job is planned, it may be accomplished through management. To manage is to direct and motivate people, resources, and processes to accomplish organizational goals and objectives. A manager uses leadership skills continuously in collaborating with others on a day-to-day basis.

Figure 6-1 A leader is a person who has earned the trust of others, and who is able to inspire others to contribute their best efforts.

© StockMediaSeller/Shutterstock.com.

Figure 6-2 Agribusinesses rely on team efforts to achieve goals. Team members work together to find effective solutions.

© Ju Photostocker/Shutterstock.com.

Figure 6-3 Participation in class discussions is the first step toward developing the qualities of leadership.
Photo courtesy of the National FFA Organization.

A landscaper is an example of a leader who plans, plants, builds, or maintains outdoor ornamental plants and landscape structures. When developing a plan for a customer, the landscaper demonstrates leadership by knowing the name, function, and performance of each plant. This information is used to develop an acceptable plan. Because the customer has personal ideas about landscaping, a professional must consider these ideas in the plan. It may test the landscaper's leadership skills to guide the customer toward a viable plan in which plants survive the climate and conform to acceptable landscape practices.

Agriscience personnel may also be called on as officers or members of professional organizations to give testimony before legislators or other public officials on the need for laws, regulations, or other actions that affect their work. Likewise, leaders may end the day presiding over a meeting or taking minutes at a professional or civic meeting.

For the agriscience student, the need for leadership skills is also apparent. Developing confidence to participate fully in class is important (**Figure 6-3**). To meet prospective employers and conduct a supervised agricultural experience program requires leadership skills. Functioning in the community, participating in group meetings, and forming friendships and professional associations all require strong leadership skills. Good leadership skills also enhance your individual marketability in the working world.

To be a good citizen, you must earn your way in life without infringing on the rights of others. A key element of effective citizenship is using leadership to promote the common good in society. **Citizenship** means functioning in society in a positive way.

Traits of Good Leaders

Good leaders must have **integrity** (honesty). Without it, others cannot trust an individual with the power to manage or control, even in minor things. A leader must have **knowledge**, which means familiarity, awareness, and understanding. Good leaders are dependable and have the **courage** and initiative to carry out personal and group decisions. To lead, one must also communicate. This requires effective speaking and listening skills.

In working with others, **tact**, or the skill of encouraging others in positive ways, is useful. Similarly, a sense of justice to ensure the rights of others is important. **Enthusiasm**, or the energy to do a job in a way that inspires and encourages others, is beneficial. **Selflessness** means placing the desires and welfare of others above yourself. It is an important quality for good leadership. These traits encourage **loyalty**, which leads to reliable support for an individual group, or cause. A good leader will actively **listen** to others, seeking understanding of their points of view. These and other traits are achieved through effective leadership development.

Agri-Profile: Communicating Effectively

Exceptional leaders communicate effectively. Fortunately, most people can learn to apply basic communication processes and competencies as they progress in leadership roles. The following skills and processes will help you as you learn to be an effective leader:

- **Listening**—Practice active listening by paying close attention to what others are saying. Ask probing questions to clarify what is meant, and rephrase what you hear to ensure that you understand.
- **Nonverbal skills**
  - **Eye Contact**—It is important to focus on the person as well as on the discussion.
  - **Alert to situation**—No fidgeting or slouching; genuine smile; firm handshake; no distracting devices; maintain open arms; project excitement/energy/animation; no interruptions.
- **Friendliness**—Friendly tone; nice/polite; ask a personal question; smile; personalize written messages.
- **Empathy**—Respect other points of view; express understanding through brief restatement of another's perspective to show that you are listening.
- **Concise and clear**—Say just enough, use few words, think before speaking, and demonstrate strong verbal skills; check the meaning of any terms you plan to use
- **Confidence**—Eye contact; firm yet friendly statements; no statements that sound like questions; no arrogance; no aggression.
- **Open-mindedness**—Listen, understand other viewpoints, and express opinions without contention.
- **Respect**—Listen actively; make eye contact; avoid distractions; stay focused.

Leadership Development Opportunities

Today's schools provide extensive opportunities for agriscience students to develop leadership skills. Students develop leadership in school organizations, athletics, and in classroom and laboratory situations. Some become leaders at home, on the job, or in community organizations.

4-H Clubs

The **Cooperative Extension System** is an educational agency of the U.S. Department of Agriculture (USDA) and an arm of your state university. It provides educational programs for both youth and adults. Its programs include personal, home and family, community, and agriscience resources development. The Cooperative Extension System also sponsors 4-H clubs. The **4-H** network of clubs is directed by Cooperative Extension System personnel to enhance personal development and provide skill development in many areas, including agriscience (**Figure 6-4**). The four *Hs* in 4-H stand for head, heart, hands, and health. These provide the basis for the 4-H pledge, which is, "I pledge my head to clearer thinking, my heart to greater loyalty, my hands to larger service, and my health to better living for my Club, my community, my country, and my world."

Figure 6-4 4-H club and FFA members share many of the same or similar experiences as they progress along the path toward becoming responsible adults and leaders in their communities.

Jeffrey Isaac Greenberg 13+/Alamy Stock Photo

| Hot Topics in Agriscience | The Agricultural Lobby |

The laws that are approved by Congress and state legislatures can have significant effects on agriculture. In addition, state and federal agencies write regulations to implement new laws. Farmers, ranchers, and agricultural processors are subject to both the laws and the regulations. In a time when politics affects the agricultural industry in so many ways, agricultural lobby efforts have become increasingly important.

Special leadership skills are required by those who lobby on behalf of agriculture. A lobbyist must be registered in most states before they can actively lobby a legislature or other political organization. Most lobbyists represent a particular group of farmers or processors, such as the United Dairymen or the Wheat Commission. The lobbyist is paid to negotiate, persuade, and be persistent in promoting legislation that is favorable to the segment of the industry that they represent.

Scout Organizations

Girl Scout and Boy Scout organizations provide opportunities for leadership development and skill development in agriscience and other areas. Scouts focus heavily on outdoor activities and provide excellent leadership development and natural resources skills. They provide recognition through a system of merit badges, which are earned by learning skills and obtaining experiences in many areas, including agriscience (**Figure 6-5**).

Figure 6-5 Boy Scouts of America and Girl Scouts of the United States of America provide opportunities for young people to develop skills in providing leadership to other members of their organizations.

Courtesy of DeVere Burton.

FFA

The National FFA Organization as we know it today is an outgrowth of events and influences of other organizations. The United States congress passed the Smith-Hughes Vocational Education Act in 1917. By 1920, high school enrollments in agriculture courses had reached 31,000 students. The Future Farmers of Virginia (FFV) was organized in 1926 for the purpose of encouraging young men to stay on the farm and offering "a greater opportunity for self-expression and for the development of leadership."

In 1928, thirty-three students representing eighteen states met at the Hotel Baltimore in Kansas City, Missouri during the American Royal Livestock Show where they voted to organize the Future Farmers of America (FFA). This new organization for farm boys extended the ideals of FFV to a national level with membership opened to participating states. The first national president elected to office was Leslie Applegate.

Agri-Profile — Career Area: Leadership Development

The art of public speaking is one of the powerful tools of a leader.
Courtesy of National FFA.

Leadership development may be a career area or specialty for teachers, consultants, personnel managers, coaches, and others. However, many agriscience positions require good leadership capabilities as a tool for everyday use. Auctioneers, salespersons, managers, entrepreneurs, corporate executives, politicians, and anyone who directs others or routinely meets the public must have strong leadership skills.

Leadership involves good planning, goal setting, and the ability to inspire others to work toward a common goal. Such skills as facilitation of positive committee interaction, adherence to parliamentary procedure, and clear self-expression are important leadership techniques that are developed through study and practice. These skills are used in church, civic, and community organizations as well as in workplaces. Group projects and club activities in agriscience provide excellent opportunities for leadership development.

The New Farmers of America (NFA) was organized in 1935 in Tuskegee, Alabama. It was patterned after the New Farmers of Virginia for the purpose of serving African American youth. The missions of the FFA and NFA organizations proved to be compatible, and the two organizations merged in 1965.

A significant issue for the FFA was female membership. Several states recognized an equity issue and allowed female membership in the state organizations. However, the issue was finally resolved nationally with female membership approved in 1969, and nearly half of all FFA members today are female.

The National FFA Organization is the name by which the organization is known today. It recognizes the concept that agriculture includes all of the affiliated professions that support and sustain production agriculture and natural resource management.

The FFA is a youth-oriented organization that was developed specifically to expand the opportunities in leadership and agriscience skill development for students in public schools. Only students under the age of 21 who are enrolled in a systematic program of agricultural education are eligible for membership in FFA.

Aim and Purposes

The FFA is part of the agriscience curriculum in most schools where agriscience programs are offered. It is an important teaching tool. It serves as a laboratory for developing leadership and citizenship skills. These, in turn, are helpful in learning agriscience skills. The primary aim of the FFA is the development of agriscience leadership, cooperation, and citizenship. The specific purposes of the FFA may be paraphrased as follows:

- to develop competent and assertive agriscience knowledge and leadership skills;
- to develop awareness of the global importance of agriscience and its contribution to our well-being;
- to strengthen the confidence of agriscience students in themselves and their work;
- to promote the intelligent choice and establishment of an agriscience career;
- to stimulate development and to encourage achievement in individual agriscience experience programs;
- to improve the economic, environmental, recreational, and human resources of the community;
- to develop competencies in communications, human relations, and social skills;
- to develop character, train for useful citizenship, and foster patriotism;
- to build cooperative attitudes among agriscience students;
- to encourage wise use and management of resources;
- to encourage improvement in scholarship; and
- to provide organized recreational activities for agriscience students.

The Emblem

The FFA emblem contains five major symbols that help demonstrate the structure of the organization (**Figure 6-6**). They are as follows:

Eagle—The emblem is topped by the eagle and other items of our national seal. The eagle was placed in the emblem to represent the national scope of the organization. It could also represent the natural resources in agriscience.

Corn—Corn is grown in every state in the United States. It reminds us of our common interest in agriscience, regardless of where we live.

Owl—The owl represents knowledge and wisdom. Use of this symbol in the emblem recognizes the fact that people in agriscience need a good education and that education must be tempered with experience to be of greatest usefulness.

Plow—The plow has been used to represent the work of tilling the soil. Hard work and sustained effort are essential to achieving positive outcomes in agriscience.

Rising Sun—The rising sun is a symbol of the progressive nature of agriscience. It is symbolic of the need for workers in agriscience to work collaboratively toward common goals.

Source: Reprinted by permission of The National FFA Organization.

Figure 6-6 The FFA emblem contains five meaningful symbols that are important to the organization.

Courtesy of National FFA; FFA #148.

The Colors

The official FFA colors are blue and gold. The shade of blue is national blue. The shade of gold is the yellow color of corn. Therefore, the colors are called national blue and corn gold.

Source: Reprinted by permission of The National FFA Organization.

Motto

The FFA motto contains phrases that describe the philosophy of learning and development in agriscience. The motto is:

- Learning to Do
- Doing to Learn
- Earning to Live
- Living to Serve

Source: Reprinted by permission of The National FFA Organization.

"Learning to Do" emphasizes the practical reasons for study and experience in agriscience. It also suggests ambition and willingness to productively use the hands as well as the mind. "Doing to Learn" describes procedures used in agriscience instruction at the doing level. Experiencing results from doing is the most permanent result of learning. "Earning to Live" suggests that FFA members intend to develop their skills and support themselves in life. Finally, "Living to Serve" indicates an intention to help others through personal and community service.

Creed

The creed is a statement of beliefs of the members of the National FFA Organization. It is studied and memorized by agriculture students during their first year of enrollment in an agriculture program. The memorized creed is repeated word-for-word at the first public speaking event in which most students participate.

THE FFA CREED

I believe in the future of agriculture, with a faith born not of words but of deeds—achievements won by the present and past generations of agriculturists; in the promise of better days through better ways, even as the better things we now enjoy have come to us from the struggles of former years.

I believe that to live and work on a good farm, or to be engaged in other agricultural pursuits, is pleasant as well as challenging; for I know the joys and discomforts of agricultural life and hold an inborn fondness for those associations which, even in hours of discouragement, I cannot deny.

I believe in leadership from ourselves and respect from others. I believe in my own ability to work efficiently and think clearly, with such knowledge and skill as I can secure, and in the ability of progressive agriculturists to serve our own and the public interest in producing and marketing the product of our toil.

I believe in less dependence on begging and more power in bargaining; in the life abundant and enough honest wealth to help make it so—for others as well as myself; in less need for charity and more of it when needed; in being happy myself and playing square with those whose happiness depends upon me.

I believe that American agriculture can and will hold true to the best traditions of our national life and that I can exert an influence in my home and community which will stand solid for my part in that inspiring task.

The creed was written by E. M. Tiffany, and adopted at the 3rd National Convention of the FFA. It was revised at the 38th Convention and the 63rd Convention.

Source: Reprinted by permission of The National FFA Organization.

Salute

The Pledge of Allegiance to the American flag is the official FFA salute. The words of the pledge are as follows: "I pledge allegiance to the flag of the United States of America and to the Republic for which it stands, one nation under God, indivisible, with liberty and justice for all."

Source: Reprinted by permission of The National FFA Organization.

Degree Requirements

The FFA awards four degrees, each of which indicates a member's progress. These degree levels are: Greenhand, Chapter, State, and American. The Greenhand and Chapter FFA degrees are awarded by the FFA chapter in the agriscience department of the school. The State FFA degree is awarded by the State Association. The American FFA degree is awarded by the National FFA.

The Greenhand degree is so named to indicate that the member is in learning mode. As such, the member is developing basic skills through FFA participation and by studying the principles of agriscience. To earn the Greenhand degree, the member must meet the requirements spelled out in the current FFA *Official Manual*. In general, the requirements for the Greenhand degree are to:

- be enrolled in an agricultural education course;
- have satisfactory plans for a supervised agricultural experience program;
- recite the FFA creed, motto, and salute;
- describe the FFA emblem, colors, and symbols;
- explain the FFA Code of Ethics and proper use of the FFA jacket;
- have satisfactory knowledge of the history of the organization and of the Chapter Constitution and Program of Activities;
- know the duties and responsibilities of members;
- own or have access to a copy of the *Official Manual* and *FFA Student Handbook*; and
- submit a written application for the Greenhand degree.

Requirements for the other three degrees help the member to learn and grow professionally from one level to the next in the organization. The *Official Manual* states the exact requirements for all degrees, provides details of membership, and explains chapter operation for the FFA.

Career Development Events (CDE)

The FFA sponsors competitive career development events for a wide range of career interests. The first level is the local chapter at the school. The second level is district or regional within the state. The third level is the state association l, and the fourth is national. Local FFA advisors counsel students in their programs to select events that will support instruction in the classroom. The competitions that are conducted at each level should reflect the content of the instructional programs.

The purpose of FFA career development events is to encourage agriscience students to develop technical and leadership skills and to practice these skills in friendly competition with other FFA members. These events include the following:

Agriculture Communications	Food and Science Technology
Agriculture Sales	Forestry
Agricultural Technology & Mechanical Systems	Horse Evaluation
Agronomy	Job Interview
Dairy Cattle Evaluation & Management	Livestock Evaluation
Dairy Cattle Handlers Activity	Marketing Plan
Environmental & Natural Resources	Meats Evaluation and Technology
Farm & Agribusiness Management	Milk Quality & Products
Nursery and Landscape	Poultry Evaluation
Floriculture	Veterinary Science

Some FFA career development events require contestants to know how to grade agricultural products, such as eggs, meats, poultry, fruits, and vegetables. Other events require students to evaluate live animals, such as beef and dairy cattle, horses, poultry, sheep, and swine. Some require mechanical abilities, such as welding, plumbing, electronics, irrigation, surveying, engine troubleshooting and repair, painting, woodworking, and general tool use (**Figure 6-7**).

Figure 6-7 Students enjoy learning by participating in career development events.
© 3D_creation/Shutterstock.com

Certain events, such as Floriculture, require knowledge of the art and science of floral arrangement. Other career development events include Forestry and Nursery/Landscape competitions, which require students to identify plants, plant materials, insects, and diseases. Land Judging (conducted by soil and water districts) involves evaluating the soil and land and recommending appropriate management practices.

Leadership Development Events (LDE)

The first point on the list of the FFA Aims and Purposes, the identification of competent and assertive agriscience knowledge, underscores the importance of effective leadership. The following leadership events have been developed for this purpose:

Agricultural Issues Forum
Conduct of Chapter Meetings
Creed Speaking
Employment Skills

Extemporaneous Public Speaking
Parliamentary Procedure
Prepared Public Speaking

Students in the Agricultural Issues Forum research a current agricultural issue and present their findings. The pros and cons associated with the issue are presented to a panel of judges. Students learn to evaluate both sides of issues as they make choices.

Parliamentary Procedure event teams demonstrate their ability to conduct meetings using their knowledge of correct parliamentary practices. These procedures are used to open meetings, conduct business, close meetings, and write minutes according to best practices in the real world.

The National FFA Organization sponsors three different types of speech competitions: the Creed Speaking, Prepared Public Speaking, and Extemporaneous Public Speaking contests. Each event provides opportunities for students to stand before an audience and deliver a speech in a public setting.

Many students begin to develop speaking skills by reciting the FFA Creed with their classmates as the audience. The next step might be to prepare and deliver a speech or debate a parliamentary issue. The ability to speak extemporaneously is enhanced by learning to deliver prepared speeches.

Agri-Profile: Be All You Can Be

Figure 6-8 FFA members participate as delegates to state and national meetings, where they introduce, debate, and ultimately approve or disapprove proposals from members to change the way the organization is run.
Courtesy of National FFA.

On the day after Christmas in 1960, a 12-year-old boy with polio was delivered to a ranch for needy boys, together with his two brothers. Their broken home was without heat, and food was scarce. After five years of growing up on the ranch, the boy enrolled in a high school agriculture program and joined an FFA chapter.

Although a love of animals drew him to the FFA, it was recognition for his successes on the parliamentary team that spurred him on. "It was the first time I had ever won anything," he later observed. From the parliamentary team, he advanced to area FFA president and on to state president and, eventually, he was a delegate to the national FFA convention (**Figure 6-8**). It was there that he presented the historic motion to open FFA membership to female members.

After graduation from college, he followed in the footsteps of his FFA advisor, teaching agriculture and inspiring young people to be all they could be. As FFA advisor, he insisted that his students could prepare themselves for any career through leadership training, talking on their feet, developing responsibility, learning to manage money, experiencing teamwork, learning respect, setting goals, and planning ahead. These experiences help the individual win in life.

His experiences in FFA, teaching, and advising helped him develop the confidence and skills he needed for a career in public life. After teaching for a while, he served eight years in the Texas Senate before going on to the U.S. House of Representatives. There, his leadership skills were soon recognized; he was elected president of the Freshman Class of Congressmen and eventually became a valued member of the House Agriculture Committee. Today he is a noted author, motivational speaker, and CEO.

What advice does this highly successful statesman give to young people pursuing careers in agriscience? The Honorable Congressman Bill Sarpalius suggests the following:

1. Be in the right frame of mind. Don't dwell on your handicaps or lack of ability like in public speaking or running. Forget, "I can't."
2. Avoid negatives.
3. Stay physically and mentally sharp. Don't let yourself get lazy.
4. Develop a religious background.
5. Concentrate on doing for others—not yourself.

In summary, he asserts, "To achieve all that is possible—we must attempt the impossible. To be as much as we can be—we must dream of being more!"

Extemporaneous speaking is delivering a speech with little or no time for preparation beyond developing a file containing information and facts about current agriscience issues that can be reviewed for a brief time once the speech topic is known. Extemporaneous speaking is a real-life skill that is used daily in agriscience careers. Research and study skills are needed to organize and deliver a speech on an assigned topic with limited preparation time.

The capstone leadership event is to experience a successful job interview. This event is an extension of extemporaneous and prepared public speaking, because it requires a candidate to think on their feet and give thoughtful responses to interview questions without knowing exactly what they might be asked.

Each of the leadership skill events is organized in such a manner that participants compete both as individuals and in teams of three or four.

Public Speaking

Oral communication skills are important for good leadership. Effective leaders must acquire and use these skills as they speak with individuals, committees, small groups, and in large forums. The ability to relax, speak clearly, and state what is pertinent to the subject at hand is a very useful life skill. These skills are developed by applying basic principles of speech preparation and organization. The remainder of this chapter presents specific information and suggestions that will prove helpful as you develop your public speaking skills.

Creed Speaking

The Creed Speaking event is for students who are enrolled in an agriscience class for the first time. It requires the speaker to repeat the FFA Creed from memory. The emphasis is on accuracy and delivery. When the speaker is finished, a specific statement from the creed is cited by the judges. The speaker explains what the statement means to them. Each contestant responds to the same statement.

The Creed Speaking event gets students started in speaking without undue concern over the content of the speech. It is a good way to help students succeed before they are expected to prepare the written content of a speech. Once confidence is gained, students often become motivated to participate in other speech contests.

Prepared Speaking

The Prepared Speaking contest provides an opportunity for a student to research an agricultural topic and develop their own ideas. The content of the speech must be original, not copied from someone else. The length of the speech should be six to eight minutes, and points are deducted if the speech is too long or too short.

Part of the scoring of this competition is based on the quality, effectiveness, and accuracy of the written manuscript. The neatly typed manuscript (double-spaced) is given to the judges ahead of time. Follow-up questions are developed from the manuscript by the panel of judges. Oral presentation also influences the score. Following the oral delivery of their speech, each contestant responds to the questions for five minutes.

Extemporaneous Speaking

Extemporaneous speaking is a valuable skill for life. This skill is used every day by most agriscience professionals. Salespeople use extemporaneous speaking skills to negotiate sales. Agricultural educators use these skills to teach classes or to teach individuals how to deal with problems and issues. Agricultural executives and administrators use extemporaneous speaking skills to convince their employees and stockholders to support their leadership and business plans. Virtually everyone can benefit from learning extemporaneous speaking skills.

Extemporaneous Speaking competitions require students to gather original documents and materials in a notebook or file, but no written preparation of a manuscript may be done before the competition. Each competitor draws one of several possible agricultural topics. They are then allowed to prepare for 30 minutes using only the materials that were assembled earlier. The speech runs five to eight minutes, after which the judges are allowed five minutes to ask questions.

Planning the Speech

A speech should have at least three sections: the introduction, the body, and the conclusion. The plan should clearly identify each section.

The Introduction

The introduction indicates the need for and importance of the speech. It should be carefully planned and presented with confidence. The introduction may be in the form of statements or questions that engage the audience's interest in the topic. If the introduction is not to the point, does not fit the occasion, or is not delivered with expression, the audience may not listen to the rest of the speech. The introduction might be only a few lines long, but it must capture the attention of the audience and make them want to know more.

The Body

The body of the speech contains the most important information. It should consist of several key points that support one central idea or objective. Each key point must be supported by additional information that explains, illustrates, or clarifies that point. It is best to write the body of the speech in outline form (**Figure 6-9**).

SPEECH OUTLINE

Introduction

Honorable judges, instructors, and fellow students
1. Rabbits, cows, plants, and plows—we still need them!
2. When the family farm goes under, a piece of America goes under with it

Body

The 2017 census revealed . . .
1. The midsize farm is likely to be the true family farm
   - Owned and operated by the family
   - Family receives benefit from their work
2. Some believe the family farm is a relic of the past!
   a. Press fascination with bankruptcy sales
   b. Farms not in view from interstate highways
   c. Small population on farms
   d. Animal rights groups and unionizing efforts
3. The family farm endures in the United States
   a. Better managed farm businesses buy the weaker farms
   b. Disease epidemics threaten specialized operations
   c. Farm retailing is on the rise
   d. Small farms are becoming legitimate
4. Farm bankruptcy must be minimized
   a. Farm failure affects general businesses
   b. We will miss fresh farm produce
   c. We will see more pollution
   d. We will depend more on imports
5. Are there remedies? Yes!
   a. Remove politics from marketing
   b. Recognize farmers as astute businesspeople
   c. See the total industry of agriculture
   d. Tax farm and for farm use
   e. Increase agriculture land-preservation programs

Conclusion

Indeed . . .
1. General George Washington nearly lost the continental army at Valley Forge from lack of food, shelter, and clothing.
2. It can happen to us if we don't maintain a healthy farm situation in the United States today
3. Don't give away our most valuable resource—the ability to feed, clothe, and shelter ourselves!

Figure 6-9 An outline of a winning speech on the importance of family farms.

An outline will help guide your written ideas and oral delivery. After outlining the speech, a carefully worded draft may be written to help the speaker fully develop the ideas and information in the speech. However, it must be emphasized that the speech should be given from the outline to avoid the temptation to simply memorize a text. Memorization of a speech has two serious pitfalls. First, there is the danger that it will sound like someone else's words and thus lack authenticity. Second, if the line of thought is lost during delivery, the speaker may not be able to locate it in the narrative. This can negatively impact the quality of the speech.

The Conclusion

The conclusion should remind the audience of the central or key points of the speech and briefly restate them. The conclusion should leave the audience motivated to take action to implement or adopt what you have said. Some speeches call for action, whereas others call for changes in attitude or perception. The more powerful speeches move people to action. The words needed to accomplish such ambitious goals must be carefully planned.

Delivering the Speech

Delivering the speech can be fun and may provide much satisfaction (**Figure 6-10**). However, such a result does not come easily. The speaker must prepare the plan and practice well. Practicing the speech until its content becomes familiar helps speaking become fluent. Be your own audience, the first several times you give the speech. Then, practice in front of a mirror to observe your facial expressions, posture, and gestures. Finally, practice giving the speech in front of others and invite them to make suggestions to improve your delivery (**Figure 6-11**).

Figure 6-10 Speaking in public can be a pleasant and exciting experience once you have learned the proper way to organize and present your thoughts. Speaking is a skill that is learned only through practice as you stand and speak in front of an audience.

Photo courtesy of the National FFA Organization.

National Public Speaking Contest — Judge's Score Sheet

PART I. FOR SCORING CONTENT AND COMPOSITION

Items To Be Scored	Points Allowed	Points Awarded Contestant												
		1	2	3	4	5	6	7	8	9	10	11	12	13
Content of Manuscript	200													
Composition of Manuscript	100													
Score on Written Production	300													

PART II. FOR SCORING DELIVERY OF THE PRODUCTION

Items To Be Scored	Points Allowed	Points Awarded Contestant												
		1	2	3	4	5	6	7	8	9	10	11	12	13
Voice	100													
Stage Presence	100													
Power of Expression	200													
Response to Questions	200													
General Effect	100													
Score on Delivery	700													

PART III. FOR COMPUTING THE RESULTS OF THE CONTEST

Items To Be Scored	Points Allowed	Points Awarded Contestant												
		1	2	3	4	5	6	7	8	9	10	11	12	13
Score on Written Production	300													
Score on Delivery	700													
TOTALS	1000													
* Less Overtime Deductions, for each minute or major fraction thereof	20													
* Less Undertime Deductions, for each minute or major fraction thereof	20													
GRAND TOTALS														
Numerical or Final Placing														

* From the Timekeeper's record.

Explanation of Score Sheet Points

Part I—For Scoring Content and Composition

1. *Content of the manuscript* includes:
   Importance and appropriateness of the subject
   Suitability of the material used
   Accuracy of statements included
   Evidence of purpose
   Completeness and accuracy of bibliography

2. *Composition of the manuscript* includes:
   Organization of the content
   Unity of thought
   Logical development
   Language used
   Sentence structure
   Accomplishment of purpose-conclusions

Part II—For Scoring Delivery of Production

1. *Voice* includes:
   Quality
   Pitch
   Articulation
   Pronunciation
   Force

2. *Stage Presence* includes:
   Personal appearance
   Poise and body posture
   Attitude
   Confidence
   Personality
   Ease before audience

3. *Power of expression* includes:
   Fluency
   Emphasis
   Directness
   Sincerity
   Communicative ability
   Conveyance of thought and meaning

4. *Response to questions* includes:
   *Ability to answer satisfactorily the questions on the speech which are asked by the judges indicating originality, familiarity with subject and ability to think quickly.

5. *General effect* includes:
   Extent to which the speech was interesting, understandable, convincing, pleasing, and held attention.

*NOTE: Judges should meet prior to the contest to prepare and clarify the questions asked.

Figure 6-11 A score sheet for evaluating speeches.

Courtesy of National FFA.

Entire books have been devoted to developing techniques that enhance the delivery of a speech. However, for the beginner, a few basic and time-tested procedures for effective speaking should suffice. Here are some suggestions:

- Have your teacher read and make suggestions about the content of your written speech.
- Learn the content thoroughly through repeated practice and self-evaluation.
- Record the speech and observe the volume, pace, and expressiveness of your voice. Make adjustments to improve the delivery.
- Practice the speech in front of a mirror to observe posture, hand gestures, and facial expressions. Your posture should be erect and natural, with hands mostly at your sides or resting lightly on the edges of the podium. Do use gestures occasionally to emphasize a point, show direction, or indicate count.
- Ask your teacher for a sample score sheet for judging speeches. Deliver the speech in front of a trusted person who can check your delivery against the score sheet and provide suggestions for improvements. This may be a friend, relative, or teacher.
- Deliver the speech in front of your class for experience and suggestions.
- Anticipate possible questions the judges may ask and prepare for them.
- Include some statements in your speech that you are well prepared to defend. They may lead to questions from the judges that you will be ready to answer because of your preparation.
- Ask your teacher to critique your speech for final approval.
- Deliver your speech in front of civic groups and/or in FFA public-speaking competitions.
- Record a video of your speech for later critique and review.

Parliamentary Procedure

What is parliamentary procedure? Why is it important? Why should agriscience students be interested in learning parliamentary procedure? **Parliamentary procedure** is a system of guidelines or rules for conducting meetings. Most Americans who are influential in their communities are familiar with parliamentary procedure.

Parliamentary procedure is used to guide the meetings conducted by city councils, school boards, church groups, commissions, professional organizations, and civic organizations, such as Lions, Rotary, and Ruritan clubs. Agriscience students should be interested in learning this procedure so they can have their opinions heard and influence decisions that affect their lives. Parliamentary procedure is important because it permits a group to:

- discuss one thing at a time,
- hear everyone's opinion in a courteous atmosphere,
- protect the rights of minorities, and
- make decisions according to the wishes of the majority of the group.

Requirements for a Good Business Meeting

A good **business meeting** is a gathering of people working together to make wise decisions. Wrong decisions can cause conflict, loss of income, inefficiency in business and social activities, injury, and other problems. Poorly run business meetings are a waste of time and accomplish little. Meetings run by groups of individuals who know and follow parliamentary procedure are smooth, efficient, orderly, and focused, and such meetings accomplish much more than poorly organized meetings (**Figure 6-12**). Some requirements for a good business meeting are as follows:

Figure 6-12 Membership in the FFA provides opportunities for students to lead class discussions and to learn to conduct the business of an organization by taking turns as the presiding officer.

Photo courtesy of the National FFA Organization.

Effective Presiding Officer

A **presiding officer** is a president, vice president, or chairperson who is designated to lead a business meeting. They should be committed to the goals of the organization and should want to lead the group in making good group decisions. A good presiding officer must know and use proper parliamentary procedure.

Competent Secretary

A secretary is a person elected or appointed to take notes and prepare minutes of the meeting. **Minutes** is the name of the official written record of a business meeting. Minutes should include the date, time, place, presiding officer, attendance, and motions discussed at the meeting. They should be written clearly, include all actions taken by the group, and be kept in a permanent secretary's book.

Informed Members

Informed members are members who are active in the organization and want to be part of the group. They give previous thought to issues to be discussed and gather useful information about the issues. They share these thoughts with others in the meeting. This permits everyone to have the benefit of the best thinking in the group and permits the best decisions to be made. Effective members know and use parliamentary procedure to bring out important points of discussion and to advance the agenda of the meeting in an orderly manner.

A Comfortable Meeting Room

The meeting place must be comfortable and free from distractions. A moderate temperature and good lighting are essential. Members should be seated so they can hear and see each other. Seating at a table or in a circle works well for small groups. Large groups must rely on a good sound system for the presiding officer to be heard, and members must speak clearly with good volume to be heard. Good public speaking skills and thorough knowledge of parliamentary procedure help the members conduct effective meetings.

Conducting Meetings

The Order of Business

Whether you lead meetings or are taking part in a meeting for the very first time, be sure to learn and follow parliamentary practices. The **order of business** refers to the items and sequence of agenda items for a meeting. The order of business is usually made up by the secretary. This generally grows out of an executive meeting. An executive meeting is a meeting of the officers to conduct the business of the organization between regular meetings. They may also consider what needs to be discussed by the total membership at the regular meeting. The essential items in an order of business are:

- call to order,
- reading and approval of minutes of the previous meeting,
- treasurer's report,
- reports of other officers and committees,
- old business,
- new business, and
- adjournment.

Other items or activities that are frequently included in orders of business are programs, speakers, or entertainment.

Use of the Gavel

The **gavel** is a wooden mallet used by the presiding officer to direct a meeting (**Figure 6-13**). It is used to call the meeting to order, announce the result of votes, and adjourn the meeting. It is also used to signal the members to stand, sit down, or reduce the noise level of the group. The gavel is a symbol of the authority of the office of president or chairperson, and it should be respected by all attending the meeting.

In some organizations, such as the FFA, a system of taps is used to signal the audience to do certain things. In FFA meetings, the gavel is used as follows:

- **One tap**—the outcome of or decision about the item under consideration has been announced by the presiding officer.
- **Two taps**—the meeting will come to order, members should sit down if standing, or members should be quiet except when recognized.
- **Three taps**—members should stand up.

Figure 6-13 The presiding officer and each member should take time to learn how to use the gavel as a tool to conduct the business of an organization.

© Kiratiya Kumkaew/Shutterstock.com.

Obtaining Recognition and Permission to Speak

For a meeting to be orderly, members must speak one at a time and in some logical and fair sequence. The presiding officer is regarded as the "traffic controller" and calls on members as they request to be recognized according to certain rules. To be recognized, the member should raise a hand to get the presiding officer's attention. The presiding officer should call the member by name; then the person should stand and address the presiding officer as Madam or Mr. Chairperson, or Madam or Mr. President. The individual should then proceed to speak. Both the presiding officer and the members should understand the correct classification of motions (**Figures 6-14** and **6-15**).

SUMMARY OF MOTIONS

Motion	Debatable	Amendable	Vote Required	Second	Reconsider
PRIVILEGED					
Fix time which to adjourn	No	Yes	Majority	Yes	Yes
Adjourn	No	No	Majority	Yes	No
Recess	No	Yes	Majority	Yes	No
Question of privilege	No	No	None	None	Yes
Call for orders of the day	No	No	None/2/3	None	No
INCIDENTAL					
Appeal	Yes	No	Majority	Yes	Yes
Point of order	No	No	None	No	No
Parliamentary inquiry	No	No	None	No	No
Suspend the rules	No	No	2/3	Yes	No
Withdraw a motion	No	No	Usually none	No	No
Object consideration of question Negative vote only	No	No	2/3	No	Yes
Division of the question	No	Yes	Majority	Yes	No
Division of the assembly	No	No	No	No	No
SUBSIDIARY					
Lay on table	No	No	Majority	Yes	No
Previous question before vote	No	No	2/3	Yes	Yes
Limit debate	No	Yes	2/3	Yes	Yes
Postpone definitely	Yes	Yes	Majority	Yes	Yes
Refer to committee	Yes	Yes	Majority	Yes	Yes
Amend	Yes	Yes	Majority	Yes	Yes
Postpone indefinitely	Yes	No	Majority	Yes	Yes vote only
Main motion	Yes	Yes	Majority	Yes	Yes
UNCLASSIFIED					
Take from table	No	No	Majority	Yes	No
Reconsider	No	No	Majority	Yes	Negative vote only

For more details on parliamentary procedure, see a parliamentary procedure book such as *Robert's Rules of Order*.

Figure 6-14 Parliamentary skills are useful in FFA and other organizations.
Courtesy of National FFA.

Presenting a Motion

A **motion** is a proposal, presented in a meeting, that is to be acted upon by the group. To present a motion, the member raises a hand and is recognized by the presiding officer. Then the member states, "Madam/Mr. President, I move that …" (and continues with the rest of the motion). The words "I move" are important to say when beginning the motion. Otherwise, you will be regarded as incorrect in your usage of good parliamentary procedure. For a motion to be discussed by the group, at least one other member must be willing to have the motion discussed. That second individual expresses this willingness by saying, "Madam/Mr. President, I second the motion."

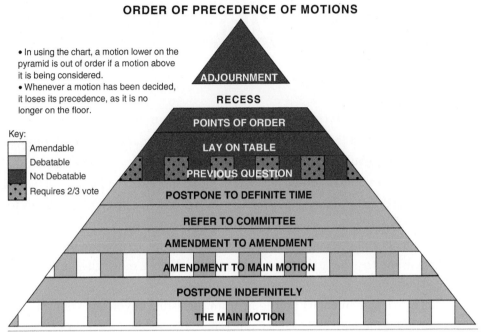

Figure 6-15 Correct order of precedence of motions.

Courtesy of National FFA.

Some Useful Motions

There are dozens of motions, but a few basic motions are generally known and widely used. These include the following:

- **Main motion**—a basic motion used to present a proposal for the first time. The way to state it is to obtain recognition from the chairman and then say, "I move …"
- **Amend**—a type of motion used to add to, subtract from, or strike out words in a main motion. The way to present an amendment is to say, "I move to amend the motion by …"
- **Refer**—a motion used to refer to a committee or person for finding more information and/or taking action on the motion. The way to state a referral is to say, "I move to refer this motion to …"
- **Lay on the table**—a motion used to stop discussion on a motion until the next meeting. The way to table a motion is to say, "I move to table the motion."
- **Point of order**—a procedure used to object to some item in or about the meeting that is not being presented properly. The procedure to use is to stand up and say, "Madam/Mr. President, I rise to a point of order!" The presiding officer should then recognize the member by saying, "State your point." The member then explains what has been done incorrectly.
- **Adjourn**—a motion used to close a meeting. The procedure is to say, "I move to adjourn."

The FFA *Student Handbook* provides a listing of additional motions and explains how to use parliamentary procedure for more effective meetings. Agriscience students are encouraged to develop effective leadership skills. The ability to work effectively as a member of a group is essential for all. The ability to function as a chairperson or officer creates more opportunities to serve and influence the communities in which we live. The development of self-confidence is essential. Self-confidence is a product of knowledge and skill. Therefore, each student should strive to learn to speak well, to function in groups through parliamentary procedure, and to use the opportunities in FFA for personal growth and development.

Chapter Review

Student Activities

1. Write the **Key Terms** and their meanings in your notebook.
2. Make a list of the many ways that you exercise leadership in your family, school, and community.
3. Develop a bulletin board showing the symbols of the FFA emblem.
4. Develop a bulletin board illustrating the purposes of the FFA. Include the FFA colors.
5. Write down five career development events and five proficiency awards for which you would like to try out. Discuss these with your classmates and teacher.
6. Prepare and present a three-minute speech on an agriscience topic to your class.
7. Ask your teacher to let you use a gavel and direct a mock class or FFA meeting to learn parliamentary skills.
8. Form a parliamentary procedure team consisting of you and your classmates, and then demonstrate various parliamentary skills to the class.
9. **SAE Connection:** Under the supervision of your teacher, develop an appropriate solution for a school or district issue. Nominate and appoint a small committee to present your idea, in the form of a main motion, to the principal or school board.
10. Consider running for a leadership position in your FFA chapter. Find which offices are available and make a list of the requirements.

What steps can you take today to develop your leadership skills?
© StockMediaSeller/Shutterstock.com

National FFA Connection: Attend an FFA Event

Courtesy of National FFA; FFA #148

FFA events connect members locally, statewide, and nationally.
Photo by Preston Keres/USDA

The FFA offers many high-interest events and activities, each dedicated to growing the next generation of leaders in agriculture. Attend an FFA activity you have never participated in. This could be a local chapter, district, state, regional, or national activity or convention. The activities range from job shadowing and mock interviews to community service and community events. Visit the FFA website to research activities and events and their locations. With your classmates, select the events or activities that most interest you. Be sure to check the location; click on the "Chapter Locator" link to find events nearest you. Then meet with your teacher to plan the next steps for participation.

Check Your Progress

Can you . . .

- define leader and leadership?
- explain why effective leadership is needed in agriscience?
- list some characteristics of good leaders?
- give examples of opportunities for leadership development in FFA?

If you answered "no" to any of these questions, review the chapter for those topics.

© Africa Studio/Shutterstock.com.

Unit 3
Natural Resources Management

Hikers, bikers, birders, prospectors, hunters, anglers, farmers, foresters, ranchers, caretakers, and occupants all must be good stewards of the land. All are dependent on the land and rely on the soil, air, water, wildlife, and other natural resources around them. Farmers, ranchers, and foresters rely on the land to grow the crops and animals of their businesses. Similarly, hunters, hikers, and other recreational users of the land and water rely on the habitat to grow and sustain the plants and wildlife. All enjoy wildlife for sport and recreation.

Land for farming and ranching is typically owned by the families who occupy the land, and they have definite property rights to grant or deny others access to their property for hunting, fishing, or other recreational pursuits. However, the good-citizen hunter or angler seeks permission of the owner to access private property and strives to protect or enhance the fish and wildlife population and habitat through legal behaviors and good stewardship practices. Both owners and good-citizen users share a love for animals and a respect for crops, pasture, and woodland. Both want to conserve the quality of land, air, and water.

Farmers and ranchers can do much to encourage growth of food, cover, and habitat for wildlife. They interact with wildlife biologists, game specialists, game officers, and other public authorities to nurture game populations and enforce game laws. Hunters and anglers (including farmers and ranchers), wildlife specialists, and the general public all help keep wildlife populations in check by harvesting excess game birds and animals. Game hunting limits are set by wildlife specialists. Scientific methods should be used in efforts to permit and encourage the removal of excess wild animals by hunting. Those animals that are not removed by hunting compete with young game animals for food and shelter. As a result, the young often perish by disease, predation, and starvation. This is nature's way of keeping animal and plant populations in balance.

Public lands are owned by federal, state, or local governments, and people who are educated in many specialized occupations are employed to care for them. Foresters, biologists, fish and game managers, game officers, park rangers, horticulturists, scientists, technicians, and others all contribute to the upkeep and improvement of public lands.

To be considered as good stewards of the land, the owners, managers, and users of both private and public lands must all cooperate in the use and conservation of soil, water, trees, crops, wild plants, livestock, wildlife, and wildlife habitats.

Chapter 7
Maintaining Air Quality

Objective
To determine major sources of air pollution and identify procedures for maintaining and improving air quality.

Competencies to Be Developed
After studying this chapter, you should be able to:
- define the term *air* and identify its major components.
- analyze the importance of air to humans and other living organisms.
- determine the characteristics of clean air.
- describe common threats to air quality.
- describe important relationships between plant life and air quality.
- discuss the greenhouse effect and global warming.
- list practices that lead to improved air quality.

Key Terms
air
ozone (O_3)
water
soil
habitat
sulfur
hydrocarbon
nitrous oxide
tetraethyl lead
carbon monoxide
radon
radioactive material
chlorofluorocarbon (CFC)
particulates
pesticide
asbestos
greenhouse effect
respiration
photosynthesis

Life on Earth requires a certain balance of unpolluted air, water, and soil. **Air** is a colorless, odorless, tasteless mixture of gases. It occurs in the atmosphere surrounding Earth, and is composed of approximately 78 percent nitrogen; 21 percent oxygen; and a 1 percent mixture of argon, carbon dioxide, neon, helium, and other gases (**Figure 7-1**).

Earth's atmosphere is made up of several layers. **Ozone (O_3)** is a molecule that exists in relatively low quantities in the lower atmosphere but in relatively greater quantities in a protective layer approximately 15 miles above Earth's surface. This layer filters out harmful ultraviolet rays from the sun. **Water** is a clear, colorless, tasteless, and nearly odorless liquid. Its chemical makeup is two parts hydrogen to one part oxygen. **Soil** is the top layer of Earth's surface that is suitable for the growth of plant life.

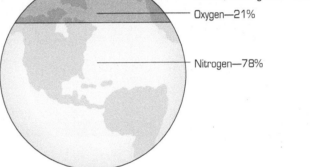

Figure 7-1 The atmosphere of Earth is composed mostly of nitrogen.

Figure 7-2 Clean air, clean water, and productive soil are necessary for a good habitat for plants, animals, and humans.

Snehit Photo/Shutterstock.com.

Air Quality

Without a reasonable balance of air, water, and soil, most organisms would perish. Slight changes in the composition of air or water may favor some organisms and cause others to diminish in number or in health. Unfavorable soil conditions usually mean inadequate food, water, shelter, and other unfavorable factors related to **habitat**, the area or type of environment in which an organism or biological population normally lives (**Figure 7-2**).

Threats to Air Quality

The mixture of gases known as air is essential for life. The air humans breathe should be healthful and life supporting. Air must contain approximately 21 percent oxygen for human survival. If a human stops breathing and no life-supporting equipment or procedures are used, the brain will die in approximately four to six minutes. Air may contain poisonous materials or organisms that can decrease the body's efficiency, lead to disease, or cause death through poisoning.

Even though Earth's circumference at the equator is 24,902 miles, the abuse of the atmosphere in one area frequently damages the environment in distant parts of the world. Air currents flow in somewhat constant patterns, and air pollution moves with them. However, when warm and cold air meet, the exact air movement is determined by the differences in temperature, the terrain, and other factors. Because humans cannot control the wind, they have an obligation to keep the air clean for their own benefit as well as that of society at large.

Science Connection: Environmental Success Story: Elmhurst Park District

The Elmhurst Park District manages parks and other recreational facilities in the city of Elmhurst, Illinois. The park district has twice been honored by "Clean Air Counts," an organization that promotes air quality improvements in the Chicago land area. Smog-forming emissions were reduced in the region by expanding the use of native landscaping in some parks, implementing the use of B-20 biodiesel in vehicles and equipment, installing energy-efficient lighting, and using cleaning supplies and paints that are low in volatile organic compounds (VOCs). The environmental committee responsible for these improvements is known as the "Green Team." The team uses several methods, including an energy audit checklist, to assess district parks' environmental needs and to develop strategies and policies to address them.

Implementing science-based practices can reduce pollution of air and other basic resources.
iStock/Chutima Chokkij

There are major worldwide threats to air quality, including sulfur compounds, hydrocarbons, nitrous oxides, lead, carbon monoxide, radon gas, radioactive dust, industrial chemicals, and pesticide spray materials, among others. Most of these products not only are poisonous to breathe but also have other damaging effects.

Sulfur

Sulfur is a pale-yellow element that commonly occurs in nature. It is present in coal and crude oil. When it combines with these and other fuels in the presence of oxygen, it forms harmful gases such as sulfur dioxide. Most smoke and exhaust from homes, factories, or motor vehicles contain some of these harmful sulfur compounds unless special equipment is used to remove them. Once these invisible gases are in the air, they combine with moisture to form sulfuric acid, which falls as *acid rain*. Some regions along the eastern and western coasts of North America have sustained considerable damage from acid rain. Acid rain damages and kills trees and other plants. It also pollutes waterways and has a corrosive effect on metals.

Hydrocarbons

As the twentieth century proceeded, **hydrocarbons** became serious problems as the numbers of factories and motor vehicles increased (**Figure 7-3**). In the United States, hydrocarbon emissions (by-products of combustion or burning) are held in check by (1) special emission-control equipment on automobiles and (2) special equipment called stack scrubbers in large industrial plants. Hydrocarbon output is also reduced in automobiles by crankcase ventilation, exhaust gas recirculation, air injection, and such engine refinements as four valves per cylinder. Without this equipment, air pollution would be much more intense in major cities and in heavy stop-and-go traffic on major highways.

Figure 7-3 Automobiles, trucks, homes, and factories burn gasoline, oil, coal, and wood, which release products of combustion that pollute the air.
Zacarias Pereira Da Mata/Dreamstime.com.

Nitrous Oxides and Lead

Nitrous oxides are compounds that contain nitrogen and oxygen. They constitute approximately 5 percent of the pollutants in automobile exhaust. Although this seems like a small amount, they are damaging to the atmosphere and current laws require that they must be removed from exhaust gases. This group of chemicals is the most difficult and perhaps the costliest pollutant to remove from automobile emissions. Scientists and engineers have partly solved the problem by developing and installing catalytic converters in automotive exhaust systems.

Hot exhaust gases from the combustion engine flow through a honeycomb of platinum metal in the catalytic converter. The reaction converts the nitrous oxides into harmless gases. Before 1986, all gasoline contained **tetraethyl lead**, a colorless, poisonous, oily liquid that improved the burning qualities of gasoline and helped reduce engine knocking. However, tetraethyl lead ruined catalytic converters. Other substitutes for tetraethyl lead were found in the 1970s, but they have all been displaced in automotive gasoline by adding up to 10% ethanol to gasoline.

Science Connection | Natural Selection as a Result of Pollution

Air pollution affects many aspects of life on Earth. One example was observed in Britain during the Industrial Revolution. In the early 1800s, coal was largely used as fuel to power factories. Burning coal polluted the air with dark soot, some of which would settle and fall to Earth. The effect of such coal pollution on small-winged moths provides a dramatic example of the phenomenon of natural selection. Small-winged moths spend daylight hours on the bark of trees hiding from predators, primarily birds. Their wings are camouflaged to match the trees on which they hide. In the 1800s, however, industrial soot from coal darkened the bark of the trees. Most of the moths at that time had lighter colored wings, so they became easy targets for predators. The few moths that had slightly darker wings gained a competitive advantage; they were able to survive and reproduce because of the darker tree bark. Eventually, most of the moth population adapted so that most had dark wings. Later, in the 1900s, the pollution from soot was reduced. The tree bark returned to its original color, as did the wing color of the moths.

Tetraethyl lead is still used in some developing countries, and lead and nitrous oxides are still major pollutants of the atmosphere in those areas. The levels of pollution, however, have been reduced as the use of lead in gasoline has declined. Today in Western nations, only fuel for small piston-powered airplanes contains tetraethyl lead.

Carbon Monoxide

Carbon monoxide is one automotive gas that cannot be removed with current technologies. This colorless, odorless, poisonous gas kills people in automobiles with leaking exhaust systems or when engines are operated in closed areas without adequate ventilation. Victims lose consciousness and die. Carbon monoxide emissions may be reduced by keeping engines in good repair and properly tuned.

Radon

Radon has become a hazard in homes in many parts of the United States. This colorless, radioactive gas is formed by the decay of radium. It moves up through the soil and flows into the atmosphere at low and usually harmless rates.

However, a hazardous condition can develop if a house or other building is constructed over an area where radon gas is being emitted. The gas can accumulate in buildings that have cracks in basement floors or walls, or it can enter through sump holes. The problem can be prevented by tightly sealing all cracks and/or providing continuous ventilation either below the basement floor or throughout the building (**Figure 7-4**).

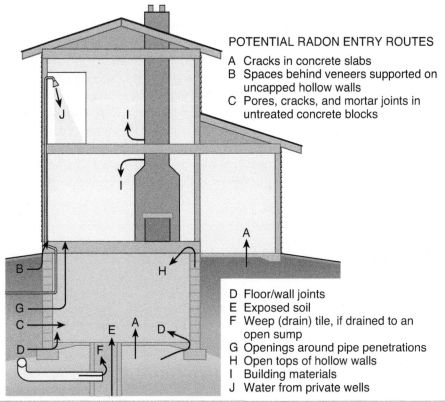

Figure 7-4 Ventilation systems must be correctly designed and carefully maintained to keep interior areas free of radon gas pollution.

Radioactive Dust and Materials

Radioactive material is matter that emits radiation. Of growing environmental concern are dust from an atomic explosion or other nuclear reaction and materials contaminated by atomic accidents or wastes. The damage from radioactivity ranges from skin burns to sickness to hereditary damage to death. Controversy over the possibility of worldwide contamination and other hazards from serious nuclear accidents has led to a reduction in the construction of atomic-powered electric-generation plants.

Chlorofluorocarbons and a Thinning Ozone Layer

Chlorofluorocarbons (CFCs) make up a group of molecular compounds consisting of chlorine, fluorine, carbon, and hydrogen. They were originally used as aerosol propellants and refrigeration gas. These materials are highly stable. Once released from an aerosol can or cooling system, they bounce around in the air and eventually float upward into the upper atmosphere. It is believed that CFCs will survive in the upper atmosphere for about 100 years. Their chlorine atoms destroy ozone molecules without themselves being destroyed. In newer equipment and consumer products, CFCs are being replaced by less damaging materials such as hydrofluoroalkane (HFA). Some aerosol products also use compressed gases, such as nitrogen and CO_2 as propellants.

Beginning in the 1970s, scientists discovered that the ozone layer over the South Pole was less dense, or thinner, than in the past. Scientists identified the cause of the thinning as the buildup of CFCs. A thinner ozone layer lets more of the damaging ultraviolet rays through. This thinning became known as a "hole in the ozone layers." The term "hole" is not literal. No place is without ozone, and concentrations of ozone vary over the course of a year. But the annual thinning has become more concerning. In 2021, researchers from the Copernicus Atmosphere Monitoring Service found an area of thinness that was larger than Antarctica.

Increasingly, international efforts are being made to address this and other environmental issues. International treaties have been negotiated to reduce the production of CFCs. Continuous efforts will be required by all nations to correct this problem.

Particulates

Small particles that become suspended in air are known as **particulates**. These tiny particles appear as smoke or dust clouds. The eruption of a volcano releases massive amounts of particulate matter to the atmosphere, Wildfires also add to the particulate load. However, a large share of particulate pollution comes from motor vehicles and industrial emissions throughout the world.

Agri-Profile Career Area: Air Quality Control and Related Fields

Air quality scientists continuously monitor air quality with instruments placed at strategic locations.
Courtesy of USDA/ARS #K-2228-11.

Careers in air quality control are available with the weather services of local, state, and national agencies. The media also employ weather reporters and meteorology forecasters. Air quality technicians collect air samples taken from various places in the atmosphere, buildings, and homes. These samples are analyzed in laboratories for pollutants known to jeopardize health. Employees of environmental protection agencies and environmental advocacy groups are important links in our efforts to maintain a healthful environment.

Air quality specialists guide industry in reducing harmful emissions from motor vehicles and industrial smokestacks. Many such specialists are engineers who design or modify air pollution-control equipment. Entomologists monitor the winds for signs of invasive insects, and plant pathologists watch for airborne disease organisms. Today, scientists also use statistical analysis and computer modeling. Salaries range widely, from $35,000 to over $100,000 depending on skill level. Education requirements include a bachelor's degree in environmental science, or environmental, chemical, or civil engineering, or a related science. A master's degree is recommended.

Particulates eventually will settle out of the air because of gravity, but they are so light in weight that the slightest breeze keeps them suspended. They are especially harmful to people who suffer from respiratory diseases such as asthma or emphysema. Most particulate matter coming from industrial processes can now be removed from gas emissions by a process called *scrubbing*.

Pesticides

A pest is a living organism that acts as a nuisance or spreads diseases. Examples include houseflies, cockroaches, fleas, and mosquitoes. A **pesticide** is a material used to control pests. Many pesticides are chemicals mixed with water so they can be sprayed on plants, animals, soil, or water to kill or otherwise control diseases, insects, weeds, rodents, and other pests. Spray materials are pollutants if they carry toxic (poisonous) materials or are harmful to more than the target organism. Such sprayed materials are generally harmful to the air if they are not used exactly as specified by the government and the manufacturer. Poisons may be thinned out or diluted by air movement, but excessive toxic materials can overburden the ability of the atmosphere to cleanse itself. The abuse of chemicals to control pests is an area of growing concern in maintaining air quality.

Asbestos

Asbestos is a heat- and friction-resistant material. In the past, it was used extensively in vehicle brakes and clutch linings, shingles for house siding, steam and hot-water pipe insulation, ceiling panels, floor tiles, and other products.

Unfortunately, asbestos fibers are damaging to the lungs and cause disease and death. State and federal laws and codes now require the removal of asbestos from public buildings, industrial settings, and general use.

One serious aspect of asbestos pollution is that pollutants are carried as dust particles by the wind to other areas. This results in damage to the environments of wild animals and fish, particularly those near large cities.

Although asbestos use has declined in automotive replacement parts, older vehicles are still on the road. As brakes and clutches wear down, the asbestos pads become dust particles that can easily become airborne. Technicians who perform such repairs are at risk of inhaling these dangerous particles. Part of the solution to this problem is to remove as much of such pollutants as possible before they are released to the environment.

Atmospheric Effects

One factor that makes life possible on Earth is our planet's proximity to the sun. If the distance between the sun and Earth were any less, the planet's surface would become too hot to sustain life. If the distance were any greater, the temperature would be too cold to support many life forms. The buildup of air pollutants in the atmosphere is widely believed to contribute to global climate change.

Besides being life-sustaining, sunlight has other effects on living organisms. For example, excessive heat in summer and drought are frequently accompanied by parched and withered crops. Medical science has also found a definitive link between extensive exposure to the sun and the occurrence of melanoma, a usually deadly form of skin cancer. Scientific evidence suggests that skin-damaging and life-threatening ultraviolet rays from the sun now reach Earth's surface with greater intensity than in the past.

Scientists and governments of the world are trying to find ways to make continued improvements in living conditions without experiencing additional reductions in air quality and other environmental factors.

The Greenhouse Effect

When sunlight passes through a clear object, such as a glass window, it heats the air on the opposite side of that object. If the warmed air is not cooled or flushed out, heat builds up under the glass—for example, under skylights, auto windshields, and in greenhouses. The glass also absorbs some of the energy from the light and radiates it as heat to the interior area. The overall result is a buildup of heat that causes the interior to become warmer than the outside. This heat buildup from the rays passing through the clear object and the resulting heat being trapped inside is known as the **greenhouse effect**. Gases that are known to contribute to the greenhouse effect include carbon dioxide, water vapor, ozone, methane, halocarbons, and nitrous oxide. They are often referred to as *greenhouse gases*. The combined volume of these gases is approximately one percent of all gases in the atmosphere.

The sun's rays include many different colors of light and types of rays. Some of the more familiar rays are ultraviolet and infrared. Ultraviolet rays are known for their potential to cause extensive skin damage and other life-threatening effects from overexposure. Infrared rays are emitted from any warm object, such as a hot stove, glass warmed by the passage of ultraviolet rays, or an open fire.

The gases in Earth's atmosphere serve as a relatively clear object through which sunlight passes. The crust of Earth absorbs, radiates, and reflects heat back into the air above it. The atmosphere encircles Earth and creates the greenhouse effect (**Figure 7-5**). This increased warming trend is believed by many people to be a serious threat to the environment and to life itself. Scientists contend that the greenhouse effect must be stabilized or reduced to protect air quality (**Figure 7-6**).

Global Warming—A Growing Consensus

Some scientists around the world still debate the issue of global warming versus climate change. Consensus has been building, however, that average temperatures are increasing within the global "greenhouse." Many researchers insist that a recent upward trend in the temperature of Earth's surface is evidence of dangerous warming of the global environment, assuming that current trends continue. They cite human activities as major factors contributing to the intensity of the greenhouse effect and to global warming.

The 2021 report by the Intergovernmental Panel on Climate Change cites the following evidence for rapid climate change:

- Global temperature rise – The average surface temperature of the planet surface has risen 2.12 degrees Fahrenheit since late in the 19th century.
- Warming oceans – Since 1969, the surface water (top 328 feet) of the oceans has increased by 0.6 degrees Fahrenheit.
- Shrinking ice sheets – Since 1969, 279 billion tons of ice has been lost per year from Greenland; 124 billion tons of ice lost from Antarctica.
- Glacial retreat – Glaciers have retreated in nearly every mountain range in the world.
- Snow cover decrease – Satellite measurements of the Northern Hemisphere reveal that spring snow covers have decreased.
- Sea level rise – Global sea levels rose by 8 inches over the last century and the rate of increase continues to accelerate.
- Extreme events – United States record high temperatures have increased and record low temperatures have decreased since 1950.

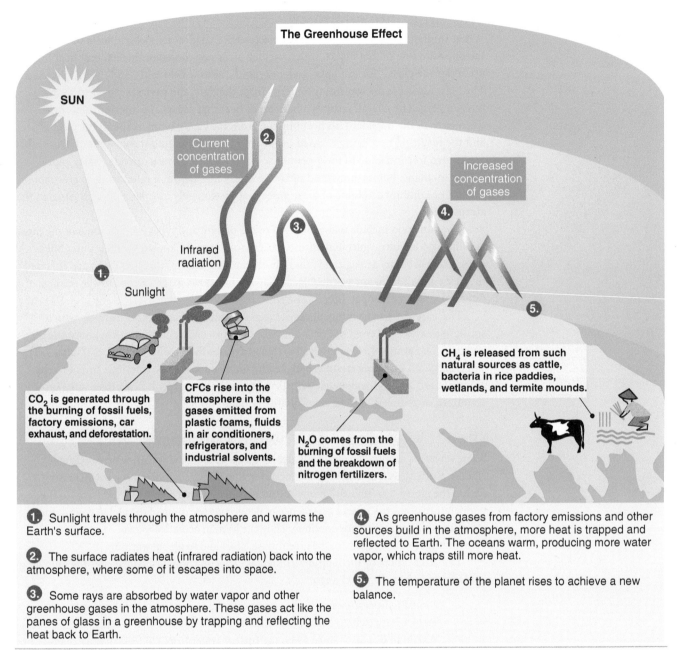

Figure 7-5 The greenhouse effect on land, sea, and air.

The Intergovernmental Panel on Climate Change concluded that:

- "Scientific evidence for warming of the climate system is unequivocal."
- "The evidence for rapid climate change is compelling."

Other scientists believe that the trend toward a rise in global temperatures may be partly the result of long-term climatic cycles and weather patterns. Daily temperatures have been recorded for a little more than 100 years, and we do not have actual historical records from which to draw conclusions on climatic cycles and weather patterns of longer duration.

Physical Factors	Recent Global Changes	Predictions
Temperature	+1.4°F since 1880	Accelerating temperature rise
Sea Level	+3.3 mm/year since 1993	+7.1 to 23 inches in this 21st century
Precipitation	+7% to +15%	Varied precipitation patterns across the planet
Glacial Retreat	Significant	Accelerating
Atmospheric carbon dioxide concentration	280 ppm - previous century 385 ppm - current data	Increasing concentration

Sources: Intergovernmental Panel on Climate Change; NASA's Global Climate Change website; and Schneider, S. (1990) Prudent planning for a warmer planet, New Scientist, 128(1743)

Figure 7-6 Evidence of global climate change.

So, are we really experiencing climate change? Geologic evidence suggests that there have been many periods in the history of planet Earth when the climate has changed significantly. Most climate changes occur slowly over hundreds or thousands of years. Since the beginning of the industrial age approximately 150 years have passed. The change in average global temperature has increased consistently since then. In terms of Earth's climate history, that is a relatively short period of time. In the short term, we are likely to be experiencing global warming, but over time the rising global temperature will surely drive climate change.

Perhaps it is time to shift the conversation from "causes" to "solutions." The more relevant issue in the climate change versus global warming debate seems to be whether nations or their citizens have the willpower and the ability to significantly slow or reverse rising Earth temperatures.

Carbon Dioxide (CO_2)

Carbon dioxide is a major product of combustion. Human's robust appetites for food, clothing, consumer goods, heated and air-conditioned spaces, and transportation have led to a cultural lifestyle that requires huge volumes of fuels to be burned to raise food, manufacture goods, and move vehicles. Highways in and around large cities are clogged with automobiles. Interstate highways are loaded with trucks, and rivers and other waterways carry heavy shipping traffic. Homes and commercial buildings use extensive volumes of electricity for light and temperature control, and farms and factories have huge machines and heating devices—all consuming fuel that expels carbon dioxide and other pollutants into the air.

Even small-engine lawn and garden equipment is now charged with contributing to pollution, and changes are needed to address this problem. Experts now observe that a gasoline lawn mower can cause 50 times more pollution per horsepower than a modern truck engine (**Figure 7-7**). Similarly, a lawn mower running for just 1 hour may create as much pollution as an automobile traveling 240 miles. A chain saw running for 2 hours emits as much pollution from hydrocarbons as a new car running from coast to coast across the United States (**Figure 7-8**).

Figure 7-7 A lawn mower can cause 50 times more pollution per horsepower than a modern truck engine.
© Tim Evans/Shutterstock.com.

Figure 7-8 A chain saw is powered by a two-cycle engine that emits large amounts of air pollutants.
Courtesy of National FFA.

Green plants use some carbon dioxide from the air and convert it into plant food, oxygen, and water. However, in most parts of the world, the mass of green plants is being reduced even as pollutants increase dramatically. The practice of slash-and-burn agriculture in the jungle regions of the world threatens to remove vast areas and volumes of plant growth with little hope of replacement. Increased carbon dioxide in the atmosphere is expected to contribute to global warming in the foreseeable future.

Methane (CH_4)

Most atmospheric methane originates from oil and coal extraction and naturally decaying plant materials such as leaves and debris on the soil surfaces of forests, swamps, and jungles. Some of it is a product of decaying organic matter, such as human and animal waste. Methane rises from piles, pits, and other accumulations of decaying animal manure, peat bogs, and sewage. Some large farms are now using carefully engineered systems to capture the gas from large manure-holding areas. The methane is then used as fuel for engines driving generators to provide electricity for farm use and for sale. Pollution of the air due to methane gas has leveled off in recent years.

Nitrous Oxide (N_2O)

Nitrous oxide has long been a troublesome pollutant. In equal amounts, it is nearly 300 times more destructive to the ozone layer than carbon dioxide, but it is much less abundant in the atmosphere. One source of nitrous oxide is exhaust fumes from gasoline and diesel engines. The ever-increasing number of automobiles, trucks, tractors, heavy equipment, aircraft, chain saws, lawn motors, boats, and other engine-driven applications has offset the tremendous improvements made in emission reduction from individual engines. Much of the recent increase in nitrous oxide emissions appears to be related to increased use of nitrogen fertilizers. Spring application of nitrogen to soil makes it available to other processes until it is absorbed by plants. Soil microbes convert some of it to nitrous oxide gas, which leaches from the soil into the atmosphere.

Science Connection: Overcoming the Effects of Air Pollution

A biological aide uses a porometer to measure the impact of CO_2 enrichment of the air on the transpiration rate and stomata activity of a soybean leaf.
Courtesy of USDA/ARS #K-3204-1.

U.S. Department of Agriculture (USDA) scientists are tackling what could be the toughest conflict of the century: the battle to breathe. Although ozone depletion is a serious problem in the upper atmosphere, ozone as a product of combustion hovers just above Earth's surface as a pollutant that decreases air quality. A recent USDA study found that in the continental United States ground-level ozone has reduced soybean yields by 5 percent and corn yields by 10 percent over the last 30 years.

Researchers have estimated that U.S. farmers experience at least $1 billion per year in lost crop yields because of air pollution. Research indicates that cutting the level of ozone in the air by 40 percent would mean an extra $2.78 billion for agricultural producers.

USDA and University of Maryland scientists discovered that treatment of plants with the growth hormone ethylenediura (EDU) can reduce damage by ozone. EDU alters enzyme and membrane activity within the leaf cells where photosynthesis takes place. A single drenching of soil with EDU effectively protected some plants from damage and reduced the sensitivity of others to excessive levels of ozone in the air. Similarly, injection of EDU in the stems of shade trees in highly polluted areas could protect them from damage.

Other major air pollutants that are damaging to crops include peroxyacetyl nitrate, oxides of nitrogen, sulfur dioxide, fluorides, agricultural chemicals, and ethylene. The task of reducing the amounts of pollutants in the air and finding ways to reduce the effects of pollutants on living organisms will continue to challenge future generations.

Air and Living Organisms

Oxygen in the air is consumed by plants and animals during a process called **respiration** (**Figure 7-9**). Animals, as well as humans, use oxygen to convert food into energy and nutrients for the body. Animals breathe in or inhale to obtain oxygen. They exhale (breathe out) carbon dioxide gas. Plants release oxygen during the day. They create oxygen through the process of **photosynthesis** (a process in which chlorophyll in green plants enables them to use light, carbon dioxide, and water to make food and release oxygen) (**Figure 7-10**).

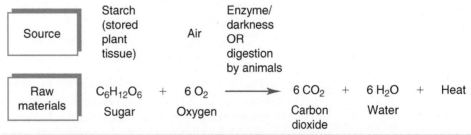

Figure 7-9 Respiration is the process by which plant tissues are broken down to produce heat, water, and carbon dioxide.

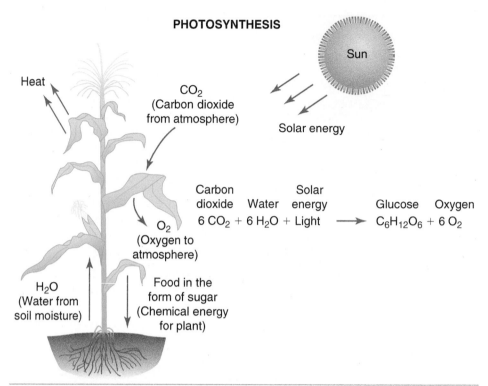

Figure 7-10 Photosynthesis is the process by which carbon dioxide is combined with water to store energy obtained from sunlight. Chlorophyll supports this reaction in which sugar and oxygen are produced.

Maintaining and Improving Air Quality

Air quality can be improved by reducing or avoiding the release of pollutants into the air and by removing existing pollution. Specific practices that can help reduce air pollution include:

- eliminating the use of aerosol products that contain CFCs.
- providing adequate ventilation in tightly constructed and heavily insulated buildings.
- checking buildings for the presence of radon gas.
- using exhaust fans to remove cooking oils, odors, solvents, and sprays from interior areas.
- regularly cleaning and servicing furnaces, air conditioners, and ventilation systems.
- maintaining all systems that remove sawdust, wood chips, paint spray, welding fumes, and dust to ensure that they function most efficiently.
- keeping gasoline and diesel engines properly tuned and serviced.
- keeping all emissions systems in place and properly serviced on motor vehicles.
- observing all codes and laws regarding outdoor burning.
- reporting any suspicious toxic materials or conditions to the police or other appropriate authorities.
- reducing the use of pesticide sprays as much as possible.
- using pesticide spray materials strictly according to label directions.

The U.S. Congress has passed various laws to prevent the loss of air quality and to resolve existing air quality problems. The first significant laws were the Clean Air Act of 1963 and the Air Quality Act of 1967. These laws required reductions in releases of industrial pollutants into the atmosphere. The laws have been expanded and updated several times since then.

Hot Topics in Agriscience

Research to Improve Photosynthesis

Improving the efficiency of photosynthesis is a key area of interest in plant research. It is well documented that the processes of photosynthesis and respiration are closely related. Photosynthesis is the process by which energy from the sun is stored in molecules of sugar; respiration is a process that converts the sugars to high-energy molecules of adenosine triphosphate (ATP) during hours of darkness.

Scientists hope to be able to stimulate plant growth through more efficient photosynthesis, such as by interrupting respiration during hours of darkness in order to maintain daylight gains in plant tissues. In another study, scientists from the USDA Agricultural Research Service (ARS) simulated the 170-step process of photosynthesis and used computer models to identify potential opportunities for increasing photosynthetic efficiency.

Scientists are researching ways to modify photosynthesis to make the process more efficient.

The Clean Air Act has helped reduce air pollution. Tall smokestacks that once dispersed harmful gases over larger areas instead of eliminating their release altogether are no longer legal for pollution control. The current law requires states to develop implementation plans that describe how the state will meet the act's requirements.

The Clean Air Act Amendments of 1970 and 1990 are the most far-reaching of the air quality laws. From these versions, a series of clean air amendments were approved. New federal and state agencies were created to interpret and enforce the laws. Among these are the Environmental Protection Agency (EPA), the Office of Air Quality Planning and Standards, the Alternative Fuels Data Center, and the Commission for Environmental Cooperation. Each government office has created new regulations and standards for air quality.

Among the federal standards that have been implemented are the Clean Air Act's National Ambient Air Quality Standards, the Clean Air Act's New Source Performance Standards, the Prevention of Significant Deterioration, Air Guidance Documents from the EPA's Office of Air and Radiation, and updated air quality standards for smog (ozone) and particulate matter. Each of these air quality standards is intended to reduce air pollution and improve air quality.

As people begin to experience the effects of pollution on the atmosphere in the form of lung and skin diseases, there will be greater motivation to solve the problems created by harmful atmospheric gases. Humans probably will not do much specifically to improve air quality for animals, but animals will benefit when humans improve air quality for themselves.

Chapter Review

Student Activities

1. Write the **Key Terms** and their meanings in your notebook.
2. Make a pie chart illustrating the components of air.
3. Examine three different aerosol cans. Which products use chlorofluorocarbons for the propellant? Why is it unwise to use such products?
4. Talk with an automobile tune-up specialist about the effect of engine adjustments on the content of exhaust gases.
5. Ask your teacher to invite an air pollution specialist to your class to discuss the problems of air pollution in your town, county, or state.
6. Do a research project on the greenhouse effect and its relationship to global warming.
7. Obtain a radon test kit, and perform a test for radon in your home.
8. Create four original drawings showing the four major pollutants. Under each of the drawings, write one sentence that explains where this pollution comes from and what can be done to improve air quality.

Inquiry Activity — Farm Management Conservation

Courtesy of USDA/ARS #K-3204-1.

After watching the **Farm Management Conservation video** available on MindTap, create a T-chart with the following categories:

Farm Management Conservation	
Category	**Solution**
Fertilizer	
Irrigation	
Insects	
Erosion	
Air pollution	
Farm wildlife	

Then watch the video again, this time pausing to list at least one conservation solution for each category. Next, choose at least one technique from the chart that you believe should be standard practice on farms today. Then, write a paragraph in which you define the technique, describe how it works, and explain why you consider it important to put into practice.

National FFA Connection: Addressing Air Pollution

FFA volunteers act locally to protect environmental air quality.
© Jim West/Agefotostock.com.

Every FFA member has the opportunity to take part in activities that make life better in their communities. Some local activities can also benefit our planet Earth! Develop your own plan, with the help of your FFA advisor, to address air pollution where you live. Consider the following ideas:

- Volunteer to chair a chapter committee to improve air quality near your home.
- Engage the committee in confirming possible sources of air pollution.
- Gather air pollution data from local meteorologists and public health officials.
- Plan events that inform and engage others in environmentally friendly events:
  - Develop mini lessons for elementary school classes.
  - Conduct radon tests in homes and public buildings.
  - Plant trees on Arbor Day.
  - Give presentations to local service clubs.
  - Find out about additional activities and events.
- Write an annual report to include what was learned, activities completed, conclusions, and recommendations.

Chapter 8
Water and Soil Conservation

Objective
To determine the relationships between water and soil in the environment and the recommended practices for conserving these resources.

Competencies to Be Developed
After studying this chapter, you should be able to:
- define water, soil, and related terms.
- cite important relationships between land characteristics and water quality.
- discuss some major threats to water quality.
- describe types of soil and water and their relationships to plant growth.
- cite examples of enormous erosion problems worldwide.
- describe key factors affecting soil erosion by wind and water.
- list important soil and water conservation practices.

Key Terms

potable	gravitational water	conservation tillage
fresh water	capillary water	contour farming
tidewater	hygroscopic water	strip cropping
food chain	no-till	crop rotation
water cycle	contour	lime
irrigation	land erosion	fertilizer
watershed	land degradation	grass waterway
water table	sheet erosion	terrace
free water	mulch	aquifer

Figure 8-1 It is difficult to view land as a resource that is in short supply because the amount of farmland seems so vast. However, we are already farming most of the land suited to producing crops.
Courtesy of USDA/ARS #K5051-5. Photo by Scott Bauer.

Flying over the vast continents of the Americas, Europe, Asia, and Africa, we might feel that the landmass of Earth is an endless resource (**Figure 8-1**). The great oceans of the world, however, combine to provide a much larger area. Even so, both land and water resources have become limited, and there is genuine concern that we are rapidly depleting them. In developed countries, a safe and adequate water supply generally faces only temporary shortages and mild inconveniences, such as restricted water for nonessential tasks like washing cars and watering lawns. In developing countries, however, a safe water supply is a luxury. Although it seems that most areas of the world have a sufficient volume of water, supplies of safe water are sometimes insufficient because of misuse, poor management, waste, and pollution.

Similarly, productive land is also becoming a scarce commodity, and ownership is expensive. Individuals seek good land for homes, farms, businesses, and recreation. Industry needs land for factories and storage areas. Governments need land for roads, bridges, buildings, parks, recreation, and military facilities. Most parties look for the best land—land that is level, with deep and productive soil. Such land is in great demand because it provides a firm foundation for roads and buildings; fertile soil produces good crops; soil also supports trees and shrubs, and it can be modified to support the desired use (**Figure 8-2**).

A

B

C

D

Figure 8-2 Good land is in high demand for (A) housing developments; (B) apartments, condominiums, business, and industry; (C) roads and bridges; and (D) farmland.
FeyginFoto/Shutterstock.com. © Elmer Cooper/Cengage 2015. © Elmer Cooper/Cengage 2015. © Ruud Morijn Photographer/Shutterstock.com.

The Nature of Water and Soil

Water

Most of Earth's surface is covered with water (**Figure 8-3**), and its oceans and lakes are vast. Most people around the world, including in the United States, live near an ocean, river, lake, or stream. Those who do not live near such bodies of water must have access to water from deep wells. Like the tissues of plants and other animals, the human body is about 90 percent water, so it can survive only a few days if the supply of potable water—that is, drinkable and free from harmful microorganisms or chemicals—is cut off.

Water is essential for all plant and animal life (**Figure 8-4**). It dissolves and transports nutrients to living cells and carries away waste products. Evaporating water cools the surfaces of leaves and plants and the bodies of animals and humans when the temperature is high. Water serves so many useful functions that a sufficient supply of potable water is one of the first considerations for a healthy community.

Figure 8-3 Most of Earth's surface is covered with water.
© StudioSmart/Shutterstock.com.

Figure 8-4 Water serves many life-supporting functions, including nutrient carrier, waste transporter, coolant, home for aquatic plants and animals, oxygen for fish, and cleanser for humans and animals.
Courtesy of Wendy Troeger.

Fresh Versus Salt Water

Most of the water on Earth is saltwater, not fresh water. It is not suitable for human consumption, however, it does support and sustain living organisms that are adapted to a saltwater environment. **Fresh water** refers to water that flows from the land to oceans and contains little or no salt. The water in oceans contains heavy concentrations of salt. Similarly, bays and tidewater rivers contain too much salt for domestic or household use. **Tidewater** refers to the water that flows up the mouth of a river with rising or in-flowing ocean tides; saltwater is not fit for land-based animal consumption or for plant irrigation.

Aquatic Food Chains and Webs

A **food chain** is made up of a sequence of living organisms that eat and are eaten by other organisms living in the community (**Figure 8-5**). Each member of the chain feeds on lower-ranking members of the chain. The general organization of an aquatic food chain moves from organisms known as producers (food plants) to herbivores (plant-eating water insects). Herbivores are eaten, in turn, by carnivores (meat-eating animals).

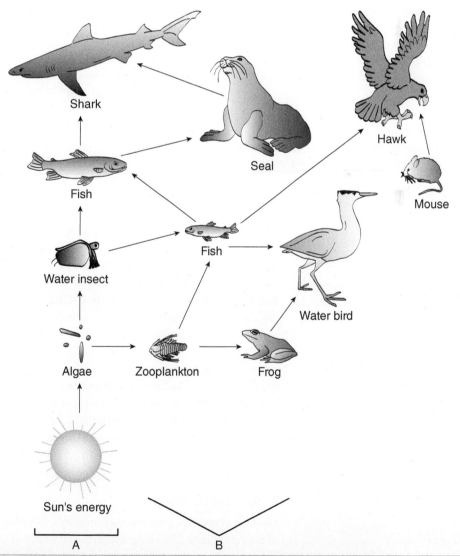

Figure 8-5 (A) A food chain model showing nutritional energy flowing in one direction. (B) A food web model showing the nutritional energy flow throughout the ecosystem.

The Universal Solvent

Water has been described as the universal solvent (a substance that dissolves or otherwise changes most other materials). Nearly every material will rust, corrode, decompose, dissolve, or otherwise yield to the presence of water. Therefore, water is seldom seen in its pure form. It generally has something in it.

Science Connection: Food Chains and Webs

Bays and oceans provide excellent support for the growth of algae, which in turn become major food sources for water-dwelling insects. Insects provide nutritional energy for shellfish. Shellfish are eaten by larger fish, which are finally consumed by top-level predators such as sharks. In this example, energy is passed from one organism to another in the form of food. This mode, referred to as a *food chain*, describes the interdependence of plants and animals for nutrition. All food chains begin with the sun, which is the primary energy source for most living organisms.

Producers at the base of the food chain capture solar energy and convert it into nutrients such as sugars and proteins. Plants and algae are good examples of producers. The next step in the food chain takes place when a consumer eats a producer.

Animals that depend on producers or other animals for food are called *consumers*. In a simple food chain, energy is passed from the sun to the producers and then to consumers in one straight line. In most ecosystems, however, feeding relationships are much more complex than a single food chain suggests. *Food webs*, which are made up of many different overlapping food chains, represent the sum of all feeding relationships in an ecosystem. Using the preceding example, it is easy to see the involvement of other organisms. For example, birds will also eat fish, as do sharks. Water insects may eat aquatic plants other than algae. Both foods chains and food webs explain the feeding relationships in given ecosystems and can be observed on land and in water.

Some minerals in water are healthful and give the water a desirable flavor. However, water sometimes carries toxic or undesirable chemicals or minerals. Water may also contain decayed plant or animal remains, disease-causing organisms, or poisons. Ocean water may be described as a thin soup. It is like human blood. Also, like blood, it gathers and transports nutrients, and it is the habitat for microorganisms. It carries life-supporting oxygen, and it neutralizes and removes wastes.

Scientists tell us that the purity levels of rivers, bays, and seas are in trouble. People are polluting air, water, and land faster than nature can cleanse and purify these resources. The sea contains the dissolved elements carried by the rivers through all Earth's history to the low places on the planet's crust. The water itself may evaporate and be carried back to the land as moisture in the clouds. Eventually, it may fall to the land again as rain or snow. However, the minerals remain behind in the ocean. Many of those substances may be toxic or otherwise threatening to organisms living in the sea. The water that falls as rain or snow would be pure if it were not contaminated by pollutants as it falls through the air.

The Water Cycle

Moisture evaporates from Earth, plant leaves, freshwater sources, and the seas to form clouds in the atmosphere. Clouds remain in the air until warm air masses meet cold air masses. This causes the water vapor to change to a liquid and fall to Earth's surface in the form of rain, sleet, or snow. Large amounts of Earth's water supply are stored for long periods of time.

Storage occurs in aquifers beneath the surface of Earth, in glaciers and polar ice caps, in the atmosphere, and in deep lakes and oceans. Sometimes this water is stored for thousands of years before it completes a single cycle. Gravity draws the water back into the ground and causes it to flow from high elevations to low elevations. The cycling of water among the water sources, atmosphere, and surface areas is called the **water cycle** (**Figure 8-6**).

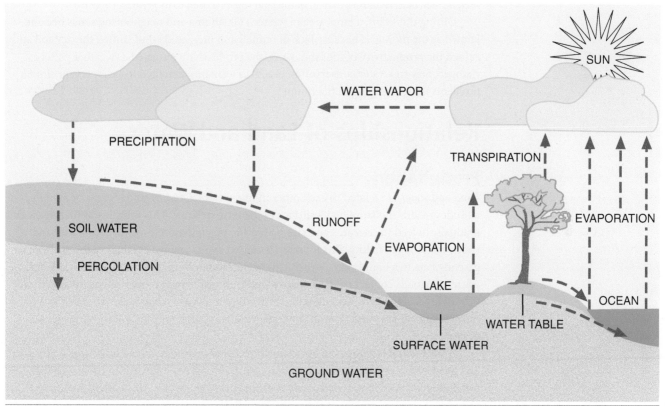

Figure 8-6 The water cycle is a natural process by which water moves in a circular flow from oceans to land and back to the oceans.

The energy that drives this cycle comes from two sources: solar energy and the force of gravity. Water is constantly recycled. A molecule of water can be used over and over again as it moves through the cycle. Solar energy trapped by the ocean is a source of heat that causes evaporation. Additional water enters the atmosphere by evaporating from soil and plant surfaces, especially in areas of hot temperatures and high precipitation.

Land

Land provides a solid foundations for buildings. It also provides nutrition and support for plants and space for work and play. Its aquifers provide storage for groundwater. Landmass also serves as a heat and compression chamber, converting organic material into coal and oil.

Soil is an important component of land. Productive soil is made up of correct proportions of soil particles and has the correct balance of nutrients. It also contains at least some organic matter and has adequate moisture to sustain the living organisms that live in a soil environment. Much of Earth's crust is too rocky or has an incorrect balance of nutrients for crop production. Much of Earth's landmass is covered by only a thin layer of productive soil or is too steep to permit cultivation. Where there is some useful soil, however, trees are often able to survive. Forest lands provide lumber, poles, paper, and other products. Forest lands also provide areas where humans can benefit from the pleasures of wildlife and recreation.

Large areas of Earth have soil with a usable balance of soil particles and minerals for plant growth, but large areas have insufficient water. Areas with continuous, severe water shortages are called deserts. Some desert areas have become productive through use of modern irrigation practices. **Irrigation** is the addition of water to the land to supplement the water provided for crop production by rain or snow. Many nations in desert regions do not have the money, technical knowledge, or water to make their deserts productive. Many nations of the world have such limited land resources that all of the land must be used to its greatest capacity.

During the 1930s, a large area of western Oklahoma and neighboring states became known as the *dust bowl* because lack of rainfall and increased wind shifted the dry soil and ruined the productivity of the land. Later, the productivity returned as weather patterns changed. Soil and water conservation practices were implemented to help prevent similar problems from recurring in the future.

Relationships of Land and Water

Precipitation

Land and water are related to each other in many ways. Land in cold regions or at high altitudes retains moisture on its surface in the form of snow. This moisture is then released gradually to feed the streams and rivers after the precipitation (moisture from rain and snow) has stopped falling. Precipitation is caused by the change of water in the air from a gaseous state to a liquid state. It then falls to the land or bodies of water. Moisture-laden, warm-air clouds contact cold-air masses in the atmosphere and the result is precipitation. Clouds are formed by water changing from a liquid to a gas when it is evaporated by air movement over land and water. *Evaporation* means changing from a liquid to a vapor or gas.

A **watershed** is a large land area in which water from rain or melting snow is absorbed and from which water drains as it emerges from springs and moves into streams, rivers, ponds, and lakes. The watershed acts as a storage system by absorbing excess water and releasing it slowly throughout the year (**Figure 8-7**).

Figure 8-7 A watershed consists of a large land mass that functions like a huge sponge. It soaks up excess water in times when a surplus exists and releases it in a somewhat uniform flow from springs and wells throughout the year.

Biletskiyevgeniy.com/Shutterstock.com.

Land as a Reservoir

In more ways than one, land serves as a container or reservoir for water. Where water soaks down into the soil, it forms a **water table**. Below that level, soil is saturated or filled with water. Water held below the water table may run out onto Earth's surface at a lower elevation in the form of springs. Springs feed streams, which in turn form rivers that flow into lakes, bays, and oceans. Since ancient times, people have been known to dig wells below the water table to extract water for human needs.

Even above the water table, excess water, held in the soil, is taken up by plant roots, from which water travels throughout the plant. Much of it evaporates from the leaves through transpiration to contribute to the moisture supply in the atmosphere.

Water supports soil tilth by improving its physical structure. It is also an essential nutrient for organisms that live in the soil, and it contributes to the soil's fertility (the amount and type of nutrients in the soil).

Types of Groundwater

Soil is saturated when water fills all the spaces or pores in the soil. If soil remains saturated for too long, plants will die from lack of air around their roots. The water that drains out of soil after it has been wetted is called **free water**, or **gravitational water**. Gravitational water feeds wells and springs. When gravitational water leaves the soil, some moisture—**capillary water**—remains, and this can be taken up by plant roots. Water that is held too tightly for plant roots to absorb is **hygroscopic water** (**Figure 8-8**).

Groundwater is easily polluted by chemicals and animal manure as polluted water seeps through the soil layers to enter the water table. In addition, constant irrigation can increase salt concentration in the soil. These and other problems are being addressed through scientific research and by soil conservation and water-management districts.

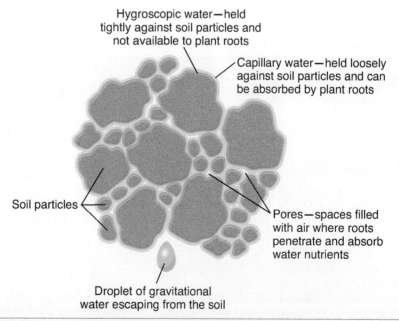

Figure 8-8 Plants use mostly capillary water. Hygroscopic water contributes to soil structure, and gravitational water is held in reserve for future use by plants and animals.

Benefits of Living Organisms

Plants break the fall of raindrops and reduce damage to the soil structure due to the impact of falling water. When plants drop their leaves, plant materials accumulate, and they provide a rain-absorbing layer on the soil surface. Worms, insects, bacteria, and other small, and even microscopic, plants and animals contribute to soil by decomposing dead plant and animal matter. As these materials decay, they add organic matter to the soil, improving its structure. Soil organisms contribute to water absorption and reduce soil erosion. Decayed plant and animal matter contributes substantially to the nutrient content of the soil. Plants assist in distributing water by taking it up through their roots and releasing it in the atmosphere. Both soil and water benefit from the living organisms that live in the soil.

Conservation and Quality Improvement of Water

How can water pollution be reduced? How can soil erosion be reduced? What is the most productive use of water and soil that does not pollute or lose these essential resources? These important questions deserve answers now. Farmers have long appreciated the value of these resources and generally have used them wisely. Economic conditions, governmental policies, production costs, farm income, personal knowledge, and other factors, however, also influence the use of conservation practices by farmers and other land users. Every citizen, business, agency, and industry influences air and water quality. Similarly, each of person has some influence on how land and water resources are used.

Improved water quality can be achieved by proper land management, careful water storage and handling, and appropriate water use. Once clean water is mixed with contaminants, it is not safe for use for other purposes until it is cleaned and purified. Purify means to remove all foreign material. Several general efforts can help improve water quality: controlling water runoff from lawns, gardens, feedlots, and fields; keeping soil covered with plants to avoid wind erosion; constructing livestock facilities so manure can be collected and spread on fields; and where feasible, using no-till or minimum tillage practices to produce crops.

Science Connection | From Manure to Methane

Technology makes it possible to convert animal manure to gas for heating and electricity.

The size of livestock enterprises has increased dramatically in recent years, with thousands of animals concentrated on large farms. Manure disposal from such farms has become challenging to many owners, particularly when they do not have enough cropland on which to spread it. Too much manure on the land is likely to contaminate the groundwater beneath the soil surface.

What to do? Treat the manure as a profitable commodity by converting it to methane gas. A manure digester is capable of providing a reliable supply of gas that can be (1) used to heat buildings on the farm, (2) used to generate electricity, or (3) sold as gas to local utilities for distribution to their customers for heating water and homes.

Some practices that help reduce water pollution are the following:

- **Save clean water.** Whenever we permit a faucet to drip, leave water running while we brush our teeth, or flush toilets excessively, we are wasting water. Taking long showers, leaving water running while washing automobiles or livestock, and using excessive water for pesticide washup wastes clean water.
- **Dispose of household products carefully.** Many products under the kitchen sink, in the basement, or in the garage are threats to clean water (**Figure 8-9**). Never pour paints, wood preservatives, brush cleaners, or solvents down the drain. Eventually, they enter the water supply, rivers, or oceans. Use all products sparingly and completely. Put solvents into closed containers and permit the suspended materials to settle out before reusing the solvent. Fill empty containers with newspaper to absorb all liquids before discarding. Then send the containers to approved disposal sites.
- **Maintain lawns, gardens, and farmland carefully.** Improve soil by adding organic matter. Mulch lawn and garden plants as shown in **Figure 8-10**. Use proper amounts and types of lime, fertilizer, and other chemicals. Leave soil untilled if it is likely to erode excessively. Cover exposed soil with a new crop immediately. Apply water to the soil only when it is excessively dry, and continue to irrigate until the soil is soaked to a depth of 4–6 inches.

Figure 8-9 Many products around the home are hazardous and require correct disposal to avoid pollution of groundwater.

iStock.com/Shank_ali

Figure 8-10 A mulching system conserves water and allows plant vegetation from lawns and gardens to be added back to the soil as a nutritious organic mulch.

Courtesy of DeVere Burton.

- **Practice sensible pest control.** Many insecticides kill all insects—both harmful and beneficial. Incorrect applications of insecticides also pollute water. Therefore, use cultural practices, such as crop rotations and resistant varieties, instead of insecticides, whenever possible. Encourage beneficial insects and insect-eating birds by improving the habitat around lawn, garden, and fields. Drain pools and containers of stagnant water to prevent mosquitoes from laying eggs resulting in larvae populations. Keep untreated wood away from soil to avoid damage from termites, other insects, and rot. Follow all pesticide label instructions exactly.

Figure 8-11 Construction of terraces on sloping land distributes water precipitation along the terrace and helps to prevent soil erosion.

Courtesy of DeVere Burton..

- **Control water runoff from lawns, gardens, feedlots, and fields.** Keep the soil surface covered with plants. Construct livestock facilities in such a manner that manure can be collected and spread on fields. Where it is feasible, use no-till cropping. **No-till** means planting crops without plowing or disking the soil. Plant alternate strips of close-growing crops (such as small grains or hay) with row crops (such as corn or soybeans). Farm on the **contour** (following the level of the land around a hill). Plant cover crops where regular crops do not protect the soil throughout the year. A cover crop is a close-growing crop planted to temporarily protect the soil surface.
- **Control soil erosion.** Reduce the volume of rainwater runoff by minimizing the amount of blacktop or concrete surface constructed. Use grass waterways in areas of fields where runoff water tends to flow. Add manure and other organic matter to soil to increase water-holding capacity. Construct terraces on long or steep slopes (**Figure 8-11**). Leave steep areas in tree production and excessive sloping areas in close-growing crops.
- **Recycle petroleum products.** Avoid spillage or dumping of petroleum products such as gasoline, fuel, or oil on the ground or in storm drains.
- **Prevent chemical spills from running or seeping away.** Do not flush chemicals away. Chemicals will damage lawns, trees, gardens, fields, and groundwater. Sprinkle spills with an absorbent material such as soil, kitty litter, or sawdust. Remove the contaminated material and place it in a strong plastic bag to be discarded according to local recommendations.
- **Follow best practices.** Properly maintain your septic system (if your home has one). Avoid using excessive water. Do not flush inappropriate materials down the toilet. Avoid planting trees where their roots might penetrate or interfere with sewer lines, septic tanks, or field drains. Do not run tractors and other heavy equipment over field drains.

Science Connection — Indicator Species

Aquatic invertebrates like this dragonfly offer clues to an environment's health.

Courtesy of DeVere Burton.

One way scientists can tell that water has been polluted is by studying *indicator species*, or sensitive organisms that show the state of an environment's overall health. The four categories of indicator species in water are fish, aquatic invertebrates (including worms, snails, and aquatic insects), algae, and aquatic plants. A pollutant in water affects these organisms in different ways. Depending on the type of pollution, the organisms can die off, have their numbers reduced, or increase their numbers.

An algal bloom is an example of how an indicator species is affected by pollution. Runoff water from farmlands on which fertilizer is too heavily applied may contain excess fertilizer. As this fertilizer enters a lake or river, it can stimulate algae growth. The result is an overgrowth called an *algal bloom*. These blooms are often seen late in the growing season and indicate to scientists that a fertilizer pollutant is in the water. Specific problems may arise from such a bloom. Algae can produce toxins that are harmful to fish, animals, humans, and other organisms. When the algae die and decompose, oxygen is consumed, and oxygen levels in the water decline, often killing other organisms that depend on the oxygen. Clean and safe water is needed for drinking, watering crops, and maintaining healthy living environments. Using indicator species helps scientists determine the safety of the water supply.

Land Erosion and Soil Conservation

Land Erosion—A Worldwide Problem

Land erosion, the process of wearing away of soil, is a serious problem worldwide (**Figure 8-12**). Both wind and water are capable of eroding soil. The food and fiber production capabilities of large nations are being compromised because of extensive damage from soil erosion. Numerous cases can be cited from history when soil erosion caused enormous problems. Yet even today the people of the world are making many of the same mistakes that caused extensive soil losses in the past.

Figure 8-12 Erosion of soil means loss of productive soil, damage to machinery, additional costs of production, and pollution of streams, rivers, lakes, bays, and oceans.

Courtesy of DeVere Burton.

Consider, for example, the port of St. Mary's City, Maryland, an early European colony established in 1634 on the eastern coast of what is now the United States. Like all ports, it was located along the shoreline, where a natural harbor made it possible for ships to load and unload cargo. By 1706, the port had been replaced by another town where the central port for the colony was located. The harbor area of the original port rapidly filled with soil eroded from the surrounding area. The water was no longer deep enough for the ships to enter safely.

Consider the massive size of the Mississippi River Delta. This land area was created from the soil deposits carried by the river. The Mississippi River Delta is more than 15,000 square miles in area. The delta was formed as the river deposited soil on the bottom of the Gulf of Mexico, continuing until it reached the surface of the water. The soil gets into the river from smaller rivers and streams receiving runoff water from surrounding land areas. Therefore, the Mississippi River Delta was built from the bottom to the surface of the Gulf of Mexico with the best topsoil in the United States. It is estimated that the amount of soil being dumped by the river into the Gulf each day would fill a freight train 150 miles long. By estimate, the amount deposited at the mouth of the Mississippi River in one year is sufficient to cover the entire state of Connecticut with soil 1 inch deep, or more than 8 feet in 100 years. Similar deltas exist at the mouth of the Nile River and other great rivers.

Agri-Profile — Career Area: Soil Science

Soil conservation professionals find employment in careers that focus on protecting soil from pollution and erosion.
Courtesy of USDA/ARS #K1918-10.

Soil science career options include a variety of opportunities to work in both indoor and outdoor environments. The seasons allow for field work to be done during the spring, summer, and fall with analysis of data and development of conservation plans, etc. done inside during inclement weather. Soil science graduates often find careers in offices of the federal Farm Service Agency, state governments, and the private sector. One may work in the field served by such offices in nearly any county in the United States.

The U.S Department of the Interior, state departments of natural resources, city and county governments, industry, and private agencies, such as golf courses, hire people with soil and water conservation expertise. These professionals manage soil and water resources for recreation, conservation, and consumption. They sometimes work as consultants, law enforcement officers, technicians, heavy equipment operators, and other emerging careers.

A soil technician or laboratory technician may begin their career by earning an associate's degree and/or working summers as a student trainee in the traditional career path. The employment entry level for soil scientists is a four-year or graduate degree from a university soil science program. The career path generally allows for advancement, depending on education and experience.

During the late 1960s and early 1970s in the People's Republic of China, much of the forested land was cleared. It soon became apparent that much of the land depended on the trees to prevent soil erosion. Due to the extent of damage from soil erosion, the country has responded with a national policy of rapid reforestation. China has more than 1.4 billion people, and all are expected to contribute to the tree-planting effort. However, it will take many years for newly planted trees to be mature enough to protect the soil again. Unfortunately, the soil that was washed away during the absence of trees can never be returned to the land. Soil scientists report that it takes 300–500 years for nature to develop one inch of topsoil from bedrock.

Another serious threat to large areas of the world is the farming system known as "slash and burn." This is used in the tropical rain forests of South America, Africa, and Indonesia. In these areas, impoverished farmers cut or slash the jungle growth and burn the plant residues. The extensive burning causes serious air pollution and destroys useful fuel. These farms produce crops for several years until the land's fertility is depleted and soil erosion takes its toll. The land is then abandoned, and the farmers move on to forested land, where the cycle is repeated. Once the land is cleared, native plant growth is not easily reestablished, and the land is often left permanently damaged. While the loss of the rain forest of the Amazon slowed from 2004 to 2012, today it declines by approximately 7,250 square miles each year. In some regions, 80 percent of the rainfall is attributed to transpiration from trees. When the trees are removed, such areas may become subject to drought.

Soil Loss on U.S. Farmland

Each year, about 5.8 tons of soil per acre are eroded away from 417 million acres of U.S. farmland and deposited into lakes, rivers, and reservoirs. One ton equals 2,000 pounds. Although some soils are deep and can tolerate a certain amount of erosion, many fragile soils cannot. The U.S. Department of Agriculture (USDA) National Resources Inventory indicates 41 million acres (or about 10 percent) of the nation's cropland are highly erodible at rates of 50 or more tons per acre per year (**Figure 8-13**).

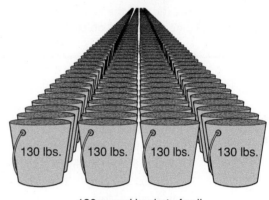

Figure 8-13 Imagine how much work it would require to replace the soil that is lost each year from one acre of highly erodible land.

Causes of Soil Erosion

Soil is formed so slowly that we consider it to be a finite natural resource; once it is gone, we cannot replace it. We must conserve the soil we have. So, what causes soil to be lost? Soil erosion processes must be understood before the loss of soil can be stopped. It is well understood that several conditions affect the rate of soil erosion:

- Intensity of water movement – Falling rain separates soil particles allowing them to be relocated by runoff water.
- Slope of the land – As slope increases, the speed of runoff water also increases, allowing exposed soil to be eroded away.
- Texture of soil – Soil structure occurs when soil particles combine with organic matter to form in clumps. Fine, sandy soil particles are easily eroded.
- Removal of forests – Tree roots stabilize soil, and the leaves form a protective cover on the soil surface. Both leaves and roots protect soil from erosion.
- Wind – The upper soil layers may be scoured off the surface by strong wind gusts, especially if plant roots and residue have been removed.

Science Connection: Land Degradation

Land degradation is the decline of soil quality or reduction in productivity. It has become one of the most important worldwide societal problems because it affects the security of the world food supply. The human population is growing and is expected to reach 8.6 billion people by the year 2030. At the same time, vast tracts of the most productive land are removed from production each year for housing and industrial development. Under these circumstances, degradation of the remaining agricultural land has become a critical problem.

Degradation of land is generally associated with human activities such as farming methods and industrial processes. For example, much of the land in the United States has been classified into land capability classes. Using land for purposes for which it is not suited contributes to the loss of soil structure, erosion, saline soils, soil crusting, compaction of soils, and pollution. Eventually, each of these problems is expected to contribute to unsustainable crop yields.

Preventing and Reducing Soil Erosion

Soil erosion can be reduced or stopped by good land management. Fortunately, management practices that reduce soil erosion also increase water retention (**Figure 8-14**). This helps stabilize the supply of fresh water available for crops, livestock, wildlife, and people. It also reduces airborne soil particles, which are threats to good air quality. Most soil conservation methods are based on (1) reducing raindrop impact, (2) reducing or slowing the speed of wind or water moving across the land, (3) securing the soil with plant roots, (4) increasing absorption of water, or (5) carrying runoff water safely away. If water is free to run down a hill, it increases in volume and speed, picks up and pushes soil particles, and carries these particles off the land. This results in sheet and gully erosion and deposits soil, nutrients, and chemicals into streams. **Sheet erosion** occurs as layers of soil are removed from the land, whereas gully erosion is severe soil loss that leaves trenches in the land surface.

A B C

Figure 8-14 Plant growth holds the key to soil and water conservation. (A) Surface mulch protects the soil surface, (B) extensive root systems of plants protect soil particles from erosion, and (C) revegetation of shorelines with willow trees helps hold streams in their courses.

Courtesy of USDA/ARS #K-3226-6, Courtesy of USDA/ARS #K-3213-1, Courtesy of USDA/ARS #K-5Z10-18.

The following practices are recommended to reduce or prevent wind and water erosion:

- **Keep soil covered with growing plants.** Plants reduce the destructive impact of raindrops, reduce the speed of wind and water movement across the land, and hold soil in place when threatened by wind or moving water.
- **Cover the soil with mulch.** Mulch is a material placed on the soil surface to break the fall of raindrops and prevent weeds from growing.
- **Manage irrigation practices to avoid flowing water across soil surfaces.**
- **Utilize conservation tillage methods.** Conservation tillage means using techniques of soil preparation, planting, and cultivation that disturb the soil the least, leaving the maximum amount of plant residue on the surface. Plant residue is the plant material that remains when a plant dies or is harvested.
- **Use contour practices in farming, nursery production, and gardening.** Contour farming means conducting all operations, such as plowing, disking, planting, cultivating, and harvesting, across the slope and on the level. This way, any ruts or ridges created by machinery occur around the slope or hill. When water flows downhill, it encounters the grooves and ridges. Because these are level, the water tends to soak into the soil, holding the water for future use.
- **Use strip cropping on hilly land.** Strip cropping means alternating strips of row crops with strips of close-growing crops, such as hay, pasture, and small grains. Examples of row crops are corn, soybeans, and most vegetables. The strips of close-growing crops intercept runoff water from the row crops and prevent it from entering streams.
- **Rotate crops.** Crop rotation is the planting of different crops in a given field every year or every several years. Crop rotation permits close-growing crops to retain water and soil, and it tends to rebuild the soil, countering losses incurred when row crops occupied the land.
- **Increase organic matter in the soil.** Organic matter is dead plant and animal tissue. Dead plant leaves, stalks, branches, bark, and roots decay and become organic matter. Similarly, animal manure, dead insects, worms, and animal carcasses decompose to make organic matter. The organic matter forms a gel-like substance that holds soil particles in absorbent granules called aggregates. An aggregated soil is a water-absorbing and nutrient-holding soil. Organic matter also releases nutrients to growing plants.
- **Provide the correct balance of lime and fertilizer.** Lime is a material that reduces the acid content of soil. It also supplies nutrients such as calcium and magnesium to improve plant growth. Fertilizer is any material that supplies nutrients for plants.
- **Establish permanent grass waterways.** A grass waterway is a strip of grass growing in an area of a field where excess water regularly flows across the soil surface (**Figure 8-15**).
- **Construct terraces.** A terrace is a soil or wall structure built across a slope to capture water and move it to areas where erosion is less likely.
- **Avoid overgrazing.** Overgrazing results when too much of a plant is eaten, reducing its ability to recover after grazing as well as the ability of the roots to hold soil in place.
- **Follow a soil conservation plan.** A conservation plan is a plan developed by soil and water conservation specialists to use land for its maximum production and water conservation without unacceptable damage to the land.

Figure 8-15 Soil that is most vulnerable to erosion should be planted to permanent vegetation.

Courtesy of DeVere Burton.

The USDA Natural Resources Conservation Service (NRCS) and Farm Service Agency (FSA) provide advice, technical assistance, and funds to assist landowners with soil and water conservation practices. The Environmental Protection Agency (EPA) monitors groundwater quality and enforces point and nonpoint source pollution laws to help protect land and water.

Of growing concern is the contamination of groundwater under large areas of the United States. Groundwater pollution emerged as a public issue in the late 1970s. The first reports documented sources of contamination associated with the disposal of manufacturing wastes. By the early 1980s, several incidences of groundwater contamination by pesticides used on field crops were confirmed. Groundwater contamination can threaten the health of large populations. For instance, in New York State, the vast aquifer, a water-bearing rock formation, that underlies Long Island, represents the only supply of drinking water for more than 3 million people. In the United States, major aquifers underlie areas that contain thousands of square miles of land, often encompassing several states. Contamination of the aquifer in one region usually results in contamination of larger areas.

There are many careers in the field of water and soil conservation. The material in this chapter discusses some of the problems encountered and/or created by humans as they use natural resources to provide food, water, shelter, recreation, and other resources for living. The health, wealth, peace, and general welfare of humanity depend on our skill in conserving these most basic resources.

Constructed wetlands are a growing trend in agriculture. Farmers understand that water is priceless. Some have chosen to set aside a fraction of their property as a wetland to help purify water and ensure the safety of this important resource.

Hot Topics in Agriscience: Constructed and Natural Wetlands

Constructed wetlands are human-made marshlands that can be used to treat polluted water. They contain plants that tolerate standing water and provide habitat for bacteria that are capable of breaking down materials that cause water pollution. As water passes slowly through the marsh, these bacteria and other microbes cleanse the water by consuming pollutants and changing them to harmless compounds. Some water plants also take up pollutants.

A natural wetland performs the same function, removing impurities from the water. Many of the natural wetlands, however, have been drained to prepare land for farming. Currently, we are beginning to recognize the value of the wetlands. Constructed wetlands usually consist of rectangular plots arranged in a series that are filled with gravel or porous soil and lined to prevent pollutants from leaching into the groundwater. Plants are added to the plots to simulate a natural marshland. Constructed marshlands do a good job of mimicking the advantages of natural marshlands. They also help create excellent wildlife habitat.

Science Connection: Safeguarding Groundwater

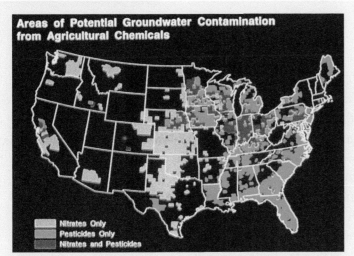

Groundwater contamination by agricultural chemicals is considered a risk in many areas.

Source: Nielsen, E. G. and Lee, L.K. October 1987. The Magnitude and Costs of Groundwater Contamination from Agricultural Chemicals: A national perspective. Resources and Technology Division, Economic Research Service, U.S. Department of Agriculture. Agricultural Economic Report No. 576. Figure 8. Retrieved from http://naldc.nal.usda.gov/download/CAT88907300/PDF

Authorities estimate that approximately one billion pounds of pesticides are applied yearly on U.S. crops. Added to this are the fertilizers that are used on farms, yards, gardens, parks, and so on, and the chemicals used to control fleas, flies, mosquitoes, ticks, termites, roaches, and other pests of livestock and humans. Although agricultural chemicals have enabled us to feed and clothe ourselves and many others in the world, pollution from these chemicals has become a threat to air, water, and land.

Applying these pesticides where they are intended and keeping them there until they change or biodegrade into harmless products is a major goal in protecting water supplies. To help safeguard surface water and groundwater from pesticide pollution, the USDA Agricultural Research Service (ARS) has developed a specific strategy that has helped shape research priorities. The ultimate goals of the plan are to (1) provide U.S. farmers with cost-effective best-management practices that will ensure ample supplies of food and fiber at a reasonable cost, while reducing pesticide movement into the groundwater; (2) identify the factors that accelerate or retard pesticide movement; and (3) provide computer models that will quickly and accurately predict contamination.

USDA water research is being concentrated in three areas to protect against contamination of groundwater: (1) agricultural watershed management, (2) irrigation and drainage management, and (3) water quality protection and management.

The computer models being used and developed are decision-enhancing tools that use a database and computer program to help select management practices. Such practices may include pesticide application procedures that are most likely to restrict pesticide movement in various soils due to climatic and other environmental conditions. These models integrate data from many sources that are frequently not available to decision makers. New technologies, such as advances in slow-release formulations, improved pesticide application scheduling, and selective placement, are also under study. Today, groundwater contamination by agricultural chemicals is considered a risk in many of the major crop and livestock production areas of the United States.

Chapter Review

Student Activities

1. Write the **Key Terms** and their meanings in your notebook.
2. Study the eating habits of one species of birds in your community. Describe the food chain that accounts for the survival of that species. What effect did the use of DDT as an insecticide have on that species of birds before DDT was banned from use in the United States?
3. Obtain a rain gauge and record the precipitation on a daily basis for several months. What variations did you observe from week to week? How do you explain the variations? What effects did these variations have on the agricultural activities of the community?
4. Conduct an experiment to determine the effect of soil water on plant growth. Obtain three inexpensive pots of healthy flowers or other plants. Each pot must be the same size and type, have the same amount and type of soil, and contain the same size and number of plants. Use the following procedure:
   - Mark the pots "1," "2," and "3." Plug the holes in the bottoms of pots 2 and 3 so water cannot drain from the pots. Leave pot 1 unplugged so it has good drainage.
   - Add water to pot 1 as needed to keep the plant healthy for several weeks. Use it as a comparison specimen (called the "control"). Add water slowly to pot 2 until water has filled the soil and the water is just level with the surface of the soil. All pores of the soil are now filled and the soil is "saturated." Weigh the pot and record the weight as "A."
   - Remove the plug from the bottom of pot 2 and permit the water to drain out. After water has stopped flowing from the drain hole, immediately weigh the pot again. Record the weight as "B." The water that flowed from pot 2 when the plug was removed is the free or gravitational water. Weight of the gravitational water, "D," should be calculated and recorded using the formula $A - B = D$. Do not add any more water to pot 2 for two weeks.
   - With the hole plugged in pot 3, add water slowly until the water is just level with the top of the soil. Do not remove the plug, and keep pot 3 filled with water to the saturation point for two weeks.
   - Keep all three pots in a good growing environment for two weeks and record all observations.
   - When the plant in pot 2 wilts badly due to lack of water, weigh the pot and record the weight as "C." Make the following calculations:

     Weight of the gravitational water $(D) = A - B$ Weight of the capillary water $(E) = B - C$

     The weight of the hygroscopic water can be determined only by driving the remaining water from the soil by heating the soil in an oven.
5. Ask your teacher to help you design and conduct a project that demonstrates some or all of the following. The effect of: the force of raindrops on soil; soil aggregation on absorption; slope on erosion; living grass or plant residue on erosion control.

National FFA Connection: Agriscience Research Proficiency Award

Earning a National FFA Research Proficiency Award will require students to engage in basic scientific research, such as on the effects of soil erosion on water quality.
Goodluz/Shutterstock.com

Research is the tool that agriscience and natural resource professionals use to measure the health or well-being of environmental systems. The National FFA Organization offers a proficiency award to students who apply research methods to environmental systems. Individuals or teams of students may do their own research, or they may collaborate with a school, experiment station, or college/university in a joint research effort. To qualify for this award, FFA members must be active participants in the following:

- Developing the experimental design and forming the hypothesis
- Collecting and interpreting the data
- Publicizing the results

Work with a classmate to locate online news articles about award-winning FFA research projects that other students have conducted in the category of environmental systems. What topics, hypotheses, and results can you find? Contact your state or local chapter of the FFA to find out about projects in your area.

Inquiry Activity: Water Conservation

After watching the *Water Conservation* video available on MindTap, create a fact sheet organized around the following elements:

- Fascinating Facts About Water
- Ways We Waste Water
- Ways We Can Conserve Water

For each category, list at least three things you learned from the video.

Next, use your notes to create and illustrate a Water Conservation poster.

Chapter 9
Soils and Hydroponics Management

Objective
To determine the origin and classification of soils and to identify effective procedures for soils and hydroponics management.

Competencies to Be Developed
After studying this chapter, you should be able to:
- define terms in soils, hydroponics, and other plant-growing media management.
- identify types of plant-growing media.
- describe the origin and composition of soils.
- discuss the principles of soil classification.
- determine appropriate amendments for soil and hydroponics media.
- discuss fundamentals of fertilizing and liming materials.
- identify requirements for hydroponics plant production.
- describe types of hydroponics systems.

Key Terms
medium
hydroponics
compost
sphagnum
peat moss
perlite
vermiculite
horizon
soil profile
residual soil
alluvial deposit
lacustrine deposit
loess deposit
colluvial deposit
glacial deposit
percolation
capability class
O horizon

A horizon	bedrock	neutral
clay	loam	fertilizer grade
silt	soil structure	active ingredient
sand	decomposer	starter solution
topsoil	soil amendment	nitrate
B horizon	pH	nitrogen fixation
subsoil	acidity	aeroponics
C horizon	alkalinity	

The roles of plants in the environment and their importance in our lives have been discussed in previous chapters. Plants are necessary to nourish the animals of the world and maintain the balance of oxygen in the atmosphere. However, they depend on a medium of soil, water, or, in some cases, air for a supportive living environment. A **medium** is a material or surrounding environment in which a living organism functions and lives.

For discussion in this chapter, the word *media* is used to mean the material that provides plants with nourishment and support through their root systems.

Plant-Growing Media

Soilless Media

Media comes in many forms. The oceans, rivers, land, and human-made mixtures of various materials are the principal types of media for plant growth. Seaweed, kelp, plankton, and many other plants depend on water for their nutrients and support. It is only recently that we have become aware of the tremendous amount of plant life in the sea. The plant life in oceans and rivers is important for feeding the animals found in water.

It has long been known that water could be used to promote new root formation on the stems of certain green plants, and it can completely support plant growth for a short time. Recently, it has been found that food crops can be grown efficiently without soil. This is done using structures where plant roots are submerged in or sprayed with solutions of water and nutrients (**Figure 9-1**). These solutions feed the plants, whereas mechanical structures provide physical support. The practice of growing plants without soil is called **hydroponics**. Hydroponics has become an important commercial method of growing some plants.

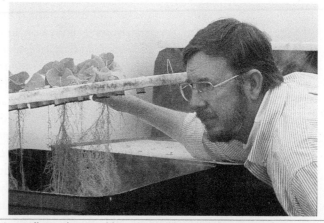

Figure 9-1 Plants can live and grow without any soil by spraying or submerging their roots in water in which the nutrients that are necessary for plants to live are dissolved.
Courtesy of USDA/ARS.

Soil

Soil is defined as the top layer of Earth's surface, which is suitable for the growth of plant life. It has long been the predominant medium for cultivated plants (**Figure 9-2**). In early years, humans accepted the soil as it existed. They planted seeds using primitive tools and did not know how to modify or enhance the soil to improve its plant-supporting performance. Ancient civilizations discovered that plant-growing conditions were improved on some land where deposits remained after river waters flowed over the land during flood season. Similarly, other land was ruined by floodwaters. Therefore, early efforts to improve plant-growing media were a matter of moving to better soil. Obviously, good soil was a valuable asset and, therefore, possession of it was the cause of intense personal disputes and wars among nations.

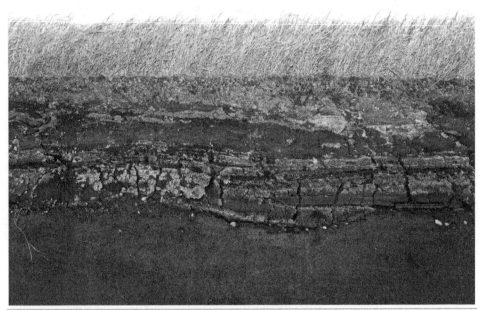

Figure 9-2 Soil is the most important plant medium for most crops.
© J. Helgason/Shutterstock.com.

Other Media

In addition to water and soil, certain other materials will retain water and support plant growth. Fortunately, some of the best non-soil and non-water plant-growing media are partially decomposed (decayed) plant materials. Hence, plants tend to improve the environment where they grow. One common material available around most homes is leaf mold and compost. Leaf mold is partially decomposed plant leaves. **Compost** is a mixture of partially decayed organic matter such as leaves, manure, and household plant wastes. Decaying plant matter should be mixed with lime and fertilizer in correct proportions to support plant growth (**Figures 9-3** and **9-4**).

There is a group of pale or ashy mosses called **sphagnum**. These are used extensively in horticulture as a medium for encouraging root growth and growing plants under certain conditions. **Peat moss** consists of partially decomposed mosses that have accumulated in waterlogged areas called bogs that are saturated with water. Both sphagnum and peat moss have excellent air- and water-holding qualities.

Many other sources of plant and animal residues may be used as plant-growing media. For instance, a fence post or a log may rot on its top and hold moisture from rainfall.

Chapter 9 Soils and Hydroponics Management   163

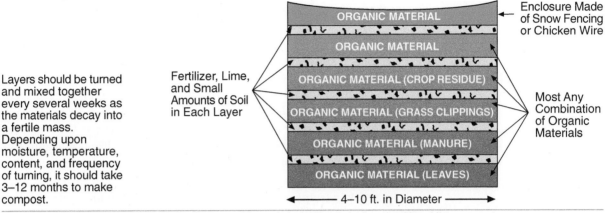

Layers should be turned and mixed together every several weeks as the materials decay into a fertile mass. Depending upon moisture, temperature, content, and frequency of turning, it should take 3–12 months to make compost.

Figure 9-3 Compost is an excellent organic matter. Most mineral soils can be improved by adding compost.

Figure 9-4 Plant materials from yards and gardens can be converted to valuable compost by placing the materials in an environment with moderate temperatures where bacteria known as decomposers have adequate moisture and oxygen.

Courtesy of DeVere Burton.

Horse manure mixed with straw is used extensively as a medium for growing mushrooms. In this instance, both animal residue (manure) and plant residue (straw) combine to make an effective medium.

Some mineral matter can also become plant-growing media. For instance, volcanic lava and ash eventually accumulate soil particles on their surface. Seeds settle into cracks, and moisture causes the seeds to germinate. Roots then penetrate and break up the volcanic residue. As time passes, the area that once was only lava and ash will be covered with plant life.

Horticulturists use certain mineral materials in plant-growing areas, too. **Perlite** is a natural volcanic glass material that has water-holding capabilities. Perlite is used extensively for starting new plants. **Vermiculite**, mineral matter from a group of mica-type materials, is also used for starting plant seeds and cuttings.

Origin and Composition of Soils

Factors Affecting Soil Formation

Productive soils develop on Earth's surface as the atmosphere, sunlight, water, and living things meet and interact with the mineral world. If soil is suitable for plant growth to a depth of 36 inches or more, the soil is regarded as "deep." Many soils of Earth are much shallower than this. Plants attach themselves to the soil by their roots, where they grow, manufacture food, and give off oxygen. Plants and animals of various sizes live on and in the soil, using carbon dioxide, oxygen, water, mineral matter, and products of decomposition.

Soils vary in temperature, organic matter, and the amount of air and water they contain. The kinds of soils formed at a specific site are determined by the forces of climate, living organisms, parent material, topography, and time (**Figure 9-5**).

Factors Affecting Soil Formation

Climate	Affects rate of weathering
Living organisms	Cause decay of organic material
Parent material	Influences fertility and texture
Topography	Affects distribution of soil particles and water
Time	Allows for the process of weathering

Figure 9-5 Soil formation depends on a number of natural factors.

Climate

Climatic factors, such as temperature and rainfall, greatly affect the rate of weathering. When temperature increases, the rate of chemical reactions increases and the growth of fungi, organisms (such as bacteria), and plants increases. The rate and amount of rainfall in a locality greatly affect the soil. In areas of high rainfall, the soils are usually leached and somewhat acidic. Leached means that certain contents have been removed from the soil by water. If the land is covered by trees, the action of high temperatures and moisture on leaf residues generally creates an acidic soil. Rainfall during cold weather has less effect on the soil than during warm or hot weather.

Living Organisms

Living organisms, such as microbes, plants, insects, animals, and humans, exert considerable influence on the formation of soil. Certain types of soil bacteria and fungi aid in soil formation by causing decay or breakdown of the plant and animal residues in the soil. Carbon dioxide and other compounds essential to soil formation are released by microbe activity. Microbes are microscopic plants and animals. Without soil microbes, organic materials would not decay.

Numerous insects, worms, and animals contribute to the formation of soil by mixing the various soil materials (**Figure 9-6**). Earthworms consume and digest certain soil substances and discharge body wastes. This aids decomposition and soil mixing. All such dead organisms add to soil organic matter.

Human activity also influences soil formation. Cultivation, bulldozing, and construction projects all disturb the surface layer. Clearing of land removes native plant life and greatly modifies soil-forming activities (**Figure 9-7**).

Figure 9-6 Living organisms like earthworms play important roles in breaking down organic matter such as leaves, stubble, and grass.

© J&L Images/Photodisc/Getty Images.

Figure 9-7 Clearing natural vegetation from land usually contributes to soil loss at a much faster rate than soil is formed.

Courtesy of National FFA Organization.

Parent Material

Parent material is the horizon of unconsolidated material from which a soil develops. **Horizon** means layer. Parent materials compose the C horizon of a typical soil profile. **Soil profile** means a cross-sectional view of soil (**Figure 9-8**).

Parent materials formed in place are called residual soils. Other soils are transported and deposited by water, wind, gravity, or ice. **Alluvial deposits** are transported by streams, and **lacustrine deposits** are left by lakes. **Loess deposits** are left by wind, **colluvial deposits** by gravity, and **glacial deposits** by ice. The composition of parent material from which soil is formed influences the natural fertility and texture of soils (**Figure 9-9**).

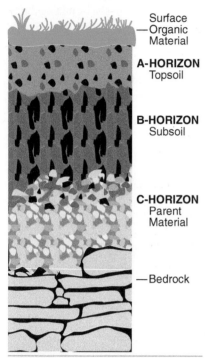

Figure 9-8 A soil profile consists of a cross-sectional view of the different layers of materials, beginning at the surface and going down to bedrock.

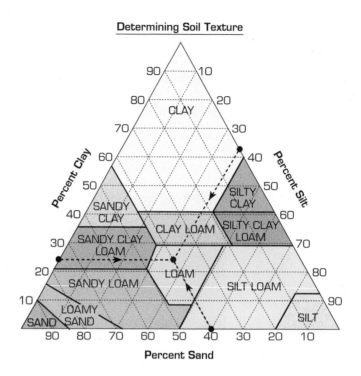

Example: Identify a soil that is 40% sand, 22% clay, and 38% silt.
1. Find 40 on the side for sand.
2. Draw a line in the direction of the arrow.
3. Do the same for clay (22%) and silt (38%).
4. The spot where the three come together is the soil texture. In this case, the soil is a loam.

A textural name may include a prefix naming the dominant sand size, as in "coarse sandy loam."

Figure 9-9 Soil texture is determined by the percentages of sand, silt, and clay found in a soil sample. The soil triangle is a tool that is used to determine texture.

Courtesy of USDA, Soil Conservation Service.

Topography

Topography, or slope of the land, affects soil erosion and drainage, thus influencing soil formation. On a steep slope, loose material is moved downward by runoff water, gravity, and movement of humans and animals. This movement not only breaks up soil materials and adds them to the lower levels, but it also exposes subsoil materials along the upper slopes. The movement of soil materials has a pulverizing effect on the material being moved as well as on the material left behind.

Slope affects the distribution of water that falls on Earth's surface. On level areas, the water soaks in and moves through the soil in a process called **percolation**. On sloping land, the water tends to run off and it moves some surface soil with it. Soils that develop on level land at low elevations tend to be poorly drained, whereas soils on gentle slopes tend to be better drained and more productive (**Figure 9-10**).

Drainage (or lack of it) affects the water table in a particular field or area. The water table has a direct bearing on soil formation, especially if it is near the surface. When a soil is saturated with water, little or no air can penetrate it. The lack of air reduces the action of fungi, bacteria, and other soil-forming activities in the soil.

Figure 9-10 Topography, or slope of the land, influences the formation of soil and how well excess water drains from it.
© Bernd Juergens/Shutterstock.com.

A wet soil is, therefore, a slow-forming soil and is usually low in productivity. Because of the lack of air, undecomposed organic matter accumulates in a wet soil. This organic matter generally causes the soil to be a blackish color. Poor drainage, accompanied by free water in the soil, impedes plant growth and affects soil formation.

Time

Soils are formed by the chemical weathering and physical weathering of parent material over time, as affected by climate, living organisms, and topography. Therefore, time itself is regarded as a factor in soil formation. Chemical weathering is the result of the chemical reactions of water, oxygen, carbon dioxide, and other substances that act on the rocks, minerals, organic matter, and life that compose the soil. The leaching action of water hastens the weathering process by removing soluble materials, and chemicals react with each other to form new chemicals in the soil. Physical weathering refers to mechanical forces caused by temperature change such as heating, cooling, freezing, and thawing. As these processes occur, rocks, minerals, organic matter, and other soil-forming materials are broken into smaller and smaller particles until soil is formed. The physical nature and chemical makeup of soils at different stages of weathering will differ widely.

Weathering causes soils to develop, mature, and age much as people do. Soils develop over time, mature, and then exhibit characteristics by which soil experts are able to identify the age of a soil. Plant nutrients are released quickly from the minerals, plant growth increases, and organic matter accumulates. Soils age more slowly during the later stages of weathering.

Eventually, nutrients in the soil are depleted. Water moving through the soil leaches away many soluble materials. At this stage, many soils are acidic because the limestone originally found in them is gone. As the supply of nutrients in the soil decreases, the amount of plant growth is reduced to the point where the organic matter decomposes faster than it is produced. When soils become acidic and have lost their native fertility, they require expensive amendments to keep them productive.

In permeable soils that permit water movement, the fine clay particles tend to move downward from the surface soil into the subsoil during the weathering process. This movement, together with further breakdown of the rock material, accounts for the fact that many soil types have a greater percentage of clay in the subsoil than in the surface soil.

Land Capability Classification System

Soil mapping and land classification have been priorities of the U.S. Department of Agriculture (USDA) throughout most of the twentieth century. Much of the original work of mapping was completed by the Soil Conservation Service (SCS). However, in recent years, this agency has been incorporated into the Natural Resources Conservation Service (NRCS). Now the NRCS can provide maps and classification information for almost any area in the United States, including small areas on individual farms. Agency personnel work with local farmers to develop farm plans, which provide recommendations for land use, cropping systems, crop production practices, and livestock systems. These plans are designed to maximize production while controlling soil erosion and enhancing productivity. Soil technicians obtain a wealth of information about the land by consulting a soils map and the accompanying material.

Capability Classes

Land capability maps indicate (1) capability class, (2) capability subclass, and (3) capability unit. **Capability classes** are the broadest classifications in the land capability classification system and are designated by Roman numerals I through VIII (**Figure 9-11**). The numerals indicate progressively greater limitations and narrower choices for practical use of land as follows.

- Class I—Soils have few limitations that restrict their use.
- Class II—Soils have moderate limitations that reduce the choice of plants or require moderate conservation practices.
- Class III—Soils have severe limitations that reduce the choice of plants, require special conservation practices, or both.
- Class IV—Soils have severe limitations that reduce the choice of plants, require careful management, or both.
- Class V—Soils are not likely to erode but have other limitations that limit their use to largely pasture, range, woodland, or wildlife food and cover.
- Class VI—Soils have severe limitations that make them generally unsuitable for cultivation and that limit their use largely to pasture, range, woodland, or wildlife food and cover.
- Class VII—Soils have very severe limitations that make them unsuitable for cultivation and that restrict their use largely to grazing, woodland, and wildlife.
- Class VIII—Soils and landforms have limitations that nearly always prevent their use for agricultural production and restrict their use to recreation, wildlife or water supply or to esthetic purposes.

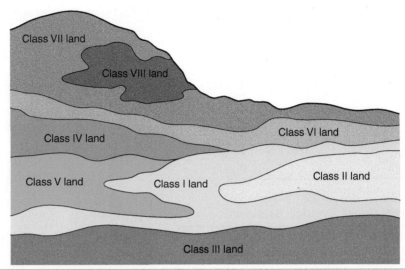

Figure 9-11 The eight land capability classes in the United States are color coded on the map. Classes I–IV include land suited to cultivation and other uses. Classes V–VIII include land limited in use and generally not suited to cultivation.

Courtesy of USDA/NRCS Soil Conservation Service.

Capability Subclasses

Capability subclasses are soil groups within one class. They are designated by adding a lowercase letter *e*, *w*, *s*, or *c* to the class numeral (e.g., IIe). The letter *e* indicates the main limitation is risk for erosion unless close-growing plant cover is maintained; *w* indicates that weather in or on the soil interferes with plant growth or cultivation; *s* indicates the soil is limited mainly because it is shallow, droughty, or stony; and *c*, used in only some parts of the United States, indicates the chief limitation is climate—too cold or too dry (**Figure 9-12**).

In Class I, there are no subclasses because the soils of this class have few limitations. Conversely, Class V contains only the subclasses indicated by *w*, *s*, or *c* because the soils in Class V are subject to little or no erosion. However, they have other limitations that restrict their use to pasture, range, woodland, wildlife habitat, or recreation.

Capability Units

The soils in one capability unit (soil groups within the subclasses) are enough alike to be suited to the same crops and pasture plants and require similar management. They have similar productivity capabilities and other responses to management. The capability unit is a convenient grouping for making many statements about the management of soil.

Capability units are generally designated by adding Arabic numerals (0–9) to the subclass symbol (e.g., IIIe4 or IIw2). Thus, in one symbol, the Roman numeral designates the capability class or degree of limitation; the lowercase letter indicates the subclass or kind of limitation; and the Arabic numeral specifically identifies the capability unit within each subclass. A map legend for each soil grouping is included with the land capability map. This type of map is also an example of how symbols can be effectively used. They make it possible to place much information in a small space on the map by use of a code.

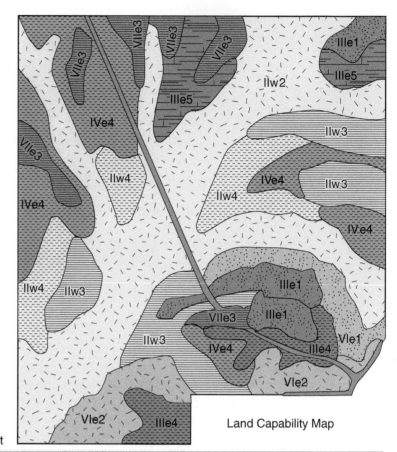

Figure 9-12 Land capability map. A land capability map is used to identify soil locations with similar characteristics. Similar land capability classes indicate that similar crops and management practices may be applied.

Courtesy of USDA/NRCS Soil Conservation Service.

Use of Maps

When the landowner and the soil conservationist start planning for the most intensive use of a farm, they need a land capability map. This will help them prepare a conservation plan or proposed land-use map. The soil conservationist and the landowner, through the use of the land capability map, discuss the kinds of soil on the farm. The current and original land uses are also discussed and noted on the map. In developing a proposed land-use map, the soil conservationist must know the personal goals or objectives of the landowner and the plans for developing the land.

Many things can be involved in reaching land-use decisions, field by field. Field boundaries may need to be changed so that all the soil in each field is suited for the same purpose and management practices. The desired balance among cropland, pasture, woodland, and other land uses needs to be considered. Appropriate livestock enterprises should also be considered to match the land's characteristics and potential. If there is a good potential for income-producing recreation enterprises in the community, an area may be used for hunting, campsites, or fishing.

The landowner and soil conservationists must consider how to treat each field to get the desired results. The NRCS conservationist can give many good suggestions, but the landowner must decide what to do, when to do it, and how to do it. As planning decisions are made, the conservationist will record them in narrative form and make them part of the plan map. This becomes the farm conservation plan. It is the guide for the farming operation in the years ahead. A conservation plan is required to obtain federal support to encourage good farming and conservation practices.

Workers from the NRCS are also available to give on-site technical assistance in applying and maintaining the farm conservation plan. They provide management advice and services to non-farm landowners, developers, strip-mining operations, and other activities where soil is involved (**Figure 9-13**).

Physical, Chemical, and Biological Characteristics

Soil Profile

Undisturbed soil will have four or more horizons in its profile. These are designated by the capital letters O, A, B, and C (**Figure 9-14**). The **O horizon** is on the surface and is composed of organic matter and a small amount of mineral matter. Organic matter originates from living sources such as plants, animals, insects, and microbes. Mineral matter is derived from non-living sources such as rock materials.

Figure 9-13 NRCS personnel advise on the many uses of land. Satellite imagery and high-altitude photography help document changing conditions on the soil surface.

Courtesy of USDA #K-5218-03.

Horizon	Name	Colors	Structure	Processes Occurring
O	Organic	Black, dark brown	Loose, crumbly, well broken up	Decomposition
A	Topsoil	Dark brown to yellow	Generally loose, crumbly, well broken up	Zone of leaching
B	Subsoil	Brown, red, yellow, or gray	Generally larger chunks, may be dense or crumbly, can be cement-like	Zone of accumulation
C	Parent material (slightly weathered material)	Variable—depending on parent material	Loose to dense	Weathering, disintegration of parent material or rock

Figure 9-14 Characteristics of soil horizons.

The **A Horizon** is located near the surface and consists of mineral matter and organic matter. It contains desirable proportions of organic matter, fine mineral particles called **clay**, medium-sized mineral particles called **silt**, and larger mineral particles called **sand**. The appropriate proportion of these soil components creates soil that is tillable, or workable with tools and equipment. With the presence of desirable plant nutrients, chemicals, and living organisms, the A horizon generally supports good plant growth. The A horizon is frequently called **topsoil**.

The B horizon is below the A horizon and is generally referred to as subsoil. The mineral content is similar to the A horizon, but the particle sizes and properties differ. Because organic matter comes from decayed plant and animal materials, the amount naturally decreases as distance from the surface increases.

The C horizon is below the B horizon and is composed mostly of parent material. The C Horizon is important for storing and releasing water to the upper layers of soil, but it does not contribute much to plant nutrition. It is likely to contain larger soil particles and may have substantial amounts of gravel and large rocks. The area below the C horizon is called bedrock (see **Figure 9-8**).

Texture

Texture refers to the proportion and size of soil particles (**Figures 9-15** and **9-16**). Texture can be determined accurately in the laboratory by mechanical analysis (**Figure 9-17**). The feel of the soil can also be used to analyze soil as follows:

1. Make a stiff mud ball.
2. Rub the mud ball between your thumb and forefinger.
3. Note the degree of coarseness and grittiness caused by the sand particles.
4. Squeeze the mud between your fingers, and then pull your thumb and fingers apart.
5. Note the degree of stickiness caused by the clay particles.
6. Make the soil slightly moister and note that the clay leaves a "slick" surface on your thumb and fingers.

The outstanding physical characteristics of the important textural grades, as determined by the "feel" of the soil, are discussed next (**Figure 9-18**).

Coarse-textured (sandy) soil is loose and single grained. The individual grains can be seen readily or felt. Squeezed in the hand when dry, it will fall apart when the pressure is released. Squeezed when moist, it will form a cast, but will crumble when touched.

Medium-textured soil is known as loam. It has a relatively even mixture of sand, silt, and clay. However, the clay content is less than 20 percent. (The characteristic properties of clay are more pronounced than those of sand.) A loam is mellow with a somewhat gritty feel, yet fairly smooth and highly plastic. Squeezed when moist, it will form a cast that can be handled quite freely without breaking.

Fine-textured (clay) soil usually forms hard lumps or clods when dry. It is usually very sticky when wet and is quite plastic. When the moist soil is pinched between the thumb and fingers, it will form a long, flexible "ribbon." A clay soil leaves a "slick" surface on the thumb and fingers when rubbed together with a long stroke and firm pressure. The clay tends to hold the thumb and fingers together because of its stickiness.

Structure

Soil structure refers to the tendency of soil particles to cluster together and function as soil units called aggregates. Aggregates, or crumbs, contain mostly clay, silt, and sand particles held together by a gel-type substance formed from organic matter.

Aggregates absorb and hold water better than individual particles. They also hold plant nutrients and influence chemical reactions in the soil. Another major benefit of a well-aggregated soil or a soil with good structure is its resistance to damage by falling raindrops. When hit by falling rain, the aggregate stays together as a water-absorbing unit, rather than separating into individual particles. When aggregates on the surface of soil dry out, they remain in a crumbly form and permit good air movement. Dispersed soil particles run together when dry and form a crust on the soil surface. The crust prevents

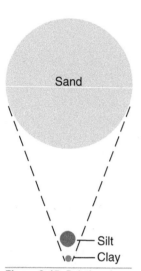

Figure 9-15 Relative size of sand, silt, and clay particles.

SIZES OF SOIL PARTICLES	
Name	Size, Diameter in Millimeters
Fine gravel	2–1
Coarse sand	1.00–0.50
Medium sand	0.50–0.25
Fine sand	0.25–0.10
Very fine sand	0.10–0.05
Silt	0.05–0.002
Clay	less than 0.002

Figure 9-16 Range of sizes of soil particles.

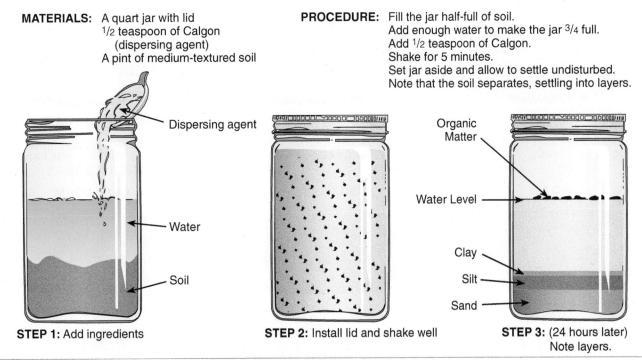

Figure 9-17 Mechanical analysis of soil to accurately determine the soil texture.

air exchange between the soil and the atmosphere and decreases plant growth. The process and benefits of aggregation apply mostly to fine- and medium-textured soils.

Organic Matter

As noted in the previous discussion, organic matter plays an important role in soil structure. Soil is a living medium with a great variety of living organisms. Some groups of organisms of the plant kingdom that are often found in soils are:

- roots of higher plants;
- algae—green, blue-green, and diatoms;
- fungi—mushroom fungi, yeasts, and molds; and
- actinomycetes of many kinds—aerobic, anaerobic, autotrophic, and heterotrophic.

Some examples of groups of organisms from the animal kingdom that are prevalent in soils are:

- those that subsist largely on plant material—small mammals, insects, millipedes, sow bugs (wood lice), mites, slugs, snails, earthworms;
- those that are largely predators—snakes, moles, insects, mites, centipedes, and spiders; and
- microanimals that are predatory, parasitic, and live on plant tissues—nematodes, protozoans, and rotifers.

Living Organisms

Living organisms excrete cell or body wastes that become part of the organic content of soil. The microbes of the soil and the remains of larger plants and animals decompose or decay into soil-building materials and nutrients.

SOIL TEXTURAL CLASSES

SAND
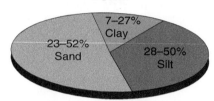

Dry: loose and single grained; feels gritty.
Moist: will form a ball that crumbles very easily.

LOAMY SAND
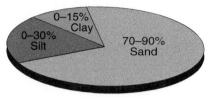

Dry: silt and clay may mask sand; feels loose, gritty.
Moist: feels gritty; forms a ball that crumbles easily; stains fingers slightly.

SANDY LOAM
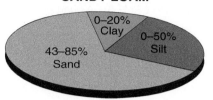

Dry: clods easily broken; sand can be seen and felt.
Moist: moderately gritty; forms a ball that can stand careful handling; definitely stains fingers.

LOAM
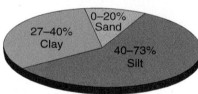

Dry: clods are moderately difficult to break; mellow, somewhat gritty.
Moist: neither very gritty nor very smooth; forms a firm ball; stains fingers.

SILT LOAM
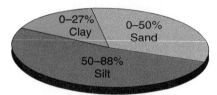

Dry: clods are difficult to break; feels smooth, soft, and floury when pulverized; shows fingerprints.
Moist: has smooth or slick "buttery" or "velvety" feel; stains fingers.

CLAY LOAM
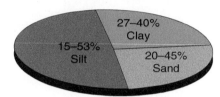

Dry: clods very difficult to break with fingers.
Moist: has slightly gritty feel; stains fingers; ribbons fairly well.

SILTY CLAY LOAM

Same as **CLAY LOAM** but very smooth.

SANDY CLAY LOAM
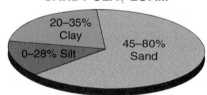

Same as **CLAY LOAM.**

CLAY
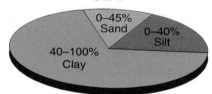

Dry: clods cannot be broken with fingers without extreme pressure.
Moist: quite plastic and usually sticky when wet; stains fingers. (A silty clay feels smooth; a sandy clay feels gritty.)

Figure 9-18 Major soil textural classes.

People who grow indoor plants at home, raise gardens, farm, produce greenhouse crops, or grow nursery stock generally find it useful to add organic materials to the soil. Popular sources of organic matter for soil amendments are peat moss, leaf mold, compost, livestock manure, and sawdust.

Some important benefits or functions of organic matter in soil include:
- making the soil porous;
- supplying nitrogen and other nutrients to the growing plant;
- holding water for future plant use;
- aiding in managing soil moisture content;
- furnishing food for soil organisms;
- serving as a storehouse for nutrients;
- minimizing leaching;
- serving as a source of nitrogen and growth-promoting substance; and
- stabilizing soil structure.

Science Connection: Digging Life

Soil is the most diverse ecosystem on Earth. The number of organisms per acre in soil far exceeds the concentration of organisms of any other place in the world. Most of us do not think about the abundant life under our feet as we cross a lawn. But just one square inch of this soil is teeming with busy creatures, most of which cannot be seen. Gardens and fields consist of fertile soil filled with living organisms. One gram of this soil contains approximately the following organisms:

3,000,000–500,000,000	Bacteria—A group of one-celled microscopic organisms
1,000,000–20,000,000	Actinomycetes—Microscopic organisms that resemble both fungi and bacteria
5,000–1,000,000	Fungi—Nonmicroscopic organisms that get their food from dead material
1,000–100,000	Yeast—Single-celled fungi. Many are used in producing food.
1,000–500,000	Protozoa—Small single-celled organisms. An example is an amoeba.
1,000–500,000	Algae—One-celled organisms that contain chlorophyll
1–500	Nematodes—Nonsegmented roundworms

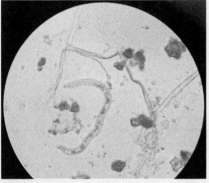
A nematode is a microscopic worm found in the soil.
© Damian Herde/Shutterstock.com

Large numbers of slime molds, viruses, insects, and earthworms are also present. Some of these organisms are harmful to plants and animals. Most are decomposers. A **decomposer** is an organism that breaks down material that was once living. They change things that are dead into the rich organic substances that add to the fertility of the soil.

The role these tiny creatures play in an ecosystem is irreplaceable. Imagine a world without decomposers. Leaves, dead animal carcasses, and huge amounts of other dead organic matter would pile up. In a short period, dead and undecomposed material would crowd out all life. Soil is a unique substance. It provides living space for billions of organisms. It is the medium in which plants and other producers grow. Soil is also a place where once-living things are broken down and changed into the fertile organic material of the soil.

Other Properties of Soils

Soil scientists, managers, technicians, and operators must be aware of numerous other properties of soils in their work. Some of these are external factors such as land position, slope, and stoniness. Soil color, depth, drainage, permeability, and erosion are important considerations.

Amendments to Plant-Growing Media

The term **soil amendment** is used here to mean addition to or change. Most soil amendments are made to add organic matter, add specific nutrients, or modify soil pH. The **pH** is a measure of the degree of acidity or alkalinity. **Acidity** is sometimes referred to as sourness, and **alkalinity** is referred to as sweetness. The pH scale ranges from 0 (maximum acidity) to 14 (maximum alkalinity). The midpoint of the scale is 7, which is **neutral**, meaning neither acidic nor alkaline (**Figure 9-19**).

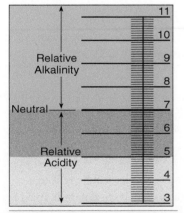
Figure 9-19 The pH is neutral in the middle. It gets progressively more acidic from 7 down to 3 and progressively more alkaline from 7 up to 11.

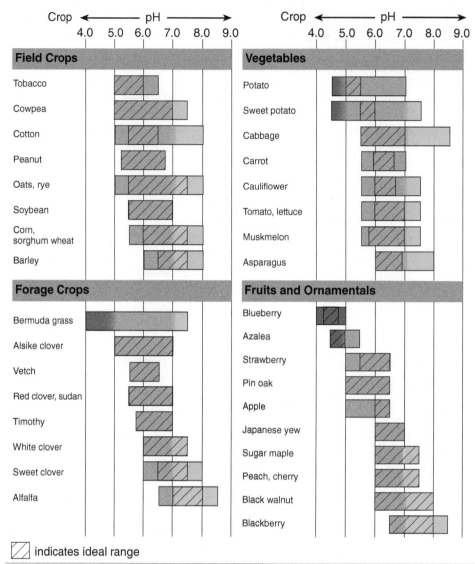

Figure 9-20 Plants grow in pH ranges from approximately 4 (very acidic) to 8.5 (very alkaline), but most plants function best within a specific pH range.

Crops grow best in media with a narrow pH range unique to that plant species (**Figure 9-20**). Most plants require a pH somewhere between 5.0 and 7.5. Some crops, such as potatoes, prefer a soil pH around 5.5. Alfalfa responds best to a pH of 7.0 to 8.0.

Liming

Areas that were historically covered by trees, such as the northeastern, western, and northwestern parts of the United States, develop acidic soils. When cleared of trees, such soils require additional lime to increase the pH for the efficient production of most farm crops.

A pH test can be performed using a pH test kit, or soil samples can be sent to a university or commercial laboratory for analysis. Laboratory analyses generally include an analysis of phosphorus, potash, and magnesium, as well as pH. Liming and fertilizing recommendations may also be provided by testing laboratories and universities (**Figure 9-21**).

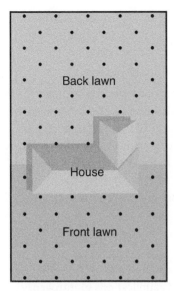
(A) Divide the area into sampling areas.

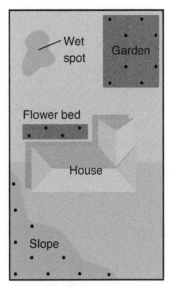
(B) Avoid wet or bare spots. Sample each different type of soil separately.

(C) Remove a 1-inch slice from one side of a V-shaped cut.

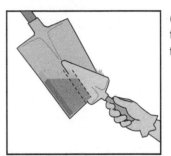
(D) Take a 1-inch strip from the middle of this slice.

Figure 9-21 A procedure for collecting a dependable soil sample is illustrated in these views.

How to Take a Soil Sample in Your Field, Lawn, or Garden

1. Select an appropriate sampling tool (spade, auger, or soil tube) (**Figure 9-22**).
2. Make a sketch dividing the area into sampling areas—for example, front lawn, garden, flower bed, slope, and back lawn. Appropriately label each area (see **Figure 9-21A**).
3. When taking samples, avoid wet or bare spots (see **Figure 9-21B**). Soils that are substantially different in plant growth or past treatment should be sampled separately, provided their size and nature make it feasible to fertilize or lime each area separately.
4. After removing surface litter, take a sample from the correct depth. This is 2 inches for established lawns and about 6 inches for gardens, flower beds, farm crop land, and other areas to be tilled.
5. Submit a separate composite sample for each significantly different area—for example, front lawn, back lawn, and flower bed. Your composite sample for each area should include a small amount of soil taken from each of 10 to 20 randomly selected locations in the area represented by each sample.
6. When using a spade, first make a V-shaped cut. Then remove a 1-inch slice from one side of the cut (see **Figure 9-21C**). Then take a 1-inch strip from the middle of this slice (see **Figure 9-21D**). This represents the soil from one spot in the sample.

Figure 9-22 A soil sample that is collected with a soil probe collects a sample that is uniform in volume at any given soil depth.

Courtesy of USDA/ARS #K8392-1. Photo by Keith Weller.

7. Air-dry the soil; do not use heat. Mix the soil from a composite in a clean bucket. Place about 1 pint of this mixture into the sample box. Use a separate box for each composite. Fill in the blanks on the box or information sheet for each box.
8. Send soil sample(s) and information sheet(s) to the soil test laboratory.

Soil samples are collected routinely to help determine appropriate management of soil amendments. The same soil sampling procedure that was described for lawns and gardens is used for farm fields, golf courses, and greenhouses. It is critical that adequate amounts of fertilizers and other soil amendments are provided to a crop without applying excessive amounts. Adding too much of a chemical to the soil is not only expensive but it also may leach into the groundwater and surface water.

Note: The Cooperative Extension System office in most counties and/or cities can arrange for soil testing.

The Petiole Test

Some crops are expensive to raise because they require large amounts of soil amendments. For example, potatoes require relatively high applications of nitrogen. When the nitrogen supply in the soil becomes depleted, the newest leaves on the plant will also have a nitrogen deficiency. By sampling and testing the petiole (young stem and leaves) of the potato plant, the deficiency can be identified and corrected before the crop yield is affected. The petiole test is effective when it is used to manage soil nutrients during the growing season.

Soil Chemistry

One characteristic of soil is that texture, depth, fertility, nutrient deficiencies, and other features tend to show considerable variation even within the same field. Because differences exist in close proximity, it becomes very difficult to manage crop growth and maturity. Productivity across a field is seldom uniform, and adding fertilizer based on average production within the field will result in nutrient deficiencies in highly productive areas and a buildup of excess nutrients in low producing sectors. Neither of these conditions is good, and both conditions are economically unsound.

The use of precision farming equipment makes it possible to adjust fertilizer applications based on past production and soil characteristics. For example, a rocky sector within a field will seldom require as much fertilizer as a highly productive loam soil. Using global positioning and computer technologies, it is now possible to apply fertilizers and other soil treatments at appropriate rates within different sectors of a field.

Soil chemistry provides the information that is needed to make this system work well. Soils cannot be precisely managed for fertility, or any other input, until we know the nutrient content in each field sector at the end of the crop production cycle. Soil testing kits to determine soil chemistry are available to individuals and school laboratories where soil testing and experimentation are integral parts of the soil science curriculum. Commercial labs are also available in most areas to provide testing services that measure fertility as well as pH and other soil characteristics.

The pH Test

The amount of agricultural lime required to raise soil to the desired pH level is indicated by a pH test. The same pH value may require different amounts of agricultural lime. This is because soils contain varying amounts of organic matter, clay, silt, and sand. The greater the organic matter and clay content, the greater the amount of lime required to correct the acidity.

Even though all the lime required by your soil is applied, do not expect the pH to increase quickly. It will generally require 2 to 3 years for all of the lime to be used and the desired pH to be reached.

Mix Well

The soil on 1 acre that is 6 inches in depth weighs about 2 million pounds. Therefore, it takes a lot of mixing to distribute a relatively small amount of lime with the soil. Liming recommendations are based on a specific plowing depth, such as 6 or 7 inches. Application rates may need to be adjusted for deep or shallow tilling.

Standard Ground Limestone

Lime recommendations are based on standard ground limestone, which should contain a minimum of 50 percent lime oxides (calcium oxide plus magnesium oxide). About 98 percent should pass through a 20-mesh sieve, with a minimum of 40 percent passing through a 100-mesh sieve. The higher the percentage of limestone passing through a 100-mesh screen, the faster this limestone will correct soil acidity. The best limestone will have the greatest calcium and/or magnesium content and will be ground to a small particle size. It is more important to finely grind a high-magnesium stone than a high calcium stone. A high-magnesium or dolomitic limestone should always be used when a magnesium deficiency is indicated by a soil test.

One ton (2,000 lbs.) standard ground limestone is approximately equivalent to 1,500 lbs. of hydrated lime or 1,100 lbs. of ground burned limestone.

Correcting Excessive Alkalinity

A reduction of soil alkalinity is desirable where soils have a high pH and the alkaline condition causes unsatisfactory crop growth. In most cases, this condition will exist where heavy applications of lime are made at one time or where lime has been applied to a soil with a high pH.

Gypsum is a soil amendment that is often used to reduce the alkalinity of agricultural soils. It is added when soil pH is too high for the intended crop to grow and thrive. In most instances, it should be added in the fall to allow time for it to be effective. For best results, it should be distributed in the upper layer of topsoil within the area where the roots of the crop are concentrated.

Another method that may be used to decrease the pH value of soil is to use sulfur or aluminum sulfate. Sulfur at 1.5 lbs. per 100 square feet or aluminum sulfate at 5 lbs. per 100 square feet will decrease the alkalinity, under most conditions, by 0.5 pH. Aluminum sulfate acts rapidly, producing acidification in 10 to 14 days. Sulfur requires 3 to 6 months in some soils before it forms into compounds the plants can use. Broadcast the material over the surface and thoroughly work it into the soil. For full benefit, sulfur should be applied in the fall, after garden crops are harvested. Aluminum sulfate may be applied in early spring.

Use chemicals cautiously to decrease alkalinity. Before these measures are taken, you should consider other factors that may be responsible for poor growth (such as drought, insect and disease injury, and fertilizer burning).

Fertilizers and Fertilizing

Essential plant nutrients are discussed in Chapter 16, "Plant Physiology." However, it should be noted that nitrogen, phosphorus, and potassium are known as the three primary nutrients. These three ingredients must be present for a fertilizer to be called a complete fertilizer.

The proportions of nitrogen, phosphorus, and potassium are known as the **fertilizer grade**, expressed on a fertilizer container as percentages of the contents of the container by weight. Therefore, a 100-lb. container of fertilizer with a grade of 10-10-10 contains 10 percent nitrogen, 10 percent phosphorus, and 10 percent potassium.

If the total amount of fertilizer (100 lbs.) is multiplied by the percentage of each ingredient, the pounds of each ingredient may be calculated. Therefore, 0.10 (percent of nitrogen) × 100 (total lb. of fertilizer) = 10 lbs. nitrogen. The amount of phosphorus and potassium is also 10 lbs. each. The other 70 lbs. consist of inert filler material.

Fertilizer is frequently shipped in 80-lbs. bags. Therefore, one bag of 10-10-10 fertilizer would have 8 lbs. nitrogen, 8 lbs. phosphorus, and 8 lbs. potassium for a total of 24 lbs. of **active ingredients** (components that achieve one or more purposes of the mixture).

Some popular grades of fertilizer are 5-10-5, 5-10-10, 10-10-10, 6-10-4, 0-15-30, 0-20-20, 8-16-8, and 8-24-8. These grades are formulated to meet the needs of a variety of crops on various soils. The amount and grade of fertilizer to apply are determined by (1) the specific crop to be grown, (2) the potential yield or performance of the crop, (3) fertility of the soil, (4) physical properties of the soil, (5) previous crop, and (6) type and amount of manure applied.

Therefore, decisions on rate of application must be made on a local basis. Soil tests, tissue tests, and plant observations are useful techniques for determining fertility needs (**Figure 9-23**).

Organic Fertilizers

Organic fertilizers include animal manures and compost made with plant or animal products. Organic commercial fertilizers include dried and pulverized manures, bone meal, slaughterhouse tankage, blood meal, dried and ground sewage sludge, cottonseed meal, and soybean meal.

Organic fertilizers have certain definite characteristics. First, nitrogen is usually the predominant nutrient, with lesser quantities of phosphorus and potassium. One exception to this is bone meal, in which phosphorus predominates and nitrogen is a minor ingredient. Second, the nutrients are only made available to plants as the material decays in the soil, so they are slow acting and long lasting. Third, organic materials alone are not balanced sources of plant nutrients, and their analysis in terms of the three major nutrients is generally low.

Inorganic Fertilizers

Various mineral salts, which contain plant nutrients in combination with other elements, are called inorganic fertilizers. Their characteristics are different from organic fertilizers. First, the nutrients are in soluble form and are quickly available to plants. Second, the soluble nutrients make them caustic to growing plants and can cause injury. Care must be used in applying inorganic fertilizers to growing plants. They should not contact the roots or remain on plant foliage for any length of time. Third, the analysis of chemical fertilizers is relatively high in terms of the nutrients they contain.

Fertilizers are needed to replenish mineral nutrients depleted from a soil by crop removal or by such natural means as leaching. Some soils with high fertility may need only nitrogen or manure. Use of fertilizers that also contain small amounts of copper, zinc, manganese, boron, and other minor elements is not considered necessary for most soils but may be needed in certain soils and for certain crops.

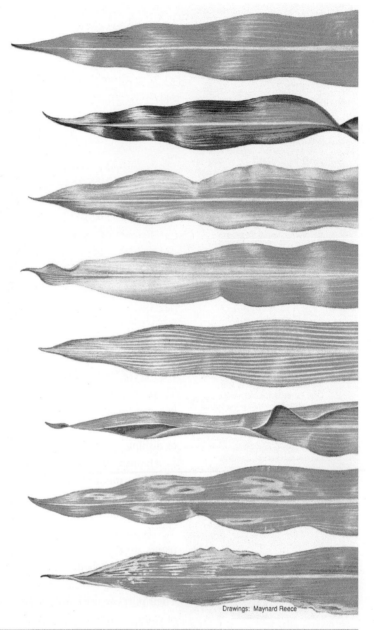

HEALTHY leaves shine with a rich, dark green color when adequately fed.

PHOSPHOROUS (phosphate) shortage marks leaves with reddish purple, particularly on young plants.

POTASSIUM (potash) deficiency appears as a firing or drying along the tips and edges of lowest leaves.

NITROGEN hunger sign is yellowing that starts at tip and moves along middle of leaf.

MAGNESIUM deficiency causes whitish stripes along the veins and often a purplish color on the underside of the lower leaves.

DROUGHT causes corn plants to have a grayish green color; leaves may roll up to the size of a pencil.

DISEASE helminthosporium blight, starts in small spots, gradually spreads across leaf.

CHEMICALS may sometimes burn tips, edges of leaves, and at other contacts. Tissue dies, leaf becomes whitecap.

Drawings: Maynard Reece

Figure 9-23 Corn plants are good indicators of nutrient deficiencies and other factors that impact plant health and productivity.

There are also unmixed fertilizers that carry only one element (**Figure 9-24**). Most important of these unmixed materials are the nitrogen and phosphate carriers. Nitrogen carriers vary from 16 to 45 percent nitrogen. Be careful in using nitrogen materials. Too much may cause excessive and soft growth.

Superphosphate fertilizers carry only phosphorus. Phosphorus promotes flower, fruit, and seed development. It also firms up stem growth and stimulates root growth. Superphosphate may be added to manure to give a better balance of nutrients for plants (100 lbs. to each ton of horse or cow manure and 100 lbs. to each half-ton of sheep manure).

Materials	Nitrogen N%	Phosphorus P$_2$O$_5$%	Potassium K$_2$O%	Calcium %	Magnesium %	Sulfur %
Ammonium Nitrate	33.5	—	—	—	—	—
Ammonium Sulfate	21	—	—	—	0.3	23.7
Urea	45	—	—	1.5	0.7	0.02
Sodium Nitrate	16	—	0.2	0.1	0.05	0.07
Calcium Nitrate	15	—	—	19.4	1.5	0.02
Calcium Cyanamide	21	—	—	38.5	0.06	0.3
Anhydrous Ammonia	82	—	—	—	—	—
Superphosphate	—	20	0.2	20.4	0.2	11.9
Liquid Phosphoric Acid	—	58	—	—	—	—
Ammonium Phosphate	11	48	0.2	1.1	0.3	2.2
Potassium Chloride	—	—	62	—	0.1	0.1
Potassium Sulfate	—	—	53	0.5	0.7	17.6

Figure 9-24 Plant nutrients in common fertilizer materials.

Fertilizer Applications

There are many ways to apply fertilizers. For the home lawn, the most likely method is broadcasting (spreading evenly over the entire surface). In the case of gardens or fields, broadcasted fertilizer may be incorporated or mixed into the soil by spading, tilling, plowing, or disking.

Band application places fertilizers about 2 inches to one side of and slightly below the seed. This method is used extensively for row crops in gardens and fields.

Side-dressing is done by placing fertilizer in bands about 8 inches from the row of growing plants. This method is popular for row crops such as corn and soybeans.

Top-dressing is a procedure where fertilizer is broadcast lightly over close-growing plants. Top-dressing is used for adding nitrogen to small grain, hay, and turf crops after they are established.

Starter solutions are diluted mixtures of fertilizer used when plants are transplanted. Their purpose is to provide small amounts of nutrients that will not burn the tender roots of young plants.

Other methods of applying fertilizers include the application of foliar sprays directly onto the leaves of plants and knife application of anhydrous ammonia gas into the soil. The practice of adding liquid fertilizer to irrigation water is also used extensively in the United States.

Using Manure

Animal manure is a valuable product when handled properly. Its content of plant nutrients makes it a valuable fertilizer material. In addition, the organic matter aids in developing and maintaining structure in soils (**Figure 9-25**).

To obtain the most nutrient value from manures, the following practices should be followed:

- Use adequate bedding to absorb all of the liquids.
- Balance the phosphorus in cow manure by adding superphosphate to the fields.
- Spread manure evenly over fields. An 8- to 12-ton application per acre is recommended. (Fifty bushels of manure and litter weigh about 1 ton.)

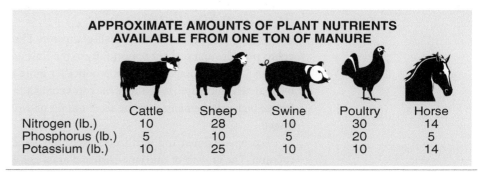

Figure 9-25 Animal manure is excellent fertilizer for most crops, and it also improves the soil by adding organic matter to the topsoil.

	N NITROGEN	**P** PHOSPHORUS	**K** POTASSIUM
Suppose the fertilizer recommendations per acre for a certain crop are:	150 lbs.	100 lbs.	100 lbs.
If well-managed cow manure is applied at the rate of 10 tons per acre, it can be determined from Figure 9-31 that the 10-ton application would provide:	100 lbs.	50 lbs.	100 lbs.
Therefore, the amount of nutrients that must be provided per acre through commercial fertilizer is:	50 lbs.	50 lbs.	0 lbs.
The shortfall of ingredients listed above could be made up by applying 500 lbs. of 10-10-0 commercial fertilizer per acre. NOTE: Caution must be exercised when estimating the nutrient values of manure due to variations in liquid and solid content captured from the animal, amount and type of bedding, and procedures used to handle and store the manure.			

Figure 9-26 When manure is applied to soil, the amount of commercial fertilizer used should be adjusted downward, considering the nutrients in the manure.

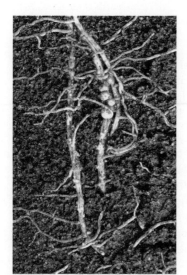

Figure 9-27 Nitrogen-fixing nodules on the roots of legume plants.

Courtesy of USDA/ARS #K5244-6. Photo by Keith Weller.

- When possible, incorporate manure into the soil immediately after spreading.
- Do not spread on steep slopes when the ground is frozen.
- When storing manure, keep it compact and under cover.
- Prevent liquid runoff from escaping from manure holding areas.
- Apply manure to crops that will give best response, such as corn, sorghums, potatoes, and tobacco.

When applying manure, the amount of commercial fertilizer should be reduced accordingly (**Figure 9-26**).

Legume Crops

Legumes are plants in which specialized bacteria use nitrogen gas from the air and convert it to **nitrate** (the form of nitrogen used by plants). Some examples of legumes are beans, peas, clovers, and alfalfa. The process of converting nitrogen gas to nitrates by bacteria in the roots of legumes is called **nitrogen fixation** (**Figure 9-27**).

Nitrogen fixation reduces or eliminates the need for adding expensive nitrogen fertilizer to legume crops. When the roots of legume plants decay, large amounts of nitrates are left behind for the next crop. Crops such as corn, which requires large amounts of nitrogen, should follow alfalfa or clover in a field because the legumes leave unused nitrogen behind.

Rotation Fertilization

Many crop rotations start with a small-grain crop. The preparation of the seedbed for small grains provides an excellent opportunity to put lime and fertilizer into the feeding zone for roots of perennial forage crops that produce for more than a year. Lime and phosphorus move slowly in the soil, and unless the roots contact these elements, they cannot provide for the high requirements of some seedlings for phosphorus in the critical first year of growth.

A properly limed and fertilized forage crop is the backbone of a successful crop rotation. The growth of nutrient-enriched grass and legume roots throughout the soil provides a most favorable medium for natural soil-building processes. The addition of organic matter, the movement of fertilizer elements by the roots into the soil, and the production of root channels by sod roots produce soils that absorb water better, erode less, are easier to work, and produce higher-yielding and better-quality crops.

Phosphorus and potassium added to sod will benefit the sod and, in addition, will be placed in the best location and be in the best form to supply these elements to the long-season row crops that follow. Nitrogen produced by the legumes or added to grass will be present in organic form and will be released in the best possible form for the row crops and small-grain crops.

The composition of soils is complex and depends on many factors. Soil is dynamic and changing all the time. Nature has provided many cycles to help provide for soil renewal. The scientific management of soils makes them more productive and helps to ensure that productive soils are capable of efficient production for future generations.

Hydroponics

The term *hydroponics* refers to a number of systems used for growing plants without soil (**Figure 9-28**). Some major systems are:

- aggregate culture—in which a material such as sand, gravel, or marbles supports the plant roots;

Science Connection | **Soil as a Water Reservoir**

A key function of soil is to hold water and dissolved nutrients in the root zone where plants can use them. The capacity of a particular soil to do this depends on several soil characteristics. Organic matter in soil increases water-holding capacity because water is easily absorbed by decaying plant material. Soil aggregates composed of individual soil particles of sand, silt, and clay, bound together in clumps, have high capacities for holding water in the root zone.

Water-holding capacity of a soil is highly influenced by soil texture. The texture of soil is determined by the proportion of sand, silt, and clay particles that characterize the soil. Sand particles have limited capacity to hold water; their function is to drain excess water away from the root zone. Clay soil consists of very small particles that attract water molecules; however, a heavy clay soil can also form a layer through which water movement is difficult. Silt is composed of medium-size soil particles with moderate water-holding capacity. A loam soil is a combination of sand, silt, and clay particles. Loam soils have the greatest capacity for water and dissolved plant nutrients. In combination with organic matter, a loam soil provides the ideal soil medium for most agricultural crops.

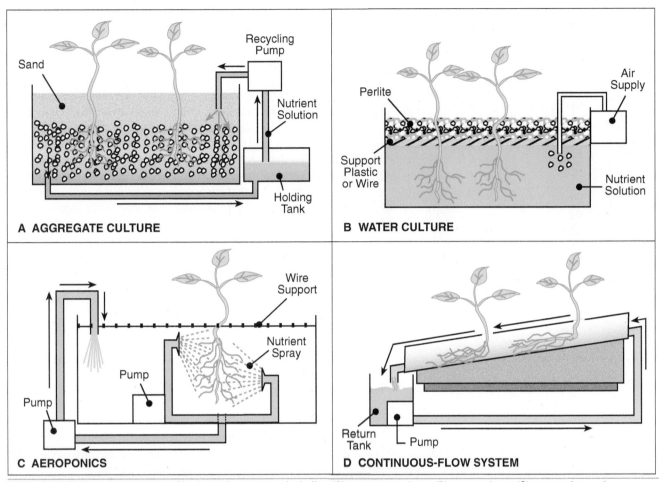

Figure 9-28 There are many types of hydroponics systems, including (A) aggregate culture, (B) water culture, (C) aeroponics, and (D) continuous-flow systems.

Redrawn based on material by Keith Staley, Middletown High School, Middletown, MD.

- water culture, solution culture, or nutriculture—the plant roots are immersed in water containing dissolved nutrients;
- **aeroponic**—in which the plant roots hang in the air and are misted regularly with a nutrient solution; and
- continuous-flow systems—in which the nutrient solution flows constantly over the plant roots. This system is the one most commonly used for commercial production.

Hydroponics is growing in importance as a means of producing vegetables and other high-income plants (**Figure 9-29**). In areas where soil is lacking or unsuitable for growth, hydroponics offers an alternative production system. Equally good crops can generally be produced in a greenhouse in conventional soil or bench systems.

When plants are grown hydroponically, their roots are either immersed in or coated with a carefully controlled nutrient solution. The nutrients and water are supplied by the solution alone and not by aggregates or other inert materials that support the roots. Inert means inactive. Without the presence of soil to absorb and release nutrients, the nutrients must be carefully controlled in a hydroponic system.

When hydroponic systems are adapted to commercial production of plants such as tomatoes, the tomatoes are referred to as *hothouse* tomatoes.

Figure 9-29 High-value crops such as tomatoes are ideal for hydroponic production methods.

© manfredxy/Shutterstock.com.

Hot Topics in Agriscience: Producing "Hothouse" Vegetables

The production of "hothouse" vegetables is done in a controlled greenhouse environment. In many cases, these vegetable production units are located near natural hot water springs to take advantage of the heat source during the winter months. The hot water is often piped into the greenhouse where it is an inexpensive source of supplemental heat. Most of these vegetable farms use some variation of hydroponics rather than using soil as a medium for plant growth. The plants are supported as they grow by tying them up using wires and strings. Tomatoes and other vegetable products are harvested frequently at their peak of quality and are shipped immediately to markets. Most "hothouse" vegetables are produced during seasons when field production is limited or impossible because of climate restrictions. As a result, these vegetables usually command high prices in the markets.

Commercial production of vegetables is extended into winter months using hydroponic practices in heated greenhouses.
© Roy Navarre, USDA

Plant Growth Requirements

Hydroponically grown plants have the same general requirements for good growth as soil-grown plants. The major difference is the method by which the plants are supported and the source of nutrients that are supplied for growth and development.

Water

Providing plants with an adequate amount of water is not difficult in a water culture system. During the hot summer months, a large tomato plant may use one-half gallon of water per day. However, quality can be a problem. Water with excessive alkalinity or salt content can result in a nutrient imbalance and poor plant growth. Softened water may contain harmful amounts of sodium. Water that tests above 320 parts per million of salts is likely to cause an imbalance of nutrients.

Oxygen

Plants require oxygen for respiration to carry out their functions. Under field and normal greenhouse conditions, oxygen is usually adequate as provided by the soil. When plant roots grow in water, however, the supply of dissolved oxygen is soon depleted, and damage or death soon follows unless supplemental oxygen is provided. Where supplemental oxygen is needed, a common method of supplying oxygen is to bubble air through the water. It is not usually necessary to provide supplementary oxygen in aeroponic or continuous-flow systems.

Mineral Nutrients

Green plants must absorb certain minerals through their roots to survive. These minerals are supplied by soil, organic matter, or soil solutions. The elements needed in large quantities are nitrogen, phosphorus, potassium, calcium, magnesium, and sulfur.

The nutrients needed in small amounts are iron, manganese, boron, zinc, copper, molybdenum, and chlorine. An oversupply of any nutrient is toxic or detrimental to plants. In hydroponic systems, all nutrients normally supplied by soil must be included in the water to form the solution or media.

Light

All vegetable plants and many flowers require large amounts of sunlight. Hydroponically grown vegetables, such as those grown in a garden, need at least 8 to 10 hours of direct sunlight each day to produce well. Electrical lighting is a poor substitute for sunshine because most indoor lights do not provide enough intensity to produce a crop. Incandescent lamps supplemented with sunshine or special plant growth lamps can be used to grow transplants, but they are not adequate to grow the crop to maturity. High-intensity lamps, such as high-pressure sodium lamps, can provide more than 1,000 foot-candles of light. They may be used successfully in small areas where sunlight is inadequate. However, these lights are too expensive for most commercial operations.

Spacing

Adequate spacing between plants will permit each plant to receive sufficient light in the greenhouse. Tomato plants, pruned to a single stem, should be allowed 4 square feet per plant. European seedless cucumbers should be allowed 7 to 9 square feet, and seeded cucumbers need about 7 square feet. Leaf lettuce plants should be spaced 7 to 9 inches apart within the row and 9 inches between rows. Most other vegetables and flowers should be grown at the same spacing as recommended for a garden.

Greenhouse vegetables will not do as well during the winter as in the summer. Shorter days and cloudy weather reduce the light intensity, thus limiting production.

Temperature

Plants grow well only within a limited temperature range. Temperatures that are too high or too low will result in abnormal development and reduced production. Warm-season vegetables and most flowers grow best between 60° F and 80° F. Cool-season vegetables, such as lettuce and spinach, should be grown between 50° F and 70° F.

Support

In the garden or field, plants are supported by roots anchored in soil. A hydroponically grown plant must be artificially supported with string, stakes, or other means.

Hydroponics in the Classroom and Laboratory

Hydroponics has become an important teaching area in agriscience programs. The use of common plastic bottles and low-cost aquarium supplies permits students and teachers to set up and perform numerous research and demonstration projects in the classroom or laboratory.

According to Dr. David R. Hershey (1994), biology education consultant and author, a solution culture system (**Figure 9-30**) may be constructed as follows:

1. Fill two dark 2-liter plastic soda bottles with hot water to loosen the glued label and base. Remove the labels from both bottles and the base from one bottle; keep the base for future use.

2. Using a fine-point felt-tip marker, draw a line around the bottle 9 inches up from the bottom.

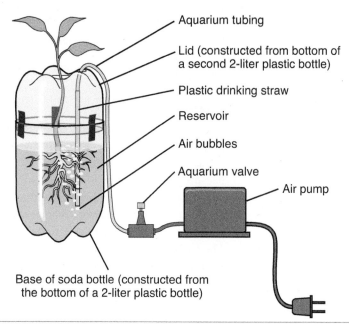

Figure 9-30 A static solution culture system built from two plastic soda bottles.

3. Using short-bladed scissors, cut on the line made in Step 2 and remove the upper portion of the bottle. This will be the reservoir.

4. Using a cork-borer, drill, or scissors, make a hole in the center of the bottle base approximately 0.5 inches in diameter to accommodate a plant stem. Close all except one of the existing (prepunched) holes in the base with black vinyl electrician's tape to prevent light passage. Then place the base on the reservoir as a dome-shaped lid.

5. Insert about 8 inches of 2-foot-long aquarium tubing into the open, prepunched hole. To keep the tubing in the reservoir rigid, place a plastic drinking straw over the tubing. Attach the free end of the aquarium tubing to an aquarium air pump with an air control valve.

Nutrient Solutions

There are many nutrient-solution recipes, but Hoagland Solution No. 1 is used widely and can be modified to create nutrient deficiencies (**Figure 9-31**). Often, Hoagland solution is used at less than full strength. Nutrient stock solutions are prepared using a balance to weigh out the salts and a graduated cylinder or volumetric flask to measure the water. Plastic soda bottles can be used to store stock solutions. When stored in a dark place at room temperature, stock solutions should last for many years. To prepare the nutrient solution, add measured volumes of stock solutions to a measured volume of water. Stock solutions can be dispensed using pipettes or graduated cylinders. Complete hydroponic salt mixtures can be purchased, and they greatly simplify nutrient solution preparation.

Aeration

Solution cultures are typically aerated using an aquarium air pump and aquarium air tubing. In a soda bottle system, the aquarium air tubing is inserted into the reservoir through one

Stock Solution				milliliters of stock solution per liter of nutrient solution							
Formula	Name	grams/liter	Complete	N	P	K	Ca	Mg	S	Fe
$Ca(NO_3)_2 \cdot 4H_2O$	Calcium nitrate, 4-hydrate	236	5	–	4	5	–	4	4	5
KNO_3	Potassium nitrate	101	5	–	6	–	5	6	6	5
KH_2PO_4	Monopotassium phosphate	136	1	–	–	–	1	1	1	1
$MgSO_4 \cdot 7H_2O$	Magnesium sulfate, 7-hydrate	246	2	2	2	2	2	–	–	2
FeNa EDTA[y]	Iron EDTA	18.4	1	1	1	1	1	1	1	–
Micronutrients[z]		–	1	1	1	1	1	1	1	1
K_2SO_4	Potassium sulfate	87	–	5	–	–	–	3	–	–
$CaCl_2 \cdot 2H_2O$	Calcium chloride, dihydrate	147	–	2	–	–	–	–	–	–
$Ca(H_2PO_4)_2 \cdot H_2O$	Calcium phosphate, monobasic	12.6	–	10	–	10	–	–	–	–
$Mg(NO_3)_2 \cdot 6H_2O$	Magnesium nitrate, 6-hydrate	256	–	–	–	–	–	–	2	–

[y]Ferric-sodium salt of ethylene-diamine tetraacetic acid. Differs from original Hoagland recipe, which used Fe tartrate.
[z]Contains the following in grams/liter:

2.86 H_3BO_3, (boric acid)
1.81 $MnCl_2 \cdot 4H_2O$, (manganese chloride, 4-hydrate)
0.22 $ZnSO_4 \cdot 7H_2O$, (zinc sulfate, 7-hydrate)
0.08 $CuSO_4 \cdot 5H_2O$, (copper sulfate, 5-hydrate)
0.02 $H_2MoO_4 \cdot H_2O$, (85% molybdic acid).

Note: For Mn-, B-, Cu-, Zn-, and Mo-deficient solutions, substitute micronutrient stock solutions missing one of the five salts in the regular micronutrient stock solution. For Cl-deficient micronutrient stock solution, substitute 1.55 $MnSO_4 \cdot H_2O$ (manganese sulfate, monohydrate) for 1.81 $MnCl_2 \cdot 2H_2O$.

Figure 9-31 Preparation of modified Hoagland nutrient solutions for nutrient deficiency system development.
Source: Hoagland, D. R. and Arnon, D. I. 1950. The Water-Culture Method for Growing Plants Without Soil. California Agricultural Experiment Station Circular 347, revised 1950.

of the prepunched holes, and the tubing in the reservoir is made rigid by slipping a plastic soda straw onto the tubing. To prevent clogging, a piece of cotton or aquarium filter floss is placed in the aquarium tubing as it exits the pump. The filter is necessary to trap either dirt from the air or pieces of the pump diaphragm. Adjust the aeration to a gentle rate of one to three bubbles per second. Some plants do not benefit from aeration but most do. An alternative to pump aeration is to let the top third of the root system remain uncovered by solution in the humid air in the top of the reservoir.

Maintenance

Water must be added to static solution cultures every few days or so to replace water lost by transpiration and evaporation. Nutrient solutions are typically replaced with fresh solutions on a schedule, such as every 1 or 2 weeks. Frequency of solution replenishment depends on the rate of plant growth relative to the volume of nutrient solution. An electrical conductivity (EC) meter is useful for determining when a nutrient solution has been depleted. To prevent interruption of growth, replace the solution when the solution EC is about half the original value.

Plants

Beans, corn, sunflowers, and tomatoes are often used for teaching hydroponics because they grow rapidly from seed; but they are difficult to handle because of their large size, high light requirements, and need for staking. Wisconsin Fast Plants, which are becoming a standard plant for teaching use, are excellent for solution culture. Fast plants go from seed to flower in 2 weeks under a bank of six 4-foot, 40-watt fluorescent lamps and remain less than a foot in height.

Houseplants, such as piggyback (*Tolmiea menziesii*), inch plant (*Tradescantia* species), evergreen euonymus (*Euonymus japonica*), coleus, common philodendron, and pothos (*Epipremnum aureum*), are also excellent in teaching hydroponics. They thrive with low light levels, are readily available, and root quickly in solution. Radish and lettuce are easily grown from seed. A carrot placed in tap water in the 2-liter soda bottle system described earlier will sprout lateral roots, produce a shoot, and flower. Pineapple fruit tops easily root in solution.

Germination

Seeds for hydroponics are conveniently germinated in containers of perlite, which is inert and is easily removed from roots before transferring the plants to solution. Small seeds, such as fast plants or coleus, can be germinated on a paper towel in a seed germinator built from a 2-liter soda bottle. The base is removed from a bottle and a 4-inch bottom section of the bottle is inverted over the base to complete the germinator. A wet paper towel is used to line the domed section, and small seeds are pressed into the towel. The seeds remain stuck until the seed germinates and the seedling is then transplanted (**Figure 9-32**).

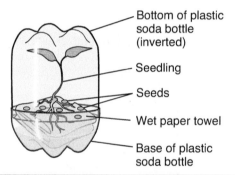

Figure 9-32 A germination chamber may be constructed with the bottom cut from a 2-liter soda bottle inverted over a soda bottle base. The dome section is lined with a wet paper towel, and seeds are pushed against the towel to absorb moisture and germinate before transplanting.

Future of Hydroponics

Hydroponics is increasing in use commercially and undoubtedly will become more important in the future. Research is expanding, and new techniques are being developed. The use of nutrient solutions as media for growing plants will be an important part of agriscience in the future.

Agri-Profile: Career Areas: Agronomy/Hydroponics

Soil scientist Ronald Schnabert (left) and hydrologist technician Earl Jacoby study natural riparian zone processes that lessen the impact of upstream agriculture on water quality.
Courtesy of USDA/ARS #K-5050-3.

Developing expert knowledge of agronomy and hydroponics opens many career options. Agronomy is the science and economics of managing land and field crops. An agronomist, or crop scientist, studies soil-based crops, just as the hydroponics scientist studies the practice of growing plants without soil. Career options in soil and hydroponics management overlap those in soil and water conservation to a certain extent, and additional opportunities occur in management of farms and hydroponics greenhouses.

The practice of hydroponics is not new, but hydroponics for commercial production has captured the imagination of the world. The recent popularity of hydroponics operations provides new career opportunities, especially in urban areas.

Soil and water management specialists are in demand on the global scene. Progressive developing countries need help in policy development, education, and project management. They hope to leap from their primitive agriculture of the past to the agriscience of the present in a few short years.

Electronic devices have become tools of the trade for soil and water research and management. For example, meteorological measuring stations like the one pictured are used to record humidity, precipitation, soil temperature, and other data. Areas of particular interest are riparian zones, which are lands that occur along the edges of rivers, streams, and other bodies of water, because of their importance to plants and wildlife. Hydrological processes affect soil quality just as upstream agriculture affects water quality.

Check Your Progress

Can you . . .

- define terms in soils, hydroponics, and other plant-growing media management?
- identify types of plant-growing media?
- describe the origin and composition of soils?
- discuss the principles of soil classification?
- determine appropriate amendments for soil and hydroponics media?
- discuss fundamentals of fertilizing and liming materials?
- identify requirements for hydroponics plant production?
- describe types of hydroponics systems?

If you answered "no" to any of these questions, review the chapter for those topics.

Chapter Review

Student Activities

1. Write the **Key Terms** and their meanings in your notebook.
2. Observe soils and other media used in flowerpots, trays of vegetable seedlings, greenhouses, gardens, road banks, construction sites, and other places.
3. Examine the living organisms in a shovelful of soil taken from an outdoor area that is damp, moist, and high in organic matter.
4. Invite a Natural Resources Conservation Service (NRCS) professional to the classroom to discuss soil mapping and land-use planning.
5. Obtain a land-use map from a Farm Service Agency (FSA) for your home, farm, or other area. Study the material and report your findings to the class.
6. SAE Connection: Observe a soil profile at a cut in a road bank, stream bank, or hole. Identify the O, A, B, and C horizons.
7. Do a mechanical analysis of a sample of soil using a fruit jar, water, and a dispersing agent such as Calgon (see Figure 9-17).
8. Observe the feel of some of the soil used in Activity 7.
9. Obtain a pH test kit and test samples of soil from around your school or neighborhood.
10. Do research on models and procedures for a home or school hydroponics unit. Plan, build, and use the unit to experiment with hydroponic production of various plants.
11. Build and use a composting structure.
12. Conduct the following soil test:
    - Add 1 teaspoon of lime to a quart of distilled water and mix well. Be sure to wear gloves and protective eyewear.
    - Find the pH of this solution by using a pH test kit.
    - Compare the pH of the sample to the pH of the soil taken in Activity 9.
    - Write a short explanation describing the pH differences and what impact lime may have on soil.
13. Learn more about one type of crop grown without soil using the Internet or other agriscience resources. In one or two paragraphs, tell which crop you learned about, in which non-soil media it was grown, and what procedure the grower used to produce crops.

National FFA Connection: Agriscience Fair: Natural Resource Systems—Soils

In one popular FFA event, students compete to identify soil types.
© Kichigin/Shutterstock.com.

The National FFA Organization sponsors state and national science fairs in which FFA members in grades 7–12 are eligible to compete. Students may compete as individuals or in teams of two. Competition is in three age divisions: grades 7–8, grades 9–10, and grades 11–12. State science fair winners are eligible to advance to the national agriscience fair in the appropriate age division.

Examples of science fair projects related to soils could include:

- Demonstrate downward movement of water through soils of different structural classes.
- Create a display of soil profiles and identify the soil horizons.
- Select any soil-related topic for your agriscience fair project.

Participate in a local or state agriscience fair. To get started, generate ideas for projects that interest you, using a T-chart like the following:

Ideas for State Fair Projects	
Individual Projects	**Projects for a Team of Two**

Then share your ideas as a class to help you launch the planning process.

Chapter 10

Forest Management

Objective
To determine the relationship of forests to the environment and the recommended practices for using forest resources.

Competencies to Be Developed
After studying this chapter, you should be able to:
- define forest terms.
- describe the forest regions of the United States.
- discuss important relationships among forests, wildlife, and water resources.
- identify important types and species of trees.
- describe how a tree grows.
- discuss important properties of wood.
- apply principles of good woodlot management.
- describe procedures for seasoning lumber.

Key Terms

forest land	deciduous	phloem
timberland	hardwood	inner bark
forest	pulpwood	heartwood
tree	clear-cut	hardness
shrub	plywood	shrinkage
lumber	veneer	warp
board foot	cambium	woodlot
evergreen	annual ring	silviculture
conifer	xylem	arboriculture
softwood	sapwood	seedling

According to the U.S. Environmental Protection Agency (EPA), the United States has more than 520 million acres of timberland and 246 million acres of other forest land, for a total of over 740 million acres. This represents about one-third of the total land in the United States. **Forest land** is defined by the U.S. Department of Agriculture (USDA) as land at least 10 percent stocked by forest trees of any size. **Timberland** is defined as forest land that is capable of producing in excess of 20 cubic feet per acre per year of industrial wood and that has not been withdrawn from timber utilization by statute or administrative regulation.

A **forest** is a complex association of trees, shrubs, and plants that all contribute to the life of the community (**Figure 10-1**). A **tree** is a woody perennial plant with a single stem that develops many branches. Trees vary greatly in size, but by definition, they grow to more than 10 feet in height. A **shrub** is a woody plant with a bushy growth pattern and multiple stems. A productive forest is one that is growing trees for lumber or other wood products on a continuous basis. **Lumber** consists of boards that are sawed from trees. It is bought and sold by the board foot. A **board foot** is a unit of measurement for lumber that is equal to $1 \times 12 \times 12$ inches (**Figure 10-2**).

Forest land may include parks, wilderness land, national monuments, game refuges, and some areas where harvesting of trees is not permitted. When you consider that there are approximately 1,400 species of trees in North America, it is evident that forestry is an important part of the economy. Forestry is the management of forests. Our citizens depend on a great variety of products that come from trees (**Figure 10-3**).

Figure 10-1 A forest contains trees, shrubs, and other plants as well as animal life.

© Thomas La Mela/Shutterstock.com.

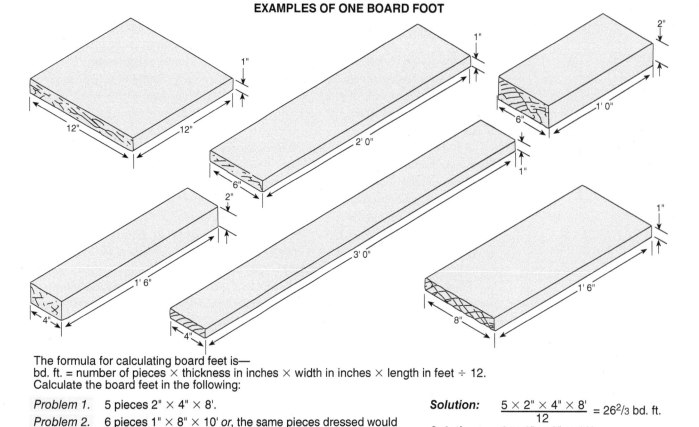

Figure 10-2 Lumber is bought and sold by the board foot. One board foot has a volume of 144 cubic inches. Any combination of dimensions that equals 144 cubic inches is 1 board foot.

Trees in the forests of the United States are divided into two general classifications: evergreen and deciduous. **Evergreen** trees do not shed their leaves on a yearly basis. Evergreen trees of commercial importance are mostly conifers. **Conifers** are evergreen trees that produce seeds in cones, have needle-like leaves, and produce lumber called **softwood**. **Deciduous** trees shed their needles or leaves every year and produce lumber called **hardwood**.

Forest Regions of North America

Eight major forest regions are generally recognized by forestry educators and other forestry professionals (**Figure 10-4**). They include the Northern Coniferous Forest, Northern Hardwoods Forest, Central Broad-Leaved Forest, Southern Forest, Bottomland Hardwoods Forest, Pacific Coast Forest, Rocky Mountain Forest, and Tropical Forest. In addition, some experts refer to the Wet Forest and the Dry Forest in Hawaii (**Figure 10-5**).

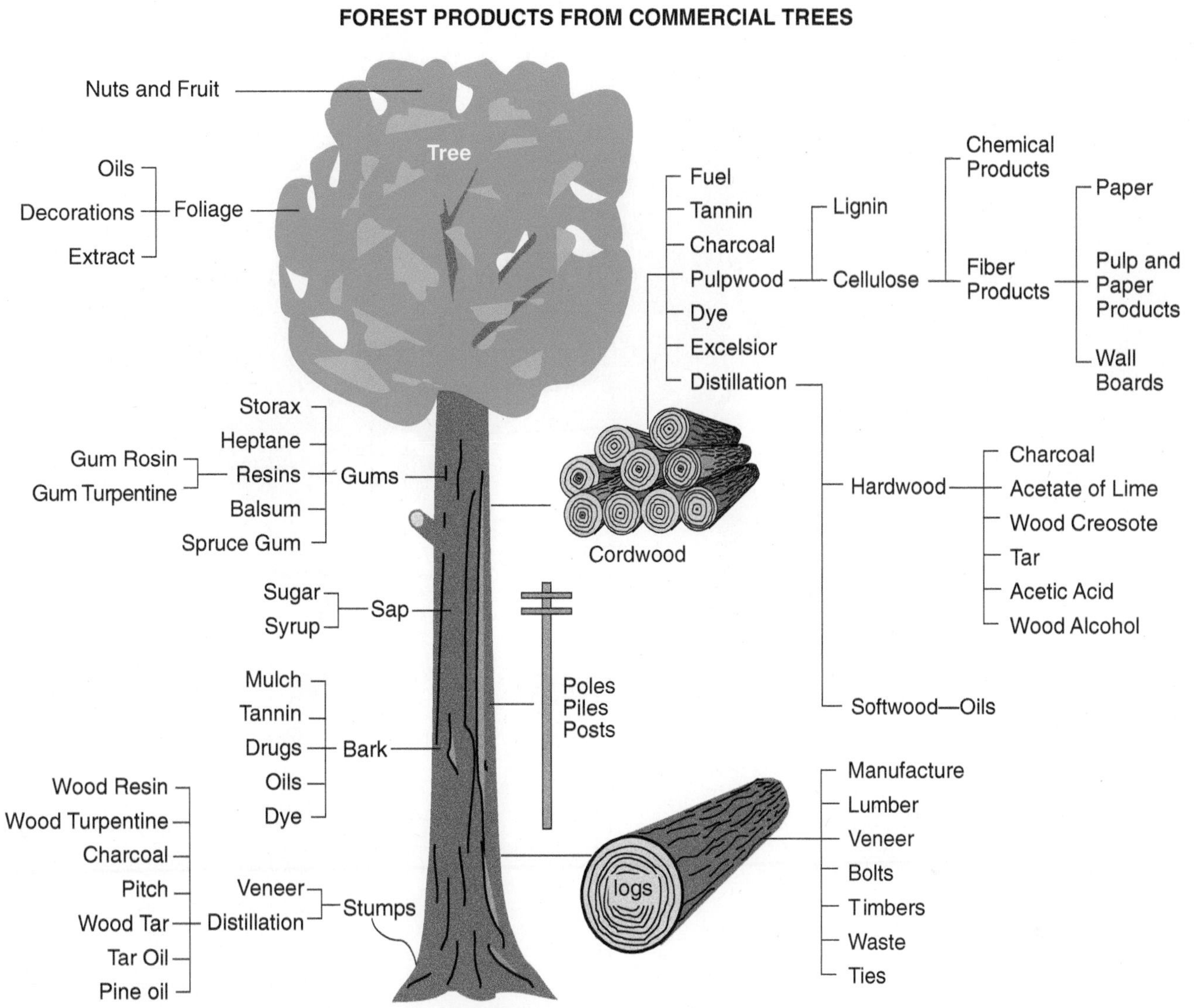

FOREST PRODUCTS FROM COMMERCIAL TREES

Figure 10-3 Many kinds of commercial products are obtained from trees and wood by-products.

Northern Coniferous Forest

The Northern Coniferous Forest region contains vast regions of softwoods. Some areas along the border between Canada and the United States contain a mixture of softwoods and hardwoods. The region is characterized by swamps, marshes, rivers, and lakes. The climate is cold. This forest is the largest region in North America, extending across Canada and Alaska. The most dominant type of tree is the evergreen, and large amounts of **pulpwood** are harvested in this region. The most important species include the white spruce, Sitka spruce, black spruce, jack pine, black pine, tamarack, and western hemlock.

Northern Hardwoods Forest

The Northern Hardwoods Forest region reaches from southeastern Canada through New England to the northern Appalachian Mountains. It blends with the Northern Coniferous Forest on the northern border and the Central Broad-Leaved Forest on the south. The region extends westward beyond the Great Lakes region, and it is populated by a number of important hardwood species, including beech, maple, hemlock, and birch trees.

Figure 10-4 Forest regions of North America.

Drawn based on information from U.S. Forest Service.

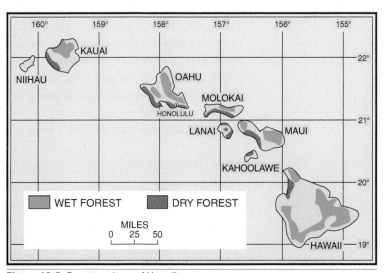

Figure 10-5 Forest regions of Hawaii.

Drawn based on information from U.S. Forest Service.

Central Broad-Leaved Forest

The Central Broad-Leaved Forest region is located mostly east of the Mississippi River and south of the Northern Hardwoods Forest. It consists of an arbitrary grouping of several distinctly different forest subgroups. It is a farming region in which most of the land has been cleared to produce cultivated crops. Little of the forested area is owned by the federal government, in contrast with some other regions. High-quality wood is produced in this region, and much of it is used to construct high-quality furniture. Hardwoods of lesser quality are used for construction and to make industrial pallets.

This forest contains more varieties and species of trees than any other forest region. It is composed mostly of hardwood trees. Hardwoods of commercial importance in the Central Broad-Leaved Forest region include oak, hickory, beech, maple, poplar, gum, walnut, cherry, ash, cottonwood, and sycamore. The conifers that are of economic value in this region include Virginia pine, pitch pine, shortleaf pine, red cedar, and hemlock.

Southern Forest

The Southern Forest region is located in the southeastern part of the United States. It extends south from Delaware to Florida, and west to Texas and Oklahoma. It is the forest region with the most potential for meeting the future lumber and pulpwood needs of the United States.

The most important trees in the Southern Forest are conifers. They include Virginia, longleaf, loblolly, shortleaf, and slash pines. Oak, poplar, maple, and walnut are hardwood trees of economic importance.

Bottomland Hardwoods Forest

The Bottomland Hardwoods Forest region occurs mostly along the Mississippi River. It contains mostly hardwood trees and is often among the most productive of the U.S. forests. This is because of the high fertility of the soil in this area. Oak, gum, tupelo, and cypress are the major hardwood species found here.

Pacific Coast Forest

The Pacific Coast Forest region is located in Northern California, Oregon, and Washington. It is the most productive of the forest regions in the United States and has some of the largest trees in the world. Approximately 48 million acres of Pacific Coast Forest provide more than 25 percent of the annual lumber production in the United States. About 19 percent of the pulpwood and 75 percent of the plywood produced in the United States comes from trees grown in the Pacific Coast Forest region.

Trees in the Pacific Coast forests include 300-foot-tall redwoods and giant Sequoias that may be as much as 30 feet in diameter. Douglas fir, ponderosa pine, hemlock, western red cedar, Sitka spruce, sugar pine, lodgepole pine, noble fir, and white fir are conifers that are important in this region. Important hardwood species include oak, cottonwood, maple, and alder.

Rocky Mountain Forest

The Rocky Mountain Forest region is much less productive than the Pacific Coast Forest region. This region is divided into many small areas and extends from Canada to Mexico. About 27 percent of the lumber produced in the United States comes from the 73 million acres of this forest region.

Most of the trees of commercial value in the Rocky Mountain Forest region are the western pines: western white pine, ponderosa pine, and lodgepole pine. Spruce, fir, larch, western red cedar, and hemlock also grow there in small quantities. Aspen is the only hardwood of commercial importance in the Rocky Mountain Forest region.

Tropical Forest

The Tropical Forest Region of the continental United States are located in southern Florida and in southeastern Texas. These compose the smallest forest region in the United States. The major trees in this region are mahogany, mangrove, and bay, which are unimportant commercially. However, they are very important ecologically.

Hawaiian Forest

The Wet Forest region of Hawaii produces ohia, boa, tree fern, kukui, tropical ash, mamani, and eucalyptus. Most of these woods are used in the production of furniture and novelties.

The Dry Forest region of Hawaii produces koa, haole, algaroba, monkey pod, and wiliwili. None of these is of commercial value.

Relationships Between Forests and Other National Resources

The relationships between forests and other natural resources, such as water and wildlife, are important to the overall well-being of the ecological system.

Forests play important roles in the water cycle. As water circulates between the oceans and the land areas, forests reduce the impact of falling rain on the soils and serve as storage areas for vast amounts of water. The stored water is released slowly from a forest watershed, allowing streams to flow throughout the year. In this manner, a forest regulates the flow of water, making it possible for fish and other aquatic plants and animals to survive.

Forests filter rain as it falls and help reduce erosion of the soil. They trap soil sediment and help maintain water quality. Trees and shrubs of the forest are also instrumental in removing many pollutant materials from the air. Similarly, the roots of trees filter excess dissolved nutrients from ground and surface water sources.

The relationships between forests and many types of wildlife are numerous. Algae, fungi, mosses, and numerous other plant and animal forms make their homes in forests. Forests provide food, shelter, protection, and nesting sites for many species of birds and animals. They also help maintain the quality of streams so that fish and other types of aquatic life can live and thrive. The shade provided by forests helps maintain proper water temperatures for the growth and reproduction of aquatic life.

The wildlife found in the different types of forests varies considerably (**Figure 10-6**). Some species must have the open areas that occur naturally or that have been clear-cut. A forest that has been clear-cut has had all or most of the marketable trees removed from it. Other wild birds and animals depend to a greater extent on mature forests to provide for their needs in life. As a forest changes naturally or as a result of harvesting practices, the types of wildlife that inhabit it also change.

Important Types and Species of Trees in the United States

Trees may be described in terms of the lumber they produce. Characteristics such as hardness, weight, tendency to shrink and warp, nail- and paint-holding capacity, decay resistance, strength, and surface qualities are used to evaluate the usefulness of lumber.

Softwoods

The softwoods, or needle-type evergreens, that are important commercially in the United States include Douglas fir, balsam fir, hemlock, white pine, cedar, southern pine, ponderosa pine, and Sitka spruce (**Figure 10-7**).

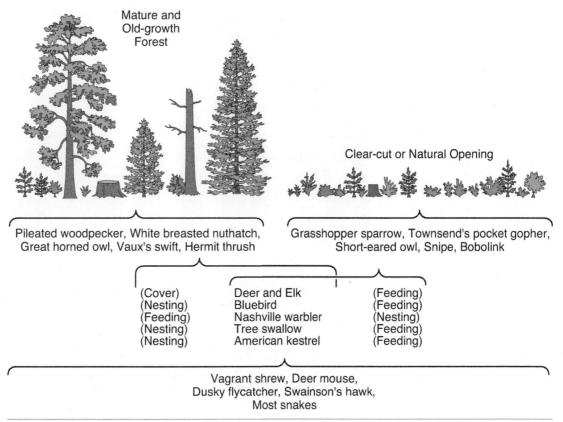

Figure 10-6 Preference of 20 wildlife species for different forest habitats.

Douglas Fir

Douglas fir is probably the most important species of tree in the United States today. It grows to a height of more than 300 feet and a diameter of more than 10 feet. About 20 percent of the timber harvested in the United States each year is Douglas fir. One-hundred-year-old stands of Douglas fir can produce 170,000 board feet of lumber per acre. This is five to six times the production of most other softwood species. Douglas fir is popular as construction lumber and for the manufacture of plywood. **Plywood** is a construction material made of thin layers of wood glued together.

Balsam Fir

Found in the forests of the Northeast, the lumber from balsam fir is used mostly for framing buildings. Balsam fir trees have soft, dark green needles and a classic triangular shape when grown at low densities. They are often used as Christmas trees.

Hemlock: Eastern and Western

Eastern hemlock is strong and is often used for building material. However, at times, it can be brittle and difficult to work with. Eastern hemlock grows over most of the Northern Coniferous Forest region.

Western hemlock grows in the Pacific Coast Forest region, where yearly rainfall averages 70 inches. Western hemlock lumber is very strong and is one of the most important sources of construction-grade lumber. It is also important for pulpwood.

Figure 10-7 Some species of softwood trees that are commercially important for wood products.

© Christopher Barrett/Shutterstock.com. © Helder Almeida/Shutterstock.com. © R.L. Hausdorf/Shutterstock.com. © Noah Strycker/Shutterstock.com. © IrinaK/Shutterstock.com. © Tom Grundy/Shutterstock.com Candia Baxter/Shutterstock.com. Courtesy of DeVere Burton. Pancaketom/Dreamstime.com.

Cedar: Eastern Red, Eastern White, and Western Red

Eastern red cedar is used for fence posts because it is resistant to decay. It is also used for lining chests and closets because its odor repels many insects. White cedar is a swamp tree with decay-resistant wood that is often used for shingles and log homes. Western red cedar resembles redwood. It is used where decay resistance, rather than strength, is important.

White Pine

White pine lumber is soft, light, and straight-grained. It has less strength than spruce or hemlock, and it is more popular as a wood for cabinetmaking. Eastern white pine grows from Maine to Georgia. Western white pine is found in the Rocky Mountain Forest region.

Southern Pine

Included in the category of southern pine is longleaf pine, shortleaf pine, loblolly pine, and slash pine. The southern pines grow in the southern and south Atlantic states. Lumber from southern pine is used for construction, pulpwood, and plywood.

Ponderosa Pine

The ponderosa pine is a large tree that grows up to 130 feet in height and 4 feet in diameter. It is widely distributed in the western United States. The wood is heavy, and it can be brittle; however, it is reasonably free of knots and other defects. Its most valuable use is for construction of wooden windowpanes and doors.

Sitka Spruce

Sitka spruce trees grow from California to Alaska, attaining a height of 300 feet and a diameter of 18 feet. Lumber from Sitka spruce is of very high quality. It is strong, straight, and even-grained. Sitka spruce is also used in large quantities for pulpwood.

Hardwoods

Hardwoods come from deciduous trees. Commercially important species of hardwoods in the United States include birch, maple, poplar, sweetgum, oak, aspen, ash, beech, cherry, hickory, sycamore, walnut, and willow (**Figure 10-8**).

Birch

Recognized by their white bark, birch trees grow in areas where summer temperatures are relatively cool. Birch lumber is dense and fine-textured. It is used for furniture, plywood, paneling, boxes, baskets, veneer, and small novelty items. **Veneer** is a thin sheet of wood glued to a cheaper species of wood that is used in paneling and furniture making.

Maple

Maple lumber is classified as both hard and soft. Hard maple lumber is heavy, and strong. It is used for butcher blocks, workbench tops, flooring, veneer, and furniture. Soft maple is only about 60 percent as strong, and it is used in many of the same applications. Some species, such as the sugar maple, produce sweet sap that is made into maple syrup.

Poplar

Poplar grows over most of the eastern United States. It is classified as a hardwood because of its deciduous structure. However, lumber from poplar trees is soft, light, and usually knot-free. Poplar lumber may be white, yellow, green, or purple, and can be stained to resemble fine hardwoods. It is used for furniture, baskets, boxes, pallets, and building timbers.

A Paper Birch

B Sugar Maple

C Yellow Poplar

D Sweetgum

E White Oak

F Quaking Aspen

Figure 10-8 Some species of hardwood trees that are commercially important for wood products.

© John A. Anderson/Shutterstock.com. © Melinda Fawver/Shutterstock.com. © Dave Barnard/Shutterstock.com. © islavicek/Shutterstock.com. © Cheryl E. Davis/Shutterstock.com. © Patricia A. Phillips/Shutterstock.com.

G White Ash
H American Beech
I Black Cherry
J Bitternut Hickory
K American Sycamore
L Black Walnut
M Black Willow

Figure 10-8 Continued

Sweetgum

The sweetgum tree is easily recognized by its star-shaped leaves and distinctive ball-shaped fruit. Sweetgum trees grow to as much as 120 feet in height and 3–5 feet in diameter. Lumber from sweetgum trees has interlocking grain and is used for house trim, furniture, pallets, railroad ties, boxes, and crates. The gum that comes from wounds in trees can be used as natural chewing gum or as a flavoring or perfume.

Oak: White and Red

The two general types of oak in the United States are white and red. White oak lumber is hard, heavy, and strong. Its pores are plugged with membranes that make it nearly waterproof. It is used for structural timbers, flooring, furniture, fencing, pallets, and other uses where wood strength is a necessity.

Red oak is similar to white oak, except that it is very porous. It is not very resistant to decay and must be treated with wood preservatives when used outside. Chief uses of red oak include furniture, veneer, and flooring.

Aspen

Aspen trees grow in the Northeast, Great Lake states, and the Rocky Mountains. Aspen is rapid growing, but the lumber tends to be weaker than most construction-grade timber. It is also used for pulpwood.

Ash

Ash lumber is heavy, hard, stiff, and has a high resistance to shock. It also has excellent bending qualities. It is popular for use in handles, baseball bats, boat oars, and furniture. It resembles oak in appearance.

Beech

Beech is grown in the eastern United States. It is heavy and hard, and it is noted for its shock resistance. It is hard to work with and is prone to decay. Beech is used in veneer for plywood and for flooring, handles, and containers.

Cherry

Cherry can be found from southern Canada through the eastern United States. Cherry wood is dense and stable after drying. It is desirable and popular in the production of fine-quality furniture. It is expensive and in limited supply; therefore, it is used mostly for veneer and paneling. It may, however, be used for other woodworking purposes.

Hickory

Hickory grows best in the eastern United States. Hickory lumber is hard, heavy, tough, and strong. It is somewhat stronger than Douglas fir when used as construction lumber. Other uses for hickory include handles, dowel rods, and poles. Hickory is also popular for use as firewood and for smoking meat.

Sycamore

Sycamore wood is used for flooring in barns, trucks, and wagons because of its strength and shock resistance. It grows from Maine to Florida and west to Texas and Nebraska. Boxes, pallets, baskets, and paneling are other uses for sycamore.

Black Walnut

A premier wood for the manufacture of fine furniture, black walnut grows from Vermont to Texas. The wood has straight grain and is easily machined with woodworking tools. Because walnut is slow-growing and demand for it is high, it is often made into veneer to get more use from its chocolate brown heartwood. It is also the source of black walnut nutmeats.

Black Willow

Most of the black willow of commercial value is grown in the Mississippi River Valley. It is soft and light and has a uniform texture. Willow is used mostly in construction for subflooring, sheathing, and studs. Some willow is also used for pallets and for interior components of furniture. Black willow is sometimes a low-cost substitute for black walnut because it has a similar brown appearance when finished. Several other domestic softwoods and hardwoods grow in the United States and many are important in local areas. The types discussed here are but a sampling, rather than a definitive list, of the commercially important trees in the United States.

Agri-Profile | Career Areas: Dendrology/Silviculture/Forestry

Many challenging careers are available to college graduates in the forest industry.
iStock.com/Snajpo

Forestry is known as a career area for rugged individuals who prefer the outdoors and like to work in relative isolation. However, many jobs in forestry are in urban areas and involve much indoor work. The United States Forest Service hires large numbers of forestry technicians and managers. Many forestry jobs do involve an extensive amount of outdoor work, but most jobs provide a desirable mix of outdoor and indoor work.

Forestry includes the work of the dendrologist engaging in the study of trees, the silviculturist specializing in the care of trees, the forestry consultant advising private forest landowners, lumber industry workers, government foresters, loggers, national and state forest rangers, and firefighters. A relatively new position is that of the urban forester, who is responsible for the health and well-being of the millions of trees found in parks, along streets, and in other areas of our cities.

The arborist is an urban forester whose work may include planting, transplanting, pruning, fertilizing, or tree removal.

Tree Growth and Physiology

Trees use carbon dioxide (carbon and oxygen) from the air and water (hydrogen and oxygen) from the soil to manufacture simple sugars in their leaves. The leaves then use additional carbon, hydrogen, and oxygen to convert simple sugars into complex sugars and starches. Nitrogen and minerals from the soil are then used to manufacture proteins, the building blocks for growth and reproduction.

A tree typically starts from a seed. For instance, oak trees grow from seeds called acorns, pine trees start from seeds in pinecones, and beech trees grow from beech seeds. Trees may also sprout and grow from stumps or other tree parts. When a seed germinates, a shoot grows upward to form the top growth, and roots grow downward and outward to form the root system.

Both roots and shoots extend themselves by growth at the tips through cell division and elongation. At the same time, tree roots, stems, and trunks grow in diameter by adding cell layers near their outer surfaces (**Figure 10-9**). This growth layer in a tree root, trunk, or limb is called the cambium. The outward growth of the cambium in 1 year creates an annual ring, as seen in the cross section of a root, trunk, or limb. Water and minerals are taken in by the roots and transported up to the leaves through a layer of cells called the xylem layer, or sapwood, which is just inside the cambium layer. Just outside the cambium is another layer of cells called the phloem, or inner bark, which carries food manufactured in the leaves to the stems, trunk, and roots.

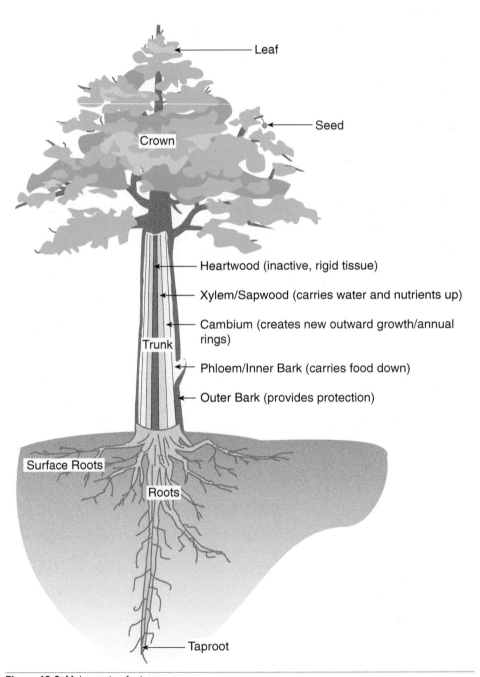

Figure 10-9 Major parts of a tree.

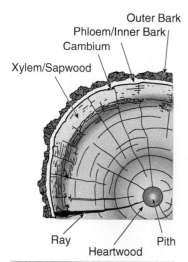

Figure 10-10 The structures of a tree illustrated in a cross section of a log or stump.

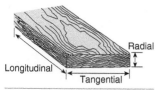

Figure 10-11 Dimensions of wood shrinkage.

Each year, the tree grows new cambium, xylem, and phloem tissues, and the older sapwood becomes heartwood. Heartwood is the inactive core that gives a tree strength and rigidity (**Figure 10-10**).

Properties of Wood

The various properties of wood determine the uses for which it is best adapted. It should be noted that there are wide variations within a specific species of trees, and the properties discussed here are general in nature.

Hardness

The property of hardness refers to a wood's resistance to compression. Hardness determines how well it wears. It is also a factor in determining the ease of working the wood with tools. Splitting and difficulty in nailing are problems that occur when wood is too hard.

Weight

The weight of wood is a good indication of its strength. In general, heavy wood is stronger than light wood. The moisture content affects the weight of wood; therefore, comparisons should always be made between woods with similar amounts of moisture.

Shrinkage

The change in dimensions of a piece of wood as it reacts to reduced humidity or temperature is referred to as shrinkage. The amount of expansion and contraction of wood greatly affects techniques used in building construction or manufacturing items from wood (**Figure 10-11**).

Warp

Warp refers to the tendency of wood to bend permanently because of moisture change. The tendency to warp is more of a problem in some types of wood than in others. Warping is caused by uneven drying of wood across its three dimensions. Warp is controlled by carefully stacking fresh lumber to allow air movement across surfaces and drying it under pressure.

Ease of Working

In wood, ease of working refers to the level of difficulty in cutting, shaping, nailing, and finishing the wood. It is influenced by the hardness of the wood and the characteristics of the grain. The grain of wood is formed by alternating hard and soft layers of new wood resulting in annual growth rings (**Figure 10-12**). In general, soft wood is easier to work with than hard, dense wood.

Paint Holding

The ability of wood to hold paint may be determined by the type of paint, surface conditions, and methods of application as well as characteristics of the wood itself. Moisture content, amount of pitch in the wood, and the presence of knots all affect how well paint adheres to wood.

Nail Holding

Nail-holding capacity refers to the resistance of wood to the removal of nails. In general, the harder and denser the wood, the better it will hold nails.

Figure 10-12 The annual rings in the cross section of a log or stump reflect the conditions of growth experienced by the tree on a year-to-year basis. These conditions influence the appearance of the grain.
Courtesy of DeVere Burton.

Decay Resistance

The resistance of wood to microorganisms that cause decay is a chief factor in selecting wood. Natural resistance to insects that live in, tunnel through, or devour wood also affects the choice of species for a given application.

Bending Strength

The ability of wood to carry a load without breaking is a measure of bending strength. Bending strength is important when determining types and sizes of lumber to use for rafters, beams, joists, and other building applications.

Stiffness

The resistance of wood to bending under a load is referred to as stiffness. When wood used in construction is not sufficiently stiff, ceilings and walls may flex and buckle, and the wallboards will crack under the load.

Toughness

The ability of wood to withstand blows is called toughness. Hardwoods are much more resistant to shock and blows than softwoods. This characteristic is particularly important in woods used for tool handles.

Surface Characteristics

The appearance of the wood's surface is sometimes the determining factor in its use. The pattern of the grain greatly influences the staining characteristics and beauty of the finished product. The number and type of knots and pitch pockets are surface characteristics that need to be considered when selecting lumber for a project.

Woodlot Management

The proper management of a wooded area or woodlot involves more than just the harvest of trees and the removal of unwanted species. A **woodlot** is a small, privately owned forest. The production of trees for harvest is a long-term investment, and mistakes in management take a long time to correct.

Some of the factors that need to be considered in the management of woodlots include soil, water, light, type of trees, condition of trees, available markets, methods of harvesting, and replanting. Using scientific methods in the management of forests is called **silviculture**. The care and management of trees for ornamental purposes is called **arboriculture**.

Restocking a Woodlot

The least expensive method of replacing trees harvested from a woodlot is natural seeding. Sources of seed for the desired species must be available in the forest, and conditions must be right for seed germination for natural seeding to take place. If seed from natural sources is not available, seeds from other sources may be planted on the forest site.

A surer method of restocking a woodlot is to plant trees of the desired species, rather than relying on seeds to do the reforestation. In most cases, **seedlings** (young trees started from seeds) are planted during late winter and early spring, before the new season's growth begins. Woodlots can be planted with one species of tree or a mixture of several compatible species (**Figure 10-13**).

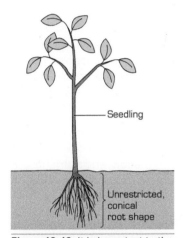

Figure 10-13 It is important to the survival of the seedling to maintain a conical root shape as the young tree is transplanted into the soil.

Management of a Growing Woodlot

Management of a woodlot is much more involved than just sitting back and watching it grow. Proper care and management are important if the forestry enterprise is to be successful.

Trees that are of no commercial value should be removed as soon as possible to eliminate competition for light, moisture, and nutrients. Because these "weed" trees are removed when they are small, there is seldom a market for them, and they are left on the woodlot floor to decay. If weed trees are of sufficient size, however, they may be used for firewood. When all the trees of a woodlot are nearly the same age (typically 15–30 years old), they often need to be thinned. Trees should be thinned any time that the crowns or branches occupy less than one-third of the height of the trees. Usually, about one-fourth to one-third of the trees in a woodlot are removed during thinning.

Trees that are being grown for lumber are often pruned of side branches to produce a better-quality log. Prompt removal of side branches helps keep logs free of knots. Only rapidly growing trees should be pruned. Branches should be pruned flush with the trunk of the tree. Pruning is usually done during the fall and winter when the trees are dormant.

Planning a Harvest Cutting

A woodlot should never be harvested without a plan. A harvesting plan can maximize the income from the woodlot over many years. The use of a forester's services is usually wise in developing a harvesting plan. A forester is a person who studies and manages forests.

Several systems can be used to harvest a woodlot. They include clear-cutting, seed-tree cutting, shelterwood cutting, diameter-limit cutting, and selection cutting.

Science Connection: Conserving Our Biodiversity

Rain forests are our richest source of plant and animal species and biodiversity.
© Dr. Morley Read/Shutterstock.com.

Our biodiversity is at risk. Not far from the shores of Florida and Texas, the tropical rain forests of Central and South America, as well as parts of Africa and Southeast Asia, are disappearing. Commonly called a jungle, a tropical rain forest is a hot, wet, green place characterized by an enormous diversity of life and a huge mass of living matter. Here, biologists estimate that more than half of the plant and animal species of the world are found. Unfortunately, many of these species are not found anywhere except in the tropical rain forests. The rain forests occupy only 6 percent of Earth's land area, and their total area is diminishing at an alarming rate.

Abundant moisture and warm temperatures encourage tremendous plant growth year round. Plants and moisture then provide an excellent environment for animals and microorganisms to grow and develop. Rain forests are covered by a dense canopy formed by the crowns of trees up to 150 feet in height. Under the shade of this canopy are smaller trees and shrubs creating a second layer called the understory. The two layers are typically woven together by strong woody vines, which may exceed 700 feet in length.

The third layer is the forest floor. Palms, herbs, and ferns dot the forest floor. The jungle floor, with its shrubs and trees, is home to a myriad of insects and animals. Annual rainfall may reach 400 inches, and the daily rate exceeds the rate of evaporation. Life is plentiful, however, and the soil beneath the jungle is shallow, and nutrients from decaying leaves are quickly used up, leaving few residual nutrients. As human populations expand into the rain forests, the foliage is cut and burned to make way for logging, grazing, and cropping.

Sadly, the land will generally support grazing or crop production for only a few years until the nutrients are gone and the soil is depleted of minerals. The occupants then typically move on to virgin jungle land to slash and burn a new tract and repeat the cycle. Once the jungle is destroyed, crops extract the scant nutrients, the soil is exposed to heavy rainfall and is soon eroded, and the land has too few nutrients to support the resurgence of jungle. Therefore, once cleared, the jungle area is often lost forever. It is estimated that one-half of the world's original tropical rain forests no longer exist.

Because the rain forests are the storehouse for more than half of the genetic diversity in the entire world, species are being lost at an alarming rate. Many species are becoming extinct before they are discovered and recorded. It is estimated that scientists have discovered and named only a sixth of all the plant and animal species in rain forests. Earth's biodiversity is indeed threatened by destruction of the tropical rain forests.

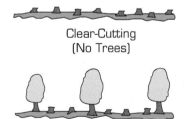

Figure 10-14 Clear-cutting is generally done in small patches, but seed-tree cutting is an economical way to harvest and reseed large areas.

Clear-Cutting

Clear-cutting is a system of cutting timber where all the trees in an area are removed. It was used extensively in cutting the virgin forests (those that have never been harvested) of the United States when there was little thought of reforestation of the cut area.

Clear-cutting is usually done in small patches, ranging in size from one-half acre to 50 acres. Prompt reforestation with desirable species of trees on the clear-cut areas is essential. This prevents erosion and the loss of fragile forest soil.

Seed-Tree Cutting

Harvesting trees by the seed-tree cutting method is very similar to clear-cutting. The primary difference is that enough seed-bearing trees are left uncut to provide seeds for reforestation (**Figure 10-14**). The trees left to produce seeds should be representative of the desired species and free from insect, disease, or mechanical damage.

The removal of the seed trees usually takes place after the cut area is repopulated with desirable seedlings.

Shelterwood Cutting

Shelterwood cutting is a modified seed-tree cutting. Enough trees are left standing after harvesting to provide for reseeding of the woodlot, and to provide shade to the tender, young seedlings. They will protect the area until the young trees are well established. After the young trees become established, the residual trees are harvested.

Diameter-Limit Cutting

When harvesting trees by the diameter limit method, all trees above a certain diameter are cut. Slow-growing or diseased trees are often left standing when harvesting is done using this method. Rapid-growing trees of desirable species may be cut before they reach their production potential, which leaves a woodlot with only slow-growing trees that may be undesirable. Diameter-limit cutting is used advantageously to remove trees left from previous harvesting, which reduces competition with younger trees.

Selection Cutting

Selection cutting is when the trees in a woodlot are of different ages. Selecting the trees to harvest can be a difficult task for an uninformed person, and obtaining the services of an experienced forester is usually wise. This system of harvesting allows for fairly frequent income and maintains the woodlot's aesthetic value.

Salvage Harvesting

Natural disasters occur regularly in forests, and salvage harvesting often follows. Trees that are dying from insect damage and diseases can still be used for lumber products if harvested before they are completely dead. Physical damage sometimes occurs to trees. Examples include charred trunks from forest fires and broken trunks and limbs caused by high winds. Once damage has occurred to the trees, they need to be harvested as soon as possible. Waiting longer than 2 or 3 years to do so will result in poor-quality lumber.

Protecting a Woodlot

A woodlot must be protected from fire, pests, and domestic animals if it is to yield a consistent harvest. More than 13 billion board feet of timber are destroyed each year by pests. This represents nearly 25 percent of the estimated net growth of forests and woodlots each year.

Fire

Millions of dollars worth of timber are destroyed by fire each year (**Figure 10-15**). In addition to actually killing trees, fire slows the growth of others and damages some so that insects and diseases may destroy them. Fire also burns the organic matter on the forest floor, which takes nutrients away from the trees and exposes the soil to erosion.

To help prevent fires, debris should be removed from around trees. Weeds, brush, and other trash around the edges of a woodlot should be removed. Conducting controlled burns at regular intervals under the supervision of professional fire personnel is an acceptable practice when it is correctly done. Limiting human use of the area during dry periods may also reduce the potential for fire. The construction of permanent firebreaks is useful for fire control.

Figure 10-15 Fire can cause major damage to timber resources when it gets out of control. It is important to plan for fire protection, especially in high-value forests.

Courtesy of Boise National Forest.

Figure 10-16 Insects and diseases sometimes become epidemic in a forest, killing many trees during periods of drought or stress.
Courtesy of DeVere Burton.

A plan for dealing with fire is important in minimizing damage should a fire occur. State and county foresters can assist in developing fire prevention and control plans. These strategies include planning for water storage and setting procedures for notifying proper authorities and obtaining appropriate equipment and help.

Pests

Insects and diseases cause more damage to existing forests than fires (**Figure 10-16**). Pests cause trees to be weak and deformed. They consume leaves, damage bark, and retard the growth of trees. They kill billions of trees each year. However, control of diseases and insects in forest lands is difficult and expensive.

Removal of dead, damaged, or weak trees can reduce disease and insect problems. Not only are weak trees attacked, but healthy ones also are favorite targets of pests. Severe drought makes trees vulnerable to pests. Prompt action in dealing with insect infestations and diseases will help ensure a profitable harvest.

Domestic Animals

The grazing of woodlots by cattle and sheep usually results in the destruction of all small seedlings in the forest. It also eliminates some of the woodlot floor coverage. Livestock may also strip the bark from trees when grazing is inadequate, causing the trees to die. Woodlots do not provide much food for grazing animals, so it is usually wise to exclude livestock from them.

Harvesting a Woodlot

Before timber is harvested, a market for it must be found. Various methods of marketing the timber should be explored to determine the most profitable alternatives. There are many uses for forest products, and trees should be marketed to maximize their value.

Hot Topics in Agriscience: To Fight or Not to Fight Fire

Once a decision has been made to fight a fire, resources, such as fire-retardant chemicals, are often used to gain control and keep it from spreading.
Courtesy of Boise Interagency Fire Center.

Before humans began managing forests, natural processes prevailed. Fire was a natural occurrence. When it erupted, it would burn underbrush, ground litter, and trees. Often, these fires would not kill the larger mature trees. Fire would kill the smaller trees that were in competition with these older trees for nutrients. The giant sequoia or ancient redwoods that stand today in some parts of the country would not have grown as tall or as thick as they have if fire had not thinned competing trees from the population.

Some types of trees need fire to reproduce. The seeds of the lodgepole pine tree cannot germinate without first going through the intense heat of fire. In the early 1900s, people in the United States began fighting forest fires. For almost 100 years, few forest fires have been allowed to burn. The result is thicker, more densely populated forests than were present a century ago. There are more trees, and they are much closer together in today's forests. Now, when fires start, the damage is much more severe. Fires burn hotter, faster, and with more intensity than in the past.

Forest fires demolish forests, kill wildlife, destroy ecosystems, burn homes, and threaten animal and human life. Although fire was once nature's way of cleaning and thinning forests, it now can wipe out all life in its path. The job of keeping our forests healthy now belongs to the forest industry. Some types of logging practices thin trees; the wood is then sold for construction, papermaking, and many other uses.

Some people and organizations are strongly opposed to logging and fighting fire. They would prefer that whole forests burn and lives of forest creatures be lost than fight fires or thin crowded trees by logging. More moderate conservationists recognize that we must thin trees and remove forest debris using all of the proven practices for doing so. This may include use of controlled burns, logging dead or diseased trees, and clear-cutting small areas to remove the most dangerous fuels. Individuals who are truly concerned with the health of our forests understand that thinning our forests by logging and seeking to control dangerous forest fires is the best way to preserve this natural resource for generations to come.

Hot Topics in Agriscience: Integrated Pest Management

Wood-boring insects cause greater economic damage to forests than any other group of insects.
Courtesy of Boise National Forest.

An insect management plan that uses a variety of control methods is the most acceptable form of pest control because it does minimal damage to insect species for which the control method is not intended. Integrated pest management is a concept for controlling harmful insects or other pests, while providing protection for useful organisms. It involves the use of some chemical pesticides in emergency situations, but it also relies on natural enemies and other biological control strategies to control harmful pests. Integrated pest management is a control program that does not attempt to kill all of the harmful insects because insect control of this kind also kills the natural enemies of the pest. To survive, natural insect enemies must have a small population of the harmful pest upon which they can depend for food.

The harvest of a woodlot can be carried out by the owner or by a contractor skilled in timber harvest. Another alternative is to sell the standing timber to a company that will harvest and market it. Regardless of how the timber is harvested, it is extremely important that contracts covering all pertinent harvest details be drawn up and signed by all parties involved before the harvest begins.

Seasoning Lumber

The proper seasoning of lumber is essential to protect it from damage. As soon as a tree is cut, it starts to lose moisture. If the wood dries too slowly, it may be subject to rot, stain, or insect damage. If it is allowed to dry too quickly, lumber may twist and warp or split. This may make it unsuitable for most uses.

Wood that is sawed for lumber should be stacked immediately after sawing to allow for even drying (**Figure 10-17**). The stacking should take place on a level, well-drained location. Wood should be off the ground and stacked so that air can circulate freely around each board. The stacked lumber should also be protected from weather. The amount of time required for lumber to dry depends on the thickness of the lumber and the species of tree from which it was cut. Drying times usually range from 30 to 200 days.

Lumber can be seasoned quickly by placing the stacked lumber in a heated kiln for a few days to dry it out. This procedure will keep the lumber from becoming warped as it dries. Under the controlled conditions that exist inside a heated kiln, most of the moisture is removed from the wood within a few hours.

The forestry industry in the United States produces about 25 billion cubic feet of standing timber each year, but harvests significantly less. However, demand for wood products, such as lumber, pulp wood, posts, and pilings, is approaching 20 billion cubic feet per year, and it continues to increase. Only by carefully managing our forest resources and replacing our trees can we fulfill our need for wood and wood products in the future.

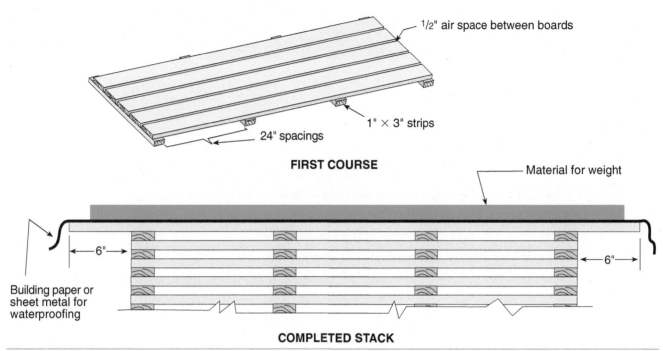

Figure 10-17 Freshly sawed lumber must be properly stacked to prevent warping and end splitting as the lumber dries.

Chapter Review

Student Activities

1. Write the **Key Terms** and their meanings in your notebook.
2. Make a collection of forest products.
3. Make a collection of tree leaves, bark, or twigs that are of economic importance in your area.
4. Make a bulletin board showing the many uses of forests and forest products.
5. Write a report on a species of trees or a type of forest product of interest to you.
6. Visit a local forest and identify the types of trees growing there.
7. Have a forester visit the class and speak about forest management.
8. **SAE Connection:** Visit a wood-processing plant operation to learn how trees are processed into other products.
9. Write a report on a career in the forest or wood products industry.
10. Study Figure 10-2 and learn how to calculate board feet (BF). Determine the number of board feet in the following:
    a. a board that is 1" × 12" × 12' = ___ BF
    b. a plank that is 2" × 8" × 16' = ___ BF
    c. a 4" × 4" × 8' board = ___ BF
    d. six 2" × 4" × 8' boards = ___ BF
    e. twenty 2" × 10" × 16' boards = ___ BF
11. What is the total board feet in Activity 10?
12. If the items in Activity 10 were rough-sawed lumber and were priced at $0.20 per board foot, what would be the total cost of the order? $ ___
13. Look up "rain forests" on the Internet or using the school library. Find one organism that is used to benefit humans (besides building products).

    Write a one-page report or develop an oral presentation. Describe:
    - the location where the organism is found
    - its importance to humans
    - its survival expectations if forests are destroyed.

    Draw an illustration or display an online image of the organism.
14. Obtain a coring tool from your teacher and find a large tree. Core the tree by following your teacher's instructions. Then, count the rings in the core to determine the age of the tree.

National FFA Connection: Forestry Career Development Event

A forester uses a digital topological map to zoom in on specific sections of forest land.
© Kaspars Grinvalds/Shutterstock.com

The National FFA Organization sponsors a Career Development Event (CDE) for FFA members who express interests in forestry careers. This CDE is conducted in a forest with students demonstrating skills that are needed in management of forest resources. They use the tools of the trade to evaluate forest health, measurement, harvest selection, and other basic skills related to silviculture and forest environments. To learn more about the Forestry CDE and to watch a short video about this event, visit the FFA website and search the term "Forestry."

Check Your Progress

Can you . . .

- give examples and definitions of forest terms?
- describe the forest regions of the United States?
- discuss important relationships among forests, wildlife, and water resources?
- identify important types and species of trees?
- describe how a tree grows?
- identify important properties of wood?
- summarize principles of good woodlot management?
- describe procedures for seasoning lumber?

If you answered "no" to any of these questions, review the chapter for those topics.

Chapter 11
Wildlife Management

Objective
To determine the relationship between wildlife and the environment and approved practices in managing wildlife enterprises.

Competencies to Be Developed
After studying this chapter, you should be able to:
- define wildlife terms.
- identify characteristics of wildlife.
- describe relationships between types of wildlife.
- understand the relationships between wildlife and humans.
- describe classifications of wildlife management.
- identify approved practices in wildlife management.
- discuss the future of wildlife in the United States.

Key Terms
wildlife
vertebrate
predator
prey
parasitism
mutualism
predation
commensalism
competition
wetlands

Figure 11-1 Some wildlife populations declined in the early 1900s, but management efforts helped restore many of them, including several species of wild sheep.

Courtesy of U.S. Fish and Wildlife Service.

Wildlife has been part of the life of humans since the beginning of time. **Wildlife** includes animals that are adapted to live in a natural environment without the help of humans. Early humans followed herds of wild animals and killed what they needed to survive. They observed what the animals did and what they ate to determine what was safe for human consumption. Early humans also used wildlife as models for their artwork and in many of their ceremonial rites.

As settlers came to North America and moved westward, wildlife often provided the bulk of the available food until food production systems could be developed. Supplies of wildlife seemed to be inexhaustible as the skies were blackened with the flight of millions of passenger pigeons, and herds of bison created dust storms as they migrated on the vast prairie (**Figure 11-1**). Unfortunately, supplies of wildlife were not, and are not, unlimited. Human activities have damaged or destroyed wildlife habitat (the area where a plant or animal normally lives and grows). Humans have polluted the air and water supplies, killed wildlife in tremendous numbers, and in some instances, generally disregarded the needs of wildlife. As a result, many species of wildlife in the United States now require some degree of management.

Fortunately, humans also have the ability to restore wildlife habitat and manage many wildlife species successfully. The Great Plains may never sustain vast herds of roving bison, but the bison have been brought back from the brink of extinction. Populations of large game animals such as deer, elk, moose, and black bear have expanded to fill the available habitat. Many smaller species, such as eagles, have also experienced population growth. A few species, however, such as the whooping crane, have not responded well to management efforts. They are now among the rarest birds in North America, with a current population of just over 500. These species of wild creatures require our best efforts to accommodate their survival needs.

Characteristics of Wildlife

All **vertebrate** animals (animals with backbones), except humans, are included in the classification of wildlife. They have many of the same characteristics as humans. Growth processes, laws of heredity, and general cell structure are common to both humans and animals. When populations become too dense, disease outbreaks occur, populations suffer from starvation, and disposal of waste becomes a problem.

With few exceptions, wildlife species live in environments over which they have no control. Wildlife must be able to adapt to whatever they are presented in terms of food and environment, or they will perish. They must also possess natural senses that allow them to avoid predators and other dangers. A **predator** is an animal that feeds on other animals. The animal that is eaten by the predator is the **prey**.

The ability to avoid overpopulation is a characteristic of many groups of wildlife. Establishing and defending territories is one way that wildlife may naturally avoid overpopulation. The stress of overpopulation causes some animals to slow their rate of reproduction or completely stop reproducing.

The wildness of an animal itself is a characteristic that allows the animal to survive without interference or help from humans. The animal's wildness often contributes to the interest that humans have in wildlife. Characteristics identified as wildness are what attract hunters to hunting and fishermen to fishing. Bird-watching and wildlife photography would be far less fascinating if wildlife were less wild and wary of humans (**Figure 11-2**).

Figure 11-2 The nature of many wild animals is to avoid humans. It is human nature to be curious about wild animals, but we must make sure that our interests do not intrude on their needs.
© Wildnerdpix/Shutterstock.com.

Wildlife Relationships

Every type of wildlife is part of a community of plants and animals where all individuals are dependent on others. Any attempt to manage wildlife must consider the natural relationships that exist. This is because relationships within the wildlife community are constantly changing, and it is difficult to set standard procedures for their management.

The balance of nature is actually a myth because wildlife communities are seldom in a state of equilibrium. The numbers of various species of wildlife are constantly increasing and decreasing in response to each other and to many external factors such as natural disasters. These include fires, droughts, and disease outbreaks. Interference of humans often upsets sensitive relationships in nature. Some of the natural relationships that exist in the wildlife community include the following: parasitism, mutualism, predation, commensalism, and competition.

Parasitism

The relationship between two organisms, either plants or animals, in which one feeds on the other without killing it is called parasitism. Parasites may be either internal or external. An example of a parasitic relationship is the wood tick, which lives on almost any species of warm-blooded animal. A warm-blooded animal has the ability to regulate its body temperature.

Mutualism

Mutualism refers to two types of animals that live together for mutual benefit. There are many examples of mutualism in the wildlife community. Tick pickers are birds that remove and eat ticks from many of the wild animals in Africa, to the mutual benefit of both. The wild animals have parasites removed from them, and the birds receive nourishment from the ticks.

A moth that lives only on a certain plant may also be the only pollinator of that plant. Some plant seeds will germinate only after having passed through the digestive tract of a specific bird or animal.

Predation

When one animal eats another animal, the relationship is called **predation** (**Figure 11-3**). Predators are often very important in controlling populations of wildlife. Foxes are necessary to keep populations of rodents and other small animals under control. Populations of predators and prey tend to fluctuate widely. When predators are in abundance, prey becomes scarce because of overfeeding. When prey becomes scarce, predators may starve or move to other areas. This permits the population of the prey species to increase again.

Commensalism

Commensalism refers to a plant or animal that lives in, on, or with another, sharing its food, but not helping or harming it. One species is helped, but the other is neither helped nor harmed. Vultures waiting to feed on the leftovers from a cougar's kill is an example of commensalism.

Competition

When different species of wildlife compete for the same food supply, cover, nesting sites, or breeding sites, **competition** exists. Competition may exist between two or more species that share the same resources. It also exists among members of the same species, especially when food or shelter is in short supply or during the mating season (**Figure 11-4**). When competition exists, one species may increase in number, whereas the other declines. Often, the numbers of both species decrease as a result of competition. For example, owls and foxes compete for the available supply of rodents and other small animals.

The various relationships that exist among species of wildlife make it necessary to consider more than just one species any time that management is contemplated. Understanding the relationships that exist in the entire wildlife community is essential if wildlife management programs are to be successful.

Figure 11-3 Predatory animals play an important role in nature by keeping populations of rodents, birds, and other animals from expanding beyond the capacity of their environments to provide food and shelter for them.

Courtesy of U.S. Fish and Wildlife Service; photo by Pedro Ramirez, Jr.

Figure 11-4 Competition among wildlife species helps keep animal populations in balance. Deer have been known to battle to the death of one or both combatants.

Courtesy of U.S. Fish and Wildlife Service; photo by John D. Wendler.

Relationships Between Humans and Wildlife

Relationships between humans and wildlife may be biological, ecological, or economic. Biological relationships exist because humans are similar to wildlife in the biological processes that control life. Relationships may be ecological because humans are but one species among nearly 1 million species of creatures that inhabit Earth, sharing its resources and environments (**Figure 11-5**).

A B C

Figure 11-5 Ecology is the branch of biology that describes relationships between living organisms and their living environments. Shown in their natural habitats are the (A) blue heron, (B) eastern cottontail, and (C) Arctic hare.
Courtesy of Chesapeake Bay Foundation.
Courtesy of U.S. Fish and Wildlife Service; Photo by William Janus.
Courtesy of U.S. Fish and Wildlife Service.

Figure 11-6 Humans are the highest ranking predators of wildlife.
Courtesy of U.S. Fish and Wildlife Service; Photo by Richard Baldes.

The economic relationship that exists between humans and wildlife is important. Originally, humans were dependent on wildlife for food, clothing, and shelter (**Figure 11-6**). Today, wildlife relationships with humans have six positive values: commercial, recreational, biological, aesthetic, scientific, and social.

The harvesting and sale of wildlife and/or wildlife products is an example of the commercial relationship between humans and wildlife. Raising wild animals for use in hunting, fishing, or other purposes also falls into this category.

Hunting and fishing, as well as watching and photographing wildlife, are examples of recreational relationships. The role of humans in preventing excessive increases in the population of a wildlife species through hunting presents humans in another relationship, that of predators. Although it is estimated that more than $2 billion are spent each year on hunting and fishing and at least another $2 billion on other recreational uses of wildlife, many of the recreational values of wildlife are intangible.

The value of the biological relationship between humans and wildlife is difficult to measure. Examples include pollination of crops, soil improvement, water conservation, and control of harmful diseases and parasites (**Figure 11-7**).

Figure 11-7 Biological values of wildlife and human relationships include the pollination of crops by honeybees.

Courtesy of USDA/ARS #K-4716-1.

Aesthetic value refers to beauty. Watching a butterfly sipping nectar from a flower, a fawn grazing beside its mother, or a trout rising to a hatch of mayflies are all examples of the aesthetic value of wildlife. Wildlife also provides the inspiration for much artwork. Even though the aesthetic value of wildlife is not easily measured in economic terms, wildlife can contribute greatly to the mental well-being of the human race.

Using wildlife for scientific studies often benefits humans. The scientific relationship between humans and wildlife has existed from the beginning of time when early humans watched wild animals to determine which plants and berries were safe to eat.

The value of the social relationship between humans and wildlife is also difficult to measure. However, wildlife species have an ability to enhance the value of their surroundings simply by their presence (**Figures 11-8** and **11-9**). They provide humans the opportunity for variety in outdoor recreation, hobbies, and adventure. They also make leisure time much more enjoyable for humans.

Figure 11-8 The presence of birds can be encouraged by providing them with feeding stations.

© gracious_tiger/Shutterstock.com.

Figure 11-9 Wildlife sanctuaries are created to provide safe habitats for animals. They are also attractive to humans who go there to observe and enjoy the birds and other animals.

Courtesy of U.S. Fish and Wildlife Service. Photo by Bob Ballou.

Classifications of Wildlife Management

Wildlife management can be divided into several classifications for ease in developing management plans. Techniques for management vary tremendously according to classification. Some wildlife management classifications are farm, forest, wetlands, streams, and lakes and ponds. The management of farm wildlife is probably the most visible wildlife management classification. The development of fence rows, minimum tillage practices, food plots, improvement of woodlots, and controlled hunting are all techniques that have long been used to manage farm wildlife. Rabbits, quails, pheasants, doves, and deer are the types of wildlife that are normally managed in this category. In some instances, farm management focuses on developing trophy animals for commercial hunting. Wild game animals, such as deer, are attracted to the area by improving habitat and supplementing the food supply.

Forest wildlife is difficult to manage. Plans should be developed so that timber and wildlife can exist in populations large enough to be sustained and possibly harvested. Management of forest wildlife may include population controls to prevent destruction of habitat. Deer, grouse, squirrels, and rabbits are wildlife species that are usually included in forestry wildlife management programs (**Figure 11-10**).

Figure 11-10 Management of wildlife in forest environments is often difficult because many species of plants and animals occupy the living environment. Interactions among all of the organisms in the environment must be considered in developing the management plan.
Courtesy of National FFA.

Figure 11-11 Canada geese populations have benefited from man-made wetlands that provide resting areas during migration and favorable habitats during other seasons.
© Glenn Young/Shutterstock.com.

The most productive wildlife management areas are wetlands. **Wetlands** include all areas between dry upland and open water. Marshes, swamps, and bogs are all wetland areas. Because these areas are sensitive to changes in environmental conditions, careful management is essential. The wetlands provide homes to ducks, geese, beaver, muskrats, raccoons, deer, pheasants, grouse, woodcock, fish, frogs, and many other species (**Figure 11-11**).

The management of running water or streams is often difficult. Water pollution competes with the need for clean water for a growing human population. Potential damage to the wildlife in streams from chemical pollution, the building of dams and roads, home construction, and the drainage of swampland are critical considerations for the stream wildlife manager.

Management of wildlife in lakes and ponds is somewhat easier because water is standing. Population levels of wildlife, oxygen levels, pollutants, and the availability of food resources are all concerns of the pond and lakes wildlife manager.

Science Connection: Population Dynamics

Northwest wildlife managers are studying Bighorn sheep to identify causes for their declining population and ways to reverse it.
© Christopher Kolaczan/Shutterstock.com.

A population is a group of individuals of a species that live in a specific area. A wolf pack in Yellowstone National Park is an example of a population. "Dynamics" refers to the factors that cause changes in a population. Biologists are interested in population dynamics that affect animal behavior and the number of animals in a population. Wildlife biologists spend a lot of time observing, counting, and calculating the information they gather from studying a population. The goal of this work is to predict what changes may take place within a population over time. Factors that can affect a population's size include changes in the number of births or the number of deaths. The number of animals that move into and out of the group affects the population, as do such factors as predation, disease, and competition for resources.

Another factor that must be considered by scientists is the carrying capacity of the environment. Carrying capacity is the maximum number of individuals the resources of an ecosystem can support. This is critical information. If the carrying capacity of the area is not determined, then it would be difficult to see what effect the other factors may have. The size of the Bighorn Sheep population in the Northwest has been on the decline in recent years. Wildlife biologists have been trying to determine what has been causing the sheep to die. Through observations, counting, blood sampling, and using radio collars, they hope to identify what is responsible for the decline. Once these things are understood, they hope to be able to predict the future for the Bighorn Sheep more accurately.

Hot Topics in Agriscience: Supplemental Feeding of Wildlife

Supplemental feeding enables more elk to survive during the cold winter months.
© Irene Pearcey/shutterstock.com.

Every year, as snow falls in big game country, a major controversy begins. In difficult winters, it is common for elk, deer, and other big game animals to die of starvation. The question of feeding the animals is debated each time large numbers of animals are found dead or dying. In years of milder winters, most big game animals survive to the next season.

Wildlife managers are concerned with stabilizing animal populations. When a dramatic decline occurs during the winter, managers become concerned that the animals may not recover on their own. Sometimes, the decision is made to feed these animals during periods of extreme food shortages.

Some individuals believe that when humans become involved in this way, the animals are no longer self-sufficient, and are therefore less wild. This is a concern of the Fish and Game Departments in some states. It becomes a balancing act to keep wildlife wild and to keep wildlife alive. For this reason, many factors must be considered before a decision is made to provide supplemental feed for big game animals: Are excess animals a threat to public safety and private property? Will excess deaths affect recovery of the animal population? Will saving the animals create a bigger problem later because of a limited or unavailable supply of food? Although wildlife managers strive to limit supplemental feeding practices, the controversy continues. Will supplemental feeding take the "wild" out of "wildlife"?

Approved Practices in Wildlife Management

Farm Wildlife

Management of wildlife on most farms is usually a by-product of farming or ranching. It is often given little attention by the farmer or rancher, except when wild animals cause crop damage and financial loss. However, there are many wildlife proponents in the agricultural sector. Many of them plant food plots on their property to support deer and other wildlife populations.

Much of the management of farm wildlife involves providing a suitable habitat for living, growth, and reproduction. This may involve leaving some unharvested areas in the corners of fields, planting fence rows with shrubs and grasses that provide winter feed and cover, or leaving brush piles when harvesting wood lots.

The timing of various farming operations is also important in a farm wildlife management program. Crop residues that are left standing over the winter will provide food and cover. Planting crops that are attractive to wildlife on areas that are less desirable as cropland is an excellent farm wildlife management practice. Providing water supplies for wildlife during dry periods is often necessary to maximize the numbers of farm wildlife on the area being managed (**Figure 11-12**).

Harvesting farm and ranch wildlife by hunting has been shown, by extensive research, to have little impact on spring breeding populations. Excess populations of farm and ranch wildlife that are not harvested by humans usually die during the winter. Even heavy hunting pressures seldom result in severe damage to wildlife populations. The sale of hunting rights to hunters is a way to increase the income of many farms and ranches. In addition, it often means the difference between profit and loss in the farming enterprise.

Management of wildlife on game preserves or farms set up specifically for hunting often differs drastically from other wildlife management programs. Species of animals and birds that are not native to the area are sometimes raised and released on the preserve. It is important to note, however, that introduction of nonnative species may violate laws

Figure 11-12 Developing clean water sources for livestock also serves wildlife, especially during periods of drought.

Courtesy of Wendy Troeger.

and regulations of state wildlife agencies, and special permits are usually required. In some instances, native wildlife species are also raised in pens and released to the farm or preserve expressly for harvest by hunters.

Science Connection: Lead Poisoning in Waterfowl

Lead shot used for hunting is known to poison waterfowl long after the hunting season is over.
© Yu Lan/Shutterstock.com.

Hunters of ducks, geese, and other waterfowl do much to promote wetlands and waterfowl populations. For many years ago, however, lead shot used in shotgun shells was killing waterfowl long after the hunting season was over. Waterfowl cannot distinguish lead shot from the small rocks they must eat to grind the seeds and plant materials that make up their food. When lead shot is eaten, it is gradually absorbed into the body, causing fatal lead poisoning. Replacing lead shot with steel shot eliminates this problem. Steel shot can be eaten by waterfowl without poisoning them.

In 2017, the United States Fish and Wildlife Service (USFWS) ordered state officials to "replace all lead-based gear" used on agency-managed property by 2022. The Lead Endangers Animals Daily (LEAD) Act of 2022 would direct the Department of the Interior (DOI) to establish and update annually a list of non-lead ammunition.

Forest Wildlife

The types and numbers of forest wildlife in any specific woodland are dependent on many factors. These include type and age of the trees in the forest, density of the trees, natural forest openings, types of vegetation on the forest floor, and the presence of natural predators.

Management of forest wildlife is usually geared toward establishing and maintaining stable populations of desired species of wildlife. If desired populations of wildlife are present, the management goal is usually to sustain those populations. Sometimes, numbers of certain species of forest wildlife increase to the point where destruction of habitat occurs. When this happens, control measures may have to be instituted to restore proper balance. The steps in developing a forest wildlife management plan should include taking an inventory of the types and numbers of wildlife living in the forest area to be managed. Goals for the use of the forest and the wildlife living in it need to be developed. The third step in the development of a forest wildlife management plan is determining the types and populations of wildlife that the forest area can support and how best to manage the forest so that required habitat is available.

The requirements for forest wildlife include food, water, and cover. These necessities must be readily available to the desired species of forest wildlife at all times. Management practices that meet these requirements include making clearings in the forest so that new growth will make twigs available for deer to feed on. Another practice is selective harvesting so that trees of various ages exist in the forest to make a more suitable habitat for squirrels and many other species of forest wildlife. Leaving piles of brush for food and cover is also a management practice that leads to increased production of forest wildlife. Care in managing harvests of forest products so that existing supplies of water are not contaminated is also important in good wildlife management.

Agri-Profile: Career Areas: Wildlife Biologist/Manager/Officer

Conducting game counts and doing habitat analyses provide information for wildlife managers.
Courtesy of USDA/ARS #K-5213-3. Photo by Scott Bauer.

Wildlife biologists work with fish and game species living in habitats such as land, freshwater streams and lakes, tidal marshes, bays, seas, and oceans. Wildlife biologists generally have master's or doctorate degrees in biology. They use the basic sciences in their work.

Wildlife managers typically have associate's or bachelor's degrees. They work in government agencies, advising landowners and managing game populations on public lands. Their work frequently requires the use of helicopters, small planes, snowmobiles, all-terrain vehicles, horses, and Land Rovers, as well as time in the wild on foot or horseback.

Wildlife officers interact continuously with the hunting and fishing community. They advise governments in establishing fish and game laws and programs for habitat improvement. Wildlife officers have the backing of strict laws and stiff penalties for offenders. However, much of their time is spent educating the public and obtaining private assistance in improving habitats and maintaining game populations.

Deer, grouse, squirrels, and rabbits are the forest wildlife species that are usually targeted for management because they are valuable for recreational purposes, especially hunting (**Figures 11-13** and **11-14**). They may also be managed to prevent the destruction of valuable forest trees and other products.

Notably, during times of overpopulation of forest animals, especially deer, it is seldom a good idea to provide supplemental food. Natural losses should nearly always be allowed to occur, including starvation of excess animals or allowance of heavier-than-normal hunting pressures. Artificial feeding of wildlife populations may result in further population increases and an expansion of the problem.

Figure 11-13 Grouse are popular gamebirds that are managed by fish and game agencies for recreational purposes.
Courtesy of DeVere Burton.

Figure 11-14 Deer management is a key activity of fish and game agencies. Management units are studied carefully to determine the size of the deer population and appropriate hunting regulations.
© Alan and Sandy Carey/Photodisc/Getty Images.

Wetlands Wildlife

No area of U.S. land is more important to wildlife than the wetlands. Wetlands include any land that is poorly drained—swamps, bogs, marshes, and even shallow areas of standing water (**Figure 11-15**). The wetlands are constantly changing as wet areas fill in with mud and decaying vegetation. They eventually become dry land that contains forests.

Figure 11-15 No area of U.S. land is more important to wildlife than the wetlands.
Courtesy of Chesapeake Bay Foundation.

Wetlands provide food, nesting sites, and cover for many species of wildlife. Ducks and geese are probably the most economically important type of wildlife that depends on the wetlands for survival (**Figure 11-16**). Other types of wildlife found in the wetlands include woodcock, pheasants, deer, bears, mink, beaver, muskrats, raccoons, and many other species.

Figure 11-16 Wetlands are important resting areas for migratory birds. They also provide nesting areas for millions of ducks and geese in the United States and Canada.
Courtesy of U.S. Fish and Wildlife Service.

Management of wetlands for wildlife may include impounding water in ponds and reservoirs for immediate or future use. Open water areas should occupy about one-third of the wetlands for optimum use by wildlife. The most useful wetlands include shallow, standing water, not more than about 18 inches in depth. This allows for growth of reeds and other aquatic plants. Shallow water also provides feeding areas for shore birds.

The management of plant life in the wetlands is also important. This may include cutting trees to create open wetland areas. Many species of wildlife require large, open areas in order to thrive. Care must be taken not to remove hollow trees that are used as nesting sites for some species of wildlife. Open wetland areas can also be created by killing excess trees rather than cutting them. This provides resting areas for many types of wetlands wildlife.

Establishing open, grassy areas around wetlands and planting millet, wild rice, and other aquatic plants in the wetlands also helps attract many types of wildlife to the area.

One hazard to wildlife in wetland habitats is excessive pollution. Polluted water flowing into the wetlands may come from agriculture, industry, or the disposal of domestic wastes. Because pollutants such as heavy metals are trapped in the mud and silt of the wetlands, the effects of these pollutants are often long-term. However, wetlands are sometimes constructed precisely for the purpose of removing such pollutants as nitrates and phosphates from the water. Wetlands tend to recycle some substances that would otherwise pollute the surface water.

In areas lacking natural nesting sites, populations of some wildlife species can be greatly increased by providing artificial nesting sites. Wood duck boxes, platforms, and islands surrounded by open water provide safe nesting sites for many species of wetlands wildlife (**Figure 11-17**).

Figure 11-17 Artificial nesting sites benefit many kinds of birds.
Courtesy of Cameron Waite. Photo taken by Claire Waite.

Raising certain species of ducks in captivity and later releasing them in wetlands areas has helped maintain viable duck populations. This has been important as more and more natural duck nesting areas have been destroyed to meet the needs of people.

Stream Wildlife

Stream wildlife are divided into two general categories: warm-water and cold-water. These categories are based on the water temperatures at which the wildlife, primarily fish, can best grow and thrive. There is little or no difference in the management practices of these two categories of stream wildlife. In general, fish are the type of stream wildlife for which management plans are developed, although many other types of wildlife also depend on streams for their existence (**Figure 11-18**).

Figure 11-18 Cool, rapid-flowing brooks and mountain streams provide appropriate habitats for trout.

Courtesy of U.S. Fish and Wildlife Service.

As land is developed, forests are harvested, civilization is expanded, and streams and their wildlife populations come under increasing pressure. Because we cannot build new streams, it is essential that existing ones be managed properly.

Management practices for streams include preventing stream banks from being overgrazed by livestock. Fencing the stream to limit access by livestock is also wise to reduce pollution and the destruction of stream banks (**Figure 11-19**).

Figure 11-19 Riparian areas, which include stream banks, can be protected by fencing livestock out. Such practices protect stream environments and the animal and plant life that is found there.
Courtesy of DeVere Burton.

Effective erosion-control practices on lands surrounding streams are important to help maintain clear, clean water. It is also important to prevent silt or chemical pollutants from entering streams (**Figure 11-20**). Maintaining stream-side forestation is important in regulating water temperatures during the warm summer months. Some species of fish stop feeding and may even die when stream temperatures become too high. The amount of dissolved oxygen in warm water is also much lower than it is in cold water. Without adequate oxygen, aquatic animals die.

Figure 11-20 Soil erosion causes siltation of streams, damaging the habitat for desirable fish species.
Courtesy of Chesapeake Bay Foundation.

Figure 11-21 Fish adapt to a wide range of environmental conditions. Some fish, such as trout, require cold, clear water to survive. Others, such as catfish, are well adapted to warm water conditions in ponds or slow-flowing streams.
Courtesy of U.S. Fish and Wildlife Service.

Anything that impedes the flow of a stream also serves to change its course, to the detriment of many species of stream wildlife and to the benefit of other species. Trout must have swift-moving cool water in which to thrive, whereas catfish are adapted to sluggish streams (**Figure 11-21**).

Care must be taken to maintain desirable species of wildlife in the stream. Introducing new species of wildlife in the stream may result in the reduction of native wildlife already living there. State fish and game agencies routinely kill all of the fish in waters that are over-populated with aggressive, non-native species. This allows desirable species to be returned to safe waters where they can successfully compete.

The maintenance of population levels of stream wildlife that are in balance with the available food supply is important. Too many fish for the available food supply normally results in stunted fish that are of little value to fishermen. This situation does provide an increased food supply for some types of birds and animals that use streams for their food supply. Overfishing of predatory species of fish, such as bass or northern pike, may allow perch or sunfish to overpopulate the stream and become stunted. Often, the only way to restore streams to a desired mix of fish species is to remove the unwanted species. This is accomplished by netting, poisoning, or electric shocking. These techniques are legal only for authorized officials and should be done only by specially trained personnel.

The artificial rearing and stocking of desired species of stream wildlife is a management practice that is important in many areas. Typically, game species of fish are stocked in streams for fishermen to catch and remove. Often few or no fish survive to reproduce, and stocking must take place each year.

The regulation of sport fishing is often necessary to maintain desirable populations of game fish. This may include closed seasons, minimum size limits, creel limits, and restricted methods of catching fish.

Lakes and Ponds Wildlife

The management practices for wildlife in lakes and ponds are usually very similar to those for managing stream wildlife. Pollution must be controlled. Wildlife populations must be managed to maintain desired mixes of species. Harvest and use must also be controlled to ensure wildlife for the future.

However, there are some differences between management of wildlife in streams and management of wildlife in ponds and lakes. Because the water in lakes and ponds is normally standing, the amount of oxygen available for aquatic life sometimes becomes critical in the hot summer months. In small ponds, artificial means of incorporating oxygen into the water may be used to prevent fish deaths.

Water temperatures at different depths in lakes and ponds are more variable than in streams. This means that different species of fish are usually dominant in ponds and lakes. In many ponds and lakes, fish populations are predominantly largemouth bass and sunfish (**Figure 11-22**).

Figure 11-22 Fish species such as sunfish and bluegill provide both sport fishing and food for predatory species such as bass.
© Rob Hainer/Shutterstock.com.

Science Profile Alaskan Adventure

Summertime fieldwork provides valuable experiences for future wildlife biologists or range scientists.
Courtesy of National FFA.

Sierra Stoneberg of Hinsdale, Montana, had developed extensive knowledge and experience of the outdoors by the time she graduated from high school. As early as the seventh grade, she became interested in botany through the Montana Range Days program. She experienced firsthand the elements of botany, biology, and other range sciences. Her summers during high school were filled with FFA activities and experiences with the Soil Conservation Service.

The summer after high school graduation provided Sierra with an Alaskan adventure studying moose, mountain sheep, and other major wildlife species. As a volunteer at the Kenai Lake Work Center, she, together with others at the center, worked 40 hours a week for lodging, boarding, and plane fare.

The setting was awe-inspiring. Huge green trees and mosses filled the high alpine country. Tiny plants and lichens on the trees and rocks contributed a unique hue to the surroundings.

Scientific observations and data collection were important facets of Sierra's many treks into the wilderness. These included monitoring of fertilizing results on rangeland for sheep and range grass response to management practices. Scouting for eagle nests and moose habitat created an interesting and stimulating blend of mental and physical activities. Even a close encounter with a black bear provided valuable experience for a future career in wildlife biology and range science. Along the way, Sierra became a National FFA Proficiency Award winner.

Based on: Bolstad, P. (April–May 1993). Alaskan Adventure: FFA prepared this member to live and work in our northernmost state. *FFA New Horizons*.

When it is necessary or desirable to rid a pond or lake of unwanted species of fish such as carp, it is often much easier to do because the water is contained. Sometimes it is possible to drain the body of water to remove the unwanted species of wildlife. More often, the body of water is simply poisoned so that all fish and other species of pond wildlife are killed. The pond is then restocked with the desired wildlife species.

The management of wildlife is an imprecise business. Often, specific species of wildlife are managed to the detriment of others. It is reasonably clear that any human interference in the wildlife community results in changes that are not always to the benefit of some species of that community.

Hot Topics in Agriscience — Environmental Cleanup: The Power of Microbes

Products made up of oil-ingesting microbes are used to restore soil and water environments that have been contaminated with oil.
© iStockphoto/Danny Hooks.

Spills from ships and tanker trucks have caused serious environmental damage, especially to marine animals. One of the remarkable scientific developments in recent years is identification of microbes for special purposes, one of which is in products that clean up such accidental spills of crude oil or other petroleum products. Products made from collections of microbes—tiny organisms made up of bacteria and fungi— break down oil-based materials to carbon dioxide and other harmless products. As a result, they are becoming important tools in restoring oil-damage marine environments. Microbial products can also be used to clean oil spills from driveways, contaminated soil, and construction materials. In addition, they are useful in oil production to separate oil from tar sands in oil fields.

The Future of Wildlife in the United States

A bright future for all of the wildlife species is not guaranteed in the United States because the needs of the human population continue to compete with those of the wildlife community. The outlook for wildlife in the future is not all bleak, however. Humans have recognized that it is possible to satisfy the needs of wildlife and the demands of humans if careful management practices are instituted (**Figure 11-23**).

Careful studies of how humans and wildlife can peacefully coexist are continuing as wildlife management plans are refined to include new scientific findings. Sincere attempts are being made to reduce the pollution of the environment. Polluted areas are being cleaned up. More extensive testing of new chemicals and other pesticides and their environmental effects is being conducted. The effects of new construction on wildlife habitat are also studied before and during the construction process.

Establishing large acreages in national parks and wildlife refuges is also important for the future well-being of the wildlife in the United States. Emphasis is being placed on management of wildlife resources, rather than simply exploiting them. This also bodes well for the future of the country's wildlife. Fish and game laws are important elements in preserving and enhancing wildlife.

Figure 11-23 Ponds and lakes can accommodate fish and wildlife and provide recreation for people.
Courtesy of Chesapeake Bay Foundation.

Realistically, some species of wildlife will decline and cease to exist in the future, while other species will proliferate. This has been the case in the past, and it seems likely to continue to be so in the future. The human population can choose to help the process or to interfere with it.

FFA members can apply the FFA motto in defense of wildlife populations at risk of becoming threatened or endangered. The first step is "Learning to do," a concept that requires in-depth study of ways to sustain wildlife populations. Step two is "Doing to learn" by organizing other FFA members or joining other interested citizens as you identify needs of wild animals where you live. Actions on behalf of vulnerable wildlife species should focus on their most critical survival needs.

Wild animals have always attracted human attention. As a source of food and clothing, they have been essential to human existence. Farm animals, domesticated from their wild condition, now fill these roles in most societies. Sadly, many wild species have become extinct at the hands of mankind. Fortunately, humans also have the ability to restore wildlife habitat and take protective action to preserve and restore the wild species that remain.

Chapter Review

Student Activities

1. Write the **Key Terms** and their meanings in your notebook.
2. Visit a wildlife area and identify the species of wildlife that you see.
3. Create a multimedia presentation or a bulletin board showing the species of wildlife that are important to hunters and fishermen in your area.
4. Have a wildlife manager visit your class and explain management practices that are used in the area that they manage.
5. Write a report on the effects of pollution on wildlife in a particular area of the United States.
6. Participate in a New Year's Day bird count.
7. Plant a feed patch area for wildlife.
8. SAE Connection: Participate in a stream or other wildlife area cleanup program.
9. Visit a zoo and list the species of wildlife housed there that are endangered.
10. Develop a list of endangered species of wildlife in your area and what is being done to prevent their extinction.
11. Write a report on a species of wildlife that interests you.
12. Have a local farmer or rancher speak to the class on what measures they are taking to enhance the environment for wildlife.
13. Invite a bird-watcher or wildlife photographer to discuss their given hobby or profession with the class.
14. Using the plans in Appendix D, construct and place birdhouses and nesting boxes.
15. Spend 1 hour outside class observing an animal in the wild. **Note:** Maintain a safe distance from the animal. Determine what activities the animal is involved in and how much time it spends on each activity. Write a one-page summary of your findings.

Check Your Progress

Can you . . .

- give examples and definitions of wildlife terms?
- identify characteristics of wildlife?
- describe relationships between types of wildlife?
- describe classifications of wildlife management?
- identify approved practices in wildlife management?

If you answered "no" to any of these questions, review the chapter for those topics.

National FFA Connection — U.S. Youth Conservation Corps

Work experience gained through the Youth Conservation Corps qualifies as a conservation Supervised Agriscience Experience.
Natural History Archive/Alamy.

Here's an exciting way that FFA members can learn valuable skills and prepare for future careers in wildlife management. Imagine a summer job working in the outdoors at a national park or a wildlife refuge. Are you interested in participating in a supervised work experience program that meshes with the FFA motto: "Learning to Do; Doing to Learn"? Such opportunities exist for youth who are 15–18 years of age to work as members of the United States Youth Conservation Corps (YCC). Those who are selected will work 40 hours per week for 8–10 weeks earning the minimum wage. In the interest of safety, candidates must have clean criminal records and no history of antisocial behavior. For more information, visit the National Parks Service website at nps.gov and search *Youth Conservation Corps*.

Inquiry Activity — Forest and Wildlife Management

Courtesy of DeVere Burton.

After watching the *Forest and Wildlife Management* video available on MindTap, answer the following questions:

1. The purpose of forest, range, and wildlife management is to balance two important goals. What are those two goals?
2. What is reduced impact logging?
3. What are some ways of managing forest or rangeland that help protect habitat? Identify at least three.
4. According to the video, how does hunting relate to wildlife management conservation? Give examples.
5. What are some ways that habitat management is accomplished? Identify at least two.

Chapter 12
Aquaculture

Objective
To recognize the biological requirements necessary for the production of aquatic plants and animals.

Competencies to Be Developed
After studying this chapter, you should be able to:
- describe the food chain in a freshwater pond.
- discuss water quality and list eight measurable factors.
- identify three major aquaculture production systems.

Key Terms

aquaculturist
gradient
amphibian
estuary
saltwater marsh
brackish water
fry
spawn

osmosis
adaptation
gills
shellfish
crustacean
molt
ppm
buffer

turbidity
ammonia/nitrite/nitrate
TAN
toxin
salmonid
ppt
larvae

Water covers three quarters of the surface of Earth. This resource produces both plants and animals that are used to feed the world. Aquaculture is the management of this and other water environments to increase the harvest of usable plant and animal products. An **aquaculturist** is a person trained in production of plants and animals in water environments. They must understand where and how organisms live, eat, grow, and reproduce in water. The manipulation of these factors determines the success of the aquaculture system. Aquaculture systems are becoming important to our food system because natural fisheries have reached their maximum capacities to produce more fish (**Figure 12-1**).

Aquaculture production systems are part of an integrated industry that requires specialized products and services. Aquaculturists include nutritionists, feed mill operators, pathologists, fish hatchery managers, processing managers, researchers, and growers. Their services are used to produce fresh and processed seafoods, freshwater fish, shellfish, and ornamental fish and plants. Natural fisheries are fish production areas that occur in nature without human intervention. They include the oceans, continental shelves, reefs, bays, lakes, and rivers. These fisheries are currently fished by sophisticated fishing fleets that are so efficient that yearly catches (yields) have leveled off or are decreasing because of insufficient supplies of fish. The population of the world continues to increase; therefore, aquaculture must be used to produce more aquatic plants and animals for food. Understanding the aquatic environment, the biology of the organisms, and how to control the production of aquatic plants and animals is essential.

Figure 12-1 Natural fisheries have reached their capacities to produce more fish.
Courtesy of Chesapeake Bay Foundation.

The Aquatic Environment

The oceans represent the largest expanse of natural resources in the world (**Figure 12-2**). They are filled with water containing salts and other soluble nutrients and materials washed from the land. Over time, the evaporation of water into the atmosphere has increased the concentration of these salts and minerals until the salinity and mineral content of the water is high. We call this seawater. The concentrations of the salts become so high that land plants and animals are unable to use this water.

Figure 12-2 The oceans represent some of our most abundant and most promising future resources.
© Allen Furmanski/Shutterstock.com.

The Salinity Gradient

Rain contains only small amounts of salts. Therefore, accumulation of water on land and its flow into the oceans generates a **gradient**, or measurable change over time or distance, in salinity. The water cycle is the means by which salinity has increased in ocean water (**Figure 12-3**). This affects the types of organisms that can flourish in an ocean environment.

Freshwater wetlands, such as marshes, ponds, and streams, are generated by rainwater (**Figure 12-4**). Here, natural rainfall accumulates and provides aquatic environments that represent the transition between aquatic (water) and terrestrial (land) plants and animals. **Amphibians**, such as frogs, turtles, and reptiles, are animals that live part of their lives in freshwater and the remaining period on land. Several plant species, such as cattails (*Typha* sp.), watercress (*Nasturtium officinale*), water spinach (*Ipomoea reptans*), and rice, also require a transitional period of flooding and drainage to flourish.

Figure 12-3 The natural water cycle operates by evaporating water from the oceans and returning it to the landmasses in the form of precipitation. The water then gradually returns to the ocean through springs, streams, lakes, and rivers.

Figure 12-4 Freshwater marshes serve as transition areas where some species of animals and plants are able to adjust while they migrate or make transitions between the aquatic and terrestrial phases in their life cycles.
Courtesy of DeVere Burton.

As water flows from fields and urban areas into large streams and lakes, this runoff water accumulates salts and soluble nutrients. The profiles of plants and animals change as other organisms that are more adapted to this changed environment displace the original residents. Flows accumulate into rivers that empty into bays, estuaries, and saltwater marshes. *Bays* are open waters along coastlines where freshwater and saltwater mix. Estuaries are ecological systems influenced by brackish or salty water. The saltwater marshes are lowland areas that are covered by seawater when the tide comes in.

Coastal areas near locations where freshwater rivers and streams flow into the ocean are different from either aquatic or terrestrial environments. In those areas, freshwater mixes with seawater to create brackish water, a mixture of freshwater and saltwater that fluctuates with the tide, flow of the rivers, and weather conditions. These areas provide unique environments for wildlife and aquatic plants.

Several types of seaweeds are commercially produced in seawater in East Asia. These include types of red, green, and brown algae. Similar plant growth of algae must be maintained in intensive fish systems to feed newly hatched fish called fry.

The migration of saltwater salmon to upstream locations to spawn (lay eggs) in freshwater streams illustrates the gradient effect on the life cycle of aquatic organisms. The body of the salmon must adjust from expelling salts and other minerals to retaining salts and expelling excess water that builds up in its tissues in the freshwater environment. At the same time, its body adjusts from drinking large amounts of sea water to drinking almost no fresh water. When the salmon entered the ocean as a smolt, these body processes were exactly opposite to those described earlier. Oceangoing fish and other aquatic animals are indeed able to accommodate tremendous changes in their living environments.

Saltwater and Freshwater Fish

Osmosis is the process by which water moves from an area of high concentration to an area of low concentration through a selectively permeable membrane. A selectively permeable membrane will allow some molecules to pass through, while other molecules (e.g., salts) cannot. **Figure 12-5** illustrates water movement from side A (pure water), which has a greater water concentration, to side B (sugar solution), which has a lower water concentration. This happens because water flows across membranes to make both sides of the membrane equal in concentration.

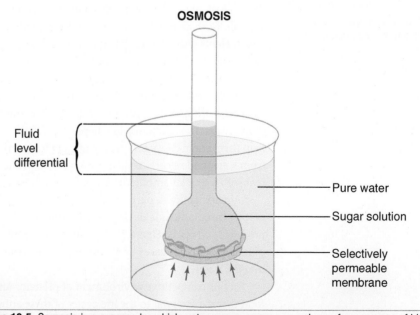

Figure 12-5 Osmosis is a process by which water moves across a membrane from an area of high water concentration to one of lower water concentration.

The skin of a fish and the cells of its gills can act as selectively permeable membranes that control water movement. In a marine environment, water would naturally move from the inside of the fish to the outside of the fish in an attempt to dilute the saltwater outside the body. If this took place without the membranes of the fish reacting to control water loss, the fish would become dehydrated and die. Conversely, in a freshwater environment, where there is more salt in a fish's body than outside it, water would move inside in an attempt to dilute the salt concentration in the body. This would result in a waterlogged fish, which would also cause severe problems and death.

Adaptation occurs when heritable traits that increase an organism's chances for survival are passed from one generation to the next. One adaptation that fish have made over time is the ability to live in freshwater, saltwater (marine), or both environments. Marine fish drink water to make up for water loss. They also have specialized "chloride" cells that rid the body of salt. Freshwater fish, in contrast, have no need to drink water. They also have chloride cells in their gills. In freshwater fish, these cells work conversely to those in their marine relatives. Their chloride cells absorb salt from the water. They also excrete a large amount of very diluted urine. Some fish, such as salmon, have adapted even further and the function of their chloride cells changes as they move between freshwater and marine environments.

The Aquatic Food Chain

The aquatic environment constantly changes to maintain a balance of organisms that function in the food chain or system. A simple illustration is the makeup of a freshwater pond. The food chain is fueled by sunlight. Green plants and algae use this energy to grow and use nutrients they absorb from the water. These plants are eaten by fish and other animals that become, in turn, the prey of larger animals (fish, reptiles, and others) and sometimes humans. These large animals return nutrients to the water as waste, or carrion, that is reabsorbed by plants for growth. The maintenance of this food chain supports life (**Figure 12-6**).

The aquaculturist must understand the effect of any management activity on this cycle and make adjustments using technology or design to maintain an aquaculture production system.

Figure 12-6 The aquatic food chain in a freshwater pond.

General Biology of Aquatic Plants and Animals

The biology of aquatic plants and animals is similar to that of terrestrial plants and animals. Both assimilate nutrients, grow, reproduce, and interact with the environment. Green plants harvest energy from the sun through photosynthesis and absorb nutrients from water to manufacture carbohydrates, proteins, fats, and cellulose. These plants serve as the waste recyclers in the aquatic environment by constantly absorbing waste products (nutrients) and contributing to the food chain. Like any land plant, they respond to fertilization, shading, competition, insects, disease, and weather. Aquatic plants are composed of many parts, as discussed in Chapter 15, "Plant Structures and Taxonomy." Certain parts help them compete within the aquatic environment. Green plants absorb carbon dioxide from water and release oxygen during photosynthesis. During the hours of darkness, plants reabsorb a smaller amount of oxygen and release carbon dioxide.

Aquatic animals, particularly fish, complement the relationship with plants by generating carbon dioxide during respiration and releasing soluble nutrients through waste products and decay. **Figure 12-7** illustrates the anatomy of a fish and shows the specialized gills that exchange gases by absorbing oxygen from water and releasing carbon dioxide into the water.

In this competitive environment of predator and prey, plants provide the shelter and food that are essential to the life cycles of these animals.

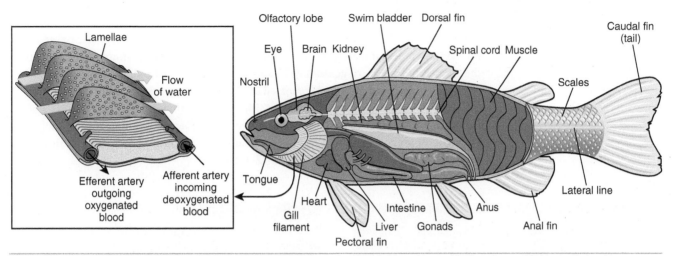

Figure 12-7 Fish are equipped with specialized organs such as gills and swim bladders that enable them to survive in a water environment.

Shellfish are aquatic animals with a shell or shell-like extensions. Some adult shellfish are nonmotile—that is, they cannot move. They include clams, mussels, oysters, and others, and they occupy a unique niche in the aquatic environment. Located on the bottom of a body of water, these organisms have developed an efficient pumping mechanism that filters great quantities of water. It filters out edible microscopic plants and animals known as *plankton*.

Crustaceans are a group of aquatic organisms with exoskeletons. These organisms molt, or replace, their outer shells as they grow. Saltwater lobsters (*Homarus* sp.), crawfish (*Procambarus* sp.), and the various crabs, shrimps, and prawns are important crustaceans. These mobile organisms are characterized by hard exoskeletons that must soften and split as the animals molt and secrete larger shells.

Aquaculture Production

The aquaculturist, in an attempt to increase the production of any aquatic organism, must monitor and maintain the optimum water quality. Water quality has several chemical and physical characteristics that interact within the water. These characteristics must be measured and maintained within a narrow range to promote growth and development of aquatic plants and animals (**Figure 12-8**).

Figure 12-8 Water quality must be monitored closely to maintain an aquatic environment.
© Goodluz/Shutterstock.com.

Several different water-quality test kits are available to test the characteristics discussed in the following paragraphs.

The concentration of dissolved oxygen (oxygen in water) depends on the temperature and pressure of the water and the concentration of atmospheric oxygen. As water is cooled, the pressure increases. The more contact with atmospheric oxygen, the greater and faster oxygen can be dissolved in water. Measured by oxygen probes or chemical tests, the results are reported as 0 to 10 **ppm** (parts per million). Water at 85° F (30° C) is saturated at about 8 ppm. Most fish can survive at levels as low as 3 ppm but quickly become stressed and succumb to other problems. Rainbow trout must have excellent, or high, levels of dissolved oxygen and can be cultured only in water that is saturated with oxygen (**Figures 12-9** and **12-10**).

Figure 12-9 Oxygen content of water is increased by churning and mixing air through water.
Courtesy of Elmer Cooper.

Figure 12-10 Trout require high levels of dissolved oxygen and a constant environment of cold water to thrive in an aquaculture setting. The water is also controlled to eliminate waste.
© iStockphoto/PPAMPicture.

The measurement of acidity or alkalinity in water is the pH. This factor affects the toxicity of soluble nutrients in the water. This measurement is recorded, using a pH meter or litmus paper tape, as a number from 1 to 14. Readings less than 7 are acidic solutions, 7 is neutral, and numbers greater than 7 are alkaline. Most aquatic plants and animals grow best in water with a pH between 7 and 8.

Water hardness is measured by chemical analysis and is expressed as ppm calcium. This element is essential in the development of the exoskeletons of shellfish and crustaceans. It also serves as a chemical **buffer** that stabilizes rapid shifts in pH.

Turbidity in water is caused by the presence of suspended matter. High turbidity limits photosynthesis and visibility. A simple method for estimating pond turbidity uses a white Secchi disc that is lowered into the water. When visibility is impaired, the depth is recorded. The greater the depth, the less turbid the water (**Figure 12-11**).

Temperature limits the adaptive range of almost all aquatic organisms. Sunlight warms the upper surface of open water but does not penetrate it. Deep waters, cool-region currents, and melting winter ice and snow can affect water temperature.

Ammonia/nitrite/nitrate compose a group of nitrogen compounds generated by aquatic animals, first as urea and ammonia. These waste products are converted first to nitrite by microscopic organisms in the water, and then to nitrate. They are ultimately converted to nitrogen gas or absorbed by plants. The accumulation of both ammonia and nitrite is toxic to fish and often limits commercial production. Total ammoniacal nitrogen,

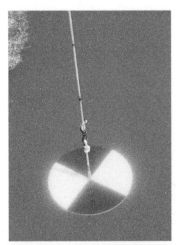

Figure 12-11 The turbidity of water is measured with a Secchi disc. It is lowered in the water until the depth at which visibility is impaired.
Courtesy of Elmer Cooper.

or **TAN**, is recorded by chemical assay in ppm. This does not reflect the toxicity of the measured amount because the toxicity of ammonia is dependent on the pH. Generally, levels of TAN are maintained at less than 1 ppm.

Toxins represent a host of materials that act as poisons, adversely affecting the growth and development of aquatic plants and animals. These include agricultural chemicals, pesticides, municipal wastes, and industrial sludges. Chemical analyses are difficult and often inconclusive.

Science Connection | **The Gradient Effect**

Several species of fish migrate from their freshwater place of birth to saltwater, where they grow to maturity. During their outward migration, their bodies adapt to thrive in a saltwater environment. Eventually, they migrate back to freshwater to spawn. This is necessary because the young fry must hatch in low-salinity water. For most fish that cross the saltwater gradient, the spawning migration is a suicide mission because their bodies are no longer adapted to survive in freshwater. Among the fish that live on both sides of the saltwater gradient are eels, salmon, steelhead trout, and some species of bass.

Selection of Aquaculture Crops

The actual selection of aquatic crops that may be grown is dependent on the resources and experience of the aquaculturist. Like terrestrial crops, each species of aquatic crop has a particular set of water-quality standards to ensure survival and reproduction. A discussion of a few of the well-known aquatic crops should indicate the diversity of this commercial industry. Characteristics will also vary between species. Information on ideal conditions should be requested from your local county extension office and local aquaculturists. Trout and **salmonids** (other related species) are high-quality fish products in high demand (**Figure 12-12**). The trout flourishes in high-quality water. Dissolved oxygen must be kept at greater than 5 ppm. Salmonids also require low salinity, cool temperature (60° F/15° C), and low turbidity. The TAN must be maintained at less than 0.1 ppm.

Catfish (*Ictalurus* sp.) farming represents one of the fastest-growing aquaculture industries in the United States. Current figures project that more than 150,000 acres of ponds are producing catfish annually. Catfish thrive at 75° F to 79° F (24° C to 26° C), a pH between 6.6 and 7.5, a water hardness of 10 ppm, and dissolved oxygen greater than 4 ppm.

Crawfish (*Procambarus* sp.) thrive in freshwater lakes and streams. Good growth occurs at 70° F to 84° F (21° C to 29° C), a pH close to neutral, a water hardness of 50 to 200 ppm, salinity up to 6 **ppt** (parts per thousand), and dissolved oxygen greater than 3 ppm.

Clams, crabs, and oysters are cultivated in bays and estuaries that are subject to tidal flows (**Figure 12-13**). Good growing conditions vary greatly with species but include approximately 6 to 20 ppt salinity, 15° C to 30° C, dissolved oxygen greater than 1 ppm, and adequate amounts of microscopic organisms. Production should improve on cultivated sites when improved spat(young oysters), are developed for stocking.

Shrimp and prawn are cultured in brackish-water ponds and estuaries. Conditions for good growth are temperatures greater than 25° C, a salinity of 20 percent, high levels of microorganisms, and dissolved oxygen greater than 4 ppm. Hatchery techniques are complex and involve several distinct growing stages.

Figure 12-12 Trout are produced on fish farms in large numbers. A high-quality source of cool water is required to raise them successfully.
© iStockphoto/Dorling_Kindersley.

Figure 12-13 The highly prized blue crab of the Chesapeake Bay.
© APaterson/Shutterstock.com.

Production Systems

The cultivation of any aquatic organism by the aquaculturist integrates the necessary cultural requirements with existing resources. This blend has developed three general production programs.

Open ponds, rivers, and bays are stocked with natural or cultured young organisms and are maintained with densities that are balanced with the existing ecosystem stock. Competing species are controlled, and natural recycling techniques are encouraged. This form of aquaculture can use both natural and constructed ponds (**Figure 12-14**). Some care must be taken to prevent any peaks in TAN. Over-fertilization stimulates rapid algal blooms, with high levels of dissolved oxygen during photosynthesis, but very low levels during early morning hours. The stress that results often leads to high levels of mortality. The aquaculturist must be careful to monitor incoming water for toxins and other suspended materials. Harvesting must involve draining or seining the entire production area. Seining is the removal of fish with nets.

Figure 12-14 Constructed ponds made of concrete are used by aquaculturists to produce large numbers of fish and other freshwater species in controlled water environments.
Courtesy of R. O. Parker.

Caged Culture

Caged culture represents a more capital-intensive program. Aquatic animals or plants are contained in a small area, and waste products are removed by the flushing action of flowing waters (**Figure 12-15**). Confining growing fish in floating pen cages or shellfish on suspended float tables is a technique for managing increasing densities of aquatic organisms. As is true on a cattle feedlot, the aquaculturist confines the animals in a limited area and provides the necessary feed and cultural management. The natural ebb and flow of the water removes waste products and replenishes dissolved oxygen.

Figure 12-15 Some aquaculture businesses raise fish and other aquatic animals in pens or cages that sit on the ocean bottom or are suspended in water.
© iStockphoto/Scott Hailstone.

Cage culture can be designed for both natural waters and newly constructed ponds. Aquaculturists have a better idea of growth rates and can adjust feeding ratios more economically. Some growers have reported problems when a pond rolls over (i.e., changes in water quality occur suddenly during certain weather conditions, bringing the less-oxygenated water from the lower levels of a pond or lake to the surface). Fish in cages are unable to move and can be stressed or killed. In this intensive production system, the aquaculturist must ensure adequate nutrition, disease control, predator control, and physical maintenance. The young stock must be legally caught from natural waters or produced in controlled hatcheries. Successful operations include the production of Atlantic salmon off the coast of Norway, Nova Scotia, and Maine. Hybrids of striped bass have been cultured in cages in Maryland and California. Most trout are cultured in concrete raceways, but net pens suspended in mountain streams and ponds are also used.

Shellfish growers in Japan and the United States have demonstrated the production of oysters, clams, and mussels on suspended float tables or ropes.

Recirculating Tanks

Many areas of the world lack sufficient water resources to maintain a viable aquaculture industry. Recirculating systems must circulate the wastewater through a biological purifier and return it to the growing tank (**Figure 12-16**). This complicated process is similar to that in a city waste-treatment plant. The system must remove the solid fish wastes, soluble ammonia/nitrite/nitrates, and carbon dioxide, and it must replace depleted oxygen. The pH must be maintained and integrated into the biological needs of the bacteria that populate the biological filters. As in any pond, if any single parameter is ignored, the organisms become stressed and production is decreased.

Figure 12-16 New technology is making fish production profitable in tanks located in sheds, warehouses, greenhouses, and other enclosures.
© iStockphoto/defun.

Hatcheries

The development of more intensive aquaculture systems will depend on a constant supply of high-quality, young organisms (**Figures 12-17A** and **B**). The natural fisheries are threatened by overfishing, pollution, and habitat destruction. Hatcheries are investigating the parameters that affect fish breeding habits, induced spawning, and fry or larvae production. **Larvae** are mobile aquatic organisms that develop into nonmotile adults. These advances in fish and shellfish management allow the industry to develop improved breeds and hybrids to support improved production. Improved genetic lines of trout, catfish, and salmon are already commercially available.

Aquaculture and Resource Management

The demand for aquaculture products will continue to increase. This demand will stimulate a tremendous growth in commercial aquaculture. During the expansion of the commercial aquaculture industry, sources of clean, pure water will have to be located or developed. Conflicts for scarce natural resources and clean water, the impact on recreational areas, and the potential pollution effects will need to be resolved by informed specialists.

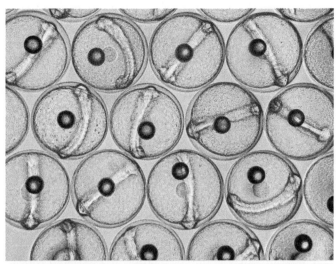

Figure 12-17A Healthy fertilized eggs are required in order for aquatic animals such as fish to be raised in captive living environments.
© Gorshkov25/Shutterstock.com.

Figure 12-17B Young fry require living environments within constant temperature ranges and adequate levels of dissolved oxygen.
Courtesy of U.S. Fish and Wildlife Service.

Science Connection | Whirling Disease

A tiny parasite is responsible for a serious disease in trout and salmon.
Courtesy of Sascha Hallett, OSU.

A major problem for trout and salmon is a small parasite called *Myxobolus cerebralis*. This harmful organism is native to Europe and Asia, but it was accidentally transported to the United States in 1955. It was not until the early 1990s that scientists discovered that a major epidemic was occurring in the waters of many states. Both wild fish and hatchery fish were affected.

The parasite is virtually indestructible. The spores can live in water systems for as many as 30 years. Eventually, a common aquatic worm ingests these spores. Inside the worm, the second life stage develops. This is the free-floating stage that moves through the water until it comes in contact with a young fish called a fry. The parasite then attaches to the host fish and burrows into the head and spinal cartilage of the fry. Here, it begins to multiply very quickly. This puts pressure on the fry's brain, causing the fish to swim in circles. Other symptoms of this disease include a deformed jaw and spine and a tail that shrivels and turns black. The fish appear to be whirling or swimming in circles. The fry that are affected have a hard time eating and difficulty avoiding predators.

"Whirling disease," as it is now called, is a concern of sportsmen, aquaculturists, fish farmers, and scientists. Whirling disease is a major focus of research in aquaculture as attempts are made to control it.

Bio-Tech Connection: Aquaculture Research Is Big Time!

Activity (left) and solids (right) tests are among the many tests needed periodically to keep fish growing in this 7400-gallon recycled water aquaculture system.
Courtesy of USDA/ARS #K-4248-2.

The Aquaculture Research Center in Baltimore, Maryland, was established to provide combined research facilities for the University of Maryland Center of Marine Biotechnology, the $160 million Columbus Center for Marine Sciences, and the National Aquarium in Baltimore. The center was established by converting a shipbuilding structure located beside the harbor into a state-of-the-art aquaculture research facility.

Scientists are growing high densities of valuable fish, such as striped bass, in closed systems using biological filters for cleaning the water to make the facility environmentally friendly. Tools of biotechnology are being used to acquire knowledge about native species and to develop novel solutions to problems associated with commercial finfish and shellfish culture. Studies in the environmental, hormonal, and molecular regulation of spawning will enable growers to have access to seedstocks whenever they need them, rather than being dependent on the normal spawning cycles.

The study of fish genes and hormones that control fish growth and the study of fish nutrition will decrease the time and cost of raising fish to market weight. Of primary importance is the study of devastating diseases that have nearly eliminated the commercial oyster population and certain other species in many natural waters. The worldwide demand for seafood is projected to double by 2030. Yet the world's natural fisheries are already stressed beyond sustainable limits. To meet the increased demand for seafood with decreasing harvests from natural waters, it is believed that aquaculture will have to increase its output three- to fourfold over the next 25 years.

As a working, participatory science laboratory with public spaces for observation, the Aquaculture Research Center is an international focal point for scientists. It also includes an educational center.

Many other institutions in the United States also conduct aquaculture research. For example, extensive research in aquaculture has been done at the University of Wisconsin with yellow perch; the University of Maryland with striped bass, trout, and tilapia; the Walt Disney World EPCOT Center with numerous species; Mississippi State University with catfish; and the University of Idaho with trout.

Agri-Profile: Career Areas: Aquaculture Researcher

Gentle pressure is applied to cause a female catfish to release her eggs for artificial fertilization.
Courtesy of USDA/ARS #KI-5325.

Catfish farming and other aquaculture enterprises are big businesses in many nations of the world. Similarly, catfish farming is one of the fastest growing food production enterprises in the United States. Cultured seafood production is rapidly approaching the volume of that taken from natural waters.

The rapid growth in aquaculture has spurred research-and-development activities. These, in turn, have stimulated career opportunities in animal science, nutrition, genetics, physiology, aquaculture construction, facility maintenance, pollution control, fish management, harvesting, marketing, and other areas.

As productive land becomes scarce among the global resources, food production will become more concentrated on the ocean environment. Aquaculture is expected to play an increasing role in providing high-quality food at affordable prices.

Science Connection: Aquaculture—A Source of Fish and Seafood

It is well known that the harvest of fish and other seafood from the oceans of the world is not likely to increase greater than current production levels. At the same time, the demand for seafood such as fish and shellfish is continuing to increase. These two trends have created ideal economic conditions for expanding the aquaculture industry. The key to success will be to produce and market high-quality fish and other seafood products economically to take advantage of this opportunity. Obtaining access to enough pure water to support aquaculture production is the first challenge in establishing this kind of farming enterprise. The second big challenge is to return the water to streams and rivers without polluting them. Aquaculture appears destined to be a growth industry to support the demand for fish and other seafood products.

Aquaculture is the practice of raising fish in captivity. Methods have been developed for conducting aquaculture in both freshwater and saltwater environments.
© leungchopan/Shutterstock.com. © Vladislav Gajic/Shutterstock.com.

Chapter Review

Student Activities

1. Write the **Key Terms** and their meanings in your notebook.
2. Visit a local supermarket or seafood market and list the seafood products. Classify them as fish, shellfish, or crustaceans; freshwater or saltwater products; or imported products.
3. Describe why some of the seafood in Activity 2 might not be produced in your area.
4. **SAE Connection:** Locate and interview local aquaculturists about their production systems. What are their greatest operational challenges and how do they address them?
5. Visit a local pet store. Describe the various parts of a freshwater aquarium. How is the water quality maintained in this recirculating system?
6. Make a multimedia presentation or bulletin board illustrating the food chain of a freshwater pond.
7. Set up a class aquarium and discuss the balance between plants and animals in the system.
8. Make saltwater and place a fresh cucumber in the water overnight. Observe any changes. In a few sentences, explain what changes you observed.
9. Research three aquaculture careers that interest you. In a short paragraph, describe the jobs and the training they require.
10. In your own words, describe the eight factors that affect water quality.
11. Here is a familiar diagram—without the labels! With a classmate, see how many parts you can identify. Then check your answers against Figure 12-7.

THE ANATOMY OF A FISH

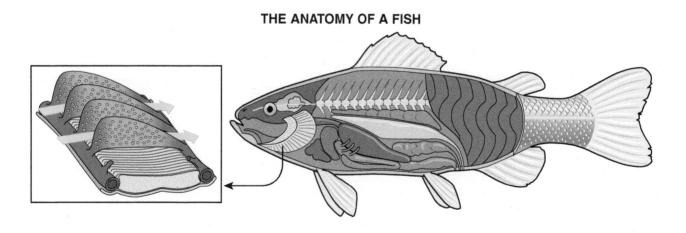

National FFA Connection — Internships and Work Placements in Aquaculture

Students can gain experience in aquaculture by developing aquarium management skills.
© M-Production/Shutterstock.com.

Successful agricultural education programs include three important components: classroom instruction, FFA, and supervised agricultural experiences (SAE). Students who have an interest in aquaculture may not have the resources to own a business in aquaculture. However, they may have an opportunity for work placement or an internship at a local school or aquaculture facility. A community member may have an indoor aquarium as a hobby.

Maintaining an aquarium requires similar management practices to those of a much larger aquaculture system. Work-based experience will open the door to many new opportunities and awards sponsored by the National FFA and their supporting industries. One of these is a natural resources proficiency award.

Start by contacting an aquaculture business and requesting a meaningful work experience. Ask your agriculture teacher to assist you in finding a work placement or internship opportunity in a related career field that interests you. If the aquaculture business or research center is located from your area, have your teacher help you set up a remote meeting online, so you can interview experts about their careers.

Check Your Progress

Can you . . .

- describe the food chain in a freshwater pond?
- discuss water quality and name several measurable factors?
- identify three major aquaculture production systems?

If you answered "no" to any of these questions, review the chapter for those topics.

Unit 4

Integrated Pest Management

Honeybees are a major player in the production of food, fiber, flowers, and ornamentals in the United States. We rely on them as the only method of pollinating certain plants and count on them to do some of the pollination of nearly all species of plants. Bees enter flowers to gather nectar and pollen for their own food and nourishment of their young. Their service to humans and animals in pollinating plants, and thereby producing seeds and fruit, is one we cannot do without. Most plants could not reproduce and survive without producing seeds. Keeping bees safe from pesticides used to control harmful insects is always at the top of the agenda for entomologists. Honeybees are numbered among the many insects that are beneficial to humans. Though they can sting if threatened, honeybees in the United States are of the European type (Photo: bee on the right) and are predictable. Except for the inexperienced person approaching a beehive, honeybee stings are generally a single sting by a single bee. And, except for the relatively few individuals who have life-threatening allergic reactions to bee stings, honeybees pose little threat because they act individually when provoked to sting.

Enter the Africanized honeybee (Photo: bee on the left)—a hybrid cousin of the domestic pollinator, famous for its aggressiveness and dubbed "killer bee"!

Africanized honeybees resulted when bees were imported from Africa to Brazil by a Brazilian scientist in 1957. The plan was to crossbreed them with the European bees prevalent in the Americas to develop a better strain of honeybees for the tropics. Unfortunately, some African bees were inadvertently released in the countryside and promptly interbred with the domestic bees. The new hybrids and their descendants known as Africanized honeybees migrated as far south as Argentina and as far north as the United States.

On October 15, 1990, the first Africanized honeybee swarm to migrate naturally to the United States was identified by entomologists near Brownsville, Texas. The swarm was promptly destroyed using standard procedure. By 2004, Africanized bees had been detected and verified in at least 10 states. Unfortunately, the dispositions of Africanized honeybees are different from the domestic bees of the United States. They tend to defend their colonies more vigorously, stinging in greater numbers and with less provocation. One bee is likely to inflict many stings. Therefore, there is greater danger in an encounter with the Africanized bees. U.S. agriculture and the beekeeping industry fear that domestic bees interbred with Africanized bees may become harder to manage as pollinators of crops and may not be as efficient as honey producers.

The challenge of observing, detecting, and stopping the northward migration of Africanized honeybees will be a top priority of government inspectors, entomologists, beekeepers, farmers, and citizens at large. At the same time, animal behaviorists will study the bees' habits and look for ways to manage them. Geneticists will study the bees' genes and look for ways to genetically engineer future bees so as to decrease their objectionable habits and enhance their abilities as pollinators and honey makers.

Chapter 13

Biological, Cultural, and Chemical Control of Pests

Objective
To develop an understanding of the major pest groups and some elements of effective pest management programs.

Competencies to Be Developed
After studying this chapter, you should be able to:
- define pest, disease, insect, weed, biological, cultural, chemical, and other terms associated with integrated pest management.
- explain how the major pest groups adversely affect agriscience activities.
- describe weeds based on their life cycles.
- describe both the beneficial and detrimental roles that insects play.
- recognize the major components and the causal agents of disease.
- understand and explain the concept of integrated pest management.

Key Terms

vector	perennial weed	noxious weed
arachnid	rhizome	exoskeleton
pathogen	node	metamorphosis
annual weed	stolon	instars
biennial weed	meristematic tissue	entomophagous

disease triangle
abiotic (nonliving) disease
biotic disease
fungi
hyphae
mycelium
bacteria

virus
mosaic
nematode
economic threshold level
quarantine
targeted pest
eradication

pheromone
cultivar
biological control
cultural control
chemical control
pesticide resistance
pest resurgence

The ability to control pests by chemical, cultural, or biological control methods has afforded people in the United States an unprecedented standard of living. We often take for granted an unlimited food supply, good health, a stable economy, and an aesthetically pleasing environment. Without effective pest control strategies, our standard of living would decrease.

Good pest management practices have resulted in dramatic yield increases for every major crop. A single U.S. farmer in 1850 could only support themselves and four people; currently, a farmer can provide food and fiber for approximately 167 people. The ability to control plant and animal diseases or disorders vectored by insects and arachnids has reduced the incidence of malaria, typhus, West Nile virus, and Rocky Mountain spotted fever. A **vector** is a living organism that transmits or carries a disease organism. An insect is a six-legged animal, such as a mosquito, with three body segments. An **arachnid** is an eight-legged animal, such as a spider or a mite.

The impact of pest management in maintaining a stable economy can be seen on a regional and national basis. The regional economy suffered shortly after the cotton boll weevil's introduction into the United States in 1892. The weevil devastated much of the cotton crop in the early 1900s (**Figure 13-1**). Similarly, the potato blight disease in Ireland caused famine and mass migration of Irish people to other parts of the world in 1845. Today, blights are still serious threats to our crops (**Figure 13-2**).

Figure 13-1 A sliced-open cotton boll showing a pink bollworm and the damage it has done.
Courtesy of USDA/ARS #K-2886-13.

Figure 13-2 Pear fruits yellowed and shriveled when the fire blight disease cut off the flow of nutrients from the tree to the fruit.
Courtesy of USDA/ARS #K-5300-1.

Types of Pests

The word *pest* is a general name for any organism that may adversely affect human activities. We may think of an agricultural pest as one that competes with crops for nutrients and water, tends to defoliate plants by eating the leaves, or carries and transmits plant or animal diseases. The major agricultural pests include weeds, insects, nematodes, and plant diseases. However, other types of pests exist. Some examples, and the classes of pesticides or chemicals used for killing them, are included in the following list:

Type of Pest	Class of Pesticide
mites, ticks	acaricide
birds	avicide
fungi	fungicide
weeds	herbicide
insects	insecticide
nematodes	nematicide
rodents	rodenticide

Damage by pests to agricultural crops in the United States has been estimated to be one-third of the total crop-production potential. Therefore, an understanding of the major pest groups and their biology is required to ensure success in reducing crop losses caused by pests.

Weeds

Weeds are plants that are considered to be growing out of place. Such plants are undesirable because they interfere with plants grown for crops. The word *weed* is therefore a relative term. Corn plants growing in a soybean field or white clover growing in a field of turfgrass are examples of weeds, just as crabgrass is considered a weed when it grows in a yard or garden.

Weeds can be considered undesirable for any of the following reasons:
- They compete for water, nutrients, light, and space, resulting in reduced crop yields.
- They decrease crop quality.
- They reduce aesthetic value.
- They interfere with maintenance along rights-of-way.
- They harbor insects and disease **pathogens** (organisms that cause disease).

Weeds can be divided into three categories—annual, biennial, and perennial—based on their life spans and their periods of vegetative and reproductive growth.

Annual Weeds

An **annual weed** is a plant that completes its life cycle within 1 year (**Figure 13-3**). Two types of annual weeds occur, depending on the time of year they germinate. A summer annual germinates in the late spring, with vigorous growth during the summer months. Seeds are produced by late summer, and the plant will die during periods of low temperatures and frost. Examples of summer annuals are crabgrass, spotted spurge, and fall panicum (**Figure 13-4A**).

A winter annual germinates in the fall and actively grows until late spring. It will then produce seed and die during periods of heat and drought stress. Examples of winter annuals are chickweed, henbit, and yellow rocket (**Figure 13-4B**).

Figure 13-3 An annual weed is a plant that completes its life cycle within 1 year. Yellow mustard is a common example of this kind of weed.

© Karin Hildebrand Lau/Shutterstock.com.

A Crabgrass (Summer Annual)
© Bhupinder Bagga/Shutterstock.com.

C Johnson Grass (Perennial)
© Cheryl Casey/Shutterstock.com.

B Henbit (Winter Annual)
© kenji/Shutterstock.com.

D Mullein (Biennial)
© Martin Fowler/Shutterstock.com.

Figure 13-4 A plant growing out of place is considered to be a weed. Weeds can be divided into three categories: annual (A, B), perennial (C), and biennial (D).

Biennial Weeds

A **biennial weed** is a plant that lives for 2 years (**Figures 13-4D and 13-5**). In the first year, the plant produces only vegetative growth, such as leaf, stem, and root tissue. By the end of the second year, the plant will produce flowers and seeds. This is referred to as reproductive growth. After the seed is produced, the plant will die. Only a few plants are considered biennials. Some examples are bull thistle, burdock, and wild carrot.

Perennial Weeds

A **perennial weed** can live for more than 2 years and may reproduce by seed and/or vegetative growth (**Figures 13-4C and 13-6**). By producing rhizomes, stolons, and an extensive rootstock, perennial plants reproduce vegetatively. A **rhizome** is a stem that runs underground and gives rise to new plants at each joint, or **node**. A **stolon** is a stem that runs on the surface of the ground and gives rise to new plants at each node. These plant parts have **meristematic tissue** (tissue capable of starting new plant growth). Examples of perennial weeds are dandelion, Bermuda grass, Canada thistle, and nutsedge.

Figure 13-5 A biennial weed is a plant that completes its life cycle in 2 years. The bull thistle is a widespread example of this type of weed.

Courtesy of DeVere Burton.

Figure 13-6 A perennial weed can live for more than 2 years and may reproduce by seed and/or vegetative growth. An example of a perennial weed is hoary cress, also known as whitetop.

© avoferten/Shutterstock.com.

Noxious Weeds

A **noxious weed** is a plant that causes great harm to other organisms by weakening those around it. Most states have developed lists of noxious weeds, and great effort is directed to control or eradicate them. Most noxious weeds are difficult to control, and they require extended periods of treatment followed by close monitoring. Noxious weeds should be handled carefully to avoid spreading seeds to unaffected areas.

Science Connection: A Prolific Weed

Cheatgrass has a competitive advantage over most other plants in its native range because it germinates early and uses up much of the available moisture before other plants begin to compete for it.
Courtesy of DeVere Burton.

Bromus tectorum is a noxious weed that was introduced to the United States from Europe in the 1850s. It has subsequently invaded every state in the country. Early pioneers nicknamed the weed *cheatgrass* because they believed the weed was cheating them out of greater wheat yields. Other common names include downy brome, downy chess, bronco grass, cheat, and 6-weeks grass. In early spring, range animals take advantage of it as a nutritious source of food. Once the plant dies in late spring, its prickly seeds become a potential danger to range animals by irritating the mouth and throat, resulting in sliver-like painful sores.

B. tectorum is a successful weed. A single plant can produce up to 5,000 seeds under favorable growing conditions. Wind, animal fur, and human clothing easily transport the seeds. Most people who spend time outdoors have collected these seeds in their shoes and socks. Cheatgrass germinates earlier than most plants and quickly uses most of the available water in the soil. When other plants begin to germinate, they often die because of lack of water. This is why cheatgrass is so harmful to crops. Cheatgrass and other noxious weeds must be aggressively dealt with using proven integrated pest management techniques.

Noxious weeds are often spread when seeds become airborne, fall into flowing streams, become attached to the hair of an animal or to human clothing, or are eaten and distributed intact by birds.

Insects

Insects have successfully adapted to nearly every environment on Earth. There are more species of insects than any other class of organism. Part of their success is due to the large numbers of offspring they are capable of producing and the short time they require to reach physical maturity. The human race is dependent on insects in many ways, and insects provide great service to us. Some of the most beneficial insects are ladybugs, praying mantises, parasitic wasps, and honeybees.

Insect Pests

Insects can vector plant and animal diseases, inflict painful stings or bites, and act as nuisance pests. They also can cause great loses to crops, livestock, and humans by injuring them or infecting them with diseases or parasites. Insects can cause economic loss by feeding on forests, cultivated crops, and stored products (**Figure 13-7**). When compared with the total number of insect species, relatively few species cause economic loss. However, it is estimated that in the Florida citrus industry alone, pest control costs $300 million annually and makes up approximately 40 percent of the cost of production.

Insect Anatomy

Insects are considered to be one of the most successful groups of animals present on Earth. Their success in numbers and species is attributed to several characteristics, including their anatomy, reproductive potential, and developmental diversity.

Figure 13-7 Some insects injure or kill plants by feeding on the leaves and stems. Such insects are called defoliators.
Courtesy of Boise National Forest.

Insects are in the class Insecta and are characterized by the following similarities (**Figure 13-8**):

- Each insect has an exoskeleton, which is the body wall of the insect. It provides protection and support for the insect.
- The exoskeleton is divided into three regions: head, thorax, and abdomen.
- The segmented appendages on the head are called antennae and act as sensory organs.
- Three pairs of legs are attached to the thorax of the body.
- Wings are present (one or two pairs) in the majority of species. This permits mobility and greater use of habitat.

Feeding Damage

Insects have either chewing or sucking mouth parts. Damage symptoms caused by chewing insects are leaf defoliation, leaf mining, stem boring, and root feeding. Insects with sucking mouth parts produce distorted plant growth, leaf spotting, and leaf burn (**Figure 13-9**).

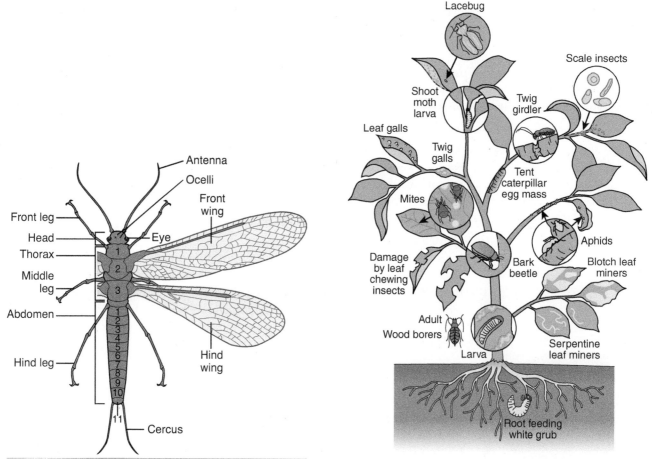

Figure 13-8 Diagram of an adult insect.

Figure 13-9 The different types of insect damage.

Development

As an insect matures from an egg to an adult, it passes through several growth stages. This growth process is known as metamorphosis. The two types of metamorphosis are gradual and complete.

Gradual metamorphosis consists of three life stages: egg, nymph, and adult (**Figure 13-10**). As a nymph, the insect will grow and pass through several instars (the stage of the insect between molts). Each time the insect sheds its exoskeleton, or molts, it passes into the next instar phase. For example, chinch bugs have five instars before they reach adult form but will vary in size, color, wing formation, and reproductive ability. When the insect reaches the adult stage, no further growth will occur.

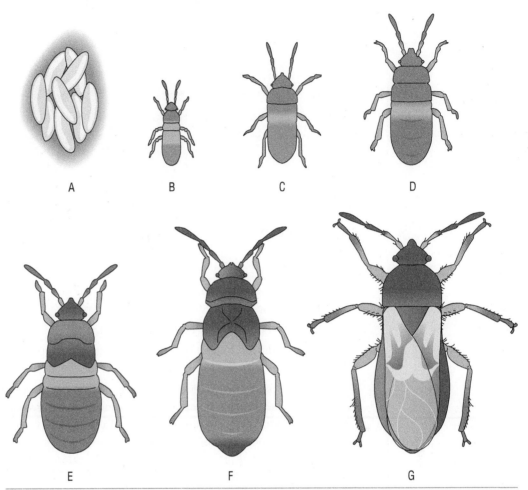

Figure 13-10 Gradual metamorphosis of the chinch bug: (A) egg, (B–F) first to fifth instars, and (G) adult.

Science Connection | Beneficial Insects

Honeybees' service to humans and animals in pollinating plants is invaluable.
© rtbilder/Shutterstock.com.

Scientists estimate that more than 1 million species of insects inhabit the Earth. A majority of them are beneficial to humans. For example, insects are necessary for plant pollination. The Natural Resources Defense Council estimates that bees pollinate more than $15 billion worth of fruit, vegetable, and legume crops per year. Honey, beeswax, shellac, silk, and dyes are just a few of the commercial products produced by insects. Many insects are entomophagous and help in the natural control of their insect species.

Entomophagous insects feed on other insects. Insects that inhabit the soil, act as scavengers, or feed on undesirable plants all play important roles. These insects increase soil tilth, contribute to nutrient recycling, and act as biological weed-control agents. Insects are at the lower levels of the food chain. Thus, they support higher life forms, such as fish, birds, animals, and humans.

Complete metamorphosis consists of four life stages: egg, larva, pupa, and adult (**Figure 13-11**). The larval stage is the period when the insect grows. As larvae molt, they pass to the next larval instar phase. A Japanese beetle will have three larval instars before developing to the pupa stage. The pupa is a resting period. It is also a transitional stage of dramatic morphological change from larva to adult.

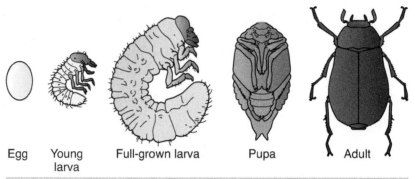

Figure 13-11 Complete metamorphosis of the June beetle.

Plant Diseases

A plant disease is any abnormal plant growth. The occurrence and severity of plant disease is based on the following three factors:

1. A susceptible plant or host must be present.
2. The pathogenic organism, or causal agent, must be present. A causal agent is an organism that produces a disease.
3. Environmental conditions conducive to support of the causal agent must occur.

The relationship of these three factors is known as the **disease triangle** (**Figure 13-12**). Disease-control programs are designed to affect each or all of these three factors. For example, if crop irrigation is increased, a less favorable environment may exist for a particular disease organism. Breeding programs have introduced disease resistance into some new plant lines for many different crops. Pesticides may also be used to suppress and control disease organisms.

Causal Agents for Plant Disease

Diseases may be incited by either abiotic factors or biotic agents. **Abiotic (nonliving) diseases** are caused by environmental or human-made stress. Examples of abiotic diseases are nutrient deficiencies, salt damage, air pollution, chemical damage, and temperature and moisture extremes.

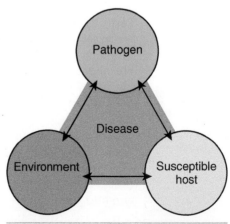

Figure 13-12 Components of the disease triangle.

Figure 13-13 Biotic diseases occur when living organisms such as fungi, bacteria, viruses, nematodes, and others cause damage to growing plants. In this instance, the roots have been killed.

Courtesy of DeVere Burton.

Figure 13-14 Powdery mildew is a fungal disease that spreads from plant to plant through airborne spores.

© Copit/Shutterstock.com.

Biotic means living. **Biotic diseases** are caused by living organisms (**Figure 13-13**). Examples of causal agents or organisms are fungi, bacteria, viruses, nematodes, and parasitic plants. Organisms are parasites if they derive their nutrients from other living organisms. Examples and discussions of causal agents for plant diseases follow.

Fungi

Fungi (plural for fungus) are the principal causes of plant diseases. **Fungi** are plants that lack chlorophyll. Their bodies consist of threadlike vegetative structures known as **hyphae** (**Figure 13-14**). When hyphae are grouped together, they are called **mycelium**. Fungi can

---

Agri-Profile   Career Area: Entomology/Plant Pathology

Plant pathologists observe a tree damaged by fire blight disease (foreground) and healthy, fire blight-resistant trees (background).
Courtesy of USDA/ARS #K-5310-1.

The work of entomologists and plant pathologists is never done. They do battle in the laboratory and in the field against insects and diseases that consume or ruin much of what we produce. The advice and service of a specialist is sought to control diseases and damaging insects, as well as to encourage beneficial ones.

Entomologists and plant pathologists attempt to control or reduce the buildup of damaging insect and disease populations. Such work may include assessing damage; attracting, trapping, counting, and observing insects; and advising, directing, and assisting those who attempt to control insects and plant diseases. The mysteries of some insects are so great that scientists must specialize in just a few insects to be truly knowledgeable about them. Chemicals no longer can be our only means of controlling insects and diseases. Rather, we now use integrated pest management.

Career opportunities exist for field and laboratory technicians as well as for degree-holding specialists. Outdoor jobs may include termite control, scouting, spraying, crop dusting, inspecting, monitoring, selling, and managing field research projects.

reproduce and cause disease by producing spores or mycelia. Spores can be produced asexually or sexually by the fungus. For example, a mushroom produces millions of sexual spores under its cap. These spores can be dispersed by wind, water, insects, and humans.

Bacteria

Bacteria are one-celled or unicellular microscopic plants. Relatively few bacteria are considered plant pathogens. Being unicellular, bacteria are among the smallest living organisms. Bacteria can enter a plant only through wounds or natural openings. Bacteria can be scattered in ways similar to fungi. Some important bacterial diseases are fire blight of apples and pears and bacterial soft rot of vegetables.

Viruses

Plant viruses are pathogenic, or disease-causing, organisms. Viruses are composed of nucleic acids surrounded by protein sheaths. They are capable of altering a plant's metabolism by affecting protein synthesis. Plant viruses are transmitted by seeds, insects, nematodes, fungi, grafting, and mechanical means, including sap contact. Viral diseases produce several well-known symptoms. A symptom is the visible change to the host caused by a disease. These symptoms are ring spots, stunting, malformations, and mosaics. A mosaic symptom is a light- and dark-green leaf pattern.

Nematodes

Nematodes are tiny roundworms that live in the soil or water, within insects, or as parasites of plants or animals. Plant parasitic nematodes are quite small, often less than a quarter inch (4 mm). They produce damage to plants by feeding on stem or leaf tissue (**Figure 13-15**). The main symptom of nematode damage is poor plant growth, resulting from nematodes feeding on the roots. The major plant parasitic nematodes are included in one of three groups: root-knot, stunt, or root-lesion.

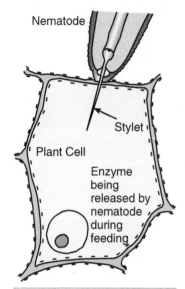

Figure 13-15 A nematode feeding on a plant cell. The stylet is a tiny needle-like feeding structure.

Integrated Pest Management

History

Integrated pest management (IPM) is a pest-control strategy that relies on multiple control practices. It establishes the amount of damage that will be tolerated before control actions are taken. The concept of integrated control is not new. Entomologists had developed an array of cultural and natural controls for the boll weevil and other insect pests by the early 1900s.

However, our approach to pest management during the period from 1940 to 1972 moved to a major reliance on chemical pesticides. Alternate control strategies were largely deemphasized because chemical control gave excellent results at a low cost.

A gradual shift toward reduced use of chemicals for pest control was triggered by a book published in 1962 entitled *Silent Spring*, by biologist Rachel Carson. After 1962, heavy reliance on chemical pest management began to be questioned. Carson's book created a public awareness of the environmental pollution that results from the overuse of pesticides. Adverse effects from misuse and/or overuse of pesticides were beginning to occur as well. These effects included pest resurgence, resistance to pesticides, and concern over human health from exposure to pesticides.

It was not until 1972 that a major change in policy occurred in the United States to encourage other pest-control strategies. Natural, biological, and cultural control programs began to be introduced as alternatives to chemical pest control.

Since then, great strides have been made in the development and implementation of IPM programs. The end result has been to reduce dependency on chemical use, while still achieving acceptable pest control.

Principles and Concepts of Integrated Pest Management

The following concepts or principles are important in understanding how IPM programs should function: key pests, crop and biology ecosystems, ecosystem manipulation, threshold levels, and monitoring.

Key Pests

A key pest is one that occurs on a regular basis for a given crop or plant type, and which causes an unacceptable level of damage that requires a control action. These pests are known to cause widespread damage to crops, trees, or humans and animals. Grasshoppers, locusts, and crickets are capable of crop destruction. The result can be devastating to a population's food supply and result in famine (**Figure 13-16**). In North America, the American grasshopper destroys crops, such as corn, cotton, oats, and peanuts. Asian longhorn beetles and Emerald ash bore can account for the death of many trees every year (**Figure 13-17**). Around the world, humans and animals are infected with insect-borne diseases, such as Zika virus, malaria, and West Nile virus. Some ticks spread Lyme disease.

Figure 13-16 Grasshoppers, locusts, and crickets are capable of crop destruction to such an extent that resulting famines have afflicted humans throughout recorded history.
Courtesy of DeVere Burton.

Figure 13-17 Vast forests are afflicted with severe insect damage throughout North America causing the death of many trees every year. Damage to the branch of an evergreen tree is shown here.

Damage	Pests	
Crops	weevils	aphids
	beetles	leafhoppers
	grasshoppers	crickets
	moths	locusts
	nematodes	cockroaches
Trees	sawflies	cone beetles
	cone maggots	seed bugs
	cone worms	cone borers
	leafminers	needleminers
	moths	mites
	aphids	psyllids
	carpenter worms	termites
Animals and Humans	flies	mosquitos
	ticks	keds
	lice	mites
	fleas	screwworms

Crop and Biology Ecosystem

The integrated pest manager must learn the biology of the crop and its ecosystem. The ecosystem of the crop consists of the biotic and abiotic influences in the living environment of the crop. The biotic components of the ecosystem are the living organisms, such as plants and animals. The abiotic components are nonliving factors, such as soil and water. Examples of human-managed ecosystems include a field of soybeans, turfgrass on a golf course, a poultry production operation, and many other similar environments.

Economic Threshold Levels

With IPM, the attempt is made to understand the influence of ecosystem manipulation on reducing pest populations (**Figure 13-18**). To illustrate this concept, the manager must ask, "What would happen to the pest population equilibrium if a disease-resistant plant were introduced?" Pest population equilibrium occurs when the number of pests stabilizes or remains steady. The introduction of disease-resistant plants should decrease the pest population to less than the economic threshold level. The **economic threshold level** is the point where pest damage is great enough to justify the cost of additional pest-control measures. Until the pest population increases to a high enough level that the cost of controlling the pest is less than the cost of the losses that the pest causes, no control actions are taken.

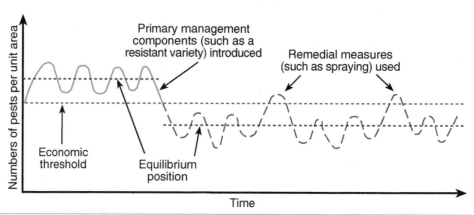

Figure 13-18 The effect of lowering the equilibrium position of a pest.

Adapted from material from Entomology Department, University of Maryland.

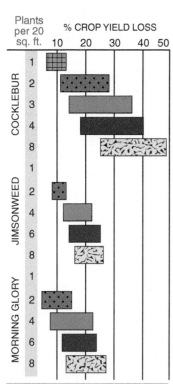

Figure 13-19 The pest-damage index for several weeds in soybeans.

The level of a pest population is important. For instance, the mere presence of a pest may not warrant any control measures. But, at some point, the damage created by insects may be great enough to warrant control measures. Various economic threshold levels are developed to determine if and when a control measure should be implemented. This prevents excessive economic loss of plants to pest damage, while minimizing the use of pesticides. Economic threshold levels are determined by first developing a pest-damage index (**Figure 13-19**). It is crucial in the decision-making process to know the level of pest infestation that will cause a given yield reduction. Pest populations are measured in several different ways. They can be counted in number of pests per plant or plant part, number of pests per crop row, or number of pests per sweep with a net above the crop (**Figure 13-20**).

Economic Threshold for Alfalfa Weevil Larvae from 30-Stem Sample

How to use table below:
1. Use plant height category that fits the field.
2. Estimate the value of crop in dollars per ton of hay equivalent and the cost to spray an acre.
3. From monitoring the field, find the number of alfalfa weevil larvae from a sample of 30 stems.
4. The number in each small box indicates the number of larvae per 30-stem sample that is required before a spray application would be profitable under these conditions.

EXAMPLE:

Plants in the field are 20 inches high (use Category II), hay is valued at $80 per ton, cost to spray is $8 per acre, and you collected 40 larvae from the sample of 30 stems. The number in the box common to $80 and $8 is 75. This means that under these conditions, 75 larvae are needed before a spray would be profitable. Since you collected only 40 larvae, a spray at this time will not be profitable.

Value of hay per ton	Category I plant height 12 to 18 inches						Category II plant height 18 to 24 inches						Category III plant height 24 to 30 inches					
$60	91	114	137	160	183	225	99	124	149	174	199	240	104	130	156	182	209	260
$80	68	85	102	119	136	171	75	94	113	131	150	186	78	97	117	137	157	195
$100	54	68	81	95	108	137	62	75	90	105	120	149	63	78	94	110	126	156
$120	45	57	68	79	91	114	50	62	75	87	100	124	52	65	78	91	105	130
$140	39	49	59	68	77	99	43	54	64	75	86	107	45	56	67	78	90	112
$160	34	43	51	60	68	86	37	47	56	65	75	93	39	49	58	68	79	98
	8	10	12	14	16	20	8	10	12	14	16	20	8	10	12	14	16	20

Cost in U.S. dollars of insecticide application per acre

Figure 13-20 A chart to determine economic threshold level for the alfalfa weevil.

Source: Tooker, J. (May 2009). Entomological Notes, Alfalfa Weevil. [Insect Fact Sheet]. Penn State College of Agricultural Sciences, Cooperative Extension, Department of Entomology. Retrieved from http://ento.psu.edu/extension/factsheets/pdf/Alfalfa%20Weevil.pdf.

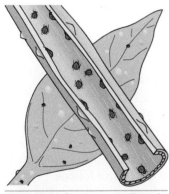

Figure 13-21 Random sampling of a plant stem and leaf to determine pest populations and damage.

Monitoring

For IPM to be successful, a monitoring or scouting procedure must be performed. Different sampling procedures have been developed for various crops and pest problems (**Figure 13-21**). The presence or absence of the pest, amount of damage, and stage of development of the pest are several visual estimates a scout must make. Scouts are monitor fields to determine pest activity. They must be well trained in entomology, pathology, agronomy, and horticulture. The method used must be speedy and accurate.

Pest-Control Strategies

In order to control pests, it is important to know their biological characteristics. The weak link in each pest's biology must be found if control of the pest is to be successful. Pest-control programs can be grouped into several broad categories: regulatory, biological, cultural, physical/mechanical, and chemical.

Regulatory Control

Federal and state governments have created laws that prevent the entry or spread of known pests into uninfested areas. Regulatory agencies also attempt to contain or eradicate certain types of pest infestations. The Plant Quarantine Act of 1912 provides for inspection at ports of entry. Plant or animal quarantines are implemented if shipments are infested with targeted pests. A **quarantine** is the isolation of pest-infested material. A **targeted pest** is a pest that, if introduced in a new area, poses a major economic threat.

If a targeted pest becomes established, an eradication program will be started. **Eradication** means total removal or destruction of a pest. This type of pest control is extremely difficult and expensive to administer. In California, the Mediterranean fruit fly, also known as the medfly, was eradicated at a cost of $100 million in 1982, only to recur several times since then. Today, it is "considered the most important agricultural pest in the world," according to the USDA Animal and Plant Health Inspection Service (APHIS). The cost is high each time the medfly is eradicated. This program relies on chemical spraying, sanitation, sterile male releases, and pheromone traps to ensure complete eradication. A **pheromone** is a chemical secreted by an organism to cause a specific reaction by other organisms of the same species, for example, seeking mates.

Pest Resistance

The development of plants having pest resistance is an extremely effective control practice. New genetic strains of plants and animals have gradually developed genetic resistance to some diseases and insects. More recently, genetic engineering techniques have made it possible to develop resistant varieties much more rapidly. Some advantages of resistant varieties are as follows:

- minimal cost of pest control,
- no adverse effect on the environment,
- a significant reduction in pest damage, and
- ability to fit into any IPM program.

Breeding programs have focused on identification and selection of plants with pest resistance. Currently, new plant cultivars with improved resistance to pests are released annually. A **cultivar** is a plant developed by humans, as distinguished from a variety as it is found in nature.

Biological Control

Biological control means control by natural agents. Such agents include predators, parasites, and pathogens. A predator is an animal that feeds on a smaller or weaker

Science Connection

IPM: Biological Pest Control in National Parks

National parks manage pest control issues with a philosophy that "less is best." This philosophy is in line with the science-based decision making of IPM. Park managers consider IPM to be the most appropriate and cost-effective approach to control specific pests. For example, the invasive weed, purple loosestrife, is known to completely choke meadows, lakes, and ponds with its aggressive foliage.

Under the IPM model, such a weed infestation is approached in different ways. When a small, isolated plant population is located, it may be sprayed with a pesticide to eradicate it before it spreads. However, a well-established population of this invasive weed is more likely to be controlled by establishing a population of hungry beetles that feed almost exclusively on purple loosestrife.

The USDA has approved beetle species to control this pest by eating the foliage to provide biological control over this particular plant pest. *Galerucella calmariensis* and *G. pusilla* feed on every part of the plant. This which prevents the seeds of the plant from dispersing. A species of root-eating weevil destroys the roots of purple loosestrife plants. The plants die before they can continue spreading.

Beetles provide biological control of purple loosestrife.
© ArjaKo's/Shutterstock.com

organism. An example of a predator is the lady beetle. Aphids are the lady beetle's principal prey. Parasites are organisms that live in or on another organism. The braconid wasp is parasitic on the caterpillars of many moths and butterflies. Pathogens are organisms that produce diseases within their hosts. For example, the bacterium *Bacillus popilliae* is a pathogen because it causes the milky spore disease in Japanese beetle grubs.

Successful biological control programs reduce pest populations to less than economic threshold levels and keep the pests in check. Such programs require a thorough understanding of the biology and ecology of the beneficial organism along with that of the pest. Careful research can match desirable plant pathogens against undesirable weeds (**Figure 13-22**).

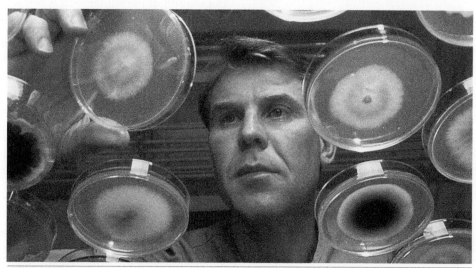

Figure 13-22 A plant pathologist examines fungi that may be used for biological control of weeds.
Courtesy of USDA/ARS #K4652-1. Photo by Scott Bauer.

Cultural Control

Cultural control is achieved when the crop environment is altered to prevent or reduce pest damage. It may include such agricultural practices as soil tillage, crop rotation, adjustment of harvest or planting dates, and irrigation practices. Other practices that are considered cultural control are clean culture and trap crops. Clean culture refers to any practice that removes breeding or overwintering sites of a pest. This may include removal of crop leaves and stems, destruction of alternate hosts, or pruning of infested parts. A trap crop is a susceptible crop planted to attract a pest to a localized area. The trap crop is then either destroyed or treated with a pesticide.

Physical and Mechanical Control

Physical and mechanical control programs use direct measures to destroy pests. Examples of such practices are insect-proof containers, steam sterilization, hand removal of pests, cold storage, and light traps. Implementation of these control practices is costly and provides pest-control results that are not always reliable.

Bio-Tech Connection: Insect Control Using Sterile Male Insects

Scientists have tried for decades to control the mosquito because they carry disease to humans.
Courtesy of USDA/ARS #K-4158-7.

Imagine a laboratory that raises millions of insects that are harmful to agricultural crops. Once the insects are mature, they are irradiated. The treatment does not kill them, but it eliminates their ability to reproduce. The treated, sterile insects are released in huge numbers into areas where the same species of insect pest is found. Because the treated insects vastly outnumber the untreated insects, the odds of treated insects mating with the untreated population are high, but no offspring are produced from these matings. By repeating the releases of sterile insects over a 2- or 3-year period, the untreated population of harmful insects is reduced to manageable levels or, in some instances, it is completely eradicated.

Now imagine a laboratory that bioengineers mosquitoes to prevent insect-borne disease. Mosquitoes transfer life-threatening yellow fever or malaria-causing organisms to humans when they "bite" them to obtain a blood meal. Scientists have bioengineered sterilized non-malaria mosquitoes that replace malaria carriers. Such mosquitoes could thin out or replace their disease-laden cousins and reduce or eliminate the malaria, Zika, Dengue fever, yellow fever, and West Nile diseases in humans and other animals. The process involves infecting male mosquitoes with the bacterium *Wolbachia pipientis* that eliminates the insect's ability to reproduce offspring. It is not known to be harmful to humans, animals, or other insects.

Chemical Control

Chemical control is the use of pesticides to reduce pest populations. Chemical-control programs have become very cost effective. However, various problems occur if this practice is misused or overused. Problems that can develop are environmental pollution, pesticide resistance, and pest resurgence. **Pesticide resistance** is the ability of an organism to tolerate a lethal level of a pesticide. **Pest resurgence** refers to a pest's ability to repopulate after control measures have been eliminated or reduced.

Integrated pest management seems to be the best defense against pests. Biological and cultural controls are favored when they are effective. However, we cannot control certain pests without the use of chemical pesticides. Under such circumstances, it is important to use minimal effective amounts of chemical pesticides.

Chapter Review

Student Activities

1. Write the **Key Terms** and their meanings in your notebook.
2. Use reference materials to find and list 10 examples of beneficial insects and 10 examples of insect pests.
3. Name five insects that are beneficial some of the time but are pests at other times.
4. Make an insect collection, with the insects properly identified and named.
5. Make a drawing of an insect and label the various body parts, appendages, and mouth parts.
6. **SAE Connection:** Research a major crop in your area, its key pests, and pest control methods.
7. Develop a collage showing pests of plants in your community.
8. Make a weed collection and identify each sample using a weed identification key.
9. Using the Internet or other sources, prepare a two-minute presentation on a local pest. Include the type of harm the pest causes and the IPM techniques used to control it.
10. Make an outline of the chapter. Phrases and words that are bold should be included together with a brief description of key concepts.

National FFA Connection — Prepared Public Speaking

While still in college, FFA award-winning public speaker Breanna Holbert was elected to a term as president of the organization.
USDA Photo by Lance Cheung.

The Pennsylvania FFA Organization partnered with Farm Credit to develop the PA FFA Prepared (Conversation) Public Speaking LDEs. These LDEs promote interest in leadership and citizenship. Members participate in agriculture public speaking activities and develop agricultural leadership and communication skills.

Two levels of participation are available to Pennsylvania FFA members:
- Junior Prepared Public Speaking — Participants write and deliver a three- to five-minute speech about a current agricultural subject.
- Senior Prepared Public Speaking — Participants write and deliver a six- to eight-minute speech about a current agricultural subject. Participants are rated based on the written speech, their delivery of the speech delivery, and their answers to judges' questions.

Note: Your state may not use the two-level format. Each state sets standards for leadership development events, but all state winners must follow National FFA rules to represent their state at the national finals.

Chapter 14
Safe Use of Pesticides

Objective
To determine the nature of chemicals used to control pests, to know important terms regarding chemical safety, and to practice the safe use of pesticides.

Competencies to Be Developed
After studying this chapter, you should be able to:
- describe the previous and current trends of pesticide use in the United States.
- recognize some popular classes of chemicals used for pest management and their roles in pest control.
- read and interpret information on pesticide labels.
- state the components of protective clothing for individuals handling pesticides.
- describe the environmental and health concerns relating to pesticide use.

Key Terms
element
compound
inorganic compound
organic compound
contact herbicide
systemic herbicide
phloem
preemergence herbicide
postemergence herbicide
photodecomposition
insecticide
protectant fungicide
eradicant fungicide
formulation
signal word
symbol
LD_{50}
LC_{50}
toxicity
acute toxicity
chronic toxicity
carcinogen

The development and use of pesticides has provided many benefits for both the producer of agricultural commodities and the consumer.

Some benefits of pesticide use are summarized as follows:
- increased yields of food and fiber;
- reduced loss of stored products;
- increased crop quality;
- economic stability;
- better health; and
- environmental protection and conservation.

Many factors affect crop yields, but it is well known that pesticides, properly applied, contribute to both quality and total yields of farm commodities (**Figure 14-1**). It has been estimated that the average total income spent on food in the United States would increase to 30 percent or more without the protection offered by pesticides. Benefits to human health are best illustrated where insecticides are used to control insects that carry and spread malaria and typhus diseases.

However, there are also risks associated with pesticide use. These risks are reduced with proper pesticide application, storage, and disposal. Improper use of pesticides will increase the risk of environmental contamination and adverse effects on human health.

Pesticide use is a controversial issue in the United States. It is important to objectively balance the benefits and the risks associated with pesticide use. The Environmental Protection Agency (EPA) conducts benefit-risk assessments on each pesticide that is registered. According to the EPA, a pesticide is any substance used for preventing, destroying, repelling, or mitigating any pest. When a pesticide is registered for use and is applied according to label directions, the benefits greatly outweigh the risks.

Figure 14-1 An agricultural drone targets an application of pesticide on a sugarcane field.
© Engineer studio/Shutterstock.com

History of Pesticide Use

The use of chemicals to control pests is not new. Elements such as sulfur and arsenic were among the first chemicals used for this purpose. An **element** is a uniform substance that cannot be further decomposed by ordinary chemical means. Homer, in about 1000 B.C.E., wrote that sulfur could be used for pest-control purposes. The Chinese, in about A.D. 900, discovered the insecticide potential of arsenic sulfide, a chemical **compound** that is composed of more than one element.

Until the late 1930s, pest-control chemicals or pesticides were mainly limited to **inorganic compounds** (any compounds that do not contain carbon). Examples of other inorganic pesticides include mixtures of mercury and Bordeaux. Bordeaux mixture is a combination of copper sulfate and lime. It is used for plant disease control. A majority of currently used pesticides are synthetically produced **organic compounds** (compounds that contain carbon).

Many companies throughout the world are continually researching new and existing chemicals to determine what characteristics they may exhibit, such as pesticide qualities, toxicity levels, persistence in the environment, risk to crops and animal life, and effectiveness in controlling pests. A tremendous amount of research is required to develop and gain government approval for a single pesticide. The cost is high, but the reward for developing an effective pesticide product is also high. A single product that is safe and effective has the potential to earn huge profits for the manufacturer. For this reason, chemical companies are willing to pay for the research to develop and gain government approval for a new product.

The organic chemistry involved in pesticide production is often complex and extremely diverse. However, a classification system for pesticides that is based on the type of pest to be treated is useful (**Figure 14-2**). The major pesticide groups are herbicides, insecticides, and fungicides.

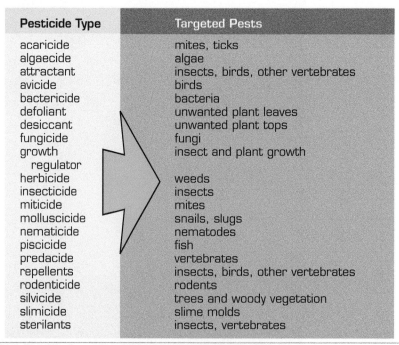

Pesticide Type	Targeted Pests
acaricide	mites, ticks
algaecide	algae
attractant	insects, birds, other vertebrates
avicide	birds
bactericide	bacteria
defoliant	unwanted plant leaves
desiccant	unwanted plant tops
fungicide	fungi
growth regulator	insect and plant growth
herbicide	weeds
insecticide	insects
miticide	mites
molluscicide	snails, slugs
nematicide	nematodes
piscicide	fish
predacide	vertebrates
repellents	insects, birds, other vertebrates
rodenticide	rodents
silvicide	trees and woody vegetation
slimicide	slime molds
sterilants	insects, vertebrates

Figure 14-2 Classifications of pesticides based on the target pests.

Herbicides

Herbicides are grouped into several major categories based on application method, type of control, and chemical structure. The terminology for herbicide use, type of control, and chemical family follows.

Selective Herbicides

A selective herbicide kills or affects only a certain type or group of plants. The selectivity of an herbicide can be caused by many different factors. Some of these factors include:
- differences in herbicide chemistry, formulation, and concentration;
- differences in plant age, morphology, growth rate, and plant physiology; and
- environmental differences such as temperature, rainfall, and soil type.

Nonselective Herbicides

A nonselective herbicide controls or kills all plants. These herbicides are used for many different purposes. Examples of their use are for railroad and highway rights-of-way, industrial areas, fence rows, irrigation and drainage ditch banks, and renovation programs.

Contact Herbicides

Contact herbicides will not move or translocate within the plant. They affect only the part of the plant with which they come in contact. In addition, they are often used for controlling annual weeds.

Systemic Herbicides

A systemic herbicide is absorbed by the plant and is then translocated in either xylem or phloem tissue to other parts of the plant. Xylem tissue transports water and dissolved minerals from the roots to the leaves. Phloem tissue is responsible for transporting carbohydrates from the leaves and stems to the plant roots.

Preemergence and Postemergence Herbicides

A preemergence herbicide is applied before weed or crop seeds germinate. A postemergence herbicide is applied after the weed or crop is actively growing.

Chemical Families of Herbicides

Herbicides are chemicals that are used to control weeds. These compounds can affect plant growth in many different ways. Chemical families of herbicides have been developed, each with a unique chemical structure and a different site or mode of action. *Mode of action* is a phrase used to describe the way herbicides affect plant growth. Several of the more important chemical herbicide families and their characteristics are described in this section.

Acetanilides

Acetanilides interfere with cell division and protein synthesis. They are applied either as preemergence or as preplant applications for control of annual grasses and some annual broadleaf weeds. Popular herbicides within this category are atrazine, Preen®, Snapshot®, Dimension®, and Barricade®. They are used for weed control in row crops grown in fields and gardens.

Bio-Tech Connection: Pest Resistance Using Genetically Engineered Plants

The juice in the foliage of beetle-resistant plants is deadly to the Colorado potato beetle.
Courtesy of DeVere Burton.

Scientists have identified the genes that make some plants resistant to a number of specific pests. Once such a gene is located on the chromosome of a pest-resistant plant, it can be isolated and transferred to other plants. This process is used to develop new plant cultivars (varieties produced by selective breeding) that are resistant to specific pests. For example, new potato plants have been developed that have genetically engineered resistance to the Colorado potato beetle. This beetle is capable of destroying entire fields of potatoes by eating all of the leaves and part of the stems of potato plants. It is a devastating pest to potato plants. The beetle-resistant potato plants have a new gene that produces a natural selective pesticide in the juices of the plant. Potato beetles that eat the foliage of these plants are poisoned without causing risk to other insects or the human population who eat the potato tubers.

Dinitroanilines

Dinitroanilines act on root tissue, preventing root development in seedling plants. They are preemergence herbicides applied to prevent weed germination and should be incorporated into the soil. The dinitroanilines are deactivated quickly by **photodecomposition** (chemical breakdown caused by exposure to light), volatilization (changing to gases), and other chemical processes. Examples of these herbicides are Ferti-lome®, Milestone®, triclopyr, and Trimec®. They are used for control of annual grasses and broadleaf weeds in many different crops.

Phenoxys

Phenoxy herbicides affect plants by overstimulating growth. They perform best when applied as postemergence foliar sprays. They are selective herbicides that affect broadleaf weeds in grass crops. The herbicide 2, 4-D was first used in 1942 and is still widely used. The herbicide 2, 4, 5-T is another product that is widely used.

Triazines

Triazines are photosynthetic inhibitors that interfere with the process of photosynthesis. They are preemergence herbicides used to control both annual and broadleaf weeds in grass crops. Metribuzin, atrazine, simazine, and cyanazine are examples of these herbicides. They are primarily used in control programs for grasses.

Agri-Profile — Career Area: Pesticide Applicator/Pesticide Specialist/Chemist

Aerial applicators are important members of the teams that apply chemical and biological pesticides to large crop, range, and forest areas.
Courtesy of USDA/ARS #K-3663-15.

Tens of thousands of chemical formulations have been developed in recent years to control insects, diseases, weeds, nematodes, rodents, and other pests. While there are benefits to using chemical control, these materials are likely to be hazardous to the operator, plants, animals, or the environment if not properly used.

Pesticide applicators are used by farmers and ranchers, lawn service companies, farm and garden supply firms, termite control companies, highway departments, and railroads as well as self-employed individuals. Special training and licensing are required for handling most pesticides.

State agencies such as state departments of agriculture are responsible for developing the regulations for pesticide consultants and applicators. In most states, people who apply for licenses must attend classes and seminars to learn how to handle pesticides properly. The training usually includes pesticide storage, mixing, application, and disposal procedures. Those who attend these classes and seminars may be required to show evidence of attendance before they are allowed to take the licensing tests that are given. Many states use these tests to establish that the candidates are qualified to prescribe and use pesticides.

Pesticide applicators may work in homes, buildings, fields, forests, and even the holds of ships. They may dust, spray, bait, or fumigate, depending on the setting and the pest. Pesticide applicators must always use protective clothing and safety devices to protect against accidental poisoning. Their tools may be as simple as aerosol cans or as complex as orchard spray rigs or specially equipped helicopters.

Glyphosate

Glyphosate is the most widely used herbicide in the United States and in the world. It is a postemergence herbicide that is sprayed on the foliage of an undesirable plant. The herbicide moves from the foliage to the roots and other live tissues of plants, killing them above and below the ground. Glyphosate is the primary ingredient of Roundup®. Several variants are also available.

A number of food crops, known as Roundup Ready crops, have been genetically engineered to be glysophate resistant. They include field corn, soybeans, cotton, and sugar beets. When Roundup is applied to the foliage of these crops, only the weeds are killed, and cultivation for weed control is not needed.

There have been concerns that glysophate is linked to serious health problems such as a form of cancer, and consumers filed lawsuits on this premise. As of May 2022, the manufacture of Roundup has settled over 100,000 lawsuits and paid out about $11 billion. As a result, Roundup for residential use was reformulated without glysophate and became available to consumers in 2023.

Insecticides

Insecticides are chemicals used to control insects. They can affect the insect in many different ways. The classification system of insecticides may be based on chemical structure and/or mode of action, as shown in **Figure 14-3**.

The botanical, inorganic, and oil insecticides are some of the original chemicals used for insect control. Sulfur, for example, may be used to control mites. Its effectiveness as an insecticide was discovered thousands of years ago.

Chemical Group	Mode of Action	Common Names
Inorganics	Protoplasmic Poisons Physical Poisons	Sulfur Silica Aerogel
Oils	Physical Poisons	Superior Oils
Botanicals	Metabolic Inhibitors	Pyrethrum Rotenone
Synthetic Organics Chlorinated Hydrocarbons	Nerve Poison	Lindane Endosulfan-Thiodan
Organophosphates	Nerve Poison	Diazinon-Spectracide Parathion-Thiophos
Carbamates	Nerve Poison	Carbofuran-Furadan Carbaryl-Sevin
Pyrethroids	Nerve Poison	Permethrin-Ambush Fluvalinate-Mavrik
Insect Growth Regulators (IGRs)	Alter Insect Growth	Methoprene-Altosid Kinoprene-Enstar
Biorational/Microbial Insecticides	Wide Range of Activity	*Bacillus thuringiensis*

Figure 14-3 Insecticide classification system.

Rotenone is a chemical present in the roots of certain species of legume plants. It affects insects by inhibiting respiratory metabolism and nerve transmission. Rotenone, first used in 1848, is a botanical insecticide still used today. It is a contact- and stomach-poison insecticide. Rotenone is principally used in vegetables for controlling fleas, beetles, loopers, Japanese beetles, and other insects (**Figure 14-4**).

Superior oils are highly refined oils and are applied to ornamentals and citrus crops to control scale insects, mites, and other soft-bodied insects. Oils act by excluding oxygen from the insect, thus causing suffocation. It is also believed that oils may destroy cell membrane function. Oils were first used in the early 1900s to control San Jose scale in fruit orchards (**Figure 14-5**).

The synthetic organic group of insecticides was principally developed after 1940. This group includes the chlorinated hydrocarbons, organophosphates, carbamates, pyrethroids, insect growth regulators, and other minor classes. A majority of these insecticides adversely affect nerve transmission. The chlorinated hydrocarbons were the first to be synthesized (human-made). Released in 1940, DDT (dichlorodiphenyltrichloroethane) was the first and most popular chlorinated hydrocarbon. This chemical had excellent insecticidal properties and was used extensively. However, environmental and health problems were linked to DDT and other insecticides within this group. Because of these risks, chlorinated hydrocarbons (such as DDT, aldrin, dieldrin, and chlordane) were banned from use in the United States in 1979.

Figure 14-4 Rotenone is a chemical insecticide used to control Mexican bean beetles and other insects.
© Panu Sahongkunvorakul/Shutterstock.com.

Figure 14-5 Superior oils are insecticides that block the oxygen supplies of insect pests.
Courtesy of DeVere Burton.

The organophosphate and carbamate insecticides are the principal insecticides currently used to control insects. Approximately 60 percent of all insecticides produced in the United States are from these two groups. They include Spectracide® Malathion, Onslaught®, acephate, and Bayer Permectrin® II. They control insects by affecting the nervous system. The potential for pesticide poisoning of people and livestock is high for these insecticides. They will affect the nervous systems of humans and animals in a manner similar to the way they affect the insects they are designed to kill. The dose, or the amount, of insecticide applied is the discriminating factor with respect to insect control or human poisoning.

Fungicides

Fungicides are chemicals used to control plant diseases caused by fungi. Chemical control of plant diseases is more difficult than it is for weed and insect control. Fungi are plants without chlorophyll. They are parasites of other plants. The fungicide must be selective enough to control the fungus but must not adversely affect the host. Also, fungi have many generations in each growing season. Therefore, reapplication of the fungicide is required to provide effective control.

Protectant Fungicide

A **protectant fungicide** is applied before disease infection. This will provide a chemical barrier between the host and the germinating spores. However, this barrier will be broken down by environmental weathering of the fungicide. Rainfall, sunlight, temperature, and plant growth are the major causes of this breakdown (**Figure 14-6**). The fungicide must be reapplied if adequate disease control is expected.

Eradicant Fungicide

An **eradicant fungicide** can be applied after disease infection has occurred. These fungicides act systemically and are translocated by the plant to the site of infection. They offer a longer control period than do protectant fungicides because they are not so prone to environmental weathering.

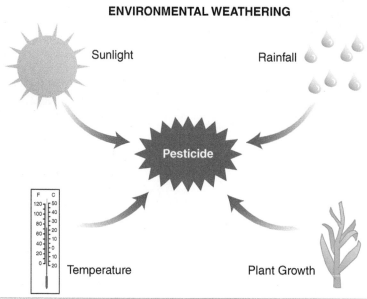

Figure 14-6 Pesticides are broken down by environmental weathering, which is caused by sunlight, rainfall, temperature, and plant growth.

Chemical Structure

Fungicides, like insecticides and herbicides, can also be classified into different chemical families. **Figure 14-7** lists some of the major groups of fungicides.

Chemical Family	Fungicide Activity	Common Name
Benzimidazoles	Eradicant	Benomyl-Tersan 1991
Dicarboximides	Protectant	Captan-Captane
Dithiocarbamates	Protectant	Mancozeb-Dithane M45
Oxathiins	Eradicant	Carboxin-Vitavax

Figure 14-7 Fungicide classification system.

Inorganic Fungicides

The elements sulfur, copper, mercury, and cadmium, or mixtures of them are some of the oldest pesticides used by humans. They protect many ornamentals and turfgrasses from diseases and are examples of inorganic pesticides.

Organic Fungicides

Organic fungicides are a newer group of fungicides. These are used both as protectants and as eradicants. These products are used for effective control of fungi with minimal environmental effects.

Pesticide Labels

The information on a pesticide container label instructs the user on the correct procedures for application, storage, and disposal of the pesticide. The pesticide label is a summary of information gathered by the pesticide manufacturer and is required for product registration. The registration process is estimated to take 8 to 10 years and costs millions of dollars per pesticide. The multiple studies required to meet federal standards are summarized as follows:

- Chemical and physical properties: water solubility, volatility, movement in soils, stability to heat and light, and other factors affecting environmental stability
- Toxicology studies: determine acute oral, dermal, and inhalation toxicity to various animals; evaluate chronic toxicity, including any effect on reproduction or the tendency of the pesticide to be a carcinogen
- Residue analysis: determines the amount of pesticide residues at the time of harvest; develops safe tolerance levels for any pesticide residue
- Metabolism studies: determine application exposure, determine consumer exposure to pesticide residues, and establish safety practices that minimize exposure

The label is a legal document indicating proper and safe use of the product. Pesticide use that differs from that specified on the label is a misuse. Such use is illegal, and the offender can be charged with civil or criminal penalties. When improperly applied, a pesticide can pose danger to the applicator, the environment, animals, and other people. Therefore, it is important to read, understand, and follow the information on the pesticide label.

The pesticide label and other labeling information must meet federal standards. The label consists of a front panel and a back panel on the product (**Figures 14-8** and **14-9**). If there is insufficient room on the panels, additional labeling information will be attached, in booklet form, to the product. The following sections discuss the parts of a label.

Use Classification

Pesticides are classified as either general use or restricted use. A general-use pesticide poses minimal risk when applied according to label directions.

Restricted-use pesticides pose a greater risk to humans and the environment. Therefore, anyone applying a restricted-use pesticide must be properly trained and certified. Applicator certification is administered by each state. An applicator must meet a minimum set of standards and is usually evaluated by testing to be certified.

Trade Name

A trade name is the manufacturer's name for its product. It appears on the label as the most conspicuous item. The manufacturer will use the name in all of its promotional campaigns. The same chemical may have several different trade names, depending on the type of formulation and patent rights.

Formulation

Formulation refers to the physical properties of the pesticide. The pesticide chemical or active ingredient will often have to be modified to allow for field use. These modifications may include adding inert ingredients, such as solvents, wetting agents, powders, or granules, to the pesticide. This will result in different formulations. Some examples of the different types of formulations and their abbreviated label names are as follows:

    Granules—G

    Solutions—S

    Wettable Powders—WP or W

    Soluble Powders—SP

    Dry Flowables—DF

    Emulsifiable Concentrates—EC or E

    Dusts—D

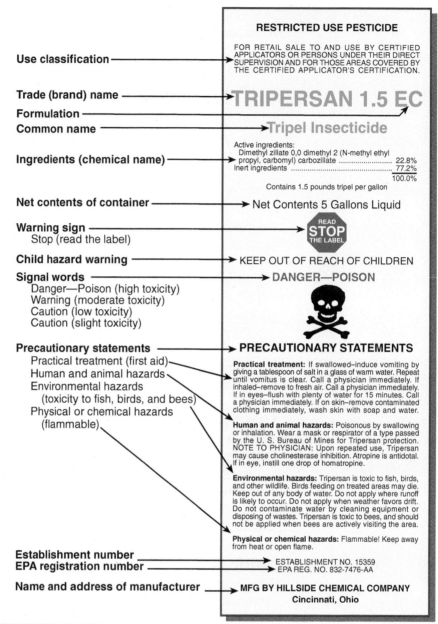

Figure 14-8 Front panel of a sample pesticide label.

Adapted from material from Vo-Ag Services, University of Illinois.

Common Name

The common name is given to a pesticide by a recognized authority on pesticide nomenclature. A pesticide is identified by a trade name, common name, and chemical name. It may have several trade names but will have only one common and one chemical name. The common name, or generic name, identifies the active pesticide ingredient and can be used for comparison shopping.

Ingredients

The percentages by weight of both the active and inert ingredients are stated on the label. The active ingredient is identified by its chemical name. This is the name of the chemical structure of that pesticide. The inert, or inactive, ingredients do not have to be listed by their chemical names. The label must state only the total percentage of inert material.

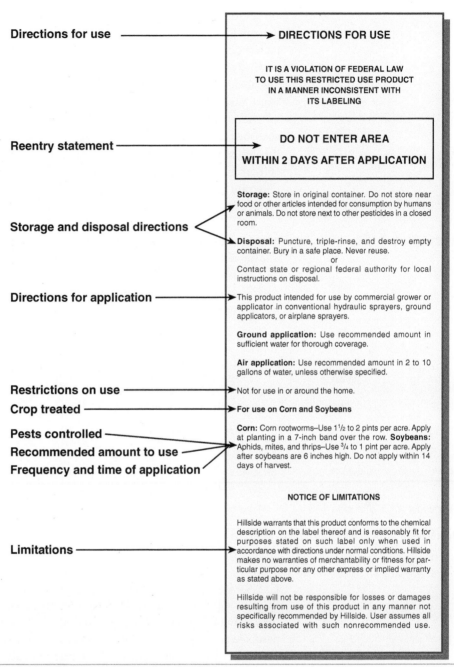

Figure 14-9 Back panel of a sample pesticide label.
Adapted from material from Vo-Ag Services, University of Illinois.

Net Contents
The label will state the amount of product in the container. This is referred to as net contents. This quantity can be expressed in gallons, quarts, pints, or pounds.

Signal Words and Symbols
The **signal words** and **symbols** describe acute toxicity of the pesticide. The different categories are based on LD_{50} (lethal dose), LC_{50} (lethal concentration) values, and on skin and eye irritation. The signal words used on pesticide labels are (1) danger—poison, (2) warning, and (3) caution. These words are used to alert the person handling or using the pesticide to the poisoning effect of contact with the chemical.

Pesticides with high toxicity—only a few drops to 1 teaspoon will kill a 150-lb. person—are labeled "DANGER—POISON." Pesticides with moderate toxicity—1 to 2 tablespoons will kill a 150-lb. person—are labeled "WARNING." Those restricted-use pesticides requiring more than 2 tablespoons of the chemical to kill a 150-lb. person are labeled "CAUTION." Obviously, even the pesticides with "CAUTION" as the signal word must be handled with extreme care.

Acute toxicity is measured by determining LD_{50} values when the pesticide is absorbed through the skin or is ingested orally. These values are determined by inhalation studies. LD_{50} is the amount or dose of the pesticide that is required to kill 50 percent of test populations. It is expressed in milligrams (mg) of pesticide per kilogram (kg) of body weight. The lower the LD_{50} value, the more toxic a pesticide is rated.

LC_{50} is the lethal concentration of the pesticide in the air that is required to kill 50 percent of test populations. It is expressed in micrograms per liter (µg/l) or in parts per million (ppm). The lower the LC_{50} value, the more toxic the pesticide is rated.

All labels must bear the statement, "KEEP OUT OF REACH OF CHILDREN," regardless of pesticide toxicity. **Figure 14-10** shows the toxicity ratings for the various signal words.

Toxicity Rating	Signal Words	Lethal Oral Dose, 150 lb. Person	Oral LD_{50} (mg/kg)	Dermal LD_{50} (mg/kg)	Inhalation LC_{50} (µg/l or ppm)
High	Danger–Poison	few drops to 1 teaspoon	0–50	0–200	0–2,000
Moderate	Warning	1 teaspoon to 1 ounce (2 tablespoons)	50–500	200–2,000	2,000–20,000
Low	Caution	1 ounce to 1 pint+ or 2 lbs.	500–5,000	2,000–20,000	200,000+
Very low	Caution	1 pint+ or 2 lbs.+	5,000+	20,000+	—

Figure 14-10 Toxicity ratings and signal words on a pesticide label warn the applicator of the degree of danger from exposure to the product.

Precautionary Statements

Precautionary statements on the label will list any known hazards to humans, animals, and the environment. They will advise the user how to minimize any adverse effect that the pesticide may have. The categories that are normally listed are as follows:
- Hazards to Humans and Domestic Animals
- Statement of Practical Treatment
- Environmental Hazards
- Physical and Chemical Hazards

Establishment and EPA Registration Numbers

The establishment number identifies the manufacturer. The EPA registration number indicates that the product has passed the review process imposed by the EPA.

Name and Address of Manufacturer

All pesticide labels must contain the name and address of the company that manufactures and distributes the pesticide.

Directions for Use

The correct amount, timing, and mixing of the pesticide are given under the "Directions for Use" section of the label. The label will also list the different pests that are controlled, the application technique, and any other specific directions for optimum control.

Misuse Statement

The misuse statement appears on the label to remind the user to apply the pesticide according to label directions. Problems associated with pesticides, whether they involve environmental pollution or human poisoning, usually occur because of pesticide misuse.

Bio-Tech Connection: Substituting Biological Control for Chemical Pesticides

Biological control of pests includes the use of predatory insects and microbes that destroy the target pest during a phase of its life cycle when it is most vulnerable. The eggs that are attached to this pest will hatch, and the larvae will eat their host.
© Stephen Bonk/Shutterstock.com.

The sound of a roaring cyclone sprayer or the sight of an airplane or agricultural drone fogging a field or orchard creates an image of chemical pesticide application in the minds of most people. However, more and more such sprays are the new, user-friendly, environmentally safe, biological-control pesticides. These materials generally contain bacteria, fungi, viruses, or other microbes that attack the targeted host but are harmless to other living organisms and the environment.

The newer, safer pesticides increase the chances of eradicating or permanently subduing some of our most troubling and costly pests. For instance, the efforts made to control cotton bollworms and tobacco budworms on the Mississippi Delta cost an estimated $50 million per year using chemical pesticides. Marion Bell, formerly an Agriculture Research Service entomologist at Stoneville, Mississippi, believed he had a better way—the *Heliothis* nuclear polyhedrosis virus. *Heliothis* has been found to be very specific to the bollworm and tobacco budworm because it works only on insect larvae that have an alkaline midgut.

Bell specialized in microbial control of insects. He and other authorities determined that treating the entire 4.7 million acres of the Mississippi Delta with *Heliothis* would cost about $7 million. However, the team was aware of the natural fears of the populace with such a widespread, general spraying program. Farmers are concerned about the effect of any general spraying program on their crops, aquaculturalists are concerned that such material could contaminate their catfish ponds, and the general public is concerned about any spray that may be damaging to the environment.

Therefore, they opted for an extensive educational program coupled with a spray program on small areas until the system was proven and more widely accepted by the public. The educational program included informational releases and face-to-face contacts with farmers, congressional representatives, extension personnel, environmental and health officials, physicians, civic clubs, and private interest groups. The message was, "The approach is solid, and the virus is harmless to every living thing except the target pests."

Before a test site was sprayed, brochures were given to the people who were affected, and catfish farmers were contacted to get permission to spray the virus around their ponds. The first year, spray planes were used to apply the virus mixture at the rate of 2 ounces of virus mix in 2 gallons of water per acre. Spray trucks were used to spray certain areas not accessible to aircraft. The researchers found that the virus application does work, and they have fine-tuned the procedure in subsequent years.

Reentry Information

Specific directions on reentering a treated area appear under this heading. Only a few pesticides require reentry times of more than a day after application. However, even if the pesticide label does not contain specific restrictions, no one should ever be allowed to enter a treated area until the pesticide has dried.

Storage and Disposal Directions

This section describes the proper storage and disposal of the pesticide. It is recommended that you purchase only the amount of pesticides needed for the current season. Stockpiling them will only increase storage risks and, ultimately, the problem of pesticide disposal, if they no longer can be used.

Many states have enacted laws and established regulations that control the disposal of unused chemicals and chemical containers. It is never wise to bury these kinds of materials because they can ultimately pollute our groundwater and soil. It may also be illegal. The best choice is to take the containers and chemicals to designated sites where hazardous materials are collected so that they can be disposed of properly (**Figure 14-11**).

Figure 14-11 Leftover pesticides should be taken to a hazardous waste station designated for the collection of toxic waste preparatory to disposing of it in safe and legal ways.
Courtesy of DeVere Burton.

Notice of Limitations

The manufacturer guarantees that the product will perform as the label states. The company also conveys inherent risks to the user if the pesticide is applied in a manner inconsistent with the label. The manufacturer will limit its liability in case of lawsuits stemming from misuse of the pesticide.

Risk Assessment and Management

Risk Measurement

An experienced pesticide applicator who understands the hazards of pesticides will take steps to reduce risks. The risk, or hazard, of pesticide applications has been expressed as the following:

Risk (Hazard) = Toxicity × Exposure

The hazard, or risk, is the relationship between the toxicity of the pesticide and the length and degree of exposure while using the pesticide. **Toxicity** is a measure of how poisonous a chemical is. These data may be expressed in several ways. **Acute toxicity** describes the immediate effects (within 24 hours) of a single exposure to a chemical. Acute toxicity data based on dermal (skin), oral (by mouth), and inhalation (breathing) exposure routes have been determined. Signal words on pesticide labels indicate acute toxicity values. **Chronic toxicity** measures the effect of a chemical over a long period. To determine this information, the chemical is administered at low levels, with repeated exposures to the test animals. The effect of the chemical on reproduction or as a potential carcinogen and any other adverse effects are evaluated.

Limiting Exposure

A pesticide's toxicity cannot be changed, but risk can be managed by addressing the exposure component of the formula. Many things can be done to reduce exposure. Examples of practices that can reduce exposure include the following:

- Select a pesticide formulation with a lower exposure potential. For example, granule formulations have a much lower exposure potential than emulsifiable concentrates.
- Use protective clothing and other safety equipment during the time of pesticide mixing, application, and disposal.
- Apply pesticides during weather conditions that will not cause pesticide drift. Focus on application methods that provide for the most effective control.
- Check all application equipment for proper working condition before applying pesticides.
- Store pesticides and application equipment properly.

Special Gear

The pesticide label will provide guidance concerning acceptable protective clothing or gear for the application of the pesticide (**Figure 14-12**). Recommended protective clothing and gear will vary according to the toxicity of the pesticide. Even if no special gear is required, it is best to minimize your exposure to all pesticides by selecting appropriate clothing.

Appropriate clothing includes long pants, a long-sleeved shirt, nonabsorbent shoes, and socks. Avoid the use of any leather clothing, particularly shoes, because leather absorbs pesticides. Use of heavy denim clothing provides good repellency to any pesticide and it can be washed to remove any pesticide residue.

Figure 14-12 Proper clothing items and safety equipment are essential in the safe use of pesticides.
Courtesy of USDA.

Gloves and Boots

Unlined rubber or neoprene gloves and boots significantly reduce pesticide exposure. Any type of cloth-lined or leather boot or glove will only increase exposure if the lining becomes contaminated. Gloves should be tucked inside sleeves if you are working below the waist. They should be left outside your sleeves if you are working above the waist. Pants legs should be placed over boots. By following these rules, you can prevent material from entering the inside of protective clothing or gear.

Hat and Coveralls

Absorption of pesticides through the skin and into the body is greatest in the scrotal area, ear canal, forehead, and scalp (**Figure 14-13**). The use of an appropriate hat and coveralls minimizes exposure to pesticides in these areas. Lightweight, one-piece, repellent coveralls with hoods are available, and they provide excellent protection.

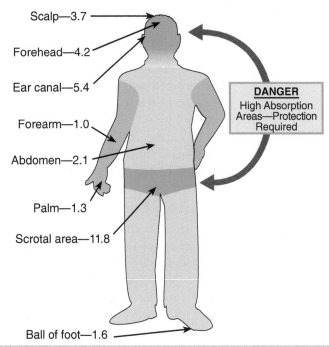

Figure 14-13 Dermal absorption sites and rates of pesticide absorption into the body. Comparison is based on forearm absorption rates.

Apron

During the mixing and loading operations, the applicator is exposed to pesticide concentrates. The use of a rubber or neoprene apron at this time will prevent pesticide concentrates from splashing onto the chest, waist, and legs of the applicator.

Goggles and Face Shield

Eyes are extremely sensitive to many pesticides. The use of goggles and face shields is recommended when mixing pesticide concentrates or working with a spray, dust, or fog.

Respirators

Respirators reduce the inhalation of pesticide fumes and/or dust. A recommended respirator for pesticide applications is a cartridge respirator that will absorb toxic fumes and vapors and filter any dust particles in the air. These respirators are often used during the mixing and application of a pesticide.

Personal Hygiene

Dermal exposure occurs through the skin. It is the principal means of pesticide entry for the applicator. Personal hygiene can drastically reduce this type of exposure. Washing or showering at the end of the workday will remove any pesticide residue on the body. In case of a pesticide spill or splash at the work site, water can be used to immediately remove the material from the skin. After pesticide use, washing your hands before eating or using the restroom will further decrease pesticide exposure. Cleaning protective gear and clothing should also be done to prevent any residual exposure to pesticides on these objects.

Pesticide Storage

Improper storage of pesticides can pose as much danger to the applicator, other people, animals, and the environment as misapplication of pesticides. Some important considerations in selecting a storage facility are the site location and building specifications.

Ideally, the site should be separate from other equipment or material storage facilities. This will reduce risk by decreasing exposure of individuals not involved in pesticide applications. The building should not be located on a floodplain where flooding will introduce pesticides into surface water. It should be built to prevent any runoff or drainage from the site onto sensitive areas. Spill and drainage containment for large storage facilities is highly recommended. Containment systems would trap the pesticides and aid in emergency situations to minimize any environmental damage if the pesticides were to move from the site.

A well-planned storage building should be well ventilated, have a source of heat and water, be fireproof, have a secure locking system, and have sufficient storage area. The storage area should be well marked with placards indicating the presence of pesticides (**Figure 14-14**). The arrangement of the pesticides within the storage area should allow for ease in handling and safety. Tips to provide good storage conditions are the following:

- Separate each pesticide class for storage on its own shelf.
- Keep products off the floor.
- Store containers so the labels remain in good condition and the containers remain orderly.
- Practice good housekeeping.

Figure 14-14 Pesticides must be stored in secure storage units approved for chemical storage.

Health and Environmental Concerns

The current use pattern of pesticides has caused a heightened awareness of their risks. Human health and environmental quality are the major issues in assessing the hazards of pesticide use. The EPA must conduct a benefit-risk assessment for each pesticide that is registered or reregistered for use. This is an extremely controversial issue. The EPA presently defines acceptable risk to the public at one death per million caused by pesticide exposure.

Environmental Concerns

After a pesticide is applied, not all of the pesticide reaches or remains in the target area (**Figure 14-15**). When this happens, the pesticide is often considered an environmental pollutant.

The movement of a pesticide from the designated area may occur in several ways. Drift, soil leaching, runoff, improper disposal and storage, and improper application are some of the major causes of a pesticide becoming an environmental pollutant. Natural resources that can be contaminated are groundwater, surface water, soil, air, fish, and wildlife. Surveys have shown that more than 50 percent of the counties in the United States have potential groundwater contamination from agrichemicals. The three main factors affecting groundwater contamination by agrichemicals are:

- soil type and other geological characteristics;
- the pesticide's persistence and mobility within the soil; and
- the production and application methods of pesticide users.

Pesticide drift is a major cause of soil and air contamination. Drift is the movement of a pesticide through the air to nontarget sites. It will occur at the time of pesticide application when small spray particles are moved by air currents to nontargeted areas. Also, vapor drift of a pesticide may occur after an application. Vapor drift is movement of pesticide vapors because of chemical volatilization of the product.

The adverse effects of pesticides on fish and wildlife may directly result in animal deaths (mortality). Pesticides may also indirectly influence animal feeding or reproduction. Pesticide labeling will indicate any potential harm to wildlife, and this information should be heeded to minimize risk. Fish, birds, bees, and other animals will be affected when pesticides reach them or their habitats.

Environmental contamination by agrichemicals can be decreased through several management practices. The use of integrated pest management (IPM) programs is expected to reduce pesticide use. IPM may also reduce the rates at which pests develop resistance to pesticides. When pesticides are used, proper mixing, application, storage, and disposal must be performed. These practices will decrease any adverse effect on the environment. An attempt must also be made to minimize any effect that temperature, soil type, rainfall, and wind patterns may have on a pesticide becoming an environmental pollutant.

Several strategies have been developed to reduce the amount of chemical that is required to control pests. One of these methods is to inject high-pressure air at the spray nozzle, causing a low-volume chemical mix to separate into tiny droplets. The air pressure also carries the droplets to the area where the chemical is intended to go. This application method has been proven effective in reducing the amount of chemical that is needed (**Figure 14-16**). Another method for reducing the amount of chemical that is needed is to induce opposite electrical charges in the chemical and on the plants to which it is being applied. This causes the chemical droplets to be attracted to the plants to which they are being applied. The net result is that a higher percentage of the chemical reaches the targeted area than it does when traditional application methods are used.

Human Health

The number of lethal pesticide-related poisonings in the United States has decreased over the years. In one year, the total number of accidental deaths from all causes in the United States was 92,000, whereas deaths attributed to pesticide poisoning numbered only 27. However, more than 100,000 nonfatal pesticide poisoning cases per year have

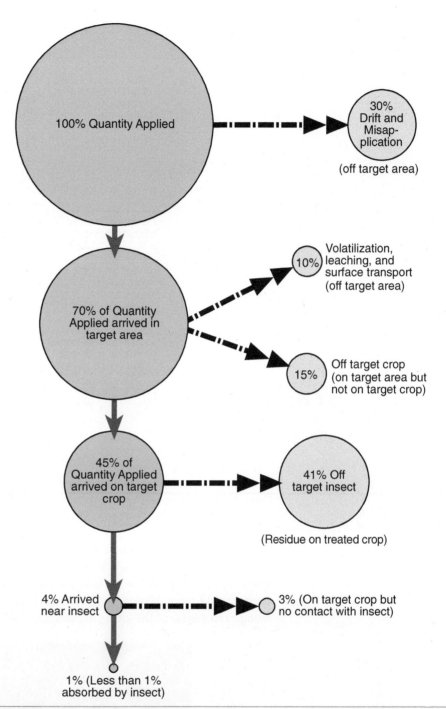

Figure 14-15 Aerial applications of pesticides result in very small amounts actually being absorbed by the insect that is targeted.

Adapted from Flint, M. L. and Van den Bosch, R. *Introduction to Integrated Pest Management*, 1981. Plenum, New York, 240 pp.

been estimated. A majority of the reported deaths were children involved in accidental ingestion of pesticides in the home. The causes have been traced to improper handling and storage practices by homeowners and other consumers.

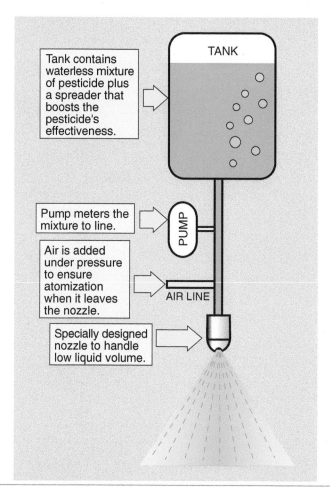

Figure 14-16 New technology permits pesticide application equipment to use less water and less pesticide to achieve equal or better control of pests.

Chemical pesticides are an important part of our food and fiber production capability. They are necessary to maintain our current standard of living in the United States and the world. Although the government does extensive research to arrive at safe pesticide-residue levels, it is important that consumers wash and handle food products to minimize the intake of pesticide residues on food (**Figure 14-17**). However,

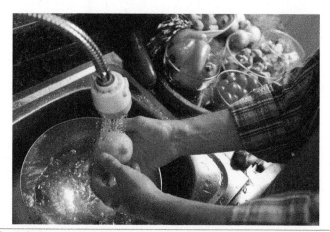

Figure 14-17 Fresh fruits and vegetables should always be carefully washed to remove any possible pesticide residue.
© Dragon Images /Shutterstock.com.

residues of a few pesticides used on food crops can pose potential health problems as carcinogens. A **carcinogen** is a material capable of producing a cancerous tumor. The National Academy of Science estimated that certain types of pesticide residues in food crops "may" cause up to 20,000 cancer cases per year. This estimate was based on a "worst-case" scenario. It assumes that exposure to the pesticide would be continuous over 70 years, with the maximum allowable pesticide residue on the food when eaten. Washing of fruits and vegetables combined with careful handling and preparation of food eliminates most of this risk.

Pesticide Safety at Home

Many people use pesticides in their homes, lawns, and gardens. People also use insect repellents and insecticides when enjoying their yards or the great outdoors. It is important to observe pesticide safety when using these chemicals (**Figure 14-18**)

Hot Topics in Agriscience — Organic Food

Organic is a labeling term that indicates the food or other agricultural product has been produced through USDA-approved methods. These methods integrate cultural, biological, and mechanical practices that foster cycling of resources, promote ecological balance, and conserve biodiversity.

The National Organic Program (NOP) regulates all organic crops, livestock, and agricultural products certified to the USDA organic standards. The USDA enforces these regulations and oversees organic certification, compliance and enforcement activities, and product labeling. To sell, label, or represent their products as organic, operations must follow all the specifications set out by the USDA organic regulations.

Any products sold as organic must have the USDA organic seal, which is obtained through a certification organization. This seal certifies that the product is organic, has 95% or more organic content, and was produced using organic methods that follow organic standards. For multi-ingredient products, the label guarantees that it is made with specified organic ingredients that are certified organic. Nowadays, organic food items can be found in most grocery stores as an alternative to conventionally-produced foods.

Pesticide Safety
Choose the right pesticide for the job.
Carefully read the label and apply the chemical according to the directions for its use.
Mix the chemical properly to assure that the correct amount is applied.
Wear appropriate protective clothing and gear.
Mix only the amount of chemical that is needed.
To avoid chemical drift, consider wind and weather conditions before making the decision to apply a chemical.
Clean the sprayer and carefully dispose of the wash water to avoid contamination to the environment.
Avoid inhaling or ingesting the chemicals and wash dangerous chemicals off the skin immediately.
Store chemicals in locked areas in the original containers away from people, animals, and feeds.
Dispose of chemical waste and empty containers according to the rules for handling hazardous waste materials.

Figure 14-18 People should observe these practices when handling pesticides at home.

Chapter Review

Student Activities

1. Write the **Key Terms** and their meanings in your notebook.
2. Ask your instructor to show examples of an approved pesticide applicator's respirator, goggles, gloves, boots, clothing, and other protective items.
3. Prepare and present a class demonstration on the proper use of one or more items of protective clothing and devices for safe handling of pesticides.
4. Do research and prepare to defend the position that pesticide residue on food in the United States is or is not a serious problem. Arrange for at least three classmates to prepare for a debate presenting various issues regarding the use of pesticides.
5. SAE Connection: Ask a professional pesticide applicator with a special interest in pesticide safety to demonstrate safe pesticide-application principles to the class.
6. Collect print or online articles on accidents involving pesticides. Study the articles and determine how each accident may have been avoided. Report your conclusions to the class. Note: Many pesticide accidents occur in or at the home, lawn, and garden. Do not overlook such cases.
7. Conduct a survey of pesticide storage-and-use practices in your home, farm, and place of employment. Correct all safety violations.
8. List the classes of chemicals used for pest management. Give a brief description of each, and explain how they work to prevent or control pests.

Inquiry Activity | Integrated Pest Management

Courtesy of USDA.

After watching the *Integrated Pest Management* video available on MindTap, discuss the following three items with a classmate.

1. Imagine that you are growing a food crop and that you are concerned about pests. What are some ways that you could use the four-tiered approach to protecting your crop? Pause the video at the 00:55 mark to help you and your classmate review each tier.
2. What is the oldest method of pest control available? What are some examples?
3. What precautions should people take to ensure the safe use of pesticides? Recall as many as you can. Then run the video again from the 02:45 mark to check that you have reviewed them all.

National FFA Connection: Research into the Safe Use of Pesticides

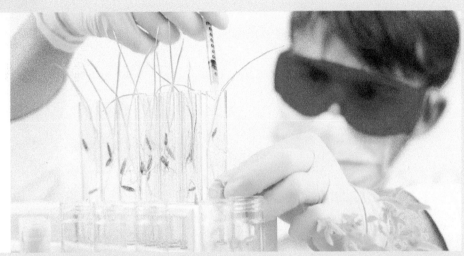

Research requires exactness in all aspects of the process, from the research design to measurement of both inputs and results.
© Monstar Studio/Shutterstock.com

One of the challenges when applying pesticides is to control the target species without disrupting other living organisms that share the environment. The FFA Natural Resource Systems Proficiency Award is focused on research into influences on the environment due to the systems that are used to fertilize crops, manage insects and diseases, and protect the shared and adjacent environment from harm.

Pesticide management is an important topic that will fit well in the framework of this FFA award. Engage your agriscience teacher and other professionals in planning and conducting the research project and reporting the results. This is a long-term project, so read all you can on the specific topic you choose and get started as soon as you can.

To get started, identify a research topic related to the safe use of pesticides. Select a topic that could qualify for the FFA proficiency award for agriscience research. Write a brief description of your topic. In your description, summarize what you know about the topic and then list the questions you would like to answer through your research.

Check Your Progress

Can you . . .

- describe the previous and current trends of pesticide use in the United States?
- cite some popular classes of chemicals used for pest management and their roles in pest control?
- read and interpret information on pesticide labels?
- state the components of protective clothing for individuals handling pesticides?
- describe the environmental and health concerns relating to pesticide use?

If you answered "no" to any of these questions, review the chapter for those topics.

Unit 5

The Quest for Better Plants

How can we grow more useful plants? How might plants provide materials for new medicines and industrial products, produce more beautiful flowers and ornamentals, and resist diseases, insects, drought, cold, or heat? Where can we find trees that produce wood faster? Such questions drive our quests in plant science.

During World War II, the USDA sent botanist Richard Schultes to South America in search of rubber trees resistant to diseases. The discovery opened the way for rubber plantations in the Western Hemisphere. These and other plants changed agriculture around the world. For example:

- Curare vines provide a muscle relaxant used in surgery.
- Ucuuba trees in Colombia have bark that provides a highly desired appetite suppressant and skin medicine.
- Rootstock from China strengthens peach trees.
- Navel oranges from Brazil have strengthened the California citrus industry.
- Durum wheats from Russia set the standard for U.S. varieties for years.
- A peanut from Peru has genes that are resistant to two major diseases that impacted the U.S. peanut industry.
- Wild oats have resulted in one of the most disease-resistant oat varieties ever developed.
- California's avocado industry was started with germ plasm from Mexico.
- Sorghum, dates, tung oil, and numerous forage grasses from around the world are now grown in the United States.
- The Annatto tree of Central and South America provides oil extract used in cosmetics and sunscreen (blocks UV rays). It is also used for yellow and orange food coloring as well as for adding mild flavoring and aroma to foods.

Over 1,600 species of plants used for medicinal purposes by indigenous people of the Colombian Amazon have yet to be studied by scientists. Such plants could carry the genetic material needed to help important species survive. They may also be the source of genetic material that is needed to develop and introduce new plant species with new uses. The need for plant collecting, cataloging, and preserving becomes more urgent each year. The genetic base for many crops we take for granted is narrow, and the encroachment of modern civilization on plants in remote places of the world threatens sources of new genes.

As samples of plant materials are brought to the United States, today's plant collectors follow these guidelines:

- Trips are organized as collaborations between the United States and the host country.
- All collected material is divided at least equally with the host country.
- All germ plasm collected with the support of the USDA is deposited in the National Plant Germ Plasm System, available to all valid users.
- All collection must "be done with a conservation ethic in mind."
- Collection must not endanger natural plant populations, and enough must be left behind so that the plant population can regenerate naturally.

Genetic diversity can be better preserved when new plants are catalogued and most are left and protected where they are found.

Chapter 15
Plant Structures and Taxonomy

Objective
To identify major parts of plants and state the important functions of each.

Competencies to Be Developed
After studying this chapter, you should be able to:
- draw and label the major parts of plants.
- describe the major functions of roots, stems, fruits, and leaves.
- draw and label the parts of a typical root, stem, flower, fruit, and leaf.
- explain some of the variations found in the structures of root systems, stems, flowers, fruits, and leaves.
- describe the relationship of plant parts to fruits, nuts, vegetables, and crops.

Key Terms

taproot	herbaceous	compound leaf
fibrous root	internode	leaf blade
adventitious root	axil	petiole
root cap	terminal bud	cuticle
xylem	vegetative bud	epidermis
root hair	flowering bud	chloroplast
bulb	phototropism	stoma
corm	margin	guard cell
tuber	simple leaf	flower

Chapter 15 Plant Structures and Taxonomy **303**

stamen	ovary	fruit
filament	ovule	taxonomy
anther	petal	genus
pollen	sepal	species
pistil	calyx	binomial
stigma	perfect flower	
style	pollination	

Plants are a basic part of the food chain. Without plants, the web of life cannot exist, and most animals and humans would die. Knowledge of plant growth is essential. To have a better understanding of plants, it is necessary to identify the parts that make up plants. The casual observer sees stems, branches, leaves, and possibly flowers and some nuts or fruits. The agriscience technician, however, sees a series of interconnected tissues and organs that depend on each other to function. The technician knows that all of the organs do not need to be present at one time but is aware that each cell or organ has an important role in the successful growth of the plant. Plant technicians and scientists are concerned with the efficient growth of plants. A plant may become stressed if one or more parts are absent or not functioning properly.

The basic industry of agriculture is dependent on the proper functioning of plants. The animal grower needs many kinds of plants to feed livestock and poultry. The plant industry also needs superior plants for feed and seed production. The horticulture industry needs plants for seeds and cuttings and for food, such as fruits, vegetables, and nuts. Plants are also needed for landscaping the inside and outside of homes and office buildings.

To be successful with plants, one must have knowledge of plant parts and how they function. Such knowledge is essential, whether you are growing, selling, or using plants.

Figure 15-1 Seeds, nuts, and fruits are plant parts commonly used for food.
Courtesy of USDA/ARS #K-3839. Photo by Keith Weller.

The Plant

Plants are composed of many parts. Each part is important in the overall life and function of a plant. The root system is normally underground and is responsible for anchoring the plant and supplying water and nutrients. The stem, or trunk, is normally above the ground and functions as a support system for the rest of the aboveground parts. Leaves constitute the food-manufacturing parts of plants. Flowers come in many sizes, colors, and shapes and function as the seed-producing parts of the plant. Healthy plants produce seeds, nuts, fruits, and vegetables. These parts are popular foods for animals and humans. They also function in reproduction of the plant (**Figure 15-1**).

Basic Necessities of Plant Life

For a plant to survive, its basic needs must be met. These needs include light, water, air, and minerals (**Figure 15-2**). Plants must also capture an energy source. Some plants can thrive in quite shady areas, whereas others need direct sunlight. The sunlight is required by plants to perform photosynthesis. In this process, plants convert the sun's energy into food. This process is discussed in Chapter 16. Another requirement for healthy plants is water. The amount of water needed varies from plant to plant. A desert cactus needs far less water than a tropical fern, but no matter the amount, all plants need it. Most water is taken in through the roots; however, many plants can absorb small quantities through the leaves. Plants also need air. Oxygen is used during plant respiration, whereas carbon dioxide is required for photosynthesis. Minerals are necessary to supply nutrients for plants. Many of these minerals are found free in the soil, but others must be supplemented by providing fertilizer.

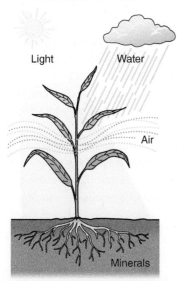

Figure 15-2 A plant draws all of its needs from the environment in which it lives.

Roots

Root Systems

The largest part of the plant is often the root system. Roots take up more space in the soil than the top part of the plant seen in the air aboveground. In fact, some roots will go down into the soil 6, 8, or even 10 feet. Some plants, such as squash, have massive root systems.

Root systems are generally either taproot or fibrous roots. The **taproot** is the main root of a plant and generally grows straight down from the stem. **Fibrous roots** are generally thing, somewhat hair-like, and numerous (**Figure 15-3**). Knowledge of these two types of root systems can be of value in caring for and handling plants.

Figure 15-3 Corn plants showing extensive fibrous root systems at an early age.
Courtesy of USDA/ARS#K5169-5.

The taproot is a heavy, thick root that does not have many side, or lateral, branches. Taproots are often used for human and livestock consumption because they are food-storage organs. Carrots (*Daucus carota*) and beets (*Beta vulgaris*) are examples (**Figure 15-4**). Some plants with taproots are used for ornamental purposes. An ornamental plant is used to improve the appearance of a structure or area.

Plants that have taproot systems have the ability to survive periods of drought. Because they grow deep into the soil and have few fine secondary roots, taproots do not stabilize the soil well.

The fibrous root system is normally shallow. Grasses, corn, and many ornamentals, such as begonia, are good examples of plants with fibrous root systems. Many small, thin-branched roots are present in this type of system. The result is that they are able to hold soil much better than taproot systems. However, fibrous root systems dry out more easily. They cannot tolerate drought conditions.

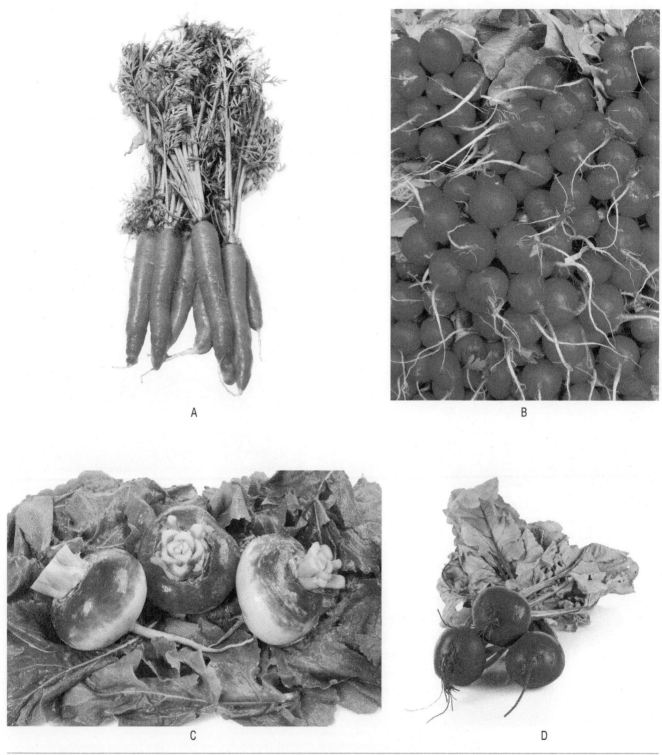

Figure 15-4 Taproots of (A) carrots, (B) radishes, (C) turnips, and (D) beets are healthy sources of food.
© Liz Van Steenburgh/Shutterstock.com. © Richard Peterson/Shutterstock.com. © JackK/Shutterstock.com. © Tinnko/Shutterstock.com.

Although many roots are in the soil, other types are seen above the ground that may not be considered roots. Some plants, such as poison ivy (*Rhus radicans*) and English ivy (*Hedera helix*), have roots that help them climb trees, walls, and sides of buildings. These are called adventitious roots. **Adventitious roots** appear where roots are not normally expected (**Figure 15-5**). Adventitious roots also prop up plants such as corn, strengthening them against the wind.

The mistletoe has roots that penetrate the bark of trees in the upper branches, or crown (**Figure 15-6**). These roots grow into the xylem and phloem tissues of the host plant and extract nutrients that originate in the soil. The dodder plant (*Cuscuta campestris*) has soil roots that die off as the plant gains a foothold in a plant. The dodder plant forms root-like attachments that penetrate the stem of the host plant and extract nutrients from that plant.

Figure 15-5 Adventitious roots are prop roots located above the soil surface. They provide a strong base to help keep a plant upright, such as when soils are wet and conditions are windy.
Courtesy of DeVere Burton.

Figure 15-6 Mistletoe is a parasitic plant with roots that grow through the bark and into the tissue of the tree branch.
Courtesy of Boise National Forest.

Root Tissues

Although there are different systems, all roots look similar when they are examined on the inside (**Figure 15-7**). The parts of the root have very specific functions in the plant. Knowledge of these parts is helpful in diagnosing diseases and other dysfunctions of plants.

Root Cap

The **root cap** is the outermost part of the root. It protects the tender growing tip as the root penetrates the soil. The root cap is a tough set of cells that is able to withstand the coarse conditions that the root encounters as it pushes its way through soil with rock and small sand particles. As the root cap wears away, the cells are replaced by more cells that develop at the root tip. This portion of the root is known as the area of cell division.

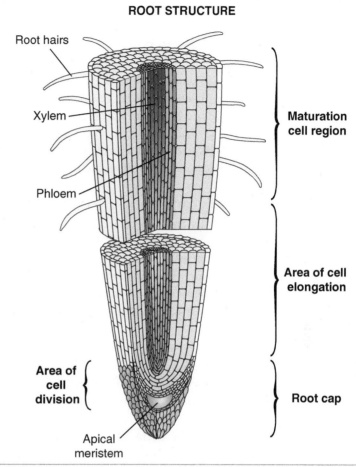

Figure 15-7 Root structure.

Agri-Profile: Plant Science Careers

Some plant science careers include working outside.
Courtesy of USDA/ARS.

Plant science is a career choice with multiple options, such as agronomist, entomologist, horticulturist, plant pathologist, weed scientist. Many plant science careers are similar to overlap with soil science careers. Each of these career opportunities requires a bachelor's degree for entry level jobs, and advancement in the career will usually require a graduate degree. These degrees require strong backgrounds in biological sciences, chemistry, and statistics at both the undergraduate and graduate levels.

Individuals working in plant science work for federal and state governments, universities, and the private sector. The job of a plant scientist includes collection of soil data, consultation, investigation, evaluation, interpretation, planning, or inspection relating to plant science. A plant scientist needs good observational skills to be able to analyze and determine the characteristics of different plants, plant diseases, and plant pests. The work environments usually involve field work outdoors and project planning, data processing, and administrative work indoors.

Area of Cell Division and Elongation

The area of cell division provides new cells that allow the root to grow longer (refer to **Figure 15-7**). The cells in this area multiply in two directions. Small, tough cells are produced on the front edge of this region. They replace cells of the root cap that are worn off or destroyed as the tip pushes its way through the soil. Small, tender cells are produced on the back of this area. They are used as the root tip grows longer. The area of cell division is actually quite thin, perhaps as thin as a strand of hair.

The next area, moving toward the base of the plant, is the area of cell elongation. In this area, the cells start to become longer and specialized. They also begin to look like the older cells and will start to perform their specific functions.

Xylem and Phloem

There are many types of cells in the root. Perhaps the most important ones are the xylem and the phloem cells. The **xylem** cells are responsible for carrying the water and nutrients from the soil to the upper portion of the plant. The phloem cells function as the pipeline to carry the manufactured food down from the leaves to other plant parts, including the roots, where it is used or stored. There are other specialized cells in the root; some are discussed later in this chapter.

Area of Cell Maturation

The area of cell maturation is where cells become fully developed (refer to **Figure 15-7**). This is also where the root hairs emerge. **Root hairs** are small, microscopic roots. They will rise from existing cells located on the surface of the root. It is the function of root hairs to absorb water and nutrients. Water and nutrients move into the root hairs, enter the xylem, and move to the upper portions of the plant. Root hairs are small and tender. They will break off very easily. This means plants must be handled very carefully when transplanting. Once the root hairs are broken off, they cannot grow again or be replaced.

Although roots are normally hidden from view, they are very important parts of the plant. Roots need the same care and consideration as the other parts for plants to grow well.

Stems

Stems are among the first things seen by the casual observer when looking at plants. Stems and branches are noticeable in the winter when the leaves are gone. They are easily seen as a plant grows. Stems support the leaves, flowers, and fruits.

Not all stems are erect structures growing above the ground. Some grow along the ground or even underground. Some stems have specialized jobs to perform. Such stems are referred to as modified stems. Examples of modified stems are bulbs, corms, rhizomes, and tubers. **Bulbs** are short stems that are surrounded by modified leaves called scales. Some examples of bulbs are Easter lilies (*Lilium longiflorum*) and onions (*Allium* sp.; **Figure 15-8**). **Corms** are thickened, compact, and fleshy stems. An example of a corm is the gladiola (*Gladiola* sp.; **Figure 15-9**). Rhizomes are thick stems that run below the ground. Johnson grass and the iris (*Iris germanica*) are examples of plants with rhizomes (**Figure 15-10**). **Tubers** are thickened, underground stems that store carbohydrates. We often eat an example of this type of stem, the Irish potato (*Solanum tuberosum*) (**Figure 15-11**).

Figure 15-8 Bulbs, such as Easter lilies and onions, are shortened stems surrounded by modified leaves called scales.
Courtesy of USDA/ARS.

Figure 15-9 Corms, such as the gladiola, are stems that are thick, compact, and fleshy.
© nimblewit/Shutterstock.com.

Figure 15-10 Rhizomes are thick stems that grow underground near the surface and give rise to new plants at each node.
© Coulanges/Shutterstock.com.

Figure 15-11 The potato is really a specialized stem called a tuber.
Courtesy of USDA/ARS #K-4016-5.

Types of Stems

Two types of stems grow above the ground: woody and herbaceous. Woody stems are composed of tough material. They often have bark around them. **Herbaceous** stems are succulent and somewhat tender. They usually do not survive the winter season in cold climates.

Some important external parts of plant stems include the node, internode, axillary bud, lenticels, and terminal bud (**Figure 15-12**). The node is the portion of the stem that is swollen or slightly enlarged where buds and leaves originate. The **internode** is the area between the nodes. The **axil** is the angle above a leaf stem or flower stem and the stalk. The axillary bud grows out of the axil. The function of the axillary bud is to develop into a leaf or branch. Lenticels are pores in the stem that allow the passage of gases in and out of the plant. The **terminal bud** is located on the tip or top of the stem or its branches. It may be either a vegetative or flowering bud. The **vegetative bud** produces the stem and leaf growth of the plant. The **flowering bud** produces flowers.

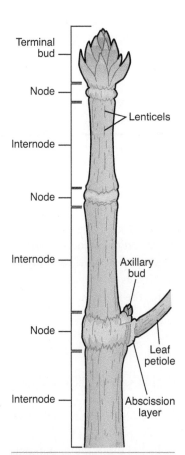

Figure 15-12 Important parts of stems.

Bio-Tech Connection: Inside a Cell: Plant Biotechnology at Work

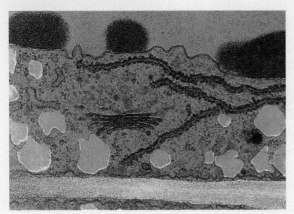

A computer-enhanced view of the interior of a soybean cell.
Courtesy of USDA/ARS.

This electron micrograph shows some of the components in a soybean seed cell. For better visibility, computer processing has colored seed storage proteins purple and stored oil yellow. Red areas are cellular compartments where synthesis of the stored protein and oil occurs. The electron micrograph was taken by Eliot Herman of the Agriculture Research Service, and the color enhancement was performed by Terry Yoo of the Science and Technology Center for Computer Graphics and Scientific Visualization at the University of North Carolina.

Through the use of standard and electron microscopes, together with other modern technology, scientists can peer into the most minute parts of plant structures. They can improve the images seen through various instruments via computer enhancement. The fields of molecular biology, cellular biology, and genetics search the mysteries of the cell and its components and seek to manipulate its biology to bring about desired changes. Larger parts of plants are observed with hand glasses or the naked eye.

Plant structures permit the plant to function as a whole. If one part becomes diseased, it will limit the performance of the rest of the plant and may lead to death of the plant. If the roots cannot obtain sufficient water and nutrients, the leaves cannot manufacture food for the plant. If the stem grows too fast without developing strength, the plant may topple and die. If seeds do not form and mature properly, the plant may not be able to reproduce and perpetuate itself. The whole is dependent on the health of its parts.

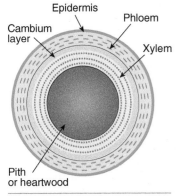

Figure 15-13 This illustration of the cross section of a stem shows the locations of xylem and phloem cells that make up the vascular system of a plant.

Parts of Stems

Stems have some of the same internal parts as roots. The xylem and the phloem continue to run the length of the stem and into all the branches of the plant. In a subclass of plants called dicotyledons, the xylem and phloem occur together in tissues called vascular bundles. In another important subclass called monocotyledons, the xylem and phloem occur in separate areas (**Figure 15-13**).

Leaves

The leaf of a plant manufactures food using light and energy in a process called photosynthesis. The plant uses this food to grow.

A plant leaf is capable of adjusting its angle of exposure to the sun. The leaves of some plants turn to allow full sunlight to shine on the leaf surfaces as the position of the sun changes during the day. The process by which this occurs is called phototropism. Without this important plant reaction to sunlight, plant growth would be reduced.

Leaf Margins

Plants may be identified by the edges, shape, and arrangement of the leaves. The leaf edges are known as **margins**. Leaf margins are named or described according to the toothed pattern on each leaf edge (**Figure 15-14**).

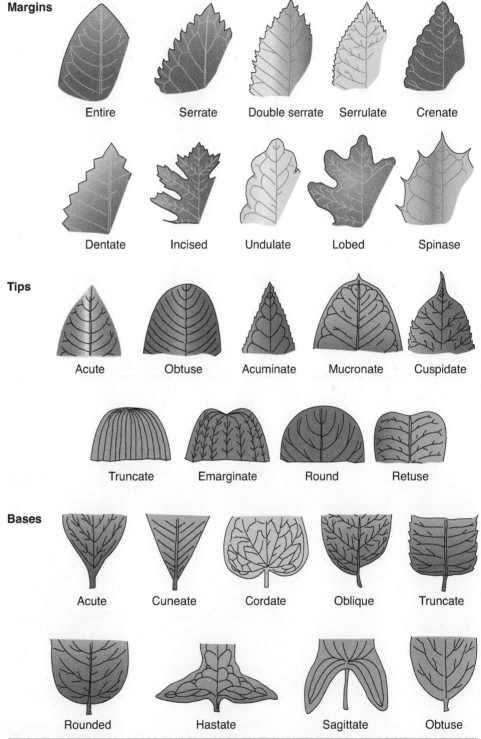

Figure 15-14 Leaf margin characteristics are helpful in identifying plants.

Leaf Shape and Form

Leaves vary in shape and form according to their species. Therefore, knowledge of the name given to each leaf shape and form is useful in identifying the plant (**Figures 15-15** and **15-16**).

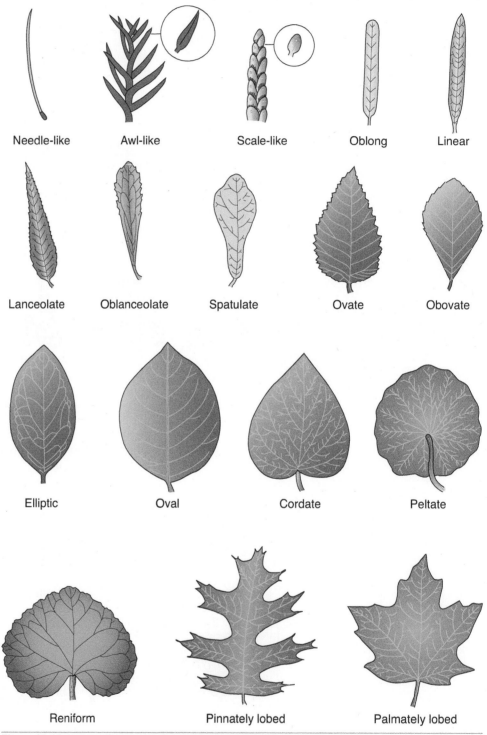

Figure 15-15 Names of various leaf shapes.

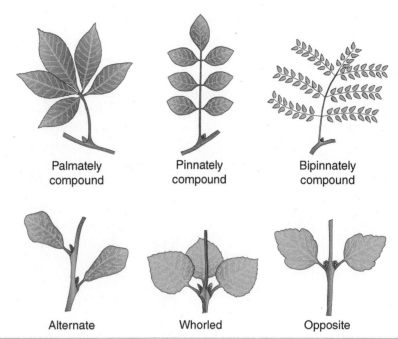

Figure 15-16 Examples of various leaf arrangements.

Types of Leaves

Leaf types vary according to the species. Therefore, leaf type is also used to identify plant species. A single leaf arising from a stem is called a **simple leaf**. Two or more leaves arising from a common point on the stem are referred to as a **compound leaf**.

Leaf Parts

A leaf consists of a petiole and blade. These are the most familiar parts of a leaf (**Figure 15-17**). The **leaf blade** is the wide portion. It may be of many shapes and sizes. The **petiole** is the stem of the leaf. It may be almost absent or may be very long.

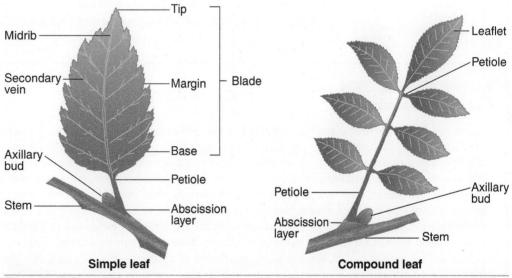

Figure 15-17 Parts of a leaf.

Internal Structure

The **cuticle** is the topmost layer of the leaf. It is waxy and functions as a protective covering for the rest of the leaf. The **epidermis** is the surface layer on the lower and upper sides of the leaf. The epidermis protects the inner leaf in many ways. Elongated, vertical palisade cells give the leaf strength and are the sites for the food-manufacturing process. These cells, as well as the lower spongy tissue layer, contain chloroplasts. A **chloroplast** is the part of the cell that contains chlorophyll, a pigment that is necessary for photosynthesis to occur. The chloroplast's lower layer is irregular and allows the veins, or vascular bundle, to extend into the leaf. The palisade cell layer and the spongy tissue are often referred to as the mesophyll (**Figure 15-18**).

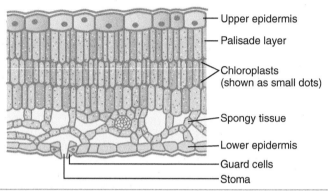

Figure 15-18 Internal structure of a leaf.

The vascular bundles contain the xylem and the phloem. These are extensions of the same tissues that are located in the root. They extend through the stem to the leaves. The xylem brings the water and minerals from the root. The phloem carries the manufactured food from the leaf to the various parts of the plant to nourish plant tissue or to be stored. The lower epidermis contains some special cells called **stomas**. These openings allow for the exchange of carbon dioxide and oxygen as well as some water. The stomas are surrounded by **guard cells** that open and close the stoma. If the plant is stressed by the lack of water or by a low light level, the guard cells will close the stoma. The result is that the plant cannot manufacture food because it will not have all the necessary ingredients.

Science Connection | The Diversity of Plants

There are four different plant types in the kingdom Plantae. Bryophytes, such as moss, are plants that lack a vascular system do not produce seeds, and depend on water for reproduction. A vascular system is a system of vessels in a plant, each of which carries water and nutrients independently from each other throughout the plant. Seedless vascular plants, such as ferns, have distinct vascular systems and need water environments to reproduce. Unlike the Bryophytes, they have true roots, leaves, veins, and stems. Gymnosperms and angiosperms produce seeds, have extensive vascular systems, and do not require water environment to reproduce. Gymnosperms bear their seeds in cones, whereas angiosperm seeds are protected in a thick tissue called fruit. Examples of gymnosperms are pine trees, ginkgo trees, and all conifers. Angiosperms, which are flowering plants, are the most abundant of all the plants. Grasses, corn, petunias, and most crop plants are angiosperms.

Flowers

The **flower** has as its primary function the production of seeds needed to continue the species. It is with this structure that the plant scientist will work to produce new and different varieties.

Not all flowers are actually flowers. The poinsettia (*Euphorbia pulcherrima*) and the flowering dogwood (*Cornus florida*), for example, have modified leaves called bracts. A bract is a modified leaf that is often brightly colored and showy. People often see the red or white bracts and call them flowers. Their function is to protect the flower parts as well as attract insects for pollination.

Flower Structure

Flowers are composed of many parts (**Figure 15-19**). The reproductive structures of the flower include the stamens, pistils, and their associated parts. The male part of the flower is the **stamen**. It consists of the filament, anther, and pollen. The **filament** supports the anther. The **anther** manufactures the pollen. The **pollen** is the male sexual reproductive cell. The female part of the flower, the **pistil**, is made up of the stigma, style, and ovary. The **stigma** receives the pollen. The pollen travels down the **style** and into the **ovary**, which contains **ovules**. Ovules are the eggs, which are the female reproductive cells. When the eggs are fertilized by the pollen, they will ripen into seeds.

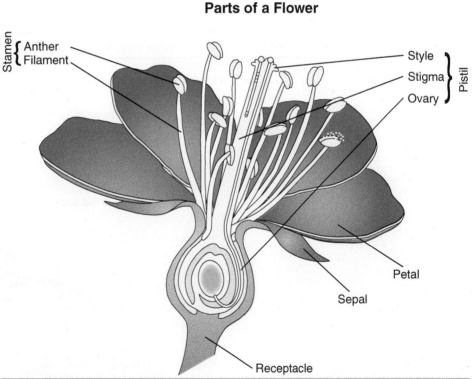

Figure 15-19 Major parts of a flower.

The outer floral parts are composed of the calyx and the corolla. The flower **petals** are collectively called the corolla. Colored petals attract insects or other natural pollinators. The **sepals** function together as a protective device for the developing flower. Collectively, the sepals compose the **calyx**.

Perfect flowers have both stamens and pistils, but they do not necessarily have sepals or petals. Imperfect flowers lack either stamens or pistils. They may or may not have sepals or petals.

In some plants, particularly the flowering plants used in horticulture, it is desirable to remove the anther sacs before the pollen ripens. This prevents pollination and stains from the pollen on the petals of the flower. **Pollination** means the union of the pollen with the stigma. In some cases, orchids, for example, the unfertilized flower will last for many months.

In plant breeding, the anther sac is removed from the plant to prevent natural pollination. It may be destroyed, or it may be used to pollinate another flower to create a new variety. Many hybrids are created in this way.

Fruits, Nuts, and Vegetables

After fertilization, the ripening seed develops in the pistil. The pistil then enlarges and becomes the **fruit** (**Figure 15-20**). The fruit may be of many different shapes and sizes (**Figure 15-21**). The true fruit consists of the seeds that carry the male and female genetic characteristics of the plant. However, the fleshy material surrounding the mature seed, and the seed itself, is commonly called the fruit. The purpose of the fleshy part of the fruit is to attract animals and humans to the seed to help spread it over wide areas. This helps in the reproduction of the plant. Entire fruits or just the seeds may be moved by wind, water, animals, or humans. Often, the fruit is eaten by animals and humans as a source of food. Then the seeds may be discarded, where they can take root and grow into new plants. Humans assist greatly in spreading seeds and starting new plants when they plant seeds for crops.

There are many kinds of fruits and vegetables. The two terms are sometimes used incorrectly. A vegetable can be any part of a plant that is grown for its edible parts. This can be a root, stem, leaf, or ripened flower. However, fruit, which is a ripened or mature ovary, is a specific plant part. A nut is also a type of fruit.

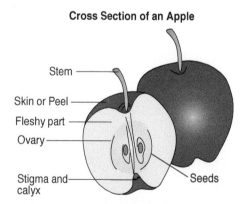

Figure 15-20 The ovary of a flower matures into a fruit that surrounds the seeds. When the fruits are eaten, the seeds are scattered to new locations.

Figure 15-21 (A) Pineapple growing in Hawaii and (B) apples in Washington.
Courtesy of USDA/ARS #K-4281. © Eric Eric/Shutterstock.com.

Plant Taxonomy

Plants can be classified in many ways. For example, agricultural classification systems depend on the methods used for production, the part of the plant consumed, or on the crops intended use (**Figure 15-22**). Other systems classify plants based on how they look, where they are grown, or how their structure compares to the structure of others.

Taxonomy is an organizational system for categorizing plants. Nomenclature is a system of assigning names to plants. The rules for assigning names to plants are established by the International Rules of Botanical Nomenclature.

Binomial System

Carl Linnaeus, a Swedish botanist, developed the binomial system of plant classification in 1753. Modern plant taxonomy is based mainly on Linnaeus's system. Prior to Linnaeus, people had tried to base classification on the leaf shape, plant size, flower color, and other features. These botanical classification systems did not prove workable.

Linnaeus's revolutionary approach was to base classification on the flowers or reproductive parts of a plant. This has proven to be the best system because flowers are the plant part least influenced by environmental changes. For this reason, knowledge of the flower and its parts is essential for anyone who is interested in plant identification.

In this system, Linnaeus gave plants a genus and species name. **Genus** is the taxonomic category between family and species. It is customarily capitalized when written together with a species name. **Species** is the subgroup under genus. A species name is generally not capitalized when written in combination with its genus. It is appropriate for both the genus and the species names to be printed in italics. For example, grain sorghum

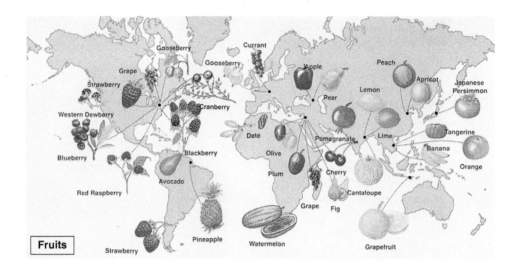
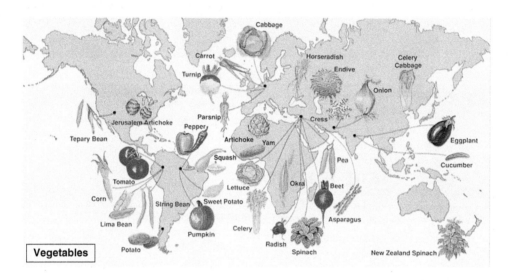
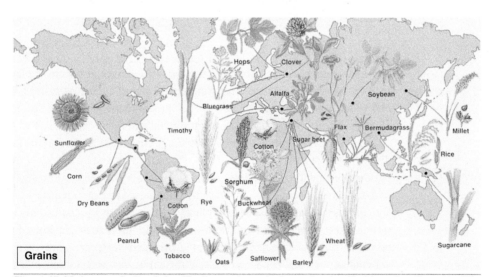

Figure 15-22 In an agricultural classification system, plants can be classified based on the part of the crop consumed, such as seeds, vegetables, fruits, or stems.

Courtesy of USDA/ARS.

is *Sorghum* (genus) *vulgare* (species). To compare the name of a plant or animal with that of a human, the species corresponds to the person's first name and the genus to the person's last name. The system of using genus and species in combination is referred to as a binomial system of classification. **Binomial** means consisting of two names. This system uses the Latin language to name plants. Latin is regarded as the universal language for the biological sciences. Having one language helps scientists and technicians around the world communicate better.

Some species are broken down into varieties (**Figure 15-23**). A variety is a subgroup of plants developed and maintained in production, as opposed to a species that originates in the wild. Variety is a rank within a species. When writing a plant variety name, it is generally capitalized. Examples are Triumph wheat and Ameristand alfalfa.

	Yellow Corn	Petunia
Common Name:	Corn	Petunia
Genus:	*Zea* (the corns)	*Petunia*
Species:	*mays* (dent corns)	*hybridea*
Variety:	Reid's yellow dent	Blue Moon

Figure 15-23 Binomial classifications for yellow corn and petunia.

Agri-Profile Career Area: Botany/Biology/Taxonomy

Knowledge of plant structure and biology is important in these career areas.
Courtesy of the USDA/ARS.

Plant scientists rely on the International Code of Botanical Nomenclature (ICBN) in order to effectively communicate about plants. Taxonomists devote their careers to identifying, classifying, and teaching others about plants. An important and exciting part of their work is that of discovering, studying, and naming new plants in their appropriate places in the classification system. Consider the excitement of discovering plants in faraway lands or even at home that have not been observed before or recognized even by the most knowledgeable specialists.

Botany is the study of plants, and biology is the study of both plants and animals. Knowledge of plant structures is critical to both. Consumers of plants for food, ornamentation, medicine, shade, wood, fuel, fiber, and other uses should have some knowledge of plant structures. Plant structures affect plant nutrition, functions, disease susceptibility, adaptation, and use.

Preparing a career in these fields requires a bachelor's degree for entry level positions and a master's degree for career advancement.

Chapter Review

Student Activities

1. Write the **Key Terms** and their meanings in your notebook.
2. Observe the plants that are commonly grown in your area. Classify them by the type of root system they have.
3. Make a chart of the plants listed in Activity 2. Indicate their common and scientific names, and discuss their responses to drought or their water maintenance requirements.
4. Make a collection of different kinds of leaves. Classify each according to its shape and type of margin.
5. Sketch the parts of roots, stems, and leaves. Label all items.
6. Make a multimedia presentation or a bulletin board showing the major parts of plants.
7. Ask your teacher to provide a microscope and slides of plant tissue. Diagram the plant parts and label the cells and other structures that you see.
8. Record the common and scientific names of plants listed in this chapter and learn the correct spelling of each.
9. List the genus and species for five plants, using the Internet, library, and other resources. Be sure to use proper punctuation.
10. **SAE Connection:** Find out more about the many career opportunities for botanists, or plant biologists. Share your findings with the class. Or invite a local botanist or plant biologist to speak to your class.

Check Your Progress

Can you . . .

- draw and label the major parts of plants?
- describe the major functions of roots, stems, fruits, and leaves?
- draw and label the parts of a typical root, stem, flower, fruit, and leaf?
- explain some of the variations found in the structures of root systems, stems, flowers, fruits, and leaves?
- describe the relationship of plant parts to fruits, nuts, vegetables, and crops?

If you answered "no" to any of these questions, review the chapter for those topics.

National FFA Connection: Plant Identification CDE

Putting your plant identification skills to the test can help you decide if a career related to plant science is right for you.
© Cassandra Lord/Shutterstock.com.

While not all FFA events are national, it's well worthwhile to learn about state-level events. For example, the Texas and Oklahoma FFA Associations sponsor a competitive team event in identification of plants that are selected because they are important habitat components for both wildlife and livestock. Teams of three or four students identify the plants, and demonstrate understanding of growth characteristics and why each is important in animal environments. A Master Plant List is provided. Sixty plants are selected from the list each time the event is conducted, and the top three scores for each team contribute to the team score. This event supports the practical application of principles learned in the classroom. It emphasizes the importance of shared environments that account for the needs of both domestic and wild animals.

Even if you live outside of Texas or Oklahoma, you can gain valuable information by practicing plant identification in your own state. Here's one way to get started. With your classmates, discuss the information you have gathered for Activity 2. Then work as a group to make flash cards that you can use to quiz each other about plants in your area. Be creative! For example, decide as a group whether to take photos or make drawings. Start with plants in your local area or go online to find a list of 10 or more plants that are common to your state. Then decide what kind of detail to include, such as the scientific and common name of the plant, a brief description, soil preference, and habitat.

Chapter 16
Plant Physiology

Objective
To determine how plants make food and to describe the relationships among air, soil, water, and essential plant nutrients for optimum plant growth.

Competencies to Be Developed
After studying this chapter, you should be able to:
- explain how plants make food.
- describe the roles of air, water, light, and media in relation to plant growth.
- trace the movement of minerals, water, and nutrients in plants.
- describe the ways that various plants store food for future use.
- compare the activity in a plant during exposure to light with periods of darkness.
- explain how plants protect themselves from disease, insects, and predators.

Key Terms

physiology
chlorophyll
glucose
light intensity
respiration
turgor
transpiration
osmosis

semipermeable membrane
pore
plant nutrition
primary nutrient
secondary nutrient
macronutrient
micronutrient
ion

anion
cation
precipitate
acid
alkaline
chlorosis

The life of a plant from its beginning to its maturity is a complex process. Many factors influence and directly control how a plant grows and what it produces. Growth, as in all living organisms, occurs by the division of cells and their enlargement as the plant increases in size (**Figure 16-1**). As the plant grows to maturity, cells are produced, divide, grow, and become specialized organs. These specialized organs are stems, leaves, roots, flowers, fruits, and seeds (**Figure 16-2**). The study of how these organs function and the complex chemical processes that permit the plant to live, grow, and reproduce is **physiology**. An understanding of the processes of germination, photosynthesis, respiration, absorption of water and nutrients, translocation, and transpiration enable the agriscience technician to maximize production of plants. The technician who works with interior and other ornamental plants must understand the environment of the plant because it is not in its native habitat.

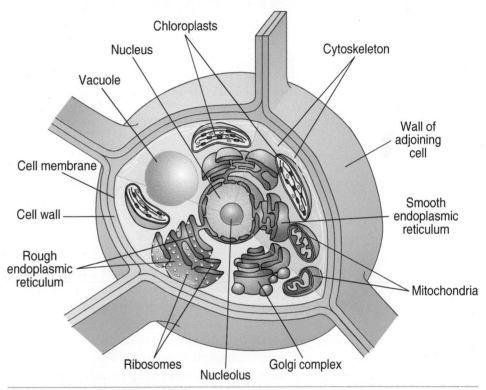

Figure 16-1 Major parts of a plant cell.

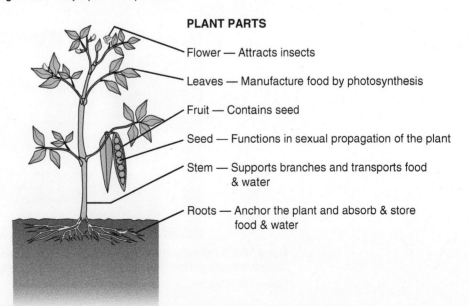

Figure 16-2 Major parts of a typical plant.

Photosynthesis

The most important life-sustaining process is photosynthesis. Without this chemical process, maintenance of life on this planet would not be possible. Plants need carbon dioxide to manufacture food. Animals need oxygen to live. The complex chemical process of photosynthesis permits both plants and animals to live and support each other.

Photosynthesis is a series of processes in which light energy is converted to a simple sugar. Chlorophyll and chloroplasts are also essential in this process. **Chlorophyll** is the green material inside the leaves and stems of the plant. It is the substance that gives the green color to plant leaves. Chloroplasts are small, membrane-bound bodies inside cells that contain the green chlorophyll pigments. The chloroplasts are located in the mesophyll of the leaf. They are the sites of the actual conversion of solar energy (light) into stored energy (simple sugars).

Photosynthesis is the conversion of carbon dioxide and water in the presence of light and chlorophyll into glucose, oxygen, and water. **Glucose** is a simple sugar and contains the building blocks for other nutrients. A simple chemical formula for the process of photosynthesis follows:

$$6CO_2 + 12H_2O \xrightarrow[\text{Light Energy}]{\text{Chlorophyll}} C_6H_{12}O_6 + 6H_2O + 6O_2$$

$$(\text{Carbon Dioxide} + \text{Water} \longrightarrow \text{Sugar} + \text{Water} + \text{Oxygen})$$

The rate at which the food-making process occurs depends on the light intensity, temperature, and concentration of carbon dioxide in the atmosphere. **Light intensity** is the quality of light or the brightness of light. Light must be present with sufficient brightness for the process to be successful. Some plants are able to adapt to various levels of light. Knowledge of the level of light required for plants to grow well is essential, particularly for indoor plant production. Temperature is also an important factor. Photosynthesis occurs best in a temperature range of 65° F to 85° F (18° C to 29° C). Extremes of temperatures slow down or may completely stop the process of photosynthesis. A lack of carbon dioxide also will affect photosynthesis. Carbon dioxide is especially important in the beginning of the process. Under normal outdoor conditions, its availability is not a problem. However, in enclosed conditions such as those found in a greenhouse, carbon dioxide shortage could be a limiting factor (**Figure 16-3**). To correct this problem, a carbon dioxide generator may be used to restore productive levels of carbon dioxide.

Figure 16-3 A technician measures the effect of carbon dioxide enrichment of the atmosphere on the transpiration rate and level of activity of the stomas that are structures in the plant leaves.

Courtesy of USDA/ARS #K3750-12. Photo by Jack Dykinga.

Science Connection: Why Trees Change Color

The pigments that are responsible for the changes in leaf color become visible due to a reduction in plant chlorophyll that occurs as temperatures decline in the fall season.
© Fedorov Oleksiy/Shutterstock.com.

In the fall of every year, trees put on a display of magnificent beauty. Leaves turn from green to orange, red, yellow, brown, crimson, purple, and scarlet. The change is a beautiful reminder that the cold temperatures of winter will soon arrive. Although the change may appear to be art, science is responsible. During the spring and summer months, the leaves of trees appear green. Chlorophyll, the green light-capturing pigment, also contains pigments called carotenoids. They are responsible for the yellow, brown, and orange colors found in bananas, rutabagas, and carrots. Anthocyanins are the red pigments that are found in the fluid within the cytoplasm of a plant cell. During the fall, anthocyanins are much more plentiful in response to sugars being trapped in the leaves. Chlorophyll is much more abundant than the other two pigments in the spring and summer and masks their colors until the fall when the day length gets shorter. As the days become shorter, plants slowly stop producing chlorophyll. This allows the other colors to appear and to produce a colorful show before winter arrives.

Respiration

All living cells carry on the process of respiration. **Respiration** is the process by which living cells (plant or animal) take in oxygen and give off carbon dioxide. Unlike photosynthesis, which occurs only in the light, respiration occurs during periods of light and darkness. It is not easily measured during the day because the presence of photosynthesis will mask or obscure the occurrence of respiration. Respiration is a breaking-down process. It uses the sugars and starches produced by photosynthesis, converting them into energy. The chemical formula for the process of respiration follows:

$$C_6H_{12}O_6 + 6H_2O \rightarrow 6O_2 \longrightarrow 6CO_2 + 12H_2O + \text{Energy}$$

(Glucose)

Figure 16-4 compares photosynthesis and respiration.

Photosynthesis	Respiration
1. Food is produced.	1. Food is used for plant energy.
2. Energy is stored.	2. Energy is released.
3. It occurs in cells that contain chloroplasts.	3. It occurs in all cells.
4. Oxygen is released.	4. Oxygen is used.
5. Water is used.	5. Water is produced.
6. Carbon dioxide is used.	6. Carbon dioxide is produced.
7. It occurs in sunlight.	7. It occurs in dark as well as light.

Figure 16-4 Comparison of the activities that occur during photosynthesis and respiration.

Transpiration

Water saturates all spaces between the cells throughout the plant. About 10 percent of the water that enters from the roots is used in chemical processes and in the plant tissues. Functions of this water include transporting minerals throughout the plant, cooling the plant, moving sugars and plant chemicals, and maintaining turgor pressure. **Turgor** is a swollen or stiffened condition in stems and leaves as a result of the plant cells being filled with liquid. When the plant does not have enough water, turgor pressure is lost, and the plant becomes wilted (**Figure 16-5**).

Figure 16-5 When a plant is unable to obtain enough moisture from the soil to replace moisture that is lost to the atmosphere, it loses turgor pressure and becomes wilted.

Courtesy of DeVere Burton.

The exchange of gases is important to the plant because air is needed for photosynthesis to occur, and water vapor must exit the plant to draw more dissolved nutrients into the roots. Both of these important functions occur through the tiny openings in the leaf called stomas. The stomas are surrounded by specialized cells called guard cells. The guard cells control the size of the opening in the surface of the leaf. The stoma openings become larger or smaller depending on the amount of water available to the plant and surrounding conditions in the environment.

Transpiration is the process by which a plant gives up water vapor to the atmosphere (**Figure 16-6**). Transpiration takes place primarily through the stoma, which open to allow water vapor and air to be exchanged by the leaf. Most plants transpire about 90 percent of the water that enters through the roots.

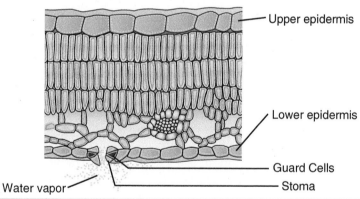

Figure 16-6 The process of transpiration occurs as water vapor and air are exchanged through the stoma.

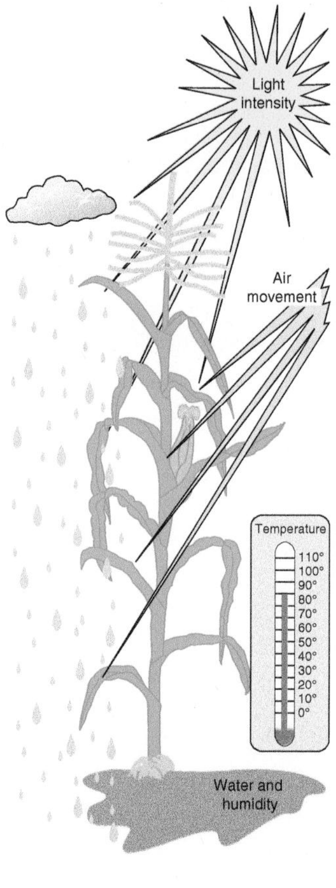

Transpiration is greatly influenced by humidity, temperature, air movement water availability, and light intensity (**Figure 16-7**). As humidity in the air around the plant increases, the rate of transpiration decreases. Conversely, as humidity decreases, the rate of transpiration increases. Increased air movement around the plant increases the rate of transpiration due to evaporation that is accelerated by air movement. Similarly, as temperature increases, the rate of transpiration increases.

Often, during dry weather or when plants are not adequately watered, transpiration causes plants to lose water faster than it can be replaced by the root system. When this occurs, the guard cells will close the stomas in the leaves, thus slowing down the rate of transpiration. This mechanism enables the plant to preserve its water content. If there is water in the soil, the plant may wilt slightly, but it will recover. However, if there is insufficient moisture in the soil, the plant may not be able to recover.

Figure 16-7 Factors that influence transpiration.

Soil

Productive soil provides a natural environment for the root zone. It provides air, water, and nutrients for the plant. Root hairs penetrate the pore spaces in the soil and absorb dissolved nutrients (**Figure 16-8**). **Osmosis** is the process by which water moves from an area of high concentration to an area of low concentration through a semipermeable membrane that separates two solutions (**Figure 16-9**). A **semipermeable membrane** will allow certain substances to pass through, whereas others cannot. The epidermis or outside cell layer of a root hair is a semipermeable membrane. The root allows substances such as water, minerals, and nutrients to enter the plant. The movement of water and dissolved minerals tends to concentrate minerals and nutrients inside the plant at greater levels than

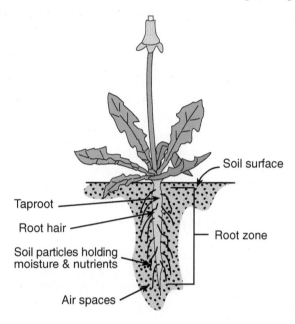

Figure 16-8 Root hairs extend from the main root into the pore spaces in the soil from which they absorb water and dissolved nutrients.

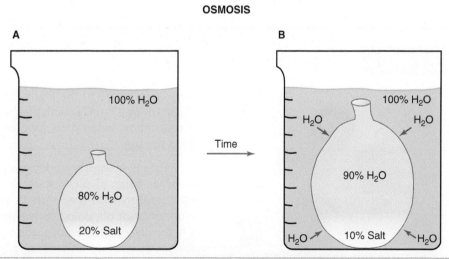

Figure 16-9 During osmosis, water moves from areas of high concentration to low concentration. (A) The water outside the balloon has a greater concentration at 100 percent than the water inside the balloon at 80 percent. (B) Over time, some of the water from outside the balloon has moved inside, crossing the selectively permeable membrane, in an attempt to balance the concentrations.

in the soil. Soil moisture containing a lower concentration of dissolved nutrients and a greater concentration of water is able to move into the root hairs. Once inside the root-hair cell, nutrients can be transported to other parts of the plant as needed. Transpiration creates a vacuum between root hairs and stomas that draws dissolved nutrients upward within the plant.

To allow the root hairs to move within the root zone, the soil must have spaces between the particles of sand, silt, and clay. Such spaces are called **pores**. The role of pore space is to store air, water, and nutrients and to permit root penetration.

Air

The air or atmosphere that surrounds the portion of the plant that is above the ground must supply carbon dioxide as well as oxygen. Generally, this is not a problem when plants grow outdoors. When plants are transplanted into artificial or unnatural environments, consideration must be given to the quality of the air surrounding the plant. In a greenhouse or other enclosed system, the quality of the atmosphere should be monitored. The presence and levels of carbon dioxide and pollutants must be understood for maximum production to occur in a greenhouse environment. With certain crops in greenhouses, such as carnations and roses, the addition of some carbon dioxide might be desirable to increase crop production. In areas where crops are growing near industrial plants or cities, or along major highways, the technician must be aware of the many types of pollutants that might affect production or severely damage the plants.

Water

A consistent supply of pure water is absolutely necessary for growth of plants. Nutrients in the soil must first be dissolved in water before they can be absorbed through the plant roots. Within the plant, water carries the nutrients to the leaves. These nutrients chemically combine with water in the process of photosynthesis. Sugars and other plant foods manufactured in the leaves are then transported throughout the plant by water. Water helps control the temperatures in and around plants through transpiration. Water gives the plant support by maintaining rigidity in the cells. It is important, therefore, that water which is used for plant production be of good quality and in adequate supply.

Agri-Profile Career Area: Plant Physiology

Physiology refers to the many functions that occur inside plants. These include familiar activities such as osmosis, nutrient uptake, translocation, respiration, photosynthesis, food movement, and food storage.

Plant physiologists work closely with technicians and scientists in other fields of plant science. They may be consultants to or collaborators with specialists in agronomy and horticulture. The work of plant physiologists is typically done by college or university faculty, employees of state or national research institutes, or specialists with agriscience corporations developing and selling seeds and plant materials.

Plant physiologists Marcia Holden and Douglas Luster inspect tomato plants for iron deficiency.
Courtesy of USDA/ARS #K4191-5.

Plant Nutrition

Plant nutrition is often confused with plant fertilization. There is a difference. **Plant nutrition** refers to availability and type of basic chemical elements in the plant. Plant fertilization is the process of adding nutrients to the soil or leaves so they become available in the growing environment of the plant (**Figure 16-10**). Before chemicals that are supplied as fertilizer can be taken up and used by plants, they generally undergo changes.

Figure 16-10 Soil and nutrient management specialists do cooperative research to determine the best rates of fertilizer to optimize corn growth.
Courtesy of USDA/ARS #K3694-5.

Essential Nutrients

Eighteen elements are essential for normal plant growth. They are required in various amounts by plants and must be available in the relative proportions needed if the plants are to produce well. Three elements are structural. They are used in huge amounts and are obtained from the atmosphere and water around the plant. They are carbon (C), hydrogen (H), and oxygen (O). Three more elements are used in relatively large amounts, and they are known as **primary nutrients**. These nutrients are nitrogen (N), phosphorus (P), and potassium (K). Other nutrients that are required in smaller amounts include calcium (Ca), magnesium (Mg), and sulfur (S). They are **secondary nutrients**. The primary and secondary nutrients, considered together, are called **macronutrients**, and all of them are obtained from the soil.

An additional nine elements are used in small quantities. They are called **micronutrients** (trace elements). The micronutrients are also obtained from the soil. They are boron (B), copper (Cu), chlorine (Cl), iron (Fe), manganese (Mn), molybdenum (Mo), zinc (Zn), nickel (Ni), and cobalt (Co). For a plant to grow at maximum efficiency, it must have all the essential plant nutrients. The absence of any one of these nutrients will cause the plant to grow poorly or show some signs of poor health.

Remembering the 18 Plant Nutrients

Various schemes have been devised to help you remember the names of the 18 plant nutrients. One technique is to first learn the chemical symbols. Use the symbols to make a logical string of words that are easy to remember. One such string of words that uses the symbols of most of the nutrients is "C. Hopkin's cafe, mighty good."

By remembering this phrase, you can recall the symbols of 10 of the 18 nutrients as follows: C HOPKNS CaFe Mg (carbon, hydrogen, oxygen, phosphorus, potassium, nitrogen, sulfur, calcium, iron, and magnesium). The remaining ones are boron, copper, chlorine, manganese, molybdenum, zinc, nickel, and cobalt. Can you devise a string of words to help you remember the symbols of these micronutrients?

Ions

Plant nutrients are absorbed from the soil-water solution that surrounds the root hairs of the plant. In fact, 98 percent of the nutrients obtained from the soil are absorbed in solution, whereas the other 2 percent are extracted by the root directly from soil particles. Most of the nutrients are absorbed as charged ions. An **ion** is an atom that has an electrical charge.

Negatively charged ions are called **anions**. Positively charged ions are called **cations**. The electrical charges in the soil are paired so that the overall effect in the soil is not changed. These ions compete and interact with each other according to their relative charges. For example, nitrogen, in its nitrate form, has a negative charge and chemical formula NO_3^-. Therefore, nitrates are anions with negative charges.

Conversely, potassium has a positive charge and is an example of a cation (K^+). Potassium nitrate, $K^+NO_3^-$, is a combination of potassium and nitrate consisting of one nitrate ion and one potassium ion. Calcium nitrate, $Ca^{++}(NO_3^-)_2$, has two nitrate ions and one calcium ion because the calcium cation has two positive charges. As you might guess, this could be confusing, but it illustrates the need to understand chemistry to manipulate plant fertility if conditions are not ideal in the natural environment.

The balance of ions is important and must be carefully monitored for good plant growth. Opposite charges attract each other, but ions with similar charges compete for chemical reactions and interactions in the soil-water environment. Some ions are more active than others and might be able to compete better in the soil. Further study would be needed to thoroughly understand why soil tests may indicate the presence of a certain element in sufficient amounts for plant growth, yet the plants may show deficiency symptoms (**Figure 16-11**). Deficiency means a shortage of a given nutrient that is required for a healthy plant. A good example of a nutrient deficiency symptom is blossom-end rot of tomato. This is common in gardens and occurs when there is not enough water to dissolve and carry calcium to the plant in sufficient quantities. The end opposite the stem is called the blossom end. The calcium deficiency produces a tomato that looks good from the top, but, when picked, the bottom end is rotten (**Figure 16-12**).

Soil Acidity and Alkalinity

The chemistry of plant elements in the soil can be affected by pH. Soil pH is a measurement of acidity (sourness) and alkalinity (sweetness) (**Figure 16-13**). Many of the nutrients in soil form complex combinations and are capable of precipitating out of solution, where they are unavailable to the plant. **Precipitate** occurs when a solid separates out of solution. If the soil pH is **acid**, or extremely low, some micronutrients become too soluble and occur in concentrations great enough to harm the plants (**Figure 16-14**).

In contrast, when soil pH is high, in the **alkaline** range, many of the nutrients can be precipitated out of the nutrient solution and not be available to the plants. The pH of soils can be determined with low-cost test kits. Fortunately, soil pH can be corrected by adding lime if the pH needs to be increased or by adding sulfur or gypsum if pH needs to be decreased. Such practices are common because soil pH is often less than ideal for the crop being grown (**Figure 16-15**).

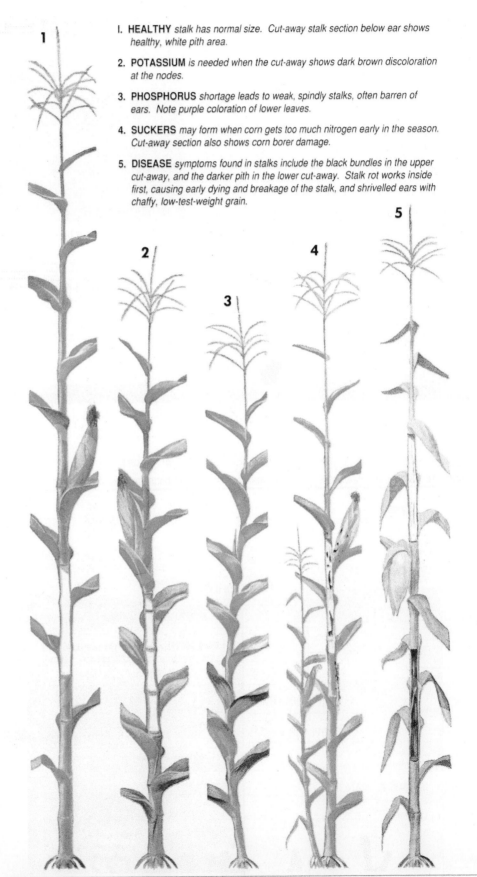

Figure 16-11 Nutrient deficiencies decrease plant health, vigor, and growth. Deficiency of specific nutrients can be determined by observing the plant.

Courtesy of International Plant Nutrition Institute, Norcross, Georgia USA.

Figure 16-12 Calcium deficiency is a cause of blossom end rot in tomatoes.
© PKDAENG/Shutterstock.com.

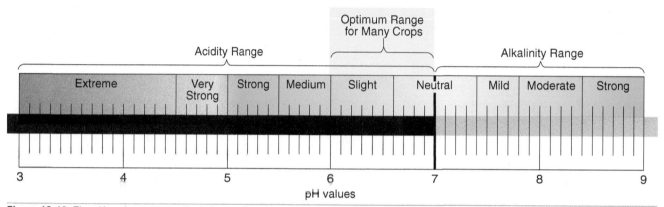

Figure 16-13 The pH scale measures the balance of positive and negative ions in a solution or in the soil.

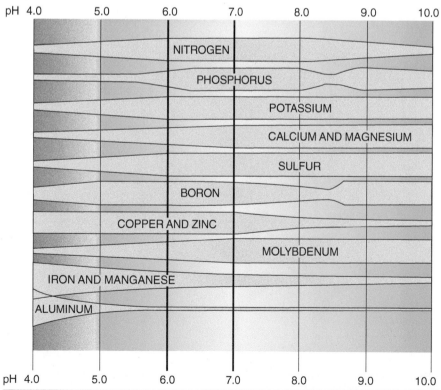

Figure 16-14 Soil pH affects the availability of plant nutrients. A wide section indicates high availability of the nutrient, and a narrow section indicates the nutrient is not available to a plant.

Figure 16-15 Soil scientist Charles Foy compares barley plants grown in soils having different pH levels.
Courtesy of USDA/ARS #K3212-1.

Plant Nutrient Functions

The importance of carbon, hydrogen, and oxygen has already been discussed under the topic of photosynthesis. The other nutrients have very specific functions and must be available in the appropriate form and correct amounts. The effects of plant nutrients may be likened to a chain—the weakest link will determine how much the chain will pull. Similarly, the nutrient in shortest supply will determine the maximum growth and production that can be expected from the plant. Deficiencies and excesses of nutrients cause predictable symptoms in plants (**Figure 16-16**).

Nutrient	Excess	Deficiency
Iron	Rare	Interveinal chlorosis, especially on young growth
Zinc	Might appear as an iron deficiency	Interveinal chlorosis, reduction in leaf size, short internodes
Molybdenum	Not known	Interveinal chlorosis on older leaves; may also affect leaves in the middle of the plant
Boron	A blackening or death of tissue between veins	Failure to set seed; death of tip buds
Copper	Might occur in low pH; will appear as an iron deficiency	New growth small and misshapen, wilted
Manganese	Brown spotting on leaves; reduced growth	Interveinal chlorosis of the leaves and brown spotting; checkered effect possible

Figure 16-16 An excess or deficiency of most nutrients will cause predictable symptoms in plants.

Nitrogen

Nitrogen is present in the atmosphere as a gas. It can be added to the soil as a single nutrient or a combination of nutrients available in commercial fertilizers. Because it exists in nature as a gas, it is easily leached (washed out of the soil). Nitrogen is responsible for the vegetative growth of the plant and its dark green color. When nitrogen is lacking, the deficiency symptoms are reduced growth and yellowing of the leaves. This yellowing is referred to as chlorosis. In contrast, excess nitrogen can cause succulent growth that is dark green, but the plants often become weak and spindly.

Phosphorus

In nature, phosphorus is present as a rock and is not easily leached out of the soil. It is important in the growth of seedlings and young plants, and it helps the plants develop good root systems. Some symptoms of a deficiency of phosphorus are reduced growth, poor root systems, and reduced flowering. Thin stems and browning or purpling of the foliage are also signs of poor phosphorus availability.

Potassium

Potassium is mined as a rock and processed into small particles to make fertilizer, but it can be dissolved in water and leached from the soil. If too much potassium is present, it can contribute to a nitrogen deficiency. A lack of potassium will appear as reduced growth or shortened internodes and sometimes as marginal burn or scorching (brown leaf edges). Dead spots in the leaf and plants that wilt easily are also indications of a potassium deficiency.

Calcium

Calcium is often supplied by adding lime to the soil. It can be leached out and does not move easily throughout the plant. Too much calcium can cause a high pH and reduce the availability of some other elements to the plant. A lack of calcium can stop bud growth and result in death of root tips, cupping of mature leaves, and blossom-end rot of many fruits. Pits on root vegetables are also signs of calcium deficiency.

Magnesium

Magnesium can be added by using high-magnesium lime. It can also be leached from soil. If magnesium is lacking, some reduction of growth and marginal chlorosis can be noticed. In some plants, even interveinal chlorosis can be seen. Cupped leaves and a reduction in seed production are also symptoms of magnesium deficiency. Foliage plants are most likely to be deficient in this nutrient.

Sulfur

Present in the atmosphere as a result of combustion, sulfur is often an impurity in fertilizers carrying other nutrients. As a result, it is rarely deficient. However, if sulfur is deficient, a yellowing of the entire plant may result.

Hot Topics in Agriscience — Better Plants, Better Yields

Agricultural crop production is becoming more efficient all the time. Yields and quality are both increasing. These improvements are partially because of the development of better plants. Scientists are continually seeking ways to improve our food crops. Most of our highly productive crop plants are resistant to one or more diseases or pests. Some of them are able to withstand drought conditions or other environmental conditions that are known to reduce crop production.

Some new plants have been modified to produce food with nutrient levels that are different from those of the parent varieties. For example, some new wheat varieties contain less protein in the grain than the same quantity of grain produced by the parent stock. This is important to the industries that make pasta and crackers. The quality of these products is improved by using low-protein wheat. New plant varieties will always be in demand because the world will always be seeking greater crop yields and higher quality.

New varieties of low-protein wheat are of great benefit to food processors that manufacture pasta products.
© Seregam/Shutterstock.com.

Food Storage

When the plant makes its food through photosynthesis, it often manufactures more than what it needs to maintain itself. This excess is stored in the plant for future use. Such food may be stored in roots, stems, seeds, or fruits.

Roots

The most common type of root that serves as a storage organ is the taproot. Some common examples of plants with extensive storage capacity are sugar beets, carrots, radishes, and turnips. The sugars and carbohydrates are transported down the phloem and into the root cells. They are held there as the root enlarges. Most of the time, this type of plant is a short-term crop that does not take long to mature. This type of root system is easy to dig or harvest.

Stems

The stems of plants usually contain cells that are necessary for support of the foliage or above-ground stems and leaves. Some specialized stems, however, are excellent food storage organs (**Figure 16-17**). The most common specialized stem used for food is the tuber. It is an enlarged portion of a stem containing all the parts of a normal stem. Nodes, internodes, and buds can be identified in tubers. The potato is an example of a tuber.

Corms and bulbs are other examples of specialized stems that contain large amounts of food manufactured by the plant. A rhizome is yet another example; however, its purpose is for propagation.

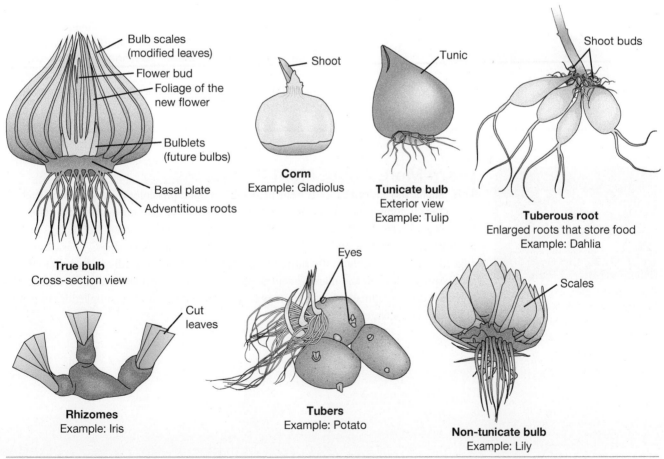

Figure 16-17 Examples of stems that are major food storage organs for the plant.

Seeds

As the ovule of a plant matures, it stores food for the young embryo to start its growth when it germinates. Seeds tend to be high in carbohydrates, fats, and oils. These are concentrated food deposits, and both humans and animals use seeds as major food sources. Humans and animals help the plant by spreading the seed to new locations, increasing the survival probability of the plant population.

Science Connection: Stimulating Photosynthesis

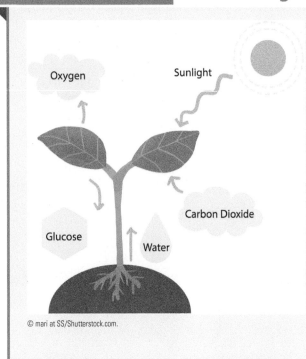

What if the process of photosynthesis could be made more efficient? What if the carbon dioxide entering plants could be modified to stimulate plant growth and production? Plant scientists are demonstrating that each of these concepts is more than just a "pie in the sky" idea. Photosynthesis captures atmospheric carbon by attaching CO_2 to organic sugar molecules using the enzyme RUBISCO as a catalyst. This enzyme is one of the most abundant proteins on Earth. Three molecules of CO_2, in tandem with RUBISCO and other enzymes, generate a single molecule of a three-carbon sugar. This process is called C3 photosynthesis. Two of these three-carbon sugar molecules then join together creating a single molecule of glucose.

Plant scientists at Cornell University developed a process whereby they engineered a RUBISCO enzyme obtained from tobacco plants, putting it inside the bacterium *Escherichia coli*. The modified bacteria provided a positive environment to speed up the process of photosynthesis.

Learn More About Plant Physiology

Figure 16-18 Missouri farmer Bill Holmes and specialists use remote sensing to examine soil fertility variations in Holmes's cropland.
Courtesy of USDA/ARS #K4914-2.

This chapter has addressed some basic principles of plant physiology. Keep in mind that physiology is complex and must be studied in great depth to gain understanding. Plant physiologists typically have master's or doctoral degrees. However, most technicians and scientists in the plant sciences will have some training in plant physiology. A basic knowledge of soils and how plants use nutrients and function in general will greatly help in the successful production and management of plants. Farmers, ranchers, and growers generally have access to sophisticated technology to help them deal with the complexities of plant production on a large scale (**Figure 16-18**).

Chapter Review

Student Activities

1. Write the **Key Terms** and their meanings in your notebook.
2. Use online diagram tools or make a bulletin board showing the cross section of a leaf. Label the various cells and leaf parts. Include the formula for photosynthesis.
3. Write an article for a newspaper or magazine that explains the importance of photosynthesis.
4. Collect plants or pictures to make a display that could be used by others to help identify nutrient deficiencies.
5. SAE Connection: Select a crop that interests you and conduct research to determine the optimum nutrient requirements. Arrange to meet with a grower to interview them and to photograph the crop.
6. Set up a demonstration to explain osmosis.
7. Make a list of all the chemical symbols of plant nutrients. Write a sentence or poem to help you remember them.
8. In your own words, explain the process of photosynthesis. Be sure to include all of the steps in the process.
9. In your own words, explain respiration. Be sure to include all of the conditions that are necessary for the steps to occur.

Check Your Progress

Can you...

- explain how plants make food?
- describe the roles of air, water, light, and media in relation to plant growth?
- trace the movement of minerals, water, and nutrients in plants?
- describe the ways that various plants store food for future use?
- explain why plants change color?
- explain how plants protect themselves from disease, insects, and predators?

If you answered "no" to any of these questions, review the chapter for those topics.

Chapter 17
Plant Reproduction

Objective
To determine the methods used by plants to reproduce themselves and to explore new propagation technology.

Competencies to Be Developed
After studying this chapter, you should be able to:
- distinguish between sexual and asexual reproduction.
- explain the relationship between reproduction and plant improvement.
- draw and label the reproductive parts of flowers and seeds.
- state the primary methods of asexual reproduction and give examples of plants typically propagated by each method.
- explain the procedures used to propagate plants via tissue culture.

Key Terms

propagation
sexual reproduction
asexual reproduction
hybrid vigor
germinate
dormant
imbibition
scarify

cutting
fungicide
rooting hormone
stem tip cutting
stem section cutting
cane cutting
heel cutting
single-eye cutting

double-eye cutting
leaf cutting
leaf petiole cutting
leaf section cutting
split-vein cutting
root cutting
simple layering
tip layering

air layering
grafting
scion
rootstock

graft union
bud grafting
T-budding
tissue culture

aseptic
agar

Plant **propagation**, or reproduction, is the process of increasing the numbers of a species or perpetuating a species. The two types of plant propagation are sexual and asexual. **Sexual reproduction** is the union of an egg (ovule) and sperm (pollen grain), resulting in a seed. Flowering plants reproduce sexually through the process of pollination. Pollination may involve one or two parent plants. **Asexual reproduction** uses a part or parts of only one parent plant. The purpose is to cause the parent plant to produce a duplicate of itself. The new plant is a clone (exact duplication) of its parent. Because this type of reproduction uses the vegetative parts of the plant, namely the stem, root, or leaf, it is known as vegetative propagation.

Sexual propagation has some distinct advantages. It is often less expensive and quicker than asexual propagation. It is the only way to obtain new varieties and to capture hybrid vigor. A hybrid is a plant obtained by crossbreeding. **Hybrid vigor** refers to the tendency of hybrid plants to be stronger and to survive better than plants of a pure variety. Additionally, sexual propagation is a good way to avoid passing on some diseases. In some plants, sexual propagation is the only way they can reproduce.

Asexual propagation has advantages as well (**Figure 17-1**). With some plant species, it is easier and less expensive to obtain the next generation of plants this way. In some species or cultivars, it is the only way they can be propagated.

PLANT PROPAGATION

Sexual Propagation Advantages	Asexual Propagation Advantages
• Less expensive • Many plants can be produced quickly • Crosses result in hybrid vigor • Avoids passing on some diseases	• Less time is required to produce a salable plant • Plants are genetically identical • The only way to reproduce some plant varieties

Figure 17-1 Two methods of plant reproduction, sexual and asexual propagation, are in common use in the plant industry. Each method has distinct advantages over the other.

Sexual Propagation

A seed is made up of the seed coat, endosperm, and embryo (**Figure 17-2**). The seed coat functions as a protector for the seed. Sometimes it is thin and soft, or it may be hard and impervious to water or moisture. The endosperm functions as a food reserve. It will supply the new plant with nourishment for the first few days of life. The embryo is the young plant itself. When a seed is fertilized and matures, it will be dormant. When it is subjected to favorable growth conditions, it will **germinate**, meaning that the seed will sprout and begin to grow.

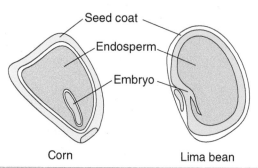

Figure 17-2 Parts of a seed.

Crop production by sexual propagation allows consideration to be given to the type of plant that is needed or the variety that is best adapted to a particular area or purpose. Hybrid plants are developed by cross-pollinating two different varieties. Many varieties in production today are the result of hybridization or crossbreeding. Seeds of hybrid plants cost more than open-pollinated varieties (**Figure 17-3**). However, the increased quality and vigor of the plants generally offset the increased cost of the seed. New varieties are constantly being developed for disease and insect resistance. In addition, some varieties have unusual cultural or product characteristics.

Figure 17-3 Hybrid seeds are produced by pollinating one variety with pollen from a different variety. Note the taller rows of corn that will pollinate the shorter variety. All of the pollen-bearing tassels will be removed from the shorter plants to ensure that cross-pollination occurs.
Courtesy of DeVere Burton.

Generally, seeds collected from plants used for commercial production will not save money in the long run. Seeds from such plants are often small. They may be poorly managed and improperly handled and stored. It is recommended that only certified seed be used. Seed saved from season to season should be stored in a sealed container at 40° F (4.4° C) and at low humidity (**Figure 17-4**).

Figure 17-4 The National Center for Genetic Resources Preservation in Fort Collins, Colorado, preserves seeds of all varieties. Seeds for cold-vault storage (0° F/ −18° C) are packaged in moisture-proof foil bags, while cryogenically stored seeds (−320° F/−196° C) are stored in polyolefin tubes.
Courtesy of USDA/ARS #K-1657-15. Photo by Jack Dykinga.

Germination

When seed is harvested, or collected, it is normally mature and in a **dormant**, or resting, state. To germinate and start to grow, it must be placed in certain favorable conditions (**Figure 17-5**). The four environmental factors that must be right for effective germination are water, air, light, and temperature.

Figure 17-5 Seeds require water, air, light, and favorable temperature to germinate and begin to grow.
© iStockphoto/Dimitris Stephanides.

Science Connection: Alternation of Generation

It takes two chromatids to make up one chromosome. One chromatid comes from the male parent and one from the female parent. A haploid cell has half the normal amount of genetic material because only a single chromatid from one of the parents is present. A diploid cell has two chromatids, one from the female parent and one from the male parent. These two chromatids, when paired, make one chromosome. The number of chromosomes in an organism depends on its species. Genes from both parents are represented in a diploid cell.

Plants sexually reproduce by a process called alternation of generation. In their life cycles, plants change back and forth between producing haploid cells and producing diploid cells. During the haploid generation, the plant produces sperm, eggs, and, in some cases, both. In flowering plants, the sperm is called pollen, and the egg is called an ovum. Once the sperm and the egg are fused, their chromatids become paired. This starts the diploid generation. After fertilization, the membrane around the ovum hardens and becomes a seed with a developing embryo inside. Under the right growing conditions, the seed will develop into a plant. At this point, the alternation of generation is complete. Once the plant matures, the cycle begins all over.

Water

Imbibition is the absorption of water. It is the first step in the germination process. The seed, in its dormant stage, contains little water. The imbibition process allows the seed to fill all its cells with water. If other conditions are favorable, the seed then breaks its dormant stage and germinates.

The quality of the germination medium is important. The medium must not be too wet or too dry. An adequate and continuous supply of water must be available. This is often difficult to control with crops directly seeded in the field. It is much easier to control in crops that are started in the greenhouse for transplanting at a later date.

Agricultural crops should be planted when soil moisture conditions are favorable for germination. However, this is not always possible, particularly when large acreages need to be planted at about the same time. In regions where irrigation is practiced, planting can be timed to follow applications of water to the fields. In many regions, however, germination is dependent on timely precipitation.

A dry period during the germination process will result in the death of the young embryo. Too much water will result in the young seed rotting. In some species, the seed coat is very hard, and water cannot penetrate to the endosperm. In these cases, it is necessary to scarify the seed, or weaken the seed coat to aid in germination.

A common way to **scarify** seed is to nick the seed coat with a knife or a file. Another method is to soak the seeds in concentrated sulfuric acid. This requires special care and experience because sulfuric acid is a dangerous material. Another technique is to place seeds in hot water—180° F to 212° F (82.2° C to 100° C)—and allow them to soak as the water cools. This process takes 12 to 24 hours. Simply placing the seeds in warm, damp containers and letting the seed coat decay over time is yet another way to scarify seeds. Commercial methods for seed scarification are used by suppliers to provide seed that is ready to plant.

Air

Respiration takes place in all viable seeds. Viable seed is alive and capable of germinating. Oxygen is required. Even in nongerminating seeds, a small amount of oxygen is required even though respiration is low. As germination starts, the respiration rate increases.

It is important that the seed be placed in good soil or media that is loose and well drained. If the oxygen supply is limited or reduced during the germination process, germination will be reduced or inhibited.

Light

Some seeds are stimulated to grow by light, while the seed growth of others is inhibited by the presence of light. Many of the agronomic crops do not require light for germination. However, some specialty crops need light to germinate, so it is necessary to have knowledge of their light requirements. Ornamental bedding plants, such as ageratum, begonia, impatiens, and petunia, are more likely to require light for germination. Lettuce also requires light for successful germination. Seeds of these plants are often deposited on the surface of the soil by nature, and the grower should follow the same procedure for successful germination.

Temperature

Heat is another important requirement for germination. The germination rate, or percentage of seed that germinates, is affected by the availability of favorable ambient (surrounding) temperatures. Some seeds will germinate over a wide range of temperatures, whereas others have more narrow limits. In the agronomic crops that are directly seeded in the field, the only way to control heat is to plant when the ground is warm. In specialty crops, particularly bedding plants and perennials, knowledge of a plant's specific heat requirements for germination should result in more efficient production. The germination requirements usually correspond to the USDA Plant Hardiness Zone Map and the recommended planting dates listed in seed catalogs (**Figure 17-6**).

Plant	Time to Seed Before Last Frost (Weeks)	Germination Time (Days)	Germination Temperature Requirements (Degrees F)	Germination Light Requirements Light (L) Dark (D)
Ageratum	8	5–10	70	L
Aster	6	5–10	70	–
Begonia	12+	10–15	70	L
Coleus	8	5–10	65	–
Cucumber	4	5–10	85	–
Eggplant	8	5–10	80	–
Marigold	6	5–10	70	–
Pepper	8	5–10	80	–
Portulaca	10	5–10	70	D
Snapdragon	10	5–10	65	L
Tomato	6	5–10	80	–
Watermelon	4	5–10	85	–
Zinnia	6	5–10	70	–

Figure 17-6 Time, temperature, and light requirements for germination of some common flower and vegetable plants.

Asexual Propagation

As stated previously, asexual propagation uses a vegetative part of a parent plant to create a new plant. The new plant is identical to its parent, which is an advantage. Other advantages are economy and time. The primary methods of asexual propagation are cuttings, layering, division, grafting, and tissue culture.

Cuttings

Herbaceous and woody plants are often propagated by **cuttings** (vegetative plant parts that are used to generate new plants) (**Figure 17-7**). Types of cuttings are named for the parts of the plant from which they are obtained, such as the stem, leaf, or root.

The procedure for taking cuttings is relatively simple. The tools needed consist of a sharp knife or a single-edge razor blade. Sharp equipment will make the job easier and reduce injury to the parent plant. To prevent the possibility of diseases spreading, the cutting tool should be immersed in bleach water made with one part bleach to nine parts water. The tool can also be dipped in rubbing alcohol.

Flowers and flower buds should be removed from all cuttings. This allows the cutting to use its energy and stored food reserves for root formation instead of flower and fruit development.

A rooting hormone containing a fungicide is used to stimulate root development. A **fungicide** is a pesticide that helps prevent diseases. **Rooting hormone** is a chemical that will react with the newly formed cells and encourage the plant to develop roots faster. The proper way to use a rooting hormone is to put a small amount in a separate container and work from that container. This procedure will ensure that the rooting hormone does not become contaminated with disease organisms. Do not put the unused hormone back in the original container.

Cuttings are inserted into a rooting medium such as coarse sand, vermiculite, soil, water, or a mixture of peat and perlite. It is best to use the correct medium for a specific plant to obtain the most efficient production in the shortest possible time. The rooting medium should always be sterile and well drained, but it should also be capable of moisture retention to prevent the medium from drying out. The medium should be moistened before inserting the cuttings. It should then be kept continuously and evenly moist while the cuttings are forming roots and new shoots. This is accomplished through regular and frequent applications of small amounts of water. Stem and leaf cuttings do best in bright, but indirect, light. However, root cuttings are often kept in the dark until new shoots are formed and start to grow.

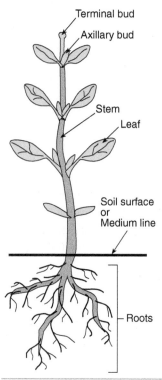

Figure 17-7 Vegetative parts of plants.

Stem Cuttings

For most plants, the most popular method of making cuttings is by stem cutting. On herbaceous plants, stem cuttings may be made almost any time of the year as long as other growing conditions are maintained. However, stem cuttings of many woody plants such as grapes or ornamental bushes are normally taken in the fall or the dormant season or both.

Stem Tip Cuttings

Stem tip cuttings normally include the terminal bud. They are taken from the end of the stem or branch. A piece of stem between 2 and 4 inches long is selected, and the cut is made just below the node, or eye. The lower leaves that would be in contact with the medium are removed. The stem is dipped in the rooting hormone and is gently tapped to remove the excess hormone. The cutting is then inserted into the rooting medium (**Figure 17-8**). The cutting should be inserted deep enough so the plant material will support itself. It is important that at least one node be located below the surface of the growth medium to encourage the growth of new roots from it.

Stem Section Cuttings

Stem section cuttings are prepared by selecting a section of the stem located in the middle or behind the tip cutting. This type of cutting is often used after the tip cuttings are removed from the plant. The cutting should be between 2 and 4 inches long, and the lower leaves should be removed. The cut should be made just beyond a node on both ends. It is then handled as a tip cutting. Make sure that the cutting is positioned with the right end up. The axial buds are always on the tops of the leaves.

Cane Cuttings

Some plants, such as the dumb cane (*Dieffenbachia* sp.), have cane-like stems. These stems are cut into sections that have one or two eyes, or nodes, to make **cane cuttings**. The ends are dusted with activated charcoal or a fungicide. It is best to allow the cane to dry in the open air for 1 or 2 hours. The cutting is then placed in a horizontal position with half of the cane above the surface of the medium. The eyes, or nodes should be facing upward. This type of cutting is usually potted when the roots and new shoots appear (**Figure 17-9**).

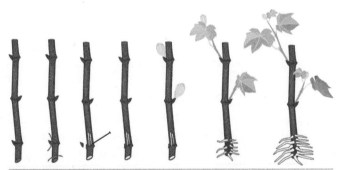

Figure 17-8 Stem tip cuttings include the terminal bud and are 2 to 4 inches in length.
© Kasakova Mariya/Shutterstock.com

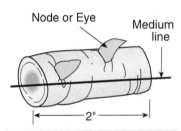

Figure 17-9 A cane cutting is made from the stem of a plant with a cane-like growth structure. One to two adjacent nodes are selected, and the cut is made to include the node or nodes that are desired.

Agri-Profile — Career Area: Plant Breeding/Crop Improvement

The plant breeder crossbreeds plants from various sources develop new varieties with desired characteristics.
Courtesy of USDA/ARS #K5146-16. Photo by Jack Dykinga.

Depending on the uses of the plants, plant breeders' objectives might include developing plants that are faster growing, more beautiful, or better flavored than current available varieties. They also may seek to develop plants that are disease resistant, drought tolerant, insect resistant, wind resistant, or frost tolerant. Much plant breeding occurs in greenhouses, but outdoor plots are often used with new crop varieties. Generally, such research is followed by field trials and seed production, which gives the plant breeder some variety in the settings where work is done.

In 2021, the U.S. Bureau of Labor Statistics (BLS) reported an annual salary range of $39,980 to $126,950 for all soil and plant scientists, which is the occupational category of a plant breeder. A bachelor's degree in plant science or a related field is the first step to becoming a plant breeder. Promotions often require advanced degrees, such as in plant genetics.

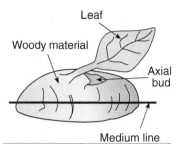

Figure 17-10 To make a heel cutting, a shield-shaped cut is made about halfway through the wood around a leaf and axial bud. The shield is placed horizontally into the growth medium with the leaf and axial bud above the medium line.

Heel Cuttings

Heel cuttings are used with woody-stem plants. A shield-shaped cut is made about halfway through the wood around the leaf and the axial bud. Rooting hormone may be used in the same manner as in the other types of cuttings. The cutting is inserted horizontally into the medium (**Figure 17-10**).

Single-Eye Cuttings

When the plant has alternate leaves, single-eye cuttings are used. The eye refers to the node. The stem is cut about a half inch above and below the same node (**Figure 17-11**). The cutting may be dipped in rooting hormone and then placed either vertically or horizontally in the medium.

Double-Eye Cuttings

When plants have opposite leaves, double-eye cutting is the preferred type of cutting. It is often used when the stock material is limited. A single node is selected, and the stem is cut a half inch above and below the node with a sharp tool. The cutting should be inserted vertically in the soil medium (**Figure 17-12**).

Leaf Cuttings

For many of the indoor herbaceous plants, a leaf cutting will produce plants quickly and efficiently. This type of cutting will not normally work for woody plants.

Leaf Petiole Cuttings

For leaf petiole cuttings, a leaf with a petiole about 0.5 to 1.5 inches long is detached from the plant. The lower end of the petiole is dipped into the rooting medium and is then placed into the medium. Several plants will form at the base of the petiole (**Figure 17-13**). These plants may be removed when they have developed their own roots. The cutting may be left in the medium to form new plants.

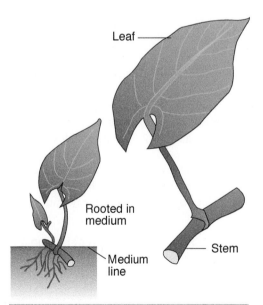

Figure 17-11 A single-eye cutting is made by cutting a section of stem above and below a single node and placing the cutting either vertically or horizontally into the growth medium.

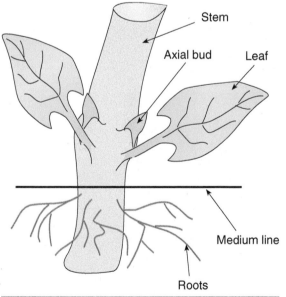

Figure 17-12 A double-eye cutting is used for propagating plants with an opposite leaf pattern. The stem is cut on both sides of a single node, and it is inserted vertically into the growth medium.

Leaf Without Petiole Cuttings

This is used for plants with sessile or petioleless leaves. Place the leaf cutting vertically into the medium. New plants will form at the base of the leaf and may be removed when they have formed their own roots (**Figure 17-14**).

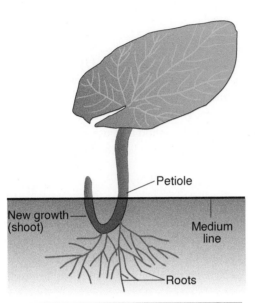

Figure 17-13 A leaf petiole cutting uses a leaf with an attached petiole that is 0.5 to 1.5 inches long. The lower end of the petiole is dipped in rooting compound and planted in the growth medium.

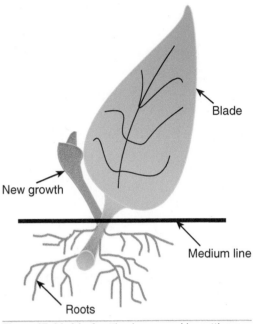

Figure 17-14 A leaf cutting is prepared by cutting a single leaf from a plant, dipping it in rooting hormone, and planting it vertically in the growth medium.

Leaf Section Cuttings

Fibrous-rooted begonias are frequently propagated using **leaf section cuttings**. The begonia leaves are cut into wedges, each containing at least one vein (**Figure 17-15**). The sections are then placed into the medium.

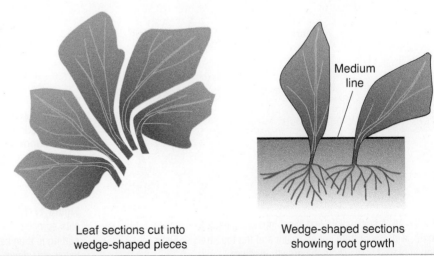

Leaf sections cut into wedge-shaped pieces

Wedge-shaped sections showing root growth

Figure 17-15 A leaf section cutting is obtained by cutting a single leaf into strips. The lower end of each leaf section is dipped in the rooting compound, followed by planting in the growth medium.

New plants will form at the vein that is in contact with the medium. A leaf section cutting is made with the snake plant (*Sansevieria* sp.). The leaf is cut into sections, 2 to 3 inches long. It is a good practice to make the bottom of the cutting on a slant and the top straight. This is done so you can tell the top from the bottom (**Figure 17-16**). The sections are placed in the medium vertically. Roots will form reasonably soon, and new plants will start to appear. These are to be cut off from the cutting as they develop root systems. The original cutting may be left in the medium for more plants to develop.

Split-Vein Cuttings

Split-vein cuttings are often used with large leaf types, such as begonias and other large-leaf plants. With split-vein cuttings, the leaf is removed from the stock plant, and the veins are slit on the lower surface of the leaf (**Figure 17-17**). The cutting is then placed on the rooting medium with the lower side down. It might be necessary to secure the leaf to make it lie flat on the surface. A good method is to use small pieces of wire, bending them like hairpins and pushing them through the leaf to hold it in place. The new plants will form at each slit in the leaf.

Root Cuttings

It is best to use plants that are at least 2 to 3 years old for making **root cuttings**. The cuttings should be made in the dormant season when the roots have a large reserve of carbohydrates. In some species, the root cuttings will develop new shoots, which, in turn, will develop root systems. In others, the root system will be produced before new shoots develop.

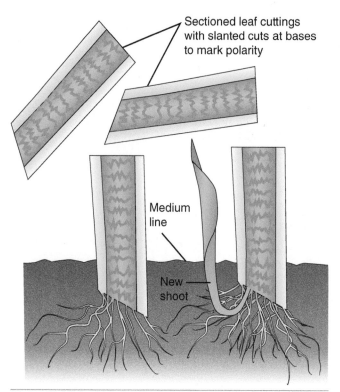

Figure 17-16 A snake plant leaf is cut into sections, 2 to 3 inches in length, with the bottom cut at a slant and the top straight. Each section is placed vertically in the growth medium.

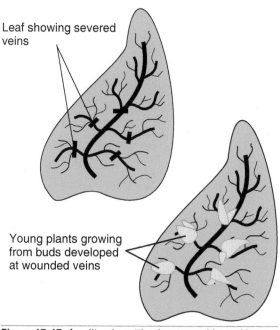

Figure 17-17 A split-vein cutting is prepared by making slits across the veins on the underside of the leaf. The cutting is then secured on the growth medium with the lower side down.

If the plant has large roots, the root section should be 4 to 6 inches long. To distinguish the top from the bottom of the root, make the top cutting a straight cut and the bottom one a slanted cut. This type of cutting should be stored for 2 to 3 weeks in moist peat moss or sand at a temperature of about 40° F (4.4° C). When removed from the storage area, the cutting is inserted into the medium in a vertical position. The slanted cut should be down, and the top straight cut just level with the top of the medium. If the plant typically has small roots, a section 1 to 2 inches long is used. The cutting is placed horizontally a half inch below the surface of the medium.

Layering

In many plants, stems will develop roots in any area that is in contact with the media while still attached to the parent plant. After roots form, shoots develop at the same point. An advantage of this type of vegetative propagation is that the plant does not experience water stress, and sufficient carbohydrates are supplied to the new plant that is forming.

Simple Layering

Simple layering is an easy method that can be used on azaleas, rhododendrons, and other plants. A stem is bent to the ground and is covered with medium. It is advantageous to wound the lower side of the stem to the cambium layer. The last 6 to 10 inches of the stem are left exposed (**Figure 17-18**).

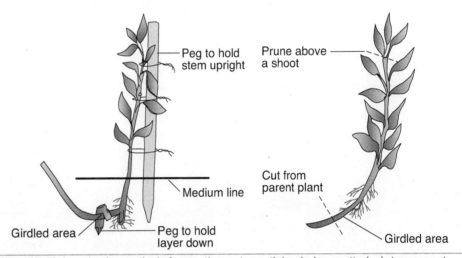

Figure 17-18 Layering is a method of promoting root growth by placing an attached stem or a cut stem beneath the soil partway along its length, with the last 6 to 10 inches of the stem exposed to sunlight.

Tip Layering

Raspberries and blackberries are propagated using **tip layering**. With this method, a hole is made in the medium, and the tip of a shoot is placed in the hole and covered. The tip will start to grow downward and will then turn to grow upward. Roots will form at the bend. When the new tip appears above the medium, a new plant is ready to be transplanted. It will be necessary to separate the new plant from the parent by cutting the stem just before it enters the medium (**Figure 17-19**).

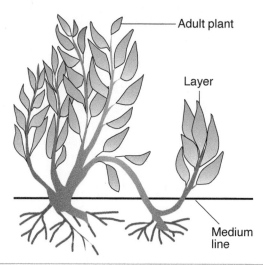

Figure 17-19 Tip layering is accomplished by making a hole in the medium next to a growing plant and burying the tip of a plant shoot in it to promote the growth of new roots and shoots.

Air Layering

Many foliage plants are propagated using **air layering**. Some ornamental trees, such as dogwood, can also be reproduced by this type of layering. The stem is girdled with two cuts about 1 inch apart. The bark is removed. The wound is dusted with a rooting hormone and is surrounded with damp sphagnum moss (**Figure 17-20**). Plastic is wrapped around the moss-packed wound and tied at both ends. In a few weeks, depending on the plant, roots will appear throughout the moss. The stem is cut just below the newly formed root ball, and the ball is planted into a well-drained potting medium.

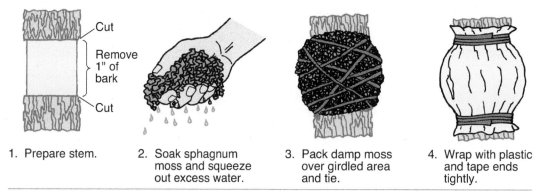

Figure 17-20 Air layering.

Division

Some plants are easily propagated by dividing or separating the main part into smaller parts. If the plant has rooted crowns, these crowns are separated by cutting or pulling them apart. The resulting clumps are planted separately. If the stems are not attached to each other, they are pulled apart. If the crowns are joined together by horizontal stems, they are cut apart with a knife (**Figure 17-21**). It is a good practice to dust the divided plants with a fungicide.

Some plants that grow from bulbs or corms form little bulblets or cormels at their bases. To produce more plants from this type, simply separate the newly formed plant part and place it in a good medium (**Figure 17-22**).

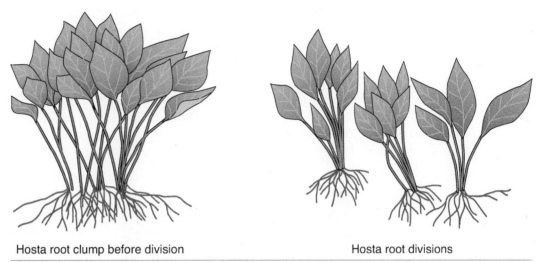

Figure 17-21 Some plants are propagated by separating the rooted crowns of the mature plant.

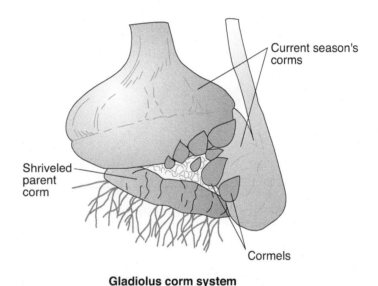

Gladiolus corm system

Figure 17-22 Plants that grow from bulbs or corms may be separated by pulling the small bulblets or cormels from the base of the mature bulb or corm.

Grafting

Grafting is a procedure for joining two plant parts together so they grow as one. This method of asexual propagation is used when plants do not root well as cuttings, or when the root system is inadequate to support the plant for good growth. Grafting will allow the production of some unusual combinations of plants. For instance, several varieties of apples can be grown on one tree. Some nut trees can be induced to grow varieties other than their own. Some unusual foliage plants can also be made by grafting. Finally, dwarf fruit trees are created by grafting regular varieties on dwarfing root stock obtained from related trees with similar, yet different genetics.

The top part of the plant that is to be propagated is called the **scion**. The **rootstock**, or stock, will provide the new plant's root system and will supply the nutrients and water. The **graft union** is where the two parts meet.

To ensure successful grafting, the following conditions are necessary: (1) the scion and the rootstock must be compatible, (2) each must be at the right stage of growth, (3) the cambium layer of each section must meet, and (4) the graft union must be protected from drying out until the wound has healed.

There are many types of grafts. Some common grafts are the whip, or tongue, graft; bark graft; cleft graft; bridge graft; and bud graft (**Figure 17-23**). Each type is used for a special purpose. The most commonly used and the easiest to perform is the bud graft.

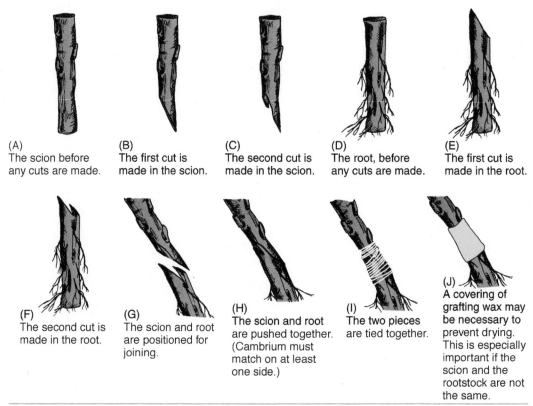

(A) The scion before any cuts are made.
(B) The first cut is made in the scion.
(C) The second cut is made in the scion.
(D) The root, before any cuts are made.
(E) The first cut is made in the root.
(F) The second cut is made in the root.
(G) The scion and root are positioned for joining.
(H) The scion and root are pushed together. (Cambrium must match on at least one side.)
(I) The two pieces are tied together.
(J) A covering of grafting wax may be necessary to prevent drying. This is especially important if the scion and the rootstock are not the same.

Figure 17-23 The process for performing a whip graft.

Bud Grafting

The union of a small piece of bark with a bud and a rootstock is called **bud grafting**. It is most useful when the scion material is in short supply. This type of grafting is faster, and it will make a stronger union than other types of grafting.

T-Budding

T-budding is a popular type of bud graft in which a vertical cut about a quarter-inch long is made on the rootstock. A horizontal cut is made at the top of the vertical cut. The result is a T-shape. The bark is loosened by twisting the point of a knife at the top of the T. A small, shield-shaped piece of the scion, including a bud, bark, and a thin section of the wood, is prepared. The bud is pushed under the loosened bark of the stock plant. The union is wrapped with a piece of rubber band called a budding rubber. The bud remains exposed. After the bud starts to grow, the remainder of the rootstock plant is cut off above the bud graft (**Figure 17-24**).

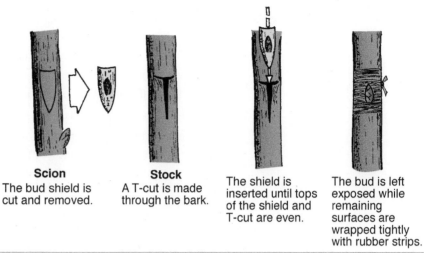

Figure 17-24 The process for performing a T-bud graft.

Scion — The bud shield is cut and removed.
Stock — A T-cut is made through the bark.
The shield is inserted until tops of the shield and T-cut are even.
The bud is left exposed while remaining surfaces are wrapped tightly with rubber strips.

Tissue Culture

A relatively new method of plant propagation is micropropagation, or **tissue culture**. Instead of using a large part of the plant, as in other types of vegetative, or asexual, propagation, a small and actively growing part of the plant is used (**Figure 17-25**). The result is that many new plantlets may be obtained from a section of a leaf. The process must be done in a very clean atmosphere, and it is not successful in the greenhouse or other traditional propagation areas. Tissue culture requires an aseptic environment. **Aseptic** means sterile or free from microorganisms. This kind of environment must be maintained for tissue culture to be successful. Currently, many commercial tissue-culture laboratories are producing a large variety of plants.

Figure 17-25 Only small amounts of tissue are needed to produce an exact clone of a plant when propagating through tissue culture.
Courtesy of USDA/ARS #K4825-1.

Advantages over Traditional Methods

The greatest advantage of tissue-culture propagation is that numerous plants can be propagated from a single disease-free plant. Plants can be propagated more efficiently and economically than with traditional methods of asexual propagation. The main disadvantage is that the work area must be very clean. All of the equipment must be sterile. In commercial production by tissue culture, there is more expense in equipment and facilities than with traditional methods of propagation.

The materials necessary for tissue culture are (1) a clean, sterile area in which to work, (2) clean plant tissue, (3) a multiplication medium, (4) a transplanting medium, (5) sterile glassware, (6) sterile tools, (7) a scalpel, razor, or craft knife, and (8) tweezers.

Preparing Sterile Media

The first step in preparing for tissue culture is to prepare the medium in which the tissue will grow. The medium is called **agar**, which is available commercially from many of the scientific supply houses. The Virginia Cooperative Extension Service offers the following formula and procedure for preparing media for experimentation in tissue culture on a small basis:

1. Use a quart jar to mix the following materials:
   - 1/8 cup of sugar
   - 1 tsp. of soluble, all-purpose fertilizer containing ammonium nitrate.
   - 1/3 tsp. of 35-0-0 soluble fertilizer

- 1 tablet (100 mg.) of inositol (myo-inositol).
- 1/4 of a pulverized tablet containing 1 to 2 mg. of thiamine
- 4 tsp. of fresh coconut milk, the source of cytokinin.
- 3 to 4 grains of a rooting hormone containing 0.1 active ingredient Indole-3-butyric acid (IBA)

2. Fill the jar with purified, distilled, or de-ionized water.
3. Shake the jar to dissolve all materials.
4. Prepare the culture tubes using either test tubes with lids or other suitable glass containers. Fill the culture tubes one quarter of the way with sterile cotton balls. Use one or two per tube. They do not need to be packed tightly. Pour the prepared medium into the culture tubes to just below the top levels of the cotton. Place the lids on loosely.
5. Sterilize the prepared tubes. Sterilization may be done in two ways: (1) heat in a pressure cooker for 30 minutes, or (2) heat in an oven for 4.5 hours at 320° F (160° C). After the culture tubes are removed, place them in a clean area and allow them to cool (**Figure 17-26**). If several days will go by before using all of the tubes, wrap them in small groups in plastic wrap or foil before sterilizing.

Hot Topics in Agriscience: Cloning a Better Potato

Tissue culture is a relatively new method for propagating large numbers of plants from a limited amount of plant material.
© iStockphoto/annedde.

Much of the seed stock that is used to produce potatoes is replaced with better cultivars over a period of a few years. Farmers who produce "seed potatoes" are constantly seeking plants that are resistant to diseases and pests. Individual potato plants that are identified in the field as being superior to other potato plants are selected as parent stock. From the materials obtained from these plants, many new potato plants are cloned. These valuable young plants are initially raised in a greenhouse environment, and the supply of "seed potatoes" is expanded in the field until an adequate supply is available for commercial plantings. The "seed potatoes" are then cut into pieces for planting. A field of potatoes is a perfect example of massive cloning efforts that have been advanced to a commercial scale of operation.

Sterilizing Equipment and Work Areas

The tools and equipment used for tissue culture must also be sterilized. This can be done as the medium is sterilized by placing the tweezers, razor blade scalpel, or craft knife in the pressure cooker or oven. After the initial sterilization, they may be cleaned by dipping them in alcohol before and after each use. The work area must be thoroughly cleaned and sterilized. Wash the area with a disinfectant. Keep a mist bottle filled with a mixture of 50 percent alcohol and sterilized water to spray work areas and tools as well as the hands and arms of the propagator.

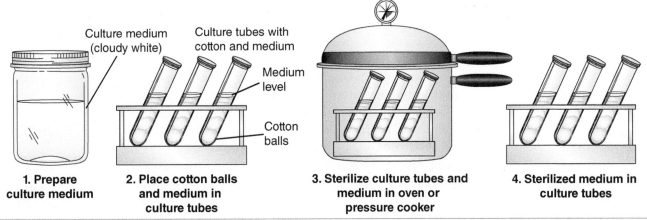

Figure 17-26 Procedures for sterilizing tissue-culturing medium.

1. Prepare culture medium
2. Place cotton balls and medium in culture tubes
3. Sterilize culture tubes and medium in oven or pressure cooker
4. Sterilized medium in culture tubes

Preparing and Placing Plant Tissue in the Culture Tube

After the growing medium is properly prepared and cooled and the work area properly cleaned, the next step is to prepare the plant tissue. Various parts of the growing plant may be tissue cultured. For the production of vigorous plantlets, use only actively growing plant parts. With some species of plants, only a small, quarter-inch-square section of the leaf is used, whereas for others, a half inch of the shoot tip is used. With ferns, a quarter inch of the tip of the rhizome is used.

Remove the part of the plant to be used and discard the excess plant material. Submerge the plant part in a solution of one part commercial bleach and nine parts water for about 10 minutes. This will disinfect the plant tissue. Remove the tissue with sterile tweezers and rinse the material in sterile water. When the plant part has been disinfected in the bleach solution, it can be handled only with sterile tweezers and must not touch any nonsterile surface.

When the plant material has been disinfected and rinsed, remove any damaged tissue with a sterilized scalpel or razor blade. Remove the lid from a properly prepared culture tube or jar and place the plant material on the agar. Take care that the plant material is not completely submerged (**Figure 17-27**). Recap quickly to avoid contamination from the air. It is best if the material is placed in front of the propagator so that work is not done over the uncapped culture tube.

Figure 17-27 Plant material in culture tube.
© iStockphoto/Sandralise.

Storing Tissue Cultures

After all plant material has been cultured, put the cultures in an evenly warmed (70° F–75° F/21.1° C–23.9° C) and well-lighted area. The plant tissue will *not* do well in direct sunlight. If any contamination has occurred, it will be evident in 48 to 96 hours as mold or rotting on the medium. If contamination occurs, remove the contaminated tubes, and then wash them for reuse.

When the plantlets have grown to a satisfactory size, take them out of the culture tubes and transplant them into a good growing medium. Remember, the plantlets are fragile, so handle them carefully. As each plant is removed from its culture tube, wash the plant thoroughly and transplant it into its own culture container (**Figure 17-28**).

When the plant is well established in its own culture container with viable roots and top growth, transplant the plant again into a suitable potting mix. Place the plant inside a protected area with high humidity. The plants are coming out of a well-protected environment with plenty of humidity and light. After they have adapted to the pot and are growing well, they may be treated like any other growing plant. This process will take about 3 to 6 weeks from the beginning to a successfully growing plant.

Figure 17-28 A young peach tree moved from the original culture tube is allowed to grow in sterile medium in a sterile container.
Courtesy of USDA/ARS.

Laboratory: Cloning of African Violets

Plant cloning by tissue culture is one of the most widely used biotechnologies. Most potatoes and many houseplants are propagated by cloning. Cloning generates multiple, genetically identical offspring from the nonsexual tissues of a parent plant.

In theory, cloning is simple: Cut a leaf off a plant, disinfect it, cut it into fragments, and then plant the fragments in nutrient agar. This may take 30 minutes.

In practice, contaminants from the air, hands, and tools quickly take over. Instead of healthy clones, you get colorful molds and bacteria. You can minimize contamination by using a simple hood and aseptic handling techniques.

Cloning African violets in the classroom is a long-term project, but it can be done within a few class periods (**Figure 17-29**). The first stage takes about 30 minutes. The violets can be transplanted to a mini-greenhouse 6 to 8 weeks later. In another few weeks, the plant will be ready for repotting.

Figure 17-29 African violets can be readily propagated through tissue culture.
© Anne Kitzman/Shutterstock.com.

IMPORTANT REMINDER!
IT IS IMPORTANT THAT ALL WORK AND THE TRANSFER OF MATERIALS BE DONE QUICKLY AND IN A CLEAN ENVIRONMENT. Scrub all areas with disinfectant and clean all tools with a disinfectant solution. Any contamination may lead to unsuccessful work. Bacteria and fungi will grow in unclean culture tubes and will overtake the new plant growth.
Courtesy of USDA/ARS #K3279-11.

Aseptic Technique

Aseptic handling is critical for successful plant tissue culture. The culture vessel is a battleground between rapidly growing microbes and slowly regenerating plant fragments. Five simple techniques of aseptic handling will minimize transfer of contaminants, and thus growth of molds and bacteria:

1. Wash hands thoroughly and scrub nails using regular soap, and then dry hands with a paper towel. Do not touch your face or other objects or put your hands in your pockets. Such practices put contaminants back on your hands and can re-contaminate your plants, equipment, work area, or medium.

2. Keep your hands from passing over open vessels.
3. Touch vessels far from the rim, neck, and similar areas. Keep the caps on when not in use.
4. Grasp tools as far from the working ends as possible.
5. Use sterile materials.

Practice Activities

1. Make thumbprints on bacteriologic nutrient agar plates before and after handwashing. Incubate overnight in a warm (not greater than 99° F/37° C) place. Bacterial colonies will appear on both plates. These bacteria are normal for humans but will hinder cloning.
2. Study violet leaf fragments using a hand lens or dissecting microscope. Note the hairs protruding from the leaf's upper surface. Now look at your fingers. Note the many ridges. Where do contaminants hide in each case? How does washing the hand and leaf change how each looks?
3. Practice aseptic handling of tools. How do you pass scissors at home? How would you do this aseptically? Why might a scalpel be better for cloning work than a single-edge razor when both cut just fine?
4. Conduct a dry run of the cloning procedure using a spinach leaf. A spinach leaf bruises as easily as a violet leaf, so it readily shows how gently it has been handled.

A Hood for Cloning

Materials:
   60-by-80-inch sheet of 2-, 3-, or 4-mil clear plastic (painter's tarp), bulldog clips (2 to 3 inches in length), support frame for hanging file folders

Assembly:

1. Place a file folder frame on the table with its arms facing you.
2. Fold the plastic sheet so it is two layers thick.
3. Drape the folded sheet with the fold line in front, so it overhangs the arms by about 2 inches.
4. Clamp the sheeting to the top of the arms to form a flat roof.
5. Straighten the sheeting to minimize creases.
6. Spray the inside of the hood and the work surface with 70 percent ethanol. Dry only the work surface, not the plastic.

The hood will look like a lean-to with a short curtain valance in front. This fits well on a student's desk. The overhanging plastic on the sides and back creates a larger work area than just the frame.

When working, stand over the hood. Do not breathe onto the cloning materials or work area.

Materials Preparation

To sterilize materials, autoclave for 15 minutes at 15 to 21 lbs. of pressure. Open only in a hood that has been surface sterilized with ethanol.

1. Mix and sterilize plant nutrient agar. One type is Murashige African violet/gloxinia multiplication medium, available from companies that supply biological products. One pack makes 30 to 40 plates. Stir one pack into 1 liter distilled water. Add 30 g. sucrose and 15 g. agar. The agar will not dissolve until it is heated. Loosely cap the flask with aluminum foil and autoclave or otherwise heat for 15 minutes. Cool in hot tap water of about 122° F (50° C). Pour 25 to 30 mL agar solution into each plastic petri dish (20 × 100 mm size). Allow gel to set up at room temperature, and then store in a refrigerator.
2. Sterilize a 100-mL beaker to hold disinfected leaves. Cap with a 4-inch square of heavy-duty aluminum foil.

3. Sterilize 150 mL tap water in a foil-capped, 250-mL Erlenmeyer flask.
4. Wrap the glass petri plate (cover and bottom assembled) in a double layer of white T-shirt rag, and then autoclave. The inside of the cover will be used as the cutting surface, and the rim of the bottom will be used to support the tools as they drain.
5. Assemble but do not sterilize the following:
   - 500-mL beaker for waste liquid
   - 20 × 150-mm test tube filled to the brim with 70 percent ethanol (support in a 250-mL Erlenmeyer flask)
   - Curved 8-inch forceps and a 6-inch scalpel handle with #11 blade
   - 200-mL beaker containing 100 mL of 70 percent ethanol to dip the leaf
   - Single-edge razor to cut the leaf from the plant
   - 100-mL beaker to hold the leaf during disinfection
6. Prepare the disinfectant. Mix a 20 percent solution of liquid chlorine bleach and add one drop dish soap per 500 mL. Swirl gently to mix; too many bubbles will inhibit wetting of the leaf surface. (Remember those little hairs.)
7. Soak the forceps and scalpel in the ethanol tube for at least 5 minutes before use. Do not store in the ethanol because the blade will rust.
8. Place the following in the hood:
   - Plant nutrient agar plate
   - Sterile H_2O flask
   - Sterile beaker
   - Sterile glass petri dish
   - 500-mL waste beaker
   - Test tube of ethanol
   - Forceps and scalpel
   - 200-mL beaker with ethanol

Cloning Procedure

As you work through the cloning procedure, review **Figure 17-30** for illustrations related to specific steps in the process.

1. Cut a young leaf so the petiole (stem) remains attached.
2. Put the leaf and petiole in the nonsterile beaker with disinfectant to remove dirt, mites, or other vermin. Leave it there for 10 minutes, but swirl it occasionally.
3. Work under the plastic cover, using the alcohol-soaked forceps to transfer the leaf, by the petiole, to the small sterile beaker. Pour the disinfectant into the waste beaker and remove the nonsterile beaker.
4. Rinse the leaf with 50 mL sterile water. Swirl and pour the water into the waste beaker. Gently hold the leaf in the small beaker with the forceps.
5. Wash again.
6. Disinfect the leaf by dipping the leaf in the ethanol leaf soak, count to 10, and remove the leaf. Meanwhile, open the glass petri plate. Use the bottom plate to drain the forceps and blade. (Resoak the forceps before cutting.)
7. Cut the leaf into fragments. Use the lid of the glass petri plate as the operating table.
   a. Cut off the petiole. Do not plant it.
   b. Cut down the midrib firmly.
   c. Cut across each half into four pieces, being sure to cut through a branching vein.
8. Plant each fragment in the plant nutrient agar so each piece is in, not just on, the agar. Plant in a spoke arrangement to minimize spread of contaminants.

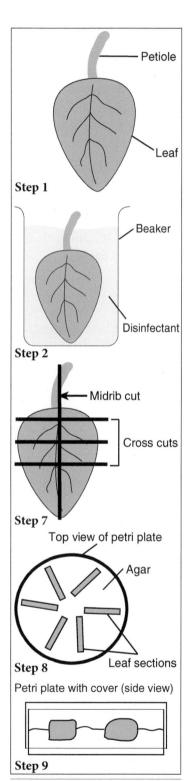

Figure 17-30 Cloning Procedure: Illustrated steps.

9. Put the lid back on and cover the plastic petri plate with plastic cling wrap to keep moisture in.
10. Store the plate in the dark for about one week, and then place it where the fragments have a daily light cycle and 68° F to 77° F (20° C to 25° C) temperatures.
11. Observe weekly. If part of a plate becomes contaminated, transfer healthy fragments to a fresh plate. Even with a commercial hood with sterile air, only about 50 percent of the plates remain completely free of contaminants.
12. Plantlets will appear on some fragments after about 8 weeks. When plantlet leaves grow to about 0.5 cm, aseptically remove the fragment. Separate its plantlets and return the plate to incubate. Gently cut the plantlets, being sure to have some root and some shoot. Place the individual plantlets in growing medium to develop under aseptic conditions.
13. After about a month, the plants should be large enough to transplant into a pot in sterile potting medium.

Science Connection — Breeding Pest-Resistant Plants

Plant geneticist Keith Schertz examines grain sorghum bred for tropical climates. Bags prevent the sorghum flowers from cross-pollinating, so the plant breeder can control which plants provide the male pollen to fertilize the female part of any given plant.
Courtesy of USDA/ARS #K2492-13

The plants we use today for food, clothing, fiber, and decoration are quite different from those found in the wild. Domestic plants or plants grown for a specific use have generally been selected or bred to survive better, grow faster, look different, or in some way perform differently from their ancestors in the wild. However, it is becoming increasingly apparent to plant breeders that we must have wild plants that are not closely related to our favorite domestic species to inject new characteristics into our favored domestic plants.

Pest resistance is an area that requires a continuous reserve of foreign genetic sources. This is to be expected because the very pests that we breed plants to resist are constantly adapting to our plants through survival of the fittest among their kind. Insects and disease-causing pathogens have an amazing capacity to adapt to and eventually break crop resistance. Resistant varieties usually become obsolete in 3 to 10 years. It generally takes 8 to 11 years to breed a new variety to resist the changing individuals of a given pest. Therefore, plant breeding programs must function on a continuous basis to maintain our current capability to feed, clothe, and otherwise supply the needs of the population.

U.S. farmers grow more than 200 varieties of wheat, 85 varieties of cotton, 200 varieties of soybeans, and many varieties of fruit, vegetables, and ornamental plants. Disease and insect resistance must be bred into each of these varieties. Many of them thrive only in specific, limited growing areas, such as one part of one state. Keeping up with known and persistent pests is a relatively manageable process, as long as our state and federal experiment stations are reasonably well funded to research evolving problems. However, the sudden appearance or introduction of a pest without resistant varieties or natural biological enemies can be devastating.

For instance, the Russian wheat aphid first appeared in the United States in 1986 and has cost wheat growers hundreds of millions of dollars since its arrival. U.S. wheat varieties have little resistance to the Russian wheat aphid. Therefore, chemical insecticides have to be used until either resistance can be bred into our domestic wheat varieties or biocontrol agents can be found or developed. Entomologist Robert Burton (1936-1993) of the Plant Science Laboratory in Stillwater, Oklahoma, believed the answer would be found by introducing selected genes from wheat varieties from Southwest Asia. Burton earned the respect of scientists throughout the world for his work in developing Russian aphid–resistant wheat and barley plants.

Chapter Review

Student Activities

1. Write the **Key Terms** and their meanings in your notebook.
2. List the crops that are grown commercially in your area. Visit a site or sites where plants are propagated and ask the grower to discuss the propagation methods used.
3. List the prevalent crops that are grown from seed in your area and the popular varieties of each. Study a seed catalog and determine the requirements for germination of each variety and why the varieties are popular in your community.
4. Plant a seed in a jar filled with medium. Place the seed near the edge of the jar so that you can see what is happening. Keep a journal of daily observations as the seed or plant changes. You may want to do this with several seeds in different jars with different amounts of water, light, air, or temperature in each jar. Note your observations and the conditions regarding each seed. Write your conclusions about what is best for maximum germination results.
5. Experiment with different kinds of plants and various kinds of cuttings. Keep a journal to determine the best type of cutting for specific plants. Keep notes on different media, temperatures, light, and rooting hormones.
6. Practice making each type of graft discussed in this chapter under the supervision of your instructor.
7. Research additional grafting methods.
8. Conduct an experiment with the tissue-culture method of propagation. Keep notes on the different kinds of plants used, the time required for root formation, and the observations regarding the benefits of propagation by tissue culture.
9. Prepare a statement about the propagation methods best suited to your purpose.
10. **SAE Connection:** With a group of classmates, develop a plan for a garden for your school or community. Sketch a garden design. Describe what the garden will grow, how space it will require, and how the plants will be used. For example, would you sell the plants to provide fresh produce for the school cafeteria, to raise money for a class project, to donate to a community organization, or for some other purpose? What would be needed to establish and maintain the garden? Provide as much relevant detail as you can. Then display or present your plan to the rest of the class.

Check Your Progress

Can you...

- distinguish between sexual and asexual reproduction?
- explain the relationship between reproduction and plant improvement?
- draw and label the reproductive parts of flowers and seeds?
- state the primary methods of asexual reproduction and give examples of plants typically propagated by each method?
- explain the procedures used to propagate plants via tissue culture?

If you answered "no" to any of these questions, review the chapter for those topics.

National FFA Connection: Nursery/Landscape Career Development Event

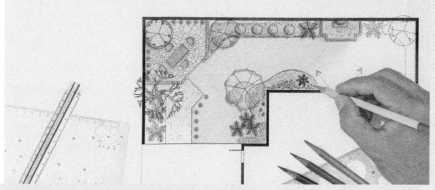

Developing skills like landscape design offers lifelong benefits.
© Toa55/Shutterstock.com

The National FFA Organization offers special events develop your skills in nursery and landscaping while learning more about the many related career paths. The Nursery/Landscape Career Development Event is designed to test and improve your skills in maintaining landscape plants, evaluation equipment and services, and performing landscape design. Find out more about by visiting the FFA website and searching the term "Nursery/Landscape Career Development Event." You will find a video that introduces the event, as well as several other related career pathway resources.

Inquiry Activity: Plant Science: Tissue Culture

Courtesy of USDA.

After watching the *Plant Science: Tissue Culture* video available on MindTap, answer the following questions:

1. What is tissue culture?
2. What is another name for tissue culture?
3. What is the advantage of tissue culture?
4. What kind of environment does tissue culture require?
5. Keeping the answer to question 4 in mind, watch the video again, this time observing the worker at the beginning of the video and then the same worker at and after the 00:28 second mark. What could the worker do to ensure that the environment *consistently* meets the requirement for tissue culture?

Unit 6

Crop Science

The growing population of Planet Earth has created a demand for food that is increasing every day. Meanwhile, the land area from which our food supply is obtained is diminishing. These two events could evolve into a problem of major proportions.

Thirty percent of the surface area of the planet is land, but only 11 percent of this resource is used to produce food. The remainder is occupied by deserts, mountains, industrial sites, and housing. Each year, our cities take in more of our most productive land, and it is unlikely that any of it will be reclaimed to raise food.

One possible solution is to focus on different resources and cultural practices to produce life-sustaining food. For example, the oceans and inland seas are resources where the plants that contribute to our supply of food could be produced. Some cultures already consume large amounts of seaweed, but most of the plants we depend on are not adapted to saltwater environments. Discovering ways to utilize saltwater for farming may be one of the most significant breakthroughs of our generation. Perhaps it will have a greater impact on producing an adequate food supply than the new technologies such as drones, satellites, precision farming, and autonomous farm machinery.

We have long described saltwater as a "soup," containing many nutrients that contribute to successful plant growth. Bridging the technology gap will require some massive engineering accomplishments if we are to shift a significant amount of our food production from land-based to seawater-based resources. Large floating structures will be needed to raise crops cultured in ocean waters. Major genetic engineering efforts will be required to develop crops that tolerate and even thrive using saltwater as a source of hydration and nutrients. Perhaps the distribution problems that have surfaced during the COVID pandemic may drive a sustained effort to develop new food sources around the world.

Progress is evident in developing salt-tolerant crops. Agrisea, a Canadian company, has modified genes in rice plants to tolerate one-third the salinity of seawater. As scientists for the company proceed with other gene sequences, they expect to have plants that can grow in the ocean without pesticides or fertilizer. The company is also working to improve salt tolerance in corn, soybeans, and kale. Other promising crops from ocean environments include sea asparagus, some forms of algae, and several varieties of edible seaweed, among others. The government of the Netherlands has reported successful trials raising vegetables on marginal land, irrigated with seawater. Crops that responded best included potatoes, carrots, red onions, broccoli, and white cabbage.

As cities require more of our prime cropland for expansion, and the demand on the supply of clean water for culinary use rises, the need for productive crop enterprises and new water sources will become critical. Perhaps the future will move some of our food production enterprises offshore where water and nutrients are abundant. We may also develop desert land into productive farms using irrigation water reclaimed from the oceans. After all, we live on the "Water Planet"!

Chapter 18
Home Gardening

Objective
To plan, plant, and manage a home garden.

Competencies to Be Developed
After studying this chapter, you should be able to:
- analyze family needs for homegrown fruits, vegetables, and flowers.
- determine the best location for a garden.
- plan a garden to meet family needs.
- establish perennial garden crops.
- prepare soil and plant annual garden crops.
- list recommended cultural practices for selected garden crops.
- protect the garden from excessive damage caused by drought and pests.
- harvest and store garden produce.
- describe the use of cold frames, hotbeds, and greenhouses for home production.

Key Terms
seasonal
loam
clod
furrow
climate
annual
biennial
perennial
cultivation
herbicide
cold frame
hotbed

Figure 18-1 Some people seem to have a natural talent for gardening. However, plant growth and the procedures for obtaining best results are based on scientific principles.
© alter-ego/Shutterstock.com.

Gardening is an activity that can be enjoyed by all members of the family. It provides fresh fruits, vegetables, and flowers for immediate use or to sell for profit. Gardening is an art and a science. It is demanding of the gardener, both in skill and creativity. A garden is alive and changing every day, presenting new challenges to the gardener (**Figure 18-1**). If gardening is to be profitable, the operator must read extensively, get help occasionally with disease and insect problems, and perform garden chores in a timely fashion. For many people, the garden is a real source of satisfaction and creates a feeling of self-sufficiency and achievement.

Analyzing a Family's Gardening Needs

First, one must decide what vegetables and flowers the family likes. It would not make sense to plant sweet potatoes, green beans, and marigolds if no one in the family cared for these vegetables and flowers. The home gardener can provide a seasonal and continuous variety of vegetables and flowers. Seasonal pertains to a certain season of the year. There are garden fruits and vegetables that are seasonal; others are available and productive despite the subtle changes that occur during the spring, summer, fall, and winter (southern latitudes) gardening seasons. Fresh vegetables are important to everyone's diet. Those who are fortunate to have access to a garden plot would do well to plan to have plenty of produce available during the growing season, with some to store for future use. Vegetables should be selected that meet the family needs, both in yield and quality (**Figure 18-2**).

Figure 18-2 The garden plan should provide for high-yielding fruits, vegetables, and flowers that meet the needs and preferences of the family.
Courtesy of USDA/ARS #K2282-7. Photo by Tim McCabe.

Agri-Profile   Community Gardens

A popular idea that has emerged in urban areas is to establish community gardens. A plot of land is prepared by measuring it into moderate-sized plots with clearly marked boundaries. Each plot is then rented to an interested person who enjoys gardening but does not have access to a garden plot. The plot owner establishes the rules that each gardener must follow. They might include weeding, observing established work times, taking water turns, and respecting the privacy of other garden plots, among others.

Some community gardens are planted and tended together with no personal ownership of the produce. At harvest time, all of the produce is divided among the shareholders who paid the membership fee and helped do their share of the work. Whatever the arrangement, community gardens are appreciated by those who enjoy working with others and eating fresh produce.

When it has been decided what vegetables and flowers the family likes, it is time to determine how much ground will be needed. It is better to have a small, well-cared-for garden than to have a garden that goes to waste because it becomes too much to handle. A good rule of thumb for four grown people is to start with a plot 10 feet wide and 26 feet long, or 260 square feet.

The Garden Plan

A prospective gardener should make a sketch on paper detailing the amount and placement of the various crops. At this stage, it is important to consider successive plantings. These crops follow each other in the season so that the ground is occupied throughout the growing season. Fall crops can follow spring and summer crops in many areas. This can be done easily by planting perennial crops and different varieties of specific annual crops. Allow adequate space between the rows for cultivation.

Locating the Garden

Depending on where you live—in the city, suburbs, or a rural area—the location of the garden is an important consideration. The garden should be convenient to the house. It should also be accessible to a water supply; be on loamy, well-drained soil; be in a sunny area; and be visible from the home, if possible.

One must visualize where to locate the garden. What happens to the proposed spot when there is a heavy rain? What happens when it is very dry? From the chosen spot, look up to determine whether trees or branches will cause problems by excessively shading the garden. When planting flower beds around the house, remember that along the south and west sides, the heat will be reflected onto these beds, so they may require extra water. Select the best site you can for both vegetable and flower gardens.

Preparing the Soil

Conditioning the Soil

Garden soil should be loose and well drained. The ideal soil type should be granular—like coffee grounds—so that water will soak in rapidly. **Loam** is granular soil with a balance of sand, silt, and clay particles. However, the soil also needs to contain enough organic matter to retain water within the root zone. Few soils are originally found this way. Soil-building practices can improve the soil over time, and the gardener must prepare the soil to achieve the best results.

If the proposed site has not been previously used as a garden, it is a good idea to add organic matter and plow, spade, or rototill it into the soil. The decayed organic matter and the freezing and thawing of the soil during the winter months will help improve the soil's physical condition. The application of fine organic matter in the spring can also improve soil condition and fertility.

Materials such as composted leaves and grasses, peat moss, composted sawdust, and sterilized, weed-free manure are good soil conditioners. Peat moss is a soil conditioner made from sphagnum moss. A good rate of application for organic matter is 1 lb. dry material per square foot of surface area.

Preparing to Plant

If the selected site has never been planted, it is best to remove the sod before tilling the soil; then spread the organic matter over the soil. When moisture conditions are favorable, turn the soil with a shovel, spade, plow, or rototiller (**Figure 18-3**). When turning the soil, it is important to break up all clods. A **clod** is a lump or mass of soil.

Figure 18-3 Soil preparation is of gardening. Soil is tilled to kill weeds, to mix materials into the soil, and to break up large chunks of soil into a granular texture.
© Savanevich Viktar/Shutterstock.com.

After the spading (turning of the soil) is completed, make the planting beds. A good procedure is to heap the soil to make raised rows with **furrows** between them. Another method is to prepare raised beds that are 4 to 8 feet wide with furrows or sunken walkways between them (**Figure 18-4**). These can deliver irrigation water or drain off excess rain.

Next, level the raised beds with a rake, but do not push the soil back into the furrows. Then, walk over the beds or tamp them with the head of a garden rake to firm them and to eliminate air pockets (**Figure 18-5**). Finally, use the hoe and the back of the garden rake to push and pull the soil to level it and to break any remaining clods into a fine, granular texture (**Figure 18-6**). The soil is now ready for planting.

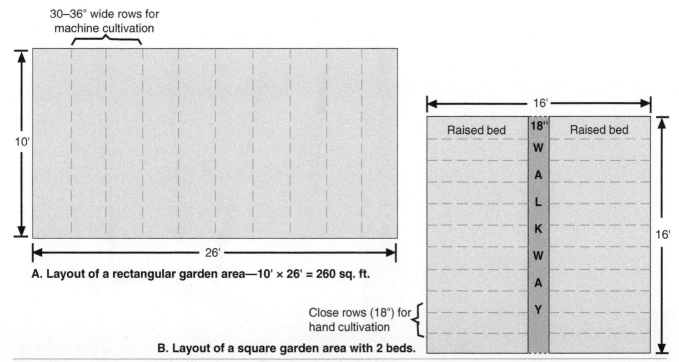

Figure 18-4 Layouts for rectangular (A) and square (B) gardens for a family of four.

Figure 18-5 Raised beds in a garden allow the soil to warm up earlier in the spring (A); they also drain away excess moisture (B).
© iStockphoto/Loretta Hostettler. © iStockphoto/Lijuan Guo.

Figure 18-6 Garden tools including the shovel, garden fork, rake, pick, and hoe are used to break up dirt clods and loosen, smooth, and level the soil.
© Mushakesa/Shutterstock.com.

Common Garden Crops and Varieties

Consider the Climate

When choosing the varieties of vegetables and flowers to plant in the home garden, consider the climate. **Climate** refers to the weather conditions of a specific region. Do not plant crops outdoors until all danger of frost is gone. Except for a few mountainous areas and the northern tier of states, almost all areas of the United States are frost-free from June through mid-September (**Figure 18-7**).

Plants grow rapidly in frost-free weather. In some areas, there is not enough time for long-season crops, such as eggplant, cantaloupe, and watermelon, to mature reliably. Some regions of the United States have intermediate growing seasons, with 5 to 6 months of frost-free weather. Long-season crops can be grown in these areas consistently. Growing seasons of 7 to 10 months are common across the mid-South, South, and low-elevation regions of the Southwest and West. In these regions, gardeners can grow many varieties of both spring and fall crops. However, these same regions are susceptible to summers that are so hot or dry that some crop varieties have difficulty surviving the intense summer heat.

Consider the Variety

There are both warm-season and cool-season crops. Certain flowers and vegetables must have continuous cool weather to do well. Heat will quickly make the plants dry up or go to seed. Where the growing season is 5 months or longer, outdoor seeding in the late summer usually results in an excellent harvest during the cool fall season. The growing season can be extended by planting varieties of the cool-season plants after other crops are harvested in late summer. Most cool-season crops can withstand light frosts with little damage.

There are two types of warm-season flowers and vegetables: those that mature quickly and those that require 4 months or more from planting to maturity. The quick-maturing types are almost always started in the garden when the soil is warm. The later-maturing types are usually started indoors and are transplanted to the garden after all danger of frost is past.

Most packets of home garden seeds are labeled to indicate the number of days that are required for the plants to mature. For example, the Blue Lake pole bean variety is labeled to mature in 60 to 65 days. Seed packet labeling is the most accurate information we have with regard to how long a plant will require to reach maturity (**Figure 18-8**). There are exceptions, however, such as the extreme northern latitudes, such as Alaska, where the sun goes down only briefly during the summer and the number of hours of daylight are much greater than in other areas of the country.

Annuals, Biennials, and Perennials

Flowers and vegetables are classified as annuals, biennials, or perennials (**Figure 18-9**). An **annual** is a plant whose life cycle is completed in one growing season. Growth is rapid. Practically all vegetables, except asparagus, rhubarb, and parsley, are annuals.

A **biennial** is a plant that takes two growing seasons or 2 years from seed to complete its life cycle. Some biennials bloom very little or not at all the first year, but they come into full bloom the second year and then go to seed.

A **perennial** is a plant that lives on from year to year. A gardener commonly treats some plants that are true perennials as either annuals or biennials. This is done because, when planted each year from seed, some perennials produce higher-quality blooms in their first year than older plants that remain from year to year. Perennial crops, such as asparagus, artichokes, rhubarb, and some herbs and flowers, should be planted in one section of the garden, separate from the annuals. By such separation, the perennials do not interfere with the cultivation of the annuals.

Vegetable Planting Guide

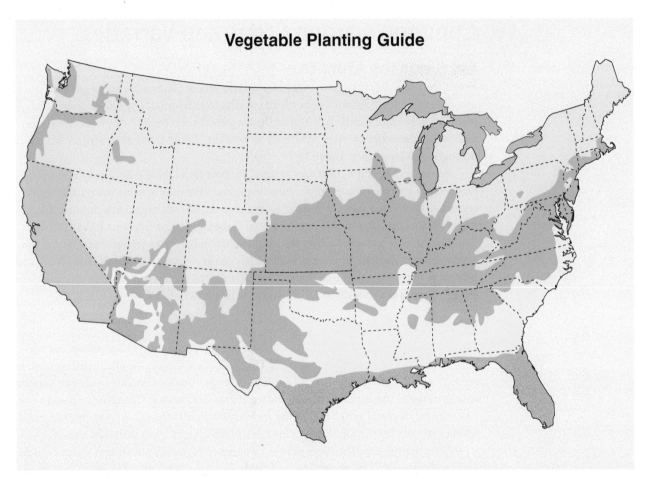

This map and accompanying chart are based largely on U.S. Department of Agriculture records showing the average dates of the last killing frosts in various parts of the country. Use this information to plan the planting and harvesting periods for your garden, but remember that these are averages, and individual conditions may vary.

	SUBTROPIC	WARM	MILD	COOL	CALIF.	PLANTING DEPTH	ROW SPACING	FIRST HARVEST	HARVEST LASTS
Beans	Apr.–Aug.	Apr.–June	May–June	May–June	Mar.–Aug.	1"	2–3 ft.	8 weeks	Until frost
Beets	Jan.–Dec.	Feb.–Oct.	Mar.–July	Apr.–July	Jan.–Dec.	1/2"	15–18"	50–78 days	6 weeks†
Broccoli	July–Oct.	Feb.–Mar	Mar.–Apr	Mar.–Apr*	Sep.–Feb.	1/4"	3 ft.	65–70 days	To frost
Brussels sprouts	Feb.–May	Feb.–Apr.	Mar.–Apr*	Mar.–Apr*	Sep.–Feb.	1/2"	3 ft.	14–20 weeks.	Past frost
Carrots	Jan.–Dec.	Jan.–Mar	Mar.–June	Apr.–June	Sep.–May	1/4"	1 1/2–2 ft.	8 weeks	8 weeks†
Chard	Jan.–Dec.	Feb.–Sep.	Mar.–Aug.	Apr.–July	Jan.–Dec.	3/4"	1 1/2–2 ft.	8 weeks	To frost
Chives	Feb.–May	Mar.–May	Mar.–May	Apr.–June	Mar.–May	1/4"	1 1/2 ft.	8 weeks	To frost
Corn	Apr.–June	Mar.–June	May–July	May–July	Mar.–Aug.	1 1/2"	3 ft.	9–12 weeks	10 days
Cucumbers	Apr.–June	Apr.–June	Apr.–June	May–June	Mar.–Aug.	1/2"	6 ft.	7 weeks	5 weeks
Eggplant	Feb.–Mar	Feb.–Apr	Mar.–May*	Apr.–May*	Mar.–May	1/4"	3–4 ft.	10–14 weeks.	To frost
Endive	July–Sep.	Aug.–Sep.	Mar.–May	Apr.–June	Jan.–Dec.	1/4"	2–3 ft.	10–12 weeks	7 weeks
Lettuce	Jan.–Dec.	Aug.–May	Mar.–June	Apr.–June	Sep.–May	1/4"	15"	6 weeks	6 weeks†
Mustard	Feb.–May	Feb.–May	Mar.–June	May–July	Mar.–Aug.	1/4"	1 1/2–2 ft.	40–50 days	2 weeks
Okra	Apr.–June	Apr.–June	Apr.–June	May–June*	Mar.–May	1/2–3/4"	3 ft.	55–60 days	To frost
Onions	Dec.–Mar	Dec.–Apr	Feb.–May	Mar.–June	Sep.–May	1/2"	18"	Variable	—
Parsley	Jan.–Dec.	Jan.–June	Feb.–June	Mar.–June	Jan.–Dec.	1/4"	1 1/2–2 ft.	10 weeks	To frost
Parsnips	Mar.–June	Feb.–June	Apr.–June	May–June	Dec.–May	1/2"	1 1/2–2 ft.	14–17 weeks.	Past frost
Peas	Jan.–May	Jan.–Apr	Feb.–May	Mar.–June	Sep.–May	1 1/2"	2–3 ft.	9 weeks	2 weeks†
Peppers	Feb.–Mar	Feb.–Apr	Mar.–May*	Mar.–May*	Mar.–May	1/4"	3 ft.	9 weeks	To frost
Potatoes	Jan.–Dec.	Feb.–Oct.	Mar.–Apr	Apr.–May	Jan.–Dec.	5"	2 1/2–3 ft.	100 days	To frost
Pumpkin	Apr.–June	Apr.–June	Apr.–June	May–June	Mar.–May	1"	8–10 ft.	14–17 weeks.	To frost
Radishes	Jan.–Dec.	Feb.–Oct.	Mar.–Aug.	Apr.–July	Sep.–May	1/4"	1–2 ft.	3–6 weeks	1–2 weeks†
Squash, Summer	Apr.–June	Apr.–June	Apr.–June	May–June	Mar.–Aug.	1"	4–6 ft.	60 days	To frost
Squash, Winter	"	"	"	"	"	1"	6–8 ft.	100 days	"
Tomatoes	Jan.–Mar	Feb.–Mar	Mar.–May*	Mar.–May*	Mar.–May	1/4"	3–4 ft.	9–12 weeks	To frost
Turnips	Feb.–Mar	Jan.–Mar	Feb.–Apr	Mar.–May	Jan.–Dec.	1/2"	1–2 ft.	6–10 weeks	3 weeks†

*Transplants recommended.
†Following harvest, space may be used for late planting of carrots, beets, or bush beans.

Figure 18-7 Planting times, planting depths, row spacing, and harvest times for common garden vegetables.

Adapted from USDA.

Hot Topics in Agriscience — Use a Plant Hardiness Map

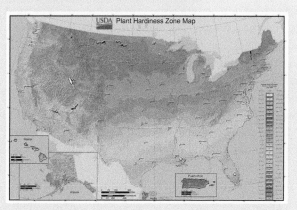

Source: USDA Plant Hardiness Zone Map. Agricultural Research Service, U.S. Department of Agriculture.

Using a plant hardiness map can help you choose the correct garden varieties for your area. The USDA interactive-GIS plant hardiness map is based on average annual minimum winter temperatures that are divided into zones for each temperature difference of 10°F. Each zone is further divided into 5°F subzones. The map is can be localized by selecting a state or zip code at www.planthardiness.ars.usda.gov.

Plant hardiness depends on the extreme minimum temperature it can tolerate. The map was developed using temperature data recorded over 30 years at each of the nation's weather stations. The plant hardiness zone map is only intended to be a guide to gardeners and farm producers in selection of plant varieties. Extreme temperatures may be recorded in the future that may exceed those on which the map is based. In addition, microclimates exist within larger zones where temperatures are affected by large bodies of water, large, paved areas, or changes in elevation.

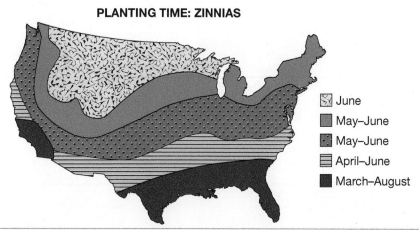

Figure 18-8 The best time to plant a specific plant can be determined by considering the length of the growing season, climatic zone maps, and the days to maturity as listed on the seed packet.

Annual	Biennial	Perennial
Lives only one year	Lives two years	Lives more than two years
Germinates, matures, and reproduces in one growing season	Germinates and grows roots, a short stem, and a cluster of leaves in the first year	Germinates, matures, and reproduces in the first year
Produces flowers, fruits, and seeds in same growing season	Produces flowers, fruits, and seeds in second growing season	Produces new shoot systems, flowers, fruits, and seeds each growing season

Figure 18-9 Characteristics of annual, biennial, and perennial plants.

Vegetables or Flowers

The kinds of vegetables and flowers to plant depend on the individual tastes of the members of your family. Certain top-ranking vegetables are popular everywhere. Others, such as okra, are popular only within certain regions of the country. The most popular vegetables are tomatoes, snap beans, onions, cucumbers, peppers, radishes, lettuce, carrots, corn, beets, cabbage, squash, and peas. Favorites of specific regions are artichokes in Louisiana and California, southern peas in warm southern climates, and melons in warm summer regions.

The gardener has a wide selection of flowers from which to choose. Generally, flower varieties are chosen because of their characteristics. Various varieties may survive well in poor soil, have a wonderful fragrance, or grow well in the shade. A few of the flowers that do well in poor soils are alyssum, cactus, cosmos, marigolds, petunias, and phlox. Flowers that are fragrant are alyssum, carnations, petunias, sweet peas, and sweet William. Some that do well in partial shade are ageratums, begonias, coleus, impatiens, and pansies.

Hot Topics in Agriscience: Master Gardener Programs

The Cooperative Extension System associated with land-grant universities has developed a program for Master Gardeners. Under this program, people who are expert gardeners are available as consultants to answer questions about home gardening problems. To find out more about programs in your state, visit the website of the American Horticultural Society, and select *Master Gardeners* from the Resources menu. People who apply for the program typically complete 30 hours of training and commit to at least 50 volunteer hours a year as a community consultant on gardening problems.

A Master Gardener is a local expert who advises the community on gardening matters.
© Alexander Raths/Shutterstock.com.

Cultural Practices for Gardens

Cultivating

Cultivation is the act of preparing and working the soil. Never put off cultivation. It is easier to do a little each day than to let weeds get ahead or the soil get too hard. Once this happens, it is difficult to get the garden back into good condition. The main purpose of cultivation is to control weeds and to loosen and aerate the soil. A sharp hoe is a necessary tool for any gardener. Buy one that can fit between plants for easy cultivation. Begin cultivation as soon as weeds or grass break through the soil. Do not wait until the weeds and grasses are tightly rooted and threaten to take over the garden. When cultivating, take care not to damage the roots of the vegetables and flowers. Use short, shallow scraping motions that cut the tops of the weeds from their roots instead of chopping deeply into the soil. Deep hoeing or cultivation will destroy desirable plant roots.

Weeding

Begin weeding as soon as weeds appear. When weeds grow near the plants, their roots can become intertwined with the crop. Therefore, take care to avoid pulling out both the weed and the plant. When weeding, select a time of day when the soil and plants are fairly dry, so the weeds wither when pulled. Be careful to shake the soil from weed roots so the weeds will die quickly. Remove persistent weeds such as purslane, crabgrasses, and Bermuda grasses from the garden area to prevent them from taking root again or reseeding a new crop of weeds. Mulches of various types provide weed barriers; using them is an important practice for controlling weeds (**Figure 18-10**).

Weed-control chemicals called **herbicides** can be used in the home garden, but they should always be handled with care. Usually, the use of herbicides is best suited to large gardens or for the purpose of controlling established stands of persistent perennial weeds. Chemicals are specific in their actions in certain types of plants, so they must be used with caution to prevent unintended injury to susceptible vegetables and flowers. Before using any herbicides, obtain detailed information from an experienced person. Always read the label on the container and use the chemical strictly in accordance with label instructions.

Figure 18-10 Plant materials, newspaper, and plastic films all make excellent mulches that help control weeds and conserve moisture.
© Alison Hancock/Shutterstock.com.

Watering

During a growing season, there will probably be some days or weeks of dry weather when the garden must be watered. Most soils require at least 1 inch of water per week, either through rain or irrigation. If water is needed, make it a practice to soak the soil to a depth of 6 inches. Frequent light watering tends to promote shallow root development and should be avoided.

When watering, run the water on the soil near the plants. A good sprinkler head, soaker hose, or water breaker is needed to ensure even distribution of water. Let the garden hose run for 15 to 30 minutes to water 100 square feet at a depth of several inches. You should apply water when the soil feels dry 1 to 2 inches below the surface.

Protection from Pests

Gardens can be damaged from a variety of insects, animals, and diseases. To help prevent severe damage, the following practices should be helpful:

- Rotate crops so that the same or a related crop does not occupy the same area every year. This helps control soil-borne diseases.
- Watch closely for insects. Pick them off by hand or knock them off with a hard stream of water. Introduce predatory insects (**Figure 18-11**).
- When using sprinklers, water early in the day so the foliage can dry before nightfall. Diseases of leaves and fruit prosper in damp conditions.
- Keep weeds out of the garden. Weeds may harbor diseases or insects and will interfere with spraying and dusting of crops.
- Use correct amounts of fertilizer and lime to promote vigorous growth.
- Protect plants from rodents, birds, and deer with battery-powered electronic devices or specially prepared repellent products.

Figure 18-11 Lady beetles feed on aphids and help keep the damaging insect population under control.
© Jamieuk/Shutterstock.com

Science Connection: When Bugs Are Beneficial

Entomologist and quarantine officer Larry Ertle at the Beneficial Insects Introduction Research Laboratory in Newark, Delaware, gently places beneficial insects in custom-designed packaging for shipment.
Courtesy of USDA/ARS #K4183-12.

Newark, Delaware, is the site of an unusual facility specializing in foreign imports—insects! The imports are flies, wasps, beetles, and other insects that parasitize, or feed on, the insects that devour our crops, devastate our gardens, cause our lawns to turn brown, and infest our houseplants. These insects have been found doing their "good deeds" in countries all over the world. Most of our recent imports have come from Argentina, Brazil, Canada, Chile, China, France, Germany, India, Indonesia, New Zealand, South Korea, Russia, and Eastern Europe.

The facility is called the U.S. Department of Agriculture (USDA) Beneficial Insects Introduction Research Laboratory. Its job is to see that insects brought into the United States are, indeed, beneficial and that the eggs, larvae, pupae, or adults of harmful insect species do not hitchhike along with the beneficial ones. The work of the entomologists at the Newark Laboratory includes picking up incoming shipments of beneficial insects at the Philadelphia airport; transporting them to the nearby Newark laboratory; taking them through a special receiving process; disinfecting all shipping materials that arrive with the insects; isolating the insects in special refrigeration units; exercising them daily; examining each insect under magnification to confirm identity; separating eggs, larvae, pupae, and adults for rearing purposes; increasing their numbers through reproduction; and shipping them under carefully controlled conditions to USDA and state research facilities throughout the United States and cooperating foreign nations.

Three sets of heavy steel doors lead into the quarantined main work area, where a large sink and an autoclave are used to clean and sterilize materials used in the laboratory. Sterilization is essential before discarding materials to avoid accidental release of both beneficial insects and "hitchhiking intruders." It is interesting to note that, just like humans, insects must receive regular exercise to remain healthy! Once a day, even on weekends and holidays, batches of insects are removed from cool storage and released in restricted, warm, and lighted areas to warm up, stretch, soak up light, and move around in the greater space.

Many innovations in packaging and handling have evolved and become standardized over the years. One of the more unusual ones is the parasite pill, or Trichocap. The parasite pill is a shipping capsule made of soft, gray cardboard and is about the size of an oyster cracker. It is used to ship insect eggs. One parasite pill holds about 500 eggs of the Mediterranean flour moth. Inside each egg, there is one future parasitic wasp used as a biocontrol against the devastating European corn borer. The discovery and introduction of parasitic wasps have helped control some of our most damaging insects, while reducing our reliance on chemical sprays.

Harvest and Storage of Garden Produce

Harvesting

Harvest the vegetables and flowers at the peak of quality or at the stage of maturity preferred by the users (**Figure 18-12**). Some vegetables, such as cucumbers, will lose their quality within a day or two, whereas others hold it for a week or more.

Vegetable	Maturity Level	Vegetable	Maturity Level
Asparagus	From seeds—3rd year. From roots—after 1st year. Cut when spears are 6" to 10" high, in spring.	Lettuce	Leaf—When tender and desired size. Head—When round and firm.
Beans, Pole	When pods are nearly full size.	Muskmelons	When stem separates easily from fruit.
Beans, Bush	When pods are nearly full size.	Mustard Greens	When large leaves are still tender.
Beans, Bush Lima	When tender, pods are nearly full size.	Okra	Cut pods when about 2" or 3" long.
Beets	When bulbs are 1¼" to 2" in diameter.	Onions	For fresh use—When ¼" to 1½" in diameter. For storage—When tops shrivel at the bulb and fall over.
Swiss Chard	Outer leaves can be harvested anytime.		
Broccoli	Before green clusters begin to open.	Parsley	Any time the outer leaves are desired size.
Brussels Sprouts	When sprouts are firm, pick from bottom up on stalks.	Parsnips	After hard frost. Can leave in ground all winter for spring use.
Cabbage	When heads are solid, before splitting.	Peas	When pods are well-filled, before seeds are largest.
Cauliflower	After blanching, when firm curds are 2" to 3" in diameter.	Peppers	When solid and nearly full size.
Carrots	When top of root is 1" to 1½" in diameter.	Pumpkins	When skin is hard and not easily punctured. Cut with some stem on.
Celery	When about ⅔ mature, harvest as needed.	Radishes	When desired size.
Collards	When leaves are large but tender. Can harvest up to winter.	Rhubarb	When stems are of desired size, twist off near base of plant and discard leaves. Do not pick more than ⅓ of plant during a season.
Corn	When kernels are filled out and milky. The silk at the tip of the ear should be dry and brown.	Spinach	When outer leaves are large enough.
Cucumbers	When slender and dark green.	Squash	Summer—When skin is soft. Winter—When skin is hard, cut with part of stem on, before frost.
Eggplant	When half grown, glossy, and bright.		
Endive	When leaves are tender and desired size. About 15" in diameter.	Tomatoes	When uniformly red and firm.
		Turnips	When 2" to 3" in diameter.
		Watermelons	When underside is yellow and thumping produces a muffled sound.
Kale	When young and tender.		
Kohlrabi	When 2" to 3" in diameter.		

Figure 18-12 Levels of maturity for harvesting vegetables for top quality.

Crop	Temperature °F	Relative Humidity Percent
Asparagus	32	85-90
Beans, snap	45-50	85-90
Beans, lima	32	85-90
Beets	32	90-95
Broccoli	32	90-95
Brussels sprouts	32	90-95
Cabbage	32	90-95
Carrots	32	90-95
Cauliflower	32	85-90
Corn	31-32	85-90
Cucumbers	45-50	85-95
Eggplants	45-50	85-90
Lettuce	32	90-95
Cantaloupes	40-45	85-90
Onions	32	70-75
Parsnips	32	90-95
Peas, green	32	85-90
Peppers, sweet	45-50	85-90
Potatoes	38-40	85-90
Pumpkins	50-55	70-75
Rhubarb	32	90-95
Rutabagas	32	90-95
Spinach	32	90-95
Squash, summer	32-40	85-95
Squash, winter	50-55	70-75
Sweet potatoes	55-60	85-90
Tomatoes ripe	50	85-90
Tomatoes mature green	55-70	85-90
Turnips	32	90-95

Figure 18-13 Recommended temperatures and relative humidity for storage of fresh vegetables.
Courtesy of University of Maryland Cooperative Extension.

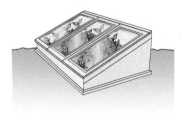

Figure 18-14 Cold frames are structures with glass or plastic covers used for starting garden plants.

For the best results with flowers, pick them in the early morning or late afternoon. Immediately place the cut flowers in lukewarm water for a short time. Display or store the flowers in a cool place.

Storage

Many vegetables that are not processed or frozen for later use can be stored successfully if proper temperature and moisture conditions are met. Only vegetables that are of good quality and at the proper stage of maturity should be stored.

Warm Storage

Vegetables that tolerate moderate storage temperatures are squash, pumpkins, and sweet potatoes. These vegetables may be stored on shelves in an upstairs storage area. Damp basement areas are not recommended. Squash and pumpkins should be kept in a heated room, with a temperature between 75°F and 85°F (23.9°C and 29.4°C) for 2 weeks to harden the shells. Long-term storage temperature should then be reduced to 50°F to 55°F (10°C to 13°C).

Cool Storage

Most vegetables require cool temperatures and relatively high humidity for successful storage. The storage area should be cool, dark, and ventilated. The room should be protected from frost, heat from a furnace, or high outdoor temperatures.

The suggested temperature and relative humidity for storage of crops are shown in **Figure 18-13**. This list will provide you with a general idea of the type of storage area that each vegetable requires.

Cold Frames, Hotbeds, and Greenhouses for Home Production

Cold Frames

Cold frames are very useful to the gardener. A **cold frame** is a bottomless wooden box with a sloping glass or clear plastic top (**Figure 18-14**). It can be constructed by using plywood or common lumber and a window sash. The size of the sash should determine the dimensions of the box. The length and width should be 2 inches smaller than the overall dimensions of the cover sash.

The following procedures should be helpful in constructing a cold frame:

1. Make the front of the box approximately 8 inches high and the back of the box about 12 inches high.
2. Cut the two sides on an angle, from 12 inches at the back to 8 inches in the front.
3. Nail the four sides together.
4. Try the sash or top for size.
5. Hinge the sash to the back of the frame for easy use. The hinge allows you to prop the cover open slightly in warm weather for ventilation.
6. Place the cold frame in a southern exposure, with good protection from the wind and with proximity to a water supply.

When the cold frame is built, it can serve three purposes:
1. It can be used as a protective home for seedlings that have been started indoors. The seedlings can continue to grow inside the cold frame and become "hardened off" before they are transplanted to the garden.
2. You can start plants in the cold frames. Seeds can be planted directly into the soil in the cold frame. After they are 4 to 6 inches tall, they can be transplanted straight into the garden.
3. Vegetables, such as lettuce and endive, grow well into the fall. Sow seeds in early autumn. Cover the cold frame with a blanket or tarp during extremely cold nights.

Science Connection: Preserving Food

The produce from a garden can be enjoyed all year when it is preserved properly.
© Poulsons Photography/Shutterstock.com.

During the growing season, a gardener spends hours weeding, watering, fertilizing, and finally harvesting. With the amount of time and energy that is put into a home garden, one would not want to see any of the produce go to waste. That is why many gardeners consider food preservation the final step in home gardening.

Within the fertile soil of a home garden, millions of microorganisms thrive. Many of these creatures are beneficial to the plants as they grow. Once eaten, however, they can cause disease. During the food preservation process, care must be taken to ensure that food is free from living microorganisms.

Common organisms that are found in improperly preserved foods include fungi and bacteria. If left in food, the food will be unsafe to eat. Fungi include molds and yeasts. Bacteria are plentiful on freshly harvested foods. Many bacteria that are commonly found in the soil can cause serious illnesses if ingested.

Local agricultural extension offices can provide material, classes, and advice on safe and germ-free food preservation. Canning, freezing, and dehydrating are all methods that can keep garden produce safe and delicious for long periods. Molds and some bacteria are destroyed between 190°F and 212°F. The boiling point of water at sea level is 212°F. By boiling canned food and blanching food that will be dehydrated or frozen, the chances that any microorganisms will remain in the food are slight. However, it takes temperatures of 240°F to kill some species of bacteria. Natural acids found in some foods can kill most bacteria. In produce with low acid levels, the food must be heated to 240°F using a pressure cooker. This will kill any remaining organisms. Although a lot of care must be taken to preserve the foods grown in a home garden, the benefits can be enjoyed all year long.

Hotbeds

A **hotbed** is simply a cold frame with a heat source. In many areas, a cold frame is usually adequate. However, in colder areas, some type of artificial heat is necessary. The hotbed should be located on well-drained land.

Electricity is a convenient means of heating hotbeds. Use either lead or plastic-coated electric heating cables or 25-watt frosted light bulbs. Temperature can be controlled with a thermostat. When using light bulbs, attach the electrical fixtures to strips of lumber and suspend the bulbs 10 to 12 inches above the soil surface. Allow one 25-watt bulb per 2 square feet of space. When using a heating cable, a standard 60-foot length will heat a 6 × 6- or a 6 × 8-foot bed. Lay cable loops about 8 inches apart for uniform heating. Lay the cable directly on the floor of the bed unless drainage material is needed.

Greenhouses

Growing flowers and vegetables in a greenhouse can be enjoyable as well as profitable. The conventional greenhouse is designed primarily to capture light and control temperature. It can be freestanding, but most often, it is attached to a building to provide convenient access, simplified construction, and a potential source of supplemental heat if needed (**Figure 18-15**). The greenhouse can provide an environment for starting plants, hardening them off, or completely growing the plants.

Figure 18-15 A home greenhouse can be attached to the house for convenience and efficiency.
© iStockphoto/Ann Taylor-Hughes.

Agri-Profile: Career Areas: Gardener/Caretaker/Market Gardener

Today, there are many ways to source fresh fruits and vegetables.
Courtesy of USDA/ARS #K4173-2.

Fruit and vegetable gardening has long been an important enterprise for providing fresh and wholesome food for rural and suburban families. Similarly, flower gardening provides a rewarding pastime and greatly increases the beauty of homes in urban, as well as rural and suburban, settings. Career opportunities exist for gardeners and caretakers at estates, institutions, colonial farms, truck farms, and in residential neighborhoods.

There are many opportunities in market gardening, too. Nationally, there is a resurgence of small farms and truck gardens where roadside stands and farmers' markets have created new interest in farm-fresh fruits and vegetables. Pick-your-own operations provide opportunities for people to harvest their own field-fresh produce and save the producer the cost of labor.

People have a new awareness of the benefits of fresh food and pay premium prices when freshness is ensured. This has created new markets for garden produce and has opened new career opportunities in production, processing, and marketing.

Science Connection: Gardens in Space

At Kennedy Space Center in Florida, scientists use a pressure chamber from a former space craft to develop a controlled ecological life support system, where plants recycle air, water, and waste to produce food.
Courtesy of NASA #89-HC-130.

As humans push back the frontiers of outer space and consider the likelihood of space travel taking days or even years, we face the age-old dilemma—how do we feed the crew? Early manned space voyages were accommodated by freeze-dried food and food in tubes. Now scientists are focusing on food production in space. Not only is there a need to produce food, but there also is a need to generate oxygen, eliminate carbon dioxide, and dispose of organic waste. The solution is a controlled ecological life support system (CELSS). This system will provide the basic components to sustain life without requiring inputs from external sources. Through photosynthesis, plants can take up nutrients from soil or water and use carbon dioxide from the air to manufacture food for themselves as well as for people and animals. They use excess CO_2 and release oxygen as a by-product of this process.

What kinds of plants and growing conditions might perform these functions? It is clear that plants will be needed that can adapt to weightlessness and light conditions of space. Researchers have identified the following plants as candidates for space gardens: thale cress, Chinese cabbage, super dwarf wheat, apogey wheat, *Brassica rapa*, rice, tulips, kalanchoe, flax, onions, peas, radishes, lettuce, garlic, cucumbers, parsley, potato, dill, basil, cabbage, *Zinnia hybrida*, mizuna lettuce, red romaine lettuce, sunflower, and *Ceratopteris richardii*. Experiments using these and other plants will surely follow.

Controlled environment facilities are gaining acceptance as a viable means of commercially growing high-value crops under scheduled production management in an environment free from harmful insects and pollution. Such systems must be refined before interplanetary space travel will be possible.

Chapter Review

Student Activities

1. Write the **Key Terms** and their meanings in your notebook.
2. Discuss with family members the types of vegetables and flowers they want in the home garden.
3. Select vegetable and flower varieties for the home garden from seed catalogs and websites.
4. Sketch a garden plot containing both flowers and vegetables that are to be grown.
5. Make a calendar indicating the dates to plant and harvest the various crops in the garden.
6. **SAE Connection:** Start and manage your own garden area. Use the flowers and vegetables at home or sell them for profit.
7. Learn to calculate area in square feet. A square foot is an area that is 1 foot long and 1 foot wide. An area that is 1 foot long and 2 feet wide contains 2 square feet. Area (A) in square feet is calculated by multiplying length (L) in feet times width (W) in feet. Therefore, $A = L \times W$, or $A = LW$.

    a. How many square feet are in a garden that is 10 ft. long × 5 ft. wide?
    A = _____ sq. ft.

    b. How many square feet are in a lawn that is 40 × 70 ft.?
    A = _____ sq. ft.

    c. What is the area of a building lot that is 109 × 150 ft.?
    A = _____ sq. ft.

    d. Suppose the lot in item c (above) is covered with lawn except for the house, which is 28 × 42 ft. What is the area of the house?
    A = _____ sq. ft.
    What is the area of the lawn?
    A = _____ sq. ft.

8. Using the produce you grew during Activity 6, dehydrate, freeze, and/or can some of it for future use. Be sure to follow approved techniques. Contact your local extension education office for more information.
9. Conduct an inventory of garden and greenhouse tools that are available to you in the school greenhouse and/or outdoor land laboratory. Perform online and textbook searches regarding the proper use of each tool. Under the direction of your teacher, demonstrate the proper use of each tool.

National FFA Connection — School and Community Garden Project

Community gardens and gardens in schools promote public health and social connection.
© Hannamariah/Shutterstock.com.

Some FFA members are fortunate to have access to garden areas at their school. Others live in communities that offer garden space through community gardens. Another gardening option is to participate in a garden cooperative. In this format, several families or groups share in obtaining a space, planning, and establishing the garden. They work together and share the produce. These options make good choices for FFA members who are motivated to establish gardens.

The ideal option would be for groups of classmates to research, plan, and establish a garden at their school, which they could then compare and synthesize as a class. Where a school garden is not available, either of the other options would be a good opportunity for individual students or groups of students to share a garden experience.

The National FFA Organization encourages the development of school garden enterprises through a grant process to provide hands-on experience correlating with classroom instruction. Visit the FFA website to find out more.

Inquiry Activity — Master Gardeners

© Alexander Raths/Shutterstock.com.

After watching the *Master Gardeners* video available on MindTap, answer the following questions. Feel free to watch the video again to locate the answers.

1. Describe the ways in which Master Gardeners pass on their knowledge and provide resources to the following members of their communities:
   - Home gardeners
   - Students
2. How does one become a Master Gardener? What does a Master Gardener need to do to maintain the title?

Chapter 19
Vegetable Production

Objective
To determine the opportunities for a career in vegetable production and marketing, and evaluate the recommended production practices for key vegetable crops.

Competencies to Be Developed
After studying this chapter, you should be able to:
- determine the benefits of vegetable production as a personal enterprise or career opportunity.
- identify vegetable crops.
- plan a vegetable production enterprise and prepare a site for planting.
- describe how to plant vegetable crops and use appropriate cultural practices.
- list appropriate procedures for harvesting and storing at least one commercial vegetable crop.

Key Terms
olericulture
market gardening
truck cropping
olericulturist
angiosperm
monocotyledon (monocot)
dicotyledon (dicot)
aeration
green manure
transplant
arid
semiarid
pre-cooling
hydrocooling

A vegetable is the edible portion of an herbaceous plant (**Figure 19-1**). The study of vegetable production is olericulture. *Herbaceous* describes a plant that has a stem that withers away at the end of each growing season. The production of vegetables can be classified into three categories: home garden, market garden, or truck crop. Home gardening, as discussed in Chapter 18, refers to the vegetable production for one family with most of the produce consumed at home. It usually does not involve any major selling of the crops. Market gardening refers to growing a wide variety of vegetables for local or roadside markets. Truck cropping refers to large-scale production of a few selected vegetable crops for wholesale markets.

Figure 19-1 Vegetables are important in the human diet because they supply basic nutrients.
Courtesy of DeVere Burton.

Vegetable Production

Home Enterprise

Growing vegetables in a home garden is enjoyed by millions of people in the United States. It has become a part of the lifestyle of many families with access to a little bit of ground. The vegetables produced can be used for fresh table consumption or can be stored for later use. Vegetable gardening not only produces nutritious food but also provides outdoor exercise from spring until fall.

The gardener who enjoys this type of activity can plant enough to provide for the family and harvest vegetables for sale near home for extra income. Fresh, homegrown vegetables are usually superior in quality to those found in the supermarkets. In addition, the gardener can grow vegetables that may be expensive to buy.

Career Opportunities

The vegetable industry is a large and complex component of the horticultural industry today. Even though millions of homeowners raise gardens, most vegetables consumed by the public are grown commercially. The commercial vegetable industry is fast moving, intensive, and competitive. It is a business that is continually changing as the demands for certain vegetables fluctuate with the tastes of consumers.

Numerous and varied career opportunities are available in the vegetable industry. These include being the owner of a small market gardening business, a member of a larger truck-crop business, a vegetable wholesaler or retailer, or a worker in a vegetable-processing plant. With adequate education and training, a person can become an olericulturist—someone who develops pest-resistant strains and new varieties of vegetables and does other specialized work. The opportunities are endless in the vegetable industry.

Identifying Vegetable Crops

Vegetables can be identified in various ways: by their botanical classifications, by their edible parts, or by their required growing season.

Botanical Classification

All vegetables belong to the division of plants known as angiosperms. These are plants with ovules and an ovary. From this division, the vegetables can be grouped into either Class I, monocotyledons (monocots) (having only one seed leaf), or Class II, dicotyledons (dicots) (having two seed leaves) (**Figures 19-2A** and **B**). A vegetable can be further grouped into a family, a genus, a species, and sometimes a variety. One of the more popular vegetable families is Cruciferae—the mustard family—which contains Brussels sprouts, cabbage, cauliflower, collards, cress, kale, turnips, mustard, watercress, and radish. Other families include Leguminosae—the pea family—which contains bush beans, lima beans, cow peas, kidney beans, peas, peanuts, soybeans, and scarlet runner beans; Cucurbitaceae—the gourd or melon family—includes pumpkins, cucumbers, cantaloupe, casaba melons, and watermelons; and Solanaceae—the nightshade family—includes eggplants, ground cherries, peppers, and tomatoes.

Edible Parts

Vegetables are grouped by the part of the vegetable that is eaten: (1) leaves, flower parts, or stems; (2) underground parts are used; and (3) fruits or seeds. **Figure 19-3** lists these categories and the vegetables in each one.

A

B

Figure 19-2 Monocots have one seed leaf (A), whereas dicots have two (B).

© iStockphoto/Clint Scholz.
© iStockphoto/Sohl.

Plants of Which the Fruits or Seeds Are Eaten		Plants of Which the Leaves, Flower Parts, or Stems Are Eaten		Plants of Which the Underground Parts Are Eaten	
Family	Vegetable	Family	Vegetable	Family	Vegetable
Grass, Gramineae	Sweet corn, *Zea mays*	Lily, *Liliaceae*	Asparagus, *Asparagus officinalis* var. *altilis*	Lily, *Liliaceae*	Garlic, *Allium sativum* Leek, *Allium porrum* Onion, *Allium cepa* Shallot, *Allium ascalonicum* Welsh onion, *Allium fistulosum*
Mallow, Malvaceae	Okra (gumbo), *Hibiscus esculentus*		Chives, *Allium schoenoprasum*		
Pea, Leguminosae	Asparagus or Yardlong bean, *Vigna sesquipedalis* Broad bean, *Vicia faba* Bush bean, *Phaseolus vulgaris* Bush Lima bean, *Phaseolus limensis* Cowpea, *Vigna sinensis* Edible podded pea, *Pisum sativum* var. *macrocarpon* Kidney bean, *Phaseolus vulgaris* Lima bean, *Phaseolus limensis* Pea (English pea), *Pisum sativum* Peanut (underground fruits), *Arachis hypogaea* Scarlet runner bean, *Phaseolus coccineus* Sieva bean, *Phaseolus lunatus* Soybean, *Glycine max* White Dutch runner bean, *Phaseolus coccineus*	Goosefoot, Chenopodiaceae	Beet, *Beta vulgaris* Chard, *Beta vulgaris* var. *cicla*		
		Orach, *Atriplex hortensis*	Spinach, *Spinacia oleracea*	Yam, Dioscoreaceae	Yam (true), *Dioscorea batatas*
		Parsley, Umbelliferae	Celery, *Apium graveolens* Chervil, *Anthriscus cerefolium* Fennel, *Foeniculum vulgare* Parsley, *Petroselinum crispum*	Goosefoot, Chenopodiaceae	Beet, *Beta vulgaris*
				Mustard, Cruciferae	Horseradish, *Armoracia rusticana* Radish, *Raphanus sativus* Rutabaga, *Brassica campestris* var. *napobrassica* Turnip, *Brassica rapa*
		Sunflower, Compositae	Artichoke, *Cynara scolymus* Cardoon, *Cynara cardunculus* Chicory, witloof, *Cichorium intybus* Dandelion, *Taraxacum officinale* Endive, *Cichorium endivia* Lettuce, *Lactuca sativa*	Morning Glory, Convolvulaceae	Sweet Potato, *Ipomoea batatas*
				Parsley, Umbelliferae	Carrot, *Daucus carota* var. *sativa* Celeriac, *Apium graveolens* var. *rapaceum* Hamburg parsley, *Petroselinum crispum* var. *radicosum* Parsnip, *Pastinaca sativa*
Parsley, Umbelliferae	Caraway, *Carum carvi* Dill, *Anethum graveolens*	Mustard, Cruciferae	Brussels sprouts, *Brassica oleracea* var. *gemmifera* Cabbage, *Brassica oleracea* var. *capitata* Cauliflower, *Brassica oleracea* var. *botrytis* Collard, *Brassica oleracea* var. *viridis* Cress, *Lepidium sativum* Kale, Borecole, *Brassica oleracea* var. *viridis* Kohlrabi, *Brassica oleracea* var. *gongylodes* Mustard leaf, *Brassica juncea* Mustard, Southern Curled, *Brassica juncea* Bok choy, Chinese cabbage, also known as Napa cabbage, *Brassica chinensis* var. *crispifolia* Pe-tsai, Chinese cabbage, *Brassica pekinensis* Seakale, *Crambe maritima* Sprouting broccoli, *Brassica oleracea* var. *italica* Turnip, Seven Top, *Brassica rapa* Upland cress, *Barbarea verna* Watercress, *Rorippa nasturtium-aquaticum*		
Martynia, Martyniaceae	Martynia, *Proboscidea louisiana*			Nightshade, Solanaceae	Potato, *Solanum tuberosum*
Nightshade, Solanaceae	Eggplant, *Solanum melongena* Groundcherry (husk tomato), *Physalis pubescens* Pepper (bell or sweet), *Capsicum frutescens* var. *grossum* Tomato, *Lycopersicon esculentum*			Sunflower, Compositae	Black salsify, *Scorzonera hispanica* Chicory, *Cichorium intybus* Jerusalem artichoke, *Helianthus tuberosus* Salsify, *Tragopogon porrifolius* Spanish salsify, *Scolymus hispanicus*
Gourd or Melon, Cucurbitaceae	Chayote, *Sechium edule* Cucumber, *Cucumis sativus* Cushaw, *Cucurbita moschata* Gherkin, *Cucumis anguria* Cantaloupe (Muskmelon), *Cucumis melo* Pumpkin, *Cucurbita pepo* Summer squash (bush pumpkin), *Cucurbita pepo* Squash, *Cucurbita maxima* Watermelon, *Citrullus lanatus* Winter melon, *Cucumis melo* var. *inodorus*				

Figure 19-3 Classification of vegetable crops by botanical family and edible parts.

Growing Seasons

There are basically two growing seasons: warm and cool. Cool-season vegetable crops grow best in cool air and can withstand a frost or two. Some of these crops, such as asparagus and rhubarb, can even endure winter freezing. This group of crops is planted early in the spring and late in the season for fall and winter harvest. Cool-season crops include mostly leaf and root crops (**Figures 19-4A** and **B**).

Warm-season, or warm-weather, crops are those that cannot withstand cold temperatures, especially frosts (**Figures 19-5A** and **B**). These vegetables and fruits require soil warmth to germinate and long days to grow to maturity. They also require warm temperatures to produce their edible parts. The edible portions of these crops are what can be picked off the standing plant or the fruit.

Cool-Season Crops		
• Asparagus	• Chinese cabbage	• Mustard
• Beet	• Chive	• Onion
• Broad bean	• Collard	• Parsley
• Broccoli	• Endive	• Parsnip
• Brussels sprouts	• Garlic	• Pea
• Cabbage	• Globe artichoke	• Potato
• Carrot	• Horseradish	• Radish
• Cauliflower	• Kale	• Rhubarb
• Celery	• Kohlrabi	• Salsify
• Chard	• Leek	• Spinach
• Chicory	• Lettuce	• Turnip

A

B

Figure 19-4 (A) Some cool-season crops. (B) Cool-season vegetable crops like cabbage and lettuce grow best in cool temperatures. They are even able to tolerate occasional freezing temperatures.

© tristan tan/Shutterstock.com.

Warm-Season Crops	
• Cowpea	• Pumpkin
• Cucumber	• Snap bean
• Eggplant	• Soybean
• Lima bean	• Squash
• Muskmelon	• Sweet corn
• New Zealand spinach	• Sweet potato
• Okra	• Tomato
• Pepper, hot	• Watermelon
• Pepper, sweet	

A

B

Figure 19-5 (A) Some warm-season crops. (B) Warm-season vegetable crops like squash respond best in summer conditions. They have little tolerance for temperatures that are near freezing.

Courtesy of DeVere Burton.

Planning a Vegetable-Production Enterprise

Before planting a vegetable garden or truck crop, the grower needs a plan. The gardener must decide where to plant, and when to plant as well as what to plant. Without some advance planning, the vegetable garden or truck farm is not likely to be successful.

Selecting the Site

Choose a site that is convenient to a water supply. The site should also be exposed to the sun a minimum of 50 percent during the day. A minimum of 8 to 10 hours of direct sunlight is needed. Also consider the type of trees that are around the proposed site. Trees can provide excessive shade. They are also likely to compete for soil nutrients that are needed by the vegetable plants. Some kinds of trees are known to produce toxins that are harmful to specific vegetables. For example, the walnut tree is toxic to the tomato. Buildings and structures cast shade that can slow down or prevent maturation of a vegetable crop. Avoid trying to grow vegetables closer than 6 to 8 feet from the northern side of a one-story structure—farther for higher structures. Both the south and west sides of a building have access to good light and often radiate heat late in the day (**Figure 19-6**).

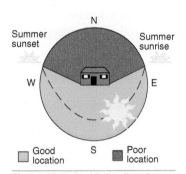

Figure 19-6 Avoid situating the garden in the shadow of a building or tree. The location of a building's shadow changes throughout the day and throughout the year.

The type of soil in the selected area is important as well. The majority of vegetables grow best in a well-drained, loamy soil. Avoid heavy clay soil that develops puddles after it rains. This is a sign of poor drainage. To test the drainage of a site, dig a trench 12 inches wide and 18 inches deep. Fill it with water and observe how long it takes for the water to drain away. If it takes 1 hour or less, the soil can be considered well drained. If the selected area has supported vegetation before, even if it has only been weeds, it will probably support a vegetable crop. If the selected site needs some alterations to its soil structure, adding organic matter can help. The organic matter can improve the drainage and allow air to move readily through the pores of the soil. If possible, the garden soil should be about 25 percent organic matter. To accomplish this, put a layer of organic matter 2 inches thick over the soil and work it in to a depth of at least 4 inches. If necessary, repeat this procedure until the final mix contains approximately 25 percent organic matter (**Figure 19-7**).

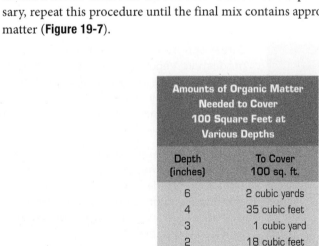

Figure 19-7 Amounts of organic matter to apply per 100 square feet of garden soil.

Scope of the Vegetable Enterprise

The size of the vegetable-growing area depends largely on the amount of ground that is available, the number of people served by the garden, and the use of the produce. A large garden that measures 50 × 100 feet will produce enough vegetables for the annual needs of a large family. **Figure 19-8** shows a suggested planting plan for a 50 × 100 feet home garden.

If this much land is not available, the fresh vegetable needs for a medium-sized family may be met with a plot that is 40 × 50 feet. A planting guide for such a garden can be found in **Figure 19-9**. When deciding on the size, it would be wise to remember that the larger the vegetable production, the more care it will require. Vegetable production is more efficient with better crops if the area is a small well-managed garden than if you take on a bigger area that is neglected and full of weeds.

Deciding What to Plant

A potential commercial gardener or truck cropper must evaluate several considerations when deciding which vegetables to plant. First, consider the maintenance that some vegetables require. The easier they are to grow and maintain, the better.

A second factor to consider is the length of the harvest time for each vegetable. If the object is fresh consumption, then vegetables that can be harvested over a long period are best. For instance, carrots can be harvested for months, whereas cabbage should be picked at maturity. Another item to consider is the expected uses for which the vegetables are planted. If they are going to be for fresh use, plant only for that purpose. Surplus vegetables can be stored in a root cellar or freezer, but plantings should match the estimated needs.

Agri-Profile — Career Areas: Vegetable Producer/Processor/Distributor/Produce Manager

The production of French fries require cooperative work of plant breeders, potato growers, food specialists, and vegetable processors, among others.
Courtesy of USDA/ARS #K4018-12.

Career opportunities in vegetable production are similar to those in fruit production. Vegetables and fruits are frequently grown, processed, and marketed by the same people. In addition to the careers mentioned in Chapter 20, many people find rewarding careers that pay well in the area of distributing and marketing produce. In this area, truckers, wholesalers, and retailers move the product from producer to consumer.

Jobs available in supermarkets include produce stockers and produce department managers. Products include fruits, vegetables, and nuts. Tasks may include inventory management, ordering, handling, stocking, and displaying produce to keep it fresh and attractive. In large cities as well as small towns, street vendors are frequently seen selling premium fruit and vegetables. Roadside stands provide opportunities for younger members of families to develop business skills while earning money for present and future needs.

Education requirements and salaries vary. For example, the average salary for a vegetable producer with a high school education is $28,792, while a produce manager with an undergraduate degree can earn between $49,033 and $62,609.

NORTH

Row	Crop				Feet between rows
1	Sweet corn		Sweet corn		3'
2	Sweet corn	1st planting	Sweet corn	2nd planting	3'
3	Sweet corn		Sweet corn		3'
4	Sweet corn		Sweet corn		3'
5	Sweet corn			3rd planting	3'
6	Sweet corn				3'
7	Tomatoes (staked)		Plant pole beans near tomato stakes in early July without disturbing the tomato plants		4'
8	Tomatoes (staked)				4'
9	Tomatoes (staked)				4'
10	Early Potatoes				3'
11	Early Potatoes				3'
12	Pepper	Eggplant		Chard (Swiss)	3'
13	Lima bean (bush)				3'
14	Lima bean (bush)				3'
15	Lima bean (bush)				3'
16	Snap beans (bush)				3'
17	Snap beans (bush)				3'
18	Broccoli				3'
19	Early cabbage				3'
20	Onion sets	This entire area may be replanted after harvest with such crops as: endive, cauliflower, Brussels sprouts, spinach, kale, beets, cabbage, broccoli, turnips, lettuce, carrots, and late potatoes.			3'
21	Onion sets				2'
22	Carrots				2'
23	Carrots				2'
24	Beets				2'
25	Beets				2'
26	Kale				2'
27	Spinach				2'
28	Peas				2'
29	Peas				2'
30	Lettuce	1st planting	Lettuce	2nd planting	Lettuce	3rd planting	2'
31	Radish		Radish		Radish		2'
32	Strawberries				3'
33	Strawberries				3'
34	Asparagus		Rhubarb		3'
35	Asparagus				3'

WEST ← → EAST 100'

←— 50' —→

SOUTH

Figure 19-8 Example of a planting plan for a large home garden 50 × 100 feet.

Rows	Crop				Feet between rows
1	Early Cabbage	(Turnips)			2'
2	Early Potatoes	(Late Cabbage)			2'
3	Peas	(Beans)	Peas	(Kale)	2'
4	Kale	(Carrots)	Turnips	(Beans)	2'
5	Parsley		Parsnips		2'
6	Onions and Radishes (same row)		Carrots	(Spinach)	2'
7	Swiss Chard	(Spinach)	Spinach	(Beets)	2'
8	Lettuce	(Beans)	Beets	(Kale)	2'
9	Beans	(Spinach)	Beans	(Lettuce)	2'
10	Peppers		Eggplant		2'
11	Lima Beans, Bush				2'
12	Tomatoes, Late				2'
13	Tomatoes, Early				2½'
14	Squash		Cucumbers		2½'
15	Corn, Sweet interplanted with Pole Lima Beans				2½'
16	Corn, Sweet interplanted with Winter Squash				2½'

(40' × 50')

NOTE: Items in parentheses are succession crops planted after the first crop is harvested or planted between the rows of mature plants to permit germination and early growth before the first crop is removed.

Figure 19-9 Example of a planting plan for a medium-sized home garden 40 × 50 feet.

The truck cropper should also consider which vegetables are in high demand in the area and which ones will produce the greatest return for the cost and labor involved. The truck-cropping enterprise is a business and must make a reasonable profit to be considered a good business. The truck cropper should also consider what vegetables can be raised in the soil and climate. Truck cropping should be a high-income business, so high quality and quantity of produce are important. Finally, the truck cropper should consider the combination of vegetables that can be grown to keep the land occupied and provide cash flow from sales throughout the season.

Preparing a Site for Planting

Preparing the Soil

Before planting vegetables, the soil must be properly prepared. This generally means adding organic matter, lime, and fertilizer. The land should be plowed if it is in sod and then left for 4 to 6 weeks for the sod to decay.

Science Connection: Encouraging the Master Tillers

Earthworms are nature's master tillers. Always a favorite for fishing bait, the lowly earthworm is now regarded as an ally of homeowners, gardeners, farmers, landscapers, and conservationists. They enhance soil tilth and crop growth by consuming and digesting organic matter and mixing it with the mineral content of the soil. The new mixtures left behind by feeding worms are called worm casts and are rich in nutrients for growing plants. Worm holes in the soil start at the surface and go as deep as the worm needs to go to escape winter cold and summer heat. This may be 3 feet deep or more in the central area of the United States. This network of holes and tunnels collects rainwater as it falls, and the enriched soil absorbs and holds water for future plant use.

The North Appalachian Experimental Watershed at Coshocton, Ohio, has been the scene of no-till experiments for more than 30 years. In a major experiment there, night crawlers (*Lumbricus terrestris*), usually 4 to 8 inches in length and ⅜ inch in diameter, and other species were studied. It was found that earthworms may be encouraged by leaving crop residues on the surface or by incorporating them into the soil. However, if crop residues are removed from the soil, the earthworms do not have an adequate food source, and their populations do not expand as much.

Most of their activity occurs in the first foot of topsoil. They make temporary burrows as they pass through the soil, consuming, digesting, and mixing soil particles and organic matter and excreting their rich casts.

Worm farming has become an important industry, and earthworms may be purchased for release for soil improvement. Worms grow and multiply the fastest where plenty of organic matter and moisture is available. Unfortunately, worms do not function as well in coarse, sandy soils as they do in soils where medium and fine particles are available. Gardeners, horticulturists, and farmers should consider practices that will encourage the work of the master tillers.

(A) Magnified image of earthworm eggs. (B) Adult earthworm—nature's master tillers.
© Henrik Larsson/Shutterstock.com. © galitson/Shutterstock.com.

Plowing

In the spring or fall, land should be plowed or spaded to a depth of 6 to 8 inches. Fall plowing has some advantages over spring plowing. First, the freezing and thawing action on the soil during the winter months improves the physical condition of the soil. Second, exposure of the soil to the weather can result in a reduced population of insects. Spring plowing should be done only a short time before planting. Take care not to plow the soil when it is too wet. Doing so tends to destroy the physical structure of the soil, resulting in hard clumps. The moisture content of loam or clay soil may be judged by pressing a handful of soil into a ball (**Figure 19-10**). If the ball crumbles easily, the soil is ready to plow. If the soil sticks together, it is too wet to plow.

Figure 19-10 Soil moisture in clay or loam may be judged by pressing a handful of soil into a ball. If the ball crumbles easily, the soil is ready to be worked.

Courtesy of DeVere Burton.

Maintaining Organic Matter

Organic matter increases the water-holding and absorption capacity of the soil. It also helps prevent erosion and promotes aeration in the soil. **Aeration** refers to the movement of air within the soil profile. Animal manure is the best material for maintaining the organic content of the soil. It is also a good source of nutrients. Manure from animals other than sheep and poultry can be turned under at the rate of 15 to 20 tons per acre, or 20 to 30 bushels per 1000 square feet. Sheep and poultry manures can be used, but at half this rate. Note: Animal manures are low in phosphorus; it may be necessary to use high-phosphorus fertilizer to balance the nutrients. For best results, apply fresh manure in the fall and well-rotted manure in the spring.

Another type of organic matter to use is green manure in the form of a cover crop, especially if it is a legume. **Green manure** is an active, growing crop that is plowed under to help build the soil. Green vegetation incorporated into the soil rots more quickly than dry material. A cover crop should be planted at the end of the growing season. A cover crop is a close-growing crop planted to prevent erosion.

Fertilizing

The amount and type of fertilizer to add can best be determined by conducting soil tests. Soil tests will help prevent application of too much or too little fertilizer and will determine the best timing. The application of commercial fertilizer is done to increase the availability of nutrients to the plants. Soils without additives are unlikely to provide the best combination of conditions for all vegetables.

The most used commercial fertilizers usually contain nitrogen, phosphoric acid, and potassium. Each vegetable crop needs varying amounts of each of these three elements. In general, fertilizer ratios used for home gardens are 5-10-10, 10-10-10, and 5-10-5 (**Figure 19-11**). The first number in a fertilizer ratio or grade represents the percentage of nitrogen, the second represents the percentage of phosphorus, and the third indicates the percentage of potassium in the fertilizer. A rough guide for home gardeners is that most vegetables need about 3 to 4 pounds of fertilizer per 100 square feet. Consult a county extension educator or master gardener for exact fertilizer recommendations based on the soil types in the area. University extension educators are specialists on local and regional conditions that affect crops and gardens.

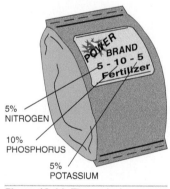

Figure 19-11 The analysis on a fertilizer bag indicates the percentages of nitrogen, phosphoric acid, and potassium in the fertilizer.

Liming

The need for lime should be determined by conducting soil tests. A pH greater than 7.0 indicates alkalinity; a pH of 7.0 is neutral; and a pH less than 7.0 indicates acidity.

If the test indicates a pH of 6.0 or less, lime should be applied. The lime should be mixed into the top 3 to 4 inches of the soil. The amount of lime to be added depends on the type of soil, the desired pH, and the form of lime used. Some plants, such as blueberries, require acidic soils. For such plants, special materials are available to make the soil acidic if the pH is too high.

Vegetable crops grow best at specific pH levels. **Figure 19-12** lists vegetables and their optimum pH ranges for maximum growth.

Planting Vegetable Crops

Vegetable crops are initially grown from seeds. Some seeds can be sown directly in the location where the plants will grow. Others need to be started indoors with the seedlings transplanted to the garden at a later date. **Transplants** are plants grown from seeds in

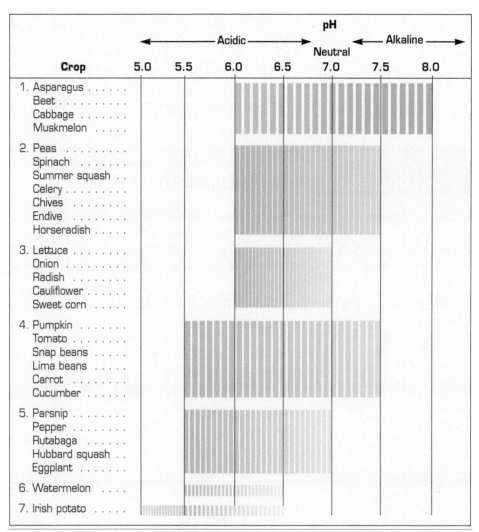

Figure 19-12 Optimum pH ranges for vegetable crops.

Figure 19-13 Some garden plants are grown in containers for later planting. This allows the gardener to extend the length of the growing season by starting frost-sensitive plants in the greenhouse or hotbed.
Courtesy of DeVere Burton.

a special environment, such as a cold frame, hotbed, or greenhouse (**Figure 19-13**). The method used depends on the climatic requirements of the plant and the germinating characteristics of the seed.

Planting Seed

Depending on the size and scope of the garden and the quantity of seed to be planted, vegetable seeds are planted (1) by hand in hills or rows, (2) through broadcasting by hand or machine, (3) with one-row hand seeders, or (4) with single or multiple-row, tractor-drawn seeders.

Broadcasting is a planting method whereby seeds are scattered on the soil surface. Regardless of the type of planting method used, the seeds should be planted at the proper depth. The soil should also be left smooth and firm over the seed. For most vegetables, seed must be fresh. Seed more than 1 year old will probably not germinate at the rate necessary to produce good yields.

Most vegetable seeds should be planted at a moderately shallow depth. It is best to plant in loamy-textured soil with adequate moisture after the danger of frost is over.

The seed will germinate best when it is planted at a depth no more than four times the diameter of the seed. To *germinate* means to sprout, grow, and produce a plant from a seed. Regardless of how deep the seeds are planted, the soil surface should be level and firmly pressed after the seed is planted. This gives the seed close contact with moist soil particles, improving germination. It also helps prevent the seed from washing away or water from puddling above the seed after a rain.

Science Connection: What Is Good About Fungi?

Fungi are plentiful organisms that are found in the soil. Despite the species of fungi that are known to be harmful to crops, some fungi are useful to crops. Mycorrhizae are fungi that growers welcome in their fields. These microorganisms do not have chlorophyll, and they cannot use the process of photosynthesis to make their own food. To survive, they attach to the roots of plants to obtain carbohydrate nutrients. Even though these fungi attach to plant roots, they are considered beneficial.

Phosphorus is a nutrient that plants have a difficult time extracting from the soil. Mycorrhizae fungi produce an enzyme that frees nutrients like phosphorous from the soil so the plants can use them. They also help plants take in water and other nutrients. Both the plants and the fungi benefit from this symbiotic relationship. Scientists have estimated that 80 percent of all plants form this kind of relationship with fungi. Fields with adequate levels of these fungi require less water and less fertilizer. Mycorrhizae fungi are commercially available, and their benefits have been demonstrated to far outweigh their costs.

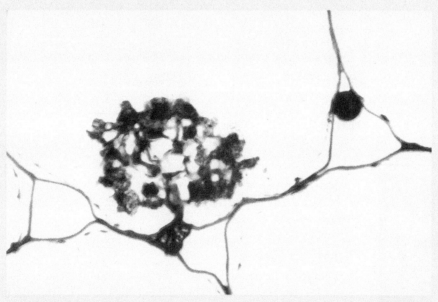

Mycorrhizae help host plants absorb nutrients from the soil and, in turn, they obtain carbohydrates from the plant.
Courtesy of USDA/ARS #K9438-1.

Transplanting Vegetable Seedlings

Three methods are used in transplanting, depending on the number of plants being transplanted: (1) hand setting, (2) hand-machine setting, or (3) riding-machine setting.

In home gardening situations, hand setting is most appropriate. Hand setting involves six steps: (1) dig a hole slightly bigger and deeper than the root ball of the plant being transplanted, (2) add some fresh soil to the hole, (3) place the plant in the soil a little deeper than it grew in its original container, trying to keep as much of the original soil as possible around the roots, (4) pull soil in and around the plant and firm it slightly, (5) add about a half pint of water and let it soak into the soil, and (6) pull in some dry soil to level off and cover the wet area, which helps prevent loss of moisture and baking of the soil.

Large commercial vegetable production practices rely on hand-machine setting or riding-machine setting methods for transplanting seedlings. Machines dig the planting furrows and pack the soil around the seedlings as workers place the seedlings. Soil conditions must be just right for the plants to survive. The soil must contain enough moisture to allow the seedling roots to become established. Stress on the seedlings can be greatly reduced by setting them out on a cloudy day or late in the afternoon. It is best to set plants just before or just after a rain.

Hot Topics in Agriscience: Robots and Artificial Intelligence (AI)

Agricultural robots lack many human-like characteristics, but in some respects, their artificial intelligence (applied computer visual recognition) and robotic arm/hand reaction skills make them ideal for the work they are programmed to do. For example, a weeding robot moves slowly through a field scanning each of the plants it encounters. The electric scanner then feeds a visual image to the computer, which determines whether each plant is a weed or a crop plant. Some robots are programmed to remove each weed, and others give each weed a tiny shot of herbicide. In either case, the weed is eliminated. As long as field conditions are favorable, the robot works tirelessly, and it never gets bored by repetitive tasks.

A tomato-harvesting robot is programmed to determine whether each tomato fruit is mature and ready to pick. Then it grasps each mature tomato with just enough pressure to remove it from the plant without crushing it, and places it gently in a container. Each harvesting robot must be carefully adjusted and adapted to specific characteristics of the crop, and new crop varieties are continually under development to make robotic harvesting more dependable.

A

B

Technology keeps changing agriculture: (A) Robots weeding a field. (B) A robot harvesting tomatoes.
© MONOPOLY919/Shutterstock.com.
© kung_tom/Shutterstock.com.

Cultural Practices

Cultivating

Cultivation, or intertillage of crops, is a proven agricultural practice. The benefits of cultivation are (1) weed control, (2) conservation of moisture, and (3) increased aeration. Cultivation increases the yield of most vegetable crops, mainly because the weeds are controlled.

All types of equipment are used to cultivate crops, from small hand tools to large tractors equipped with cultivating tools. The cultivation should be shallow and done only when there are weeds to be killed. Excessive cultivation is not beneficial and may even be damaging.

Controlling Weeds

Weeds are easier to control when they are small, using shallow cultivation. Weeds that are more established are more difficult to control. For them, deeper cultivation is required to rid the area of weeds. This can result in injury to the roots of vegetable crops (**Figure 19-14**). The key to weed control is to remove them as soon as they show up and as often as needed.

Figure 19-14 Mechanical weed control is applied regularly in the production of vegetable crops.
Courtesy of USDA/ARS # K5226-18.

Using herbicides is appropriate for large home gardens and commercial vegetable production plots. They can greatly reduce the amount of effort that is required to remove weeds by mechanical methods. The use of any herbicide depends on the registration of the herbicide by federal and state environmental protection agencies. Do not use any herbicide unless it is clearly stated on the label that it is intended for a particular crop. Always seek the latest information for use of herbicides on vegetables.

Irrigation

Nearly all commercial vegetable operations irrigate their crops. Irrigation is especially important in California and other arid and semiarid, or dry, regions of the West. Irrigation is important because rainfall is rarely uniform and adequate for high yields.

Water for irrigation can be obtained through established water rights from a stream, lake, well, spring, or stored stormwater. Definite laws and regulations guide the practice of irrigation in each state. Several types of irrigation systems are available, including sprinkler irrigation; drip irrigation; surface or furrow irrigation; and subirrigation. The selection of a system should be guided by local needs and proven practices in the region.

Mechanical sprinkler systems are versatile machines (**Figure 19-15**). They can be used in many situations, such as irrigating when soil moisture is in short supply. They also can deliver shallow and light irrigation to promote seed germination and can be used for the application of liquid fertilizers. Drip irrigation uses a system of pipes and tubes to deliver the water to individual plants rather than wetting all of the soil (**Figure 19-16**). This method is gaining in popularity because of its water-conserving benefits.

Figure 19-15 Sprinkler irrigation is a proven method for providing water to vegetable crops. The water can be applied uniformly in the proper amounts.

Courtesy of DeVere Burton.

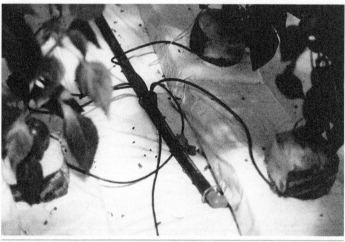

Figure 19-16 Drip or trickle irrigation delivers water to individual plants through small openings in pressurized tubes. It is the most efficient method of irrigation for vegetable crops.

Courtesy of DeVere Burton.

Figure 19-17 Mulching is the practice of adding material such as straw, leaves, or plastic film to the surface of the soil to reduce or prevent the growth of weeds.

Courtesy of DeVere Burton.

Surface or furrow irrigation has the advantage of requiring a relatively low investment. The topography of the land is important for surface irrigation. The land must be gently sloping and uniform. This method is well suited for irrigating vast areas of land, but it generally requires land leveling to work well.

Subirrigation requires large amounts of water. With this system, the water is added to the soil so that it permeates the soil from below. This method is expensive and is suited only to commercial enterprises in locations where the depth of groundwater can be manipulated.

Drip or trickle irrigation is a form of subirrigation in which water is delivered under low pressure near the plants. The water seeps through porous hose under the rows. This system is expensive, but gives excellent results, with increased yields and less water consumed than with other methods.

Mulching

Mulch is created whenever the surface of the soil is modified artificially by covering it with straw, leaves, grass clippings, paper, or polyethylene film (**Figure 19-17**). Mulching helps control weeds, regulates soil temperatures, conserves soil moisture, and provides clean, weed-free vegetables. The mulching material should be spread around the base of the plants and between the rows.

Figure 19-18 Insects such as the Colorado potato beetle (shown) and the Mexican bean beetle can strip the leaves from vegetable plants in a matter of hours.
Courtesy of USDA/ARS #K1291-2. Photo by Tim McCabe.

Plastic film is frequently used with cantaloupes, melons, cucumbers, eggplants, summer squash, and tomatoes. It is important that the plastic be laid over moist soil that has been freshly prepared. It is advisable to lay plastic mulch 1 to 2 weeks before planting the vegetables. This allows the soil to warm up underneath the mulch.

Pest Control

Good pest-control practices are essential for quality vegetables. A combination of cultural, mechanical, biological, and chemical methods can be used. Insecticides and fungicides can be effectively used by both home and commercial vegetable growers.

Insect and disease infestations bring about losses that can be devastating to vegetable crops (**Figure 19-18**). Types of losses include the following: (1) reduced yields, (2) decreased quality of produce, (3) increased costs of production and harvest, and (4) increased expenditures for materials and equipment. The use of chemical control of pests in vegetable production is sometimes restricted due to proximity of vegetables for which a particular chemical has not been approved. Government standards may apply when a crop matures before sufficient time has passed to reduce the pesticide residues to acceptable levels.

Modern biological controls are expected to play a greater role in vegetable production in future years. Integrated pest management (IPM) practices call for a combination of control measures to be used to combat pests and diseases in plants. This approach puts priority on controlling pests with the least invasive methods; however, it does not rule out the limited use of chemical pesticides when other methods of control prove ineffective.

Many measures can be taken to help vegetable growers control insects and diseases (**Figure 19-19**). Some of these are the following:

- Dispose of old crop residues that serve to shelter pests, and thus enable them to overwinter or survive from one crop to another.
- Rotate the location of individual crops.
- Choose insect- and disease-resistant vegetable varieties.
- Treat seeds with fungicides and insecticides.
- Control weeds.
- Use adequate, but not excessive, amounts of fertilizer.
- Follow recommended planting dates.

A

B

C

Figure 19-19 The following practices can help control insects and diseases. (A) Plant disease-resistant varieties. (B) Spray plants with nontoxic Bt to spread insect-fighting bacteria. (C) Introduce insects that eat pests. A spined soldier beetle eats a leaf-devouring Mexican bean beetle larva.
Courtesy of USDA/ARS #K-3461-14. Courtesy of USDA/ARS #K-2785-2. Courtesy of USDA/ARS #K1461-3.

Hydroponics

The ability to produce vegetables on a commercial scale without soil has contributed to the supply of tomatoes, cucumbers, and other fresh vegetable crops during the winter season (**Figure 19-20**). The plant most often produced using hydroponic methods is the tomato (see Chapter 9). Although actually a fruit, the tomato is used as a vegetable, and it is included in this chapter for that reason. Most commercial greenhouses that produce tomatoes plan to deliver most of the crop during the winter season when tomato prices are highest. The plants are usually rooted in inert filler material through which water and nutrients flow, or they may be placed in containers to which the nutrients and water are delivered using drip irrigation methods. Large hydroponic operations harvest the crop daily or every other day during the production cycle. (See Chapter 9 for more information on hydroponics.)

Figure 19-20 The use of hydroponic methods to produce vegetables is big business. Water and nutrients may be delivered through a channel in the floor or by drip irrigation tubes.

Courtesy of DeVere Burton.

Hot Topics in Agriscience: Organic Vegetable Farming

A farmer who raises *organic* vegetables produces them without using the chemical pesticides and fertilizers on which most vegetable farmers depend. This does not mean that the vegetable crop is produced without insect controls or fertilizers. Fertilizers are obtained from animal manure and/or green manure crops such as legumes that are plowed under when they are 6 to 12 inches in height. It simply means that an organic farmer first tries to maintain a healthy environment for the plants. This is because a healthy plant located in an environment that supports its needs is less susceptible to pests and diseases than a plant that is not well adapted to its environment.

The organic farmer may also introduce the natural enemies of pests to the environment to control their populations. In addition, pepper and garlic sprays are common methods for discouraging insect pests. As a final resort, the organic farmer may apply sprays consisting of horticultural oils, isopropyl alcohol, soaps, ammonia, baking soda, and even bug juice (a spray made with water and insects that have been pulverized in a blender). Some chemicals such as rotenone and sulfur are acceptable to organic farmers because they break down easily and are degraded quickly to materials that are no longer harmful.

Organic vegetables that are of high quality command a premium price among some consumers in comparison with the price of produce that was produced using standard farming methods that rely on chemical fertilizers and pesticides. Regulations are in place in most states that restrict the additives that may be used and the cultural practices that may be implemented to produce and market organic vegetables.

Harvesting, Marketing, and Storing Vegetables

Harvesting

For the best possible quality, vegetables should be harvested at the peak of maturity. This is possible when the vegetables will be used for home consumption, sold at a local market, or processed. The best harvesting time varies with each vegetable crop. Some will hold their quality for only a few days, whereas others can maintain quality over a period of several weeks. Frequent and timely harvest of crops is needed if the vegetable grower wishes to supply the market or the kitchen table with high-quality produce over time. Specifications for harvesting each kind of vegetable crop should be consulted to achieve the best results.

Harvesting is done by hand and through the use of machines for a few crops. Mechanical harvesting is especially useful for the commercial grower (**Figure 19-21**). When harvesting a crop, care must be taken to prevent injury or bruising to the produce due to overly aggressive machine action or poor handling methods.

A

B

Figure 19-21 (A) Mechanical harvesting of vegetable crops such as potatoes requires special machinery that is designed to separate the crop from leaves, stems, and soil. (B) Metal parts are often rubber-coated to avoid bruising the tubers.
Courtesy of DeVere Burton. Courtesy of DeVere Burton.

Science Connection: Ripening Genes—Key to Better Taste and Longer Storage

Tomatoes benefit from new gene sequences that produce firm, full-sized, flavorful fruits.
© iStockphoto/kokophoto.

The discovery of gene sequences in tomatoes that control the ripening process is coming to *fruition* in the tomato industry, and perhaps the entire fruit industry will benefit. Scientists at the U.S. Department of Agriculture and the Boyce Thompson Institute for Plant Research, Inc., at Cornell University have discovered *ripening genes*. These genes are from a family of genes known as "MADS" and consist of more than 100 gene sequences that are found in plants, animals, and microbes. They influence the developmental processes of living things, including the ripening process of fruits.

It is hoped that this discovery might soon make it possible to delay the ripening process until all fruits are fully grown and then make it possible to harvest full-sized, ripe tomatoes that are still firm enough to ship. Allowing the fruits to remain on the vine until they are ripe gives them the flavor of home garden tomatoes instead of the bland flavor of fruits that are shipped before ripening. These gene discoveries are expected to have similar ripening effects on several fruits and vegetables, including strawberries, bananas, melons, and bell peppers. Imagine, perhaps someday, these commodities may be stored and shipped without spoiling and with little or no refrigeration!

Marketing

The most important aspect of marketing fresh vegetables is to assure that the quality is high. This requires the vegetable plants to be protected from the stresses of disease and insects. Vegetables must be harvested when they are of optimum size and maturity. At this point in the process, it is important to market the produce as soon as possible.

Farmers markets are ideal outlets for marketing fresh vegetables to local consumers (**Figure 19-22**). These markets are intended to bring several farmers together in a central location with vegetables and fruits of sufficient quality and quantities to attract local residents to the site. In some instances, these markets are established on private property. They may also be located on federal, state, or city property as long as permission is obtained from the proper government authorities. The U.S. Department of Agriculture has developed rules under which these markets can be legally established in local communities. It is important for a manager to assume the duties of running the business. Good business practices and high-quality fruit and vegetable products are keys to operating a successful farmers market.

Figure 19-22 A farmers market is organized by a group of farmers who bring fresh fruits and vegetables to a single location. Such a market serves both those who sell and those who buy.
© hillaryfox/iStockphoto.com.

Storing

When storing vegetables, the proper stage of maturity is essential. There are ideal temperatures and humidity levels for storage of vegetable crops (**Figure 19-23**). Most crops need 90 to 95 percent humidity. Most homeowners find it difficult to reproduce the exact temperatures and humidity levels, but commercial growers can accomplish storage through specialized facilities (**Figure 19-24**).

Vegetables differ in the amount of time that they can be stored. Some may be stored for several months, others for only a few days. For the vegetables that store well, good storage conditions are essential to prolong the marketing period.

Fresh vegetables placed in storage should be free of skin breaks, bruises, decay, and disease. Such damage or disease will decrease storage life.

Refrigerated storage is recommended (**Figure 19-25**). This type of storage reduces respiration and other metabolic processes. It also slows ripening; slows moisture loss and wilting, reduces spoilage from bacteria, fungus, or yeast, and reduces undesirable growths, such as potato sprouts.

Vegetable Storage		
Crop	Temperature °F	Relative Humidity Percent
Asparagus	32	85-90
Beans, snap	45-50	85-90
Beans, lima	32	85-90
Beets	32	90-95
Broccoli	32	90-95
Brussels sprouts	32	90-95
Cabbage	32	90-95
Carrots	32	90-95
Cauliflower	32	85-90
Corn	31-32	85-90
Cucumbers	45-50	85-95
Eggplants	45-50	85-90
Lettuce	32	90-95
Cantaloupes	40-45	85-90
Onions	32	70-75
Parsnips	32	90-95
Peas, green	32	85-90
Potatoes	38-40	85-90

Figure 19-23 Ideal temperature and humidity levels for storing common vegetables.

Storage Conditions							
Vegetable	Temperature (°F)	Relative Humidity (%)	Storage Life	Vegetable	Temperature (°F)	Relative Humidity (%)	Storage Life
Pea, English	32	95-98	1-2 weeks	Squash, winter	50	50-70	___[4]
Pea, southern	40-41	95	6-8 days	Strawberry	32	90-95	5-7 days
Pepper, chili (dry)	32-50	60-70	6 months	Sweet corn	32	95-98	5-8 days
Pepper, sweet	45-55	90-95	2-3 weeks	Sweet potato	55-60[3]	85-90	4-7 months
Potato, early	___[1]	90-95	___[1]	Tamarillo	37-40	85-95	10 weeks
Potato, late	___[2]	90-95	5-10 months	Taro	45-50	85-90	4-5 months
Pumpkin	50-55	50-70	2-3 months	Tomato, mature green	55-70	90-95	1-3 weeks
Radish, spring	32	95-100	3-4 weeks	Tomato, firm ripe	46-50	90-95	4-7 days
Radish, winter	32	95-100	2-4 months	Turnip	32	95	4-5 months
Rhubarb	32	95-100	2-4 weeks	Turnip greens	32	95-100	10-14 days
Rutabaga	32	98-100	4-6 months	Water chestnut	32-36	98-100	1-2 months
Salsify	32	95-98	2-4 months	Watercress	32	95-100	2-3 weeks
Spinach	32	95-100	10-14 days	Yam	61	70-80	6-7 months
Squash, summer	41-50	95	1-2 weeks				

[1]Spring- or summer-harvested potatoes are usually not stored. However, they can be held 4-5 months at 40°F if cured 4 or more days at 60-70°F before storage. Potatoes for chips should be held at 70°F or conditioned for best chip quality.

[2]Fall-harvested potatoes should be cured at 50-60°F and high relative humidity for 10-14 days. Storage temperatures for table stock or seed should be lowered gradually to 38-40°F. Potatoes intended for processing should be stored at 50-55°F; those stored at lower temperatures or with a high reducing sugar content should be conditioned at 70°F for 1-4 weeks or until cooking tests are satisfactory.

[3]Sweet potatoes should be cured immediately after harvest by holding at 85°F and 90-95% relative humidity for 4-7 days.

[4]Winter squash varieties differ in storage life.

Figure 19-24 Ideal temperature, humidity level, and storage life for common vegetables.

Figure 19-25 Climate-controlled vegetable storage units control humidity, temperature, and sometimes the level of oxygen in the atmosphere. This makes it possible to extend the storage life of vegetables and fruit.

Courtesy of DeVere Burton.

One practice that can increase the successful storage of vegetables is pre-cooling. **Pre-cooling** is the process of rapidly removing the heat from freshly harvested vegetables before storage or shipment. One effective method of pre-cooling is termed **hydrocooling**. This is a method wherein vegetables are immersed in cold water where they are immersed long enough to reduce the core temperature to the desired level.

Chapter Review

Student Activities

1. Write the **Key Terms** and their meanings in your notebook.
2. Using images from digital or print seed catalogs, prepare flash cards of warm-season and cool-season vegetables.
3. Bring in an uncommon vegetable to show to the class. Be prepared to tell the class which part is edible.
4. Select a site for planting a vegetable crop and prepare a planting plan to scale.
5. Take a soil sample from the site selected in Activity 4. Submit the sample to the local county extension service for analysis and lime and fertilizer recommendations for the vegetables planned.
6. Plant lima bean seeds in containers to observe the growth of the stem and leaves. Determine whether the plant is a monocotyledon or a dicotyledon.

How would you classify each of these plant types? Why?

7. Plant tomato seeds in a greenhouse for later transplanting.
8. Develop a chart with pictures illustrating the various insects that attack vegetable plants. Find pictures of insects in vegetable catalogs or extension service bulletins.
9. Store vegetables under different conditions (light/dark, dry/humid, warm/cool). Record what happens under each set of conditions. Note which conditions are best for each vegetable.
10. Visit a grocery store and count the number of vegetables offered to consumers. Notice that a number of varieties of each vegetable are usually available. For example, how many varieties of lettuce are there? In your opinion, what are the advantages of producing more than one crop? What are the disadvantages?
11. **SAE Connection:** As a class, create a business plan for a virtual vegetable production enterprise. The plan should include all estimated costs, such as for fuel, seed, hired help, and so on. The size of the property and the type of crop to be planted should be given. Assume that the business will have a fair yield and the crop will sell at the average market value. Find the estimated profit by subtracting the income from the costs. As a class, discuss the business plan.

National FFA Connection — Vegetable Production Proficiency Award

The FFA promotes proficiency in management practices for the production of vegetables.
© Somchai_Stock/Shutterstock.com

The National FFA Organization offers separate proficiency awards for vegetable production. Entrepreneurship and Placement proficiency awards can be earned at the local, state, and national levels. A student who establishes ownership of a business producing and marketing vegetables is eligible to compete for the entrepreneurship award. A student who works for a business that produces and markets vegetables may compete for the placement award.

Each of these proficiency awards requires students to practice "best available management practices" in both production and marketing of vegetable crops. Local agriscience teachers can provide guidance as you become involved in vegetable production and as you compete for the proficiency award for which you are best qualified.

Check Your Progress

Can you…

- name some of the benefits of vegetable production as a personal enterprise or career opportunity?
- explain how to prepare a site for planting?
- describe how to plant vegetable crops?
- give examples of proven agricultural practices?
- recall appropriate procedures for harvesting and storing at least one commercial vegetable crop?

If you answered "no" to any of these questions, review the chapter for those topics.

Chapter 20

Fruit and Nut Production

Objective

To determine the opportunities for a career in fruit and nut production, and evaluate the recommended production practices for key fruit and nut crops.

Competencies to Be Developed

After studying this chapter, you should be able to:

- determine the benefits of fruit and/or nut production as a personal enterprise or career opportunity.
- identify fruit and nut crops.
- plan a fruit or nut production enterprise and prepare a site for planting.
- describe how to plant fruit and nut trees and use appropriate cultural practices in fruit and nut production.
- list appropriate procedures for harvesting and storing at least one commercial fruit or nut crop.

Key Terms

pomologist
semidwarf
dwarf
pome
drupe
nursery
terminal
cane

Home fruit operations provide high-quality and bountiful varieties of fruit and nuts. Fruits grown at home can be enjoyed at the peak of ripeness. This quality is difficult to obtain from supermarket fruits. Home fruit and nut operations can be enjoyed as hobbies or turned into profitable commercial enterprises.

Career Opportunities in Fruit and Nut Production

Like other agricultural enterprises today, the production of fruit and nut crops is more a business than a way of life. The small-fruit grower has to be a good financial manager as well as knowledgeable about science and farming. Producing fruit and nuts, however, can be an enjoyable and rewarding profession.

Career Descriptions

The production, harvest, and marketing of fruits is all part of the large fruit industry. The United States accounts for a sizable part of the combined world crops of apples, pears, peaches, plums, prunes, and oranges, grapefruit, limes, lemons, and other citrus fruits.

Career opportunities exist at all levels of the industry. One can be an owner/grower, orchardist, plant manager, foreman, or technician. People who work in this field must be able to propagate fruit/nut trees and vines as well as plant, transplant, prune, thin, train, and fertilize them. They must also be able to provide protection from diseases and insects. A **pomologist** is a fruit grower or fruit scientist. There is currently a shortage of educated plant science graduates in the fruit industry, including pomologists (**Figure 20-1**).

Figure 20-1 Research into methods of production conducted by pomologists and other scientists has led to improved varieties of fruits and nuts.
© Supavadee butradee/Shutterstock.com.

Identification of Fruits and Nuts

Types of Fruits

There are three broad categories of fruit. Some fruits grow on trees, some grow on canes and small bushes, and others grow on vines. The shape and form of a plant, whether it is a tree, a bush, a cane, or a vine, is known as the growth habit of the plant. Fruit is produced on woody twigs and stems from plants that exhibit each of these growth habits.

Tree Fruits

Tree fruits are popular with home gardeners as well as large producers. One drawback, however, is that some types of trees can take several years to mature before producing the first harvest. Three types of fruit trees based on growth habit are in production in the fruit industry. They are standard, semidwarf, and dwarf. These three terms refer to the size of the tree and the length of time it takes from planting until the first crop is produced (**Figures 20-2A** and **B**).

A standard tree is one that has its original rootstock, its root system and stem, and reaches normal size. A standard apple tree can grow to be 30 feet high. However, semidwarf and dwarf trees are smaller. They are standard varieties of trees, but they are grafted onto dwarfing rootstocks. Such rootstocks cause the tree to produce less annual growth. Thus, it remains smaller than a standard tree throughout its life. The semidwarf tree averages 10 to 15 feet in height, whereas dwarf trees average 4 to 10 feet. Dwarf trees are better suited to the home garden because they bear earlier and take up less space than semidwarf or standard trees (**Figure 20-3**).

There are two types of tree fruits: pome and drupe. A pome is a fruit with a core and embedded seeds. The common pome fruits are apples, quince, and pears. A drupe is a fruit with a large, hard seed called a stone. Drupe fruits include peaches, plums, apricots, and cherries.

When growing tree fruits, particular attention must be paid to the climate of the area. Peaches, plums, and cherries are especially sensitive to climatic conditions.

Small-Bush Fruits

Small-bush and cane fruits that are popular with gardeners, and commercial operations include strawberries, blueberries, red raspberries, currants, gooseberries, and thornless blackberries (**Figure 20-4**). Small-bush fruits grow either low to the ground or only 3 to 4 feet high. They require less maintenance than tree fruits or vine fruits, and they tend to bear quickly—usually between 9 months and 1 year after planting. Some of these fruits will bear a harvest in the summer and again in the fall. A factor to consider in growing bush fruits is to buy plants from a reliable nursery that stands behind their products. It is vital that the new nursery stock required to establish a fruit enterprise is disease-free and healthy.

A

B

Figure 20-2 (A) twelve-year-old semidwarf apple trees and (B) a seven-year-old dwarf apple tree.

Courtesy of USDA/ARS #K4048-17.
© ustun ibisoglu/Shutterstock.com.

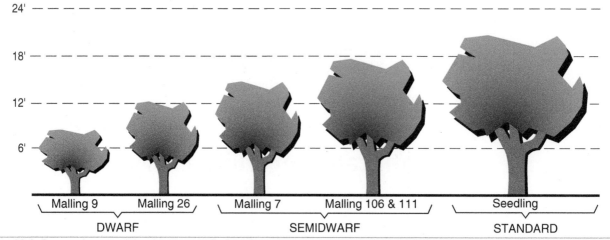

Figure 20-3 Rootstocks control the size of a tree. Each of the codes refers to a particular rootstock with known "dwarfing" characteristics.

Figure 20-4 Agricultural engineer David Peterson evaluates his spiked-drum shaker as it harvests a row of blackberry bushes.

Courtesy of USDA/ARS #K2949-6.

Vine Fruits

The best-known vine fruit is the grape. Grapes are also the most prevalent vine fruit. They are easily grown and have a wide range of varieties and flavors. Grapes will occupy the land for many years, so site selection is important. The vines need to be trained and pruned for production and yield. Grapes require a growing season that has at least 140 frost-free days. A disadvantage of grape production is the 3- to 4-year wait for the vines to reach maturity before they produce fruit (**Figure 20-5**).

Types of Nuts

Nut trees can be useful for shade, landscaping, food, and wood. Wherever trees grow, a nut tree can also be grown. Popular nut trees grown in the United States are pecan, black walnut, filbert, hickory, almond, English walnut, chestnut, and macadamia (**Figures 20-6**).

Figure 20-5 Grapes of many varieties are popular as fresh fruit, dried fruit (raisins), juice, and wine.

© Kitamin/Shutterstock.com.

Figure 20-6 The black walnut tree can be used for shade in landscaping. It also produces a good-quality, edible nut.

Courtesy of DeVere Burton.

Planning Fruit and Nut Enterprises

Selecting Fruit or Nuts to Plant

Certain varieties and cultivars of fruits or nuts are more suited to one type of climate than another. For example, citrus fruits can be grown only in southern-type climates, such as in Florida, Texas, and California. This information is a key factor when deciding which fruit to grow. Recommendations for each fruit crop for a given area can be obtained from local county agricultural extension educators.

Selecting Rootstock

The variety of the root or rootstock of a grafted tree is important because many fruit crops are propagated by grafting. Select a rootstock that is true to its variety and is hardy enough to support the tree. Most nurseries will readily provide this information about the rootstock. The rootstock will determine whether the tree is a standard, semidwarf, or dwarf tree.

Agri-Profile Career Areas: Pomology/Fruit Grower/Nut Grower/Packer

Pomology is the scientific study and cultivation of fruit, including nuts. Careers in pomology include work on farms and in nurseries, orchards, and groves. Fruit and nut specialists may be plant breeders, propagators, educators, food scientists, or consultants. Technical-level jobs are available for graders, packers, supervisors, and managers.

In the past, fruit production meant hard labor because of handpicking. However, the frequent shortages of labor for harvesting have created new opportunities for agricultural engineers, electronics technicians, plant breeders, systems managers, food scientists, and inventors to develop mechanical harvesting methods. Mechanical harvesters range from hydraulically controlled buckets, to tree shakers, to robots directed by visual sensors teamed with complex computer programs (artificial intelligence).

Development of new pecan tree varieties is monitored continuously during the growing season.
Courtesy of USDA/ARS #K3318-8.

One promising new technology is the use of harvesting robots teamed with computers that are capable of visual scanning to locate each fruit.
© asharkyu/Shutterstock.com.

Frost Susceptibility

Locate crops in areas as protected from frost as possible. Also, select varieties that are late bloomers if your area is susceptible to late frosts. Planting areas on hillsides, near ponds, or near cities will remain warmer on clear, frosty spring nights. Such locations decrease the amount of damage caused by frosts. Apricots and sweet cherries are especially susceptible to late frost damage. The home gardener can protect some fruit crops, such as strawberries, from frost by re-mulching or irrigating and by placing cloth or plastic sheets over plants. Commercial growers use irrigation sprinklers, heaters, and fans to help reduce frost damage.

Fertility

The production of high-quality fruit is dependent on the fertility of the soil. On average, soil should have a pH of 5.5 to 6.0. However, for growing blueberries, the soil should be more acidic—a pH of 5.2 or less. Fruit crops should be fertilized annually after they are planted. However, too much fertilizer can damage roots and create other problems. One pound of 10-10-10 fertilizer per 100 square feet of surface soil should give good results.

Planning for Pollination

Most fruit crops require pollination. Pollination is defined as the transfer of pollen from the male stamen to the stigma, or female part of the plant (**Figure 20-7**). Pollination is necessary for the fruit to set, or form. The gardener should choose crops that are known for having high-quality and abundant pollen. The pollen is distributed by wind or insects. Sometimes it is necessary to have several varieties that bloom at the same time for adequate cross-pollination. Similarly, it may be necessary to plant a tree just for its pollen quality and availability. Many fruit growers own or rent honeybees to help ensure good pollination (**Figure 20-8**).

When favorable conditions exist, just two varieties of a particular fruit crop are necessary for cross-pollination. Choose varieties that bloom at the same time or, at least, overlap for a period. Varieties of pears, plums, and cherries all tend to bloom at approximately the same time. However, apple varieties differ greatly in their times of bloom. Careful selection of apple varieties is important to ensure good cross-pollination.

Figure 20-7 The honeybee is one of our most important insects because it is such a good pollinator. It transfers pollen from one plant to another as it gathers nectar.

Courtesy of DeVere Burton.

Figure 20-8 Honeybees are essential for good pollination of some fruit trees. The sale of honey can provide another source of income for the fruit grower.

© Tischenko Irina/Shutterstock.com.

Growing and Fruiting Habits

The various fruit and nut crops have different growth patterns and fruiting dates. This is especially true of fruit trees. Some varieties of apple trees require as long as 8 years before they bear fruit. Newly planted peach and sour cherry trees can bear fruit in as few as 3 years. It pays to understand what to expect from the chosen varieties.

Most fruit trees go through two stages of growth: a vegetative period and a productive period. The vegetative period consists of rapid, vigorous growth of the tree. In younger trees, 4 to 5 feet of growth may occur in a single season. When the tree enters the productive period of fruit bud formation, a change is noticeable. Fruit buds are larger and plumper. On apple, pear, cherry, and plum trees, short, spur-like growths occur (**Figure 20-9**). On the peach tree, the peach buds become larger and occur in clusters, with two large ones surrounding a center one. The vigorous terminal growth slows down to only 18 to 20 inches in a season.

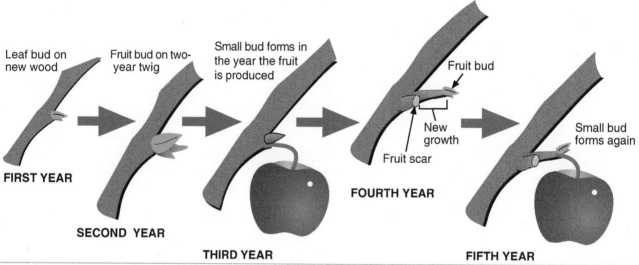

Figure 20-9 Growth of an apple-fruiting spur.

Soil and Site Preparation

Selecting the Site

Because many fruits require full sunlight for maximum production and yield, choose a site that allows for maximum exposure to the sun. A southern exposure is best. To help prevent frost damage, select land with a slight slope and good air circulation. This is important because cold air always flows downhill to the lowest location, leaving warmer air on the slope. The soil for fruit crops should be well drained and of medium texture.

Soil Preparation

A soil sample should be taken a year in advance of planting. Because many fruit and nut crops have specific pH and fertilizer requirements, a soil sample tested by a reputable testing laboratory will provide important information. Add lime to adjust the soil pH.

Fertilizer should be applied as needed. The soil should also be deeply plowed because the root systems of fruit trees and bushes grow deep into the soil. Deep plowing can be accomplished the year before the trees are planted. This permits the freezing and thawing action of the winter months to improve the texture of the soil.

Caution should be observed when planting small-bush and cane fruits in soils that have recently produced vegetable crops, such as peppers, tomatoes, and potatoes. These vegetables may leave residues in the soil that contain disease organisms to which the fruit plants are susceptible.

Planting Orchards or Small-Fruit and Nut Gardens

When to Plant

Tree stock from reputable nurseries can be planted either in the spring or fall. There are advantages to planting in each season. Fall planting is advisable if the soil will be less than desirable or wet in the spring. Planting in the fall also allows the crop to get its root system established before rapid top growth occurs in the spring. Weather conditions may also be more stable in the fall than in the spring.

Planting in the spring has some advantages:
- The crop can make considerable top and root growth before the stress of the winter season.
- The threat of damage by rodents is reduced.
- The threat of winter kill is reduced.

Laying Out the Orchard

Each fruit or nut species and type will have its own space requirements. It is important to follow the specific spacing recommendations for each variety. When planting small trees, the distance between trees seems enormous. However, the trees will grow and fill the spaces in time. For aesthetic reasons, plant the trees in a straight line. A crooked row of trees will be obvious. Orchards are often organized in a grid pattern so that tractors and equipment may control weeds between rows and between the trees within the rows.

Planting the Fruit or Nut Crop

A scion is the main stem of an immature fruit tree from which branches and leaves emerge. It is obtained from a branch of a mature fruit tree, and it will produce the same kind of fruit as the tree from which it came. A rootstock is a piece of a root from a hardy, related tree species to which a scion or bud is attached to form a graft. The grafted tree will be stronger and hardier than a tree obtained from the seed of the "mother" tree.

Each fruit or nut variety has its own specifications for planting. To plant a tree, make the hole wide enough to accommodate the roots without crowding them. The hole should be deep enough to allow the tree to be set 2 to 3 inches deeper than it was in the nursery. Make sure that the graft between the rootstock and the scion faces to the north to avoid sunburn damage. The two sides of the hole should be parallel but not smooth. This allows the roots to attach more easily. The bottom should be as wide as the top. By making the hole to these specifications, the soil can be packed uniformly around the roots to avoid leaving air spaces. Air spaces permit the roots to dry out, reducing the likelihood that the plant will survive.

When planting trees, add some prepared soil to the hole. Then spread the roots, adding more soil until all roots are covered. Move the tree up and down several times, while firming the soil around the roots. This procedure allows all roots to be in contact with the soil and avoids air pockets.

Before the hole is completely filled, add about 2 gallons of water. After the water soaks in, add the final soil. Most trees need to be staked for good support. A metal pipe or post may be driven 8 inches from the tree and the tree tied to it (**Figure 20-10**). Following planting, keep a 12-inch circular weed-free and grass-free area around the tree.

Cultural Practices in Fruit and Nut Production

Fertilization

A complete fertilizer, such as a 5-10-10 or a 10-10-10, should be used annually on fruit and nut trees. After filling the hole of a newly planted tree, apply 1 lb. of the 5-10-10 or a half lb. of 10-10-10 fertilizer around each tree. Do not place the fertilizer closer than 8 to 10 inches from the trunk. Excessive fertilization can result in damage or death to young trees. After the first year, apply fertilizer at the rate of 1 lb. times the age of the tree. After the tree is 6 to 8 years old, do not increase the rate of application. As the tree becomes larger, fertilize all of the ground area under the outer dripline of the tree limbs.

These rates are useful for all tree fruits except pears. Pears should not be fertilized because increased susceptibility to fire blight more than offsets the benefit from fertilizing.

Figure 20-10 Provide support for the newly planted tree.
© Elena Elisseeva/Shutterstock.com.

Science Connection — Saving the Fruit with Ice

Ice on fruit blossoms can trap an insulating layer of water vapor inside the flower that prevents the fruit crop from the effects of freezing.
© iStockphoto/Juliann Itter.

Fruit growers are at risk of losing money when their crops freeze. A number of techniques have been developed to reduce the risk of this happening. One successful method that is often used by fruit and vegetable growers appears to defy science. When air temperatures fall to 32°F, the irrigation sprinklers are turned on. Water falls onto the fruit trees and becomes frozen. At this point, all a nervous farmer can do is wait to see if their efforts will save the harvest.

At first thought, this method makes no sense. Why would a person add water to trees that will surely become frozen? The explanation seems to be a trick of science. As the water is misted onto fruit blossoms, it begins to freeze. As this happens, heat that is trapped inside the blossom is released. The result is that a very thin layer of water vapor becomes trapped between the blossom and the newly formed ice crystals. The water vapor acts as insulation protecting the ovule of the flower from freezing.

Fertilization requirements for the various bush and cane fruits vary considerably. Therefore, it is advisable to consult a grower's guide for each variety.

Pruning

The purpose of pruning and training young fruit trees is to establish a strong framework of branches that will support the fruit. This framework will resemble the shape best suited for the tree and for the fruit it will bear. Each fruit tree has a specific way that it should be pruned. For example, peach trees are pruned for an open-center, V-shape (**Figure 20-11**). However, apple trees should have a strong central leader and be pruned into a Christmas-tree shape (**Figure 20-12**).

Pruning a grapevine is based on its fruiting habit. Next year's grape clusters are on shoots produced during the current growing season. Therefore, this new growth should be carefully pruned. When the shoots have fruited and the leaves have fallen, the shoots are called **canes**. The canes will produce the new shoots and next year's crop.

Figure 20-11 A peach tree should be pruned to an open-center, V-shape in the center of the tree.

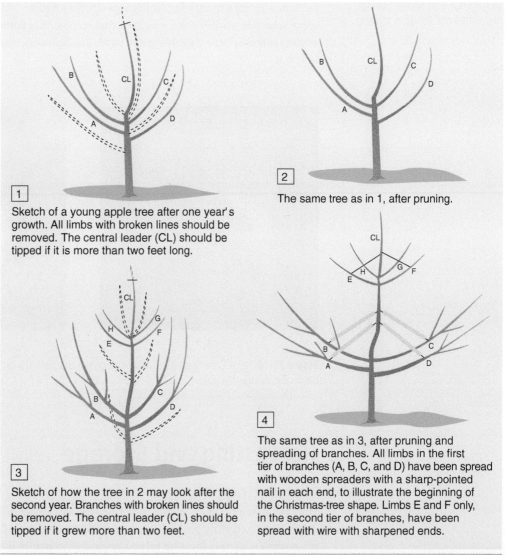

1 Sketch of a young apple tree after one year's growth. All limbs with broken lines should be removed. The central leader (CL) should be tipped if it is more than two feet long.

2 The same tree as in 1, after pruning.

3 Sketch of how the tree in 2 may look after the second year. Branches with broken lines should be removed. The central leader (CL) should be tipped if it grew more than two feet.

4 The same tree as in 3, after pruning and spreading of branches. All limbs in the first tier of branches (A, B, C, and D) have been spread with wooden spreaders with a sharp-pointed nail in each end, to illustrate the beginning of the Christmas-tree shape. Limbs E and F only, in the second tier of branches, have been spread with wire with sharpened ends.

Figure 20-12 A young apple tree should be pruned to maintain a strong central leader with strong, wide-angled main branches to support the weight of the fruit.

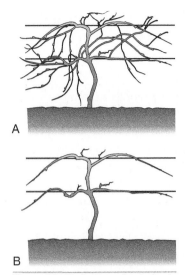

Figure 20-13 A grape vine trained to the Four-Arm Kniffin System (A) before and (B) after pruning.

A common method of training grapevines is the Four-Arm Kniffin System. In this system, the vine is pruned to form a double-T on a two-wire trellis system. Two arms originating at the two wires are trained to grow in opposite directions from the trunk. The vines are pruned to remove all canes except those that are saved for fruiting and renewal growth (**Figure 20-13**).

Pruning and thinning small-bush and cane fruits should be accomplished according to the specific standards for each fruit. Pruning, thinning, and training these smaller fruits are essential practices for high-quality and high-volume yields. These practices also help prevent pests and diseases.

Disease and Pest Control

A great deal of the time and money in fruit and nut production is spent controlling diseases, insects, and other pests. Control of diseases and insects is a major component of fruit and nut production (**Figure 20-14**). The type of pesticide needed and how often it is applied varies tremendously among all the species of fruits and nuts.

Each tree fruit, nut, and small-fruit variety must be treated individually. What may work on one variety may not be appropriate for another. Most orchardists will have a spray schedule worked out for each of their crops. Tree fruits require more pesticide than any other fruits. Before applying any pesticide, however, read the label thoroughly and follow it.

A

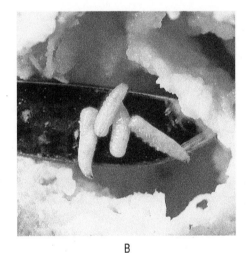
B

Figure 20-14 (A) A Caribbean fruit fly laying eggs in the peel of a grapefruit. (B) Fruit fly larvae then hatch and devour the pulp.

Courtesy of USDA/ARS #K-898-8. Courtesy of USDA/ARS #K-1674-7.

Harvesting and Storage

Harvesting

Harvesting fruits can be a daily practice. The advantage of homegrown fruit is that it can be picked at its peak of ripeness (**Figures 20-15A, B, C, and D**). However, fruits for commercial uses must be picked several days before they are ripened so they will withstand shipping and handling.

Figure 20-15 Harvest fruits at the peak of ripeness for home use. (A) Apples. (B) Orange. (C) Peaches. (D) Plums.
© zimmytws/Shutterstock.com. © Arkady/Shutterstock.com. © Tomas Pavelka/Shutterstock.com. © Protasov A&N/Shutterstock.com.

For the best quality of fruit, harvest apples, pears, and quince when they begin to drop, soften, and become fully colored. Some varieties will ripen over a 2-week period, requiring picking each day for best quality. Other varieties will ripen all at once.

Peaches, plums, apricots, and cherries should be harvested when the green disappears from the surface skin of the fruit. A yellow under-color should be developed by this time. The fruit should be soft when pressed lightly in a cupped hand (**Figure 20-16**).

The nut varieties ripen from August to November. Harvest nuts immediately after they fall from the tree. Most of the nuts that do not fall can be knocked off the tree with poles or mechanical harvesters.

Figure 20-16 Cherries ripen earlier in the season than most tree fruits.
Courtesy of USDA/ARS #K6012-20. Photo by Brian Prechtel.

The nut varieties of pecan, hickory, chestnut, and Persian walnut will lose their husks when ripe. However, the husks of black walnuts and other types of walnuts need to be removed.

Citrus fruits are handpicked, and the size of the fruits is the key in determining the time to harvest. Some citrus regulations set minimum size standards in order to market the fruit. For example, fruit size, for legal purposes, is determined by measuring individual fruits with standard-sized rings. Size 96 requires 96 fruits to fill a standard 1.6-bushel box, size 80 requires 80 fruits, and so forth.

Oranges are usually ring-picked only at the beginning of the harvest season. Smaller fruits are left on the tree, allowing them to grow to marketable sizes. As the season progresses, the increased size of oranges makes it more efficient to pick without using the rings, and fruits that are too small are discarded at the processing plant. The target size for most grapefruit harvests is 96, and ring picking is common practice.

Grapes should be harvested only when fully ripe. The best way to judge when the grape is at full maturity is to sample an occasional grape. Harvesting is by hand or by using machinery. Hand harvesting is usually practiced with a small number of vines, but it may be employed with larger, high-value crops. Most large vineyards are harvested by machines. Clusters of grapes should be cut from the vine with part of the stem intact.

Small-bush and cane fruits, such as strawberries and raspberries, should be harvested when they are fully ripe. The best indicator is when the fruit is fully colored. When these fruits and berries reach maturity, they should be picked every day. When harvesting, pick when the fruit is dry to avoid mildew and mold damage.

Storage

Fruit storing is limited to those varieties that mature late in the fall or those that can be purchased at the market during the winter months. The length of time fruits can be stored depends on the variety, stage of maturity, and soundness of the fruit at harvest (**Figure 20-17**).

Commodity	Freezing Point °F	Storage Conditions		Length of Storage Period
		Temperature °F	Humidity	
Apples	29°	32°	Moderate Moisture	Fall/Winter
Grapefruit	29.8°	32°	Moderate Moisture	4 to 6 weeks
Grapes	28.1°	32°	Moderate Moisture	1 to 2 months
Oranges	30.5°	32°	Moderate Moisture	4 to 6 weeks
Pears	29.2°	32°	Moderate Moisture	4 to 6 weeks

Figure 20-17 Fruit storage temperatures as adapted from the U.S. Department of Agriculture.

Source: USDA, ARS. (Draft-Revised April 2004). The Commercial Storage of Fruits, Vegetables, and Florist and Nursery Stocks. Gross, K C.; Wang, C. Y., and Saltveit, M. (Eds.). Agriculture Handbook 66. Retrieved from http://www.ba.ars.usda.gov/hb66/contents.html.

For long-term storage of apples, the temperature should be as close to 32°F as possible. Apples can be stored in many ways, but they should be protected from freezing. A cellar or other area below ground level that is cooled by night air is a good place for apple storage (**Figure 20-18**). The storage area should have moderate humidity. Pears have the same storage requirements as apples.

Nuts should be air dried before they are stored. Nuts, especially pecans, keep longer if they are left in the shell and refrigerated at 35°F. They can also be frozen if they are kept in their shells. Chestnuts have special requirements. They should be stored at 35°F to 40°F and high humidity, shortly after harvest.

Small-bush and cane fruits, including grapes, are not suitable for storage. These fruits perish too quickly to keep very long and are best used as soon as they are harvested.

Figure 20-18 A fruit storage facility in which temperature and humidity are controlled.
Courtesy of DeVere Burton.

Chapter Review

Student Activities

1. Write the **Key Terms** and their meanings in your notebook.
2. SAE Connection: Develop a multimedia presentation or a bulletin board display explaining opportunities in fruit- and nut-related occupations. Include those that can be found on the local, state, and national levels.
3. Bring a specimen of some type of uncommon fruit or nut to class. Discuss its origin and the techniques for growing the fruit or nut.
4. Prepare fruit and nut flash cards using pictures from seed catalogs.
5. Prepare a planting plan for fruits and nuts, drawn to scale. Indicate the types, quantity, and location of plants to be grown.
6. Many fruits and nuts consumed in the United States are imported from other countries. How does the U.S. government ensure that the produce is safe? Research the process that is followed before imported crops are allowed into the United States. Why is this process necessary?

Inquiry Activity | Modern Crop Production

Courtesy of DeVere Burton.

After watching the *Modern Crop Production* video available on MindTap, discuss with a partner how the innovations in modern crop production presented in the video can be applied to fruit and nut production. Use the following questions to guide the discussion:

- How can the innovations presented in the video improve fruit and nut production?
- Which innovation do you think would be the most beneficial to fruit and nut production? Why?
- When developing new plant varieties, what traits are considered desirable by scientists?
- Could hydroponics help address the world's food needs? Explain.

Check Your Progress

Can you...

- name some of the benefits of fruit or nut production as a personal enterprise or career opportunity?
- identify some examples of fruit and nut crops?
- explain how to prepare a site for planting?
- describe how to plant fruit and nut trees?
- give examples of proven agricultural practices?
- recall appropriate procedures for harvesting and storing at least one commercial fruit or nut crop?

If you answered "no" to any of these questions, review the chapter for those topics.

How are fruits and nuts included in your regular diet? Which fruit and/or nut trees would you like to grow and/or learn more about? Why?

Chapter 21

Grain, Oil, and Specialty Field-Crop Production

Objective

To determine recommended approved practices for production of grain, oil, and specialty field crops.

Competencies to Be Developed

After studying this chapter, you should be able to:
- define important terms used in crop production.
- identify major crops grown for grain, oil, and special purposes.
- classify field crops according to use and thermal requirements.
- describe how to select field crops, varieties, and seed.
- prepare proper seedbeds for grain, oil, and specialty crops.
- plant field crops.
- describe current irrigation practices for field crops to meet their water needs.
- control pests in field crops.
- harvest and store field crops.

Key Terms

field crop
malting
forage
oilseed crop
linen
linseed oil
ginning
cash crop
cereal crop
seed legume crop
root crop
stimulant crop
thermal requirement
conventional tillage
minimum tillage
sprinkler irrigation
surface irrigation
drip irrigation
mechanical pest control
genetic control

Anthropologists tell us that the cultivation of the land and the growing of crops began about 10,000 years ago in Africa. The need to produce food for the animals that humans had captured and begun to domesticate caused early humans to change from hunters to farmers. There were no guidelines to follow in selecting plants. Early agriculturists had to rely on observing what the animals were eating to decide which plants to grow. Trial and error and thousands of years of selection have led to the crops that are grown today. New types, varieties, and uses of plants continue to be developed in response to current needs and in anticipation of future demands for food and plant fiber by an ever-increasing world population.

In the United States, the production of grain, oil, and specialty crops occupies more than 450 million acres. These crops are called **field crops**. This acreage represents nearly 20 percent of the landmass in the United States. U.S. agriculturists are among the most efficient in the world, producing enough food for themselves and people in many other countries. As a result, less than 2 percent of U.S. workers are engaged in the production of food and fiber. The efficiency of the U.S. agriculturist allows the U.S. population to spend less of its income on food than citizens of most other nations. It also allows for sizable exports of food crops all over the world. This has helped maintain a favorable balance of trade for the United States in recent years. Today, we have so many products from agriculture that we no longer are aware of the plant and animal origins of many common nonfood items (**Figure 21-1**).

Major Field Crops in the United States

Grain Crops

There are seven major grain crops in the United States. Grain crops are grasses that are grown for their edible seeds. These crops are corn, wheat, barley, oats, rye, rice, and grain sorghum. Visit the website of the USDA Economic Research Service to see statistics and information about crops that are featured in this chapter.

Figure 21-1 New products from plants and animals through agriscience: (top row) plastics from vegetable oils and industrial uses for corn starch; (second row) cocoa butter substitute, improved cotton processing; (third row) tallow-based soap, detergents with improved surfactants, more uses for corn starch; and (fourth row) starch-based rubber products, plastics from vegetable oil, starch-based rubber.

Courtesy of USDA/ARS #K-4796-20. Photo by Keith Weller.

Corn

Corn is the most important field crop grown in the United States. It is well adapted, high yielding, and is one of the most important crops in many states. About 35 to 40 percent of the corn produced annually in this country is grown in the Midwestern states, commonly called the Corn Belt. These states include Iowa, Illinois, Nebraska, Minnesota, Indiana, Kansas, South Dakota, and Ohio, in descending order of production (2020). The United States accounts for more than 1/3 of the corn produced in the world (**Figure 21-2**).

Figure 21-2 Corn is a plant that produces high-quality grain for humans and animals and stored feed for livestock.
© Zoran Karapancev/Shutterstock.com. © JIANG HONGYAN/Shutterstock.com.

Originating in Central America, corn served as a major part of the diet of the native peoples (**Figure 21-3**). Corn was unknown to the rest of the world until European explorers explorers observed Native Americans growing corn.

Figure 21-3 Corn is a Native American plant known as maize, or "Indian corn."
Courtesy of DeVere Burton.

Less than 10 percent of the corn grown in the United States is for human consumption. The rest is used for livestock feed, alcohol production, and hundreds of other products.

The major types or classifications of corn are dent corn, flint corn, popcorn, sweet corn, flour or soft corn, and pod corn. Most of the corn grown in the United States is dent corn.

A general discussion of cultural practices for crops appears later in this unit. However, the following approved practices are useful for growing corn:

- Select a hybrid that will mature during the growing season.
- Obtain a current soil test.
- Lime and fertilize according to the soil test and desired production.
- Select a field with deep, rich, well-drained soil.
- Select the proper tillage method.
- Prepare a firm seedbed or use a proper no-till planter.
- Plant corn when soil temperature is 50° F or warmer.
- Calibrate the planter for the proper plant population.
- Match the plant population to soil-yield potential.
- Adjust the planter for proper depth of seed placement.
- Calibrate the sprayer for proper application of materials.
- Use selected herbicides for controlling problem weeds.
- Cultivate to help control weeds.
- Use insecticides as necessary for proper insect control.
- Apply fungicides when needed.
- Make a yield check.
- Check moisture content during the harvest season.
- Harvest the crop as soon as moisture permits, to reduce losses.
- Take a forage analysis on ensiled corn.
- Store ensiled or high-moisture corn in sound structures.
- Track market trends and habits.
- Market the crop at an optimum time.
- Keep accurate enterprise records.
- Summarize and analyze records.

Hot Topics in Agriscience — Native Corn: American Maize

Maize, or corn, was an important food source for Native American cultures.
© argonaut/Shutterstock.com

Long before the birth of the United States, and before corn became the number-one grain crop produced here, corn was known as maize. It was the most important grain crop for Native Americans. Maize was cultivated by native tribes as their principal food source for many years before the arrival of European settlers. Plant breeders have modified the maize plant, also known as "Indian corn," as they have developed the plant for specific uses. For example, we now have popcorn, sweet corn, grain corn, and silage corn in many varieties. All these corn plants can be traced back to the Native American maize plants.

Corn is now used in many ways besides providing food for people and animals. It is used to make products that we use each day, such as cooking oil, biodegradable plastic bags, and ethanol (a component of gasohol). Large amounts of corn are exported throughout the world, which helps sustain the U.S. balance of trade. Corn also remains the most important feed grain for livestock in the United States.

Wheat

Wheat is one of the most important grain crops in the world (**Figures 21-4A** and **B**). In the United States, wheat ranks second to corn in bushels produced. Leading states in the production of wheat are Kansas, North Dakota, Washington, Montana, Oklahoma, and Idaho.

Figure 21-4 Worldwide, wheat ranks second to rice as the most important grain crop.
Courtesy of USDA/ARS #K3597-18. © Zeljko Radojko/Shutterstock.com.

Wheat is used primarily for human consumption. Wheat is ground into flour, which is then made into products such as bread, cakes, cereal, crackers, and pasta. Other uses of wheat include the manufacture of alcohol and livestock feed.

Types of wheat grown in this country include common, durum, club, Poulard, Polish, Emmer, and spelt. Most of this wheat is the common type. Classes of common wheat include hard red spring, hard red winter, soft red winter, and white.

Barley

The leading states in the production of barley in the United States include Idaho, Montana, North Dakota, Colorado, and Washington (**Figures 21-5A** and **B**). Barley ranks fourth among the grain crops produced in the United States.

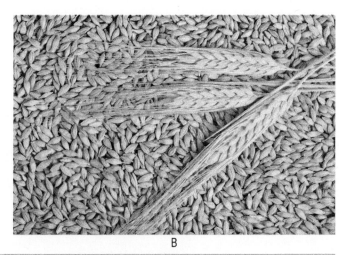

Figure 21-5 (A) Most of the barley grown in the United States is used for livestock feed. (B) Barley is important as a feed grain, and it is also used in the production of malt.
© Fedor A. Sidorov/Shutterstock.com. © Madlen/Shutterstock.com.

Hot Topics in Agriscience: Biofuels Versus Feed Grains

Is it possible to have the best of both worlds, to have your cake and eat it, too? The world faces difficult choices as corn and other grains are diverted in ever-increasing amounts to develop renewable energy sources. Ethanol and biodiesel are just two of the alternative energy products to come from feed grains and oilseed crops. Within the past decade, the monetary value of these crops has skyrocketed in comparison with the marginal grain prices that farmers historically received.

Now we face a new challenge: The price of food is rising, driven up by competition in the world markets. The *Law of Supply and Demand* is proving itself in dramatic fashion. Cereal grain foods and animal products cost more in the grocery stores than at any time in recent memory.

Livestock farmers operate at considerable risk as feed prices escalate, but they have also benefited from higher prices for their products. However, with higher prices come higher risks. Even a short delay in covering production costs through the sale of milk, meat, and eggs can drive a livestock producer out of business.

So what is the best result that can come from these apparently conflicting interests? Perhaps, in the final analysis, consumers will one day depend on renewable energy sources, and the value of agricultural products will be equitable with other sectors of the economy.

A

B

Corn and other grains have found new markets. (A) Ethanol production plant in South Dakota. (B) Ethanol refinery setup.
© Jim Parkin/Shutterstock.com. © AZP Worldwide/Shutterstock.com.

Most barley is used for livestock feed. It has slightly less food value than corn, but it can be grown in less favorable climates. The production of barley for malting is also important. **Malting** is the process of preparing grain for the production of malt, an ingredient in beer and other alcoholic beverages.

Cultural practices for growing wheat, barley, oats, and rye are similar. The following list of approved practices for growing small grains should be useful for growing any of these crops:

- Select a variety adapted to your conditions.
- Plant certified seed.
- Obtain a current soil test.
- Add lime and fertilize according to soil test results and desired production.
- Select a field suitable for your grain choice.
- Select the proper tillage method.
- Prepare a firm and smooth seedbed.
- Calibrate the drill for the proper plant population.

- Match the plant population to soil-yield potential.
- Adjust the drill for proper depth of seed placement.
- Calibrate the sprayer for proper application of materials.
- Use selective herbicides for controlling problem weeds.
- Use insecticides as necessary for adequate control of insects.
- Apply fungicides as needed.
- Make a yield check.
- Adjust the combine for proper threshing tolerances.
- Monitor the grain moisture content during the harvest season.
- Harvest the crop as soon as the moisture level permits.
- Store grains in disinfected grain storage tanks.
- Bale straw as soon as moisture levels allow.
- Track market trends and habits.
- Market the crop at an optimum time.
- Keep accurate enterprise records.
- Summarize and analyze records.
- Make business decisions based on records.

Oats

Oats as a grain crop are fourth in acres produced in the United States (**Figure 21-6**). Major oat-producing states are South Dakota, Minnesota, Wisconsin, North Dakota, and Iowa.

The value of oats in adding bulk and protein to the diets of livestock is well documented. However, about 5 percent of the oats produced in the United States are made into oatmeal and cookies. Oats are also used in the production of plastics, pesticides, and preservatives. In addition, they are important in the paper and brewing industries.

Figure 21-6 Oats provide bulk and protein to the diets of animals as well as food for humans.
Courtesy of DeVere Burton.

Rye

Although rye is grown in nearly every state in the United States, it is the least economically important grain crop. It does, however, have many uses (**Figure 21-7**). Most rye is grown in North Dakota, Pennsylvania, and Wisconsin.

Figure 21-7 Rye is used as a livestock feed and for the production of specialty flour. Economically, it is the least important grain crop produced in the United States.
Courtesy of DeVere Burton.

About 25 to 35 percent of rye acreage is used for grain. The rest is used for forage, as a cover crop, or as a green manure crop. **Forage** is hay or grass grown for animal feed. Cover crops are planted to protect the soil from erosion. Green manure crops are grown to be plowed under to add nutrients and organic matter to the soil. The rye grown for grain is used for livestock feed, flour, whiskey, and alcohol production.

Rice

Rice is the most important grain crop grown for human food in the world. Most of the rice in the United States is grown in Arkansas, California, Louisiana, Missouri, and Texas. Rice is the only commercially grown grain crop that can grow and thrive in standing water. The types of rice grown in the United States are short grain, medium grain, and long grain varieties (**Figure 21-8**).

The majority of the rice grown in the United States is used for human consumption. The excess that is produced is exported to other countries of the world.

Sorghum

Sorghum is grown in the United States primarily for livestock feed. It is about equal to corn in food value. Other uses of sorghum include forage, the manufacture of syrup or sugar, and the making of brooms (**Figure 21-9**). Leading states in the production of sorghum include Kansas, Texas, Colorado, South Dakota, and Nebraska. Based on acres harvested, sorghum is the third most important grain crop in the United States.

The five types of sorghum are grouped according to use. They are grain, forage, syrup, grass, and broomcorn.

Figure 21-8 Rice is the most important grain crop in the world. (A) Rice is the only commercially grown grain crop that will grow in standing water. (B) Most of the rice grown in the United States is for human consumption.

Courtesy of USDA/ARS #K-3444-1. © Mikus, Jo./Shutterstock.com.

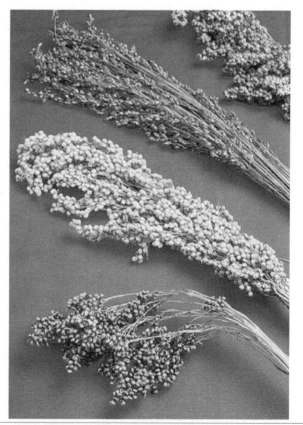

Figure 21-9 Sorghum is a grain crop that is produced in a few Midwestern states. It is mostly used for livestock feed, either as grain or as forage.

Courtesy of USDA/ARS #K9371-7. Photo by Peggy Greb.

Oilseed Crops

Crops that are grown to produce oil from their seeds are called **oilseed crops**. These crops are increasing in importance each year as people in the United States rely more and more heavily on vegetable oils and less on animal fats in their diets. Important crops grown for the oil extracted from their seeds are soybeans, peanuts, corn, cottonseed, canola, safflower, flax, and sunflowers (**Figure 21-10**). Corn and cottonseed are discussed in other sections of this chapter.

Figure 21-10 The rape, or canola, plant is getting new attention as an oil-producing crop. This plant belongs to the mustard family, as evidenced by the bright yellow flowers.
Courtesy of DeVere Burton.

Peanuts

The peanut is actually a pea rather than a nut, despite its nut-like taste and shell. It is grown primarily in the South, where warm temperatures and a long-growing season are keys to success. Leading peanut-producing states are Georgia, Alabama, Florida, Texas, and North Carolina.

One ton of peanuts in the shell will yield about 500 lbs. of peanut oil and 800 lbs. of peanut-oil meal. The remaining 700 lbs. are mostly shells. The oil meal is used for livestock feed and serves as a good protein source in human diets. Other food products produced from peanuts include peanut butter and dry-roasted peanuts (**Figure 21-11**).

Figure 21-11 George Washington Carver, the famous agricultural research pioneer, developed more than 300 products from peanuts.
Courtesy of USDA/ARS #K4297-14. Photo by Jack Dykinga.

Soybeans

Approximately 75 million acres of soybeans are grown in the United States each year. With an average yield of close to 52 bushels per acre, the production of soybeans grosses more than $40.9 billion each year (**Figures 21-12** and **21-13**).

Oil and grain products are the major uses of soybeans. The meal resulting from the extraction of oil from soybeans is an important source of protein in livestock feeds. Soybeans are also harvested for hay, pasture, and other forage crops. Research has led to the development of hundreds of other uses for soybeans.

Figure 21-12 The late Edgar E. Hartwig devoted half a century to soybean research. Sometimes called the Soybean Doctor, this world-renowned agronomist developed productive plants with built-in resistance to insects, nematodes, and diseases.

Courtesy of USDA/ARS #K5272-1. Photo by Keith Weller.

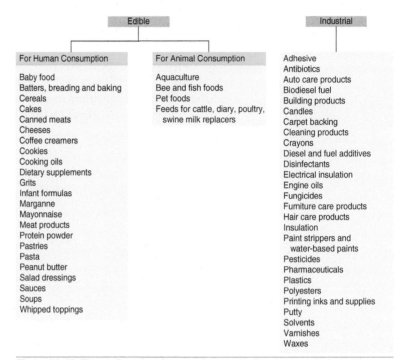

Figure 21-13 The soybean has become famous for its many products and is growing in importance in world markets.

Large centers of production include the Midwestern states of the Corn Belt. Major soybean-producing states include Illinois, Iowa, Minnesota, and Nebraska. Internationally, Brazil has become a major producer of soybeans. Approved practices for raising soybeans are similar to those for raising corn, except that soybeans and other legume seeds should be treated with the proper inoculants (bacteria to ensure good nitrogen fixation).

Safflower

Production of safflower for oil occurs mainly in California, Montana, and Utah. Safflower plants grow 2 to 5 feet in height and have flower heads that resemble Canadian thistles (**Figure 21-14**). The oil comes from the wedge-shaped seeds that the plant produces. The seeds contain from 20 to 35 percent oil.

Figure 21-14 Safflower is a drought-tolerant plant that looks much like a thistle. The seeds contain up to 35 percent oil, which is used for cooking and industrial purposes.

© getcloser/Shutterstock.com.

Safflower oil is used in the production of paint and other industrial products. It is also used as cooking oil and in low-cholesterol diets.

Flax

Originally, the production of flax was mostly for fiber. Flax fibers from the stems of plants are used to produce **linen** (a cloth product).

The oil produced from the seed of the flax plant is called **linseed oil**. It is an important part of many types of paint and has hundreds of uses in industry. The linseed-oil meal that is left after extracting the oil is an excellent source of protein for animal feeds.

Most flax is grown in North Dakota, Montana, South Dakota, and Minnesota.

Sunflowers

Figure 21-15 The sunflower is the state flower of Kansas, the "Sunflower State." A field of sunflowers is beautiful and a source of valuable oil-rich seed.

© formiktopus/Shutterstock.com.

Production of oil-type sunflower seed has been important in the United States in recent years (**Figure 21-15**). Most of the sunflower production is located in North Dakota, South Dakota, and Minnesota.

Two types of sunflowers are grown commercially in the United States: oil-type and non–oil-type. About 90 percent of sunflower production is of the oil-type. Oil-type sunflower seeds contain 49 to 53 percent oil. The meal that remains after the oil has been removed is high in bulk and contains 14 to 19 percent protein. It is used for livestock feed. The oil is used for margarine and cooking oil. Sunflower oil can also be processed into biodiesel as a substitute for petroleum-based diesel fuel in tractors and trucks (**Figure 21-16**).

Figure 21-16 Some vegetable oils, such as rapeseed oil and sunflower oil, can be used to produce fuels for diesel tractors. This kind of fuel is known as biodiesel.

Courtesy of DeVere Burton.

Specialty Crops

Fiber crops, sugar crops, and stimulant crops are grouped into the category called specialty crops. Specialty crops grown in the United States include cotton, sugar beets, sugarcane, and tobacco.

Cotton

Cotton originated in Central and South America. It has been an important crop in the South since colonial days. The cotton plant requires warm temperatures and a long growing season to reach maturity. The leading cotton-growing states include Texas, Georgia, Arkansas, and Mississippi. Arizona's extensive irrigated acreages of cotton crops, producing multiple crops per year on the same land, also place it among the leading cotton-producing states.

About 17.3 million bales of cotton are harvested per year in the United States. Approximately 3.6 million bales are needed by the United States textile industry, and the rest is exported to other countries of the world. (**Figure 21-17**).

Cotton seeds must be removed from cotton after it is harvested. This process is called ginning. The cotton seed is then processed to remove its oil, which is a major contributor to the vegetable-oil needs of the United States. Following the extraction of the oil, the seed is ground into a high-protein animal feed.

Figure 21-17 Cotton has been an important crop since colonial times, but care must be taken to protect cotton plants from insects and diseases.
© Sherry Yates Sowell/Shutterstock.com.

Sugar Beets

The production of sugar beets for sugar accounts for about 55 percent of the refined sugar produced in the United States (**Figure 21-18**). This crop is grown for its thick, fleshy storage root in which sugars are accumulated. The leading states for sugar beet production are Minnesota, North Dakota, Idaho, Michigan, and California.

Figure 21-18 Sugar cane and sugar beets are the sources of nearly all of our domestic sugar supply. Large factories and refineries process the raw product into syrup and then into the familiar sugar crystals we use to sweeten foods.
© Marie C Fields/Shutterstock.com.

Sugarcane

Sugarcane production in the United States is concentrated in subtropical areas of Florida, the Gulf Coast states, and Hawaii. Sugarcane accounts for about 45 percent of the sugar refined in the United States.

This crop, which is a grass, is grown from sections of stalk called seed pieces, rather than from seed. It takes about 2 years for sugarcane to reach the harvesting stage in Hawaii. Compared with production in Hawaii, in the southern states, sugarcane is harvested about 7 months after planting, with a corresponding loss of yield. The same field of sugarcane can be harvested several times before it must be replanted because the plant regenerates its top growth (**Figures 21-19A** and **B**).

Tobacco

Tobacco is an original North American product that was used by American Indians in religious rites. It is produced as a cash crop in the southeastern states, predominantly in

Figure 21-19 (A) Forty-five percent of U.S. sugar production comes from sugar cane produced in the southern states and in Hawaii. (B) Sugar beet production accounts for about 55 percent of the domestic sugar supply.

Courtesy of Elmer Cooper. © Bits and Splits/Shutterstock.com.

North Carolina, Kentucky, Virginia, Tennessee, Georgia, and South Carolina. Tobacco is considered to be a **cash crop** because it is sold for its commercial value instead of using it on the farm for livestock feed. The United States ranks fourth among tobacco-producing countries behind China, India, and Brazil.

Tobacco production requires large amounts of labor and is therefore best adapted to small farming operations. Warm temperatures and plenty of rainfall are required for optimum production of high-quality tobacco.

Hot Topics in Agriscience — Biodiesel—A Renewable Resource

Biodiesel can be manufactured from vegetable oils as well as animal fats and used cooking oil.

© Florian Augustin/Shutterstock.com.

One of the great concerns about world dependency on fossil fuels has been that these fuels are nonrenewable resources. Once the oil reserves are used up, this important resource will be lost to the human race. Biodiesel is different because this product is produced by plants. It is obtained from oil seeds such as rapeseed and sunflower seeds.

Plant scientists and agricultural engineers have pooled their talents to develop ways that plant oils can be used to replace diesel fuel. Vegetable oils have shown great promise as fuels in diesel engines, but except when diesel prices are high, biodiesel has been expensive in comparison with the cost of petroleum-based diesel fuel. As the cost of crude oil rises, the cost to produce biodiesel also increases—mostly due to the increased cost of vegetable oils. Concurrently, consumers have tended to blame the increasing cost of food on the use of grains and oil crops to produce ethanol and biodiesel.

Despite the cost, global consumers will eventually run short of crude oil, and a renewable source of energy for fuel will become a necessity. A less expensive approach to producing biodiesel is to produce it from recycled vegetable cooking oils and from animal fat. These oils and fats form chemical compounds called esters when they are combined with methyl alcohol under alkaline conditions. This type of biodiesel is an improvement over earlier biodiesels because this fuel product works better in cold weather than the early biodiesel products.

Seed Crops

The production of good-quality field crops depends on good seeds. An entire seed industry has emerged to supply the need for high-quality seeds for many varieties of field crops. Seeds must come from healthy plants that are free of diseases. Many of the vegetable crops that we produce are hybrid plants (see Chapter 17) that come from different strains of parent stock. Seed crops are usually raised in areas that are isolated from related commercial crops. This helps maintain the purity of the genetic makeup of the seeds because stray pollen from undesirable sources is minimized.

Seeds for most field crops are produced in similar climates and growing conditions that will be encountered when the seeds are used to produce commercial crops. This assures that they are adapted to the conditions under which they are expected to grow and produce. Care must be taken, however, to ensure that the seed crops do not become infected by diseases and destructive organisms such as fungi. Such an infection of a seed crop could spread the infection wherever the seed is planted. In all cases, seed crops must be subjected to inspection and testing to ensure that they are pure and that the quality of the seed is high.

Most vegetable seeds are produced in isolation from commercial crops. An arid climate is desirable in the production of most vegetable seed crops because such a climate is not favorable to fungi, bacteria, and many other organisms that cause damage to crops. As a result, most vegetable seeds are produced on irrigated farms at isolated locations in the western desert region of the United States (**Figure 21-20**). Seeds are high-value crops that must be protected from the stresses that are often found in the environment. The seeds must have favorable conditions that allow them to mature completely before they are harvested; otherwise, the germination rate will not be acceptable.

After the harvest, seeds are placed in storage under carefully controlled climatic conditions to preserve their ability to germinate. Seed samples are routinely submitted to seed laboratories for certification. The seed laboratories test the germination rates of different seed samples, and they carefully inspect the seed samples for contamination with weed seeds and diseases. Seed crops must meet strict standards to be labeled as certified seed.

Figure 21-20 Most vegetable seed crops are raised in arid climates at isolated locations, such as Melba, Idaho, to avoid contamination from the pollen of related field crops and to minimize exposure to fungal and bacterial diseases that thrive in damp environments.

Courtesy of Elmer Cooper.

Classification of Field Crops

Field crops can be classified in several ways. Three of these classifications are (1) use, (2) thermal requirements, and (3) life span.

The classification of field crops according to use is as follows:
- **Cereal crops** are grasses grown for their edible seeds. They include corn, wheat, barley, oats, rice, rye, and sorghum.
- **Seed legume crops** are nitrogen-fixing crops that produce edible seeds. Included in this class are soybeans, peanuts, field peas, field beans, and cowpeas.
- **Root crops** are grown for their thick, fleshy storage roots. Beets, turnips, sweet potatoes, and rutabagas are root crops.
- Forage crops are grown for hay, silage, or pastures for livestock feed. Examples of forage crops include alfalfa, clover, timothy, orchard grass, and many other crops that are harvested for the feed value of their stems and leaves.
- Sugar crops are grown for their ability to concentrate and store sugars in their stems or roots. They include sugarcane, sugar beets, and sorghum. Corn is also used to produce sugar.
- Oil crops are produced for the oil content of their seeds. Examples are soybeans, peanuts, cottonseed, flax, rapeseed, sunflower seed, safflower seed, and castor beans.
- Tuber crops are grown for their thickened, underground storage stems. Potatoes and Jerusalem artichokes are examples of tuber crops.
- **Stimulant crops** are grown for their ability to stimulate the senses of the user. Examples of stimulant crops are tobacco, marijuana, coffee, and tea.

Crops can also be classified according to their thermal requirements. **Thermal requirement** refers to the heat requirement or length and characteristics of the growing season that are required for the crop to mature. Two major thermal groups are warm-season and cool-season crops.

Warm-season crops must have warm temperatures to live and grow. They are adapted best to areas where freezing or frost seldom occurs. They also normally require longer growing seasons than cool-season crops. Examples of warm-season crops are cotton, tobacco, and citrus fruits.

Cool-season crops are normally grown in the northern half of the United States, where temperatures below freezing are normal. These crops often need a period of cool weather to attain maximum production. Most of the grains, tubers, and apples are cool-season crops.

Crops can also be classified according to their life spans. They may be annual, biennial, or perennial. An example of an annual crop is corn; red clover is a biennial crop; and alfalfa is a perennial crop.

Growing Field Crops

A number of factors must be considered when selecting which field crops to grow:
- Select crops that will grow and produce the desired yields under the type of climate available. Be sure to consider length of growing season, average yearly rainfall, average daily temperature, humidity, and prevailing wind.
- Crops must be adapted to the type of soil available. Consider soil pH, soil type, soil depth, and soil response to fertilizers.
- Consider demand and availability of markets for the crop to be produced.
- Assess labor requirements and availability of labor for the crop.
- Identify machinery and equipment that will be needed to grow the crop.
- Consider the availability of enough land to justify production of the crop.
- Identify potential pest-control problems.
- Estimate yields.
- Anticipate production costs. Can a reasonable profit be expected?

Agri-Profile — Career Areas: Agriscience Support Industry

Plant technicians and research specialists plan and conduct research projects leading to improved plant varieties, such as the pearl millet pictured here.
Courtesy of USDA/ARS #K3882-2.

Corn, wheat and other small grains, sugarcane, soybeans, sugar beets, and other specialty crops are grown on large acreages in the United States, Canada, and many nations of the world. In the United States, grain, oil, and specialty crops account for large amounts of exports and do much to help maintain our balance of payments in foreign trade. Grain brokers, futures brokers, market reporters, grain elevator operators, crop forecasters, farmers, and others owe their jobs to these crop enterprises.

Custom combine operators, truck drivers, maintenance crews, cooks, and other service workers follow the grain harvest from Mexico to Canada. At season's end, the crews return south and prepare for the next season, when the cycle is repeated. At the same time, small farm owners and operators grow, harvest, and market millions of acres of field crops as part, or all, of the farm operations.

Along with crop enterprises, jobs are available in building and storage construction, systems engineering, machinery sales and service, welding/repair, irrigation, custom spraying, hardware sales, agricultural finance, chemical sales, and seed distribution.

Seedbed Preparation for Field Crops

The purpose of seedbed preparation for field crops is to provide conditions that are favorable for the germination and growth of the seed. Not only does the seedbed need to be prepared for seed germination but the area under the seedbed must also be prepared for the root growth of the crop (**Figure 21-21**).

Figure 21-21 A good seedbed consists of crumbly, mellow soil that is free of clods, rocks, and debris.
Courtesy of DeVere Burton.

Eliminating competition from weeds and crop residues is a consideration when preparing a seedbed for planting. Proper seedbed preparation can also increase the availability of soil nutrients to plants.

Seedbeds should not be overworked. The texture of the soil should be porous and allow for free movement of air and water. Small seeds require a seedbed with a finer texture than larger seeds require. The seedbed should contain enough fertility to encourage germination and growth until additional fertilizer can be applied.

Several methods can be used to properly prepare seedbeds for field crops. They can be divided into three general categories: conventional tillage; reduced, or minimum tillage; and no-tillage.

In **conventional tillage**, the land is plowed with a moldboard or disk plow, turning under all the residue from the previous crop. The soil is then worked with tillage machinery to smooth and further pulverize the soil for the seedbed (**Figure 21-22**).

Reduced or **minimum tillage** is a system of seedbed preparation that works the soil only enough so that the seed can make and maintain close contact with the soil improving germination. A chisel plow is frequently used (**Figure 21-23**). Minimum-tillage systems usually combine several operations into one pass across the field. This method reduces the amount of soil compaction, conserves soil moisture, and usually provides less opportunity for soil erosion (**Figure 21-24**).

No-till preparation of seedbeds involves planting seeds directly into the residue of the previous crop, without exposing the soil. Seed is usually planted in a narrow track opened by the seed planter. It is extremely important that good management practices be used when using the no-till method of seedbed preparation. In addition to coordinating planting with soil moisture conditions, practices such as controlling weeds, insects, and diseases should be implemented. Competition from previous crop residues must be managed (**Figure 21-25**). For example, as crop residue breaks down, it competes with a growing crop for soil nitrogen. Extra nitrogen may be needed to assure good crop yields.

Figure 21-22 With conventional tillage, the soil is turned using a moldboard plow so that all crop residues, livestock manure, lime, and fertilizer are mixed through the plow layer.

Courtesy of Elmer Cooper.

Figure 21-23 A chisel plow loosens the soil but leaves the crop residue on the surface.

Courtesy of Elmer Cooper.

Figure 21-24 A tillage tandem disc/chisel-point tiller cuts crop residue and loosens surface soil via the discs, loosens the deep soil via chisel points, and leaves a fine seedbed on the surface.

Courtesy of Elmer Cooper.

Figure 21-25 No-till planting retains crop residues on the surface for better erosion control.

Courtesy of Dr. Allen Hammer.

Planting Field Crops

The invention of the seed planter was one of the most important events for U.S. agriculture. Three general types of planters are used in planting field crops today. They are row crop planters, drill planters, and broadcast planters.

Row crop planters plant seeds in precise rows and with even spacing within the rows. Three types of row crop planters are drill planters, hill-drop planters, and checkrow planters. Hill-drop planters drop two or three seeds together in rows. Checkrow planters plant several seeds together in a checkered pattern in the field to permit cross-cultivation. Row crop planters are used to plant corn, beans, sugar beets, soybeans, sorghum, and cotton.

Science Connection — Designer Foods

Golden rice has been genetically modified to address vitamin A deficiency, a leading cause of child blindness in developing nations.
© Alan49/Shutterstock.com.

In the 1940s, food products were being fortified with vitamins and minerals. Before this time, many people experienced illness that was a result of vitamin and mineral deficiencies. Once the causes of such sicknesses were discovered, foods such as breads and cereals were supplemented with the needed nutrients. Currently, the back of any breakfast cereal box shows that fortification is still practiced. Biotechnologists are certain that there is another way the food we eat can improve our health. For example, a variety of bioengineered rice is available in the United States and Canada. Golden rice has been genetically altered to produce vitamin A.

Crops such as golden rice are created using a process called recombinant DNA technology. This process begins when a desired gene is found and cut out of the DNA of the cells from another plant. These cells are called donor cells. This is done using proteins called restriction enzymes. These enzymes cut DNA at a specific location. Next, the donor gene is put into a plasmid. A plasmid is a circular piece of DNA that can be found in some bacteria. The bacteria are then placed in a petri dish, where they begin to replicate. Many copies of the donor gene are made. Through bacterial action, the desired gene can be put into the cells of the target plant. The resulting plants reproduce themselves with the new donor gene in place.

In many developing countries, undernourished children experience childhood blindness. This condition is caused by vitamin A deficiencies in young children. Rice is often the only food source available to poor families in these countries. When golden rice is made available to the citizens of these nations, the number of children affected by childhood blindness can be demonstrated to decrease dramatically. Bioengineered foods have the potential to do for this generation what food fortification has done for generations in the recent past.

Drill planters plant seeds in narrow rows at high population rates. They drop seeds individually in a row at set distances apart (**Figure 21-26**). Drills are available in many row spacings and planter widths. Seeding rates are less accurate than with row crop planters. Among the field crops that are planted with drill planters are wheat, oats, barley, rye, forage legumes, and many grasses. Fertilizer and pesticides may be applied at the same time the seed is planted with drill planters.

Broadcast planters scatter the seeds in a random pattern on top of the seedbed. The seeding accuracy using this type of planter is the poorest of the planting methods. Broadcast seeders cover wide areas and usually plant seeds much faster than other methods. They are sometimes used when weather conditions make it difficult to get machinery into the fields. Airplanes are sometimes used in combination with broadcast planters, especially on difficult terrain following a fire. Knapsack seeders and spinners are other types of broadcast seeders.

One disadvantage of using broadcast seeders is that some kinds of seed need to be covered to germinate and to protect them from loss. This means that a second trip must often be made over the field to cover the seed. Small grains, grasses, and legumes such as clover and alfalfa are sometimes planted by broadcasting.

There are other considerations in planting field crops. These include the date to plant, germination rate of seeds, uniformity of seed, weather conditions, and insect- and disease-control problems.

Figure 21-26 Drill seeders are used to plant small seeds, such as small grains, grasses, alfalfa, and clover, at high population rates.
© Alexey Fursov/Shutterstock.com.

Meeting Water Needs of Crops

The soil in which plants grow acts as a storage vat for the water needed by the plant. Under ideal soil conditions, approximately half of a soil's pore space is filled with water. About half of this water is available for use by plants. Unfortunately, this ideal condition seldom exists; often, much less water is available for plant use than is needed. Factors that affect water availability for crops include the type of soil, natural rainfall, water-table levels, and prevailing winds.

When conditions are such that sufficient water is not available for the crop, irrigation may be the answer to obtaining profitable yields. Irrigation is a means of providing adequate amounts of water to crops.

Irrigation of crops has been practiced for more than 5000 years. The Nile River was used by the Egyptians to irrigate crops grown in the fertile deserts of the area. The Chinese diverted the water from many rivers to irrigate their rice fields. American Indians of the American West used irrigation to ensure the production of corn in arid areas.

The major methods of supplying irrigation water to crops are sprinkler systems, surface irrigation, and drip irrigation.

Sprinkler irrigation is one of the most efficient uses of supplemental water because it places the water where it is needed at the time it is needed. The amount of water that is applied is easily controlled by the length of time that the irrigation system is allowed to operate on the same setting. Unlike flood irrigation, which tends to over-irrigate part of the field while under-irrigating other areas, sprinkler irrigation delivers a uniform supply of water to all areas of the field (**Figure 21-27**).

Surface irrigation water is delivered to the crop by gravity, flowing over the surface of the soil or in ditches or furrows. It is an inexpensive means of providing water to crops. However, the cropland may need to be leveled before it can be irrigated.

Drip irrigation supplies water to the roots of crops in a uniform manner. Use of tubes located either above or beneath the soil to deliver the water to the crops makes a drip irrigation system expensive to set up. It has the advantage of low operating costs when in operation, however, and it permits the most efficient use of water (**Figure 21-28**).

Figure 21-27 Modern sprinkler irrigation systems are capable of delivering a uniform supply of water throughout the field.
© Brenda Carson/Shutterstock.com.

Figure 21-28 Cucumbers growing in a greenhouse are irrigated by a drip irrigation system that distributes water and nutrients.
Courtesy of DeVere Burton.

Pest Control in Field Crops

The control of pests in field crops is often the factor that determines whether a crop is profitable or not. Pests of field crops include diseases, weeds, insects, and animals. They may destroy the seed before it germinates, attack the growing crop, or render the harvested crop to be unusable and unfit for humans to eat. Economic losses from plant pests total billions of dollars each year (**Figure 21-29**).

There are three main categories of losses from plant pests. They are reduced yields, reduced quality, and storage losses.

Reduced yields occur when weeds germinate and grow more vigorously than the crop plants. Weeds compete successfully with the crop for moisture and nutrients, often causing the crop to be unhealthy. Parasitic plants draw nourishment from the host plants and may cause them to shrivel and die.

Damage from insects also causes the crop to yield less than expected. The damage usually occurs as adult insects feed, or during the growth stages as immature insects feed on the crop. Reduction of yield may also occur as insects spread diseases from plant to plant.

Diseases can reduce yields by interfering with the ability of the plant to manufacture food. They can also cause other plant processes to be changed, affecting the health of the plant.

Figure 21-29 A few lesser Grain Borers left unchecked in stored grain soon reproduce to become a devastating horde of insects that is capable of ruining an entire crop of grain.
© Tomasz Keljdysz/Shutterstock.com.

Figure 21-30 The ridge tillage machine for cultivation is capable of reducing or eliminating pre-plant tillage under most soil conditions.
Courtesy of USDA/ARS #K3239-1.

Figure 21-31 Parasitic wasps provide biological control of some pests by laying their eggs within or on the larvae of specific pests.
Courtesy of USDA/ARS #K5035-20.

Figure 21-32 Plant scientists are able to transfer genetic-resistant genes from one species to another. Potatoes that are genetically resistant to the Colorado potato beetle no longer require pesticides to control this insect.
Courtesy of DeVere Burton.

The quality of a crop is sometimes reduced from such things as weed seeds or rodent hairs and droppings intermixed with the crop. Foreign materials are usually unsanitary and may cause flavors that are objectionable to consumers. Foreign materials must be removed before the crop can be used.

Damage from insects and diseases can make the crop less desirable in appearance and can increase processing costs tremendously. Food crops may be deemed unsuitable for human consumption and result in a total loss if they are too severely infested with insects and diseases.

Spoilage of crops sometimes results when weeds hinder the drying process. Insects may also cause stored crops to overheat and mold, rendering them unfit for use.

Methods of controlling pests in field crops include mechanical control, cultural control, biological control, genetic control, and chemical control.

Mechanical pest control refers to anything that affects the environment of the pest or the pest itself. Cultivation is the normal mechanical control of weeds (**Figure 21-30**). Cultivation of the soil may also expose insects and soilborne diseases to the air. The effect of abrupt changes in temperatures as a result of exposure to the sun and air often proves fatal to some pests. Other types of mechanical control of pests include pulling or mowing weeds and the use of screens, barriers, traps, and electricity to exclude them from the crop.

Cultural control refers to adapting or changing farming practices to control pests. Cultural controls include timing farming operations to eliminate pests, rotating crops, planting resistant varieties, and planting trap crops that are more attractive to insects than the primary crop.

Biological control of plant pests involves the use of natural predators or diseases as the control mechanisms. The release of sterile male insects and the use of baits and repellents are also examples of this type of pest control. When using insects or diseases to control crop pests, it is important that the control be specific to the intended pest (**Figure 21-31**).

The development of crop varieties that are resistant to pests is called **genetic control** (**Figure 21-32**). This may involve making the crop less attractive to the pest because of taste, shape, or blooming time. Developing more rapidly growing crops that crowd out weeds is also an example of genetic pest control. Crops with resistance to diseases also fall into this category.

Chemical control of plant pests involves the use of pesticides to control pests of field crops. Excellent management practices must be exercised when using chemicals to control pests. Care should be taken to correctly identify the pest to be controlled and the chemical to be used. Dosage, runoff, and pesticide residues need to be carefully monitored (**Figure 21-33**).

Harvesting and Storing Field Crops

Harvesting field crops at the proper stage of maturity is a key to maximizing profits. The harvest of the crop is the culmination of a growing season of work and anticipation of the rewards for a job well done.

Development of mechanical harvesting equipment allows field-crop producers to harvest thousands of bushels of grain daily and with far less labor than previously required. This allows for tremendous increases in the amount of food available for people and animals and a greatly improved standard of living.

The primary harvesting machine for field crops is the combine (**Figure 21-34**). It performs the tasks of cutting the crop, threshing it, separating it from the straw, and cleaning it. Threshing refers to the separation of grain from the rest of the plant materials. Many types of combines have been adapted to the harvest of specific crops.

Science Connection: New Crops for New Uses

Checking lesquerella for seed set. New crops, such as lesquerella and wormwood, promise advances in fabrics, plastics, lubricants, corrosion inhibitors, cosmetics, and medicines.
Courtesy of USDA/ARS #K4692-2. Photo by Jack Dykinga.

As the medical profession scrambles to find new drugs and treatments for new and old medical problems, agriscience is becoming an increasingly important tool. At the Southern Weed Science Laboratory in Stoneville, Mississippi, scientists are developing a weed-control system to ensure the supply of a medicine for patients with malaria. Annual wormwood has been used for 2000 years in China to treat malaria. Now a new, refined drug derived from the plant is in demand worldwide. The World Health Organization is seeking ways to increase the production of wormwood sufficiently to meet the growing need.

Annual wormwood grows throughout the United States as a weed. The woody-stemmed, cone-shaped plant reaches a height of 4 to 8 feet and could become a cash crop for farmers in the future. Wormwood is harvested with a sickle-type machine similar to that used to cut sugarcane. Cut plants must be protected from prolonged exposure to the sun because ultraviolet light breaks down artemisinin—the antimalarial substance.

Artemisinin is found in the leaves of growing tips, and it takes 2.2 lbs. of the leaves to process 1 gram of artemisinin. Authorities indicate that it would take thousands of acres of wormwood to meet the world demand for the drug. However, before commercial production of wormwood could occur, farming methods, including weed control, must be developed. According to the World Health Organization, there were about 240 million cases of malaria in 2021.

Another newly emerging crop is lesquerella. Lesquerella has been developed from *Lesquerella fendleri*, a wild plant native to Arizona, New Mexico, Texas, and Oklahoma. The University of Arizona and two commercial firms joined with the U.S. Department of Agriculture to develop the crop. Machinery used for other grain and oil seed crops is suitable for lesquerella.

Cosmetic manufacturers and other companies are creating demand for the oils extracted from lesquerella seed. The oils can be used in resins, waxes, nylons, plastics, high-performance lubricants, corrosion inhibitors, coatings, and cosmetics such as hand soap and lipstick. The oil meal that remains after the oils are extracted is suitable as a high-protein livestock feed.

Figure 21-33 Environmentally friendly, ultra-low-volume herbicide application methods can significantly reduce the amount of agricultural chemicals used.

Courtesy of USDA/ARS #K5287-4. Photo by Keith Weller.

Figure 21-34 Late corn harvest using a combine near College Park, Maryland.

Courtesy of DeVere Burton.

The proper storage of crops after harvesting is important. Threats to the quality of stored crops include heat, moisture, fungi, insects, and rodents (**Figure 21-35**).

Drying grain to reduce moisture and heat is important for successful long-term storage. Much grain is harvested with a moisture content that is much too high. If stored without drying, the grain may heat up and encourage the growth of fungi, causing the grain to spoil. Foreign materials, such as high-moisture weed seeds, may also cause stored crops to spoil.

Stored crops must be protected from insects and rodents if quality is to be maintained. Rodent droppings, hair, and urine, as well as insect parts, render crops unfit for human consumption. Reduction of food value and spoilage are other hazards of stored crops when insects and rodents are not controlled.

The production of field crops generates more income for U.S. agriculturists than any other production enterprise. Nearly one-half billion acres of the land in the United States is currently being used for growing crops, and we are fortunate to harvest much more than we need each year. The excess production contributes to a more favorable balance of trade. Crop production will surely remain our most vital national product.

Figure 21-35 Modern weather- and rodent-proof grain storage facilities provide optimum storage and conditions to minimize insect and mold damage in stored grain.

© Denton Rumsey/Shutterstock.com.

Chapter Review

Student Activities

1. Write the **Key Terms** and their meanings in your notebook.
2. Compile a list of the field crops grown in your area.
3. Write a report on a field crop of interest to you.
4. Select a field crop and determine as many uses for it as possible.
5. **SAE Connection:** Visit a local crop farm and talk to the operator about the advantages and disadvantages of growing a particular crop.
6. Prepare an advertisement to promote a field crop.
7. Create a multimedia presentation or a bulletin board about field crops and products made from them.
8. Visit a farm-machinery dealer. Make a list of all the equipment sold there that is used in the production and harvesting of field crops.
9. Make a collection of as many different field crops as you can.
10. Do a germination test on the seeds of several types of field crops. Observe the number of days required for germination to take place and the percentage of the seeds that germinate. Also conduct germination tests under a variety of environmental conditions, such as warm and cool, wet and dry, or with and without light. Compare the results to determine optimum conditions for the germination of crop seeds.
11. Make sketches of each of the field crops (and their seeds) provided by your teacher for identification. Label each drawing with its name. Write any notes that will help you identify the plant in the future. Review this activity from time to time to ensure that you remember how to identify each field crop.

Inquiry Activity | Oilseed Processing

Courtesy of DeVere Burton.

After watching the *Oilseed Processing* video available on MindTap, take turns with a classmate to briefly summarize each of the following steps of oilseed processing:

- Cleaning
- Drying
- Dehulling and flaking
- Reducing size
- Tempering and cooking
- Pressing

National FFA Connection — Diversified Crop Proficiency Award: Placement

Ready to test your knowledge of forage production?
© IoanaB/Shutterstock.com.

FFA members with interests in and employment with a farming business that practices "best management practices" may become eligible for local, state, and national awards in diversified crop production. The farm business for which the member works must produce and market crops in at least two of the following proficiency categories:

- Vegetable production
- Fruit production
- Grain production
- Fiber and oil crop production
- Forage production
- Specialty crop production (excluding floriculture production)

The National FFA website provides procedures and instructions that will help chapter advisors work with their members to prepare appropriate applications for this award.

Check Your Progress

Can you...

- define important terms used in crop production?
- identify major crops grown for grain, oil, and special purposes?
- classify field crops according to use and thermal requirements?
- describe how to select field crops, varieties, and seed?
- discuss how to prepare proper seedbeds for grain, oil, and specialty crops?
- describe examples of current irrigation practices for field crops?

If you answered "no" to any of these questions, review the chapter for those topics.

Chapter 22
Forage and Pasture Management

Objective
To determine the nature of and approved practices recommended for forage and pasture production and management.

Competencies to Be Developed
After studying this chapter, you should be able to:
- define important terms used in forage and pasture production and management.
- identify major crops grown for forage and pasture.
- select plant varieties for forage and pasture.
- prepare proper seedbeds for forage and pasture crops.
- plant forage crops and renovate pastures.
- control pests in forage crops and pastures.
- harvest and store forage crops.

Key Terms
hay
silage
pasture
nurse crop
overseeding
carrying capacity
tiller
haylage
silo

Forages are crop plants that are produced for their vegetative growth. Forage and pasture crops rank first in total acres among crops grown in the United States. There are approximately 655 million acres of pasture and rangeland. Another 50.7 million acres are used to produce hay. The importance of forages is evident when you consider that half of the pasture and rangeland in the United States is not suited for the production of cultivated field crops.

Forage production is divided into three general categories: hay, silage, and pasture. **Hay** is forage that has been cut and dried until its moisture content is reduced to safe storage levels. **Silage** is green, chopped forage that has been allowed to ferment in the absence of air. **Pasture** is forage that is harvested by livestock as they graze. Forages are generally planted or maintained with one of these uses in mind.

Forage and Pasture Crops

Alfalfa

The most important forage crop in the United States is alfalfa (**Figure 22-1**). It is often called the "queen of the forages." Alfalfa is a legume that adds nitrogen to the soil. It is high in protein and other nutrients and is productive in fertile soils (**Figure 22-2**). It is also one of the oldest cultivated forage crops; it was mentioned in the Bible and in other early writings.

When properly harvested and stored, alfalfa is the most economical source of nutrients for ruminant animals. Including alfalfa and alfalfa mixtures together, annual alfalfa production ranks third to corn and soybeans in dollar value.

The North Central states account for about two-thirds of the yearly alfalfa production in the United States. However, alfalfa can be grown in nearly every state. California ranks number one in alfalfa production, followed by Idaho, Montana, and South Dakota.

Figure 22-1 Alfalfa is a forage crop that is high in protein and other nutrients that are needed by livestock. It is the most important forage crop in the United States.

Courtesy of DeVere Burton

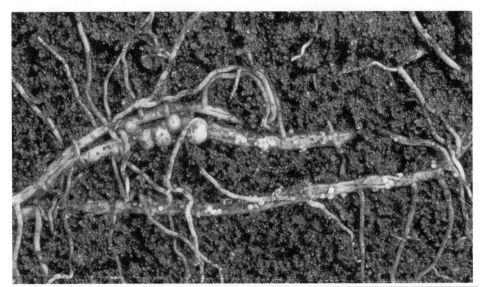

Figure 22-2 Bacteria in the nodules on alfalfa roots are capable of converting nitrogen gas from the atmosphere into nitrates, a required plant nutrient.
Courtesy of USDA/ARS.

True Clovers

True clovers include about 300 species; however, only about 25 have agricultural importance. Clovers of greatest economic importance in the United States include red clover, white clover, crimson clover, ladino clover, and alsike clover (**Figure 22-3**).

Clover ranks second among hay crops grown in this country. It is a legume with the ability to add nitrogen to the soil. Clover has the disadvantage of smaller yields than alfalfa under similar growing conditions. It is popular as hay, pasture, and silage and grows well in combination with many forage grasses. The Northeast and North Central states produce much of the clover grown in the United States.

Figure 22-3 Clover is an important hay and pasture crop in the United States.
Courtesy of Sharon Rounds.

Sweet Clover

Sweet clover is used most often in areas that are hot, drought stricken, or both, where its ability to survive and produce a crop is unsurpassed. It is used as hay, pasture, and sometimes as a green manure crop. Sweet clover is also used in Texas in rotation with cotton to help control a cotton-root disease. It is also an excellent source of nectar, which honeybees make into honey.

Three species of sweet clover are grown in the United States: biennial yellow, biennial white, and annual. Biennial white sweet clover yields more than the other species, although biennial yellow sweet clover is usually of better quality. Sweet clover will grow in all areas of the United States. The most favorable production areas are concentrated in the deep South from Georgia to Texas.

Bird's-Foot Trefoil

Bird's-foot trefoil is a comparatively new crop in the United States. It originated in Europe, where it has been a forage crop for 300 years. Approximately 2 million acres are in production in the United States. Bird's-foot trefoil is used as pasture in most cases, though some is also grown for hay (**Figure 22-4**).

The food value of bird's-foot trefoil is about equal to that of alfalfa, but because of smaller yields, bird's-foot trefoil is unlikely to seriously challenge alfalfa in importance. It does have the advantage over true clovers in being longer-lived. States that report large acreages of bird's-foot trefoil are Pennsylvania and New York.

Figure 22-4 Bird's-foot trefoil is a legume containing tannin, which reduces the danger of life-threatening bloat to grazing cattle.
Courtesy of USDA/ARS #K2610-10.

Lespedeza

This legume is grown primarily in the South, where 1 to 2 million acres are harvested as pasture and hay each year. It can grow and thrive in soils that are low in fertility. Because lespedeza has a lower nutritive value than true clovers and alfalfa, it is recommended as feed for beef cattle but not for dairy cattle.

Types of lespedeza include annual and perennial varieties. Most of the lespedeza grown in the United States is of the annual type (**Figure 22-5**).

Figure 22-5 Lespedeza is a forage plant that provides pasture and hay for cattle. Most of the roughage from lespedeza is fed to beef cattle.

Courtesy of DeVere Burton.

Peanut Hay

Peanut hay is a by-product of peanut production and is used in the southeastern region of the United States. Although it lacks the protein content of forage legumes, the addition of nitrogen compounds such as urea improves its quality. The extra nitrogen in the hay can be converted to protein by ruminant animals such as cattle and sheep.

Bromegrass

Bromegrass is an important forage grass throughout the northern half of the United States. It is extremely hardy and grows to a height of 2 to 3 feet (**Figure 22-6**). Because bromegrass produces many rhizomes, thin stands rapidly thicken with age. Rhizomes are horizontal, underground stems from which new plants arise. Fertile soil is required for the production of bromegrass.

Orchard Grass

Orchard grass is known for its rapid germination and early-spring growth (**Figure 22-7**). It also recovers quickly after being harvested. Timing of the harvest of orchard grass is important because quality decreases rapidly after the plant reaches maturity. Orchard grass makes excellent pasture and is often grown for hay in combination with legumes.

Figure 22-6 Bromegrass is grown throughout the northern half of the United States.
Courtesy of DeVere Burton.

Figure 22-7 Orchard grass is known for its rapid, early-spring growth.
Courtesy of DeVere Burton.

Timothy

Timothy is a cool-season grass that grows best when temperatures are between 65° F and 72° F (**Figure 22-8**). The use of timothy as a forage grass has declined in recent years because bromegrass and orchard grass out-yield it in those areas where all three species are adapted. Most timothy is grown in the northeastern part of the United States.

Figure 22-8 Timothy grows best when temperatures are between 65° F and 72° F.
© Tyler Olson/Shutterstock.com.

Reed Canary Grass

In areas that are wet or poorly drained, reed canary grass is often the only answer to producing a forage crop (**Figure 22-9**). It can produce more than 4 tons of forage per acre in the cool, damp areas where it thrives. This grass grows as tall as 7 feet and can produce high-quality feed if harvested before reaching maturity. Mature grass becomes coarse with high fiber content. In pellet form, it is sometimes used in combination with wood pellets as fuel in wood-burning stoves. Most reed canary grass is grown in the Northwestern and Northcentral regions.

Figure 22-9 Reed canary grass grows in areas that are too wet for the production of other grasses.
Courtesy of DeVere Burton.

Kentucky Bluegrass

Kentucky bluegrass is raised over much of the United States, even though it is best adapted to the Northeast. It is the major grass of many pastures, lawns, and even golf courses in areas where summers remain moderately cool. Kentucky bluegrass is a good native, permanent pasture grass, but it is out-yielded by other grasses, and it is not practical to use for hay due to low yields. It also goes dormant during hot weather, making it necessary to use bluegrass in combination with other grasses such as ryegrass to maintain the availability of forages during hot summer months (**Figure 22-10**).

Fescue

Tall fescue and meadow fescue are perennial grasses that are used for pasture and hay, usually in combination with other grasses and legumes. Fescue is best adapted to the Southeast, where 10 to 20 million acres are grown each year. The quality and palatability of the fescues are less than most other forage grasses.

Bermuda Grass

Bermuda grass is a warm-season grass that is dormant during cool weather. It is adapted to pastures and lawns because it grows only 6 to 12 inches tall. Common Bermuda grass is considered to be a weed in some areas because it tends to crowd out other grasses. However, improved varieties of Bermuda grass make good pastures in the Southeast (**Figure 22-11**).

Figure 22-10 Kentucky bluegrass is a cool-season grass. Its primary uses are (A) lawn and golf course turf when mowed regularly and (B) pasture and forage crops for livestock.
© iStockphoto/Jean Assell. Courtesy of DeVere Burton.

Figure 22-11 Improved varieties of Bermuda grass make good pastures in the Southeast.
© Torian/Shutterstock.com.

Hot Topics in Agriscience | Rangeland Forage

Range forage is a valued resource to owners of cattle and sheep. These ruminant animals are able to convert the forage to valuable meat for human consumption.
Courtesy of DeVere Burton.

In many parts of the country, ranchers use rangelands to graze their animals. Some of these lands are privately owned, while the government owns others. Some people are opposed to grazing on public land, citing the fear that the animals will destroy the environment. However, most public rangeland is managed under the multiple-use system, allowing shared use of the resource. This includes grazing, mining, and recreation.

On rangelands, the forage is generally provided by nature, in the form of grass or edible wild forage. In native areas, forage management is much different compared with forage production in pasture settings. Animals must be carefully monitored to avoid overgrazing. Overgrazing occurs when animals are allowed to feed in an area for too long. The result is damage to the soil and loss of valuable native plants. When this occurs, the area may not produce enough forage in subsequent years. Ranchers who rely on rangelands as feed for their animals must carefully manage grazing practices to maintain the health of rangelands.

Some of the common techniques that are used to prevent overgrazing are fencing, rest rotation, deferred rotation, supplementation, and herding. Fencing off areas of the rangeland helps keep the animals where they are supposed to be. Nutrient supplements such as salt are also used to entice animals to stay where they are wanted. Fencing also plays a part in both rest rotation and deferred rotation. Rangelands can be divided into pastures; for example, a rancher may divide his or her range into four different pastures. Rest rotation is when one pasture is not grazed for a given period. This gives plants a chance to complete at least one life cycle. Ideally, this should be done every other year. Deferred rotation occurs when the normal rotation between pastures is changed.

Range riders and herders are people who take care of range animals. They move them from place to place, making sure that the animals have access to sufficient forage and that the land and plant cover are preserved. When ranchers properly apply these methods, both the animals and the environment benefit.

Dallis Grass
Dallis grass grows 2 to 4 feet tall in the warm areas of the South. It cannot stand continuous use as pasture because it needs to be able to recover from close grazing. However, it is productive earlier in the spring than other warm-season grasses.

Corn and Oats as Forages
Alfalfa and clovers are often planted with a nurse crop of oats. A **nurse crop** consists of plants that are planted with other plants that require protection during early stages of development. The oats protect the tiny legume plants during the early stages of growth. They are then cut for silage while the leaves are still nutritious and the grain is immature. Oats make excellent forage when they are harvested at this stage of maturity. The new alfalfa or clover crop responds to the reduction in competition, often producing a hay crop in the first growing season.

Corn is produced as a forage crop on many farms, where it is used for making silage. Approximately 4.25 million acres are used for this purpose in the United States. Corn silage is a highly nutritious feed for beef and dairy cattle, and high yields are possible when adequate water and fertilizer are available. The corn plant is chopped and ensiled before the first frost when the grain is in the full dent stage of maturity. The acids that are produced as the silage cures are highly digestible, and the nutritional value of the forage is maintained at a high level. It is a high-moisture feed; therefore, it is not recommended as a large part of the diet for young cattle. They are unable to eat enough silage to satisfy their nutritional needs. Dairy cows require dry hay in combination with silage to maintain high levels of milk production.

Growing Forages

Many considerations must be made when selecting forages for hay, silage, and pasture:
- Intended use of the forage
- Expected yield
- Nutrient value of the crop
- Climatic conditions under which the forage will be grown—warm season versus cool season, summer and winter temperatures, humidity, soil type, anticipated rainfall/access to irrigation water, nutrient level, and length of growing season
- Pest-control measures
- Methods of establishment
- Compatibility with other forages when grown in mixtures
- Expected life span of the crop
- Care and maintenance
- Equipment and labor necessary for growing, harvesting, and storing the crop

Agri-Profile — Career Areas: Agronomist/Plant Physiologist/Forage Manager/Range Manager

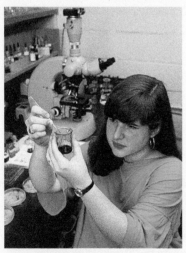

Summer employee Aimee Crago stains an alfalfa cotyledon for microscopic examination.
Courtesy of USDA/ARS #K4264-8.

Forage and pasture provide a variety of career options, ranging from farm and ranch duties to plant breeding and physiology. Grains such as silage corn provide enormous volumes of feed for dairy and beef cattle. Crops harvested for silage, as well as those cut for hay, are numbered among the forages. Similarly, pasture and range plants are forages. Those who specialize in the science of forage growth and use are called agronomists.

Forage specialists are hired by universities and agricultural research centers as well as by large farms and businesses. In certain parts of the United States, hay businesses are on the increase as more people engage in part-time farming, especially those with pleasure horses. In addition, hay and straw are needed by large commercial dairies, beef feedlots, and racetracks, creating strong markets for these forages in many areas.

Hay business managers, truckers, and dealers also conduct thriving businesses distributing hay and straw. Large concentrations of dairy cattle and pleasure horses have created markets for cross-country shipping of hay, a practice once regarded as too expensive to be worthwhile. Large amounts of hay are also shipped during periods of drought.

Forage crops that are adapted to the production of hay include alfalfa, clover, bromegrass, orchard grass, timothy, and fescue. Forages used for pastures include clover, lespedeza, Kentucky bluegrass, and Bermuda grass. Almost any legume or grass crop can be used for silage. In addition, corn and most small-grain crops make excellent silage when harvested when the grain is nearing maturity and before the leaves and stalks lose their nutritional value.

Seedbed Preparation

In general, preparing a seedbed for forages is much the same as it is for producing- field crops. Residues from previous crops must be incorporated in the soil. The soil must be prepared for the planting of seed or vegetative pieces used in the propagation of some grasses. The soil also must be amended so that its pH and fertility are suitable for the germination and growth of the intended forage crop. With some crops, the moisture level of the seedbed may need to be regulated to ensure germination.

One difference between preparing a seedbed for most forages and one for field crops is that the texture of the soil in the seedbed must be finer and the seedbed somewhat firmer for forage crops than is necessary for most field crops. This is because most forage plants have very small seeds that have difficulty making firm contact with the soil in seedbeds unless the soil is finely textured.

Conventional tillage methods are used when preparing seedbeds for starting grass and legume crops for forage. The residues from previous crops are plowed down or shredded. The soil is then pulverized using modern tillage machinery, followed by smoothing and leveling. For very small seeds, further preparation may be necessary. Final preparation of the seedbed should occur immediately before planting the seed. New seeding equipment makes it possible to plant new grass seed into established sod. This practice improves pastures and greatly reduces soil losses while providing protective vegetation until the seedlings become established

Planting Forages and Renovating Pastures

Forage crops are usually planted by drilling or by broadcasting. The same grain drills that are used for planting small grains are often equipped with metering devices for small seeds, allowing them to be used to plant forages. A separate seedbox on the drill is needed for this purpose. A nurse crop provides shade and is used to protect another crop while it becomes established. Forage seed is often planted along with a grain crop. Many grasses are planted in this way. The grain crop germinates faster than the forage and acts as a nurse crop until the forage plants can become established.

Many forage crops are also planted separately, using either drills or broadcast planters. With some forages, the seed must be covered after it is broadcast into the soil by broadcast planters. This is done to protect the seed from pests, such as birds, and to reduce the drying effects of sun and wind. Covering the seed may be accomplished by a light harrowing of the soil surface or by the use of the corrugated wheels of a cultipacker, which presses the seed into the seedbed (**Figure 22-12**). Some types of forage seeds are broadcast into growing crops. In these instances, they must be able to germinate fairly easily because contact with the seedbed is often minimal. Red clover may be overseeded into stands of forage grasses to make mixed hay. **Overseeding** is the practice of seeding a second crop into one that is already growing. This is usually done during late winter and early spring, when freezing and thawing of the soil help provide contact between the seedbed and the seeds.

Forages may be planted with no-till planters through crop residues (**Figure 22-13**). The no-till planter disc opens a narrow furrow in which the seed is dropped. A packing wheel follows the disc, compacting soil particles over the seed to assist germination. Advantages of a no-till planter include fewer trips across the field with heavy tractors and tillage equipment.

Figure 22-12 The packing wheels of this modern small seed drill planter create a firm seedbed, bringing the seeds in close contact with the soil particles.
Courtesy of DeVere Burton.

Figure 22-13 The no-till grassland drill places seed, fertilizer, and pesticides in one pass across the field.
Courtesy of Elmer Cooper.

This means less compaction of the soil and little exposure of the seedbed to erosion. Care must be taken to control pests in no-till plantings.

There are several methods of renovating pastures in the United States. The existing pasture can be killed with herbicide followed by preparation of the soil, forming a fine seedbed. The pasture is then reseeded with the desired types of grasses or legumes. The pasture is usually treated for insects and diseases at this time if necessary. This is also an ideal time to apply fertilizers and lime to the soil.

Another method of renovating a pasture involves using selective herbicides to kill unwanted species of plants. The pasture is then tilled to break up the existing sod. This allows

Figure 22-14 Grasshoppers are serious pests in forage crops because they eat the leaves and growing tips of the plants. This grasshopper represents only one of more than 600 species of grasshoppers in the United States.
Courtesy of USDA/ARS #K1066-2.

easier entry of moisture and nutrients into the soil. The pasture may or may not be overseeded with desirable species to improve yield. It is also fertilized and limed according to soil test recommendations.

No-till grassland seeders may be used to place new seed in existing sod, permitting new plants to augment the existing plant population. These seeders can also introduce new, improved, and more aggressive varieties of forage plants.

Pest Control in Forage Crops

The control of weeds, insects, diseases, and rodents in forage crops helps promote optimum yields. Because many forage crops grow for more than one growing season, pest control is usually an ongoing part of forage crop management (**Figure 22-14**).

The proper identification of the pest or pests affecting a crop is important. To that end, personal experience and the advice of trained professionals are often necessary.

The actual methods of pest control are many, and they are discussed in previous units. Chemicals, cultural practices, biological control, genetic control, and the timing of crops are all tools to be used in controlling the pests of forage crops. Not to be overlooked in pest control management is the control of pests in stored forages. When chemicals are used to control pests, care must be taken to ensure that the pesticide is properly applied at the recommended rates. Timing the application of the pesticide for maximum effect is also essential. Care must also be taken to ensure that overspray and runoff of pesticide materials do not adversely affect organisms other than the intended pest. Measures to ensure that pesticide residues do not end up in food sources are critical.

Harvesting and Storage of Forage Crops

The proper harvesting and storage of forage crops is extremely important in the management of the forage enterprise. For maximum profits, special care must be taken to control moisture levels in the harvested crop. It is also important for the forage to be removed from the field as soon as possible. Once it is in storage, forage quality is much easier to maintain.

Too much moisture in ensiled crops causes nutrients to be leached away in the liquids that drain from the storage area. Too much moisture in hay causes it to become moldy, sometimes resulting in fires ignited by spontaneous combustion due to a buildup of heat.

Pastures

The harvesting of pastures involves several factors. One of the most important is that the **carrying capacity** of the pasture must be determined to figure out how many animals can be fed by the pasture that is available. The carrying capacity of a pasture equals the number of animals the pasture is capable of feeding.

Pastures also need time to recover from the ravages of the animals that graze them. The rotation of animals using the pastures is important if the pastures are to meet their potential yields. Pasture rotation breaks the life cycles of parasites that afflict the grazing animals.

Hay

Harvesting hay involves several operations that must be accurately timed if high-quality hay is to be produced. The hay crop must be cut at the optimum time, with an eye on the weather forecast for the next several days. Hay is normally dried in the field by the sun. Nothing ruins hay faster than rain on the crop after it starts to dry and cure. The maturity of the forage being harvested for hay is also critical (**Figure 22-15**). The more mature the forage becomes, the lower will be the quality and nutritional value of the hay.

Type of Hay	When to Harvest
Alfalfa	Pre-bud to bud stage
Clover	¼ to ½ bloom stage
Bird's-foot trefoil	¼ bloom stage
Sweet clover	Start of the bloom stage
Brome grass	Medium head stage
Timothy	Boot stage to early bloom stage
Lespedeza	Early bloom stage
Orchard grass	Full head but before blooming
Reed canary grass	When the first head appears

Figure 22-15 Maturity levels at which selected forages should be harvested for hay.

Hot Topics in Agriscience: Producing High-Protein Alfalfa Hay

The alfalfa plant is a great forage plant because it is a succulent with a lot of leaves. Just before it begins to bloom, it reaches the peak of quality. If it is harvested much sooner, valuable tonnage is lost. Later harvesting results in a smaller percentage of protein. At just about the same time, young **tillers**, or shoots, begin to grow at the base of the plant. These tillers are the new stems for the next hay crop. If the harvest is delayed beyond this stage of maturity, the plant goes into seed production. The stems become thickened, and the proportion of stems to leaves will increase. This reduces the percentage of protein in the hay.

High-quality alfalfa hay can contain more than 20 percent protein when it is harvested at the proper time. Hay of this quality commands a premium price as dairy hay, because dairy cows require more protein than other farm animals. The best high-protein alfalfa hay should be fed to dairy cows to ensure that they consume enough nutrients to sustain a high level of milk production.

Another key element in producing high-quality hay is to process the hay immediately and move it into a covered stack to protect it against precipitation. Hay that is rained on between cutting and baling never is high-quality hay. This is because nutrients and color are leached out, and leaves are usually lost during the extra days that are required to get it dry enough to bale.

Most of the forages grown on U.S. farms and ranches are fed to livestock maintained on the same farm. Forages are usually the least expensive sources of nutrients available for cattle, horses, and sheep.

Terms that describe forage crop maturity include boot stage (immature seed development), pre-bud (early stage of flower formation) and 1/4 to 1/2 bloom (1 in 4 or 1 in 2 flower buds have opened).

Most hay is cut with either a sickle-bar mower or a rotary mower. The sickle-bar mower cuts with the same action as a pair of scissors. These types of mowers are best adapted for forages that are standing upright.

The rotary mower cuts with blades that move in a rapid circular motion parallel to the ground. This type of mower will cut any type of forage, although greater amounts of horsepower are required when the stands of forage are extremely thick and heavy.

A swather, also known as a mower/conditioner, is commonly used for cutting hay. It performs two processes in a single pass through the field. It cuts the forage plant off just above the ground, and it places the cut forage in a windrow for convenience in the baling or chopping operations.

Crushing and mashing of the stems is also performed by conditioners that are built into most swathers to reduce the length of time required to dry the foliage. In areas where rain is frequent, the hay may have to be cut with a mowing machine and allowed to lie in the swath where it falls. This spreads the hay sufficiently for the sun to dry it quickly. In areas of high humidity, chemicals to speed the drying of the forage are sometimes sprayed on the forage as it is cut.

Hay is often raked at least once before it is baled to allow it to dry more evenly and more rapidly. Legume-type hay should be raked during the part of the day when humidity is highest, to minimize the loss of leaves. Too much raking reduces hay quality because the high-protein leaves fall off. Raking hay also puts it into windrows so that it can be baled.

When hay has dried to the desired moisture level, it is baled or chopped and removed from the field. In some cases, it may be cubed or pelleted, particularly if it is being exported or transported long distances. Most hay is baled and hauled with automated hay wagons that pick up the bales from the fields and transport them to the storage area, where they are stacked in place.

Chapter 22 Forage and Pasture Management **461**

Most hay is harvested by forming it into rectangular bales or into round bales of various sizes (**Figure 22-16**). The baler gathers the dried hay from the windrow, compresses and shapes it into a bale, and ties twine or wire around the bale to hold it together, allowing for handling. The bale is then expelled from the baler. A recent trend has been to bale hay with balers that make large bales weighing up to 1 ton each. These bales are preferred for shipping long distances (**Figure 22-17**).

Machines designed to move baled hay from the field to the storage site include trucks, wagons, bale loaders, bale handlers, and bale stackers (**Figure 22-18**). Hay may also be harvested with machines called stack wagons. These machines gather forages from the field and form dense stacks or loaves of weather-resistant hay. With special handling equipment, stacks can be moved with few problems.

Figure 22-16 Hay is harvested by mowing, conditioning, air drying, baling, and hauling to storage areas.

© Prixel Creative/Shutterstock.com.

Figure 22-17 Large, rectangular bales are preferred for shipping hay by truck because they can be loaded by machine, they are easier to secure on the truck, and the load is not easily shifted.

Courtesy of DeVere Burton.

Figure 22-18 Bale stackers have done much to replace manual labor with machine labor. One person operating a bale wagon can haul more hay in a day than a crew of four using flatbed wagons.

Courtesy of USDA.

Hay cubes were developed to allow for full mechanization and ease of feeding hay. Cubes are also favored for the export market due to their high density and the ease of shipping. The cubing machine gathers dry forage and compresses it into cubes about 1.25 × 1.25 × 2 to 3 inches. Normally, only legumes are cubed because of difficulty in getting grass cubes to bind together. Artificial binders are sometimes used.

Hay may be stored in buildings or in the open field. Regardless of the method of storage, care must be taken to ensure that quality is maintained. Hay that is stored in buildings must be kept free from pests, particularly rodents. Care must also be taken to ensure that the moisture level of stored hay is low enough that spoilage does not occur.

When hay is stored outdoors, it should be placed on a site that is well drained and sloped to drain water away from the hay. Stacks of baled hay stored outside should be covered with plastic or other materials so that little moisture can penetrate them. Protecting hay from adverse weather conditions will preserve its quality until it is needed (**Figure 22-19**).

Figure 22-19 Protective covers prevent baled hay from spoiling in the stack due to moisture entering the baled hay.
Courtesy of DeVere Burton.

Silage

Harvesting forages for silage is easier than making hay. Less equipment and labor are required, and the entire harvesting operation is usually completed in minimal passes over the field. A forage harvester cuts the green forage, chops it into small pieces, and deposits it into a wagon or truck to be hauled to the storage facility. In instances where the crop is unusually succulent, the forage may need to be cut and allowed to dry for a few hours in the field to reduce the moisture content before it is ensiled. Forages used for **haylage** are cut and allowed to dry to a moisture content of 40 to 55 percent before they are placed in a silo. A **silo** is an airtight storage facility for silage or haylage. Haylage has the advantage of being lower in moisture than silage. This results in a product with enough dry matter to allow it to be fed to young cattle, and it requires less storage space than silage.

Silage may be stored in upright silos or in trenches or bunkers. It may also be stored in piles on top of the ground and sealed with plastic or other materials to make it airtight. The production of silage is dependent on sealed storage in the absence of air or oxygen. The fermentation that occurs in the absence of air preserves the silage. Spoilage occurs when green forages are stored in the presence of air (**Figure 22-20**). The chopped forage must be tightly packed to remove the air.

Figure 22-20 Forage silos must restrict the movement of air to preserve and store silage.
Courtesy of DeVere Burton.

The Fermentation Process

Silage-making is a fermentation process whereby bacteria break down sugars and starches to produce the acids that preserve the forage. Silage-making and silage-keeping require that several conditions be met and maintained. First, the crop must be chopped into short pieces and have sufficient moisture to pack tightly enough to exclude most of the air. Second, the crop must contain a sufficient amount of sugars and starches for the fermentation process to occur. Third, the silo or silage container must be airtight so air cannot reenter the silage after it has fermented.

What happens in the silo to change the taste, texture, and chemical content of the forage? If the ensiled forage contains grain, such as corn, sorghum, wheat, barley, oats, rye, or other seeds, it will ferment when packed in an airtight container or silo with about 60 percent or more moisture. If the mixture does not have sufficient sugars and starches, or if the moisture content is not within the acceptable range, then grains or preservatives must be added to the forage to make good silage.

When the moisture content is less than 60 percent and the silo is sufficiently tight, a low-moisture silage known as haylage may be made with the addition of preservatives.

Immediately after suitable forage has been chopped and packed in an airtight space, bacteria that live in the presence of oxygen in the air that fills the pores of the forage start to multiply. They feed on the sugars and starches in the forage and consume the oxygen in the process. They feed and multiply until the oxygen is consumed, so long as there are sufficient sugars and starches in the forage to keep them feeding. The bacteria make lactic and acetic acids, which smell good and give the forage a pleasant taste for livestock.

The frantic burst of bacterial activity in freshly packed forage causes the temperature of the forage to increase within hours after the process starts. This continues until the oxygen is consumed and the bacteria die. The resulting level of acid in the forage, now called silage, prevents any other bacterial action so long as oxygen-laden air does not reenter the silage. Without bacterial action, the silage cools down and is preserved until air enters the silage again. In practice, air may get into silage and cause spoilage around poorly fitted silo doors and holes left in the plastic film used to seal silos. Air movement through poorly packed or excessively dry silage also leads to spoilage. Spoiled silage can be detected by the presence of mold or the foul odor of butyric acid, which forms when undesirable fermentation takes place.

Approved Practices for Forage Crops

Many crops are usable as forages. Forages are found in the form of grazing of young plants by livestock, cut and dried as hay, or harvested and stored in silos as medium- or high-moisture haylage or silage. Some approved practices for the production of forages are as follows:

- Select a perennial species for long-term stands.
- Select a perennial or annual species for short-term stands.
- Plant certified seed of the selected variety.
- Obtain a current soil test.
- Apply lime and fertilizer according to the soil test results and desired production.
- Select a field with soil suitable for your forage choice.
- Select the proper tillage method to prepare a firm and smooth seedbed.
- Calibrate the drill for a proper plant population.
- Match the plant population to the soil-yield potential.
- Adjust the drill for proper depth of seed placement.
- Decide if forage may best be grown with a companion crop.
- Calibrate the sprayer for proper application of materials.
- Use selected herbicides for controlling problem weeds.
- Use insecticides as necessary.
- Make sure chemicals have been approved on forage crops.
- Decide the stage of maturity to harvest the crop.
- Make a yield check.
- Keep harvesting equipment in good repair to save downtime.
- Check the moisture content of harvested forage.
- Harvest the crop as soon as the moisture level permits.
- Store hay in a clean, dry structure.
- Store silage in an airtight structure.
- Obtain a forage analysis.

- Track market trends and habits.
- Market the crop at the optimum time.
- Keep accurate enterprise records.
- Summarize and analyze records.

The production of forages in the United States accounts for about half of all the land in agricultural use. Much of this land is grazing land that is unsuitable for production of other crops. With the production of hay ranking third only to corn and soybeans, forages are extremely important to U.S. agriculture. Because forages are the least expensive sources of nutrients for cattle, sheep, and horses, they are likely to remain essential for years to come.

Science Connection: Kenaf: Super Forage?

Kenaf, a bamboo-like plant with high-protein forage, shows promise to become a super foliage crop of the future.
Courtesy of USDA/ARS #K2975-9.

Kenaf was viewed for several years as a possible new source of fiber for newsprint, rope, and other fiber-based products. However, it appears that its best use may be as a roughage and protein source for sheep and cattle. Research at the Agricultural Research Service Forage and Livestock unit at El Reno, Oklahoma, indicates that animals will eat the leaves and stems, and the digestibility of certain parts of the plant is about the same as that of alfalfa. This combination of high fiber and high protein has attracted interest in the plant as cubes or pellets for cattle. The leaves are of growing interest as a small-animal feed.

Kenaf grows rapidly, and a 30-day harvest rotation seems to work best. Kenaf leaves contain 20 to 29 percent crude protein when harvested at this maturity stage. Stalks must be cut at least eight inches above the ground, however, to ensure good survival. When cut, two or more shoots spring from each stalk to produce new plants. The potential is to grow multiple forage crops in a single season. Being an annual, the crop is easy to establish and has few or no insect and disease pests. Production in the United States is increasing as research continues on kenaf as a forage crop.

Chapter Review

Student Activities

1. Write the **Key Terms** and their meanings in your notebook.
2. Put fresh-cut forage in each of two quart jars. Pack the forage tightly into the jars. Seal one jar and leave the other open to the air. Compare the contents of the two jars after 10 days.
3. Make a collection of as many different forage seeds as you can.
4. Write a report on a forage crop of economic importance in your area.
5. Make a collection of local forages.
6. Visit a machinery dealership to learn about the types of seedbed-preparation, planting, and harvesting equipment for forages.
7. Create a multimedia presentation or a bulletin board about forages.
8. Obtain samples of hay and silage, and compare them for firmness of leaves and stems, odor, and color.
9. Ask your instructor to obtain samples of lactic acid (found in good silage) and butyric acid (found in spoiled silage) from the science laboratory, and compare the odors produced by these acids. Do not smell the vapors of these acids directly from the container.
10. A cow eats about 90 lbs. of feed each day and drinks 35 to 50 gallons of water. Hay makes up about 30 lbs. of its total daily feed intake. Determine how many pounds of hay a cow will eat in one year in a confined feedlot with no supplemental grazing.
11. **SAE Connection:** Determine the cost for 1 ton of hay in your area by calling an area hay farmer. Determine how much it will cost for hay in the ration of one cow for one year, using the answer you calculated in Activity 10.

Inquiry Activity Forage Crops

Courtesy of DeVere Burton.

After watching the *Forage Crops* video available on MindTap, recall what you learned from just one viewing. Make a list for each of the following topics:

- Considerations when deciding what forage plant to grow
- Reasons alfalfa is the most important forage crop in the United States
- Steps of the harvesting and storing process of forage crops

Compare your list with that of a partner, and add to your lists as needed. Then, watch the video again. Review your lists and add information gleaned from the second viewing.

Unit 7

Ornamental Use of Plants

The words "fastest growing segment of U.S. agriculture" may evoke an image of great fields of corn "as high as an elephant's eye" or feedlots of cattle as far as the eye can see. Few would picture acres of greenhouses filled with flowers, golf courses with gorgeous turf, or beautifully landscaped neighborhoods. But flowers, potted plants, and landscape trees and shrubs are some of the hottest agricultural products in periods of economic prosperity. They beautify our surroundings and help purify the air.

According to the U.S. Floral Crops Department of Agriculture (USDA), the total value of sales across the U.S. floral industry hit $6.4 billion in 2021. The value of production is growing at the rate of 2.5 to 5 percent per year

We tend to assume our decorative plants are simply gifts from nature. In reality, years of painstaking research and development account for most ornamental plants, and research must continue to stay ahead of pests and to meet the public appetite for new decorative plants. For the most part, we seldom see plants from the wild. Nearly all the plants in common use are products of plant-improvement efforts.

For example, research and genetic improvements to the poinsettia have contributed greatly to this plant's phenomenal increase in commercial value over a 2- to 30-year period. During this time, the poinsettia has often been ranked the number one potted plant in the nation based on numbers sold—despite an annual sales period of only 6 weeks. About 35 million pots are sold each year during the holiday season. In 2020 alone, the retail value of poinsettias totaled $157 million.

Most of the USDA's work on the poinsettia and other ornamentals takes place the National Arboretum in Washington, DC, and the Agricultural Research Service's Florist and Nursery Crops Laboratory in nearby Beltsville, Maryland. Scientists there have developed many new ornamentals and are improving others for commercial use. In one project, scientists are investigating the biochemical and physical bases for color in plants and people's perceptions of color. Such research could lead to new kinds of flowers and provide a scientific basis for variety identification.

As applied research continues to answer the immediate needs of growers, such as problems impacting commercial ornamental enterprises, long-range research aims to stay well ahead of current knowledge and technology. Thanks to both efforts, ornamental plants will continue to play an important role in our health and well-being.

Chapter 23

Indoor Plants

Objective
To use indoor plants for beautification, air quality, and pollution control.

Competencies to Be Developed
After studying this chapter, you should be able to:
- identify plants that grow well indoors.
- select plants for various indoor uses.
- grow indoor foliage plants.
- grow indoor flowering plants.
- describe elements of design for indoor plantscapes.
- describe career opportunities in indoor plantscaping.

Key Terms
floriculture
succulent
foliage
variegated
herb
foot-candle
phototropism
rootbound
plantscaping
basic color
accent color
texture
accent
sequence
balance
formal
symmetrical
informal
asymmetrical
scale

The world of indoor plants is a fascinating one. There are plants that have magnificent blooms, unusual shapes, fancy foliage, and fragrant smells. If selected and cared for properly, they can last for years and may even be passed from generation to generation. The industry of indoor plants is the floriculture part of ornamental horticulture. **Floriculture** involves the production and distribution of cut flowers, potted plants, greenery, and flowering herbaceous plants. Indoor plants present a worthwhile challenge to people who wish to add nature's living color to homes, workplaces, and public spaces. (**Figure 23-1**).

Figure 23-1 Indoor plants add beauty and atmosphere to homes and public areas.
© Ru Bai Le/Shutterstock.com.

Almost all plants can be grown indoors. There are plants, however, that favor indoor conditions. While many trees and shrubs do better when grown outdoors, small succulent plants are best for indoor use. **Succulent** means having thick, fleshy leaves or stems that store moisture. A wide variety of shapes and sizes of indoor plants are available. These plants can be divided into two major groups—those that flower, and those that are grown only for their foliage. **Foliage** consists of stems and leaves.

Popular and Common Indoor Flowering Plants

African Violets

One of the most popular and common indoor flowering plants is the African violet (*Saintpaulia ionantha*). This plant can be recognized by its small size and hairy leaves. The leaves are oval-shaped, dark green, and covered with soft, short hairs. The flowers contain four to five petals arranged in a clover pattern. They vary in color from deep purple to brilliant white (**Figure 23-2**).

Figure 23-2 African violet (*Saintpaulia ionantha*).
© Roger Swift/Shutterstock.com.

Figure 23-3 Fuchsia (*Fuchsia triphylla*).
© irakite/Shutterstock.com.

Fuchsias

Fuchsias (*Fuchsia triphylla*) are plants with colorful flowers that cascade from the plant. Most have flowers that are two-tone pinks and reds. The foliage is dark green, and the leaves tend to be long and oval in shape, with a bronzy hint of color (**Figure 23-3**).

Gardenias

Gardenias (*Gardenia jasminoides*) are particularly fragrant, flowering, indoor plants that have deep-green shiny foliage and pure white flowers. The leaves are in clusters of three and are pointed (**Figure 23-4**).

Figure 23-4 Gardenia (*Gardenia jasminoides*).
Courtesy of H. Edward Reiley.

Geraniums

One of the most versatile flowering indoor plants is the geranium (*Pelargonium zonale*). These indoor plants are among the oldest. The leaves are rounded, yellowish green with scalloped edges. The flower is borne on a stem and consists of many petals in a cluster shaped like a ball. The flower color ranges from the most popular red, to white and pink (**Figure 23-5**).

Figure 23-5 Geranium (*Pelargonium zonale*).
Courtesy of H. Edward Reiley.

Impatiens

If an indoor plant with many blooms is desired, the impatiens (*Impatiens walleriana* or *Impatiens sultanii*) is a good choice. The flowers are small and rounded, with five petals. One petal is shaped like a tube that protrudes from the underside of the flower. Flower colors include white, pink, salmon, coral, lavender, purple, and red. The leaves are lance-shaped and have succulent stems (**Figure 23-6**).

Figure 23-6 Impatiens (*Impatiens walleriana*).
Courtesy of H. Edward Reiley.

Hot Topics in Agriscience — A Truly Popular Potted Plant: Poinsettia

The poinsettia is often the sales leader for potted plants in the United States.
© Scott Prokop/Shutterstock.com.

The poinsettia (*Euphorbia pulcherrima*) is native to Mexico and Central America. The Ancient Aztecs, who used its sap to treat fevers, called it Cuetlaxochitl. In Mexico, it is known as Flor de Noche Buena, or Christmas Eve flower. It was introduced to the United States in 1825 as Poinsettia in honor of Joel Roberts Poinsett, a botanist who was then the first U.S. Ambassador to Mexico. Over time, it became one the most popular potted plants sold in this country, often ranking number one.

The poinsettia far outsells other potted plants and is typically sold only in November and December. The brilliant red leaves are often confused for flowers. However, the red petal-like foliage is actually a form of specialized leaves called bracts. Its true flowers are small and unassuming.

Additional Indoor Flowering Plants

Many other varieties of plants can be grown indoors for their flowers. Some flowering plants are both indoor and outdoor plants, such as wax begonias, ageratums, verbenas, and petunias. Any other flowering plant capable of withstanding the rigors of an indoor environment may be considered here.

Popular and Common Indoor Foliage Plants

Indoor foliage plants can be divided into five groups for easy identification. These groups are ferns, indoor trees, vines, cacti/succulents, and specimen plants. A foliage plant is grown for the appearance of the leaves and stems (**Figure 23-7**).

Figure 23-7 Business and public areas with foliage plants are attractive and appealing.
© Igor Stepovik/Shutterstock.com.

Ferns

Ferns come in a variety of types. Some of the most popular indoor ferns are Boston fern (*Nephrolepis exaltata*) (**Figure 23-8**), asparagus fern (*Asparagus sprengeri*; **Figure 23-9**), maidenhair fern (*Adiantum capillusveneris*), sword fern (*Nephrolepis cordifolia*), rabbit's foot fern (*Davallia canariensis*), and staghorn fern (*Platycerium bifurcatum*). Ferns are categorized by their long and often multicut leaves. Most ferns are feathery in appearance. Some ferns are used extensively as greens in floral arrangements.

Figure 23-8 Boston fern (*Nephrolepis exaltata*).
Courtesy of H. Edward Reiley.

Figure 23-9 Asparagus fern (*Asparagus sprengeri*).
Courtesy of H. Edward Reiley.

Indoor Trees

An indoor tree can provide an excellent accent to a room or an attractive addition to a patio or hallway. Trees can grow to be 6 to 7 feet tall or higher indoors. Some of the more popular indoor trees are Norfolk Island pine (*Araucaria excelsa*; **Figure 23-10**), fiddle leaf fig (*Ficus lyrata*), umbrella plant (*Schefflera actinophylla*; **Figure 23-11**), rubber plant (*Ficus elastica*; **Figure 23-12**), fragrant dracaena (*Dracaena fragrans*), weeping fig (*Ficus benjamina*; **Figure 23-13**), and croton (*Codiaeum variegatum*). Indoor trees vary in the type and size of their foliage, but all have woody-type stems.

Figure 23-10 Norfolk Island pine (*Araucaria excelsa*).
Courtesy of H. Edward Reiley.

Figure 23-11 Umbrella (*Schefflera actinophylla*).
© Melica/Shutterstock.com.

Figure 23-12 Rubber plant (*Ficus elastica*).
© Imageman/Shutterstock.com.

Figure 23-13 Weeping fig (*Ficus benjamina*).
Courtesy of H. Edward Reiley. Courtesy of DeVere Burton.

Vines

Vines are characterized by their habit of climbing or draping from the sides of the pot. Some of the more widely recognized vine-type indoor plants are philodendrons (*Philodendron* sp.; **Figure 23-14**), inch plant, (*Tradescantia fluminensis* "Variegata"; **Figure 23-15**), grape ivy (*Rhoicissus rhomboidea*), and English ivies (*Hedera* sp.). These plants can be trained to climb up a piece of wood, around a planter box, or up a wall.

Figure 23-14 Philodendron (*Philodendron sp.*).
Courtesy of H. Edward Reiley.

Figure 23-15 Inch plant (*Tradescantia fluminensis*).
© Stephen VanHorn/Shutterstock.com.

Cacti and Succulents

Cacti and succulents make up a unique group of indoor plants that originated in deserts. They are among the easiest plants to grow indoors. These plants tend to hold water within their stems and leaves and can survive dry heat, low humidity, and varying temperatures. Cacti usually have some type of prickly needles, whereas succulents do not. One of the most popular succulents is the jade plant (*Crassula arborescens*; **Figure 23-16**).

Figure 23-16 Jade plant (*Crassula arborescens*).
Courtesy of H. Edward Reiley.

Specimen Plants

There are numerous other types of indoor plants that do not fit into the categories of ferns, indoor trees, vines, cacti, and succulents. These indoor plants tend to have special characteristics or features that people like to display; thus, they are called specimen plants. A few examples of popular specimens are peperomia (*Peperomia caperata*), which has pale pink to red stems with deeply grooved heart-shaped leaves (**Figure 23-17**); the spider plant (*Chlorophytum elatum vittatum*), which shoots off baby plants or "spiders" (**Figure 23-18**); purple passion plant (*Gynura aurantiaca*;), which has rich royal purple leaves that are covered with velvet-type hair; bromeliads, one variety of which has a deep pink color at the center; and snake plant (*Sansevieria trifasciata*), which has long, spike-like, thick leaves that are variegated with gold (**Figure 23-19**). **Variegated** means having streaks, marks, or patches of color.

Figure 23-17 Peperomia (*Peperomia caperata*).
© matka_Wariatka/Shutterstock.com.

Figure 23-18 Spider plant (*Chlorophytum elatum vittatum*).
Courtesy of H. Edward Reiley.

Figure 23-19 Snake plant (*Sansevieria trifasciata*).
Courtesy of H. Edward Reiley.

Selecting Plants for Indoor Use

Two general rules should be followed when selecting plants for use indoors. The first is to be selective when you purchase the plant. The second is to choose the right plant for the growing conditions available in the location you want the plant to live. Careful attention to these two rules greatly increases the chance for success with indoor plants.

Purchasing an Indoor Plant

When buying an indoor plant, look for one that appears to be healthy. Look closely at the plant for insects, being careful to check the undersides of the leaves where many insects hide or lay their eggs. Select plants with even green color, no yellowing leaves, and without any spots or blotches on the leaves. Avoid plants that have spindly growth or which appear wilted. If possible, purchase the plant during its growing season. Finally, look for new growth, such as leaf or flower buds.

Choosing the Right Plant

Before selecting an indoor plant, decide where the plant will be placed. Note particularly the light intensity and duration, as well as the temperature of the location. Each species of indoor plant has specific conditions for optimum growth. The plant should match the location for best results.

Light

The amount of available light for a plant is a factor that is difficult to control. A shadow test can help determine this. To conduct this test, hold a piece of paper up to the light (window or lamp) and note the shadow it makes. A sharp shadow means you have bright or good light. However, if there is barely a shadow visible, the light is dim or poor. It is important to know how much light the plant needs. If a plant needs direct or full sun, then exposure of the sun is needed for at least half of the daylight hours. When a plant needs indirect or partial sun, the light should be filtered through a curtain or slats. Plants that fit into this category are Chinese evergreen (*Aglaonema modestum*; **Figure 23-20**) and dracaena (*Dracaena fragrans*; **Figure 23-21**). Even for plants preferring no direct sunlight, the room should be bright and well lit. Plants that need shade should be kept in a well-shaded part of the room. In summary, too much or too little light can greatly affect the health of the plant.

Figure 23-20 Chinese evergreen (*Aglaonema modestum*).
Courtesy of H. Edward Reiley.

Figure 23-21 Dracaena (*Dracaena fragrans*).
Courtesy of H. Edward Reiley.

Temperature

Plants are adapted to specific temperature ranges, and they do not do well when these temperatures are exceeded. Indoor plants are grouped into three temperature categories: cool, moderate, and warm. The cool temperature range is 50° to 60° F, with temperatures not falling to less than 45° F. The moderate temperature range is from 60° to 70° F, with a minimum of 50° F. The warm temperature range is from 70° to 80° F, with a minimum limit of 60° F.

Uses of Indoor Plants

Uses of indoor plants vary. Plants can be used as room dividers or to brighten up dull spots in the kitchen or bathroom. They can make a workplace more inviting. They may be placed in containers with special watering devices, or several plants may be grouped together to form a natural barrier. Containers are available in several sizes and shapes, some with castors on the bottom so they can be moved easily. Indoor trees and climbing plants are best used as living-area dividers (**Figure 23-22**).

Specimen or showy plants, such as the Boston fern or date palm (*Phoenix dactylifera*), are good for brightening up dull areas (**Figure 23-23**). Plants may be placed on ornate plant stands or in decorative pots. A series of wall shelves may also be used to display interior plants. Plants on shelves should be compact, and perhaps trailing, for a more dramatic effect.

Hanging baskets filled with indoor plants are useful when space is at a premium. Baskets can be easily suspended from the ceiling to utilize the available space above head level. Care should be used when selecting plant containers so that water does not drip on the furniture or the floor. Select baskets that are designed with drip trays attached to the bases.

Interesting areas for hanging baskets include stairwells, offices, and hallways. An important element that is unique to hanging baskets is the hook securing the basket to the ceiling. It should be secured firmly enough to support the basket when the soil is wet and when the plant is being cared for (**Figure 23-24**).

Bathrooms are excellent places for indoor plants. These rooms are humid and warm, which provides a good atmosphere for plants. They can be placed on windowsills or around the tub. Shelves can also be installed in a bathroom to display indoor plants. Plants will add a touch of color to a bathroom because some bathrooms are expanding to hold a hot tub or Jacuzzi (**Figure 23-25**).

Figure 23-22 Decorative plants help define areas.
© Michel Cramer/Shutterstock.com.

Figure 23-23 Date palm (*Phoenix dactylifera*).
Courtesy of H. Edward Reiley.

Figure 23-24 Hanging baskets are versatile and attractive.
© Nadia Zagainova/Shutterstock.com.

Figure 23-25 The humidity of a well-lit bathroom is ideal for plants.
Courtesy of Elmer Cooper.

The kitchen is another room where plants can add a beautiful touch. Refrigerators, other appliances, and kitchen windowsills all make good places for plants that require cooler temperatures. Most plants in the kitchen should be relatively small because space is usually at a premium. Several plants suitable for the kitchen are aloes (*Aloe variegata*; **Figure 23-26**), maidenhair ferns, peperomias, and even a variety of herbs. **Herbs** are plants kept for aroma, medicinal purposes, or seasoning.

Growing Indoor Plants

Growing plants indoors is much like growing them outdoors. Consideration needs to be given to the environment in each instance. Aspects of the plant environment that must be considered are light, temperature, water, drainage, and nutrients. Although indoor plants are not harvested, they need to be maintained. Some aspects of maintenance that need to be addressed are grooming, repotting, and propagation.

Figure 23-26 Aloe (*Aloe variegate*).
Courtesy of H. Edward Reiley.

Light

Light is one of the most crucial factors to consider when growing indoor plants. Light is measured in foot-candles. A **foot-candle** is the amount of light found one foot from a burning standard candle, known as a candela. Two aspects of light that need to be determined are intensity and duration. The intensity and duration of indoor light vary during different seasons, at different times of the day, and through different windows. A plant near a window with southern exposure receives more intense sun, and for a longer period, than a plant in a window with northern exposure. Pulling shades across a window can reduce the intensity of light, and a tree outside a window can reduce both the intensity and duration of light that enters the window. Summer sun is much more intense than winter sun because the angle of the sun to the Earth is near perpendicular during the summer.

Too little light will limit photosynthesis causing some indoor plants to grow tall and spindly. They lose leaves because the long, thin, weak stems can no longer support the

Figure 23-27 Plants generally must be moved or rotated periodically to provide correct lighting to all parts of the plant.

© Elena Elisseeva/Shutterstock.com.

plants. Conversely, too much light causes indoor plants to wilt and lose their vibrant, green colors. The youngest leaves on plants are affected first by unfavorable conditions. Another item concerning light is the tendency of plants to grow toward the strongest source of light. If you notice that a plant is leaning toward the light, you may need to rotate the plant container periodically so that it maintains a balanced shape (**Figure 23-27**).

There are five basic light categories for indoor plants. These are full sun, some direct sun, bright indirect light, partial shade, and shade. Full sun is a location that receives at least 5 hours of direct sun a day. This amount of sun can be found in areas that have southern exposure. Some direct sun occurs in areas that are brightly lit but receive less than 5 hours of direct sun a day. This usually occurs in windows facing east and west.

Bright indirect light describes areas that receive a considerable amount of light but no direct sun. An example might be an area 5 feet away from a window that receives full sunlight. Partial shade refers to areas that receive indirect light of various intensities and durations. Areas 5 to 8 feet away from windows that receive direct sun are partial-shade areas. Shade refers to poorly lit sections away from windows that receive direct sun. Few plants can survive low-intensity light.

Temperature

Plants survive best at constant temperatures. Temperatures that fluctuate up and down are not ideal for plant growth. Temperature interacts with light, humidity, and air circulation. It is best to maintain temperatures in a range from 60° to 68° F for optimum indoor plant growth. As temperatures increase beyond this range, the air tends to become warmer, and the available moisture in the air decreases. Although the thermostat in a house reads one temperature, each room usually varies as much as 5 degrees in one direction or the other. Because plants respond differently to various temperature ranges, it is advisable to place plants in rooms that match their specific temperature requirements.

Science Connection — Phototropism in Plants

Plant leaves respond to light by turning to face the light source. Sometimes an entire plant will grow toward light, requiring a potted plant to be rotated regularly.

Courtesy of DeVere Burton.

The tendency of plant parts to move or turn toward or away from a light source is known as **phototropism**. The term comes from two ancient Greek words, *photo*, meaning "light" and *tropos*, meaning "a turning." This phenomenon occurs most often when the leaves of a plant turn toward the sun to maximize the sun's intensity on the leaf surface. It is a survival mechanism that allows photosynthesis to occur in the leaf at the maximum rate.

Phototropism is also observed in flowers that turn to face the sun each morning, following the track of the sun across the sky until it sets in the evening. Sometimes the petals of flowers close at night and open the next morning when the sun comes up. Each of these adaptations by plants to the sun or to other light sources is an example of phototropism.

You can observe phototropism for yourself by using a potted plant that is growing on or near a window. Simply turn the planter so that the plant is bending away from the direction of sunlight. Within 8 to 12 hours the plant will have bent toward the sunlight again.

Water

Water is an essential ingredient for growth of any living organism. The amount of water needed by indoor plants is usually not as much as one would think. The most likely problem with unhealthy indoor plants is too much water. More plants die from overwatering than from any other cause. As with other environmental factors such as light and temperature, each plant varies in its requirements for water. Unless you know the particular needs of the plant, it is best to water when the soil around the plant is a little dry. Most plants do best if allowed to dry out between applications of water. When indoor plants are watered properly, the roots remain more active than if the soil becomes waterlogged or excessively wet.

There are numerous ways to water plants. The basic rule about watering is to use water that is neither hot nor cold. Water should be tepid, or a moderate temperature. One way to water is to soak the pot in a bucket of warm water for half an hour, remove the pot, and drain it. A second way is to pour the water on top of the soil slowly, filling the pot to the top with water. Allow it to absorb the water until the excess drains from the hole in the bottom of the pot. Do not water again until the soil becomes dry to the touch (**Figure 23-28**).

Figure 23-28 Plants need to receive water on a regular basis, but too much water causes most plants to do poorly or even die.
© StockLite/Shutterstock.com.

Besides water in the media around their roots, plants also need moisture in the air. The moisture content in the air is referred to as humidity. It is expressed as relative humidity (a relationship between the amount of water vapor in the air compared with the maximum moisture the air will hold at a given temperature). Almost all indoor plants prefer 50 percent relative humidity. The simplest method for humidifying the air around plants is to set pots in trays filled with gravel and add water to just cover the gravel. Misting is a good way to add humidity, but it should be done several times each day to be effective. The humidity around plants can also be increased by grouping them together. However, plants should have enough space around them to allow for adequate air circulation.

Drainage

Good drainage is achieved by using pots with porous materials in the bottom and drainage holes in the pot. Good drainage is essential for indoor plants. Adding coarse material such as sand or perlite to the soil will improve soil drainage. The addition of gravel or bits of broken clay pot material to the bottom of a pot before adding the soil is a substantial aid to good drainage.

Fertilizing

Plants need nutrients on a regular basis for good health. A balanced fertilizer, such as a 5-5-5, should be applied at regular intervals. Fertilizers with a greater proportion of nitrogen than of phosphorus and potash are often used to keep foliage plants green and healthy looking. Plants should be fertilized at 2- to 6-week intervals, depending on the type of fertilizing material. Slow-release–type fertilizers dissolve slowly and release nutrients evenly over weeks or months. In contrast, liquid fertilizers suitable for foliar application are used by the plant within a few days. They may also be applied at weekly or biweekly intervals (**Figure 23-29**).

Plants should be fertilized while they are actively growing. Fertilization should be discontinued when the plant is dormant, or in a resting stage. Flowering plants need more fertilizer. Addition of fertilizer once every 2 weeks is recommended from the time flower buds first appear until the plants stop blooming. Avoid fertilizing plants when the soil is excessively dry. Under such conditions, the fertilizer solution is likely to be too highly concentrated and may cause burned leaves or roots.

Figure 23-29 Plants need to be fertilized with materials that correspond to their specific needs.
Courtesy of DeVere Burton.

Grooming

Grooming plants is important even with the best combination of light, temperature, humidity, water, and drainage. Grooming should be done weekly. This task includes removing wilted or withering leaves, flowers, and stems using sharp scissors or shears (**Figure 23-30**). The plant should be observed to determine if it is getting spindly or thin.

Figure 23-30 Plants need to be examined and groomed weekly.
© auremar/Shutterstock.com.

Figure 23-31 Repotting at appropriate times prevents the plant from becoming rootbound and promotes good plant health.

© Christina Richards/Shutterstock.com.

If so, pinch out new growth to force the plant to branch out. To avoid crooked stems, stake plants when they are young. Climbing or trailing plants need a stake made of bark to enable them to climb and cling to the stake.

Once a week, the plant should be dusted. Dust accumulates on the leaves and blocks the stoma so that the plant cannot breathe or transpire as well as it should. The soil around the base of the plant should be loosened with a fork or small spade to allow air to enter the soil and water to percolate through. To help control insects, mist the infested plants with a diluted solution of mild dishwashing soap and water.

Repotting

In time, plants develop root systems that are restricted by the pot or container. When this happens, the plant is said to be **rootbound**. The roots have no place to continue to grow, so repotting is necessary for plant health. In general, repotting is done in the spring or the fall. A good rule for determining pot size is to use one with a diameter at the top of the rim equal to one-third to one-half the height of the plant. This rule does not apply to plants whose growth habit is tall and slender, as opposed to a balanced top growth. When repotting, it is not desirable to move an established plant to a new pot that is more than 2 inches wider than the original pot. Excessively large pots result in wasted soil, water, and nutrients (**Figure 23-31**).

Flowering plants are best repotted after the flowers have faded. During repotting, check the roots for insects and root damage. Remove any roots that look or feel unhealthy.

Propagation

Propagating most indoor plants is relatively easy. The method chosen depends on the type of plant (whether it is herbaceous or woody, flowering or foliar). Both sexual and asexual methods of propagation are used.

In sexual propagation, seeds may be started in containers with a good potting soil, plenty of moisture, and adequate air circulation. Keeping pots in warmer areas of the house increases the speed of seed germination. Covering the pots with glass or plastic held up by stakes will also aid in the germination process.

Some popular methods of asexual propagation of indoor plants are leaf and stem cuttings, removal of plantlets from parent plants, and air layering. Each of these procedures results in the multiplication of the plant.

Flowering Plants

Indoor flowering plants may take some extra, special care to ensure blossoming. These plants are more sensitive to the availability of light, so some artificial lighting may be needed. They are also more sensitive to temperature changes. Flowering plants have particular seasons in which they flower. It is important to know when to expect the plants to bloom. Some examples of plants with specific bloom times are the Christmas cactus (*Schlumbergera x buckleyi*), which is expected to bloom between October and late January, and the Easter cactus (*Rhipsalidopsis gaertneri*), which blooms in April or May. These two plants are similar in their leaf types and flowers, but they bloom in opposite seasons of the year. Indoor plants may flower in the winter, spring, summer, or year-round.

Hot Topics in Agriscience: Genetically Engineered Flower: Blue Rose

The world's first blue rose was developed by splicing a pansy gene into the DNA of a rose.
© Veniamin Kraskov/Shutterstock.com.

The world's first blue rose was genetically engineered by introducing a gene for blue pigment from a pansy into a rose plant. After 20 years of research, scientists at Suntory Flowers of Japan made headlines around the world in 2009 with the release of the new rose. On one hand, the new rose proved capable of producing a blue pigment named *delphinidin*, which is not found naturally in any rose. On the other, the new rose was not true blue, but somewhat violet in color.

In 2019, research scientists in Tianjin University in China used genetic engineering to synthesize indigoidine, an amino acid in rose petals that produces true blue pigmentation. Unfortunately, the blue of such a rose fades quickly and it would not be possible to breed new flowers from the modified rose. However, the Chinese research team continues to work with other genetic techniques in hopes of yielding a true blue rose that will keep its color and which could be used to grow new roses.

Roses have been cultivated for more than 5,000 years, and they are among the most revered flowers on the planet. Their common colors are red, pink, yellow, and white, but a true blue rose would mark one of the biggest floral breakthroughs of all time. It would also introduce a lucrative product into a global market for roses that's already worth $5 billion.

Foliage Plants

Foliage plants are nonflowering plants grown and sold for their attractive leaves. Many shopping malls, offices, and other public buildings prefer foliage plants over flowering or fruiting plants because they are very easy to maintain and add a nice decorative element to interiors. In the early 1970s, wholesale customers purchased an average of $29 million worth of foliage plants in the United States. In 2005, that number soared to $721 million. This increase was not only spurred by an increase in popularity, it also was driven by a technique called *new tissue culture technology*. This new method speeds the rate of propagation, allowing growers to significantly increase production.

A wide variety of foliage plants is grown in homes and offices, and specific instructions for their care compose entire books. However, the correct management of light is probably the greatest single factor for growing foliage plants. Light is the source of energy for the process of photosynthesis, whereby the leaves produce sugars and starches to feed all parts of the plant. Different foliage plants have different light requirements. It is recommended that a good reference book on indoor plants be used to determine the particular requirements for any given plant. Regardless of the amount and duration of light, it is desirable to rotate plants so that each side receives the same amount of light over time.

Interior Landscaping or Plantscaping

Plantscaping is the design and arrangement of plants and structures in indoor areas. This design and arrangement is an art. Interior plantscaping is an activity that is fun. However, it should be approached seriously, just as the arrangement of furniture or other interior decorations must be done carefully. The indoor plants should be used to complement people-oriented spaces. The more creative you are in designing, the more distinctive indoor plantscapes will become. There is no right or wrong way to design with indoor plants, but there are elements that enhance the interior plantscaping technique.

Design

Before beginning a plantscape design, the designer must know the purpose or intent of the plants. Are the plants to be used as a space divider? Are they being used to accent existing furnishings? Will they fill empty space? Answers to these questions will result in an organized design, rather than a happenstance.

Another question to consider is the function or functions of the plants. Are they to create a specific shape, emphasize a specific area, or support a specific architectural feature of the room? If the function of the plantscape is not considered, the end result is not likely to be successful.

The physical characteristics of color, form, and texture should be considered when selecting plants for a plantscape. These are determined in concert with the perceptual characteristics of accent, sequence, balance, and scale. These characteristics are the basic tools a designer uses to create an interior plantscape (**Figure 23-32**).

Figure 23-32 Interior plantscaping is an important element in shopping malls and other public areas.
© Philip Lange/Shutterstock.com.

Color

Color is the most important physical characteristic of a plantscape. It can influence emotions, set a specific mood, and subtly or dramatically change the look of the environment. Two types of color must be considered by the designer of an interior plantscape. First is the basic, or background, color. **Basic color** is the color of the walls, ceiling, and floor. These colors should influence the selection of the flowering characteristics and foliage of plant material. **Accent color**, the second type, is the color of the plants or other attention-getting objects.

Form

Form refers to shape. There are different forms that plants possess naturally. The most common shapes are round; oval; weeping (or drooping); upright; spiky; and spreading or horizontal. These shapes can be used individually or grouped to form an artificially sculpted shape and form. Each plant shape or cluster shape adds its own particular feature to the design.

Science Connection: Guarding Against Whiteflies

A threat to ornamentals and non-ornamentals alike, whiteflies settle on plants like the season's first snow.
© iStockphoto/alohaspirit.

Even as adults, whiteflies are not much larger than the head of a pin, yet they can suck the life from plants, including fruits, vegetables, flowers, shrubs, and field crops. Whiteflies use needle-like mouth parts to suck sap from the phloem, the food-conducting tissues in plants and leaves. The costs associated with control of these pests and the damage they cause run in the millions of dollars. Greenhouse growers, nurseries, and retail outlets wage a continuous battle against whiteflies in an effort to save crops from damage and infestations that would prevent the sale of these products. Similarly, homeowners and interior plantscapers must be on a constant vigil, because unobserved eggs can hatch into an infestation of devastating proportions. Like the season's first snowfall, the eggs may suddenly appear as white specks all over the plant. When the plant is moved or the insects disturbed, the masses of adult whiteflies take flight and create the appearance of a snowstorm in the air.

First observed in Florida in 1986, an especially viral strain of the whitefly spread rapidly from Florida to California. It is challenging scientists and growers alike as it spreads nearly unchecked by either pesticides or natural enemies. Across the country, researchers are showing the textile industry how to wash off the sticky material that the insects leave on cotton fiber. Researchers are also discovering ways to deal with plant viruses left in the wake of the troublesome, testing such environmentally friendly controls as insect parasites, predators, fungi, plant extracts, and dish detergent/vegetable oil mixtures.

Scientists have identified more than 30 predators and 25 parasites as allies in the fight against whiteflies. One of the most promising is the big-eyed beetle. It secretes a sticky substance from its mouth to stick the whitefly to the leaf. Then it slowly devours the fly. Scientists have released some big-eyed beetles on a pilot basis. Another tiny but potentially useful insect that has been identified devours an estimated 10,000 whitefly eggs or 700 nymphs in its 6- to 9-week lifetime. However, not enough is yet known about this insect to release it. Similarly, a number of parasitic wasps are under examination.

As for spray materials, one of the best options is a bio-soap, which provides some control over whiteflies as well as other damaging insects. It controls those insects on which it is directly sprayed, so thorough applications are important.

Texture

Plant **texture** refers to the visual or surface quality of the plant or plants. Texture is influenced by the arrangement and size of leaves, stems, and branches. It is described in terms of coarseness or fineness, roughness or smoothness, heaviness or lightness, and thickness or thinness. Coarse-textured plants, such as fiddle leaf fig or prickly pear cactus, should be used in large spaces, whereas a maidenhair fern should be placed in small spaces, such as on shelves.

Accent and Sequence

Accent means a distinctive feature or quality. It captures the attention of the viewer and has a dramatic effect on the visual appearance of the room or part of the room. You can create an accent with the use of color, form, or texture. It can also be created through the use of a sequence. **Sequence** refers to a related or continuous series. Sequence is created with plant material by repeating the same color, texture, or form. The overuse of accents, however, will detract from their function.

Balance and Scale

Balance is the state of equality and calm that is created between items in a design. In a design, two types of balance are used: formal, or symmetrical; and informal, or asymmetrical. **Formal**, or **symmetrical**, balance occurs when the items are equal in number, size, or texture on both sides of the center of the design. **Informal** or **asymmetrical** balance occurs when the items are not equal in number, size, or texture on the two sides of the center. **Scale** refers to the size of items. A large patio with African violets as accents would be out of scale. Plant materials need to complement the size and, therefore, the scale of the room. A better selection for a patio would be Norfolk Island pines and dracaenas.

A Design Process

There are various types of design processes. A particularly effective one to use has three phases. The three-phase process emphasizes the how and why of using plants for the particular space.

Phase One

The first phase of the process is preplanning, which includes three steps. The first step is to develop the design objectives by determining the purpose of the plantscape. The second step is to determine the space capacities, which include the habitat and the circulation of people. Circulation is how people tend to move in the area. The amount of light, temperature, and humidity in the room is referred to as habitat. The third step is to determine the development limitations. Development limitations include the amount of money available to implement the design; room and space characteristics; and habitat limitations created by light, temperature, people, pets, and other factors.

Phase Two

The second phase is to develop the plan, including the basic design and arrangement of plants and materials. The physical characteristics of the plants are important in this phase, as tentative plant selections and visual placement in the room are planned. The planning is done on paper, and alternatives are examined. The finished plan includes the final selection of all the plants that will be used.

Phase Three

Implementation is the third phase. This is where the design comes to life. This phase has three steps. The first step is the preparation of the documents. If any construction is involved, such as platforms, decks, or planter boxes, drawings for these objects must be completed. Drawings and specifications for the installation and maintenance of water lines and fixtures, electrical devices, and plants should also be finalized at this point. This is important, especially if you are designing for someone other than yourself. The second step is the installation of all physical modifications. This includes the selection of plant containers and the planting of the indoor plants. The plants are then put in place. The final step is evaluating the project. A thorough look at the plants, their containers, and their position in the room or area must be done. If everything looks balanced, no changes need to be made. If not, minor adjustments may be made. The interior plantscape is now complete.

Plantscape Maintenance

Maintenance of the plantscape with attention to light, moisture, humidity, and grooming is important. These factors are discussed in preceding parts of this unit. The nursery and greenhouse industries often provide critical expertise in developing and maintaining plantscapes (**Figure 23-33**).

Figure 23-33 Greenhouses, nurseries, florists, and suppliers provide a constant supply of plant materials.
© R. Gino Santa Maria/Shutterstock.com.

Agri-Profile — Career Area: A Bumper Crop of Plant-Related Opportunities

The florist industry provides cut flowers and arrangements for all occasions. Plants grown in greenhouses, nurseries, and fields become part of interior plantscapes.
Courtesy of DeVere Burton.

From florals to foliage to indoor and outdoor plantscapes, the popularity of plants and gardens has created an abundance of opportunities for career seekers. The widespread popularity of foliage plants certainly offers "green" potential for horticultural career seekers. From small trees to potted greenery, foliage plants are extensively used as décor in such interior public areas as shopping malls, office buildings, institutions, and housing complexes.

Youth in suburban settings can also find new career and employment opportunities caring for plants, installing interior plantscapes, rotating plants between growing areas and display areas, and contracting to maintain interior plantscapes. These early experiences frequently lead to lifelong careers. Home gardeners have also been known to turn their personal interest into a career by seeking out local opportunities to both learn on the job and pursue formal training part-time.

Several of these aspects of horticulture offer individuals the chance to establish their own businesses with relatively low investment. For example, one interesting occupation in the field of plantscaping is the plant rental business. Many office and building managers want the benefits of plantscaping but cannot spend the time required for maintenance. Plant rentals, which include regular maintenance, give building managers the best of both worlds.

With more experience and education, one may become a plantscape designer, business owner–operator, grower, wholesaler, plant doctor, or extension specialist. Ornamental horticulture specialists are hired by universities, research institutes, and business firms to develop improved plants and horticultural practices. The florist industry is also a source of jobs in design, arranging, care, and delivery.

The education required for a career in indoor plantscaping varies widely. Persons with high school technical experience in horticulture, two-year college degrees, or four-year college degrees can be successful in this career. Many career seekers also take college-level or certification courses in floriculture, business management, and marketing. Indeed, since many careers in this field combine technological skills, scientific knowledge, and aesthetics, or an eye for beauty and design, the possibilities for lifelong learning are virtually unlimited.

Chapter Review

Student Activities

1. Write the **Key Terms** and their meanings in your notebook.
2. Take an inventory of the species, type, and number of indoor plants in your home.
3. Select indoor plants that are suitable for your home from reference books and seed or plant catalogs.
4. Develop a plan for an interior plantscape of a room in your home, the school office, or another location.
5. Repot an indoor plant.
6. Propagate indoor plants using one or more methods.
7. Perform a shadow test at several windows in your home or classroom to determine the intensity of the light available.
8. Prepare hanging baskets for use in the home or office.
9. **SAE Connection:** Check with a local greenhouse or flower shop for any discarded plants you could try to save. Try to rehabilitate the plants using the knowledge you gained from this unit about fertilizer, watering, and light requirements. Donate any nice-looking plants to a nursing home, school, or community organization.

National FFA Connection — Proficiency Award: Diversified Horticulture

The FFA offers opportunities to develop a well-rounded knowledge of horticulture.
© Zoran Pucarevic/Shutterstock.com

The National FFA offers separate proficiency awards for supervised horticulture experience. "Entrepreneurship" and "Placement" proficiency awards can be earned at the local, state, and national levels. Experience is required in two or more of the following proficiency areas: landscape management, nursery operations, turfgrass management, and floriculture production. A floriculture production project plus experience in design and/or sale of floral products such as corsages, boutonnieres, floral arrangements, etc. may also qualify.

A member who owns a business where best management practices are followed to produce and market plants is eligible to compete for the entrepreneurship award. A student who works for a business that produces and markets plants may compete for the placement award.

Each of these proficiency awards requires students to follow best available management practices in both production and marketing. Local agriscience teachers can provide guidance as you become involved in horticulture production and as you compete for the proficiency award for which you are best qualified.

Inquiry Activity: Improving Indoor Spaces

Courtesy of H. Edward Reilley

After watching the *Improving Indoor Spaces* video available on MindTap, imagine you are proposing a plan to improve the school lobby by adding plants. Using information from the video, describe how you will assess the space and what you will need to consider as you devise your plan.

Consider how a professional in the plant industry might present such a plan to potential clients. For example, you could give an oral presentation to your class. Include visuals in the presentation, such as sketches of your proposed design and a digital slide show of the types of plants you would use. Invite your audience to ask questions and offer feedback.

Check Your Progress

Can you . . .

- identify plants that grow well indoors?
- discuss how to select plants for various indoor uses?
- explain how to grow indoor foliage plants?
- explain how to grow indoor flowering plants?
- describe elements of design for indoor plantscapes?
- give examples of career opportunities in indoor plantscaping?

If you answered "no" to any of these questions, review the chapter for those topics.

Chapter 24
Turfgrass Use and Management

Objective
To understand growth and development of turfgrasses and the establishment and cultural practices involved in managing these plants.

Competencies to Be Developed
After studying this chapter, you should be able to:
- identify and describe careers available in the turfgrass industry.
- identify turfgrass plant parts.
- select turfgrass species for various purposes and locations.
- state the basic cultural practices for turfgrass production and maintenance.
- list the basic steps for turfgrass establishment.

Key Terms
turfgrass
seminal root
crown
extravaginal growth
vegetative reproduction
intravaginal growth
seed culm
inflorescence
induction
sheath
blade
ligule
collar
recuperative potential
fungal endophyte
thatch
syringing
seed blend
seed mixture
sodding
sprigging
stolonizing
plugging

Who would have expected 50 years ago that farmers would one day raise fields of grass for a purpose other than feeding livestock? Yet, a huge industry has risen for the purpose of growing "instant lawns" that are transported from the fields to the front yards where green grass suddenly appears and thrives. It is a classic case of the instant gratification we have come to expect in modern society.

The Turfgrass Industry

It has been estimated that the annual economic impact of the turfgrass industry in the United States is approximately $40 billion. This industry is large, diverse, and still growing. Career opportunities in this field continue to expand and are projected to continue growing in the years ahead. **Turfgrass** includes all the grasses that are mowed frequently to maintain a short and even appearance.

Golf course superintendents and other athletic-field managers maintain turfgrass at a certain level of playability. Playability level means suitability for the intended use. It is determined by the type of sporting or recreational event. For example, putting greens are areas used for playing golf, and the turfgrass is very short. These areas are maintained to provide a surface for consistent, yet adequate, putting speeds (**Figure 24-1**). A football field may be managed to offer secure footing and sufficient turfgrass resiliency for player safety.

Figure 24-1 (A) Sports turf requires maintenance practices that provide an acceptable level of playability. (B) Soils near golf greens tend to become compacted due to heavy use. Aeration involves removing small plugs of soil to restore infiltration of air and water in the root zone.
Courtesy of Bob Emmons. © iStockphoto/Michael Braun.

The maintenance of turfgrasses at large government, apartment, university, commercial, and private complexes requires personnel trained in turfgrass management. Lawn care services offer many job positions and are among the largest employers within the industry.

Sod-production farms and landscaping businesses are involved in the establishment and installation of turfgrass for lawns (**Figure 24-2**). Turfgrass specialists are also needed for these segments of the industry (**Figure 24-3**).

Figure 24-2 Turfgrass is widely used to establish new lawns around homes, businesses, and public buildings.
Courtesy of Elmer Cooper.

Figure 24-3 Turfgrass is grown in large fields from which it is harvested for the purpose of creating "instant lawns."
Courtesy of USDA.

The turfgrass industry supports a substantial sales force. Companies that produce or distribute seed, fertilizer, pesticides, and turfgrass equipment require an extensive support and sales staff (**Figure 24-4**).

Federal and state governments and private companies hire turf specialists and scientists with advanced degrees. Career opportunities in these areas offer challenging positions in research, teaching, and extension.

Figure 24-4 Turfgrass requires regular maintenance, such as fertilizing and mowing.
© iStockphoto/BanksPhotos.

Turfgrass Growth and Development

Turfgrasses are plants grouped into the Poaceae family. These grasses differ from other grass plants because they can withstand mowing at low heights. They can also tolerate vehicle and foot traffic. Turfgrasses are frequently used to hold the soil and as ornamental plants. These traits have made turfgrasses the most widely used ornamental crop in the United States. To properly maintain turfgrasses, an understanding of their growth and development is required.

The Turfgrass Plant

The grass plant can be divided into two broad areas known as the root and shoot systems. The root system consists of adventitious and seminal roots. The shoot system includes the stem and leaves of the plant.

Root System

Seminal roots develop from the seed during seed germination. They initially anchor the seed into the soil. The seminal root system will be active for 6 to 8 weeks. The adventitious roots develop from the nodes of stem tissue. They usually compose the entire root system of a mature turfgrass stand.

Turfgrass roots are multibranching and fibrous. Their function is nutrient and water absorption. They also prevent soil erosion by effectively stabilizing and anchoring the soil particles in place.

Seasonal changes in root growth are dependent on soil temperature and moisture. Active root growth for warm-season turfgrasses occurs in the summer. A warm-season turfgrass is one of a group of grasses adapted to the southern region of the United States. A cool-season turfgrass is a plant adapted to the northern region of the United States. These turfgrasses have active root growth in the fall and early spring. Optimum growth occurs at temperatures from 60° to 75° F.

Agri-Profile — **Career Area: Turfgrass Grower/Groundskeeper/Landscape Maintenance Technician**

Turfgrass production, establishment, and maintenance provide many good jobs and create attractive surroundings for homes, businesses, and recreation.
© cappi thompson/Shutterstock.com.

Turfgrass production and management have become big business in the United States. In some states, turfgrass ranks as the number one crop, based on total acres in production. Growth and development of golf courses has stimulated the turfgrass industry with high-paying salaries for golf course superintendents and other turfgrass specialists.

Currently, career opportunities in turfgrass production, management, service, supervision, research, and consultation are extensive. Turfgrass technicians and specialists generally work in attractive and appealing surroundings. Many work outdoors in sunny weather and indoors when the weather is bad. While most annual salaries range between $30,000 and $56,000, demand for experienced professionals in some areas has led to salaries as high as $122,000.

Educational programs in turfgrass science and management are available in high school technical programs, technical schools, colleges, and universities. With the movement toward urbanization, interest in open spaces, concern for the environment, and increasing population, the outlook for careers in turfgrass production and management is excellent.

Rooting depth is affected by plant species, soil factors, and cultural or maintenance practices. Average rooting depth for turfgrasses is 6 to 12 inches. The warm-season grasses have deeper root systems than the cool-season grasses. Well-drained, sandy-loam soils with neutral soil pH are best adapted for turfgrass roots.

Cultural practices that influence rooting depth and growth include mowing, fertilization, and irrigation. Frequent mowing and low mowing heights reduce rooting depth. In addition, fertilization programs that emphasize only shoot growth are likely to impair root growth. Light, frequent irrigation results in shallow-root grasses. Heavy but infrequent irrigation is more conducive to deep-root penetration and growth.

Turfgrass maintenance practices should attempt to optimize the rooting potential of turfgrass plants. An extensive root system allows a plant to recover from drought and other stress conditions more rapidly.

Shoot System

The shoot system consists of stems, leaves, and seed head, or inflorescence. These plant parts are involved in capturing solar energy through photosynthesis and storing it in forms that the plant can use for growth and seed production.

Stems

Turfgrass stems include the crown, tillers, rhizomes, stolons, and seed culms. The crown is the meristem tissue of the grass plant from which new growth occurs. The **crown** is a stem with the nodes stacked on top of each other (**Figure 24-5**). All root, leaf, and other shoot growth originates from this area. The crown is located at the base of the grass plant in the soil surface area.

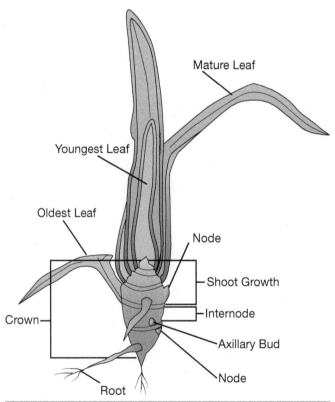

Figure 24-5 The crown of a turfgrass plant is the meristem tissue from which new growth occurs.

Rhizomes and stolons are horizontal stems. A rhizome is a creeping underground stem, whereas a stolon is an aboveground stem (**Figure 24-6**). Each originates from an axillary bud on the crown and will penetrate through the lower leaf sheath. This type

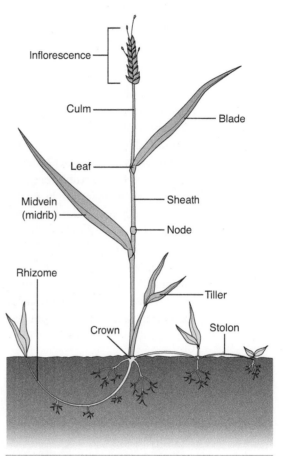

Figure 24-6 The major parts of a grass plant.

of growth is referred to as **extravaginal growth**. Rhizome and stolon growth allows for vegetative spreading of turfgrasses. **Vegetative reproduction** is reproduction from plant parts other than seeds.

Tiller

Tillers are new shoots of a grass plant that develop at the axillary bud of the crown. They form within the lower leaf sheath of the plant. This type of growth is referred to as **intravaginal growth**. Increased tillering will enhance turfgrass density. All turfgrasses produce tillers. Optimum tillering for the cool-season grasses occurs in the spring and fall months. Warm-season grasses have optimum tillering during the summer. Under low-moisture and high-temperature stress conditions, tiller, rhizome, and stolon development is reduced.

Seed Culm and Inflorescence

The **seed culm**, or seed stem, supports the inflorescence of the plant. The seed culm originates at the top of the crown. **Inflorescence** is the arrangement of the flowering parts of a grass plant. Cool-season grasses produce their inflorescence in the spring. Warm-season turfgrasses produce their inflorescence in the late summer.

Flower **induction** or initiation is caused by several environmental conditions. Temperature and photoperiod are the two induction processes for grasses. When seed heads form, they will cause a decrease in playability and appearance of the turfgrass stand. Mowing will remove the seed head and thus improve turfgrass quality.

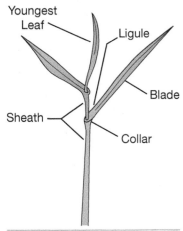

Figure 24-7 The leaf blade and leaf sheath of a grass plant.

Leaf

The turfgrass leaf consists of the sheath and blade (**Figure 24-7**). The **sheath** is the lower portion of the leaf and may be rolled or folded over the shoot system. The **blade** is the upper portion of the leaf. At the junction of the blade and sheath are the collar and ligule. The **ligule** is located on the inside of the leaf and is a membranous or hairy structure. The **collar** can be found on the outside of the leaf and is a light green or white-banded area (**Figure 24-8**). These two features are important vegetative traits for turfgrass identification.

Turfgrass growth is dependent on the production and use of carbohydrates. The turfgrass leaf is responsible for photosynthesis and, ultimately, carbohydrate production. Reserve carbohydrates will be stored in crown, rhizome, and stolon tissue.

During unfavorable growth conditions, the plant will go into dormancy. All leaf tissue will die. However, when favorable environmental conditions recur, carbohydrate reserves from the crown and other stem tissue will be used for plant re-growth.

Turfgrass maintenance programs attempt to optimize root growth and carbohydrate accumulation. Greater recuperative potential and plant persistency occur when these two basic concepts of growth are understood and managed. **Recuperative potential** is the ability of a plant to recover from drought or damage.

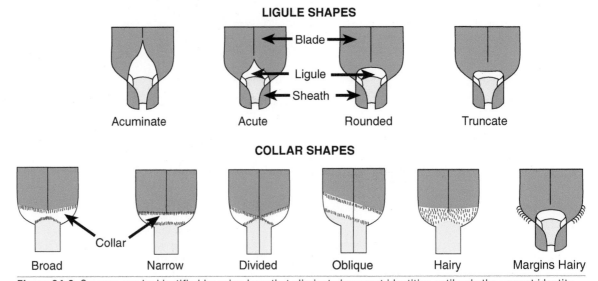

Figure 24-8 Grasses can be identified by using keys that eliminate incorrect identities until only the correct identity remains. The shapes of ligules and collars are used to describe plant characteristics included in the identification key.

Turfgrass Varieties

Approximately 7,500 plants are classified as grasses. Only a few dozen are considered useful for turfgrass. Turfgrasses are divided into two major groups based on climatic adaptation. The two groups are cool-season and warm-season grasses. The United States may be divided into a number of climatic zones based on temperature and moisture conditions. It is helpful to be aware of these zones when selecting turfgrass (**Figure 24-9**).

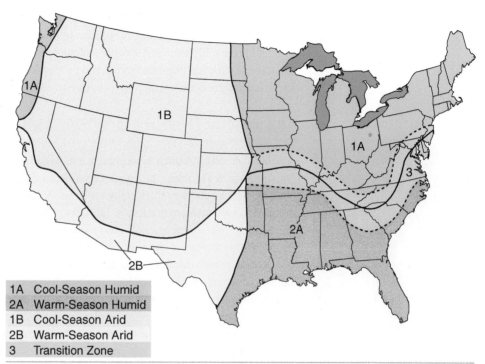

1A Cool-Season Humid
2A Warm-Season Humid
1B Cool-Season Arid
2B Warm-Season Arid
3   Transition Zone

Figure 24-9 The major turfgrass adaptation zones.

Cool-Season Turfgrasses

Cool-season turfgrasses originated in Europe and Asia. They have optimum growth at temperatures from 60° to 75° F. They predominate in the northern and central regions of the United States. Species adaptation within the cool-season group is determined by rainfall, soil fertility, and turf use. Major cool-season turfgrasses in the United States include Kentucky bluegrass, tall fescue, red fescue, perennial rye grass, creeping bentgrass, and crested wheatgrass.

Kentucky Bluegrass (*Poa pratensis* L)

Kentucky bluegrass (**Figure 24-10**) is used extensively in residential and commercial lawns, in recreational facilities, and along highway rights-of-way. It performs well with moderate levels of maintenance. The plant has a medium leaf texture and an extensive rhizome system. Texture refers to leaf width. Fine-texture turf contains grasses with narrow blades, whereas coarse-texture turf consists of wide-blade grasses.

Figure 24-10 Kentucky bluegrass.
© Joshua Boman/Shutterstock.com.

Kentucky bluegrass grows best with full sun, moist and fertile soil, and a mowing height of 1.5 to 2.5 inches. More than 100 cultivars of Kentucky bluegrass have been developed for different geographic areas and specific maintenance conditions. A cultivar is a plant of the same species that has been discovered and propagated because of its unique characteristics. Cultivar differences exist with respect to disease tolerance, leaf width, color, and other traits.

Tall Fescue (*Festuca arundinacea* Schreb)

Tall fescue (**Figure 24-11**) is a coarse-textured, bunch-type grass used in home lawns or as a utility-type turf. A bunch-type grass grows in clumps rather than spreading evenly over the soil surface. However, the bunching tendency of tall fescue can be overcome by heavy seeding, which produces thick stands of grass. A utility-type grass refers to a turfgrass adapted to low maintenance levels. Tall fescues have extensive root systems and are among the most drought-tolerant, cool-season species. However, they are prone to winter injury in the northern range of the cool-season zone.

Recent genetic developments have introduced many new and improved cultivars of tall fescue. These new cultivars have medium leaf texture with more aggressive rhizome development. This turfgrass will not tolerate low mowing heights and should be mowed at 2.5 to 3 inches. Insect resistance and plant persistency are excellent for tall fescues infected with fungal endophytes. A **fungal endophyte** is a microscopic plant growing within a plant. They tend to improve turfgrasses.

Figure 24-11 Tall fescue.
© Abine sh/Shutterstock.com.

Red Fescue (*Festuca rubra* L)

Red fescue is a fine-textured turfgrass well adapted to shady, dry locations. It has excellent drought tolerance and can persist on rather infertile soils and under acidic soil conditions (pH 5.5 to 6.0).

Red fescue is often seeded in mixtures with Kentucky bluegrass for lawn turf. It is not used on athletic fields as a permanent turf because it has poor recuperative potential. However, in the South, it may be overseeded into dormant Bermuda grass greens to provide winter play endurance and color.

Red fescue functions satisfactorily if it is mowed at 1.5 to 2.5 inches in height and is provided with minimal levels of nitrogen fertilizer and water. The major pest problem of red fescue is leaf spot disease.

Perennial Ryegrass (*Lolium perenne* L)

Perennial ryegrass (**Figure 24-12**) has a medium leaf texture and is most often used in seed mixes with other turfgrasses. It has a rapid germination and excellent seedling vigor. It is often used in seed mixes to provide soil stabilization during the establishment period.

Figure 24-12 Perennial ryegrass.
© 3DMIShutterstock.com.

The perennial ryegrasses are used extensively for recreational turf. They have good wear tolerance and can be rapidly established if the turfgrass stand is damaged. They are also used in winter overseeding programs with warm-season turfgrasses.

Perennial ryegrasses require moderate levels of maintenance to form attractive turf. As a group, they have poor disease resistance. Ryegrasses express improved plant persistence and insect resistance in turf plantings that have been inoculated with fungal endophyte. A symbiotic relationship exists between ryegrass and fungal endophyte.

Creeping Bentgrass (*Agrostis palustris* Huds)

Used in close-cut, high-maintenance areas, creeping bentgrass (**Figure 24-13**) is an extremely attractive turfgrass and fits well when used on putting greens or bowling greens mowed at five-sixths of an inch. It can be maintained at higher mowing heights (0.5–0.75 inch) for use on football fields, golf course fairways and tees, lawns, and formal gardens.

The plant has extensive stolon growth and is a fine-textured turfgrass. It is best adapted to slightly acidic soil with a pH of 5.5 to 6.0 that has good internal drainage and is not prone to compaction. Creeping bentgrass has excellent cold tolerance but poor heat tolerance.

Figure 24-13 Creeping bentgrass.
© Funbee/Shutterstock.com.

Creeping bentgrass requires a high level of maintenance to produce a quality turf. It requires proper irrigation, disease control, mowing, and cultivation practices when grown as a sports turf. Because of these high-maintenance requirements, creeping bentgrass is not recommended as a lawn turf.

Crested Wheatgrass (*Agropyron cristatum*)

Crested wheatgrass is also known as fairway crested wheatgrass and is used on nonirrigated lawns and fairways in dry, cold regions. It is a coarse-textured, noncreeping, bunch grass. Crested wheatgrass is not considered a high-quality turf, but it is durable in semiarid, northern areas of the Great Plains. Its extensive, deep root system enables the plant to survive lengthy drought conditions without irrigation.

Recommended mowing height for crested wheatgrass is 1.5 to 2.5 inches, and low-to-moderate fertility is required. Heavy watering will stress the plant; therefore, irrigation should be minimal.

Warm-Season Turfgrasses

Warm-season turfgrasses originated in Africa, North and South America, and Southeast Asia, but they are adapted to the southern United States (**Figure 24-14 A–D**). They make optimum growth at temperatures of 80° to 95° F. These grasses go into winter dormancy as temperatures decrease to lower than 50° F. Fourteen species-of-interest of warm-season turfgrasses are found throughout the world.

Some major warm-season turfgrasses in the United States include Bermuda grass, zoysia grass, St. Augustine grass, and buffalo grass.

A

B

C

D

Figure 24-14 Warm-season turfgrasses: (A) Bermuda grass; (B) zoysia grass; (C) buffalo grass; (D) St. Augustine grass.
© pyzata/Shutterstock.com. © Tatchaphol/Shutterstock.com. © Shuang Li/Shutterstock.com. © tammykayphoto/Shutterstock.com.

Bermuda Grass (*Cynodon dactylon* L)

Bermuda grass is considered the most important and widely used warm-season turfgrass in the United States. It is principally used as a lawn turf and sports turf. Improved breeding lines have provided fine-textured Bermuda grasses capable of being used on putting greens and fairways. The common type of Bermuda grass is a medium-textured turfgrass used for airport runways, rights-of-way, and other low-maintenance areas.

The plant spreads by both stolon and rhizome growth. It has excellent wear resistance, recuperative potential, and drought tolerance. Bermuda grass can persist on a wide range of soil types and soil pH. However, it has poor shade tolerance and low winter hardiness.

The improved types of Bermuda grass require a high level of maintenance and must be vegetatively established. Recommended mowing heights for these grasses range from 0.25 to 1 inch. The common type of Bermuda grass is established by seed, requires a greater mowing height than the improved cultivars, and in some situations is considered a weed.

Zoysia Grass (*Zoysia japonica* Steud.)

Zoysia grass has excellent low-temperature hardiness and is found in home lawns as far north as New Jersey. It can also be found on golf course fairways and tees within the transition zone. The transition zone is a geographic area of the United States where the warm-season and cool-season adaptation zones overlap.

Zoysia grass has good drought and shade tolerance. It can survive in a wide range of soil types. However, this plant will not perform well in poorly drained soils that remain waterlogged. Though zoysia grass has excellent wear resistance, it has a low recuperative potential. It is not as aggressive as Bermuda grass.

Zoysia grass requires a low-to-moderate level of maintenance. It is vegetatively established because seed germination is poor; however, new strains show promise for seed propagation. This grass should be mowed at a height between 1 and 2 inches. Two major pest problems of zoysia are nematodes and billbugs.

Buffalo Grass (*Buchloe dactyloides*)

Buffalo grass is adapted to the dry, semiarid Great Plains region of the United States. It is native to Oklahoma, Texas, Arizona, Kansas, Colorado, North Dakota, South Dakota, and Montana. Regarded as the most drought-tolerant turfgrass in the United States, it survives both high and low temperature extremes and persists through droughts without irrigation. However, its shade tolerance is poor.

Buffalo grass makes a gray–green turf with fine texture. Its vertical growth is slow, but it spreads by stolons, creating a continuous turf. It tolerates alkaline soil and prefers soils of the heavier, fine-textured type. It is used on nonirrigated lawns and golf fairways with cutting heights ranging from 0.5 to 2 inches. Buffalo grass is also used along roads and other low-maintenance areas for erosion control. Little or no fertilizer is necessary for survival, but turf quality may be improved by light applications of fertilizer. The grass is propagated by seed or plugs. It is virtually pest-free if not overfertilized or overirrigated.

St. Augustine Grass (*Stenotaphrum secundatum*)

St. Augustine grass is a major lawn grass in the Deep South and is used in the warmest area of the subtropical zone. It is adaptable to many soil conditions but does best on moist, well-drained sandy soils. Drought resistance is only fair, so irrigation is required in dry weather. The grass has thick stolons and produces a coarse turf of good color, medium density, and excellent shade tolerance.

St. Augustine grass is propagated with sprigs or sod and has a good establishment rate. The grass has a vigorous growth rate with moderate maintenance requirements. Medium fertility is required, and acceptable cutting heights range from 0.5 to 3 inches. Thatch buildup, chinch bugs, and the St. Augustine decline virus are some of the problems encountered with this grass.

Turfgrass Cultural Practices

Mowing, fertilization, and irrigation are the most common and most important cultural practices performed to maintain turfgrass stands. These practices have tremendous effects on turfgrass quality and persistency. Improper mowing, fertilization, and irrigation are major causes of poor lawns (**Figure 24-15**). Once a choice of turfgrass has been made, it is important to understand and address the unique requirements of the variety or mixture.

Mowing

Mowing will influence the functional use, persistency, and aesthetic value of a turfgrass. Grasses used for recreational purposes must be playable. The playability of an athletic-field turf is principally determined by its mowing height. A uniform turf surface, fine-leaf texture, and freedom from weed encroachment require proper mowing height and frequency.

Turfgrasses are capable of being mowed because the crown, their growing point, is located just below or at the soil surface. However, mowing does have several adverse effects on the grass plant. Reduced rooting depth and decreased carbohydrate reserves occur in mowed turfs. Improper mowing practices only accentuate these and many other adverse effects, causing a decline in turfgrass quality (**Figure 24-16**).

Causes of Poor-Quality Lawns

- Using the wrong turfgrass species or cultivars
- Using poor-quality seed
- Mowing the lawn too closely
- Permitting excessive growth between mowings
- Using too little or too much lime or fertilizer
- Improper watering
- Too much shade
- Droughty or poorly drained soils
- Too much traffic
- Damage by insects or disease
- Improper use of chemicals

Figure 24-15 Major causes of poor-quality lawns.

Figure 24-16 The cutting deck of a home lawnmower should be set for the proper depth of cut to ensure a healthy lawn.

© Zoom Team/Shutterstock.com.

Science Connection: Washington, DC, National Mall: Tough Turf

At the National Mall, the turf is a mixture of zoysia grass and tall fescue.
Courtesy of the D.C. Committee to Promote Washington.

Each summer, millions of people visit the museums, monuments, of the National Mall in Washington, DC. The daily foot traffic of local Washingtonians, as well as the continuous flow of tourists, means many thousands of feet pounding the turfgrass sections throughout the summer months. Furthermore, the seasonal use of the Mall for national concerts, inaugurations, and other events pounds and punishes the turfgrass and soil like no other place. The end of summer generally leaves the turfgrass thin and vulnerable at best and bare, muddy, and void at worst. The best of traditional turfgrasses have not been up to the task, and the most heavily traveled areas have been converted to gravel and concrete.

Turfgrass is still the preferred surface for large areas from the standpoint of both beauty and function. The unique needs of the mall have stimulated the development of a turfgrass mixture that fills the need for toughness, sustainability, and year-round green color. However, this is a tall order in the Washington, DC climate, which may swing from less than 0° F in the winter to more than 100° F in summer.

Agronomists from the U.S. Department of Agriculture (USDA) got to work on a solution. Why not mix two high-quality turfgrasses, one with exceptional durability in hot weather and one with exceptional performance in cool-to-cold weather. Would the mix result in a high-quality, tough, year-round turf?

For example, zoysia grass forms a dense, tough, fine-textured, attractive, and highly competitive turf from May to October. Unfortunately, it turns brown with October's first frost and stays brown until late April or early May. It is usually propagated with plugs.

In contrast, the new, improved, tall fescue turfgrasses shine in the cooler weather of fall and spring and generally stay green through Washington's winter months. They are propagated with seed. In fact, the tall fescue seed germinates with exceptional ease, and the seedlings become established and grow in competitive environments.

Both grasses are tough competitors and tend to crowd out other plants. Could new strains, varieties, or cultivars of each species be found or bred that would survive together and combat the common enemies—excessive heat, cold, wet, dry, and the constant pounding of foot traffic in all kinds of weather?

The agronomists screened hundreds of varieties of grass in search of a fine-bladed fescue that could hold its own against zoysia grass. Eventually, they were able to test a mixture of fescue and zoysia grass in field trials. Some 200,000 square feet, or the area of four football fields, were eventually sodded with zoysia and overseeded with tall fescue on the National Mall. This turfgrass appears to be enduring the test of time.

Meanwhile, a zoysia grass that will produce seed has been developed, and the turfgrass industry has picked up on the new products. Similarly, agronomists of the National Turfgrass Evaluation Program have established a testing program for zoysia grass and mixtures at more than 25 locations throughout the country. The greatest promise for the zoysia–fescue mixture is in the transition zone that stretches from New Jersey to Georgia and as far west as Kansas. Perhaps one plus one can equal one: One warm-season turfgrass plus one cool-season turfgrass equals one exceptional year-round turfgrass!

Recommended Mowing Heights	
Species	Mowing Height Range in Inches
Bahia grass	2–4
Bermuda grass Common	0.5–1.5
Hybrids	0.25–1.0
Carpet grass	1–2
Centipede grass	1–2
St. Augustine grass	1.5–3.0
Zoysia grass	0.5–2.0
Creeping bent grass	0.2–0.5
Colonial bent grass	0.5–1.0
Fine fescue	1.5–2.5
Kentucky bluegrass	1.5–2.5
Perennial ryegrass	1.5–2.5
Tall fescue	1.5–3.0
Crested wheatgrass	1.5–2.5
Buffalo grass	0.7–2.0
Blue grama	2.0–2.5

Figure 24-17 Recommended mowing heights for different turfgrasses.

Mowing Height

The recommended ranges for mowing height are listed in **Figure 24-17**. If one mows below or above these ranges, various problems will occur. Mowing below the desired range will reduce photosynthesis, thus preventing carbohydrate production within the plant. This causes a reduction in rooting and decreases the recuperative potential of the plant.

Mowing above the recommended height will increase thatch buildup and leaf texture. It will also decrease turf density and appearance. **Thatch** is the buildup of organic matter on the soil around the turfgrass plants. Excessive thatch causes poor water infiltration, increased disease activity, and decreased rooting depth.

Mowing Frequency

Frequency of mowing is determined by the mowing height and the growth rate of the plant. No more than one-third of the top growth should be removed per mowing (**Figure 24-18**). This allows sufficient leaf area for photosynthesis after mowing. The lower the mowing height, the more frequent the mowing if the one-third rule is observed. For example, creeping bentgrass mowed at a quarter inch will be mowed five to six times per week. However, tall fescue cut at 3 inches requires only one mowing per week. Under ideal growing conditions, grasses require more frequent mowing than they do under poor growing conditions.

Fertilization

Fertility programs attempt to supply adequate levels of plant nutrients to allow for favorable plant growth (**Figure 24-19**). Proper fertilization should increase turfgrass density and color. Lush or succulent growth caused by excessive fertilization should be avoided.

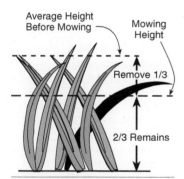

Figure 24-18 Mowing more than one-third of the top growth of turfgrasses can result in reduced rooting and less recuperative potential in the plant.

Figure 24-19 Lawns require fertilization with the proper nutrients to maintain vigor and color.
© Le Do/Shutterstock.com.

A complete fertilizer is often recommended, with the rate of application based on the amount of nitrogen required per 1,000 square feet (**Figure 24-20**). A complete fertilizer consists of nitrogen, phosphorus, and potassium. The fertilizer analysis states the percentage of nutrients by weight in the fertilizer. A 10-6-4 fertilizer consists of 10 percent nitrogen, 6 percent phosphorus, and 4 percent potassium. A formula can be used to determine fertilizer amounts based on nitrogen-rate recommendations (**Figure 24-21**). Nitrogen is the most important element in turfgrass fertilization.

Kentucky Bluegrass		
Time to Apply	Nitrogen Applied per 1,000 Square Feet	Acceptable Fertilizer per 1,000 Square Feet of Area
September	1 lb.	8 to 10 lb. 12-4-8 or 10 lb. 10-6-4 or 5 lb. 20-10-10
October	1 to 1½ lb.	8 to 12 lb. 12-4-8 or 10 to 15 lb. 10-6-4 or 5 to 8 lb. 20-10-10
November/December or February/March	1 to 1½ lb.	8 to 12 lb. 12-4-8 or 10 to 15 lb. 10-6-4 or 5 to 8 lb. 20-10-10
May–June 20	0 to ½ lb.	0 to 4 lb. 12-4-8 or 0 to 5 lb. 10-6-4 or 0 to 3 lb. 20-10-10

Figure 24-20 The fertility requirements for Kentucky bluegrass change with the season of the year. This is because its greatest growth occurs during the cool part of the growing season.

Formulas for Fertilizer Use	
Question #1:	How much nitrogen is there in a 50 lb.-bag of 10-6-4 fertilizer?
Solution:	% nutrient(s) x Weight of fertilizer = Weight of nutrient(s) or 0.10 x 50 lb. = 5 lb. of nitrogen
Question #2:	How many pounds of 10-6-4 fertilizer must be used to apply 1 pound of nitrogen per 1,000 ft^2 on a lawn that measures 5,000 ft^2?
Solution:	Pounds of nutrient needed divided by % nutrient equals pounds of fertilizer needed. or 5 lb. N divided by 0.10 = 50 lb. of 10-6-4

Figure 24-21 Math applications: Determine nutrient content and calculate fertilizer application rate.

Proper timing and application rate are the keys to a successful fertilization program. Timing of fertilization will vary depending on turfgrass species and geographic location. For cool-season turfgrasses, the fall is the desired time period. Optimum root growth occurs during this time. In contrast, fertilization of warm-season turfgrasses should be done during late spring or the summer months (**Figure 24-22**).

Bermuda Grass and Zoysia Grass		
Time to Apply	Nitrogen Applied per 1,000 Square Feet	Acceptable Fertilizer per 1,000 Square Feet of Area
April or May	1 lb.	8 lb. 12-4-8 or 10 lb. 10-6-4 or 5 lb. 20-10-10
June	1 lb.	8 lb. 12-4-8 or 10 lb. 10-6-4 or 5 lb. 20-10-10
July	1 lb.	8 lb. 12-4-8 or 10 lb. 10-6-4 or 5 lb. 20-10-10
August	1 lb.	8 lb. 12-4-8 or 10 lb. 10-6-4 or 5 lb. 20-10-10

Figure 24-22 Fertility requirements and some recommended fertilizers for Bermuda and zoysia grass.

Improper fertilization will reduce turfgrass quality. Turfgrasses receiving insufficient amounts of fertilizer will lack color, density, and recuperative potential. Excessive fertilization will reduce heat and drought tolerance, increase disease and insect damage, and cause excessive top growth. Appropriate nitrogen rates range from 0.25 to 1 lb. nitrogen per 1,000 square feet. The rate is dependent on growth and environmental conditions. Phosphorus and potassium rates should be determined by soil testing.

Science Connection: "Grasscycling" Reduces Landfills

Recycle clippings whenever possible to reduce the amount of biodegradable material entering landfills.
© iStockphoto/pagadesign.

A major ecological problem is what to do with all of the garbage that is filling up U.S. landfills. Some of what's tossed out isn't garbage at all. A single-family home can produce up to 400 lbs. in grass clippings each year per 1,000 square feet of lawn. This translates into 8 tons of clippings per acre each year. Many people bag their clippings in plastic trash bags and set them by the curb, never to be thought of again. However, with landfills maxed out all over the country, changes are being made to keep the grass out.

One solution is "grasscycling," or the practice of recycling grass. For example, clipped grass can be used as an organic fertilizer in the form of mulch. In this process, cut grass is mixed with leaves, soil, and other dead material in a plastic bin. Over time, the mixture becomes a fertile mulch that can be used anywhere commercial mulches would be used. Another option is to leave the grass as it lies on top of the lawn. When doing this, the grass must be cut into small pieces that filter through the tall grass and onto the soil surface, where it decomposes and returns nutrients to the soil. To accomplish this involves cutting grass more often or mowing with mulching mowers. Given the need to reduce landfills, "grasscycling" offers the ideal alternative to curbside grass removal.

Irrigation

The application of water to turf may accomplish several different objectives. First, sufficient soil moisture will allow for optimum plant growth. Irrigation is also used to establish turfgrass, reduce plant surface temperatures, and incorporate fertilizer and pesticide applications. **Syringing** is a light application of water mist to a turfgrass, particularly on golf putting greens. It may be used to reduce plant temperatures or to remove dew and frost from the turfgrass leaf.

The amount of irrigation needed to maintain optimum plant growth is dependent on many factors. Some of these factors are turfgrass species, geographic location, soil type, weather conditions, and turfgrass use. General recommendations are to apply 1 inch of water per week during the summer months. It takes approximately 620 gallons of water per 1,000 square feet to provide 1 inch of water. This amount of water will produce a green and actively growing turfgrass stand.

Heavy but infrequent irrigation will force root growth deep into the soil. Light and frequent irrigation will keep the surface soil moist, thus encouraging shallow rooting.

The best time to irrigate is at night when evaporation and wind are low. However, an increase in disease activity will occur at this time. Therefore, early morning watering is often selected because the effects of evaporation, wind, and disease activity are at a minimum.

Turfgrass Establishment

Turfgrasses may be established by seeding or vegetative propagation. Regardless of the method used, proper establishment practices, including site preparation, should be performed. Correct establishment practices will ensure adequate turfgrass quality and persistency.

Turfgrass Selection

Selection of the correct turfgrass species is one of the most important decisions in the establishment process. Selecting a turfgrass seed blend or seed mixture should be based on the intended use and performance data of the turfgrasses. A **seed blend** is a combination of different cultivars of the same species. A **seed mixture** is a combination of two or more species. Many state land-grant universities evaluate turfgrass species and new cultivars and provide information on their performances (**Figure 24-23**).

Kentucky Bluegrass Cultivar	Quality Rating*					
	April 1	May 4	June 8	July 15	Sept. 2	Oct. 17
Adelphi	4.0*	5.6	5.4	4.5	5.9	5.7
Baron	3.8	5.1	4.8	4.4	5.4	5.5
Bensun (A-34)	3.6	5.3	6.0	5.9	6.1	5.4
Cheri	3.2	5.2	4.7	4.9	5.8	6.0
Emmundi	4.0	4.9	4.6	4.5	5.4	5.7
Glade	4.2	5.1	5.4	5.5	6.3	6.3
Majestic	3.9	6.1	6.0	5.8	6.0	5.8
Newport	3.7	3.6	3.8	4.0	4.2	4.9
Parade	4.5	6.8	5.9	4.4	5.7	5.8
Ram 1	4.3	6.9	5.9	6.0	6.2	6.4
Sydsport	3.8	6.2	6.0	5.9	6.4	6.1
Touchdown	4.1	6.3	5.9	5.7	6.2	6.5

*It is important to note that quality ratings can vary significantly from one region or location to another. A rating of 1 = no live turf, 9 = ideal turf, >5 = acceptable quality.

Figure 24-23 Quality ratings for Kentucky bluegrass cultivars.

Site Preparation

Proper site preparation may include any or all of the following activities:
- debris removal;
- nonselective weed control;
- installation of a subsurface drainage or irrigation system, or both; and
- soil modification, tillage, and grading.

The removal of woody vegetation, leftover construction debris, and any large stones will reduce future maintenance problems. If woody debris is buried on the site rather than removed, the soil will settle and diseases such as fairy ring may result. Excessive amounts of stone and rock present in the seedbed will cause localized dry spots and interfere with future cultivation practices.

Nonselective weed control may be necessary if perennial grassy weeds or difficult-to-control weeds are present. A nonselective herbicide or a soil fumigant should be applied before seeding. The herbicide glyphosate (Roundup) is often used to provide nonselective weed control. A soil fumigant, such as metham, may also be used.

If the site is to be used as a sports turf, the installation of drainage and irrigation systems may be necessary. This should be done before final grading.

Soil modifications, followed by tillage and grading, are the last steps in site preparation (**Figure 24-24**). Soil amendments, such as organic matter and sand, may be incorporated to improve nutrient retention and/or drainage. Topsoil may also be used to provide a favorable growth medium for turfgrasses. The soil should be worked to a depth of 6 inches and then lightly tilled to provide a uniform seedbed and adequate surface drainage (**Figure 24-25**).

Figure 24-24 The site must be prepared properly by tilling and grading the soil before grass seed is planted.

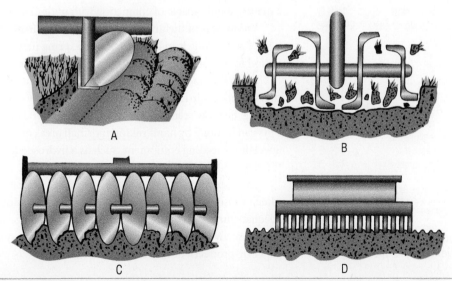

Figure 24-25 Tillage equipment frequently used for turfgrass establishment: (A) plow, (B) rototiller, (C) disc cultivator, and (D) harrow.

Incorporation of lime and fertilizer should be done during this time. Excessive tillage will destroy soil tilth. A seedbed with soil aggregates of 0.25 to 1 inch in diameter is ideal.

Additional preparation is necessary on sites where bunch grasses are selected as turfgrass crops. The lack of a strong fibrous root system makes it difficult to hold the harvested turf plugs together while they are shipped and transplanted. Netting materials are used to compensate for the lack of an extensive root system. Netting materials must be placed on the soil surface before the grass is planted. As the turfgrass matures, it becomes bound into the netting material. This allows it to be harvested and handled while causing minimal damage to the turf (**Figure 24-26**).

Planting

Turfgrasses can be established by seeding, sodding, stolonization, sprigging, and plugging. The last four methods are vegetative means of planting. Vegetative establishment is usually practiced when the selected turfgrasses produce infertile seed or have low seed yields. Also, if quick establishment is required, sod may be installed.

Seeding

Seeding is the principal method for turfgrass establishment because it is the least expensive. **Figure 24-27** provides the seeding rates for some popular grasses. These rates will vary depending on seed size and percentage of seed germination.

The seed label informs the buyer or consumer of the quality and ingredients of a seed blend or mixture. Information on purity, germination, weed seed content, inert matter, and test date is present on the label (**Figure 24-28**). Certification programs are available for turfgrass seed. Seed certification programs are administered by state governments, and they ensure that seed is true to type. They also require the seed to meet other minimum quality standards concerning percentage of germination and weed seed content.

The best time to plant turfgrass depends on the type of turfgrass that is to be established. The optimum planting time for cool-season grasses is in the fall. Ideal weather conditions and less annual weed competition are the principal reasons for fall establishment. The plant will also have sufficient time to develop an adequate root system before summer stress conditions occur. Spring seeding is less desirable because there is insufficient time to develop a mature stand capable of competing with undesirable grasses.

Warm-season turfgrasses are established during the spring and summer months. The best planting time is in late spring. Late-spring planting gives the new grass the longest period for optimum growth and development after establishment.

Different types of equipment may be used to apply grass seed. Fertilizer drop spreaders, overseeders, hydroseeders, and cultipacker seeders are some types of equipment used for seeding. Shallow seed placement (a quarter inch) is important for proper germination. This can be done by hand raking for small areas or the use of a drag mat for larger areas. Specialized seeding equipment, such as a hydroseeder or cultipacker, will place the seed at the right soil depth (**Figure 24-29**).

After seeding, the use of mulch provides a favorable environment for seed germination. Mulch conserves soil moisture and prevents soil erosion. This is extremely important because adequate surface moisture must be present for seed germination. Straw mulch is preferred and is usually applied at the rate of 2 bales per 1,000 square feet. Other mulches are wood and paper by-products, net or fabric, and peat moss.

Figure 24-26 Some grass species that lack extensive root systems require netting to be applied to the soil surface to hold the turf together when it is harvested.

Courtesy of Arva Burton.

Seeding Rates	
Species	**Pounds of Seed per 1,000 ft²**
Bahia grass	3–8
Bent grass	
Colonial	0.5–1.5
Creeping	0.5–1.5
Bermuda grass (hulled)	1–2
Bluegrass	
Kentucky	1–2
Buffalo grass	3–7*
Carpet grass	1.5–5
Centipede grass	0.25–2*
Fescue	
Fine	3–5
Tall	5–9
Grama, blue	1.5–2.5
Ryegrass	
Annual	5–9
Perennial	5–9
Wheatgrass, crested	3–6
Zoysia grass	1–3

*The higher rates are best, but lower rates are commonly used because the seed is expensive.

Figure 24-27 Seeding rates for the major turfgrass species.

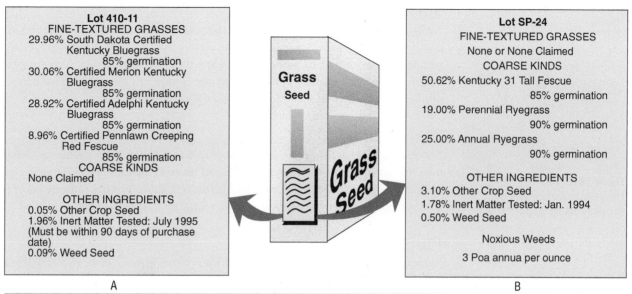

Figure 24-28 Sample labels for lawn and turf seed. (A) Label describing a recommended seed mixture of cool-season grasses. (B) Label describing a poor-quality seed mixture of cool-season grasses.

Figure 24-29 One method of seeding turfgrass is the hydroseeding technique, mixing seed, mulch, fertilizer, and water and then spraying it on the planting surface.
© iStockphoto/BanksPhotos.

Sodding

Sodding refers to removing a rectangular piece of grass and a shallow layer of the soil beneath it and moving it to another location. The grass and the soil immediately beneath are referred to as **turf**. Sodding offers the ability to establish turf at any time of the year. It also provides an instant cover. However, the cost is comparatively high, and the sod must be installed soon after harvesting. Specialized equipment for harvesting sod has been developed (**Figure 24-30**). Sod is cut into 12- to 24-inch widths and lengths of up to 3 feet or in carpet-like pieces. The depth of cut will vary from 0.3 to 0.5 inch. The cut sod is then rolled, folded, or stacked on pallets for transport and later placed on a carefully prepared seedbed. The new turf is then rolled to ensure good contact with the soil. After the sod is in place, it must be watered thoroughly.

Figure 24-30 Cutting sod for placement in a new location as turfgrass.
Courtesy of H. Edward Reiley.

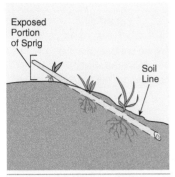

Figure 24-31 When sprigging, place two or more nodes with shoots in the soil and water well.

Sprigging

Sprigging is the planting of a section of a rhizome or stolon, referred to as a sprig. A sprig may be up to 6 to 8 inches in length. For successful establishment, a section of the sprig with several nodes must be properly placed into the soil (**Figure 24-31**). After sprigging, irrigation must be applied to prevent drying out. Equipment for sprigging has been developed and will plant sprigs in rows spaced 6 to 18 inches apart.

Stolonizing

Stolonizing is similar to sprigging, in that 6- to 8-inch sections of rhizomes or stolons are used. However, in stolonizing, sprigs are broadcast onto the soil surface. These sprigs may be lightly top-dressed with soil and rolled before irrigation. Survival rates are lower than with sprigging. Therefore, stolonizing requires greater quantities of sprigs.

Plugging

Plugging is the establishment of a turfgrass stand by using plugs, or small pieces of existing turf. Plug sizes vary, as noted in **Figure 24-32**. Spacing of plugs can be on 6- to 18-inch centers, depending on how quickly the turfgrass needs to be established. Plugging can be done by hand or with specialized equipment. It requires a longer time to cover the soil than using the other establishment techniques.

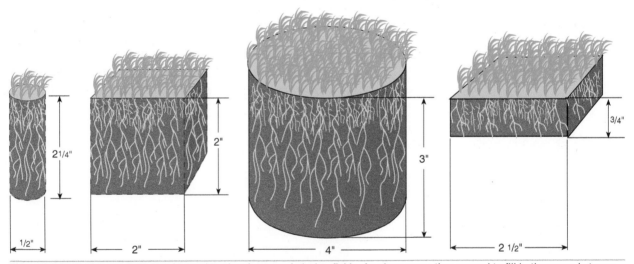

Figure 24-32 Turfgrass plugs may be used to plant lawns and playing fields. As plugs grow, they expand to fill in the space between plants.

Maintenance

After investing a lot of time and money establishing a grassy area, it would be unfortunate if a lack of maintenance ruined the lawn. Homeowners and grounds keepers need to do a number of things to keep lawns looking nice. Monitoring and adjusting water should be done throughout the growing season based on temperature and rainfall. For example, when sprinklers are set to automatically turn on every other day for 20 minutes and it rains for a week, the grass is susceptible to rot and fungus infestations.

Grassy areas need to be monitored for pests and weeds. If a problem is discovered, it should be treated quickly to avoid damage. In the late spring or early summer, older lawns should be thatched to remove excess organic build-up. Organic matter is good in moderate amounts, but it can choke grass plants at excessive levels. It is also important to reseed a lawn every few years in early spring. Reseeding is important to keep desired grasses strong and to keep undesirable grasses such as crab and quack grasses at bay.

Agri-Profile — Career Area: Golf Course Maintenance Supervisor

Golf course fairways and greens require regular early morning care, especially during periods of heavy use.
© iStockphoto/vm.

Golf is one of the most popular American pastimes. New golf courses are designed and constructed every year, and the current U.S. total is more than 15,000. All of these golf courses must be maintained. A new course requires a person who knows how to establish turfgrass plantings and keep them healthy. Likewise, even a well-established turf requires constant maintenance and care. The many tasks include controlling weeds, insects, and rodents; reseeding damaged turf; irrigation management; soil testing; fertilization; mowing; caring for trees; and establishing and maintaining other plantings. Each of these tasks requires expert knowledge.

Students who are interested in a career in golf course maintenance can obtain the required training by enrolling in a two-year technical college program in turfgrass management or by earning an undergraduate degree in plant science or a related field of study. A summer internship on a golf course would also be a valuable component. A qualified manager of golf course resources can expect to have good employment opportunities.

Chapter Review

Student Activities

1. Write the **Key Terms** and their meanings in your notebook.
2. Determine whether your community is located in a warm-season, transitional, or cool-season zone.
3. Visit or call your county cooperative extension office and obtain copies of the various publications available on turfgrass or lawn production and maintenance.
4. Determine the recommended cultural practices for establishing and maintaining a major turfgrass species or mixture for your locality. Report your findings to the class.
5. **SAE Connection:** Arrange to visit a golf course and discuss with the superintendent the duties of turfgrass workers on the golf course.
6. Obtain a map of your county and attach it to a classroom bulletin board. Place pins at the locations of all businesses, schools, government buildings, and institutions that have extensive lawn or turfgrass areas around them. Ask your classmates to help identify them. Use different colored pins to identify different institutions and post a key for the map.
7. Obtain a grass identification key from your teacher or the cooperative extension service office. Collect 10 specimens of different turfgrasses and identify them by using the identification key.
8. Make your own grass clipping mulch. Inside a compost bin or in a plastic garbage container with ventilation holes, mix grass clippings, leaves, soil, and any other plant material. Add water and stir the mixture each week to add moisture and oxygen to the compost. Soon you will have useful nutrient mulch that can be used in a flower bed or a garden plot.

> ### Check Your Progress
> Can you . . .
> - give examples of careers available in the turfgrass industry?
> - identify turfgrass plant parts?
> - summarize the basic cultural practices for turfgrass production and maintenance?
> - recall the basic steps for turfgrass establishment?
>
> If you answered "no" to any of these questions, review the chapter for those topics.

National FFA Connection: Exploring the Wide World of Turfgrass

Get to the root of turfgrass management's appeal with FFA resources and events.
Courtesy of the USDA.

From special events to interviews with experts, the National FFA Organization offers a wealth of resources for developing a "deep-rooted" knowledge of the turfgrass industry. Search the term "turfgrass" on the FFA website for instance, and you will find interviews with the experts who keep athletic fields in tiptop condition for players, coaches, and sports fans. From the Super Bowl to your local high school, the need for such expertise is great.

Wondering if a career related to turfgrass might be right for you? Search the AgExplorer on the FFA website. As you will discover, a wide range of career opportunities is available in this booming industry.

You can also find opportunities to build your knowledge at a literal grassroots level. A growing number of states offer Career Development Events (CDE) in Turfgrass Management. These events cover all aspects of the industry in producing, marketing, utilizing, and maintaining turfgrass as well as related products, equipment, and services. Each CDE is designed to stimulate a career interest, encourage proficiency development, and recognize student excellence.

For example, in Pennsylvania, individuals and teams demonstrate their ability to identify turfgrasses, weeds, and diseases common in that state. They also demonstrate knowledge of the scientific principles and skills related to propagation, growth requirements, growing techniques, and maintenance of turfgrass. Students are also tested on business skills, from communications and customer service to accurate recordkeeping and ability to understand business documents.

To find out if your state offers a Turfgrass Management CDE, gain insights from turfgrass professionals, and learn more about career opportunities, visit the FFA website.

Chapter 25
Trees and Shrubs

Objective
To use trees and shrubs for beautification, wood products, improvement of air quality, and pollution control.

Competencies to Be Developed
After studying this chapter, you should be able to:
- identify ornamental trees and shrubs.
- select trees and shrubs for appropriate landscape use.
- classify trees and shrubs according to growth habit, growth habitat needs, and other requirements.
- identify trees and shrubs using proper nomenclature.
- purchase plant material for installation in a landscape.
- plant and maintain plant material.

Key Terms
specimen plant
border planting
group planting
hardiness
nomenclature
genus

bare-root
heel in
balled and burlapped (B & B)
root pruning
containerized plant
stake

drip line
pruning
arborist

Trees and shrubs are major components of the environment. They add natural beauty and help provide oxygen to humans and animals. This need has created a demand for horticulture technicians who understand the production and care of landscape plants.

Trees and Shrubs for Landscapes

A well-designed landscape increases the value of a home (**Figure 25-1**). Residential properties are much more attractive when trees and shrubs are used to enhance their beauty and balance.

Trees and shrubs do much more than look good, however. An acre of healthy trees or shrubs will produce enough oxygen to keep 16 to 20 people alive each year. These plants also help keep our air clean by using the carbon dioxide produced by people, automobiles, and factories.

There are other benefits to such greenery, too. Trees and shrubs cut noise pollution by acting as barriers to sound. They can deflect sound as well as absorb it. When used properly in the landscape, trees and shrubs also provide shade and act as insulators to keep the house cooler in the summer and warmer in the winter.

Many urban areas depend on trees and shrubs to soften the concrete, blacktop, and steel environment. In fact, many cities now employ urban foresters to help design, install, and maintain trees and shrubs in large cities (**Figure 25-2**).

Figure 25-1 Trees and shrubs add character and value to a property.
© Susan Law Cain/Shutterstock.com.

Figure 25-2 Trees and shrubs add beauty, provide sound insulation, purify the air, and provide privacy.
© Lowe Llaguno/Shutterstock.com.

Agri-Profile

Career Area: Landscape Architect/Landscape Technician/Landscape Contractor/Plant Specialist

Plant care and management require many skills, including the diagnosis of disease and insect problems.
© raluca teodorescu/Shutterstock.com.

Career opportunities related to trees and shrubs span both the arts and the sciences. Ornamental trees and shrubs vary greatly in size, shape, temperature preference, light preference, fertility needs, and pest tolerance. This variety reflects the wide range of jobs and specialties in the area of ornamental horticulture.

Landscape architects practice the art of design in that they design and plant landscapes pleasing to their clients. They must know plant species and plant materials to develop plans that are attractive yet functional in the environment. Landscape architects generally have bachelor's degrees in landscape design and considerable experience in ornamental horticulture.

Horticulture specialists and technicians have careers centering on nursery or greenhouse management, landscape contracting, groundskeeping, wholesaling, retailing, public information, writing, and consulting. They may specialize in just trees, just shrubs, or both. Others may specialize in pruning, tree planting, pest control, or landscape maintenance. The opportunities in ornamental horticulture offer an abundance of career opportunities in many localities.

Forest Resources

Vast regions of North America produce forest products such as wood, cardboard, paper, solvents, medicines, fuels, and many other products. An entire industry depends on trees for the raw materials that are used to process these important products. Thirty-one percent of Earth's land area is forest land, and forest products are important to the economies of developed countries. Some of the trees that are used for this purpose are produced on private land. Vast tracts of public land are also devoted to timber production.

Forests are important resources in many ways besides production of wood and forest products. A forest also functions as a biological filter system that cleans the environment by removing impurities from air and water. Forest plants restore oxygen to the atmosphere and improve watersheds that contribute to consistent supplies of fresh, pure water. Forests furnish habitat to many kinds of wild animals and birds. They also provide recreation opportunities for people who enjoy outdoor activities such as camping, picnics, hiking, boating, fishing, and hunting. Forests provide many kinds of resources that are important to people.

Plant Selection

Trees and shrubs are distinct groups of plant materials. Trees are woody plants that produce a main trunk and grow to a height of 15 feet or more. Shrubs are woody plants with a low growth habit, produce many stems or shoots from the base, and do not reach more than 15 feet in height (**Figure 25-3**).

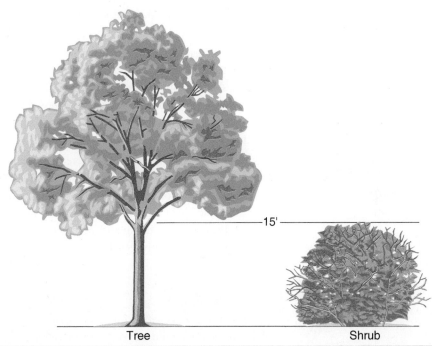

Figure 25-3 A tree has a single stem and a height of 15 feet or greater. A shrub has multiple stems and does not grow to a height greater than 15 feet.

Use

Ornamental trees and shrubs in the landscape are both beautiful and functional. They are used as specimen plants, border plantings, or groupings. A **specimen plant** is used as a single plant to highlight or provide some other special feature to the landscape. A **border planting** is used to separate some part of the landscape from another or to serve as a fence or a windbreak (**Figure 25-4**). A **group planting** consists of a number of trees or shrubs that are planted together so they point out a special feature, provide privacy, or create a small garden area.

The location of plant material in the landscape will play an important part in the selection process. Color of the leaves, texture of the plant, and color of the flowers are just some of the factors that must be considered before purchasing or planting trees or shrubs.

Figure 25-4 A border planting is used to separate one landscape feature from another.

© Nessli Orpmas/Shutterstock.com

Geographic Location

The geographic region in which a plant will be used is an important factor in the selection of a tree or shrub. Some plants lack **hardiness**—that is, they may not be able to survive or even grow properly in an area for which they are not adapted. The hardiness of a plant is affected by the intensity and duration of sunlight, length of the growing season, minimum winter temperatures, annual precipitation, summer droughts, and humidity. The U.S. Department of Agriculture (USDA) has issued a plant hardiness zone map. (See Chapter 18.) There are 13 zones and 26 subzones in the United States, including Alaska and Hawaii. Each zone represents an area of winter hardiness that is based on the average annual minimum winter temperatures. It is possible that local climates may vary from the general zone map. They may be colder or warmer than is indicated on the map. Local nursery personnel are generally willing to help in the selection of the best plant material. Most nursery catalogs list the plant for the coldest zone in which it can reasonably be expected to grow. Two examples of such catalog listings are the following:

1. *Cornus florida* (flowering dogwood)—Zones 5–9: A low-branched, flat-topped tree that has a horizontal branching habit. It will grow 20 to 30 feet high and 25 to 35 feet wide. Growth rate is slow to medium, and the texture is fine to medium. White flowers appear in April to May. Red berries appear in the fall.

2. *Pyrus calleryana* "Bradford" (Bradford pear)—Zone 5: A dense, pyramidal tree that becomes brittle with age. It will grow 30 to 50 feet high and 30 to 35 feet wide. The growth rate is medium, and the texture is medium to fine. Its use as an all-purpose tree, shade, or street tree has declined in recent years due to the tendency for the trunk to split or limbs to break when it is exposed to ice, heavy snowfall, or strong winds. Profuse white flowers one-third inch in diameter appear in late April or early May in clusters up to a diameter of 3 inches.

A plant can also be expected to live in a warmer zone than indicated if rainfall, soil, and summer conditions are comparable. In some cases, it may be necessary to adjust these conditions through irrigation, correction of soil conditions, wind protection, and alteration of shade or sun exposure. It is also possible to grow some plants in areas north of the indicated zone. Such plants may need special attention to protect them from wind or cold. Without such protection, they may not perform normally and are likely to suffer winter injury.

Site Location

When planning the location for planting trees and shrubs in a given area, many of the following factors should be given consideration:

- Fruit size and type: Many trees, such as some crabapples or cherries, drop messy fruit. Therefore, they should not be placed near an area that will be walked on or heavily used by people. Such areas include driveways, walkways, patios, and near swimming pools.
- Other structures: Do not plant trees directly in front of doors; near wells, cesspools, or field drains; or under utility lines, where interference is likely when the plant matures.
- Ornamental characteristics of the plant: Consider flowering time, shape, foliage texture, seasonal changes, pest resistance, landscape suitability, and mature size when selecting a plant.
- Flower color: Is it compatible with the house, fence, patio, and nearby plants?

Type and Growth Habit

Plants have many different types of growth habits, an important consideration when selecting plants (**Figure 25-5**). A good landscape planner or designer will be aware of the type of growth habit for each plant used in a landscape design.

Plant size in relation to the structure around which the landscape will surround is also important. A common mistake, made by homeowners and landscape workers, is planting a tree or shrub without considering what the plant will look like in 20, 50, or even 100 years. Some trees, like the cottonwood (*Populus deltoides*), cover lawns with a thick cotton-like layer of seeds. Many of these seeds germinate and result in undesirable woody shoots. The willow (*Salix babylonica*) tree is a beautiful addition to a landscape. However, its roots grow shallow and wide. Often, when planted too close to structures, willow tree roots grow into foundations of buildings. This can cause expensive structural damage. This tree is better suited for planting away from structures.

Maintenance and frequent pruning of plants can be expensive. Pruning of shrubs is frequently done without climbing, whereas most trees require the use of a hydraulic lift or they must be climbed to prune them. The use of a tree that is not in proportion to nearby structures can result in an expensive and frequent pruning program. **Figure 25-6** illustrates some common trees and their mature sizes. A tree that is too tall will require extra work to keep it in proper relationship to adjacent structures. Such a design error would cancel out the goal of property enhancement.

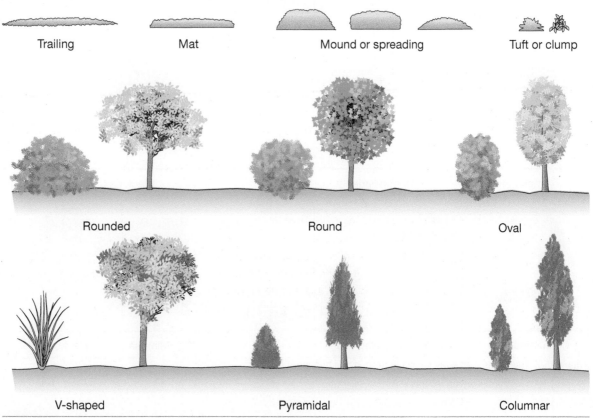

Figure 25-5 Types of growth habit.

Figure 25-6 Tree types with sizes shown in relation to a two-story house.

Shape of the Plant

Form or shape is another important factor in the selection of plant material. Common undesirable effects can be avoided by asking basic questions. Does the plant grow straight up and give little shade? Does it have a trunk that divides and spreads to cast unwanted shade? Is the plant a low-growing tree that gives no shade? Consider which

Figure 25-7 Typical shapes of landscape trees.

shapes of trees and shrubs are needed in the landscape. Then select the appropriate plant to achieve the objective (**Figure 25-7**). Before plants are selected, research is necessary to better understand the types of plants to be used in a particular area. To do this, study the types of plants that are sold and used in the locality, and note the various characteristics of each plant. Catalogs are available from nurseries that sell plants in the locality. Also, a visit to local garden centers and nurseries will pay dividends in gathering information on plant selection.

Plant Names

Trees and shrubs must be ordered by their proper names. Common names such as flowering dogwood and upright juniper are not governed by any formal code of nomenclature. Nomenclature is a systematic method of naming plants or animals. The botanical or scientific name is recognized internationally. Scientific designations are always written in Latin and consist of two names. The first name is the genus, and the second is the species. The genus name always begins with a capital letter and is a noun. The species name is usually written in all lowercase letters and is an adjective.

The genus (plural is genera) is defined as a group of closely related and definable plants composed of one or more species. The common, definable characteristics are fruit, flower, leaf type, and arrangement. The species (plural is also species) is the basic unit in the classification system whose members have similar structures and common ancestors, and that maintain their characteristics.

Variety (var.) is a subdivision of species. A variety has various heritable characteristics of form and structure that are perpetuated through both sexual and asexual propagation. The term for variety is written in lowercase letters and underlined or *italicized*. Often in catalogs and on plant labels, the abbreviation var. is used. An example of how this is used might be *Cornus florida rubra* or *Cornus florida* var. *rubra*.

A cultivar (cv.) is a group of plants within a particular species that has been cultivated and is distinguished by one or more characteristics and that, through sexual or asexual propagation, will keep these characteristics. The term is written inside single quotation marks. An example is *Pyrus calleryana* 'Bradford' or *Pyrus calleryana* cv. 'Bradford.'

It is important to become familiar with this plant-naming system because common names vary from area to area. There are no standards for creating common names. Professional horticulturists order plants using the method of naming just described. Landscape architects place scientific names on landscape drawings, and nurseries use the scientific names in their catalogs to avoid confusion by anyone who orders plant material.

Obtaining Trees and Shrubs

After determining the specific plants needed in a landscape, the plants must be purchased. Plants are normally dug and shipped bare-root, balled and burlapped (B & B), or containerized (**Figure 25-8**).

A   B   C

Figure 25-8 Methods of preparing plants for shipping and handling: (A) Bare-root, (B) Balled and burlapped (B & B), (C) Containerized.
Photo courtesy of Boise National Forest. © J. Bicking/Shutterstock.com. © Vaidas Bucys/Shutterstock.com.

Bare-Root Plants

A **bare-root** plant has been dug, and the soil has been shaken or washed from the roots. Normally, only deciduous trees, shrubs, and trees with taproots are shipped this way. They are dug while the trees are dormant. Plants ordered from nursery catalogs that can survive as bare-root stock are shipped this way because of lower transportation costs and easier handling. It is not economical to ship soil with the roots because it is so heavy.

Bare-root plants are best planted while they are dormant. In all instances, the roots must be protected to keep them from drying out. If the plant cannot be planted when it is received, the roots must be protected by putting them in a container of water, wrapping them in burlap after a good watering, or placing wet newspaper around them. If plants cannot be planted for many days, the plants may be heeled in. To **heel in** a plant, a trench is dug in the soil deep enough to hold the roots of the plant, the plant roots are placed in the trench, and then they are covered with soil (**Figure 25-9**). Then the soil around the roots is compacted with the heel of a shoe or boot. It is important to wet the soil well after heeling in. Frequently, bare-root plants will need additional pruning at planting time.

HEELING IN

1. Dig V-shaped trench in moist, shady place.

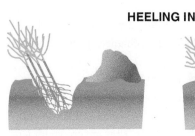
2. Break bundles and spread out evenly.

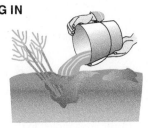
3. Fill in loose soil, and water well.

4. Complete filling in soil and firm with feet.

HANDLING SEEDLINGS IN FIELD

CORRECT
In bucket with sufficient wet moss to cover roots.

INCORRECT
Incorrect handling promotes drying of the root.

CORRECT AND INCORRECT DEPTHS

CORRECT
At same depth or 1/2" deeper than seedling grew in nursery.

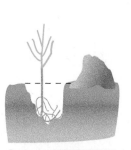
INCORRECT
Too deep and roots bent.

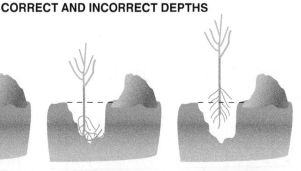
INCORRECT
Too shallow, roots exposed.

Figure 25-9 Handling and planting seedling trees.

Balled and Burlapped Plants

Balled and burlapped (B & B) plants are dug with a ball of soil remaining with the roots. This is wrapped with burlap and laced with twine. Normally, these plants have been root pruned. **Root pruning** is a process whereby roots are cut close to the trunk so that a good root system develops close to the trunk before the plant is dug. The result is that when the ball is dug, the tree or shrub will have a compact root system in the ball. This gives the plant a good chance to reestablish itself in a new environment.

It is important that transplanted trees and shrubs have every opportunity to grow after transplanting. Plants typically balled and burlapped for transplanting are deciduous trees with branching root systems. Other plants that are sold as B & B are conifers, azaleas, rhododendrons, and other plants that have fibrous root systems.

As is true with bare-root plants, it is important that B & B plants' root systems not dry out. However, because there is soil around the roots, the drying-out process is more gradual. B & B plants should be avoided if they do not have a sound ball of soil surrounding the roots. Handling tends to break the ball up and strip the hair roots from the plants. Therefore, a good root ball will help substantially to avoid damage when handling. Plant material that is shipped as B & B may be planted anytime, as long as the soil can be worked.

The mechanical tree spade is becoming more popular in the industry today. A tree spade is an expensive piece of equipment that will dig a tree in a matter of a few minutes with a very specific-sized ball. In addition, it is used to dig holes for trees and shrubs quickly and efficiently. The tree spade saves many hours in digging and planting trees. Frequently, B & B trees and shrubs are shipped in burlap with wire cages, specially prepared baskets, or other containers. The tree spade has helped make this an efficient way to ball and burlap.

These types of plants require little pruning at planting time. This makes B & B a popular way of handling plants by the mechanized professional horticulturist.

Containerized Plants

The use of container-grown plants is increasing in the nursery industry. A **containerized plant** is grown and shipped in a pot or can (**Figure 25-10**). Normally, the smaller types of plants are handled in this manner. They are grown for a reasonable period in the containers in which they are shipped and purchased.

The disadvantage of using containerized plants is that they need to be planted carefully. The roots of the plant have developed in a limited space and are generally rootbound. The roots also may have grown back around the trunk. To prevent the plant from strangling itself and also to encourage the roots to grow out of the confined area, these plants must have the "container ball" broken. This is done by placing a sharp shovel through the root mass or by breaking and spreading the root mass as it is planted in the new hole.

Figure 25-10 Nurseries package and ship many of their trees and shrubs to customers in disposable containers.
Courtesy of DeVere Burton.

Planting Trees and Shrubs

Planting may be done in spring, summer, or fall, as long as soil can be worked. A good practice for preparing the hole is to dig it about one-third wider than the ball or container. If the soil is hard, compacted, or of poor quality, it is advisable to dig the hole 4 to 5 inches deeper than the ball. This will allow peat moss or another soil conditioner to be added to the soil and placed under and around the root ball. Fill the hole with enough good soil to allow the ball to be the proper depth when it is put in the hole. The ball or container should be about 2 to 5 inches above the top of the hole. After cutting the twine that holds the ball together, peel back and remove the burlap. Backfill the hole around the plant, tamping the soil lightly to ensure removal of all air pockets. Continue backfilling the hole until it is level with the surrounding soil. Avoid over-filling with soil above the root graft. Attempt to attain the same north-south orientation as the tree occupied before transplanting. In most cases, this can be accomplished by turning the weakest side of the tree to face toward the sun. Place a ring of soil about 4 inches high around the backfill to retain water.

In the case of a container plant, remove the plant from the container and slice the root zone with a knife or shovel. Then place the plant in the hole and backfill carefully.

After backfilling, fill the ring with several inches of water to saturate the soil to the bottom of the backfill. This will settle the soil and provide moisture for the plant (**Figure 25-11**).

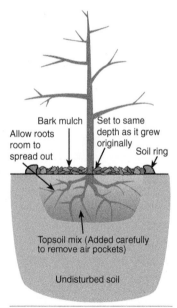

Figure 25-11 It is important to follow all of the instructions for planting trees and shrubs.

Science Connection: Biocontrol—Tamer of the Euonymus Scale?

How serious is the threat of scale insects? And what does this mean for euonymus plants, which include some of the best-known flowering plants? According to entomologist John J. Drea of the U.S. Department of Agriculture (USDA) Insect Biocontrol Laboratory in Maryland, "Insects that attack euonymus plants rank among the most insidious enemies of trees and bushes in the United States." Native to temperate Eastern Asia, *Unaspis euonymi* is commonly known as the euonymus scale. It has become established and multiplied in the United States without facing the usual risk of natural enemies.

Euonymus plants are part of a genus of flowering plants that rank among the top 20 plants used by the multibillion-dollar landscaping industry. They are used as ground covers, hedges, vines, shrubs, and trees. Yet scale insects have become such a problem that many nurseries have stopped selling them.

The female euonymus scale is brownish in color, about one-sixteenth inch in length, and shaped like an oyster shell. It is possible for the scale to have three generations in 1 year, so the buildup can occur at a rapid rate. The female forms an armor-like covering of wax, inserts her mouth parts into the host plant, and settles into uninterrupted feeding and egg laying until she dies. Pesticide sprays cannot get to her. After the eggs hatch, the flying males live about 1 day—sufficient time to mate. The fertile females then form their waxy shelters, feed, lay eggs, and start another generation. Controls based on sprays and dusts are not useful because of the brief window of insect exposure during the life cycle. Also, the use of pesticides on plants infested with scale insects tends to kill any natural enemies of the scale that might exist.

Therefore, biocontrol such as the use of natural enemies may be the only likely solution to the problem. Researchers are studying scale's natural enemies in Asia to identify those that could be introduced to the United States without becoming pests themselves. Testing such organisms for effectiveness and safety is ongoing.

Two species of predatory beetles have been identified and tested, and after years of study, they are finally available to combat this troublesome insect pest. They are the red-spotted, black Asian lady beetle, *Chilocorus kuwanae*, and the one-twenty-fifth nitidulid, *Cybocephalus prob. nipponicus.* Both lay eggs under the body of the euonymus scale and in cracks in the bark and other protected places on the host plant. When the larvae hatch, they feed on the scale insects and their offspring. The powerful jaws of Asian lady beetles enable them to burrow under protective cover and chew through the scale's armor.

The USDA and landscaping industries have successfully released and tested these and other predatory wasps to help control euonymus scale. Researchers are also evaluating other natural enemies as candidates.

A   B

(A) A red-spotted ladybug devours euonymus scale insects and (B) deposits an egg to start a new generation of predators.
Courtesy of USDA/ARS #K2848-4. Courtesy of USDA/ARS #K2848-12.

Mulching

Trees and shrubs will benefit from the addition of mulch around the planted area. Mulch in the form of shredded hardwood or pine straw will reduce evaporation and help hold the soil moisture. A newly planted tree will require about 15 to 20 gallons of water twice a week in a hot, dry environment. The conservation of water is important. Other advantages of mulch are that it helps provide a more constant soil temperature, helps control weeds, prevents erosion, and helps prevent soil compaction. Some species of plants have shallow root systems and will compete with any ground cover for nutrients. The addition of mulch reduces this competition.

The use of mulch will also prevent damage from lawn mowers or *weed whackers* by keeping grass and ground covers away from the trunk. Mulch should be applied 3 to 4 inches deep. For estimating the mulch needs for a bed of plants, use the rule that 1 cubic yard of mulch can be spread over an area of 100 square feet. Mulch will last for 1 to 3 years, depending on the type applied. Pine bark must be replaced annually, whereas shredded hardwood mulch will last about 3 years.

For added benefit, consider the use of landscape fabrics to help control weeds. Such materials are placed under the mulch. The fabric is made of fiberglass and lasts many years. It is better than sheet plastic because it allows water and fertilizer to move through it, whereas the plastic sheet blocks water and fertilizer.

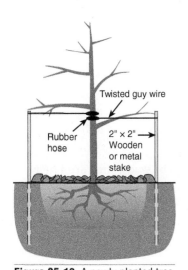

Figure 25-12 A newly planted tree should be anchored in position with guy wires.

Staking and Guying

Newly planted trees and shrubs need to be staked or guyed. This is to avoid loosening of the soil and disfiguration of the plant by the wind. To **stake** a tree, drive a wooden pole or metal post into the ground near the plant and tie the upper part of the plant to it with rope or wire. The best time to do this is before the roots are covered with soil to avoid damaging the roots. Guying is a form of staking, and it is accomplished by tying a tree to two to four stakes with wire or rope (**Figure 25-12**). Stakes and guys should be left in place for at least one growing season. When using wires to stake or guy a tree, make sure the wires do not contact the trunk. Using old water hose with the wire running through is a good practice to protect the trunk. Another good practice is to tighten the wires at the stake and not at the trunk. The best method is to use a double strand of wire twisted together for tightening. Guy wires should be checked for tightness several times during the growing season.

Fertilizing

Plants need nutrients to maintain their vigor and to make healthy new growth. If the soil is fertile, it is not necessary to add fertilizer at planting time. Trees and shrubs should not be fertilized during the first year of growth. This practice is followed to prevent the plant from developing too much top growth in relation to the root growth. Plants should be fertilized every 3 to 5 years, starting with the growing season after the first year. The best time to apply fertilizer is in the early spring. It is not a good practice to fertilize a plant after middle to late July. This practice would force new growth that will not mature enough to escape damage by the winter cold.

Unless the plants are in a bed, it is not a good practice to place fertilizer on the surface of the soil. It will wash away in the rains or will not penetrate into the root-zone area. To fertilize a tree properly, it is necessary to place the fertilizer into a hole 1 inch in diameter and 18 inches deep. Care should be taken to avoid underground utility wires and pipes, such as electric lines, telephone wires, gas pipes, and water pipes.

Holes for fertilizer are made using a 1-inch steel stake placed in concentric circles starting about halfway out from the trunk and the drip line. The **drip line** is the outer edge of the tree where the branches stop. The concentric circles should be about 24 inches apart, and the holes should be located on the circle about 24 inches apart (**Figure 25-13**). Each hole receives the appropriate amount of fertilizer and is sealed with soil or by closing the hole with a heel of the foot.

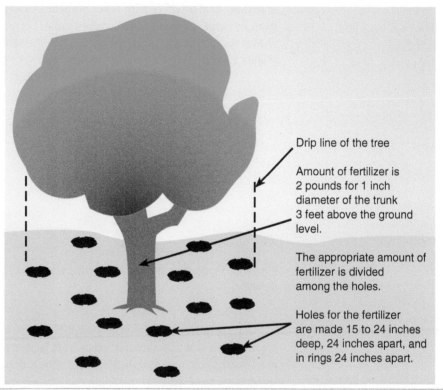

Figure 25-13 Holes for fertilizer applications should be placed within the drip line of the tree.

The amount of fertilizer needed by a tree is determined by measuring the trunk about 4.5 feet above the ground. If the trunk of the tree is less than 8 inches, multiply the number of inches of diameter by 3. If the trunk is greater than 8 inches, multiply by 6. This will give the number of pounds of fertilizer the tree needs. Divide the number of pounds of fertilizer by the number of holes to determine the amount of fertilizer to put in each hole. For example, if the tree measures 6 inches at 4.5 feet, multiply 6 × 3. The answer is 18 lbs. of fertilizer. If the tree measures 12 inches at 4.5 feet, then add 12 × 6 = 72 lbs. of fertilizer. If the tree has more than one trunk, combine the diameters of all the trunks and multiply by the appropriate factor.

A fertilizer with an analysis of 10-6-4 or 10-10-10 is generally acceptable. Never apply more than 100 lbs. of fertilizer to any given tree in a year. If the rate to be applied is more than 100 lbs., make the application over a 2-year period. Trees should be fertilized every 3 to 5 years.

Pruning

Pruning is the process of removing dead or undesirable limbs from a tree or shrub. Removing dead, broken, diseased, and insect-infested wood helps protect the plant from additional damage. Proper pruning techniques and timing are important in growing good ornamental trees and shrubs. Incorrect pruning can leave a plant in worse condition than before it was pruned.

Trees may have bad-angle crotches, branches, and waterspouts (clusters of skinny branches), that interfere with other branches, or branches that form an asymmetrical habit or shape. These need to be taken out or corrected. Sometimes it is necessary to reduce the top growth of the plant to match the root ball on a newly transplanted plant (**Figure 25-14**).

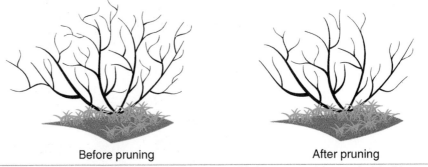

Before pruning   After pruning

Figure 25-14 Careful selection of branches for removal is important when pruning deciduous shrubs.

A plant in weak condition is susceptible to insects and diseases. Poor pruning can cause the loss of a season of flowers or fruit. In general, flowering plants should be pruned just after they bloom or produce flowers. This is done so that the next season's flowering wood is not removed.

It is easiest to prune deciduous trees and shrubs in the late fall or winter, after the leaves drop. The framework is bare and easier to see. When removing large limbs, care should be taken to avoid splitting and tearing of the limb (**Figure 25-15**). Specific plants have specific pruning requirements. Determine how each plant should be pruned before you start. Local plant specialists, university extension programs, and reference books are all reliable sources of accurate information on pruning specific plants.

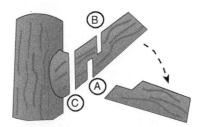

First cut partway through the branch at A, then cut it off at B. Make the final cut, just outside the branch collar, at C.

Figure 25-15 It is important to follow the recommended order of cuts when removing a large tree branch. The wrong cutting order is likely to damage the tree.

Insects and Diseases

Figure 25-16 Insect infestations can build up fast on susceptible plants.

Courtesy of USDA/ARS #K2849-1.

Most plants are subject to damage by insects and diseases. It is easier to prevent such damage than to control it, and several preventive measures are available. Pest-resistant varieties should be used in areas where pest populations are well established. Trees should be carefully managed to remain healthy and vigorous. Weakened trees and shrubs are easily overcome and damaged by insects and diseases (**Figure 25-16**).

It is important that plants are selected for the environments in which they are to be planted. Plants cannot tolerate stressful conditions continuously. They may have some ability to adapt to various conditions, but some plants adapt better than others. Some plants can withstand air pollution, drought, heat, or even wet soils. Other types of plants can tolerate infertile conditions, but not hot, dry weather. Still others may not survive in heavy shade or in full sun.

In its lifetime, any plant can be expected to experience stress from insects and diseases. When an infestation is suspected, it is important to act quickly. Often, caretakers do not know what is causing damage to a tree. State agricultural universities regularly employ a state **arborist** or tree surgeon who provides consultations concerning tree problems. Many county extension offices maintain plant experts on staff who can identify issues, offer suggestions, and help save the plant. It is wise to become aware of the plants in your area and to consult cooperative extension professionals for more specifics on the types of insects and diseases that can be expected to infect various plants in a given locality.

Science Connection: Tree Gall Infestation

Gall consists of growth of extra tissue formed by a tree in response to injury or infestation by parasitic organisms.

Courtesy of DeVere Burton.

A tree can be a host to many different kinds of organisms that do not cause any problems. Moss or lichen on a tree is rarely cause for concern, and some birds, such as hummingbirds and orioles, are pollinators that benefit trees. However, some organisms attack a tree either by eating it or by laying its eggs in it.

When this happens, the tree will form extra tissue called a gall. This abnormal growth serves as both a habitat and a food source to the maker of the gall. The parasitic organism moves into the lumpy or complicated structure to live, reproduce, and eat. Galls are believed to form in response to a plant-growth-regulating chemical produced by the attacking organism. Typically, tree galls form during the rapid growth period of new foliage and flowers in late spring. They usually remain on trees for more than one season since it is not until they are fully formed that most gardeners notice them.

The growths can be caused by many different kinds of organisms, including fungi, viruses, insects, mites, and nematodes. They can be found on leaves, branches, trunks, and roots. Galls usually do not cause serious damage to a well-established tree and are more of a bother than a serious threat. Cleaning up any leaf debris in the fall can help limit the spread.

Chapter Review

Student Activities

1. Write the **Key Terms** and their meanings in your notebook.
2. Contact a local nursery and obtain a nursery catalog.
3. Visit a garden center or nursery and compile a list of trees and shrubs that are available and recommended for your community.
4. Prepare a chart of popular ornamental trees and shrubs for your community. For each plant, list the scientific name, common name, mature size of the plant, time of flowering, color of flowers, spring leaf color, and fall leaf color.
5. Survey your home or other assigned area and make a list of the trees and shrubs that are present. Use books, nursery catalogs, and other resources to help identify the plants.
6. For a given area, determine the diameter of each tree 4.5 feet above the ground. Work up a recommended fertilizer program for the trees, including the amount of fertilizer per tree, local fertilizer prices, cost of fertilizer for each tree, time of year to fertilize, and the total cost of fertilizer for the area.
7. Prepare a pruning schedule for plants around your home or other area based on local recommendations. Include in the schedule the common and scientific names of each plant, time of pruning, and special requirements for pruning.
8. Prepare a sketch or a scale drawing of your home property or other assigned area. Locate the existing trees and shrubs on the sketch, using a circle with the initials of the scientific name to represent each plant. Add new plants that you believe will enhance the property you have surveyed.
9. **SAE Connection:** Volunteer at a local park to help plant new trees.
10. Create a game, song, or activity that will help you remember the tree names featured in the figures.

Check Your Progress

Can you...

- identify examples of ornamental trees and shrubs?
- discuss how to select trees and shrubs for appropriate landscape use?
- explain how trees and shrubs are classified and give examples?
- recall proper nomenclature for several examples of trees and shrubs?
- discuss how to purchase plant material for installation in a landscape?

If you answered "no" to any of these questions, review the chapter for those topics.

National FFA Connection: Are Trees and Shrubs on Your Career Landscape?

An interest in trees and shrubs can open up a broad range of career possibilities.
Courtesy of the USDA.

Interested in "branching out" into a career related to trees and shrubs? The National FFA Organization hosts Nursery/Landscape Career Development Events (CDE). The purpose of such an event is to promote career interest, encourage proficiency development, and recognize excellence. These CDEs include all aspects of the industry, such as producing, marketing, utilizing, and maintaining trees, shrubs, and other landscape plants.

As part of the competitive event, students evaluate classes of trees, shrubs, bedding plants. They also take an identification exam on plants and tools common in the horticulture industry, demonstrating proper transplanting technique and general horticultural knowledge.

An excellent way to start preparing for such events is to hone your plant identification skills. For example, can you name your state tree? Do you know what kinds of trees and shrubs are native to or commonly grown in your area? Several FFA chapters have posted helpful study tools online, including slideshow guides to shrubs, trees, bedding plants, vines, ground covers and other plants. Such guides may inspire you and your classmates to develop guides for your own area. A general online search of "Nursery/Landscape Career Development Event" plus "FFA" suffices to find these and other state- and chapter-level resources.

Searching "Nursery/Landscape" on the National FFA Organization website reveals more resources, including a video and a handbook about the CDE. To dig even deeper, search the FFA website using such terms as "tree" and "shrub." The search results include several interviews and articles about FFA members' achievements. Read about students who work together to improve local parks and host community giveaways of trees and shrubs. Gain insights from outstanding individuals whose horticultural skills led to national acclaim, helped grow a family business, or launched an unexpected and satisfying new career path.

Unit 8

Animal Sciences

What will animals of the future be like? As current animal science shows, researchers are answering this question through improvements in breeding, health, food production methods, and environmental sustainability. Agriscience research is the key element to maintaining the supply of food for a rapidly increasing global population.

Ovine (Sheep) Science In the mid-twentieth century, desert range sheep averaged a 75 percent lamb crop—every 100 female sheep averaged a total of 75 live lambs per year. Today the annual average is 108 live lambs. Farm flocks under intensive management tend to be even more productive. Scientists at the U.S. Sheep Experiment Station in Dubois, Idaho believe that the average crop could approach 150 percent within 50 years if the gene responsible for multiple lamb births can be isolated and bred into the genetic makeup of the most productive breeds. Current ovine research studies include:

- Pneumonia transmission between and among domestic and wild sheep
- Production efficiencies: improving reproduction rate; reducing death rate
- Improving herd genetics
- Sustainability of rangeland ecosystems.

Bovine (Dairy) Science In the 1950s, the United States had 12 million dairy cows, and the national average milk production per cow was 9.751 lbs. By 2021, 9.4 million cows were averaging 23,391 lbs. of milk per year. In 50 years that average could increase to 40,000 lbs. per year. Will such increases require larger cows? Dairy scientists at the U.S. Department of Agriculture (USDA) Dairy Forage Research Center in Madison, Wisconsin do not believe so. Instead, better feed could render the conversion of feed to milk more efficient. Dairy-based research studies include:

- Organic dairy management practices
- Precision technologies: robotic milking; cow behavior sensors
- Computer chip analysis of milk: early detection of illness; milk contamination
- Production of probiotic dairy products.

Bovine (Beef) Science Management of beef cattle has improved remarkably through research, resulting in more efficient production of cattle and better beef products. Beef-based research studies include:

- Potential for cattle-borne disease pathogens to cause human illnesses
- Improving feedlot efficiencies
- Discovering and developing new marketing opportunities
- Effects of different rangeland stocking rates
- Responses of deer and elk to intensively managed forests and rangelands.

Avian (Poultry) Science Devastating diseases, such as avian influenza and Newcastle disease, still threaten the poultry industry. Both viruses have strains that range from non-fatal to birds to those that are 100 percent lethal. Thus, when only a few birds are diagnosed with the disease, the entire flock must be destroyed to prevent widespread infection. Scientists at the USDA Southeast Poultry Research Laboratory in Athens, Georgia are studying ways to better manage poultry diseases. Poultry research also includes:

- Mortality management and litter control
- Rearing environment, ventilation, and energy
- Biogas for electrical power generation
- Modification of egg composition
- Predicting specific virus behavior in poultry.

Porcine (Swine) Science The swine industry in the United States relies on research to improve production methods and product quality. High-priority porcine-based research studies include:

- Disease control, such as for porcine epidemic diarrhea and swine dysentery
- A swine fever surveillance program
- Oral fluid testing for pathogens and antibodies
- Selecting replacement gilts for longevity.

Chapter 26

Animal Anatomy, Physiology and Nutrition

Objective
To determine the nutritional requirements of animals and learn how to satisfy those requirements.

Competencies to Be Developed
After studying this chapter, you should be able to:
- compare animal digestive systems.
- understand the basics of animal physiology.
- understand how nutrients are used by animals.
- identify classes and sources of nutrients.
- identify symptoms of nutrient deficiencies.
- explain the role of feed additives in livestock nutrition.
- compare the composition of various feedstuffs.

Key Terms
nutrition
ration
deficiency disease
vitamin
mineral
anatomy
skeletal system
muscular system
voluntary muscle
involuntary muscle
protein
circulatory system
carbohydrate
respiratory system
central nervous system
peripheral nervous system
urinary system
endocrine or hormone system

hormone
digestive system
ruminant
rumen
roughage
monogastric
concentrate

fructose
galactose
sucrose
maltose
lactose
starch
cellulose

fat
supplement
feed additive
antibiotic
dry matter
TDN

Feed is animal food. It represents the largest single-cost item in the production of livestock. Therefore, it is important to understand the complex nature of animal nutrition. **Nutrition** is the process by which animals process food and use its nutrients to live, grow, and reproduce (**Figure 26-1**).

The axiom "You are what you eat" applies to both humans and animals. This chapter explores the relationship between good nutrition and good health.

Figure 26-1 Animal health, growth, and reproduction all are directly related to nutrition.
Courtesy of DeVere Burton.

Nutrition in Human and Animal Health

The relationship between proper nutrition and health has long been recognized. Early sailors stocked their sailing vessels with limes when going to sea for long periods. This was to prevent scurvy. Scurvy is a disease of the gums and skin caused by a deficiency of vitamin C in the diet.

Proper nutrition for animals is just as important as it is for humans. Feed efficiency, rate of gain, and days-to-market weight are all uppermost in the minds of people who raise livestock for meat. Proper nutrition is of equal importance for animals that produce milk, wool, or fur. Slow growth, poor reproduction, reduced production, and poor health are generally the results of less-than-adequate animal rations. The amount and content of food eaten by an animal in 1 day is referred to as a **ration**. When the amount of feed consumed by an animal in 24 hours contains all of the needed nutrients in the proper proportions and amounts, the ration is a balanced ration.

Numerous diseases can result from imbalances or improper amounts of vitamins and minerals. Such diseases are called **deficiency diseases**. These diseases usually occur because of inadequate diets or digestive disorders. **Vitamins** are organic substances that are required in small amounts for normal metabolism. **Minerals** are elements found in nature that are essential for normal body functioning of all humans and animals. A shortage of either vitamins or minerals in the diet can lead to a deficiency disease. It should be noted, however, that not all types of animals require the same vitamins and minerals to maintain good health.

Agri-Profile: Career Area: Animal Nutritionist/Feed Formulator

Farmers use science to determine the healthiest rations for farm animals. Healthy dairy cows, such as these, are capable of producing more and higher-quality milk than cows with inadequate nutrition.
© iStockphoto/Alexandru Nika.

Animal nutrition is an interesting field of study that can lead to several types of career opportunities. Some careers are involved in basic research in which scientists investigate how an animal uses its food supply to grow or reproduce. They also determine the nutrient values of feeds, including how digestible they are. This career field extends to production, processing, and sale of feeds. Some people in this career field raise farm animals. Raising and managing farm animals requires an understanding of animal nutrition.

Elements of nutrition are basic, such as the composition of feed grains, animal by-products, and the basic nutrients needed by animals. However, the nutritive content of forages and other feedstuffs varies considerably according to stage of growth, condition, and quality. The digestive capabilities of animals vary considerably from species to species. Nutritional needs of animals vary with age, stage of development, production, and pregnancy.

Careers in animal nutrition may be predominantly in the basic sciences or may be in the application of science to the nutrient requirements of an animal species. They may focus on fish, small animals, pets, horses, poultry, livestock, dairy, or wild animals.

Animal Anatomy and Physiology

The internal functions and vital processes of animals and their organs are referred to as animal physiology. The various body systems, such as the skeletal, muscular, circulatory, respiratory, nervous, urinary, endocrine, digestive, and reproductive, must each be properly nourished and working together for the animal to be healthy and productive. To this end, proper nutrition is a must. The various organs and parts of the body are collectively known as **anatomy**.

Skeletal System

The **skeletal system** (**Figure 26-2**) is made up of bones joined together by cartilage and ligaments. The purpose of the skeletal system is to provide support for the body and protection for the brain and other soft organs of the body.

Bone is the main component of the skeletal system. It is composed of about 26 percent minerals. This mineral material is mostly calcium, phosphate, and calcium carbonate. Another 50 percent of bone is water, 20 percent is protein, and 4 percent is fat.

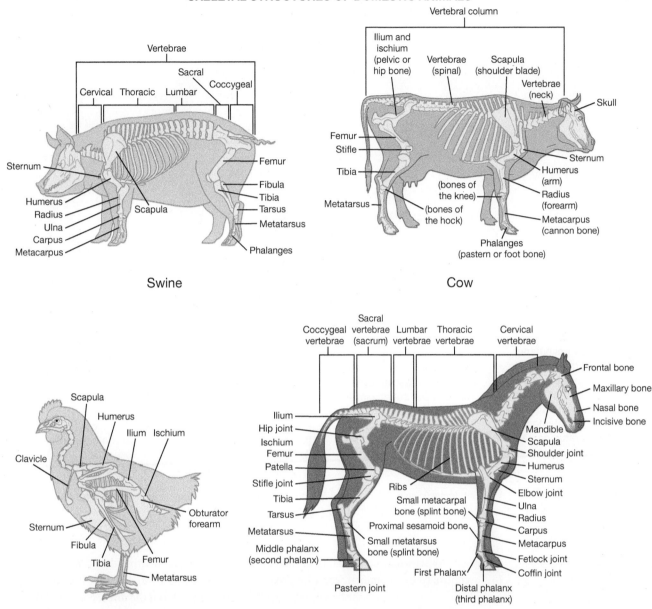

Figure 26-2 These domestic animals may have different bones and skeletal structure, but their skeletal systems all serve the same purpose.

The material inside bones is called bone marrow, and it produces the body's blood cells. The growth and strength of bones are greatly affected by the minerals and vitamins in animal rations.

Muscular System

The **muscular system** (**Figure 26-3**) is composed of muscle fibers, which are the lean meat of the animal. This is the part of the animal used to produce steaks, chops, and roasts. Muscles provide for body movement in tandem with the skeletal system, and they support life (as in the heart muscle and the diaphragm).

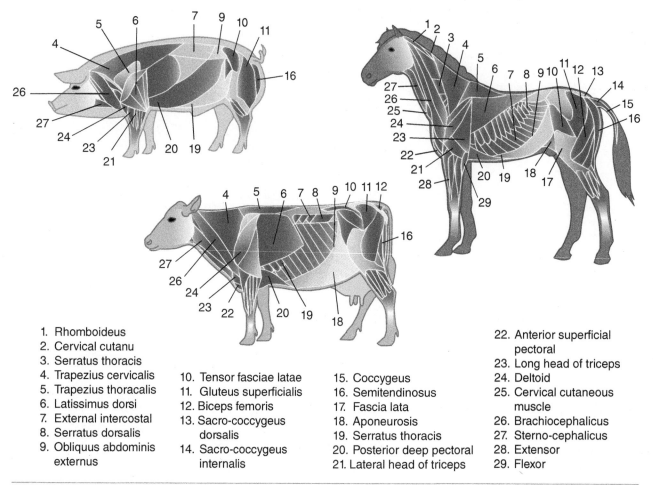

Figure 26-3 The muscular systems of domestic animals are similar in many ways; however, they vary among species depending on which traits have been priorities in selecting breeding stock over many generations.

1. Rhomboideus
2. Cervical cutanu
3. Serratus thoracis
4. Trapezius cervicalis
5. Trapezius thoracalis
6. Latissimus dorsi
7. External intercostal
8. Serratus dorsalis
9. Obliquus abdominis externus
10. Tensor fasciae latae
11. Gluteus superficialis
12. Biceps femoris
13. Sacro-coccygeus dorsalis
14. Sacro-coccygeus internalis
15. Coccygeus
16. Semitendinosus
17. Fascia lata
18. Aponeurosis
19. Serratus thoracis
20. Posterior deep pectoral
21. Lateral head of triceps
22. Anterior superficial pectoral
23. Long head of triceps
24. Deltoid
25. Cervical cutaneous muscle
26. Brachiocephalicus
27. Sterno-cephalicus
28. Extensor
29. Flexor

Muscles may be voluntary or involuntary, depending on whether they can be physically controlled by the animal. **Voluntary muscles** can be controlled to do such things as walk and eat food. **Involuntary muscles** operate in the body without the direct control of the animal, and they function even during sleep. Examples of involuntary muscles are heart and diaphragm muscles.

Muscles are composed largely of protein. These large amounts of protein are required for the maintenance of the animal and for growth and reproduction. **Proteins** are nutrients made up of amino acids, the building blocks of muscles.

Circulatory System

The heart, veins, arteries, capillaries, and lymph system compose the **circulatory system** (**Figure 26-4**). This system transports blood that contains nutrients and oxygen to the cells of the body and it filters waste materials from the body. Lymph glands secrete disease-fighting materials into the body. They are part of the lymphatic circulatory system that operates within the blood circulatory system. A key function of this system is to transport excess water from the cells of the body.

Vitamins, minerals, proteins, and carbohydrates are all essential for the smooth function of the circulatory system. **Carbohydrates** provide sugars and starches that supply energy to the animal.

CIRCULATORY SYSTEMS OF DOMESTIC ANIMALS

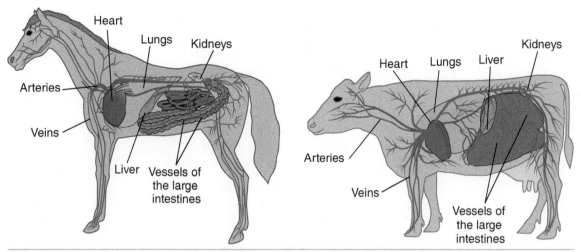

Figure 26-4 The circulatory system distributes food and oxygen that is dissolved in the blood to the cells of the body.

Respiratory System

The **respiratory system** provides oxygen to the blood of the animal and removes waste gases such as carbon dioxide from the blood. The respiratory system is composed of the nostrils, nasal cavity, pharynx, larynx, trachea, and lungs (**Figure 26-5**). This system controls breathing and uses the muscular and skeletal systems to draw air in and out of the lungs. Oxygen passes from the lungs to the blood.

RESPIRATORY SYSTEMS OF DOMESTIC ANIMALS

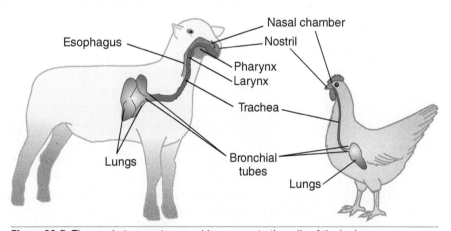

Figure 26-5 The respiratory system provides oxygen to the cells of the body.

Nervous System

The nervous system is composed of the central nervous system and the peripheral nervous system. The **central nervous system** includes the brain and the spinal cord. It is responsible for coordinating the movements of animals and also responds to each of the senses. The senses are hearing, sight, smell, touch, and taste. The **peripheral nervous system** controls the functions of the body tissues, including the organs. The nerves transmit messages to the brain from the outer parts of the body (**Figure 26-6**).

Because the nervous system is composed primarily of soft tissues, proteins are particularly important in maintaining its health.

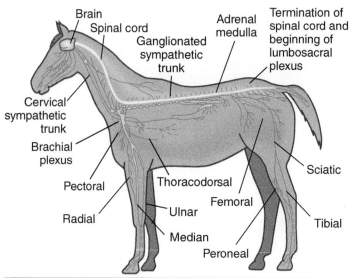

Figure 26-6 The nervous system consists of the brain, spinal cord, and the nerves that are distributed throughout the body. This system coordinates all of the other body systems.

Urinary System

The function of the urinary system is to remove waste materials from the blood. The primary parts are the kidneys, bladder, ureters, and urethra (**Figure 26-7**). The kidneys also help regulate the makeup of blood and help maintain other internal systems.

Abnormal levels of proteins fed to animals have been known to cause stress to the urinary system, which rids the body of excess protein. Greater-than-recommended levels of minerals may also cause kidney problems.

Endocrine System

The endocrine or hormone system is a group of ductless glands that release hormones into the body. Hormones are chemicals that regulate many of the activities of the body. Some of these body functions are growth, reproduction, milk production, and breathing rate. Hormones are needed in only minute amounts. For example, only 1/100,000,000 g. of oxytocin hormone will stimulate immediate letdown of milk in female animals. Oxytocin is a hormone from the hypothalamus gland.

Proper levels of all nutrients, especially minerals, are important for the proper functioning of the endocrine system.

Digestive System

The digestive system provides food for the body and for all of its systems. This system stores food temporarily, prepares food for use by the body, and removes waste products from the body. There are three basic types of digestive systems: polygastric, or ruminant; monogastric; and poultry.

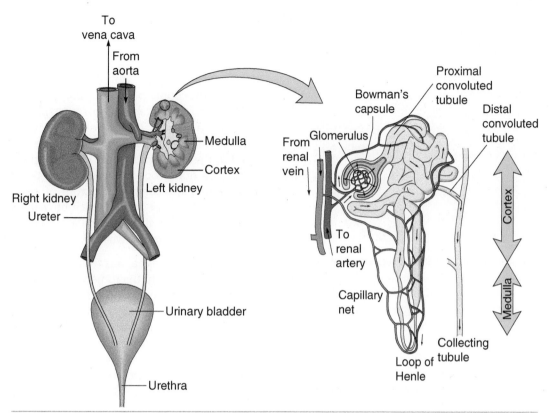

Figure 26-7 The urinary system removes waste materials from the blood.

Polygastric or Ruminant System

Ruminants are a class of animals that have stomachs with more than one compartment (**Figure 26-8**). Cattle and sheep are ruminants and have multicompartment stomachs. The largest compartment is called the **rumen**. The rumen can store large amounts of roughage. Examples of **roughage** include grass, hay, silage, or other high-fiber feed. Ruminants have the ability to break down plant fibers and to use them for food far more efficiently than nonruminants. B-complex vitamins are manufactured in the digestive systems of ruminant animals by bacteria. Such vitamins need not be added to the diets of these animals, even though they are required by the body. Notably, calves do not develop true rumens until they are several months old. Therefore, they need to be fed complete rations like nonruminant animals until their rumens develop.

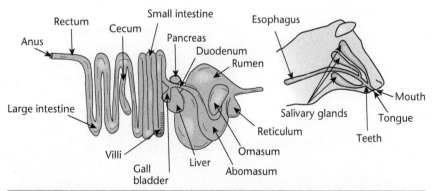

Figure 26-8 The ruminant digestive system processes large amounts of roughage.

Science Connection: A Four-Chambered Stomach

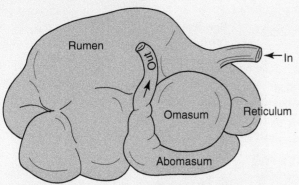

The stomach chambers of a ruminant animal include the rumen, reticulum, omasum, and abomasum.

Ruminant animals have distinct advantages over their nonruminant counterparts. They can digest roughage. The main source of energy in roughage is cellulose. Cellulose is a complex, nutrient-rich plant material that is unusable to nonruminant animals because they cannot digest it. The ruminant stomach is host to specific strains of bacteria that produce enzymes with the capability of unlocking the complex structure of cellulose and digesting it. Not only do these microbes aid in breaking down cellulose but they become a valuable source of protein for the animal as well.

The ruminant animal has four compartments that make up its stomach: the rumen, reticulum, omasum, and abomasum. The compartments work together to digest food that is high in fiber. Food can flow easily between the rumen and the reticulum. Most of the roughage is broken down there. Next, food enters the omasum, where the food particles become smaller and water is reabsorbed into the body. Next, food enters the abomasum or true stomach. Here, digestive enzymes are secreted that break down the food into basic nutrients. The nutrients then move into the small intestines where they are absorbed. Unused materials move into the large intestine where more water is absorbed, and the remaining material leaves the body as feces.

Monogastric System

The digestive systems of swine, horses, and many other animals are called monogastric (**Figure 26-9**). **Monogastric** means having a stomach with one compartment. The stomachs of swine and horses are relatively small and can store only small amounts of food at any one time. Most of the digestion takes place in the small intestines. A horse is able to digest roughage because it benefits from bacterial action on cellulose in a section of the large intestine called the caecum. Except for the horse, monogastric animals are unable to break down large amounts of roughage. Therefore, their rations must be greater in concentrates. **Concentrates** are composed mostly of grains that are low in fiber and high in total digestible nutrients. B-complex vitamins also must be included in the diets of monogastric animals because they cannot make such vitamins in their digestive systems.

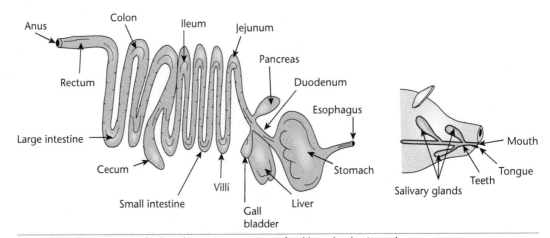

Figure 26-9 The monogastric digestive system processes food in a simple stomach.

Poultry Digestive System

Although poultry have monogastric digestive systems, their digestive systems are different enough to discuss separately (**Figure 26-10**). Chickens have no teeth and must swallow their food whole. The food is stored in the crop and passed on to the gizzard, which grinds it up. The gizzard contains pebbles and other hard objects (eaten by the bird) that crush seeds and other large food particles as the muscular gizzard contracts and agitates its contents. Food particles then pass on to the small intestine where digestion occurs. Poultry rations must be high in food value because birds have no true stomachs and little room for storage of food they have eaten.

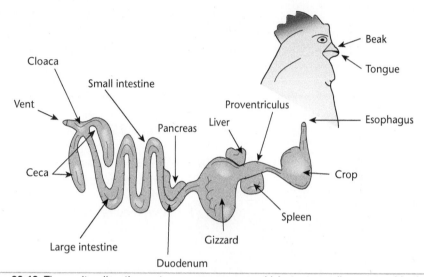

Figure 26-10 The poultry digestive system a crop, an organ which stores small amounts of feed, and a gizzard, an organ that grinds seeds and other materials.

Major Classes of Nutrients

Water

Water is the largest component of nearly all living things. The muscles and internal organs of animals contain 75 percent or more water.

Water is one of the most important requirements for living organisms. It is the solution in which all nutrients for animals are dissolved or suspended for transport throughout the body (**Figure 26-11**). Water provides rigidity to the body, allowing it to maintain its shape. The liquid solution in each cell is responsible for this rigidity. Water reacts with many chemical compounds in the body to help break down food into nutrients that can be used by the body.

Water regulates the body temperature of animals through perspiration and evaporation. Because water absorbs and transports heat, body temperatures of animals increase and decrease more slowly than would be possible otherwise.

ROLE OF WATER IN LIVING ORGANISMS

- Dissolve or suspend nutrients
- Create rigidity to maintain the shape of the organism
- Break down foods to forms the organism can use
- Regulate the body temperature of animals

Figure 26-11 The role of water in living organisms is vital to their survival. Most animals can live only a few days without access to water.

Protein

Protein is the major component of muscles and tissues. Proteins are complex materials that are made of various nitrogen compounds called amino acids. Some amino acids are essential for animals and some are not. Therefore, the quality of proteins fed to animals must be considered.

Monogastric animals need specific amino acids, so it is important that they receive high-quality proteins containing the appropriate amino acids. However, in ruminant animals, quantity of protein is more important than quality. Ruminants can convert amino acids in their rumens to different amino acids to meet their needs.

Protein is used continuously by animals to maintain the body because cells are continually dying and being replaced. In young animals, large amounts of protein are used for body growth. Protein is also important for healthful reproduction.

Science Connection: The Amino Acid Lysine

Much of the nation's corn crop is fed to pigs. The development of high-lysine corn has greatly benefited the health of pigs and the efficiency of pork production.
© iStockphoto/Carol Gering.

Animals require nutrients of the right kinds and in the right amounts to produce the proteins that make up muscles and other body tissues. When an amino acid is unavailable, a protein that requires that amino acid can no longer be manufactured by the body.

The amino acid lysine is too low in some corn varieties to support rapid growth in hogs. Lysine must be added to corn to avoid this problem. Humans who rely on diets that are high in corn are also at risk for a lysine deficiency in the diet. However, people can supplement their diets with fish, turkey, or other high-lysine foods to correct the deficiency.

Researchers have developed new corn varieties known as "high-lysine" corn that have been shown to correct the nutrient deficiency. However, the new variety is lower in grain yield than most dent-corn varieties, and fields of "high-lysine" corn must be separated from other corn fields to prevent cross-pollination. Cross-pollinated corn yields the usual "low-lysine" corn because the gene for "low-lysine" corn is dominant over the gene for "high-lysine" corn.

Carbohydrates

Carbohydrates make up a class of nutrients composed of sugars and starches. They provide energy and body heat to animals. Carbohydrates are composed primarily of the elements carbon, hydrogen, and oxygen.

The energy obtained from carbohydrates is used for growth, maintenance, work, maintaining body heat, reproduction, and lactation (milk production). Carbohydrates come in several forms, with the sugars being the simplest. Examples of simple sugars used in animal feeds are glucose, fructose, and galactose. Compound sugars include sucrose, maltose, and lactose. Complex forms of carbohydrates include starch and cellulose.

Carbohydrates make up about 75 percent of most animal rations, but there is little carbohydrate in the body at any one time. Carbohydrates in the diet that are not used quickly are converted to fat and stored in the body. **Fat** is a tissue that stores energy in a concentrated form; it contains 2.25 times as much energy per gram as carbohydrates.

Minerals

Minerals have many functions in the animal. The skeleton is composed mostly of minerals. Minerals are important parts of soft tissues and fluids in the body. The endocrine system is heavily dependent on various minerals, as are the circulatory, urinary, and nervous systems.

Fifteen minerals have been identified as being essential to the health of animals. These are calcium, phosphorus, sodium, chlorine, potassium, sulfur, iron, iodine, cobalt, copper, fluorine, manganese, molybdenum, selenium, and zinc. In the past, most of these minerals were provided naturally by feeds grown on fertile soils and by contact with the soil itself. Today, it is increasingly important to provide additional mineral matter to the diet of animals. A mineral fed as a separate feed is called a **supplement**. Supplements are especially important for animals that spend their lives in confinement.

Vitamins

Vitamins are acquired by animals in several different ways. Some are available in roughages and concentrates, some are available in feeds containing animal by-products, and some are made by the body itself.

Science Connection — Energy Content

Like humans, livestock need food for energy. But it is important to know how much energy an animal will receive from the food it eats. The calorie content of a feed source can be calculated. A *calorie* is a term used to describe available energy. One calorie has enough energy to raise 1 g. of water 1°C. Animals need a specific number of calories to maintain good health and be productive. It is a balancing act. Too much feed and an animal will gain an inappropriate amount of weight and not be as productive as it could be. In contrast, if an animal does not receive an adequate amount of calories, it will lose weight and will not reach its production potential.

Macronutrient Calorie Content

Macronutrient	Calories per Gram	Livestock Feed Example
Carbohydrate	4 calories per gram	Grain
Protein	4 calories per gram	Alfalfa
Alcohol	7 calories per gram	Silage
Fat	9 calories per gram	Plant oil additives

In the preceding table, each macronutrient is listed with the calorie content available in 1 g. of each. An example is provided, showing a feed source for each nutrient.

*Note: The feed examples listed may contain other nutrients. For example, alfalfa is also a source of carbohydrate.

The table on macronutrient calorie content shows the amount of energy (in calories) available for each macronutrient. This is useful information for individuals who need to provide proper nutrition to livestock and receive the highest levels of production possible. For example, if a grain contains 50 g. of carbohydrates, 20 g. of protein, and 10 g. of fat in a cup of feed, how many calories will each cup of feed contain?

50 g. carbohydrates × 4 calorie/gram = 200 calories from carbohydrates

20 g. protein × 4 calorie/gram = 80 calories from protein

10 g. fat × 9 calorie/gram = 90 calories from fat

200 + 80 + 90 = 370 calories per cup of feed

*With this information, a measurement of calories from feed can be calculated based on an animal's weight and energy requirements.

Vitamins are required in only minute quantities in animals. They act mostly as catalysts for other body processes. There are large variations in the necessity for vitamins in various species of animals that are important to agriscience.

Some of the specific ways that vitamins are used in animals include clotting of blood, forming bones, reproducing, keeping membranes healthy, producing milk, and preventing certain nervous system disorders.

Fat

Only small amounts of fat are required in most animal diets. The addition of fat to the diets of animals improves the palatability, flavor, texture, and energy levels of feed. The addition of small amounts of fat to the diet has also been shown to increase milk production and to aid in the fattening of meat animals. Fats are also necessary in the body as carriers of fat-soluble vitamins, A, D, E, and K.

Sources of Nutrients

The sources of nutrients for animals are many and extremely varied. Important animal feed components include roughages, concentrates, animal by-products, minerals from mineral deposits, and chemically made nutrients called synthetic nutrients.

Proteins

The major sources of protein for animals include oilseeds such as soybeans, peanuts, cottonseed, and linseed. These seeds are processed by cooking and other procedures to remove the bulk of the oil from them. The remainder of the seed content is then dried and ground up for feed. Feed consisting of ground oilseeds with the oil removed is called oil meal.

Cereal grains provide lesser amounts of protein than oil meal, but they are also important protein sources. Good-quality legume hay, such as alfalfa or clover, is another good plant source of protein for ruminant animals.

Animal protein is generally of greater quality than plant protein. More specifically, animal protein usually contains more of the essential amino acids than plant protein. Sources of animal protein include meat and bone meal, meat meal, fish meal, blood meal, skim milk, whey, feather meal, and meat products. It should be noted that most meat by-products obtained from mammals cannot be fed to ruminant animals. Some exceptions include products made from blood obtained from slaughter facilities and milk by-products.

Nonprotein nitrogen in the form of urea can be used as a substitute for some of the protein required by ruminant animals. Urea is a synthetic source of nitrogen made from atmospheric nitrogen, water, and carbon. The rumen bacteria found in ruminant animals such as cattle and sheep can convert urea and other nitrogen sources to amino acids and proteins. Rumen bacteria are also a source of protein because they are digested together with the feed. The feeding of urea should be limited to not more than 1 percent of the total dry matter in the ration. Young ruminant animals and all nonruminants are unable to digest urea.

Carbohydrates

Carbohydrates are found in all plant materials. The major sources of carbohydrates for animal feed are the cereal grains. Corn is the most important of these grains in the United States, followed by wheat, barley, oats, and rye. Other sources of carbohydrates include nonlegume hays such as orchard grass, timothy, other grasses, and molasses. Animal rations generally contain adequate levels of carbohydrates.

Fats

Because fats are needed in fairly small amounts in the diets of animals, it is seldom necessary to identify specific sources of dietary fat. Carbohydrates are converted to fats when a surplus exits in the diet, and most proteins are also sources of fat. This is especially true for the oil seeds and animal by-products.

Vitamins and Minerals

Vitamins and minerals are part of all the normal feeds for animals. Ruminants manufacture B-complex vitamins in their rumens. Exposure to sunlight allows the body to manufacture vitamin D. Contact with the soil, coupled with other feeds grown on fertile land, provides most of the mineral requirements for animals that have access to pasture and high-quality feeds. However, it is sometimes necessary to supplement, or add to, natural sources of vitamins and minerals. Commercial vitamin and mineral supplements are formulated for specific classes of animals and their special needs. Such supplements are available in the developed countries of the world.

Symptoms of Nutrient Deficiencies

Animals must be fed appropriate types and amounts of feed regularly to remain healthy and to produce milk, meat, wool, eggs, fur, work, or healthy offspring. Shortages, or deficiencies, of various nutrients will generally produce observable effects in animals. Some common symptoms of disorders caused by vitamin and mineral deficiencies are described in **Figure 26-12**.

Feed Additives

A feed additive is a nonnutritive substance that is added to feed to promote more rapid growth, to increase feed efficiency, or to maintain or improve health. Feed additives fall into two major groups: growth regulators (mostly hormones) and antibiotics. Antibiotics are substances used to help prevent or control diseases.

Common growth regulators include hormones such as progesterone, estrogen, and testosterone. They are known to increase growth rates and feed efficiency by as much as 5 percent. Use of growth regulators is controlled by government meat inspectors, who test for the presence of these substances in meat.

A wide range of antibiotics are added in low levels to the diets of animals such as swine and poultry. Antibiotics keep certain low-grade infections at bay. Antibiotics added to feed allow growing animals to gain weight at their greatest potential rate.

Controversy has arisen over including antibiotics and growth hormones in the feed of animals. Of major concern is the possibility of these substances remaining in the meat of animals slaughtered for human consumption. To reduce this possibility, the substances must be removed from the feed well before the animals are marketed.

Composition of Feeds

All feeds are composed of water and dry matter. The material left after all water has been removed from feed is dry matter. Water makes up 70 to 80 percent of most living things. Dry matter generally contains only 10 to 20 percent water.

Dry matter is made up of organic matter and ash or minerals. The organic-matter portion of animal feed consists of protein; carbohydrates, such as starch and sugar; fat; and some vitamins. The proportion of these materials varies widely among different feeds.

Disorders of Vitamin Deficiency	Disorders of Mineral Deficiency
Vitamin A 　1. Night blindness 　2. Loss of young 　3. Poor growth 　4. Nasal discharge 　5. Diarrhea **Vitamin C** 　1. Scurvy 　2. Gum inflammation 　3. Hemorrhages 　4. Slow healing of wounds **Vitamin D** 　1. Bone weakness and deformities 　2. *Rickets* and *osteomalacia* are bone diseases in young and old animals, respectively 　3. Thin egg shells 　4. Weak, deformed young 　5. Lowered milk production **Vitamin E** 　1. Reproductive failures 　2. Degeneration of certain muscles 　3. *Stiff lamb disease*, or muscle degeneration in lambs 　4. *White muscle disease*, or muscle degeneration in young calves 　5. Poor egg hatchability **Vitamin K** 　1. Poor blood clotting 　2. Internal hemorrhages **Thiamine** 　1. Poor appetite 　2. Slow growth 　3. Weakness 　4. Nervousness **Riboflavin** 　1. Slow growth 　2. *Dermatitis*, or skin disorder 　3. Eye abnormalities 　4. Diarrhea 　5. Weak legs in pigs **Niacin** 　1. Dermatitis 　2. Retarded growth 　3. Digestive troubles **Pyridoxine** 　1. *Anemia*, or low red-blood-cell count 　2. Poor growth 　3. Convulsions in pigs **Pantothenic Acid** 　1. "Goose-stepping" in pigs 　2. Unhealthy appearance 　3. Digestive problems **Biotin** 　1. Dermatitis 　2. Loss of hair 　3. Retarded growth **Choline** 　1. Poor coordination 　2. Poor health 　3. Fatty liver 　4. Poor reproduction in swine **Folic Acid** 　1. Blood disorders 　2. Poor growth **Vitamin B_{12}** 　1. Slow growth 　2. Poor coordination 　3. Poor reproduction	**Calcium** 　1. Rickets 　2. Poor growth 　3. Deformed bones 　4. Milk fever **Phosphorus** 　1. Lameness 　2. Stiff joints 　3. Rickets 　4. Poor milk production **Sodium Chloride** 　1. Lack of appetite 　2. Unhealthy appearance 　3. Slow growth **Potassium** 　1. Slow growth 　2. Joint stiffness 　3. Poor feed efficiency **Sulfur** 　1. General unthriftiness (lack of strong growth) 　2. Poor growth **Iron** 　1. Anemia 　2. Labored breathing 　3. Edema (swelling) of the head and shoulders 　4. Flabby, wrinkled skin **Iodine** 　1. *Goiter*, or enlarged thyroid gland in the neck 　2. Weak or dead offspring at birth 　3. Hairlessness 　4. Infected navels, especially in foals **Cobalt** 　1. Delayed sexual development 　2. Poor appetite 　3. Slow growth 　4. Decreased milk and wool production **Copper** 　1. Abnormal wool growth 　2. Poor muscular coordination 　3. Anemia 　4. Weakness at birth **Fluorine** 　Poor teeth **Manganese** 　1. Poor fertility 　2. Deformed young 　3. Poor growth **Molybdenum** 　Poor growth rate **Selenium** 　1. Muscular degeneration 　2. Heart failure 　3. Paralysis 　4. Poor growth **Zinc** 　1. Poor growth 　2. Unhealthy wool or hair 　3. Slow healing of wounds 　4. *Parakeratosis*, or a skin disease similar to mange

Figure 26-12 Some disorders caused by vitamin and mineral deficiencies in animals.

Classification of Feed Materials

In general, animal feed is classified into two types: concentrates and roughages. Concentrates are low in fiber and high in total digestible nutrients, abbreviated as **TDN**. TDNs include the digestible protein, soluble carbohydrates, digestible crude fiber, and 2.25 times the digestible fat contained in the ration. Conversely, roughages are high in fiber and low in TDN.

Concentrates

Included under the classification of concentrates are the feed, or cereal, grains. These include corn, wheat, oats, barley, rye, and milo as well as others. These grains make up the bulk of most concentrates.

Grain by-products, such as wheat bran, wheat middlings, brewer's grain, and distiller's grain, are concentrates. They are materials left over from the production processes used in making flour and alcohol. A by-product is a secondary product resulting from the production of a primary commodity.

The oil meals are by-products resulting from the processing of vegetable oil from oil seeds. Oil meals and sugar in the form of cane molasses and beet molasses are considered to be concentrates. They are widely used in animal feeds.

Roughages

Roughages can be divided into three categories: dry, green, and silage. The most important of the dry roughages is hay (**Figure 26-13**). Grass hays include timothy, orchard grass, bromegrass, Bermuda grass, and a few others. Types of legume hay include alfalfa, clover, lespedeza, soybean, and peanut. The hulls of cottonseed, peanuts, and rice are also included in the category of dry roughages.

Green roughages are plant materials with high moisture content, such as grasses in pastures and root plants (including sugar beets, turnips, and rutabagas). Tubers such as potatoes are also considered to be green roughage.

Silage is the feed that results from the storage and fermentation of green crops. Fermentation takes place in the absence of air. Corn silage is the most important member of this group. Other examples are grass, legume, and small-grain silages.

The successful production of animals for fun, profit, or sport requires proper animal nutrition. A knowledge of animal physiology, feed materials, and nutrition helps keep animals healthy and productive. Knowledge of nutrition-deficiency symptoms permits the animal manager to take corrective steps when an animal exhibits signs of malnutrition.

Figure 26-13 Roughages are feeds consumed in large amounts by ruminant animals. They consist of high-fiber feeds such as hay, silage, and pasture.

Courtesy of DeVere Burton.

Chapter Review

Student Activities

1. Write the **Key Terms** and their meanings in your notebook.
2. Obtain several different commercial feed tags and compare the percentages of protein, fat, TDN, and other nutrients listed on the tag. Also make note of any vitamins and mineral supplements that are part of the ingredients. Types of feed additives should also be noted. If the prices are known, try to determine what causes variations in the prices of the various feeds.
3. Compare the digestive systems of ruminants, nonruminants, and poultry. Note the parts that are alike and those that are different.
4. Dissect and compare the contents of the digestive tracts of ruminant and nonruminant animals.
5. Make an outline of this unit. Phrases and words that are bold should be included in the outline, together with a brief description of key principles.
6. Draw and label the three different digestive systems discussed in this unit.
7. **SAE Connection:** Research one of the disorders caused by a vitamin or mineral deficiency described in the unit. Write a one-page report on the disorder you have chosen. Be sure to include the symptoms, treatment, and prevention techniques associated with the illness.
8. In your own words, describe the two classifications of feed material.

Inquiry Activity — Using Science in Animal Products

Courtesy of DeVere Burton.

After watching the *Using Science in Animal Products* video available on MindTap, discuss the following two items with a classmate.

1. Recalling the video, what are some advancements that helped produce improvements in animal products?
2. Pause the video at the 1:36 mark to review the list of ethical considerations. Choose one of the questions to discuss. Draw upon what you already know and what you learned in the video to support your position.

National FFA Connection — No FFA Chapter in Your Area? No Problem!

FFA members learn about animals of all kinds, from pets to production animals.
Courtesy of FFA/Photo by Michael Wilson.

The National FFA Organization has close to 9,000 chapters, and the number keeps growing. However, if your area doesn't yet have an FFA chapter, you and your classmates can work with your instructor to start one of your own.

Simply go to the FFA website and search "Start an FFA Chapter." There you will find not only the steps to follow, but resources that make it easy to do so. For example, the FFA provides survey forms and tips to help you gauge student and community interest in forming a local chapter. And since each state is different, you will find everything you need to share your plan with FFA agricultural education leaders in your state.

Indeed, even members of a long established chapter will find it worthwhile to visit the web page, as several resources offer the opportunity for a "refresh." For example, some resources explain how to write or update your chapter's philosophy—that is, the purpose of your chapter and how it will serve your school and your local community. You will also find inspiration as you read about the activities of other FFA chapters, from an Arizona chapter that raises chickens and donates the eggs to charity, to a chapter in Vermont that hosts local pet adoptions.

Check Your Progress

Can you...

- compare two different animal digestive systems?
- discuss the basics of animal physiology?
- give examples of how nutrients are used by animals?
- identify some classes and sources of nutrients?
- name some common symptoms of nutrient deficiencies?
- explain the role of feed additives in livestock nutrition?
- compare the composition of various feedstuffs?

If you answered "no" to any of these questions, review the chapter for those topics.

Chapter 27
Animal Health

Objective
To determine the most effective strategies to maintain animal health.

Competencies to Be Developed
After studying this chapter, you should be able to:
- identify physical signs of good and poor animal health.
- identify symptoms of animal diseases and parasites.
- understand how to prevent animal health problems.
- explain various methods of treating animal health problems.

Key Terms
disinfectant
host animal
contagious
noncontagious
abortion
roundworm
fluke
protozoa

secondary host
mange
balling gun
drench
intravenous
intramuscular
subcutaneous
intradermal

intraruminal
intraperitoneal
infusion
cannula
vaccination
immune
veterinarian

Maintaining animal health is the key to a profitable and satisfying animal enterprise. Several considerations need to be addressed in dealing with the health of animals. These include knowing how to recognize signs of good and poor health, maintain a healthy environment, identify animal diseases and parasites, and treat health problems that occur.

Animal Health Indicators

The ability to recognize the signs of good health or the symptoms of health problems is the single most important factor in maintaining healthy animals. A keen sense of observation is important, as is the innate ability to know when something is not right with an animal (**Figure 27-1**). One of the best signs of good health is a contented animal. Of course, recognizing contentment in animals this requires a great deal of experience. Some specific signs of good health include:

- the chewing of the cud in ruminant animals
- shiny hair coat, bright eyes, and pink membranes
- normal body discharges of urine and feces.
- a normal body temperature, pulse rate, and respiration, or breathing, rate (**Figure 27-2**).

Figure 27-1 A shiny coat, bright eyes, and clean nostrils are indicators of good health for many production animals.
Courtesy of USDA ARS/ Photo by Peggy Greb.

Class of Livestock or Poultry	Degree F Average	Degree F Range
Cattle	101.5	100.4–102.8
Sheep	102.3	100.9–103.8
Goats	103.8	101.7–105.3
Swine	102.6	102.0–103.6
Horses	100.5	99.9–100.8
Poultry	106.0	105.0–107.0

Figure 27-2 Normal body temperatures for animals.

Signs of Poor Health

Often, it is easier to tell when an animal is sick than to tell when it is healthy. Loss of production, especially in dairy cattle, is often the first sign that the animal is not well. A rough hair coat and dull, glassy eyes are also indicators of possible poor health. Sick animals often isolate themselves from other animals and stay alone with their heads down. Such animals may be drawn up and walk slowly when forced to walk. Other signs of poor health may include:

- abnormal feces, either too hard or too soft
- discolored urine
- abnormal body temperature
- labored breathing
- rapid pulse rate.

Taking Temperatures

Taking an animal's temperature is basically the same as taking a human baby's temperature. It is usually taken in the rectum. The rectum is the last organ in the digestive tract. Animal thermometers are normally longer and heavier than those used in human medicine. Animal thermometers are manufactured with string at one end to prevent loss of the thermometer in the digestive tract of the animal.

To use an animal thermometer, coat the thermometer probe with sterile jelly to make insertion easier. Do not force the thermometer into the rectum, as resistance by the animal may result in injury. Instead, correct the conditions that are causing the resistance, and then reinsert the thermometer. Leave the thermometer in place until directed—usually by a sound—before removing it. Read the temperature on the digital screen.

Determining Pulse and Respiration Rates

The pulse rate for a large animal can be taken by holding your ear against the animal's chest or by using a stethoscope and listening to the heartbeat. The number of heartbeats in 1 minute is the pulse rate. The respiration rate of an animal can be determined by watching its rib cage move. Counting the number of breaths that the animal takes in one minute indicates the rate of respiration. Readings are compared to normal temperature and respiration rates for healthy animals of the same species.

Hot Topics in Agriscience — Modern Ways to Monitor Herd Health

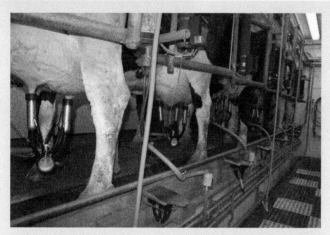

Abnormal milk temperature is a good indicator of an unhealthy cow. An electronic temperature sensor, placed in the milk pipeline at each milking station, helps identify cows that need medical attention.
© Beth Van Trees/Shutterstock.com

The earlier that a herd health problem is detected and treated, the more likely it is that a sick animal will recover. Modern electronic devices make it possible to detect health problems in animals during the early stages of infection. For example, electronic sensors are available that can take the body temperature of an animal as it walks by. These devices are placed on alleyway fences that lead to water or from milk parlors. When an animal with an abnormal body temperature approaches the sensor, a hydraulic gate sorts the animal into a holding pen to be checked by a veterinarian.

A different type of sensor is available to monitor the health of dairy cows. This sensor is placed in the milk line that carries milk from each cow to the main pipeline. The sensor measures the temperature of the milk as it comes from the cow. The sensor is useful on those dairy farms where each cow is identified by a computer chip as she enters the milking area. The sensor enters the milk temperature in the computer record for each cow, and a sorting gate is activated that sorts each cow with an abnormal milk temperature into a holding pen as she leaves the milking area. These types of devices make it possible to identify sick animals in the early stages of infection. Early detection and treatment contribute to early recovery from infections and diseases.

Healthful Environments for Animals

Maintaining a healthy environment for animals is critical. It is usually much less expensive to maintain a healthy environment for animals than to treat animals that are unhealthy due to a poor living environment (**Figure 27-3**).

Figure 27-3 A clean living environment is essential to good animal health.
© Lukas Guertler/Shutterstock.com.

Sanitation

Good sanitation is important to good health, including keeping animal facilities clean. Sanitation also requires the use of clean equipment when dealing with animals, including feed containers, milking equipment, artificial-breeding equipment, needles and syringes, and surgical equipment. A syringe is an instrument used to give injections of medicine or to draw blood and other body fluids from animals (**Figure 27-4**). Simple, on-farm surgical procedures should always be performed with the strictest sanitation possible. The liberal use of disinfectants when dealing with animals is important. A **disinfectant** is a material that kills disease organisms.

Figure 27-4 Syringes are instruments used to inject medications and vaccines into animals. They are also used to withdraw blood and other body fluids.
© photomak/Shutterstock.com.

Housing

Maintenance of proper housing is an important part of ensuring good animal health. Housing should be clean and free from cold drafts. However, good air circulation throughout the shelter is important to help decrease high temperatures in the summer and reduce humidity in the cold of winter. Proper circulation also helps reduce the spread of germs through the air. Extremely dry and dusty conditions should be avoided where possible. Proper structural maintenance is also important. In poorly maintained animal housing, loose boards, roofing materials, and rusty nails can cause injuries and pose other problems.

Handling Manure

Piles of manure are often sources of serious health problems in animals. It is important that manure not be allowed to accumulate in areas frequented by animals. Manure piles sometimes harbor diseases and parasites. They also attract flies, which spread diseases. Cages and pens that are continually soiled by animal waste products can also lower the quality of the air that animals breathe. Feedlots, areas in which large numbers of animals are grown for human consumption, must also be well maintained. Wet, poorly drained, manure-soiled feedlots usually reduce the rate of weight gain in beef cattle and swine. Foot and leg problems can often be traced to poorly maintained feedlots.

Figure 27-5 An immunologist prepares equipment for the purification of parasitic cattle grub proteins for possible use in anti-parasite vaccines.
Courtesy of USDA ARS/Photo by Scott Bauer.

Controlling Pests

The control of pests and parasites is an important consideration in the maintenance of animal health and welfare. To that end, the development of a prevention program is a wise practice. For example, regular use of disinfectants to control parasites such as lice and flies is necessary in a good disease-prevention program. Regular, close observation of animals also is necessary to determine when outbreaks of parasites occur. Many parasitic infections can be prevented through scheduled parasite testing and vaccination (**Figure 27-5**).

The control of other pests, such as birds and wild animals, is also part of a good animal-health program. Many birds carry parasites on their bodies and in their droppings. When the parasites move from infected animals to healthy ones, they sometimes carry diseases. Wild animals and pets may also cause serious health problems when allowed to roam freely around farm animals. Dogs and coyotes will sometimes chase animals and cause injuries. Bites from animals of any kind may cause infection, additional health problems, and death.

Isolation

The isolation of animals that are new to a herd is an important part of a preventive health program. New animals may be harboring diseases or parasites that are not readily apparent, so it is wise to keep a new animal isolated from other animals for a time, usually a minimum of 30 days. This gives the new owner time to observe the isolated animals closely for health problems (**Figure 27-6**).

Similarly, isolation of diseased animals is important. Animals with contagious diseases that can be spread by contact should never be allowed to remain with healthy animals. It is also difficult to treat unhealthy animals when they are living among large groups of animals. Healthy animals tend to become aggressive with unhealthy ones, making it harder for such animals to recover their health.

Figure 27-6 It is important to isolate animals that are being added to a herd for up to 30 days. This allows time for disease symptoms to develop if they are present, and thus protects the rest of herd from the risk of exposure.
© Baloncici/Shutterstock.com.

Pasture Rotation

The rotation of pastures helps maintain a healthy environment for animals and prevents health problems from "gaining ground"—literally. Many disease organisms that infect animals are harbored in the soil. The only way to control them is to prevent them from coming into contact with host animals for extended periods. A **host animal** is an animal which is infected by disease organisms and/or harmful parasites. Rotating pastures for grazing animals on a regular basis is an effective way to break the life cycles of most parasites, which helps control parasite populations (**Figure 27-7**).

Figure 27-7 Pasture rotation breaks the life cycles of diseases, insects, and internal parasites, and it pays additional dividends in greater pasture production.
Courtesy of USDA/ARS #K-3714-2.

Animal Diseases and Parasites

Diseases

Diseases are infective agents that result in declining health of living things. Animal diseases can be divided into two major classes: contagious and noncontagious. **Contagious** diseases are those that can be passed on to other animals by contact.

Noncontagious diseases are not spread through contact. It is important that animals with contagious diseases be isolated from other animals in the herd as soon as the disease is identified. In addition, some contagious animal diseases can be transmitted to humans, so care must be taken when handling infected animals. Similarly, humans who handle animals familiarize themselves with proper techniques, immunizations, and sanitary practices to avoid contracting human diseases and parasitic infections from animals.

Noncontagious diseases pose no threat to humans or other animals, except to the animals afflicted with the diseases. Therefore, there is more leeway in dealing with these animals. However, it is still a good idea to isolate such animals from the herd for their own good.

Causes

Contagious diseases are mostly caused by bacteria and viruses. They can be spread by direct contact with infected animals, from shared housing, or from contaminated feed or water. In some cases, the spread of infectious diseases takes place through intermediary hosts, such as birds, rodents, or insects. Diseases such as sore mouth and brucellosis have the potential to infect humans.

Noncontagious diseases may be caused by nutrient deficiencies or nutrient excesses. Eating poisonous plants or foreign material and developing open wounds that become infected can cause, or lead to, noncontagious disease.

Science Connection: Fighting "Fire" with "Fire"

The right bacteria can save newborn pigs from deadly *E. coli* bacteria.
© Fahroni/Shutterstock.com.

The number one cause of death in newborn pigs is an infection caused by the *Escherichia coli* bacteria. As the result of many deaths, the swine industry loses millions of dollars every year. Up to this point, antibiotics have been used to treat the sick animals. Unfortunately, current techniques are becoming less and less effective because of the tendency of bacteria to develop resistances to the antibiotics. When a bacterium has developed a resistance, it means that an antibiotic no longer kills it.

A promising alternative is a mixed culture of "good" bacteria from the digestive systems of healthy pigs. These bacteria, known as "RPCF" (recombined porcine continuous-flow), are then introduced in the digestive tracts of newborn piglets. When this is done, the good bacteria attach themselves to the limited number of attachment sites in the intestines, thereby preventing *E. coli* from colonizing and causing an intestinal infection.

Symptoms

General symptoms of disease are extremely varied and may include the following:
- Poor growth, reduced production, or both
- Reduced intake of feed
- Rough, dry hair coat
- Discharge from the nose or eyes
- Coughing or gasping for breath

- Trembling, shaking, or shivering
- Unusual discharges, such as diarrhea or blood in feces or urine
- Open sores or wounds
- Unusual swelling of the body, including lumps and knots
- **Abortion**, or the loss of a fetus before it is fully developed
- Peculiar gait, or walking pattern, or other odd movements

Some diseases have little or no external symptoms and may even progress so rapidly that death of the animal occurs before symptoms can be observed.

Parasites

Parasites may also be grouped into two general classifications. These are internal—inside the animal's body—and external—on the outside of the animal's body.

Generally out of sight and frequently microscopic in size, internal parasites are small organisms that live in the flesh or internal organs of larger animals, where they draw nutrients from the bodies of their hosts. The most important internal parasites that infest animals are **roundworms** (slender worms that are tapered on both ends) (**Figure 27-8**). Other types of internal parasites include flukes and protozoa. **Flukes** are very small flat worms, and **protozoa** are microscopic, one-celled animals. In nearly every instance, an internal parasite spends at least a part of its life cycle outside the host animal (**Figure 27-9**). It is during this period that the parasite is most easily spread to other animals. Contact with discharges from infested animals, contaminated feed, water, housing, or secondary hosts may result in the spread of internal parasites. A **secondary host** is a plant or animal that carries a disease or parasite during part of its life cycle. Some internal parasites are also spread by insects such as flies and mosquitoes.

Figure 27-8 Roundworm parasites come in many sizes. They infect the digestive tracts of animals and humans and are common in most parts of the world. The worms are round, as the name implies, and tapered on both ends.

© Oleksande Lytvynenko/Shutterstock.com.

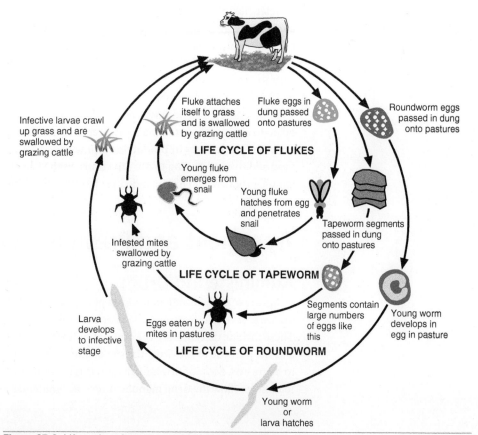

Figure 27-9 Life cycles of some common internal parasites.

The control of internal parasites in production animals is a persistent animal-health problem. Animals can become infected with parasites, and they are helpless to do anything about it when symptoms begin to appear. Fortunately, the body has some ways of keeping such intruders in check. If otherwise healthy and free of reinfection, the hosts can sometimes function fairly well even though they are infected with parasites. Sometimes they can even rid their systems of the parasites. Generally, however, the host needs treatment.

External parasites include flies, ticks, lice, mites, and fleas. They are spread from animal to animal through physical contact.

Symptoms of parasite infestation may include the following:
- Poor growth
- Weight loss
- Constant coughing and gagging
- Anemia
- Reduced production and reproduction
- Diarrhea or bloody feces
- Worms in the feces
- Swelling under the neck
- Poor stamina
- Loss of hair
- The occurrence of **mange**, a crusty skin condition caused by mites
- Visible evidence of the parasite itself

Feed Additives

Feed additives are included in animal rations to stimulate appetites, correct nutrient deficiencies, and control the incidence of low-level infections, such as those caused by parasites, in growing animals. The materials used to control infection have traditionally been antibiotics that help increase feed efficiency and rate of gain while also reducing disease. However, the consumer trend for antibiotic-free animal products has driven a shift to probiotic alternatives. Probiotics consist of living microorganisms that contribute to a healthy digestive system.

Sometimes, feed additives are used to control internal parasites. Caution must be taken to always follow the manufacturer's recommendations concerning the use of such materials particularly in production animals used for food. Failure to use these and other food additives according to the directions on the label may lead to illegal contamination of animal products that are intended for human consumption.

Treatment and Prevention

Administering Drugs

Drugs are manufactured and sold as pills, powders, pastes, or liquids and are used to treat diseases. Several factors must be considered before administering drugs to an animal. These include determining the amount or dosage to be administered, the type of drug to use, the purpose of the drug, the site of administration of the drug, and the type of animal to be treated. Most of this information can be found on the drug container. It is important to closely follow the drug manufacturer's recommendations.

Science Connection — The Unseen Harvesters

Researchers weigh a parasite-infected calf that has been treated.
Courtesy of USDA/ARS #K-4736-3.

Farmers, ranchers, veterinarians, and scientists have long known that animals with parasites grow poorly and seldom reach their full potential. No matter how much they eat, they are undersized and underachievers. Researchers confirm that internal parasites do more than just consume nutrients intended for the host.

The animal's immune system responds to invading organisms by producing chemical signals that modify the host's metabolism. These immune-response signals are small proteins called cytokines that manipulate the hormones regulating feed intake, nutrient use, and, ultimately, growth of the animals. Dairy and beef calves infected with a protozoan parasite called Sarcocystis run fevers, lose their appetites, and become emaciated. Even after an animal has been treated and the parasites have been brought under control, some calves are incapable of growing normally.

Research was conducted on calves at Beltsville, MD, to find some explanations. Before, during, and after acute infection with the parasite, the concentration of growth-regulating hormones in the blood was measured. It was found that after acute infection, the concentration of a hormone essential for growth decreased and the concentration of another hormone that blocks growth-hormone secretion increased. These hormone changes persisted in the infected calves even after the symptoms of infection were gone.

It is thought that the body implements a survival strategy in response to the invasion of the parasites. The strategy is one of growth restriction so more energy can be available to fight off the intruders. It is therefore essential that every effort be made to keep animals in surroundings that are free from parasites and to use precautionary measures whenever applicable. When symptoms appear, it is important to get a professional diagnosis and promptly administer the proper treatment.

Monitoring Medication Withdrawal

Production animal owners must consider the amount of time required for the drug to be inactivated in the animal's body. It often requires several days for drugs to leave an animal's system. If the withdrawal period has not been observed for treated animals before slaughtering one for its meat or consuming its eggs or its milk, illegal residues of the drug are likely to end up in human food. Contaminated milk must be discarded to make sure it does not enter the human food supply. Likewise, it is critical to accurately determine how long to wait before a treated animal can be slaughtered.

When humans frequently consume foods that are contaminated with drugs, they may build up a tolerance for antibiotics among disease organisms that infect humans. Such tolerances for antibiotics make medications less effective when they are needed for the treatment of animal and/or human infections. Additionally, people who are allergic to antibiotics sometimes experience serious allergic reactions to these drug residues.

Pills

The procedure for giving a pill to an animal is to restrain the animal and lift its head so that the mouth opens. Gently place the pill as far back on the tongue as possible, using a **balling gun** (a device used to place a pill in an animal's throat). Massage the animal's throat until it swallows the pill.

Liquids

Liquid drugs administered orally (by mouth) can be placed directly in the animal's stomach by drenching. To **drench** an animal, a fairly large amount of liquid medication is force-fed to an animal by mouth. A syringe or drenching gun is used. In the process, the animal is restrained, with the head held level. The upper lip of the animal is lifted, and the tube is inserted along the side of the tongue. The drug is released slowly, and the animal is allowed to swallow. Care must be taken not to get the drug into the animal's lungs.

Powders

Powdered drugs are normally mixed in the feed or water of the animal. Powders mixed in this way are effective only when the animal is well enough to consume a full dose. Powders can also be mixed with water and administered in liquid form as a drench.

Pastes

A paste is normally used for treating horses for worms. The preparation is placed on the back of the horse's tongue with a caulking gun, and the horse is stimulated to swallow. Pastes are used for horses because it is often nearly impossible to treat them for worms by any other method.

Injection

Injection is the process of administering drugs by needle and syringe. The injection of drugs into animals takes many forms, based on the location of the injection site. Injection sites include **intravenous** (in a vein), **intramuscular** (in a muscle), **subcutaneous** (under the skin), **intradermal** (between layers of skin), **intraruminal** (in the rumen), and **intraperitoneal** (in the abdominal cavity). One factor in determining where injections are made is how fast the drug needs to work. A drug injected into the blood is more readily available than one injected under the skin. Often, it is desirable for drugs to be released slowly over a long period. Growth hormones are generally administered in this way. To give an injection, use the following procedure:

1. Fill the syringe, making sure that all air is removed.
2. Restrain the animal.
3. Select the location for the injection.
4. Disinfect the area to be injected.
5. If the injection is to be made intradermally, clip the hair from the area to be injected.
6. Pop the needle into the desired area without the syringe attached (this prevents the loss of the drug if the animal jumps).
7. Attach the syringe to the needle and inject the liquid.

Infusion

Infusion is another method of getting drugs to the site of the infection. It is used most often to treat dairy animals with udder and teat problems. The udder is the milk-secreting gland of an animal. Teats are the appendages of an udder through which milk is obtained by mechanical means or by suckling offspring. To perform an infusion, a sterile **cannula** (blunt needle) is inserted into the opening of the teat, and the drug is forced into the teat canal from a syringe (**Figure 27-10**).

Chapter 27 Animal Health

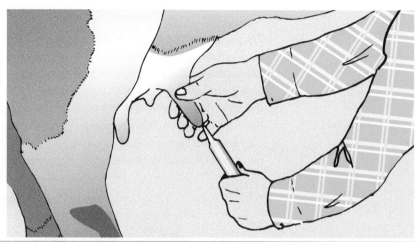

Figure 27-10 Infusion is a method of treating cows and other large animals for infections in the udder.

Dipping

Dipping is a process for treating animals, mostly cattle and sheep, for external parasites. It involves filling a vat with medicated water and forcing the animal to walk or swim through it. Dipping is used in cases where large numbers of animals must be completely covered with the medication.

Restraining Animals

There are a number of ways to safely restrain animals for observation and treatment of diseases and parasites. These include head gates, squeeze chutes, halters, twitches, nose leads, and casting harnesses. Head gates trap the heads of large animals, whereas squeeze chutes hold the whole animal. When halters are used, they are usually tied to a post or something else substantial in order to hold the animal. Twitches are used to hold the tender upper lip of a horse (**Figure 27-11**). As the name suggests, nose leads hold cattle by the nose (**Figure 27-12**). Sometimes, large animals must be lying down to be examined. An effective way to accomplish this is by using a casting harness (**Figure 27-13**). When steady pressure is applied by pulling the rope, the casting harness causes an animal to gently fall down.

Figure 27-11 A twitch is used to redirect the attention of a horse so that it is attentive to its handler instead of the person who is administering a diagnosis or medical treatment.

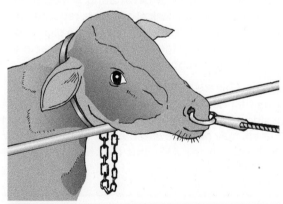

Figure 27-12 Cattle can be restrained during examination or treatment by using a nose lead to hold them in position. The nose lead distracts the animal's attention from the procedure by applying moderate pressure on the tissue between the nostrils when the animal struggles.

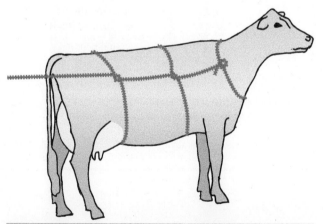

Figure 27-13 A properly used casting harness causes a large animal to lie down on its side while it is examined and treated.

Vaccination

The prevention of a disease is nearly always less expensive than treating animals once they have become infected, so a good disease-prevention program should include vaccination. **Vaccination** is the injection of a modified disease organism (the vaccine) into an animal to stimulate development of immunity against (protection from) a specific disease (**Figure 27-14**). An animal that is **immune** to a disease should no longer be vulnerable to the disease organism. Vaccination programs are among the services regularly offered by veterinarians. These programs vary by type of animal, the disease organisms involved, and the area of the country in which the program is being implemented.

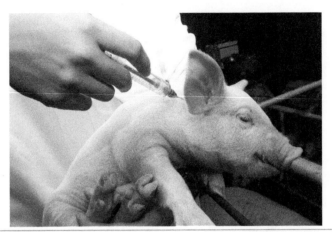

Figure 27-14 Most animal diseases can be prevented by a timely injection of a vaccine that targets a specific disease organism.
© Pattakorn Uttarasak/Shutterstock.com.

Science Connection: African Swine Fever (ASF)

Microbiologist Zhiqiang Lu uses a DNA sequencer to examine genetically engineered African swine fever viruses.
Courtesy of USDA ARS/Photo by Keith Weller.

Eradicating the African swine fever (ASF) and Classical swine fever (CSF) is a major focus of the global swine industry. The two diseases are caused by unrelated viruses, but both are considered 100 percent fatal. In 2018, a devastating outbreak of ASF in China, the global leader in pork production, reduced the world's largest swine herd by an estimated 100 million pigs according to Cable News Network. It triggered a massive research effort to prevent the disease from spreading.

Since then, Chinese scientists have reported progress on an ASF vaccine that has tested near 80 percent immunity without negative side effects. An effective vaccine, successfully tested and delivered, may help prevent the spread of the current and future outbreaks of the ASF virus that has infected much of Europe and beyond.

Veterinary Services

A veterinarian is an essential part of any good health program for animals. **Veterinarians** are professional animal doctors who have graduated from an accredited veterinary college. Their services are critical to animal health, but all owners of animals should learn a few basic veterinary skills.

Many common animal illnesses have distinct symptoms and can be treated initially in the field. Medicines for these illnesses can be purchased at a farm supply store or from a veterinarian. Some knowledge on how to provide medical care to animals can be learned through observation and experience. Other skills are learned through organized education classes and by enrolling in animal technician programs at a college or university. Agricultural extension workshops, clinics, and veterinary handbooks also teach basic veterinary skills that are needed on the farm or ranch. In addition, a lot can be learned about preventing and treating routine health issues by consulting with a local veterinarian.

It is important for animal owners to know when to call the veterinarian for help and when to deal with a problem themselves. There are no hard-and-fast rules in this regard, and it will vary greatly depending on the experience of the individual.

Some procedures that are commonly performed by animal owners are birthing assistance, castration, treating for lice or mites, minor cut repair, mastitis treatment, and vaccination, among others. Farmers and ranchers save on animal medical costs by administering all of the routine vaccinations required by their animals.

A veterinarian should be consulted when:
- Planning and executing a disease-prevention program
- An animal is experiencing reproductive problems, such as failure to conceive, abortion, or great difficulty in giving birth
- An animal suddenly dies for no apparent reason
- Animals have symptoms of a contagious disease.

Agri-Profile Career Areas: Veterinarian/Veterinary Technician

A veterinarian records data on a dairy calf that has a genetic disorder that prevents its white blood cells from fighting off infection.
Courtesy of USDA ARS/Photo by Chuck Greiner.

Animal pathologists, animal behaviorists, physiologists, biologists, zoologists, microbiologists, geneticists, nutritionists, and others must work together to understand the complexities of animals. Animals exist as pets, production animals, work animals, pleasure animals, fish, fowl, birds, wild animals, and specimen animals in zoos. The need for health services for animals varies with the species and its primary use.

The desire to become a veterinarian is a popular career choice among youths. This interest has permitted the numbers of veterinarians to increase, despite the rigor of the college curriculum and tough competition to be accepted into veterinary schools. Colleges of animal sciences offer curricula in other career areas in animal sciences, such as nutrition, breeding, education, and production.

In urban and suburban areas, veterinary clinics for pets offer many opportunities for licensed veterinarians. Animal shelters, hospitals, kennels, and pet stores provide additional career opportunities for those interested in animal health at the technician level. In rural areas, large-animal veterinarian, veterinary assistant, laboratory veterinarian, and laboratory technician are typical positions in the animal-health industry.

Chapter Review

Student Activities

1. Write the **Key Terms** and their meanings in your notebook.
2. Compare the methods of administration, dosage rates, and times of withdrawal of several drugs for animals.
3. Practice giving injections of water using discarded animal ears from a meat-processing plant.
4. Develop a complete disease-prevention program for a livestock operation.
5. **SAE Connection:** Check the housing of production animals for animal safety and proper sanitation at home, on someone else's farm, or at someone's business.
6. Visit a farm supply store or a pet center and record the names of animal medicines, animal pest controls, and disinfectants that are for sale. Also list the use(s) of each item.
7. Interview the manager of a small animal or livestock operation concerning the disease-prevention and health-maintenance practices that are used. Report your findings to the class.
8. With permission of the owner and with supervision, practice some restraint techniques on an production animal as discussed in this chapter. Determine what works best for you and for the animal.
9. Contact a local veterinarian and ask to spend a few hours observing them as they work.

Inquiry Activity Animal Health

Courtesy of DeVere Burton

After watching the *Animal Health* video available on MindTap, answer the following questions.

1. What is gene mapping?
2. How does gene mapping impact animal health?
3. What are the four types of nucleotides? What are the pairs?
4. Why is it important for scientists to discover gene markers?

National FFA Connection

Diversified Livestock Proficiency Award: Entrepreneurship or Placement

One best management practice for livestock is to take precautions to prevent disease.
©Lukas Guertler/Shutterstock.com.

FFA members with interests in ownership or employment in a livestock business that uses "best management practices" may become eligible for local, state, and national awards in diversified livestock production. The livestock business in which the member has ownership or employment must produce and market livestock in at least two of the following proficiency categories:

- Beef production
- Dairy production
- Sheep production
- Swine production
- Equine production
- Goat production
- Specialty animal production
- Poultry production
- Small animal production and care

The website of the National FFA Organization details procedures and instructions that will help chapter advisors work with their members to prepare appropriate applications for these awards.

Check Your Progress

Can you...

- give examples of physical signs of good and poor animal health?
- cite symptoms of animal diseases and parasites?
- discuss how to prevent animal health problems?
- give examples of methods of treating animal health problems?

If you answered "no" to any of these questions, review the chapter for those topics.

Chapter 28
Genetics, Breeding, and Reproduction

Objective
To determine the current roles of genetics, breeding, and reproduction in animal science.

Competencies to Be Developed
After studying this chapter, you should be able to:
- define terms associated with genetics and reproduction.
- understand the principles of genetics.
- identify systems of breeding.
- identify parts of reproductive systems.
- understand new technologies in animal reproduction.
- evaluate animals for type and production.
- understand types of testing programs.

Key Terms

ova
sperm
zygote
gene
chromosome
gamete
dominant
recessive
cell

protoplasm
nucleus
organelle
cell membrane
nuclear membrane
lipid
mitochondrion
ribosome
endoplasmic reticulum

Golgi apparatus
Centrosome
mitosis
meiosis
deoxyribonucleic acid (DNA)
homozygous
heterozygous
incomplete dominance
genotype

phenotype
sex-linked
mutation
heritability
testosterone

estrogen
estrus
progesterone
ovulation
fetus

gestation
inbreeding
semen

Some of the fastest-growing areas of technology in agriscience are in genetics and reproduction. For example, artificial insemination has allowed for more improvement in milk production in the last 50 years than had occurred in the previous 200 years. Artificial insemination (AI) is the process of placing sperm in a female animal in close proximity with female reproductive cells, called eggs or ova, by a method other than natural mating. Sperm consists of male reproductive cells. AI allows for use of a superior male animal to sire many times more offspring than would be possible naturally.

Embryo transfers have made it possible for superior female animals to produce far more offspring than would be possible otherwise. It is a process that removes fertilized eggs, or zygotes, from a female animal and places them in the reproductive tract of another female animal to be carried and nurtured until birth (**Figure 28-1**).

A geneticist is a scientist who studies genetics and heredity. Heredity is the passing on of traits or characteristics from parents to offspring. Genetic engineering makes it possible to increase resistance to diseases and improve production and efficiency of animals. Gene splicing, recombinant DNA, and biotechnology are common terms used by geneticists and reproductive technicians today.

This chapter explores some new technologies as well as the basics of animal breeding. Also notable are current directions of research in animal breeding.

Figure 28-1 Highly trained veterinarians and technicians backed up by sophisticated laboratory equipment have developed processes for producing more offspring from our best animals.
Courtesy of USDA.

Role of Breeding and Selection in Animal Improvement

English geneticist Robert Bakewell (1725–1795) s considered the father of animal husbandry. His work in the selection of Merino sheep for fine wool production and quality encouraged other farmers of his era to try to improve their livestock. Bakewell and others took care to mate the most desirable female animals with the best male animals, with the expectation that the offspring would be as good as or superior to their parents. Over the centuries, these practices have brought about major advances in animal agriscience (**Figure 28-2**).

By continually selecting animals for a specific type or characteristic, the resulting generations of animals tend to conform to the characteristics for which they were selected. For example, 200 years ago, cattle were not identified as dairy and beef types. Through careful selection of those animals with superior milk production and those types with excellent meat production, two distinct types of animals emerged from the same ancestors. Many distinct breeds of domestic animals have been developed in the same way. A breed is a population of animals having common ancestors and similar physical characteristics that are consistently passed on to their offspring. It should also be noted that selection is an extremely important part of animal agriscience today. This is especially true as consumer demands for animal products change.

Figure 28-2 Mating the most desirable female animals with the best male animals has been the traditional method used to produce modern livestock breeds.
Courtesy of USDA/ARS #K-2681-13.

Principles of Genetics

Gregor Mendel, an Austrian monk, (1822–1884) is generally credited with discovering the basic principles of genetics. Mendel, who excelled at physics and mathematics, did this through keen observation as he raised peas in his garden. These principles have become the foundation of modern genetics. They are summarized as follows:

- In every living organism, there is a set of paired genes in every cell that determines every trait in that individual. A **gene** is a unit of hereditary material located on a chromosome. A **chromosome** is the rod-like structure that functions as a carrier for genes.
- Individuals receive one gene or more genes for each trait from each parent.

- Genes are transmitted from parent to offspring as unchanging units.
- A **gamete** is a reproductive cell. In the production of gametes, gene pairs separate; only one gene from each pair is contained in each gamete.
- When different genes are present for a single genetic trait, in most instances, only one trait will be expressed. The trait that is expressed is known as the **dominant** trait, whereas the trait that is masked is called the **recessive** trait.

Science Connection — Genomes, Humans, and Cattle—a Surprising Connection

© Jorg Rose-Oberreich/Shutterstock.com

Launched in 1990 and completed in 2003, the Human Genome Project was an international research effort to determine the DNA sequence of the entire human genome. This is a history lesson worth reviewing. Before mapping a human's genome, the scientists sequenced the genomes of other organisms to perfect their techniques. They started with bacteria, which have much less DNA than a human. Researchers also worked on the cattle genome. Once the human genome was finished, comparisons could be made between organisms of different species. The results of one comparison involved the DNA of cattle and a human. This comparison showed that the two species share many of the same genes. They also found that four whole chromosomes shared identical genes.

Today, scientists can identify a gene in one organism and apply that understanding to another organism. In the relatively new field of agrigenomics, the application of genomics to agriculture, the value of the Human Genome Project has been immeasurable and has raised new possibilities. For example: Could the gene responsible for some cancers be identified and eliminated? Could the lactation gene be manipulated in such a way that a single cow could produce as much milk as 20 cows on less feed? Thanks to the historic Human Genome Project, today's agriscientists are working to achieve healthier and more productive animals.

The Cell and Its Parts

Cells are the basis of all genetic activity. A **cell** is a single unit of living material; it is the basic structure or building block of all living things. Inside the cell are distinct structures with specific functions. The cell contents are collectively called **protoplasm** (**Figure 28-3**). Cells are microscopic in size, requiring a microscope to see them.

All plant and animal life begins as a single cell. The **nucleus** of the cell contains pairs of chromosomes on which genes are located at specific locations. Each gene exerts control on a genetic trait. The gene for a specific trait is always located in the same place on the same chromosome pair in a given plant or animal species.

The structures within a cell are called **organelles**. The nucleus is typically the largest organelle in a cell, containing all the instructions needed to direct the cell's growth and activities. Cells in an organ or a tissue are separated from each other by **cell membranes**. These form protective barriers around the cell contents and control which substances enter or exit the cell. The **nuclear membrane** works similarly to guard the nucleus. The cytoplasm is a gel-like substance that gives structure to a cell and cushions the other organelles. Often, in an animal cell, **lipid** droplets can be seen through a microscope. These are stored fats, which will be used by the cell as needed.

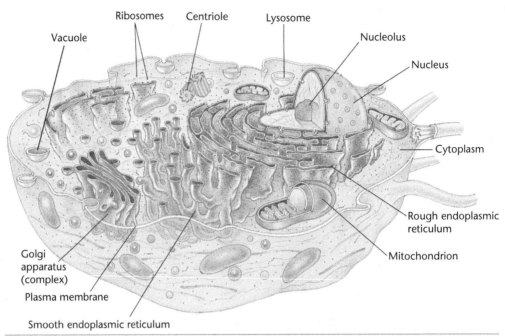

Figure 28-3 Each animal cell consists of distinct structures that carry out the functions of the cell.

Many processes take place within cells that are vital to an animal's very existence. For example, the **mitochondrion** is an organelle that plays a major part in converting an animal's food into usable energy. A **ribosome** functions as a kind of micro-machine for making proteins, which are continually needed by an organism. Some ribosomes float freely within the cell, some cluster together into groups, while others are attached to the endoplasmic reticulum. The **endoplasmic reticulum** stores proteins and facilitates their movement to other parts of the cell as needed. Similarly, the **Golgi apparatus**, also known as the Golgi body, stores and packages lipids and proteins from the endoplasmic reticulum and distributes them where needed. The last organelles discussed here are the **centrosomes**. These structures form microtubules that function like strings to pull cells apart during cell division.

Cell Division

Animal growth and reproduction take place through cell division. Simple cell division for growth is called **mitosis**. The process begins as each chromosome first divides into two parts. The wall of the nucleus disappears, and the chromosomes move to opposite sides of the cell. A new nucleus wall forms around each of the groups of chromosomes. The walls on opposite sides of the original cell then move toward each other until they divide the cell into two new cells complete with nuclei and pairs of chromosomes (**Figure 28-4**).

The cell division that occurs during reproduction resulting in the formation of gametes is called **meiosis** (**Figure 28-5**). It differs from mitosis primarily in that instead of the chromosomes dividing and moving in pairs to the opposite sides of the cell, they separate and move individually to the cell walls. When the new cells are formed, each cell contains only one of each chromosome pair rather than both chromosomes. In animals, meiosis occurs only in the reproductive organs. When an egg is fertilized, the sperm contributes one-half of each new chromosome pair; the egg contributes the other.

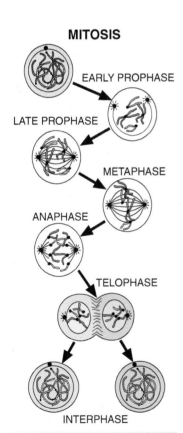

Figure 28-4 Mitosis is the process of division or duplication of a typical cell.

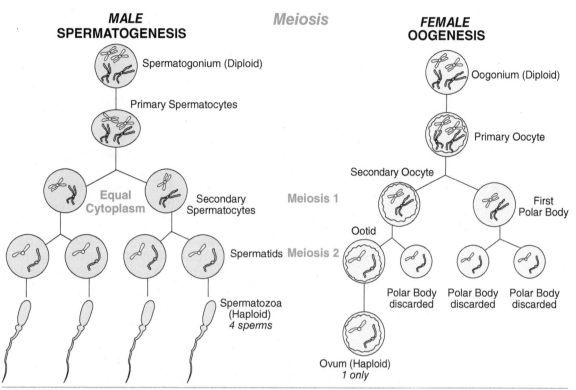

Figure 28-5 Meiosis is the division or duplication of egg and sperm cells.

Genes

Genes are the units of genetic material that are responsible for all of the traits or characteristics that an animal inherits from its parents. Genes occur at specific locations on structures called chromosomes. Chromosomes control the production of specific enzymes and proteins that influence the physical traits in animals. The chromosomes are themselves composed of a protein covering surrounding two chains of **deoxyribonucleic acid (DNA)**. This structure provides the coding mechanism for heredity.

In pairs of genes on matching chromosomes, the genes may be either alike or different. Pairs of genes that are alike are said to be **homozygous**, whereas those pairs that are different are called **heterozygous**. When the two genes in a pair are different, one gene usually expresses itself, and the expression of the other remains hidden. The gene that expresses itself is referred to as dominant. The gene that remains hidden and expresses itself only in the absence of a dominant gene is called recessive. Sometimes, neither gene of a pair expresses itself to the exclusion of the other. When this happens, the gene pair is referred to as expressing partial dominance, or **incomplete dominance**. The actual configuration of genes in an animal is called the **genotype**, whereas the physical appearance of the animal is referred to as **phenotype**. All of this is important when exploring the basics of genetics and the use of genetics in animal breeding.

Some traits are controlled by genes that are located on the chromosomes that determine the sex of the animal. These are called **sex-linked** traits. The chromosomes that determine an animal's gender are a pair: one chromosome from the mother and one from the father. The chromosome from the female that determines gender is shaped like an X. The gender-determining chromosome from the male is either an X chromosome (resulting in female offspring) or a Y chromosome (resulting in male offspring). For example, when a male passes on a Y-shaped chromosome, the offspring will be male. If the male passes on an X-shaped chromosome, the offspring will be female. Other genes are also found on these X and Y chromosomes. When a mutation occurs on one of these gender-determining chromosomes, it results in a sex-linked trait. Both males and females can express this type of mutation, but the majority of sex-linked traits occur in male offspring. Color blindness is an example of a sex-linked trait.

Genes normally duplicate themselves accurately. However, sometimes accidents or changes occur. These genetic accidents or changes in genes are called **mutations**. Sometimes these mutations result in desirable changes in animals. One such example is the polled characteristic in breeds of cattle that are normally horned. Polled means naturally or genetically hornless. In other cases, the mutation results in a lethal characteristic, which causes an animal to be born dead or to die shortly after birth.

Once an egg has been fertilized, cell divisions occur and the embryo begins to grow. Mitotic cell division takes over as cells divide, forming the organs and tissues and increasing the size of the developing fetus. Occasionally, a fertilized egg divides into two identical eggs, and identical twins result (**Figure 28-6**). It has become common practice with superior cattle to mechanically divide the growing embryo during the stage of growth when it consists of 16–32 cells. Each cell mass develops as before, and two or more embryos develop, having identical genetic make-up. They are clones, manipulated through technology to result in multiple births.

Figure 28-6 These two polled calves are identical twins that resulted when a fertilized egg was divided into two identical cells or masses of cells in the early stages of embryonic development.
Courtesy of USDA/ARS #K-4323-18.

Genetics in the Improvement of Animals

The improvement of animals through genetics can be either natural or planned. In natural selection, the "survival of the fittest" occurs. In other words, as changes in genes occur naturally, those animals experiencing gene changes that make them better adapted to their environment are most likely to survive. Popular examples include protective colorations, ability to digest certain feeds, and ability to survive in extreme heat or cold.

In planned or artificial selection, people decide which traits they want in animals. They then use the animals with the desirable traits in the breeding program. Over several generations, the offspring that result from such selection practices show more and more of the desired traits.

Heritability is the capacity of a trait to be passed down from a parent to offspring. Many of the traits for which people are selecting animals are the result of a combination of more than a single pair of genes. Because of this, few traits are 100 percent heritable from parents. For example, the extent of heritability for loin-eye size in pigs is 50 percent. A boar with a 9-inch loin-eye is crossed with a sow that has a 7-inch loin-eye. The expected average loin-eye size for the resulting offspring would be 8 inches if loin-eye size were 100 percent heritable. However, because loin-eye size is only 50 percent heritable, the offspring can only be expected to have 7.5-inch loin-eyes. The heritability rates of other traits can be found in **Figure 28-7**. These rates should be used as a guide when attempting to improve animal traits through genetic selection.

Environmental factors often play a part in the expression of genetic traits, masking to some extent the true potential of the animal. For example, an animal that is improperly fed or cared for may never reach the size or weight that its genetic potential would permit.

Trait	Estimated Percent Heritability					
	Cattle	Sheep	Swine	Poultry	Rabbits	Horses
Fertility	0–10	0–15	0–15	0–15	–	Low
Number of young weaned	10–15	10–15	10–15	–	3	–
Weight of young at weaning	15–30	15–20	15–20	–	35	–
Postweaning rate of gain	50–55	50–60	25–30	–	60	–
Postweaning gain efficiency	40–57	20–30	30–35	–	–	–
Fat thickness over loin	40–50	–	40–50	–	–	–
Loin-eye area	50–70	–	45–50	–	60	–
Percent lean cuts	40–50	–	30–40	–	60	–
Milk production (lb.)	25–30	–	–	–	–	–
Milk fat (lb.)	25–30	–	–	–	–	–
Milk solids, nonfat (lb.)	30–35	–	–	–	–	–
Total milk solids (lb.)	30–35	–	–	–	–	–
Body weight	–	–	–	35–45	40	–
Feed efficiency	–	–	–	20–25	–	–
Total egg production	–	–	–	20–30	–	–
Age at sexual maturity	–	–	–	30–40	–	–
Viability	–	–	–	5–10	–	–
Speed	–	–	–	–	–	25–50
Wither height	–	–	–	–	–	25–60
Body length	–	–	–	–	–	25
Heart girth circumference	–	–	–	–	–	34
Cannon bone circumference	–	–	–	–	–	19
Points for movement	–	–	–	–	–	40
Temperament	–	–	–	–	–	25

Figure 28-7 The rates of heritability for certain traits of domestic animals.

Genetic Engineering

Genetic engineering is the process of transferring genes from one individual to another by inserting them in the chromosomes of the new individual or organism. Geneticists have been able to link specific genes to specific traits. They have also developed procedures for removing the genes from the cells of one animal and inserting them into the cells of another animal. Genetic engineering has much potential for improving animals for human purposes.

The potential for change in animals using genetic engineering is tremendous. For example, if an animal species is genetically resistant to a certain disease, genes that make that animal resistant may be inserted into cells of an animal species that is not resistant. Because genes are passed on to offspring from parents, resulting generations of animals are also resistant to that disease.

Geneticists are also exploring how genetic engineering could achieve breakthroughs in such areas as disease resistance, cancer research, vaccines, increased growth and production, and immunology.

Reproductive Systems of Animals

Male Reproductive System

The male reproductive system functions to produce, store, and deposit sperm cells. Secondary functions include production of male sex hormones and elimination of urine from the body (**Figure 28-8**).

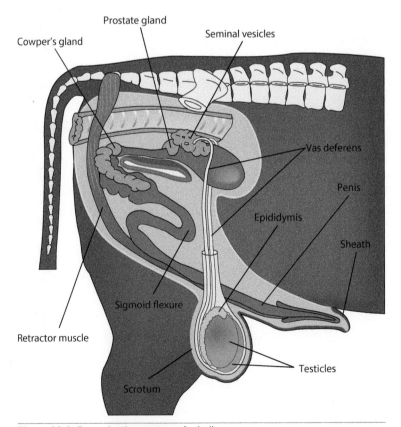

Figure 28-8 Reproductive system of a bull.

The actual structural makeup of the male reproductive system varies widely with different species of animals. The testes are the organs that produce sperm cells. They also produce **testosterone**, the male sex hormone. The testes are attached to the body by the spermatic cord and are protected by the scrotum. A coiled tube called the epididymis stores and transports the sperm cells. The vas deferens carries the sperm cells to the urethra, which extends through the penis. Other glands of the male reproductive system include the seminal vesicles and the prostate and Cowper's glands. Seminal vesicles secrete seminal fluid.

The prostate gland provides nutrition to the sperm cells, and the Cowper's gland prepares the urethra for the passage of the sperm cells. The penis serves to deposit the sperm cells into the female reproductive tract.

Female Reproductive System

The female reproductive system produces the ova, or egg, together with the female sex hormones estrogen and progesterone (**Figure 28-9**). **Estrogen** regulates the heat period, **estrus**, whereas **progesterone** prevents estrus during pregnancy and causes development of the mammary system. The mammary gland produces milk, the food of young mammals.

The parts of the female reproductive system include two ovaries, which produce ova, also referred to as eggs. The infundibulum is a funnel-shaped structure that catches the eggs during **ovulation** (the process of releasing a mature egg from the ovaries). The egg then passes to the fallopian tube, which is also called the oviduct. A fallopian tube is associated with each ovary, and they are the sites where fertilization of the eggs takes place. The fertilized egg, or embryo, then moves to the uterine horn, where it attaches to the wall of the uterus and remains until birth (also known as parturition). When

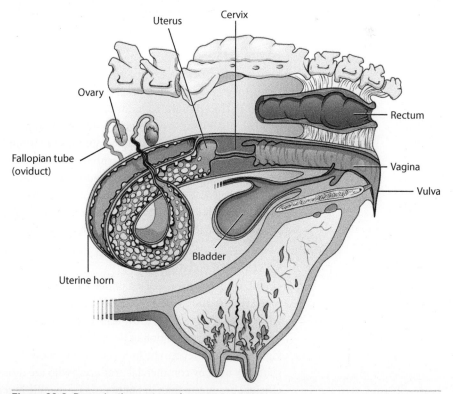

Figure 28-9 Reproductive system of a cow.

the embryo attaches itself to the uterine wall, it becomes known as a *fetus*. The time period between fertilization of the egg by a sperm cell and birth is called the *gestation* period. The vagina, which is separated from the uterine horn by the cervix, serves as the passageway for the sperm cells. The external opening, or vulva, protects the rest of the female reproductive system from outside infection.

Reproductive Problems

Sterility is the inability of an animal to reproduce. A number of conditions may result in sterility or reproductive failure. Some of these problems are physical and others are genetic. Examples of genetic reproductive problems include sterile female cattle (also some female sheep) that are born as a twin or triplet with male offspring. These female animals are often exposed to male hormones before birth that interfere with the development of their female organs. A female animal that exhibits this problem is called a freemartin. Another problem of a genetic nature includes the scrotal hernia. This is a muscle tear that is usually caused by a hereditary weakness in the muscle. Undeveloped or missing ovaries are often thought to be the result of genetic or developmental accidents.

Infections and diseases are important physical causes of sterility. Physical damage to the reproductive system and nutritional deficiencies also are known to contribute to reproductive failures.

Systems of Breeding

Several breeding systems are important to animal science. Which system or systems to use depends on several factors. Some of the considerations include type of operation, available markets, available resources, climatic conditions, size of operation, goals of the breeder, and personal preference.

Commonly recognized systems of breeding include pure-breeding, crossbreeding, grading up, inbreeding, and outcrossing (**Figure 28-10**).

Pure breeding

Pure breeding occurs when a purebred animal is bred to another purebred animal of the same breed. A purebred animal is one of a recognized breed and one whose ancestors can all be traced to the foundation stock of the breed. A registered animal is a purebred that can qualify for breed registry. Registration papers are records of ancestry. Although there are no guarantees, purebred animals are usually considered to be superior in some ways to animals that are not purebreds. They are sometimes used as show animals and are important parts of the crossbreeding and grading-up systems of breeding.

People who elect to use the pure breeding system need to have ample resources and a good knowledge of genetics. They should also be effective salespersons in order to market purebred animals with superior qualities at premium prices.

Crossbreeding

Crossbreeding is the breeding of one recognized breed of animal to another recognized breed. The resulting offspring are called hybrids. Hybrid animals have a number of advantages. They tend to grow faster, be stronger, and be capable of greater production as a result of the combination of desirable traits from the two breeds. This is called hybrid vigor. They also tend to be more fertile and more disease resistant. Crossbreeding is generally used by commercial producers who are more interested in offspring that are efficient producers than in maintaining a specific breed of animals.

	Relationship of Mates	Advantages	Disadvantages
Pure-breeding	Unrelated	1. Concentration of selected traits 2. Breed assn. to create demand	1. May result in less-desirable traits 2. Loss of hybrid vigor
Crossbreeding	Unrelated	1. Increased growth 2. Increased prod. 3. Increased hybrid vigor 4. Higher fertility 5. Disease resistance	1. Less uniformity of offspring 2. Not eligible for registry
Grading up	Unrelated	1. Herd improvement w/o purchasing purebreds 2. Develop uniformity	Slow process of improvement
Closebreeding	Sire-daughter Son-dam Brother-sister	Concentrates desirable traits	1. Concentrates undesirable traits 2. Expression of abnormal traits
Linebreeding	Not closer than half-brother to half-sister	Concentrates desirable traits of one individual	1. Concentrates desirable traits 2. May result in expression of abnormal traits
Outcrossing	Unrelated	Produces hybrid vigor within a breed	

Figure 28-10 Systems of breeding for animal improvement.

Grading Up

Livestock producers who are raising animals of mixed breeding sometimes use the grading-up breeding system to improve their herds. When a non-purebred female animal, called a *grade*, is mated with a purebred male animal, the process is called *grading up*. The idea is that the purebred male animal should be superior to the grade female animal and the resulting offspring should be superior to their mother. Succeeding generations of female animals are also mated to superior-quality purebred males of the same breed. Over several generations, the breeding herd takes on the best qualities of the breed from which the sires were selected. The purposes of grading up include improvement of quality and production in the progeny or offspring. The development of uniformity in the herd is also a reason for grading up a breeding herd.

Inbreeding

Inbreeding is the mating of animals that are genetically related. The purpose of inbreeding is to intensify the desirable characteristics of a particular animal or family of animals. Unfortunately, inbreeding also intensifies the undesirable and abnormal characteristics.

There are various degrees of inbreeding, based on how closely related the individual animals are. When a father animal, or sire, is mated with its daughter, a son is mated with its mother, or dam, or a brother is mated to its sister, the term *close breeding* is used. Linebreeding is the mating of less closely related individuals that can be traced back to one common animal ancestor within a few generations. Normally, the most closely related cross made in linebreeding is half-brother to half-sister.

Inbreeding must be used carefully because inbred animals tend to exhibit more undesirable characteristics than animals produced by other systems. Unless a breeder is willing to be highly selective among the progeny resulting from inbreeding, this system should be avoided. A few highly superior individuals may come at the cost of discarding even greater numbers of inferior animals that are the result of concentrating undesirable traits.

Outcrossing

Outcrossing is the mating of unrelated animal families within the same breed. This is probably the most popular system of breeding used in purebred breeding herds. It also has many of the advantages of crossbreeding, including increased production and improved type.

Methods of Breeding

Three general methods are used for breeding animals: natural service, AI, and in vitro. Natural service occurs when the male animal is allowed to mate directly with the female animal. Several factors impact the selection of the breeding method to be used. These factors include the type of operation the breeder currently runs or wants to run, the amount of labor available, the number of animals in the herd, the location of the herd during the breeding season, requirements of breed associations, and personal preference.

Pasture mating is a system of natural service in which the male animal is allowed to roam freely with the female animals in the herd. The male animal detects the heat period of each female in the herd and mates with her at the appropriate time. There are some disadvantages to this system. One male can mate with only a limited number of females. Also, if more than one female is in heat at the same time, all may not get bred. In addition, breeding records are difficult to keep. When this system is used, the possibility exists that a sterile male may not be detected in time, and the females may not be bred at all. Commercial beef ranchers usually modify this system by placing several bulls with a large herd of cows to make sure each of the cows has access to a fertile and healthy bull.

Hand mating is accomplished by bringing the female animal to the male animal for mating. More labor and better management are required because someone must determine when the female is in heat and get the female to the male animal so mating can take place. Advantages of hand mating include being able to keep more accurate breeding records and being able to mate more females to a single male.

Mating animals through AI has numerous advantages. There are a few disadvantages as well. AI involves collecting semen from a male animal and placing it in the reproductive tract of the female. **Semen** is composed of the sperm cells and accompanying fluids. The technique of AI has been responsible for tremendous increases in animal productivity in recent years.

Advantages of AI are that semen, collected from the male, can be used fresh or stored frozen in liquid nitrogen for later use. One ejaculate, or the amount of semen produced at one time by a male animal, can be diluted and used to breed many females. Because semen can be frozen and stored for long periods, the use of an outstanding male can be greatly extended. AI also greatly reduces the need to keep male animals and reduces the danger of having dangerous male animals around. There is less chance of injury to people and their animals, reduced spread of reproductive diseases, and improved recordkeeping.

There are some disadvantages to AI as a method of mating animals. A person trained in AI must be available when the animal is in heat. Semen collection from the male animal can be a dangerous activity requiring the services of trained technicians, special equipment and facilities, and excellent management. Finally, genetic defects of the male must be identified early on to prevent defects from spreading faster and more broadly.

In vitro mating occurs outside the animal's body. Mature eggs are flushed from the female animal and fertilized by sperm cells collected from a male. The fertilized eggs are then placed in host females for development into offspring. Although this process is very exacting in its requirements and facilities, at times this is the only way that a viable fetus can be obtained. This means of mating is often used as part of the new technology of genetic engineering.

Hot Topics in Agriscience: Embryo Cloning from Outstanding Female Animals

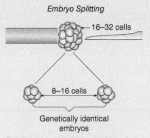

Embryo splitting makes it possible to obtain numerous offspring from a single cow.

Outstanding cows are routinely given hormone treatments to increase the number of ova they produce. This allows them to produce more than one calf per year when "extra" embryos are placed in other cows of lesser quality. In addition, the number of embryos can be expanded by splitting the cells that make up each embryo mass.

In this instance, each of the cell masses obtained from the procedure is capable of producing identical individuals, or clones. An embryo that is in the 16- to 32-cell stage of development can be divided into as many as four to eight different clones.

When these cloned embryos mature in the uterus of a cow that is not the genetic mother, they retain the genetic makeup that they inherited from the cow and the sire that produced the embryos. Each of these calves may be nourished by its surrogate mother, but it is the genetic offspring of the cow from which the embryos were collected. In this manner, numerous offspring can be obtained from a single outstanding cow.

Selection of Animals

There are several methods by which animals may be selected for a breeding herd. These methods fall into two major types: selection based on physical appearance and selection based on performance or production of either the individual or its progeny.

Selection based on physical appearance is generally used when choosing purebred animals. Often, the sole criterion for selection of an animal is how well he or she performs in the show ring. This is an acceptable means of selection when animal breeders are raising animals for show and to sell to others for the same purpose. However, this method often leaves much to be desired when the animal so chosen is expected to produce a product or perform a desired activity. Unfortunately, animals fitted to perform or look their best in the show ring often fall short of expectations in the milking parlor or in other performance settings.

The use of comparative judging helps individuals develop skills in evaluating animal appearance. In livestock-judging events, one animal is compared with another and frequently judged in groups of four (**Figure 28-11**).

Figure 28-11 Comparative judging is a good technique for studying the details of the animal, such as feet, legs, udder, head, body, and overall appearance.
© Stanislaw Tokarski/Shutterstock.com.

Selection based on production or performance is usually a more reliable means of choosing animals. If a breeder is selecting dairy cows to produce milk, it makes more sense to select cows with high-production records and high-producing relatives. In meat animals, progeny testing, or the testing of performance of the offspring, is often the only way to predict the breeding value of the parents. Other measures of production on which various animals may be selected include rate of gain, feed efficiency, butterfat production, back-fat thickness, loin-eye area, yearly egg production, and pounds of wool produced.

Sometimes, selection of an animal is based on its pedigree, a record of an animal's ancestry, and is included on registration papers for purebred animals (**Figure 28-12**). Although the consideration of pedigree can be important in the selection process, it should always be used in combination with other methods of selecting animals.

In summary, tremendous gains have occurred in the productivity of domestic animals in the last 100 years. Fewer and fewer animals are providing for the needs of an ever-growing human population. Genetics and gains in animal breeding have been responsible for much of the increase in productivity. This trend is expected to continue as technology in the field of animal agriscience continues to advance.

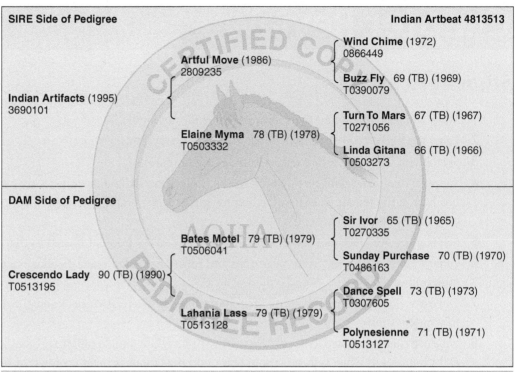

Figure 28-12 Pedigree and registration papers are valuable documents to those who buy or sell purebred animals because they document the purity of the bloodlines.

Agri-Profile Career Area: Geneticist/Genetic Engineer/Technician

A technician removes frozen semen from liquid nitrogen for artificial insemination of a farm animal.
Courtesy of USDA.

The Bureau of Labor Statistics expects the employment rate of geneticists to grow 27 percent by 2028. This is faster than the average growth rate for nearly all other U.S. occupations and is linked to rapid advances in all branches of genetics.

Genetic engineers work with the structure of DNA in animals, plants, and humans to modify organisms. Their work helps improve plant and animal production, fight diseases, improve manufacturing processes, clean up environmental disasters, and more. For example, animal geneticists develop selective breeding programs, conduct genetic research and tests, improve heritability of desirable traits, and report genetic trends. Reproductive technicians, employed mostly in livestock facilities, provide breeding services, such as managing artificial insemination, pregnancy checks, and sanitation. Lab and research technicians are needed in all branches of genetics.

Geneticists find work in livestock facilities, research labs, zoos, private companies, and government. Salaries range from $36,000 to $120,000, based on education, experience, and employer. Education requirements include an undergraduate degree in genetics and/or a science in one's specialty. For example, an animal geneticist may earn advanced degrees in animal science or dairy science.

Chapter Review

Student Activities

1. Write the **Key Terms** and their meanings in your notebook.
2. Label the parts of the male and female reproductive systems.
3. Sketch and label the various stages in mitosis and meiosis.
4. Suppose that a Hereford bull that has horns is mated to six female animals that were born without horns (polled). If the female animals were homozygous for the polled characteristic, how many would probably bear calves having no horns? Why?
5. Select a species of animal and determine the origin of the popular breeds in that species.
6. Clip pictures of animals from magazines and newspapers and make a collage of popular breeds for the bulletin board.
7. Develop a bulletin board showing the popular breeds of sheep, cattle, horses, swine, dogs, rabbits, and cats in your community.
8. Suppose that you mated a black male rabbit with a white female rabbit and that the female gave birth to eight bunnies. If white is dominant to black and the female is heterozygous for hair color, how many of the bunnies are likely to be black?
9. If a red pig with floppy ears was crossed with a white pig with erect ears and all possible characteristics were homozygous, what is the probability that the offspring will be red with erect ears? In pigs, white hair color and erect ears are dominant.
10. Dissect the reproductive organs of a female animal and identify the major parts.
11. **SAE Connection:** Visit a farm or other animal facility and observe AI or other reproductive techniques being conducted.

Inquiry Activity | Animal Husbandry

After watching the *Animal Husbandry* video available on MindTap, create graphic organizers to process the information presented on artificial insemination and embryo transfer.

- Using a T-chart, list the benefits of artificial insemination and the benefits of embryo transfer.
- Using a Venn diagram, describe how the processes of artificial insemination and embryo transfer are similar and how these processes are different.

National FFA Connection: Animal Genetics and Livestock Evaluation

Genetic improvement of farm animals has captured the interest of today's generation of FFA members.
Courtesy of the USDA.

Thanks to events and resources offered by the National FFA Organization, an increasing number of students are taking their interest in genetics beyond the classroom. For example, the Livestock Evaluation Career Development Event has proved to be a good starting point for many. During this event, students rank breeding and marketing classes of livestock by making careful observations and identifying desirable traits. Here's how two FFA members applied what they learned about livestock evaluation and genetics to launch their careers:

- Bailey, who was an FFA member from Montana when she began competing in livestock judging, become so fascinated by the science that she changed her college focus from medicine to livestock genetics. After earning advanced degrees, she traveled to Australia to conduct research as a bovine geneticist. Her work is aimed at improving cattle's disease resistance.
- Livestock evaluation skills and a knowledge of genetics benefited Jacob, too. As an FFA member in Nebraska, Jacob used what he has learned to build his cattle herd when he bid for a bred cow at a live auction. To make his selection, he read the genetic information provided for each animal and compared these predictors of desirable traits with his own observations. An SAE grant from his state FFA Foundation helped him bid for an animal with the traits of most interest to him.

To learn more about the Livestock Evaluation Career Development Event, visit the FFA website. Search the term "animal genetics" to find additional resources and interviews.

Check Your Progress

Can you…
- define terms associated with genetics and reproduction?
- identify systems of breeding?
- identify parts of reproductive systems?
- discuss new technologies in animal reproduction?
- discuss the evaluation of animals for type and production?

If you answered "no" to any of these questions, review the chapter for those topics.

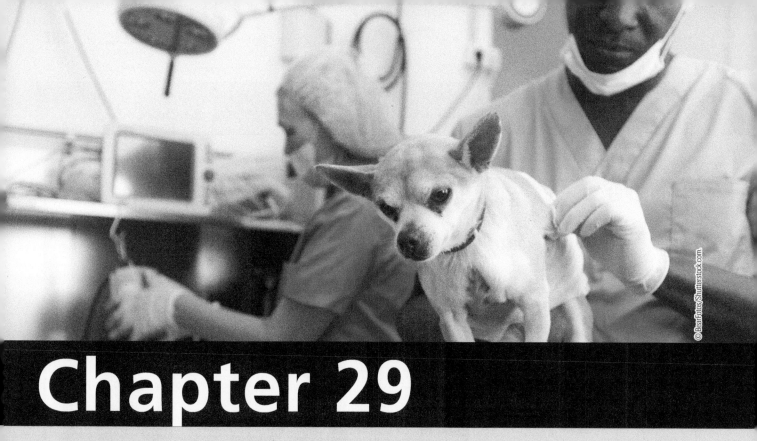

Chapter 29
Small Animal Care and Management

Objective
To determine the types, uses, care, and management of small animals.

Competencies to Be Developed
After studying this chapter, you should be able to:
- describe the domestication and history of small animals.
- determine the economic importance of various classes of small animals.
- list the types and uses of the various classes of small animals.
- describe the approved practices in feeding and caring for small animals.

Key Terms

domestic	hen	angora
jungle fowl	tom	buck
waterfowl	poult	doe
broiler	drake	apiculture
layer	duckling	apiary
chick	gander	queen
cockerel	goose	worker
rooster	gosling	drone
pullet	squab	

As our world population continues to expand and less space remains for humans and large animals to coexist, the benefits of small animals are clear. They are more efficient than large animals in converting the feed they eat into food and other products that humans use. They are less intrusive on the lives of people, more of whom today keep animals as pets (**Figure 29-1**). There are many important species of small animals. This chapter explores the topics of pet, poultry, rabbits, and honeybees and how these small animals contribute to our daily lives.

Figure 29-1 Pet ownership and the animal care industry have grown extensively in the last two decades.
Courtesy of Sandy Clark.

Poultry

History and Domestication

A **domestic** animal is one that has been tamed and cared for by humans who use such animals for labor and food. The domestication of chickens occurred about 4000 BCE in Southeast Asia. The **jungle fowl** is an ancestor of our modern chickens. Its association with humans benefited both the jungle fowl and its human neighbors. Humans made small clearings in the jungle that attracted insects and other food for the jungle fowl, which in turn provided eggs and meat for humans. Over the centuries, this association led to the domesticated chicken of today. When early settlers came to the Americas, they brought chickens, too. For example, people at the Jamestown settlement kept pens of chickens.

Turkeys are the only domesticated animals of agricultural importance to have originated in the Americas. European explorers observed native Central Americans raising domesticated turkeys as food for themselves and their animals. Although modern domestic turkeys are direct descendants of the wild turkeys of the Americas, their physical makeup has changed to such a degree that they are totally dependent on humans and cannot survive in the wild.

Poultry includes all domesticated birds (**Figure 29-2**). The various types and breeds of ducks and geese have originated from places all over the world. Ducks and geese are also known as **waterfowl**.

Figure 29-2 Poultry include a wide range of domestic fowl, including chickens, ducks, geese, and turkeys.
© Eric Isselee/Shutterstock.com.

Science Connection: Taxonomy of Farm Poultry

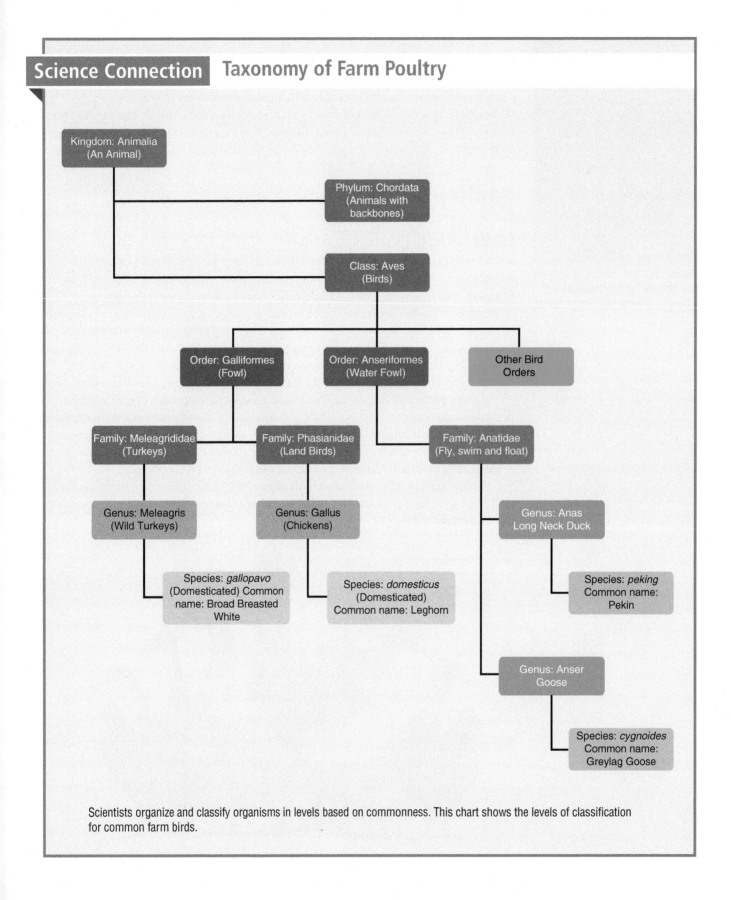

Scientists organize and classify organisms in levels based on commonness. This chart shows the levels of classification for common farm birds.

Most poultry farms are large, with thousands of birds in production. This is because of the greater potential to earn a profit when fixed costs such as buildings are divided among more animals. Despite this, it is still attractive to many people to keep a few chickens or other poultry for their own use. Chickens, ducks, geese, and turkeys can be raised in small numbers in nearly any environment, including backyards and small acreage properties. They tend to control insects, such as grasshoppers. In outdoor settings, they are also capable of finding some of their own food. Care must be taken, however, to protect them from predators such as dogs and skunks.

Economic Importance

According to the United States Department of Agriculture (USDA), the consumption of red meat decreased by 40 lbs. per person between 1970 and 2009. Negative publicity regarding fat and cholesterol influenced the declining consumption. In contrast, while average chicken consumption has more than doubled since 1970, beef consumption has fallen by more than a third. By 2014, Americans were consuming an average of 47.9 pounds of chicken a year, compared to 39.4 pounds of beef. Today, some of the largest farms in the United States are poultry operations.

Broilers are young chickens grown for meat. In 2021, the United States produced 59.2 billion lbs. of broilers. The top producers by state were Georgia, Arkansas, Alabama, Mississippi, North Carolina, and Texas. That same year, Americans consumed 96 lbs. of broiler chickens per person. People in the United States also eat about 288 eggs per person each year. Iowa, Ohio, and Indiana are the leading producers of eggs in the United States, followed by Pennsylvania and Texas.

As a result, the poultry industry will continue to be an important part of the U.S. agricultural industry.

Types and Uses of Poultry

The types of poultry can be divided into the following general groups: chickens, turkeys, ducks, geese, and captive game birds.

Chickens are usually classified as either layers or broilers, depending on their intended use. A **layer** is a chicken that is developed to produce large numbers of eggs (**Figure 29-3**). They may produce either white or brown eggs, depending on the breed. Laying chickens are also maintained to produce eggs to be hatched for the production of broiler chicks. A **chick** is a baby chicken.

Chickens produced for meat are usually classified according to age. Broilers are young-meat chickens usually not more than 8 weeks old (**Figure 29-4**). A roaster is a mature chicken used for meat. Cornish and White Rock crosses are popular breeds of chickens raised for meat.

Figure 29-3 Laying hens are bred to convert feed into eggs rather than body flesh.

Courtesy of Bill Muir, Purdue University.

Figure 29-4 Broilers are produced in large numbers, especially in the southern states.

Courtesy of USDA.

A young male chicken is called a cockerel, whereas an adult male is called a cock or rooster. These terms also apply to male pheasants.

A young female chicken is called a pullet, and an adult female chicken is called a hen. Adult female turkeys, ducks, and pheasants are also called hens.

Other classes of chicken include the bantam, or miniature chicken, and ornamental chickens, which are valued as hobby chickens.

There are more than 200 recognized breeds of chickens in the United States. However, nearly all of the layer and broiler types resulted from crossbreeding to maximize production. The foundation breed of most laying-type chickens is the White Leghorn. Most broilers and roasters can trace their ancestors back to Cornish or White Rock chickens.

Turkeys are of a single breed but of eight varieties, including Beltsville Small White, Black, Bourbon Red, Bronze, Narragansett, Royal Palm, Slate, and White Holland. Of these, only the White Holland, which includes the Broad-Breasted White (**Figure 29-5**), and the Bronze, which includes the Broad-Breasted Bronze, are used extensively for commercial production. The broad-breasted birds have been intensively bred for superior meat qualities. They are produced by artificial breeding because they are no longer capable of breeding naturally. However, these two varieties account for nearly all of the turkey meat that is produced in the United States each year. A male turkey is called a tom, a female turkey is called a hen, and a young turkey is called a poult.

Figure 29-5 The most popular commercial turkey in the United States is the Broad-Breasted White.
Courtesy of USDA.

The production of turkeys is spread over a wide area < , > and the top producing states are Minnesota, North Carolina, Arkansas, Indiana, Missouri, Virginia, Iowa, and California. In 2021, consumption of turkey meat was 15.5 lbs. per person, and approximately 224 million turkeys were produced. Of the more than 30 million ducks raised for meat in the United States each year, 53 percent (16 million) come from Long Island, New York. New York, Missouri, Iowa, South Dakota, and Minnesota are also major producers of geese. Ducks can be classified as meat producers or egg producers. The primary meat breed is the Pekin (**Figure 29-6**). Ducks reach a market weight of about 7 lbs. in 8 weeks. This makes them faster growing than broilers, which reach 4 lbs. in the same period. Other breeds of duck used for meat production are Aylesbury, Muscovy, Rouen, and Call.

Figure 29-6 Pekin ducks are the most popular breed of duck used for meat.

Courtesy of Jurgielewicz Duck Farm.

Figure 29-7 Khaki Campbell ducks are the most popular domesticated breed in the U.S.

Courtesy of John Metzer, Metzer Farms.

Egg-laying ducks are generally either Khaki Campbell or Indian Runner. The Khaki Campbell is the champion egg layer of the bird world, often averaging more than 350 eggs per year (**Figure 29-7**). This compares to an average of about 250 eggs laid per year for laying chickens. A male duck is called a **drake**, and a young duck is a **duckling**.

Geese are raised primarily for meat. There is also a limited market for geese used for weeding certain crops. The Chinese breed is popular for this use. Other breeds of goose are Toulouse, Emden, Pilgrim, and African. A male goose is called a **gander**, a female is a **goose**, and a young goose is a **gosling**.

Captive game birds include pheasants, quail, chukar partridge, and pigeons (**Figure 29-8**). The uses of game birds include meat and eggs. Some game birds are also raised to release to the wild or on game preserves for hunting.

Pigeons are often raised for sport (racing) or as a hobby, but some breeds are raised for meat purposes. Many people maintain pigeon lofts for the sheer pleasure of watching these birds. They require little space, and they can be raised almost anywhere. The adult male pigeon is a cock; an adult female is a hen. A young pigeon that has not left the nest is called a **squab**. When a pigeon is used for meat, it is processed before it learns to fly, and the meat is called *squab*. It is usually considered a *delicacy* meat served in restaurants.

A   B

FIGURE 29-8 Two types of game birds are pictured here: (A) Chinese Ring-necked Pheasant and (B) Bobwhite Quail.

Courtesy of U.S. Fish and Wildlife Service. © iStockphoto/Robert Blanchard.

Science Connection: How Food Fuels Energy

Which animal requires more energy?
© Alexander Jache/Shutterstock.com. © Makarova Viktoria (Vikarus)/Shutterstock.com.

Hunger makes humans and other animals feel tired and weak. This is one of the brain's ways of letting the body know it needs food. Food provides the basic building blocks for the body's energy supply. Inside cells, the mitochondria (the cell's power plant) works continuously to produce energy. The process whereby food is transformed into chemical energy is known as the Kreb cycle. Before the Kreb cycle can begin, another process breaks a sugar molecule called glucose in half, creating two energy molecules called adenosine triphosphate (ATP). ATP is the molecule that cells use for energy. Next, the two sugars are sent into the Kreb cycle, where they go through several chemical reactions that result in 34 more ATPs.

Animals require a lot of energy. From the beating of a heart to running, every function of the body takes energy. Which animal do you think needs more energy, a horse or a rabbit? The answer may surprise you: It is the rabbit. Small animals need more energy per ounce than larger animals. In general terms, the bigger an animal is, the less energy it needs. Eighty to 90 percent of the energy of the body is spent regulating the body's temperature. Small animals lose heat much faster than large animals, so small animals use more energy or ATP to maintain their bodies at the right temperature.

Some Approved Practices for Poultry Production

Up-to-date information on poultry and other small animals is available from suppliers of breeding stock, equipment, medicine, and building materials. Recommended or approved practices are available from your state university. Approved practices for the production of poultry include the following:

- Purchase young poultry with a specific use in mind.
- Purchase young poultry or eggs for hatching only from reputable hatcheries or breeders. A hatchery is a business that hatches young poultry from eggs.
- Purchase chicks, pullets, or poults that are immunized and free from disease.
- Purchase young poultry at the proper time for meeting a target market date. Broilers should be 7 to 8 weeks old before marketing them; ducks, 7 to 8 weeks; turkeys, 12 to 14 weeks; and geese, 12 to 14 weeks old. Layer chicks should be purchased 20 to 22 weeks before you expect them to produce eggs.
- Ensure that proper housing is available for the type and number of poultry you are planning to raise. Housing considerations include size, ventilation, ease of cleaning, lighting, heating and cooling, feed storage, and maintenance requirements.
- Secure and maintain the proper equipment for the type of poultry operation planned. Consider feeder and waterer space and brooder size.
- Feed a balanced ration designed for the type of poultry being grown.
- Develop and implement a plan to control external parasites.
- Plan and follow a flock health program.
- Plan for marketing at the optimum time.
- Properly clean and disinfect facilities before introducing a new flock of poultry.

Rabbits

History and Domestication

Much of the early history of the rabbit is obscure. It is believed that the Phoenicians brought rabbits to Spain about 1100 BCE. They are credited with having introduced rabbits to what was then the known world.

Romans kept rabbits in special enclosures. Roman women were known to have eaten large quantities of rabbit meat. They believed it enhanced their beauty.

Early monasteries produced large amounts of rabbit meat and fur. These religious institutions are said to have domesticated the rabbit. Great care was taken to produce high-quality rabbits and there was much trading of rabbits between monasteries.

Rabbit meat has long been an important component of the diets of people in densely populated countries of Europe. Rabbits are efficient converters of feed to meat. They take up relatively little space and reproduce rapidly.

Rabbits have been raised in the United States since the time of the early settlers, but serious rabbit production did not begin until the turn of the twentieth century. An intense advertising campaign was conducted for Belgian hare at that time to promote commercial production of rabbits. Rabbit production also got a boost in the United States during the two world wars. At a time when shortages and rationing of food products occurred, rabbits became an inexpensive source of lean, red meat.

Economic Importance

Rabbit production is an important agricultural enterprise in the United States. Each year, 6 to 8 million rabbits are raised. The U.S. population consumes 25 to 30 million lbs. of rabbit meat each year. Another 600,000 rabbits are used each year for biomedical teaching and research.

Rabbit production is an ideal enterprise for a young person because it can be started with limited capital (**Figure 29-9**). With only a small investment in housing and equipment, a person with one pair of rabbits can produce 50 to 60 rabbits each year to eat or to sell. Because they are small and generally accepted by people, rabbits are better adapted for production in suburban and urban areas than most other types of animals. As for its market appeal, rabbit meat is low in fat (4 percent), sodium, and cholesterol and is high in protein (25 percent).

The outlook for rabbit production in the future is variable, with a need for good market analysis and development, as well as careful production management. As competition between humans and animals for grain products increases, rabbits could play an important role in meeting the protein needs of humans.

Figure 29-9 Rabbit production can be done on a small basis with very little capital and space.
Courtesy of FFA.

Science Connection: Taxonomy of Rabbits

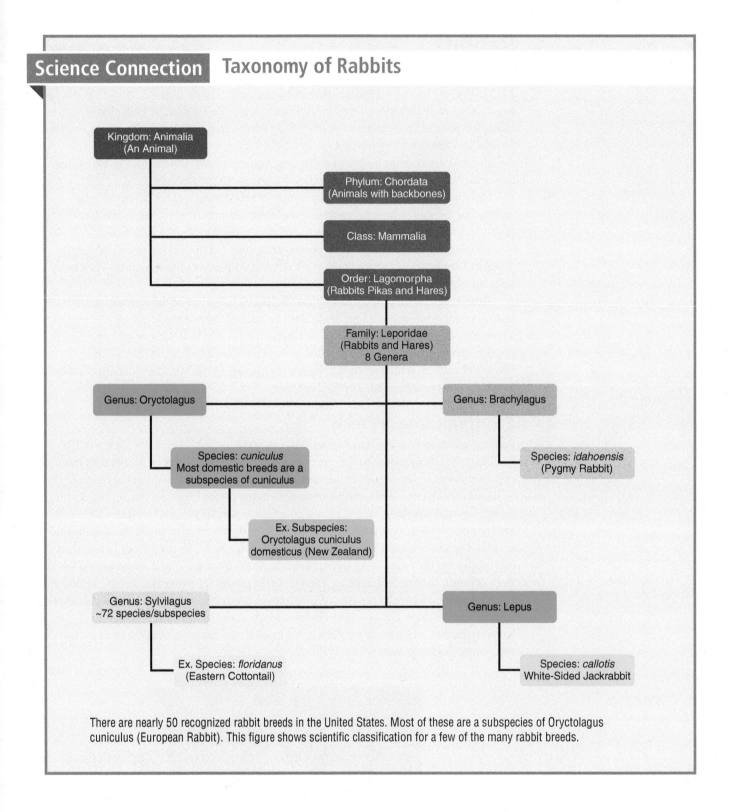

There are nearly 50 recognized rabbit breeds in the United States. Most of these are a subspecies of Oryctolagus cuniculus (European Rabbit). This figure shows scientific classification for a few of the many rabbit breeds.

Types and Uses

Rabbits come from two families and three genera, with distinct differences. These include rabbits, cottontails, and hares. Rabbits bear their young in underground burrows in the wild. The young are born blind, hairless, and completely helpless. In contrast, cottontails and hares usually give birth in nests above the ground. The young are born with their eyes open and with hair. They are able to fend for themselves shortly after birth. Hares also have larger hind legs and longer ears.

Hares include the jackrabbit, Arctic hare, and Snowshoe hare. Because they belong to different genera, cottontails, hares, and rabbits cannot interbreed.

Domestic rabbits can be divided into a number of groups based on use. These groups include meat, fur, pets, show, and laboratory use. Many breeds fall into several of these use groups.

The primary use of rabbits in the United States is for the production of meat, with pelts being a by-product. A pelt is an animal skin with the hair attached. Almost 100 million rabbit pelts are used in the United States each year. Most of these are imported from other countries because, in the United States, rabbits are slaughtered at too young an age to have desirable pelts.

Although all the breeds of rabbit will produce meat, some breeds are far more efficient in producing desirable-quality meat. The New Zealand White is the most popular breed of rabbit in the United States for meat production. (Despite the name, it originated in the United States. It was bred in 1916 by a California breeder, reportedly from wild New Zealand wild rabbits.) This rabbit comes in several colors, including white, red, black, and sometimes blue or "broken" (white with other color). It is of medium size. It may be grown into a 4-lb. rabbit at 8 weeks of age, using about 4 lbs. of feed for each pound of rabbit produced. The Californian and Champagne D'Argent are also popular breeds used for meat production (**Figure 29-10**).

Some breeds of rabbit are grown for their lustrous fur, which is used in the manufacture of fur coats and other rabbit-fur products. The Satin, Rex, and Havana are examples of rabbit breeds grown for this purpose (**Figure 29-11**).

Figure 29-10 Important meat-producing breeds of rabbit include the (A) New Zealand, (B) Californian, and (C) Champagne D'Argent.
Courtesy of American Rabbit Breeders Association. Courtesy of American Rabbit Breeders Association. Courtesy of American Rabbit Breeders Association.

Figure 29-11 Important fur-producing breeds of rabbit include the (A) Satin, (B) Rex, and (C) Havana.
Courtesy of American Rabbit Breeders Association. Courtesy of American Rabbit Breeders Association. Courtesy of American Rabbit Breeders Association.

Rabbits have also been important in laboratory work. They are used for research in the development of drugs for treating a wide range of diseases. They are also important in nutritional studies and genetic research. Breeds used for laboratory work include New Zealand White, Dutch, and Florida White. Producers who breed rabbits for laboratory work should be aware that many laboratories will use only white rabbits of medium size (**Figure 29-12**).

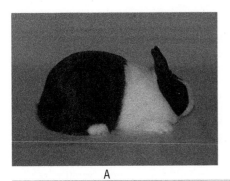  

Figure 29-12 The (A) Dutch and (B) Florida White join the (C) New Zealand White as important breeds in biomedical research and education.

Courtesy of American Rabbit Breeders Association. Courtesy of American Rabbit Breeders Association. Courtesy of American Rabbit Breeders Association.

The Angora rabbit is used strictly for the production of a wool called **angora**. The wool from Angora rabbits is sheared or pulled from the rabbit about every 10 to 12 weeks. Mature Angora bucks may produce 1 to 1.5 lbs. of wool each year. A **buck** is a male rabbit. A female rabbit is a **doe**. There are two breeds of Angora rabbit: French and English (**Figure 29-13**). All of the 40 breeds of rabbit recognized by the American Rabbit Breeders Association can be used as pets, for show, or both. They range in size from the Flemish Giant, which can weigh nearly 20 lbs., to the Netherland Dwarf, which seldom weighs more than 2 lbs. and makes a popular pet (**Figure 29-14**).

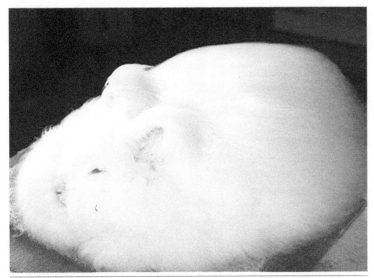

Figure 29-13 Angora rabbits, such as the Giant Angora (pictured here), are raised for the production of angora wool.

Courtesy of American Rabbit Breeders Association.

Figure 29-14 Netherland Dwarf rabbits, generally weighing in at less than 2 lb., are popular pets.

Courtesy of American Rabbit Breeders Association.

Personal preference and the availability of breeding stock usually determine what breed or breeds of rabbit to raise for pets or show.

Approved Practices for Rabbit Production

Approved practices for the production of rabbits include the following:

- Select the correct breed for the intended use.
- Use purebred stock if you plan to sell breeding stock and to maintain uniformity in your herd.
- Purchase breeding stock only from reputable breeders with accurate records.
- Build or choose a hutch of the proper size for the breed of rabbit that you are growing. A hutch is a cage or house for a rabbit. For small-sized and medium-sized breeds, provide hutches 30 inches wide × 36 inches long × 18 inches high. For large-sized breeds, provide hutches 30 inches wide × 48 inches long × 18 inches high.
- Place the hutch where the rabbit will have adequate ventilation and be protected from heat, wind, rain, sleet, and snow.
- Provide adequate feeder and waterer space. Rabbits should have access to fresh, clean water at all times.
- Provide a separate hutch or cage for each mature rabbit.
- Breed rabbits when does are 5 to 8 months old, depending on the breed, and when bucks are 6 to 7 months old. It may be wise to delay breeding of large rabbits because they are slower in reaching sexual maturity.
- Take the doe to the buck's cage for breeding and return her to her own cage immediately after breeding.
- Maintain 1 mature buck for every 10 to 25 does, depending on the breed and mating management.
- Place a nesting box in the doe's cage 25 days after mating occurs.
- Keep the handling of rabbits to a minimum to avoid injury. When handling rabbits, hold them by the skin on the back of the neck, with the other hand supporting the weight of the rabbit.
- Feed a commercial pelleted-feed free choice (feed available at all times) to does and litters (a group of young born at one time to the same parents). Feed single bucks and does 3 to 6 ounces of feed each day. Rabbits need to be fed only once a day, and preferably, they should be fed in the evening.
- Maintain accurate breeding, production, and health records for all rabbits.
- Tattoo all breeding rabbits for identification. A tattoo is a means of marking rabbits and other animals for identification. Rabbits are tattooed in the ear.
- Plan for and maintain a strict herd health program.
- Dispose of sick and dead rabbits promptly.
- Market rabbits as soon as they reach market size or weight.

Honeybees and Apiculture

History and Domestication

Honeybees (*Apis mellifera*) are an important part of human history. Early man drew pictures of bees and honey collection on cave walls. In Egypt, mummies were embalmed and stored in a liquid based on honey, the thick, sweet substance made by bees from the nectar of flowers. Jars of honey have been found in many Egyptian tombs. The Bible makes many references to honey and to the use of honey for food.

The ancient Greeks and Romans were very familiar with honeybees and honey. Pompey, an ancient Roman general, used poisoned honey to defeat his enemies in at least one battle. The ancient philosopher Aristotle wrote in great detail about bees and their production of honey. Ancient Greek athletes often ate honey before competing in Olympic games to enhance their strength and endurance. Most early civilizations considered honey to be the food of gods.

In ancient times, honey was not produced in nice, neat combs as it is today. Instead, the hive was usually destroyed in the process of removing the honey and comb. A hive is a home for honeybees. The comb is the wax foundation in which bees store honey. Early beekeepers kept their bees in hollow logs, straw hives, or even in crude clay cylinders. All of these containers had to be destroyed to remove the honey.

In the 1850s, the invention of movable combs with wax foundations encouraged bees to make neat, straight honeycombs. This changed the beekeeping industry. Honey was finally a commodity to be enjoyed by nearly everyone (**Figure 29-15**). Next came the discovery that honey could be whirled out of the comb. This led to the invention of the honey extractor. It was no longer necessary to destroy the comb to get to the honey. The comb, after being emptied of honey, could be placed back in the hive to be refilled by the bees.

Figure 29-15 The invention of moveable honeycomb made it possible to remove honey from a beehive without destroying the home of the bees.
Fernando M. Elkspera/Shutterstock.com.

Today, the production of honey in the United States is a large and profitable business. Modern beekeeping is known as apiculture. Of even greater importance than the production of honey is the work that honeybees do in the pollination of important agricultural crops.

Economic Importance

It is difficult to accurately gauge the true economic importance of honeybees. They are responsible for about 80 percent of insect pollination of plants. Without honeybees, many crops important to agriculture would simply disappear from Earth.

Pollination of orchard crops such as citrus, peaches, and apples by honeybees is so important that many beekeepers rent their bees to orchardists when these trees are in bloom. Many commercial beekeepers earn more money from bee rental than from honey. Such beekeepers use flatbed trailers to transport their hives, including from Florida to Maine and from Texas to Washington State (**Figure 29-16**).

There are as many as 300,000 beekeepers in the United States, of which most are hobby or part-time beekeepers. These 300,000 people care for about 2.5 million honeybee hives. (The numbers have declined from 6 million in the 1940s.) In a normal year, one hive will produce 100 to 150 lbs. of honey in excess of the amount needed by the bees to live (about 150 lbs.). The production of honey is big business.

Figure 29-16 Bees are responsible for up to 80 percent of all insect pollination of plants and are essential for pollination of fruit crops such as citrus, peaches, apples, and other orchard crops.
Courtesy of USDA/ARS #K-4715-1.

Approved Practices for Beekeeping

The following is a list of approved practices used in the keeping of bees:

- Check local regulations before starting a beekeeping operation.
- Secure equipment before starting an **apiary**, an area for keeping beehives.
- Purchase bees from reputable sources. It is usually far more profitable to purchase a 3-lb. package of bees with a purebred queen than to rely on a swarm to populate a new beehive. A **queen** bee is the only fertile, egg-laying female bee in each hive (**Figure 29-17**). A swarm is a group of bees complete with a queen that leaves an overcrowded hive to find a new home.
- Locate bees out of direct contact with people and neighbors' yards and gardens.
- Place hives facing away from prevailing winds. They should also be protected from hot summer sun.
- Thoroughly clean and disinfect hives before new groups of bees to use them.
- Replace queens every 2 years.
- Have a federal inspector check bees annually for contagious diseases.
- Make sure that each hive of bees has a store of at least 75 lbs. of honey for the winter. Hives that do not have enough surplus honey stored for winter should be fed a sugar–water mixture to supplement their own honey stores.
- Always be sure that bees have ample room to store the honey they produce.
- Remove surplus honey as soon as the bees have capped it over with wax.
- Extract honey from the comb as soon as possible after harvesting it. Honey stored for long periods in the comb may granulate, which makes it impossible to extract.
- Remove honey in the evening or at night when nearly all the bees are in the hive. Supers containing surplus honey can be freed from bees by blowing cool smoke over the bees and brushing them off the comb with a bee brush. A super is a box filled with a movable wax foundation that is used by the bees to store honey. You can also use a bee excluder between the honey to be removed and the hive body.
- Keep honey that has been removed from bees in an area that bees cannot get to. Otherwise, they will steal all of the honey in a short time.
- After moving a hive, put a deflector in the entrance of the hive so the bees will notice that they have been moved. Hives must be moved at least 5 miles to prevent bees from returning to the former site of their hive.

Figure 29-17 Worker bees tend the queen (marked), the hive, and the brood, as well as making and storing honey for the future.
Courtesy of USDA/ARS #K-5069-22.

- Inspect beehives at least monthly to determine the strength of the hive and the queen. Be sure to observe the number of eggs being laid by the queen. Also note whether the worker bees are building drone or queen cells in the hive. **Worker** bees are undeveloped female bees and constitute all of the working force of the hive. A **drone** is a male bee whose only purpose is to fertilize the queen once in his life span. Drone and queen cells look like peanuts and should be destroyed.
- Reduce or prevent swarming of bees by providing ample hive space for the bees and eliminating queen cells as they are found. Overcrowding often causes bees to develop a second queen. A new queen will attract a group of worker bees and leave the hive to start a new colony. This process is called swarming. Bees will not swarm without a queen because she is their only hope of survival.
- Be aware of pesticides in the area that could kill bees or be stored in the honey.
- Develop a market for your honey.

Science Connection: Bees in Biocontrol Business

Fire blight is caused by a bacterium that enters apple and pear trees by colonizing the stigma of the fruit blossom. Honeybees, carrying a fire blight–fighting bacteria, are used to colonize the stigma with beneficial bacteria as well as pollinate the fruit blossoms.
Courtesy of Paul C. Pecknold.

According to a popular saying, "If you want to get a job done, ask someone busy to do it"—and who else is "as busy as a bee"? Now entomologists are deploying busy honeybees to spread biocontrol agents to plants as they pollinate billions of dollars' worth of crops each year. The unsuspecting honeybees deliver biocontrol agents right where the agents are needed on plants. This innovative process targets two serious pests, the fire blight bacterium, which infects pear and apple trees, and the corn earworm, one of the worst insect pests of corn and cotton.

Fire blight is caused by *Erwinia amylovora*, a bacterium that first colonizes a flower's stigma, the part that receives pollen grains during pollination. The bacteria then multiply and spread quickly. The disease causes cankers on twigs and branches and causes the leaves to have a burnt appearance, hence, the name "fire blight." The disorder weakens or kills the trees. On the positive side, scientists have found that spraying blossoms with beneficial bacteria helps prevent fire blight disease. The beneficial bacteria apparently compete with the harmful bacteria by consuming the nutrient-rich stigma, and the pathogen cannot get a foothold to invade the tree.

Rather than spraying with the beneficial bacteria, entomologists and plant pathologists at Utah State University and Oregon State University are conducting research to identify the honeybees that deliver the beneficial bacteria to the point-of-entry for the fire blight pathogen—the stigma. Such a strategy maximizes the use of the biocontrol material. The procedure is being field tested, wherein numerous beehives are fitted with devices called pollen inserts. The pollen insert is a slotted passageway for bees that automatically dusts the bees with the beneficial bacterium, which is the biocontrol in this case, each time they leave the hive. A hive contains from 10,000 to 100,000 bees, each visiting perhaps 100 blossoms per hour. The beneficial bacteria do no apparent harm to the bees.

Entomologists in Tifton, Georgia, use honeybees to carry a natural virus, *Heliothis*, that destroys the larval stage of the corn earworm. The method used to load each bee involves having it use different openings in the hive to enter and exit. The exit device includes a metal tray that contains the virus material. The exiting bees carry the material on their feet, legs, and undersides and deposit it on their rounds. This technique has reduced earworm infestations in clover and should be effective on other flowering plants.

Other Small Animals

Small animals can make excellent companions and can provide specialized functions for children and adults alike. There is often a fine line between pet, companion animal, and work animal. For example, guide dogs function as essential "eyes" for the blind and must make intelligent choices for their owners when crossing streets (**Figure 29-18**). Similarly, dogs and other pets provide companionship for people and can alert them to potential intruders. Dogs may ward off attackers and provide other protection when appropriately trained. In sheep farming, dogs are essential for herding sheep and protecting them from predators.

Figure 29-18 Guide dogs serve dual purposes. When at work, they act as eyes for the blind. When at rest or play, they provide companionship.
Courtesy of Guide Dog Foundation for the Blind, Inc.

Many kinds of caged birds and reptiles are used for pets. Their needs are rather specialized, and advice for their care is usually obtained from pet shops and other suppliers. Similarly, a wide variety of fish and other aquatic creatures make attractive and challenging aquarium projects for all ages. Even some exotic animals, such as llamas and kangaroos, have found their way onto farms and into the hearts of Americans (**Figure 29-19**). Some states prohibit the possession of certain animals, so it is important to check regulations before adopting a pet.

Figure 29-19 The care of exotic animals creates new challenges for experienced pet handlers.
Courtesy of FFA.

Pet Care and Management

Dogs and cats are probably the most prevalent companion animals in the United States. Although their temperaments and behaviors vary, they have similar needs. The following is a list of approved practices for the care of dogs and cats based on the findings of Pennsylvania State University:

- Select animals that are alert and healthy.
- Select a breed adapted to your situation.
- Vaccinate for rabies at age-determined intervals.
- Prepare a clean, draft-free living area.
- Provide an adequate number of feed dishes.
- Provide a diet of high-quality food for the breed and age.
- Share a moderate daily exercise routine with your pet, such as active play for a cat, or brisk walks for a dog (**Figure 29-20**).
- Provide toys, treats, or both.
- Clean the food and water dishes daily.
- Keep clean, fresh bedding in the sleeping area.
- Provide plenty of clean, fresh water.
- Clean the pen or box and exercise area daily.
- Develop and implement a sound health plan.
- Consult with your veterinarian to prevent and control internal and external parasites.
- Vaccinate animals routinely at proper times.
- Properly bathe and groom the animal.
- Use a proper carrier for transport.
- Use proper restraint procedures.
- Maintain accurate breeding and production records.
- Select a male animal to mate with your female animal.
- Use a superior proven male animal.
- Take extra precautions for female animals in heat.
- Prepare a clean area for whelping (giving birth).
- Train properly.
- Market and sell young animals.
- Develop a private market for the animals.

Figure 29-20 Dogs and other pets need regular exercise just as people do.
© Andrey Yurlov/Shutterstock.com

- Complete a registration application if animals are purebred.
- Neuter animals that are not intended for breeding.
- Summarize and analyze records.

What one person regards as a pet, another person may regard as a business opportunity, such as to provide feed, supplies, and equipment for that pet, as well as pets for adoption (**Figure 29-21**). Raising small animals provides the opportunity for individuals with limited capital and facilities to get a start in animal agriculture. Most small animals are better adapted than larger animals to production in urban and suburban areas. As with large-animal agriculture, small animal enterprises offer opportunities for planning, raising, managing, and marketing, but without the large outlay of cash needed for the production of large animals.

Figure 29-21 Cats make great pets and provide companionship to people. They also have needs, such as food and veterinary care, that must be provided by their owners.
Courtesy of FFA.

Science Connection Animal Care Includes Disaster Preparednes

How can we keep pets and other animals safe when disaster strikes? During Hurricane Katrina in 2005, some pet owners remained in harm's way because people lacked options for evacuating with their pets. Soon after, Congress passed the Pets Evacuation and Transportation Standards (PETS) Act, and Federal Emergency Management Agency (FEMA) took on this lifesaving mission. The USDA also oversees its regulated animals when disasters strike. Regulated animals include those in zoos and research labs that have special transportation and biosecurity needs.

To prepare for the unexpected, pet owners and farm animal managers alike should have an evacuation plan in place. For example, all animals need to have durable and visible identification. Today, many animals are also micro-chipped. Likewise, people should contact the relevant local services well in advance of a disaster, such as animal shelters for pets or the county extension service for farm animals. All animal owners can contact their local emergency management office for detailed advice and to learn how local services can help in the event of an evacuation.

Agri-Profile: Animal Technician/Grower/Beekeeper/Manager

Career opportunities for small animal care and management are available in many areas. The extensive use of small animals for pets, fur, zoological parks, and use of laboratory animals for research, ensures strong job growth.

The operation of animal hospitals, kennels, grooming services, pet stores, training programs, boarding facilities, public aquariums, animal shelters, and humane societies provide opportunities in a number of fields. These include animal nutrition, facilities construction and maintenance, feeds, health services, care and management, production, breeding, and marketing.

For example:

- Small-animal veterinarians provide clinical and surgical care and disease surveillance.
- Veterinary technicians record medical histories, provide nursing care or emergency first aid, administer anesthesia before surgery, and collect lab samples for tests.
- Small-animal biotechnologists study the effects of nutrients on animal health in order to improve the reproductive and feed process.
- Animal shelter managers oversee the daily care of animals and the sanitary care of the facility.
- Small-animal nutritionists research the current physical condition and medical history of pets to determine the best diets and exercise schedules.

Educational requirements vary by job, from pon-the-job training for pet sitters to a four-year undergraduate degree and a Doctor of Veterinary Medicine degree for researchers, university-level instructors, and veterinarians. Research careers also require skill in the use of advanced statistical processes.

Compensation varies accordingly. For example, according to the U.S. Bureau of Labor Statistics in 2020, small-animal veterinarians earned a median annual salary of $99,250 , while the median for small-animal veterinary technicians was $36,800.

A

B

(A) Animal trainers teach animals to obey commands and to perform tricks for audiences.
(B) Domesticated animals require human care to provide for basic needs such as food, shelter, and health care.
© iStockphoto/Joe Brandt. © Danny E. Hooks/Shutterstock.com.

Hot Topics in Agriscience: The Role of Small Animals in Medical Research

Small animals such as rats and rabbits are used in medical research to test new vaccines and other medications before they are used to treat humans.
© Sebastian Duda/Shutterstock.com.

One of the many reasons why animals are used in medical research is that animals and humans are biologically similar. For example, mice share over 98 percent of DNA with humans.

Another reason is that, since animals are vulnerable to many of the same diseases as humans, there are mutual benefits. For example, medical researchers have identified the causes and many of the cures for a variety of diseases and infections that afflict people and animals. This has been accomplished by studying the organisms that cause these infections. Once such an organism has been identified, it is cultured, modified, weakened, and tested in animals in an effort to create a vaccine that will prevent the infection from becoming established.

For example, fertile chicken eggs are often used to culture such organisms in live tissue. These eggs are supplied to medical research facilities from farms that specialize in producing the quality of eggs that is needed for this purpose. Such vaccines would simply not be possible without access to such live organisms.

Rabbits and other small animals are used in medical research to test the effects of potential vaccines and medications on animal life. Much of the progress that medical science has made in controlling diseases and infections of various kinds has depended on the use of animals for research purposes. According to Stanford University, 95 percent of animals needed for biomedical research are mice and rats, which are specially bred for laboratory research.

Federal law prevents researchers from using human subjects until medicines and vaccines have been proved to be effective in laboratory animals.

The USDA has established strict regulations governing the care and use of laboratory animals in biomedical research. These regulations require research labs to follow detailed recommendations for animal care, including proper housing, feeding, cleanliness, ventilation and medical needs, and prevention of pain.

The regulations also require research institutions to establish an Institutional Animal Care and Use Committee to oversee all work with animals. The committees must include one veterinarian and one community representative. The law also requires researchers to justify their need for animals to the committee, which can approve or reject the research project.

Without research animals, we would not have the ability to control many of the diseases and infections that are known to afflict humans. The average life span of humans would be much shorter in the absence of research animals.

Chapter Review

Student Activities

1. Write the **Key Terms** and their meanings in your notebook.
2. Create a multimedia presentation or a bulletin board display of breeds and types of poultry and rabbits.
3. Attend a fair or show and record the names of the breeds of poultry and rabbits shown there.
4. Interview a local beekeeper about beekeeping practices in your area.
5. Compare the label from a bag of poultry feed with one from a bag of rabbit feed. Determine the differences in ingredients, percentage of protein, additives, and fiber.
6. Set up an observation beehive in the school.
7. Make a list of local crops that are of importance to agriscience and that bees pollinate.
8. Develop a crossword puzzle or word search using the **Key Terms**.
9. Take your small animal to the veterinarian for a checkup. While there, have all of the scheduled vaccinations given.
10. Ask your veterinarian to discuss with you the proper care that your animal needs. Be sure to talk about how much food and water it should be getting as well as any special care it may require.
11. Select a rabbit breed based on one of the photos in this chapter. Using online resources, identify its full scientific name from kingdom to subspecies. Then, make a chart like the one shown below. Use it to compare and contrast the rabbit you selected with another rabbit breed featured this chapter.

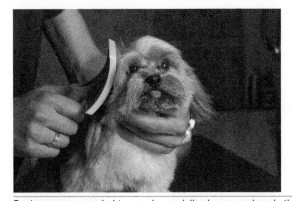 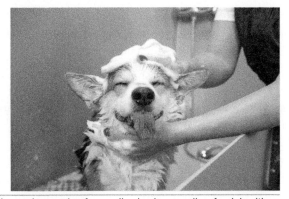

Businesses are needed to supply specialized care, such as bathing and grooming for small animals, as well as food, health supplies, and equipment.

© Serdar Tibet/Shutterstock.com

Compare-Contrast		
	RABBIT 1	RABBIT 2
Breed		
Scientific name(s)		
Physical features		
Role or use		
Other details		

National FFA Connection — Career Development Events in Small Animal Care

FFA events and resources provide opportunities to build—and test—your skills on a wide range of small animal care issues.
Courtesy of FFA/Photo by Michael Wilson.

Many chapters of the National FFA Organization hosts Career Development Events that give students opportunities to evaluate their abilities in the area of small animal care. These areas range from small animal veterinary skills to pet store management.

While events may vary by state or county, events typically include a team and individual components. For example, team members might explore how they would solve a problem with kennel management, nutritional needs, or issues with aquarium product sales. Individuals might be tested in such areas as breed identification, care and grooming, anatomy, and veterinary terms.

To find out more about such events and to read articles about FFA members' experiences with small animal care and management, visit the FFA website and search "small animals."

Check Your Progress

Can you . . .

- briefly summarize the domestication and history of one or more classes of small animals?
- explain the economic importance of one or more classes of small animals?
- give examples of types and uses of one or more classes of small animals
- give examples of approved practices in feeding and caring for small animals?

If you answered "no" to any of these questions, review the chapter for those topics.

Chapter 30
Dairy and Livestock Management

Objective
To determine the history, types, uses, care, and management of dairy and livestock.

Competencies to Be Developed
After studying this chapter, you should be able to:
- describe the history and economic importance of dairy and livestock.
- recognize major types and breeds of livestock.
- list major uses of livestock.
- understand basic approved practices in the care and management of dairy and livestock.

Key Terms

mammal	steer	lamb
veal	implant	mutton
calf	pork	ewe
colostrum	sow	ram
heifer	gilt	mohair
cow	boar	chevon
draft	farrow	lactation
bull	wool	kid

Large animals, including dairy, beef, sheep, goats, and swine, are the backbone of animal agriculture. The capturing and domestication of animals changed how humans lived. If these hunter-gathers wanted to raise animals as livestock, they would have to provide feed for them, which required staying in one place and farming feed. In doing so, hunting was no longer necessary as the people had meat to eat. Today, much of agriscience is centered on the production of animals and animal products and the production of feed for those animals.

Dairy Cattle

Origin and History

A **mammal** is an animal that produces milk, a highly nutritious white or yellowish liquid secreted by the mammary glands of animals for the purpose of feeding their young. In the wild, mammals normally produce only enough milk to feed their offspring. When early humans realized that milk was a healthful food and that some types of animals produced more milk than others, the domestication of dairy animals began. Although the cow, buffalo, camel, goat, ewe, and mare have been and are currently being used to produce milk in various parts of the world, this chapter focuses primarily on cattle with a minor emphasis on goats in dairy production.

Early historical records note the use of cattle to produce milk, and the Bible also has a number of references to milk and the production of milk. Hippocrates, the father of modern medicine, recommended milk as a medicine in his writings around 400 BCE.

While early native peoples of the Americas sourced milk from llamas, goats, sheep, and other animals, there were no dairy cows. European explorer Christopher Columbus brought dairy cows to the Americas on his second voyage in 1493. Cattle were also transported to Jamestown, the first permanent English colony in America, which was founded in 1611. The production of milk was limited to a few cows per family during the colonial period. It was not until the late 1800s that dairy farming became an important agricultural industry in the United States. Since then, intensive efforts have been directed toward breeding and developing dairy cattle (**Figure 30-1**).

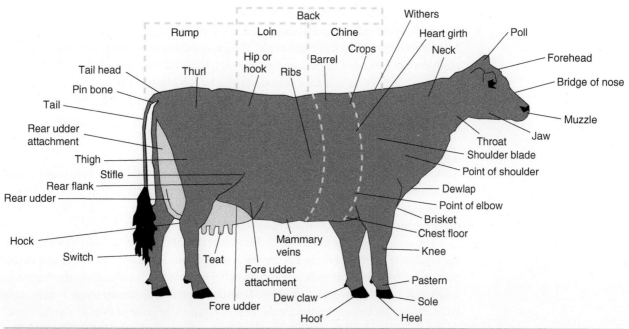

Figure 30-1 Anatomy of a dairy cow.

Economic Importance

The production of milk is the second most important animal enterprise in the United States, when sales dollars are the criterion for importance. However, consumption of milk as a beverage has been trending downward, while consumption of other dairy products such as cheese and butter continues to increase (**Figure 30-2**). According to the U.S. Department of Agriculture (USDA), in 2020 the average American consumed 655 lbs. in dairy including milk, cheese, yogurt, butter, ice cream, and other dairy products. By contrast, consumption of dairy averaged just 539 lbs. per consumer in 1975.

Figure 30-2 Consumption of milk as a beverage is declining but consumption of cheese and butter has increased.
© ifong/Shutterstock.com.

About 9.45 million cows in the United States produce about 227 billion lbs. of milk annually. As of 2021, production of milk per cow in the United States averaged 23,948 lbs.

The production of milk is not the only income-generating part of dairy production. Bull calves enter the meat production product stream as **veal** (the meat of young calves), or they are grown to market size for beef (meat from grown cattle). Similarly, cows that are no longer profitable producers of milk are sold for beef.

Most milk sales in the United States conform to Federal Milk Marketing Orders (FMMOs). FMMOs establish the provisions, or conditions, under which dairy processors purchase fresh milk from dairy farmers. The provisions also recognize different milk classes depending on how the milk will be used.

Class I milk is produced under strict standards and is intended for consumption as fluid milk. Fluid milk includes whole milk, reduced-fat milk, and cream. Class II milk can be produced under less strict standards. The intended use of this milk is to make soft dairy products like ice cream, yogurt, and sour cream. Class III milk is used to produce cheese and whey. Class IV milk is used to make butter, butter-like spreads, and dry products such as milk powder. It should be noted that some states have developed their own milk classes and market their milk outside of FMMOs.

Types and Breeds

Around 94 percent of all dairy cattle in the United States are of the Holstein breed. These familiar black and white cattle have the highest average production of milk of any breed in the country. Because of the large numbers of cattle involved, the breed also has been able to make the most genetic improvement in recent years.

The second most popular breed of dairy cattle is Jersey. In terms of size, Jersey cows are the smallest of the dairy breeds, but they rank number one in the amount of butterfat in their milk. Butterfat is the fat in milk from which butter is made. Their rich milk is also high in milk solids such as protein, from which cheese is made. Another well-known breed of dairy cattle is Guernsey. This breed is known for the yellowish tint to the color of its milk. Ayrshire and Brown Swiss round out the top five breeds of dairy cattle in the United States (**Figure 30-3**).

Figure 30-3 Major breeds of dairy cattle in the U.S. include (A) Holstein, (B) Jersey, (C) Guernsey, (D) Ayrshire, and (E) Brown Swiss.
Courtesy of the Holstein Association USA, Inc. Courtesy of the American Jersey Cattle Club. Courtesy of the American Guernsey Association. © smereka/Shutterstock.com Courtesy of the Brown Swiss Cattle Breeder's Association.

Figure 30-4 Today's female calves become tomorrow's dairy cows, so good calf care and management are top priorities on dairy farms.
© Oleks/Shutterstock.com.

Approved Practices for Raising Calves

Approved practices for raising dairy calves include the following (**Figure 30-4**):
- Ensure the newborn calf receives colostrum as its first food as soon after birth as possible, and continue to feed colostrum for at least the first 36 hours of its life. Colostrum is the milk a cow produces for a short time after calving (giving birth). It contains antibodies that protect the newborn calf from diseases until it can build up its own natural defenses.
- Feed milk or milk replacer daily at 8 to 10 percent of the calf's weight until the calf is 4 weeks old. Milk replacers are dry dairy or vegetable products that are mixed with warm water and fed to young calves instead of milk. Such products are less expensive than whole milk.
- Begin feeding calf starter, a grain mixture, free choice after about 10 days. Free choice means making feed available at all times.
- Wean calves from milk when they are eating 1.5 lbs. of calf starter per day.
- Feed calves green, leafy hay and water free choice at 2 to 9 months of age. Up to 4 lbs. of grain can be fed daily.
- Make up the bulk of the ration with forages fed free choice after 9 months of age.
- Remove horns at an early age, preferably as soon as the horns begin to develop.
- Remove extra teats at an early age.
- Identify calves with ear tags or tattoos as soon as possible after birth.
- Prevent calves from sucking the ears of other calves or licking each other.
- Keep hooves properly trimmed.
- Vaccinate for cattle diseases at the recommended times.
- Maintain the calf in clean and sanitary conditions.
- Plan for and maintain a disease and parasite-prevention and control program.
- Maintain heifers and calves in uniform groups according to size and weight. A heifer is a female that has not given birth to a calf.
- Breed heifers to calve at 20 to 24 months of age.

Approved Practices in Dairy Cow Production

Approved practices in dairy cow production includes the following:
- A cow is a female animal of the cattle family that has given birth. Cows should be bred to calve once every 12 months. Rebreed cows 45 to 60 days after calving.
- Observe cows for evidence of heat period twice daily, mornings and evenings (**Figure 30-5**). Cows noticed in "True Heat" in the morning probably have been in heat for several hours.
- Check cows 45 to 60 days after breeding to determine whether they are pregnant.
- Provide a dry period of about 60 days before calving to allow the cow to rebuild its body. The dry period refers to the time when a cow is not producing milk.
- Feed cows according to their levels of production and stages of pregnancy.
- Maintain complete health, breeding, and production records for every cow in the herd, or group (**Figure 30-6**).
- Establish and maintain a disease- and parasite-prevention and control program.
- Use the services of a veterinarian for serious or unknown health problems.
- Milk dairy cows at regular intervals each day. Two milkings each day, approximately 12 hours apart, is traditional. However, milking three times per day results in greater milk production, and is recommended in high-producing herds.
- Maintain a regular routine in handling dairy cattle for maximum production.
- Cull (remove) unprofitable dairy cows.
- Properly maintain dairy housing and milking equipment.

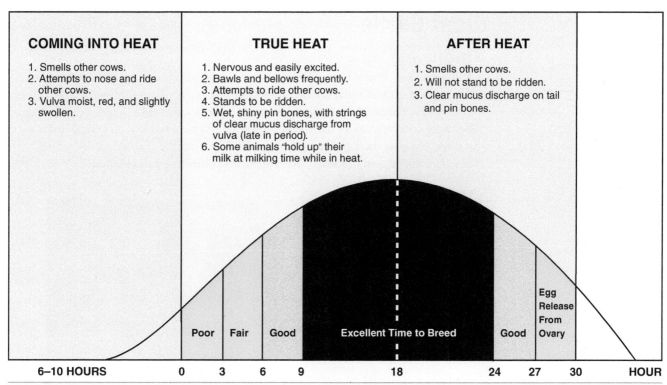

Figure 30-5 Breeding at the proper stage of the heat period is essential to maximize pregnancy rates. The diagram is based on average conditions.

Factors Determining Efficiency			
Factor	Results	Goals	State Average
1. Number of cows in herd			
2. Total lb. of milk produced			
3. Total lb. of butterfat produced			
4. Average annual milk production/cow (lb.)			
5. Average annual butterfat production/cow (lb.)			
6. Average annual percent of butterfat for student's herd			
7. Feed cost/lb. of butterfat produced			
8. Average feed cost/cow			
9. Percent of calves sold or kept until 3 months of age			
10. Profit or loss			
11. Lbs. of milk sold/hour of total labor			
12. Dollar returns/hour of self labor			
13. Production cost/lb. of milk			

Figure 30-6 Efficiency factors are good indicators of how the dairy business is operating.

Beef Cattle

Origin and History

Cattle were likely domesticated in Europe and Asia sometime during the New Stone Age. Domesticated cattle of today are likely descended from one of two wild species, *Bos taurus* or *Bos indicus* (**Figure 30-7**). Some of these wild cattle stood as tall as 7 feet at the shoulders. They were domesticated for meat, milk, and draft. **Draft** is a term for an animal that is used for work. Oxen are draft animals. Owning cattle was a symbol of wealth in early times. They were worshipped in some cultures and used as draft animals in others.

Figure 30-7 *Bos taurus* was one of the wild ancestors of today's domesticated cattle.
© Robert Frerck/Stone/Getty Images.

Cattle came to the Americas with the earliest European colonists, who were more interested in animals that could do heavy work than in those that could produce meat. As the European colonists arrived in the Americas, they also brought cattle with them. Spanish colonists brought longhorn-type cattle as a source of food for Christian missions in the Southwest. As demand increased for beef, the cattle industry developed on the frontier, where grass and the large open spaces required for cattle were abundant. Great cattle drives originated in these areas as cattlemen drove their cattle to transportation centers to market them. Today, the beef cattle industry still thrives where plenty of feed is available and production costs are minimal.

Economic Importance

The beef cattle industry is the number one red-meat production source in the United States. On average, the U.S. population eats about 58 lbs. of beef per person per year. This is part of a total consumption of about 225 lbs. of meat, poultry, and seafood. Other products are obtained from cattle. For example, cattle convert inedible grasses into food for people. In addition, cattle manure provides fertilizer for crops, and meat by-products are made into many nonfood products that people use every day (**Figure 30-8**).

By-Products from the Meat-Processing Industry

- Bone for bone china.
- Horn and bone for carving set handles.
- Hides and skins for leather goods.
- Rennet for cheese.
- Gelatin for marshmallows.
- Stearin for chewing gum and candies.
- Glycerin for explosives used in mining and blasting.
- Lanolin for cosmetics.
- Chemicals for tires that run cooler.
- Binders for asphalt paving.
- Medicines, such as various hormones and glandular extracts, insulin, pepsin, epinephrine, ACTH, cortisone, and surgical sutures.
- Intestines for violin strings.
- Animal fats for soap, non-ruminant feed, and industrial lubricants.
- Wool for clothing.
- Ear hair of camels for artists' brushes.
- Bone charcoal for high-grade steel, such as ball bearings.
- Collagen for special glues for marine plywoods, paper, matches, window shades.
- Curled hair for upholstery.

Figure 30-8 Many important by-products are derived from the meat-processing industry. These by-products are used in a variety of industries to make everything from household goods to medicine.

Types and Breeds

The general types of beef-cattle operations include purebred breeders, cow–calf operations, and slaughter cattle, or feedlot operations. In a purebred operation, only cattle of a single, pure breed are raised. Operations are geared to produce purebred **bulls** (male animal of the cattle family) for cow–calf operations and to produce animals to be sold to other purebred producers. Breeders of purebred cattle have been responsible for much of the genetic improvement in beef cattle in recent years.

Cow–calf operations produce feeder calves that are placed in feedlots owned and managed by producers of slaughter cattle. Cow–calf operations are located mostly in the upper Great Plains states and in the western range states where grass is in abundance and much of the land is unsuited to produce other crops. Calves are usually born in the spring, remain with their mothers during the summer, and are weaned in the fall. They are then sold to feedlot operations. Most cow–calf producers use purebred bulls to breed the grade cows in the herd, producing superior calves.

The feedlot operator buys calves from cow–calf operators and feeds them until they reach slaughter weight. These operations are concentrated in the Midwest, where there is an abundance of corn and other grain for feed. Large feedlot operations are located wherever an abundance of cheap feed exists.

Until about 40 years ago, there were three major breeds of beef cattle in the United States: Hereford, Angus, and Shorthorn. Although these three breeds are still important today, they have been joined by more than 50 other breeds from all over the world (**Figure 30-9**). These breeds can be divided into many classifications. However, for our purposes, they are designated as English, exotic, and American.

English breeds, which originated in Great Britain, include Hereford, Angus, Shorthorn, Galloway, Devon, and Red Poll. In general, English breeds of cattle are of medium size and are noted for the excellent quality of meat they produce.

Figure 30-9 Today's predominant beef breeds in the United States are (A) Hereford, (B) Angus, (C) Simmental, (D) Limousin, (E) Beefmaster, and (F) Charolais. More of today's predominant beef breeds in the United States are (G) Brahman and (H) Shorthorn.

Courtesy of the American Hereford Association. Courtesy of the American Angus Association. Courtesy of the American Simmental Association. Courtesy of the North American Limousin Foundation. Courtesy of the Beefmaster Breeders United. Courtesy of the American-International Charolais Association. Courtesy of the American Brahman Breeders' Association. Courtesy of the American Milking Shorthorn Society.

Figure 30-10 Many commercial producers have crossed and mixed breeds of cattle, such as the Brangus bull pictured here.
© Eduardo Garcia Furtado/Shutterstock.com.

Exotic breeds of cattle were first imported to the United States from all over the world when consumers began demanding leaner beef. Beef producers also became aware that calves from these breeds of cattle grew faster and much more efficiently than most of the English breeds. Exotic-breed bulls are often used in grade cow–calf operations to increase the weight of calves being produced. Examples of exotic breeds of cattle include Charolais, Limousin, Simmental, Salers, Wagyu, and Maine Anjou.

American breeds of beef cattle were developed from necessity. To grow beef cattle in the South and Southwest, heat tolerance and disease- and parasite-resistance were needed. To develop cattle that met these requirements and still had desirable meat quality, breeders crossed Brahman cattle from India with the English breeds. Examples of American breeds of beef cattle are Brangus, Beefmaster, Santa Gertrudis, and Barzona. Many commercial herds are of crossbred, or mixed, breeds (**Figure 30-10**).

An understanding of beef production includes knowing and using the terms that are used to describe beef cattle (**Figure 30-11**).

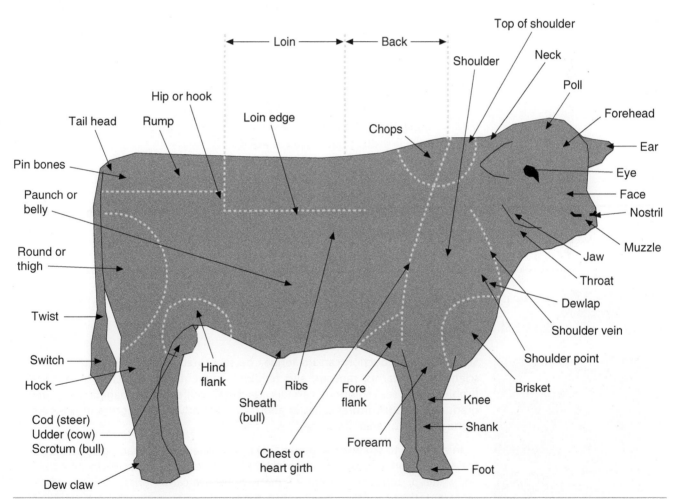

Figure 30-11 Learning the proper names of cattle parts is necessary for discussing and evaluating animals.

Approved Practices in Beef Production

Approved practices in the production of beef cattle include the following:

- Select breeds by intended use, area of the country, and personal preference.
- Buy cattle only from reputable breeders.
- Select purebred cattle by physical appearance, pedigree, and production records.
- Isolate new animals for at least 30 days to observe for diseases and parasites.
- Provide enough human contact so that beef cattle can be handled when necessary.
- Break calves to lead as soon as possible if they will be exhibited at fairs.
- Plan a complete herd-health program and follow through with it.
- Use the services of a veterinarian for serious or unknown health problems.
- Vaccinate to prevent diseases of local concern.
- Castrate males, dehorn, and permanently identify calves at an early age.
- Breed heifers to calve at 2 years of age.
- Implant **steers** (castrated males) and heifers grown for slaughter with approved growth hormones. An **implant** is a substance that is placed under the skin. It is released slowly over a long time period to stimulate efficiency and growth.
- Wean calves at 205 days of age and at 450 to 500 lbs.
- Provide supplemental nutrition when natural forages are in short supply.
- In a pasture-breeding system, allow one mature bull for every 25 to 30 cows. With pen mating, one bull can be used to service 30 to 50 cows.
- Restrict the length of breeding season so all calves are born within 30 to 60 days.
- Assure that cows calve in clean stalls or pastures.
- Group the cattle being fed for slaughter according to size and sex.
- Provide shelter from inclement weather.
- Provide access to clean, fresh water at all times.
- Feed slaughter cattle to reach market weight at 15 to 24 months of age.
- Utilize proper facilities and equipment for the type of operation.
- Market animals at the optimum time to maximize profits (**Figure 30-12**).
- Maintain complete and accurate records.

Agri-Profile — Career Area: Farm Manager/Herd Manager/Farmer/Rancher/Breed Association Representative

Farming, ranching, and feedlot management increasingly involves scientific research.
Courtesy of USDA.

Science and production can no longer be separated. Individuals and teams of scientists visit farms, ranches, and feedlots frequently to gather data, analyze problems, and seek solutions. Farmers, ranchers, and feedlot managers are usually graduates of agricultural colleges and may have advanced degrees in business, management, or animal sciences.

In addition to scientists and managers, technicians find meaningful careers in the animal production businesses of the nation. Livestock enterprises include beef, dairy, sheep, and swine. There are many non-farm career opportunities as field representatives for breed associations, feed companies, marketing cooperatives, supply companies, animal-health products, herd improvement associations, and financial-management firms.

Many individuals use their farming and ranching backgrounds in agriscience writing, publishing, and telecasting careers. Farm and ranch experience provides an excellent background for careers in agriscience teaching and extension work.

ANGUS BEEF CHART

■ Steaks and roasts suitable for broiling, panbroiling and roasting
■ Thrifty cuts requiring longer cooking methods

A 1000 pound steer yields 465 pounds of retail cuts from a 600-pound carcass.

25% are steaks
25% are roasts
25% is ground beef and stew meat
25% is made-up of fat, bone and shrinkage

Cuts labeled: ROLLED RIB ROAST, STANDING RIB ROAST, DELMONICO STEAK, CLUB STEAK, T-BONE, PORTERHOUSE, SIRLOIN, CHUCK ROAST, ARM ROAST, BRISKET Fresh or Corned Beef, SHORT RIBS, FLANK, GROUND BEEF, STEW MEAT, RUMP ROAST, ROUND

Body sections: Chuck 26%, Rib 8%, Short Loin 8%, Sirloin 9%, Rump 3%, Round 21%, Hind Shank 3%, Flank 4%, Plate 5.5%, Short Ribs 1.5%, Brisket 6%, Fore Shank 4%

VARIETY MEATS: SWEETBREADS, TONGUE, LIVER, OXTAILS, KIDNEY, HEART

Area in red represents the most desirable cuts and accounts for about 90% of the retail value of a carcass.

Figure 30-12 Many different cuts of meat are obtained from beef animals.
Courtesy of the American Angus Association.

Hot Topics in Agriscience: Finding a Better Beef in Genes

Genetic engineering techniques make it possible to control and lock in desirable hereditary traits in food animals.
Courtesy of USDA/ARS #K-4235-8.

Lean, tender beef is the product that is in demand from cattle producers. These qualities have been pursued in the past by attempts to breed cattle that will develop marbling in the meat without putting on too much fat on the outside of the carcass. Now science has identified a new approach to producing lean, tender beef. A gene has been isolated that codes for a protein called myostatin. In its active form, this gene controls the number of muscle fibers that develop in the fetus of a calf. In its inactive form, myostatin genes are paired in an animal, a condition known as double-muscling occurs.

Double muscling is considered undesirable because the extra muscle mass tends to produce stress in animals, and it has been shown to cause difficult births. When this gene is paired with another form of the gene, more extensive muscling occurs without the related stress. The animal produces a carcass with approximately 7 percent more beef and 14 percent less overall carcass fat than cattle that have active myostatin.

Swine

Origin and History

Swine were domesticated in China over a period of about 5,000 years, beginning about 4900 BCE. However, it appears that domestic swine originate from two wild stocks: the European wild boar, *Sus scrofa*, and the East Indian pig, *Sus vittatus*. Several other wild types of swine exist today. Even domesticated swine can revert quickly to the wild when the opportunity or need arises.

Swine came to the Americas in 1493 on the second voyage of Columbus. As European explorers came to what is now the United States, they brought swine as a food supply. The original 13 head of swine multiplied to more than 700 head in just 3 years and provided food for the colonists.

European colonists also brought swine with them as they settled on the East Coast of the United States. Swine could find food on their own and reproduce rapidly. Swine, in excess of local needs, were exported as pork and lard. **Pork** is the meat from swine. Lard is pork fat that has been obtained by cooking and pressing the oil from fatty tissue. This process is called *rendering*.

As westward expansion occurred, swine production followed. In time, the center of swine production settled in the Corn Belt, the Midwestern region of the United States. This area provided the large amounts of corn and other feed grains necessary for large-scale pork production.

Science Connection: A Solution to Offensive Smells

Suburban sprawl is closing in on farmland. As a result, conflicts are becoming more frequent because of the smells produced by animal wastes. Fortunately, researchers at the University of Florida have designed a piece of equipment called a fixed-film digester that may be a solution. A system of plastic pipes lines a 100,000-gallon tank. Bacteria inside the pipes transform smelly waste into compounds that do not have an odor problem.

The system can reduce offensive smells by as much as 90 percent. Besides clearing the air, the digester also produces a free fuel source for the farmer, biogas (methane). Another benefit is that water can be cleaned and recycled. While installation of the system has a high price tag, about $1,500 per cow, moderate operating costs and the added benefits make the system a reasonable farm expense.

Economic Importance

After beef, pork production is the second-leading red-meat industry in the United States, with 72.2 million hogs as of 2022. In a recent USDA report, China was the largest producer of pork, followed by the European Union (EU), and the United States. Together, China, the EU, and the United States account for 76 percent of the world's pork production. About 784 million head of swine are in the world today, more than 50 percent of them in China alone. China produced more than 50 million metric tons of pork in 2021, while U.S. production was 11 million metric tons. U.S. production of pork makes up approximately 13.2 percent of the world supply.

Over the last 50 years, the annual consumption of pork in the United States has remained mostly flat, hovering between 48 to 52 lbs. per person, up slightly from the 2011 low of 45.7 lbs.

The modern hog is vastly different from its ancestors. It is lean and trim compared with the lard-type hog that was in demand 50 years ago and even the meat-type hog of 20 years ago. Hogs are the most efficient converters of feed into meat among the large red-meat animals. It takes about 3.5 lbs. of feed to gain 1 lb.

Leading states in the United States in the production of swine include Iowa, North Carolina, Minnesota, Illinois, and Indiana. More than 31 percent of the swine production in the country takes place in Iowa alone.

Swine Operations

A **sow** is a female animal of the swine family that has given birth; a **gilt** is a young female that has not given birth. A **boar** is a male member of the swine family. A piglet is a baby pig. Once they are weaned, young pigs are called feeders.

The basic types of swine operations in the United States are purebred breeders, feeder-pig producers, and market-hog producers, though purebred breeders tend to engage to some extent in all of these types. This kind of swine operation is responsible for producing high-quality boars and gilts for feeder-pig operations and purebred stock for other purebred farms. They contribute much to the genetic improvement of swine in general.

Operations that produce feeder pigs usually maintain large herds of sows that annually produce approximately 2.1 litters of 9.5 piglets weaned per litter. Feeder pigs are often sold to other producers who feed them until they reach market weight. In many operations, specific hybrid sow lines are bred to compatible hybrid boar lines to produce offspring with superior hybrid vigor. These superior hog types have predictable, uniform meat characteristics that will command top market prices (**Figure 30-13**).

Market-hog operations normally purchase pigs at 5 to 8 weeks of age from feeder-pig producers. They feed the pigs until they reach a market weight of about 240 to 260 lbs. They are then marketed and sent to slaughter plants for processing.

Figure 30-13 Feeder-pig producers keep brood sows and boars that provide young stock for the swine industry.

Courtesy of Cole Swine Farms, Inc.

Types and Breeds

Purebred swine in the United States consist of meat-type hogs. With the decreased demand for lard and the demand for lean pork nearly devoid of fat, the lard-type hog has been bred out of existence.

Popular breeds of swine that were refined in the United States include Duroc, Hampshire, Chester White, Poland China, and Spotted Hog. The Berkshire, Yorkshire, and Tamworth breeds were developed in England, and the Landrace originated in Denmark. Duroc, Hampshire, Berkshire, and Yorkshire are the most popular breeds of swine in the United States today (**Figure 30-14**).

Figure 30-14 Major breeds of swine in the U.S. include (A) Duroc, (B) Yorkshire, (C) Hampshire, (D) Spotted, (E) Chester White, and (F) Landrace.

Photos A, B, C, and E: Courtesy of the National Swine Registry. Courtesy of Swine Genetics.

Approved Practices in Swine Production

Approved practices in swine production include the following:

- Buy pigs only from reputable producers or at certified feeder-pig sales.
- Observe newly purchased animals for signs of disease and parasites.
- Group pigs according to size in groups of not more than 20 to 25 animals.
- Feed a complete, balanced ration based on the age and weight of the animal.
- Ensure that access to an unlimited supply of fresh water is available at all times.
- Keep facilities and equipment clean and sanitary.
- Clean and disinfect all facilities and equipment after each group of animals leaves and before the next group arrives.
- Select replacement gilts for the breeding herd at an early age and raise them separately from market hogs.
- Breed gilts at 8 months of age or 250 to 300 lbs. so they farrow at approximately 1 year of age. To **farrow** means to give birth to baby pigs.
- Use a hand-mating system to breed gilts and sows. In a hand-mating system, the boar and sow are kept separate except during mating. Use a boar to check for animals in heat.
- Put bred gilts or sows in farrowing facilities or farrowing crates 3 days before they are due to farrow. Farrowing crates are specially made pens in which sows give birth. This 3-day period gives the mother pig time to adjust to new surroundings before the piglets are born.
- Perform the following to the piglets at birth:

1. Clip needle or wolf teeth.
2. Clip or tie navel cord and dip the end in iodine.
3. Provide supplemental iron.
4. Dock tails of pigs to be marketed for meat. To dock means to remove all but about 1 inch of the tail.
5. Weigh all pigs in the litter.
6. Ear notch all pigs for identification. Ear notching is a system of permanently marking animals for identification by cutting notches in their ears at specific locations.

- Provide creep feed for the baby pigs by the time they are 1 week old. Creep feed is feed provided especially for young animals to supplement the milk from their mothers.
- Castrate male pigs at an early age.
- Wean pigs at 4 to 6 weeks of age. Weaning at about 6 weeks is normal in most herds of swine.
- Rebreed sows on the first heat period after weaning the piglets. This usually occurs about 3 days after the pigs are weaned.
- Limit the feed for gestating sows to prevent them from getting too fat.
- Provide protection from heat and cold, especially heat. Swine have no sweat glands, and care must be taken to keep them cool in hot weather.
- Maintain complete health and production records for each animal in the breeding herd.
- Set realistic production goals and cull animals that do not meet the goals.

Sheep

Origin and History

The domestication of sheep occurred before the time of recorded history. Fibers of wool have been found in ruins of the earliest Swiss villages. Egyptian sculptures showing the importance of sheep date back to the period of the great dynasties. Even ancient historical texts are filled with mentions of sheep and shepherds.

Sheep were probably domesticated from wild types in Europe and Asia by early humans to use for meat, wool, pelts, and milk. **Wool** is a modified hair with superior insulating qualities. It is usually obtained from sheep, but it is also obtained from some other animals such as llamas. Wool has been an important product of international trade for many centuries, and it is still one of the most important fibers known to humans.

As civilization advanced, the production of wool became a priority in sheep production. As a result, specific wool-producing breeds were developed in Europe. Most of today's breeds can be traced back to these breeds developed 500 to 1,000 years ago.

Columbus brought sheep with him on his second voyage to the West Indies in 1493. Spanish explorer Hernando Cortéz also brought sheep when he explored Mexico in 1519. Spanish missionaries also kept sheep and taught Indigenous Peoples of the of the Southwest how to weave wool into cloth. English colonists on the East Coast also raised sheep for the production of wool. Lamb and mutton were of secondary concern. The term **lamb** refers to a young sheep as well as the meat of young sheep. **Mutton** is meat from mature sheep.

Centers of sheep population gradually moved from the Northeast to the West as populations expanded. Areas of open spaces and abundant grasses such as western rangelands have proved to be ideal for the production of sheep.

Economic Importance

Production of sheep in the United States has declined in the last 50 years, while dairy, beef, and swine production has expanded. Since the 1960s, the annual consumption of lamb and mutton in the United States has dropped from 5 lbs per person to less than a pound (0.9 lb.) per person. This amount is not expected to increase. Each person in the United States also uses about 0.8 lb. of wool each year.

In 1884, the U.S. sheep population peaked at 45 million, but by 2020 has declined to 5.2 million sheep. The production of wool yields about 17 million lbs. Leading states in the production of lamb include Texas, California, Wyoming, Colorado, and Utah.

Because sheep have the ability to survive in areas of limited feed and harsh climates, they are of more economic importance than would be expected. In many such areas, only sheep can survive, and they, therefore, constitute a major animal enterprise (**Figure 30-15**).

Figure 30-15 Sheep can use low-quality forage, graze land unsuited for other purposes, survive severe weather conditions, and produce meat and wool.

Courtesy of USDA/ARS #K-4166-5.

Types and Breeds

Sheep operations can be divided into two basic types: farm flocks and range operations. *Flock* is a term describing a group of sheep. Farm-flock operations are generally small and are often part of diversified agricultural operations. They may raise either purebred or grade sheep. These flocks usually average fewer than 150 animals. They are responsible for about one-third of the sheep and wool produced in the United States.

The other two-thirds of the sheep in the United States are produced on range operations. Many flocks contain 1,000 to 1,500 head or more. They are nearly 100 percent grade sheep. Range production is concentrated in the 12 western states. In the United States, there are five basic classifications of sheep according to wool type. They are fine wool, medium wool, long wool, crossbred wool, and fur sheep.

Fine-wool breeds of sheep produce wool that is fine in texture with a moderate staple length. The wool is dense, with a wavy texture, and is used to make fine-quality garments. Fine-wool sheep can produce up to 20 lbs. of wool per sheep per year. Breeds of fine-wool sheep all originated from the Spanish Merino breed. Fine-wool breeds in the United States include Merino, Debouillet, and Rambouillet (**Figure 30-16**).

Figure 30-16 Fine-wool breeds of sheep include (A) Rambouillet and (B) Merino.
© Lynn M. Stone/Alamy.com. © stockphoto mania/Shutterstock.com.

Medium-wool breeds of sheep were developed for meat, and little emphasis was placed on the production of wool. Popular medium-wool breeds include Suffolk, Shropshire, Dorset, Hampshire, and Southdown (**Figure 30-17**). The long-wool breeds of sheep were developed in England. They tend to be larger than most of the other breeds. Their wool tends to be long and coarse in texture. Long-wool breeds in the United States include Leicester, Lincoln, Romney, and Cotswold.

Figure 30-17 Medium-wool breeds of sheep include (A) Hampshire and (B) Suffolk.
Courtesy of the American Hampshire Sheep Association.

Crossbred-wool breeds are the result of crossing fine-wool breeds of sheep with long-wool breeds. They were developed to combine good-quality wool with good-quality meat. Because they tend to stay together as a group better than other breeds of sheep, they are said to have a *herding instinct*. They are popular in the western range states. Crossbred-wool breeds include Corriedale, Columbia, Panama, Targhee, and Polypay (**Figure 30-18**).

Figure 30-18 Crossbred wool breeds of sheep include (A) Columbia and (B) Corriedale.
Courtesy of the Columbia Sheep Breeders Association of America. Courtesy of the American Corriedale Association.

There is only one breed of fur sheep in the United States. The Karakul is grown for the pelts of its lambs. Young lambs are killed shortly after birth, and the pelts are made into expensive Persian lamb coats. The production of wool and meat is of little importance in this breed.

Approved Practices in Sheep Production

Approved practices in sheep production includes the following:
- Select lambs that are growing rapidly and are large for their age.
- Select purebred stock based on physical appearance, production, and pedigree.
- Select breeding stock with a history of multiple births.
- Provide shelter from severe weather conditions during the lambing season.
- Provide good-quality forages and unlimited fresh, clean water.
- Vaccinate for disease problems of local concern.
- Treat regularly for internal and external parasites.
- Breed ewes to lamb at no more than 2 years of age. A **ewe** is a female animal of the sheep family.
- Use marking harnesses on rams to tell when ewes have been bred. A **ram** is a male animal of the sheep family.
- Use a system of identification to distinguish ewes from one another.
- Do not unnecessarily disturb ewes during lambing. With sheep, lambing means giving birth.
- Shear ewes at least 1 month before lambing except in extreme climates. Shearing is the process of removing wool from sheep.
- Provide a clean, warm, dry environment for lambing.
- Make sure that ewes accept their newborn lambs.
- Dock (remove) lambs' tails at 7 to 10 days of age.
- Castrate ram lambs that will be marketed for meat.
- Keep the hooves of farm sheep properly trimmed.
- Maintain a complete flock disease- and parasite-prevention and control program.
- Cull ewes that do not lamb or those that have health problems.
- Maintain a complete and accurate recordkeeping system.

Goats

Origin and History

The domestication of goats probably took place in Western Asia during the Neolithic Age, between 7000 and 3000 BCE. Remains of goats have been found in Swiss lake villages of that period. Mention of the use of mohair from goats is made in the Bible. **Mohair** is hair from Angora goats that is used to make a shiny, heavy, wooly fabric.

Goats were imported to the United States from Switzerland for milk production early in the colonial period. Angora goats, to be used for mohair production, were also imported from Turkey. Approximately 97 percent of the mohair-producing goats in the United States are located in Texas. Milk or dairy-type goats can be found throughout the United States; however, there are concentrations on the East and West coasts.

Economic Importance

Goats are of relatively low economic importance in the United States. Most dairy goats are raised in small numbers by suburbanites and small farmers to produce milk and meat for their own families. There are few large herds of milk goats. Finding a processor to bottle goats' milk is often difficult.

Mohair production in The United States has significantly declined, but annual domestic production still accounts for nearly 20 percent of the mohair in the world. Texas provides more than 97 percent of the 589 thousand lbs. produced in the United States annually.

Goats provide little competition for food with cattle and sheep. They prefer to eat twigs and leaves from woody plants rather than grass.

Types and Breeds

Three types of goats are raised in the United States: hair producing, meat producing, and milk producing.

The only hair-producing goat is the Angora (**Figure 30-19**). It produces 6 to 7 lbs. of mohair per year. Mohair ranges in length from 6 to 12 inches. Angora goats are best adapted to a dry climate with moderate temperatures. In addition to producing mohair, Angora goats are used for meat and to help control weeds and brush. The current interest in Boer, Kiko, and Spanish meat goat breeds has resulted in a market for meat-producing goats (**Figure 30-20**). Goat meat is called **chevon**.

Figure 30-19 The Angora is the only commercial breed of hair-producing goat in the United States.

© Eric Isselee/Shutterstock.com.

Figure 30-20 (A) Boer and (B) New Zealand Kiko are the most widely used commercial breeds of meat-producing goats in the United States.

© Acreagemedia/dreamstime.com. © imagesbycat/Shutterstock.com.

Dairy goats are found in every state in the United States (**Figure 30-21**). Normal production per goat averages 1 gallon of milk per day during a 10-month lactation period. **Lactation** is the process of producing and releasing milk. The common breeds of dairy goat are LaMancha, Nubian, Alpine, Saanen, and Toggenburg (**Figure 30-22**).

Figure 30-21 Dairy goats are highly adaptable, thriving in mountain pasture settings to confinement operations.
Courtesy of the American Dairy Goat Association.

A　　　　　　　　　　　　　　　　　B

Figure 30-22 (A) LaMancha and (B) Nubian are widely used commercial breeds of dairy goats in the United States.
Courtesy of the American Dairy Goat Association. Courtesy of the American Dairy Goat Association.

Approved Practices in Goat Production

Approved practices in goat production includes the following:
- Select goats according to intended use.
- Use physical appearance, pedigree, and records as a basis of selection.
- Purchase replacement animals from reputable breeders.
- Provide adequate feed for hair goats on the range in winter.
- Feed dairy goats supplemental grains based on the amount of milk production.
- Breed goats in the fall to have **kids** (young goats) in the spring.
- Breed does (female goats) for the first time at 10 to 18 months of age.
- Use one male goat (buck or billy) for every 20 to 50 does.
- Shear or clip hair goats twice each year.
- Maintain clean and sanitary conditions for the production of milk.
- Castrate bucks that are not to be used for breeding at an early age.
- Maintain a herd-health program.
- Milk dairy goats twice daily.
- Dehorn dairy-type goats at an early age.
- Maintain complete and accurate records of reproduction, production, and health.

Importance of Livestock

Nearly every facet of life is affected in one way or another by animals and animal products. There are types of livestock production that are adapted to almost every locality and situation. The production of dairy and livestock in the United States is big business. When one considers the value of products and the income from jobs that are created by this sector of the United States economy, the total economic impact is approximately $289 billion annually. Although food from animals is more expensive than food from crops, animal products add variety and quality to the human diet. Similarly, animals are sources of high-quality fabrics, leather, and many other products. Therefore, animals and the production of animals will be important enterprises far into the future.

Science Connection: Radio Signals on the Ranch

Wearable transponders permit better research and management of farm animals. Devices send data to a computer that records it. Then scientists are able to analyze the data.
© attraction art/Shutterstock.com.

Transponders are tiny devices used to receive and transmit a radio signal. Transponders can be attached to the ears of cows and other livestock to identify them. As animals pass through the electronic fields of certain other devices, their identities can be noted and recorded by computer. The computer can be programmed to simply note and remember what the animal does, or it can direct other devices to provide selected kinds and amounts of feed, water, medicine, pesticide, or other treatment. This technology is becoming more common in modern milking parlors and feeding systems.

At the USDA Agricultural Research Service (ARS) Fort Keogh Livestock and Range Laboratory in southeastern Montana, scientists found transponders to be the answer to a problem. To conduct feeding, grazing, pasture-improvement, and pasture management experiments, they had to weigh their research subjects, range cattle, at frequent intervals. Animal response to treatment generally shows up in the form of weight gain or loss. At the same time, any disturbance or change in an animal's routine, such as cattle roundup, forced weighing, or catching in a head gate, will influence an animal's weight. Quality data required a system that could weigh the animals at frequent intervals without disturbing them—that is, a way was needed for the animals to weigh themselves.

The answer was found in a system using a scale, a computer, and a transponder. Cattle need water. A continuous supply of fresh water must be readily available if body functions are to be optimal. Therefore, the researchers set up a scale and guide railing so the animal had to cross over the scale to drink. An ear-tag transponder on the animal identifies the animal as it enters the range of the electronic pickup device and steps on the scale. The electronic scale sends continuous weight readings to the computer, indicating the animal's weight before, during, and after drinking. The computer records the weight data, together with time, temperature, weather, and other information monitored at the site.

Chapter Review

Student Activities

1. Write the **Key Terms** and their meanings in your notebook.
2. Develop a word search or other puzzle using the Terms to Know. Trade your puzzle with someone else in class and solve their puzzle.
3. Make a chart showing when and where the various types of animals were domesticated.
4. Participate in a class discussion on how and why certain animals were domesticated and others were not.
5. Take a class survey of all of the breeds of animals owned by students and their families.
6. Invite a breeder of purebred livestock and a breeder of commercial, or grade, livestock to class to discuss advantages and disadvantages of each type of operation.
7. Conduct a survey of your area to determine the types of livestock being raised, the breeds, and the numbers of each.
8. Make a bulletin board or slideshow presentation showing the various types of livestock in your area and the importance of each.
9. **SAE Connection:** Visit a livestock operation to determine the types of jobs that need to be performed there and what training would be necessary to perform the tasks.
10. Survey various purebred livestock magazines to determine prices of animals being sold at various purebred sales.
11. Make a notebook-cover collage showing as many breeds as possible of the particular type of livestock in which you are interested.
12. Make a bulletin board or slideshow presentation showing some items made from animal products.

Inquiry Activity | Dairy Cows

Courtesy of the Holstein Association USA, Inc.

After watching the *Dairy Cows* video available on MindTap, complete the following items.

1. A female calf's diet changes throughout its first year. Create a timeline of a female calf's diet from newborn through nine months old. Rewatch the video as needed to complete your timeline.
2. Imagine you are a dairy farmer talking to a group of students visiting your farm. In your own words, describe the practices of managing dairy cows including breeding, feeding, and milking.

National FFA Connection: Dairy Cattle Evaluation and Management

The Dairy Cattle Evaluation and Management Career Development Event of the National FFA Organization helps students gain skills in dairy cattle selection and herd management.
Courtesy of the American Guernsey Association.

Have you ever wanted to compete "until the cows come home"? Now's your chance. At one popular FFA career development event, teams of students use their skills to evaluate dairy cattle based on physical characteristics, explain milk classes, and analyze a herd record. Teams also respond to a dairy farm management scenario provided by judges. It's up to each team member to help identify problems and solutions as if they were actual professional consultants hired to advise a commercial dairy producer.

An introductory video to the competition is available on the National FFA Organization website. (Search for "Dairy Cattle Evaluation and Management Career Development Event.") View the video to become acquainted with the skills you will need to develop as you participate in this event. Local and state competitions require many of these same skills. Let your agriscience instructor know of your interest in participating.

Check Your Progress

Can you…

- briefly summarize the history and economic importance of dairy and livestock?
- identify major types and breeds of livestock?
- give examples of major uses of livestock?
- describe basic approved practices in the care and management of dairy and livestock?

If you answered "no" to any of these questions, review the chapter for those topics.

Chapter 31
Horse Management

Objective
To determine the role of horses in our society and to learn how to care for and manage them.

Competencies to Be Developed
After studying this chapter, you should be able to:
- understand the origin and history of the horse.
- determine the economic importance of the horse in the United States.
- recognize the various types and breeds of horse.
- understand approved practices for the care and management of horses.
- understand the basics of English and western riding.
- list rules of safety for handling horses.
- understand the vocabulary associated with horses.

Key Terms

light horse	jack	filly
draft horse	mare	foal
donkey	hinny	colt
equine	stallion	gelding
horse	jennet	farrier
ponies	tack	blemish
hand	gait	unsoundness
color breed	equitation	
mule	vice	

Horses have been closely associated with humans for much of recorded history. Of most significance was their role in transporting people and materials. For close to 5,000 years, they provided power for farming, and teamsters used them to haul goods and materials, much as our modern trucks move products of commerce. Horses and mules have also filled important roles in military operations, making it possible to move soldiers and supplies. Today over 3.6 million horses and ponies are in the United States. Farmers and ranchers still own and use horses, but most people use horses primarily for pleasure, training, breeding, and companionship (**Figure 31-1**).

Horses

Origin and History

The horses of today represent just one branch of a vast family tree that took root millions of years ago. Fossil remains of the horse family dating back nearly 58 million years have been found in what is now North America. The earliest horses were only as big as a medium-sized dog. It is a common belief that early horses migrated to Europe and Asia across the Bering Strait when Alaska and Siberia were connected. About 5,000 years ago, horses were domesticated in Central Asia or Persia. They were likely one of the last animals to be adapted for human use. Horses arrived in Egypt in about 1680 BCE from Asia, and from there, the Egyptians introduced horses over the known world. However, horses were not used to any great extent until AD 500 to 600. By the time of the arrival of Europeans in what is now known as the Americas, horses no longer lived here.

Figure 31-1 Horses are companions and pets to youths and adults.
© Aturner/Shutterstock.com.

The Arab horse is the ancestor of most modern light breeds of horse. **Light horses** are breeds used for riding. Draft horses probably originated from the heavy Flanders horse of Europe. **Draft horses** are used for work. At the time of domestication, they were often used in religious rites for sacrifice or for food. Donkeys were domesticated in Egypt some time before 3400 BCE when they appeared on slates of the First Dynasty. A **donkey** is a member of the horse family and has long ears and a short, erect mane. Mention of using donkeys appears in many places in the Bible and ancient Greek mythology. They were generally used as saddle animals or beasts of burden.

In AD 1493, Columbus imported horses to the Americas. When Cortez came to Mexico in 1519, he brought nearly 1,000 horses with him. The death of the explorer Hernando de Soto in the upper Mississippi region led to the abandonment of many of his horses. These horses, coupled with some of those lost or stolen from Spanish missionaries, likely formed the nucleus of the horses used by the Indigenous peoples of the Great Plains. Herds of wild horses roam the desert regions of the American West even today.

The introduction of the horse to the Indigenous peoples of the Great Plains completely changed their culture. Horses permitted easier hunting of buffalo in wider areas. This led to competition between various tribes and a nearly constant state of war in an area where peace had previously prevailed.

The European settlers on the East Coast were far less dependent on the horse than were the explorers. It was of little advantage to have a riding horse with no place to ride. As draft animals, oxen or cattle were more popular because they were stronger than horses, and they produced meat and milk as well as work.

It was only as early Americans prospered that horses became an important part of their lifestyles. Horses have had an illustrious past, and a bright future looms ahead.

Hot Topics in Agriculture: Wild Horses in the United States

Wild horse roundups and government wild horse sales make wild horses available to anyone who has the facilities and the interest to own one.
Courtesy of DeVere Burton.

In 2022, the Bureau of Land Management estimated that 64,000 wild horses and about 17,000 burros were scattered across nearly 43 million acres of wild desert rangelands in the western United States. Most of these animals live on federal lands where the ranges are shared with livestock and other wildlife.

The wild horse population is capable of doubling every four years under good conditions. Agencies managing the herds have been hard-pressed to maintain the population at a consistent level that can be supported by the available supply of forage. Legislation mandates that the herds may not be reduced by lethal means; therefore, horses are routinely rounded up, and excess animals are offered to the public for adoption.

It is generally believed that the wild herds are the remnants of horses that escaped from the Spanish conquistadors before other Europeans arrived in what is now the Americas. Wild horses are smaller than most domestic light horse breeds. Wild stallions weigh about 1,000 lbs., whereas stallions of domestic breeds generally weigh close to 1,200 lbs. Controversy has surrounded these wild animals in recent times because of the tendency of the horses to overrun the ranges as their numbers expand. Adoptions are useful in controlling the expansion of wild herds, but they are also expensive. Adopted horses are usually penned for several months, requiring feed, immunizations, and transportation.

Benefits and Economic Importance

The U.S. horse industry has a direct impact of approximately $50 billion on the economy each year and provides nearly one million jobs. Direct impact means activities occurring within the horse industry, such as care and management. These activities have a ripple effect creating additional economic activity outside of the horse industry, such as tourism spending by participants and spectators of events. Thus, the U.S. horse industry has the total economic impact of generating $122 billion and providing 1.7 million jobs a year. The racehorse industry alone generates nearly $3.5 billion of annual revenue.

Horses provide numerous benefits besides economic ones. They help young people develop a sense of responsibility. They provide physical activity, companionship, and opportunities for families to participate in outdoor activities together. Horseback riding is also therapeutic for the healing of certain injuries and disabilities (**Figure 31-2**).

Figure 31-2 Horses help young people develop a sense of responsibility.
Courtesy of DeVere Burton.

Science Connection: All in the Horse Family

The theme song of an old TV show begins "A horse is a horse, of course, of course." But what about animals that only resemble horses, such as zebras and donkeys? And where do mules fit in? This abbreviated version of the taxonomic hierarchy shows the biological classification of these animals.

Taxonomic Hierarchy
Kingdom: Animalia
Phylum: Chordata (sharing the common feature of a spinal cord)
Class: Mammalia
Order: Perissodactyla (odd-toed, hoofed)
Equidae: (sometimes known as the horse family) Horses, zebras, asses, and various extinct related species known only from fossils
Genus: (Existing species) Horses, Zebra, Donkeys, Mules*

*Unlike horses, zebras, and donkeys, mules are not a separate species, but a hybrid of two species, a female horse and a male donkey. A species is a group of organisms that can interbreed and create offspring. Two mules cannot be bred to reproduce a mule.

Anatomy

Among the first things to learn about horses are the names of their body parts (**Figure 31-3**). Differences in body parts define the different breeds. Knowing these terms also helps a person understand what horse enthusiasts are talking about when they discuss the merits of a particular horse.

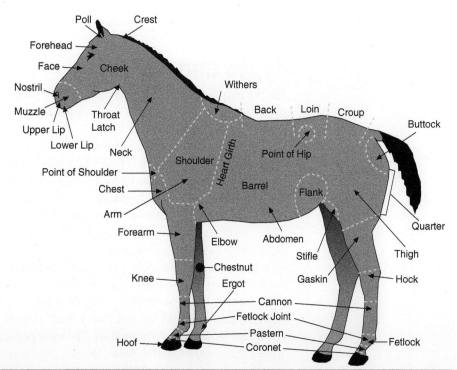

Figure 31-3 Learning the proper names of the parts of a horse aids in discussing, evaluating, and caring for horses.

Agri-Profile — Career Areas: Horse Veterinarian/Trainer/Jockey/Farrier

Maintaining working horses in peak condition is vital to the daily operations of many enterprises.
Courtesy of Barbara Lee Jensen, After Hours Farm.

Although many people raise horses as a hobby, careers in horse management are available in a variety of settings.

For example, racetracks and breeding farms depend on several categories of professionals:

- Track veterinarians ensure that racehorses are healthy to compete.
- Trainers manage the health and performance of racehorses in their care, from feeding schedules to routine vet visits. Some trainers start out as attendants and groomers, maintaining the cleanliness of stables and horses.
- Farriers are skilled workers who trim and shoe horses' hooves, often forging horseshoes by hand.
- Jockeys are trained professionals who ride horses in races and for the animals' exercise.

Tack, feed, and equipment supply centers provide business opportunities for those wishing to interface with the horse industry but preferring not to handle horses directly. For career seekers with a keen interest in horse health, soundness, breeding, behavior, there are related career openings in research, veterinarian services, and education. Horse-related jobs may be indoors or outdoors and run the spectrum from laborer to scientist. Educational requirements range from a high school degree for groomers to advanced degrees for equine medical and science careers.

The national average salary for a horse veterinarian is $84,000 for a surgeon and $105,900 for a vet at a major equestrian organization. The salaries of jockeys range from $10,049 to $271,427 with a median salary of $48,880. The typical salary range for a horse trainer or farrier is $33,000 to $42,000.

Types and Breeds

For the purpose of this discussion, members of the **equine** (horse or horse-like) family are divided into horses, ponies, donkeys, and mules. Horses are further divided into light horses, coach horses, and draft horses.

Members of the horse family that are 14.2 hands or taller are called **horses**. Animals of the same family that are less than 14.2 hands tall are **ponies**. A **hand** measures 4 inches and is used as a unit of measure for members of the horse family to determine the size of horses, mules, donkeys, and ponies.

The light breeds of horses are comprised of the most popular horses in the United States. Light breeds are further divided into true breeds, or purebreds, and color breeds. True breeds have registration papers and parents of the same breed. **Color breeds** need only be specific colors or color patterns, although other characteristics may be considered for registration. They need not have purebred parents.

Examples of the nearly 50 true breeds of horse in the United States include Arabian, Morgan, Thoroughbred, Quarter Horse, Standardbred, Tennessee Walking Horse, and American Saddle Horse (**Figure 31-4**). The most popular of the light breeds of horse in the United States is the Quarter Horse. Quarter Horses are used for riding, hunting, racing, sports, and ranch work with cattle and sheep.

Figure 31-4 The most popular breeds of light horses in the United States, based on number registered, are (A) Quarter Horse, (B) Thoroughbred, (C) Arabian, (D) Standardbred, (E) Paint Horse, and (F) Appaloosa.

Courtesy of the American Quarter Horse Association. Courtesy of the Illinois Racing News. Courtesy of Johnny Johnston. Courtesy of the U.S. Trotting Association. © Zuzule/Shutterstock.com. Courtesy of the Appaloosa Horse Club.

Some of the color breeds of horses include the Buckskin, Palomino, Appaloosa, and Pinto or Paint. Color breeds may or may not breed true to color. The uses of the color breeds are the same as those of the true breeds of light horse.

Draft, or work, horses were once the backbone of U.S. agriculture. The invention of the internal combustion engine replaced the horse as it can do the work of many horses and requires less maintenance and care. Today, draft horses are primarily used for show and recreation. Breeds of draft horse include Belgian, Clydesdale, Percheron, Shire, and Suffolk (**Figure 31-5**).

Coach horses combined the qualities of both the light horse and draft breeds. They were fast (like the light horse) and strong enough (like a draft horse) to pull heavy stagecoaches and freight wagons. When stagecoaches disappeared from roads in the United States, there was no longer a need for coach horses, so they became rare. The Cleveland Bay is the only coach horse breed in the United States today.

Ponies are smaller versions of the horse. They are used for riding and driving and as pets. Some breeds were originally developed to work in coal mines and in other pursuits, where small size is important. The common breeds of pony include Shetland, Welsh, Gotland, Appaloosa, Connemara, and Pony of the Americas. The Shetland is the most popular breed of pony in the United States (**Figure 31-6**).

Donkeys are used in the United States for work and as pack animals. Miniature donkeys also make excellent pets. The male donkey is used as the male parent of a mule. A **mule** is a cross between a jack and a mare (**Figure 31-7**). A **jack** is a male donkey or mule. **Mares** are mature female horses or ponies. The opposite cross, a stallion with a jennet, results in a **hinny**. A **stallion** is an adult male horse or pony. A **jennet** is a female donkey or mule

Mules are thought to be more intelligent than horses and can do more work than comparable-sized horses. There are several types and sizes of mule, depending on the breed or type of horse that serves as the female parent.

Figure 31-5 Draft horses are now kept primarily for historic interest, show, and recreation.

© Becky Swora/Shutterstock.com.

Figure 31-6 The Shetland pony can be trained for a variety of uses, ranging from riding to show pony.

Courtesy of Multi-World Champion Shetland Harness Pony.

Figure 31-7 The mule is a hybrid cross between a jack donkey and a horse mare.
© patti jean_images & designs by patti jean guerrero/Shutterstock.com.

Hot Topics in Agriscience: Color Genetics

The dominant gene for color in horses results in bay coloring or one of its shades.
© Alexia Khruscheva/Shutterstock.com.

The color of a horse is determined by the combination of several genes and the presence or absence of two important pigments in the hair and skin. Two primary pigments are found in the coats of mammals. These pigments are eumelanin, a black or brown pigment, and pheomelanin, which is red. All of the horse colors are combinations of these pigments. In many instances, these pigments are diluted by certain gene combinations, resulting in different degrees or shades of color. Spots of white indicate the absence of the pigments. Genes control the distribution of the pigments in the skin and hair. This explains the white and colored spots of the Appaloosa and Paint breeds of horses.

Overriding all other genes for color is the dominant gene A, which results in the coloring of bay or one of its shades. Every time the A gene is present, the horse will be bay in color. Another interesting color combination is "roan." In this instance, hair of different colors is mixed together. A "sooty" or "smutty" color occurs when black is mixed with hair of other colors.

Many different colors are possible in horses. However, once the offspring of a stallion has been observed for color, it is possible to work out the genetics that are involved. When the color genetics of both the stallion and the mare are known, the possible color combinations in the foal can be predicted with reasonable accuracy. However, it is beyond the scope of this book to expand on the specific genetic combinations and the colors they produce.

Riding Horses

The styles of riding horses can be divided into two general classifications: English and western. The style of riding and the **tack** (horse equipment) vary greatly between the two classifications. Even the **gaits** (the ways an animal moves) of the horses involved have different names.

English

Equitation is the art of riding on horseback. English riding attire includes close-fitting breeches, jodhpurs, jacket, and a hard hat or derby (**Figure 31-8**). A saddle is the padded leather seat for the rider of a horse. Stirrups are foot-supports hung from each side of the saddle. In English equitation, the saddle is smaller and lighter and has shorter stirrups than the western saddle (**Figure 31-9**).

The gaits of the English-equitation horse are walk, trot, canter, gallop, rack, and pace. The walk is a slow, four-beat gait. The trot is a fast, two-beat diagonal gait. The canter is a

Figure 31-8 Proper riding attire for English equitation riders.
© pirita/Shutterstock.com

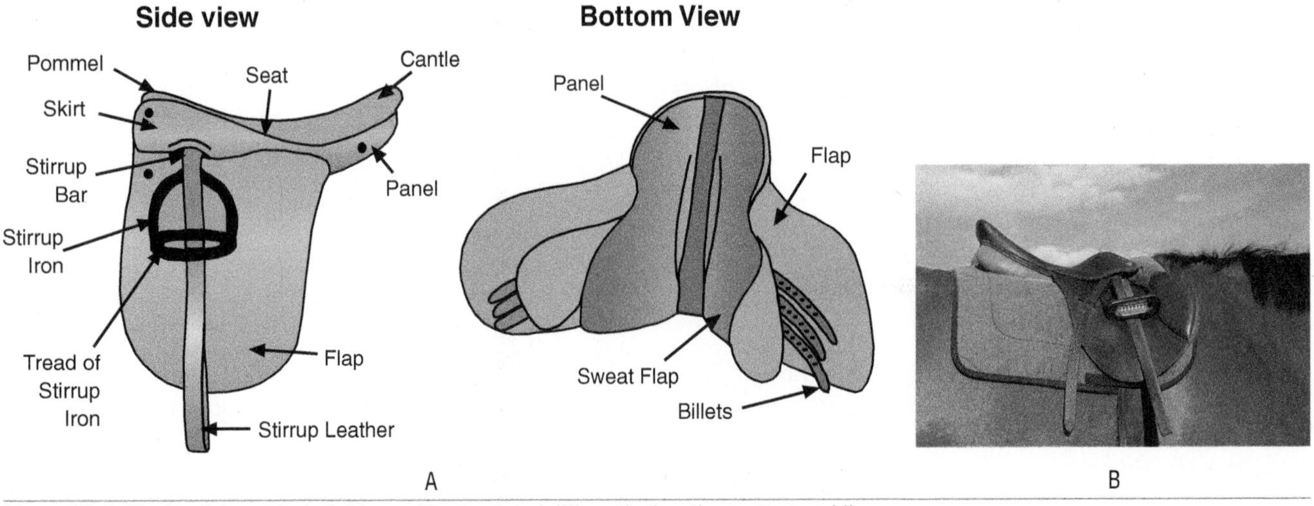

Figure 31-9 The English saddle is lighter and is structured differently than the western saddle.
© marekuliasz/Shutterstock.com

slow, three-beat gait, whereas the gallop is a fast, three-beat gait in which all four feet come off the ground at the same time. The rack is a fast, four-beat gait, and the pace is a side-to-side, two-beat gait (**Figure 31-10**).

The English-equitation horse is controlled by reins that may be held either in one or both hands. Reins are leather or rope lines attached to the bit. The bit is a metal mouthpiece used to control the horse.

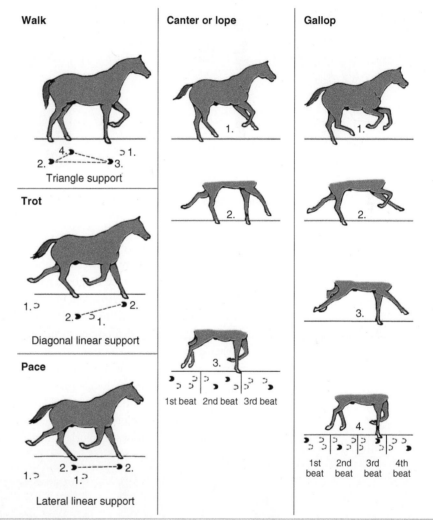

Figure 31-10 Basic gaits of English equitation include walk, trot, pace, canter, and gallop.

Figure 31-11 Proper riding attire for western riders.
© Diane Garcia/Shutterstock.com.

Western

Western riding differs from English riding. Western riding attire was designed for comfort (**Figure 31-11**). It consists of jeans, chaps, western shirt, wide-brimmed cowboy hat, and cowboy boots. The western saddle was designed for the comfort of cowboys, who spend much of their time in a saddle (**Figure 31-12**). The western saddle is heavy and has a horn that is used to secure the rope when roping livestock. This saddle has longer stirrups than the English saddle.

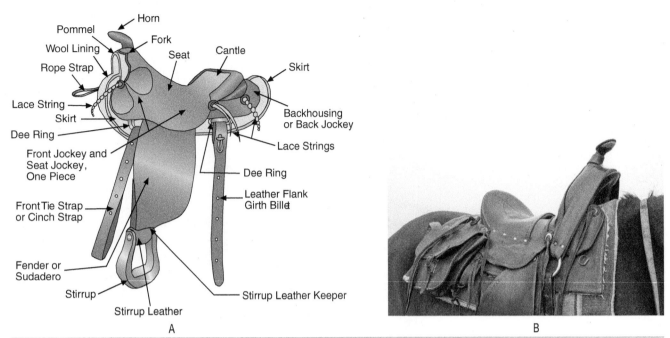

Figure 31-12 Western saddles were originally developed for cowboys who worked with cattle.
© Jan S./Shutterstock.com.

The gaits performed by the western horse include walk, jog, lope, and gallop. The jog is a slow, smooth, two-beat diagonal gait. The lope is a slow canter. The western rider controls the horse with the reins held in one hand (**Figure 31-13**). The hand holding the reins cannot be changed during competition riding.

Working horses are still found on farms and ranches (**Figure 31-14**). They are essential for managing and caring for cattle and sheep on the vast western rangelands. They are used extensively on ranches in areas where other forms of transport are difficult, impossible, or illegal (such as roadless areas). Range riders depend on their horses to get them into the canyons and sometimes difficult terrain where cattle and sheep graze. Horses are also used to check cattle in large feedlots where it is necessary to separate individual animals from the herd for treatment of illnesses or injuries. A well-trained horse is a great asset for this task.

Figure 31-13 Western riding requires control of the horse with one hand on the reins and the other hand free to handle a rope.
© iStockphoto/Rosemarie Gearhart.

Figure 31-14 Training of horse and rider is the key to success. They must coordinate their movement to control herds of livestock.
© iStockphoto/Jason Lugo.

Safety Rules

Some safety rules to observe when riding and caring for members of the horse family include:
- Approach a horse from the front-left side only.
- Do not act in a way that would startle or scare a horse as it may kick or rear up.
- Make the horse sure knows exactly what you intend to do to or with it.
- Pet the horse by putting your hands on its shoulder, not on its nose.
- Tie horses that are strangers far enough apart so they cannot fight.
- Walk beside the horse when leading it, not in front of or behind it.
- Do not wrap the lead rope around your hand when leading a horse.
- Adjust the saddle on the horse tight enough that it will not slip or slide.
- Mount the horse from the left side only (**Figure 31-15**).
- Keep the horse under control at all times.
- Do not allow the horse to misbehave without disciplining it.
- Walk the horse up and down steep slopes, on rough ground, and across paved roads.
- Reduce speed when riding on rough terrain or in wooded areas.
- Ride in single file in groups and on the right side of the road.
- Be calm and gentle when dealing with your horse.
- Wear appropriate riding attire, including hard hats when jumping obstacles.
- Do not tease the horse.
- Walk the horse to and from the barn or holding area to prevent it from developing a habit of riding to the area when it comes into sight.
- Know the temperament and **vices**, or bad habits, of the horse.
- Be especially aware of barking dogs and other things that may startle or frighten the horse.
- Never allow other people to ride your horse unsupervised.

Figure 31-15 Always mount a horse from the left side.

© DDCoral/Shutterstock.com.

Approved Practices

Breeding Horses

Some of the approved practices for breeding horses are the following:
- Breed fillies so they will foal for the first time at 3 to 4 years of age. A **filly** is a young female horse or pony. A **foal** is a newborn horse or pony. To foal means to give birth in the horse family.
- Breed mares in April, May, or June when their fertility is normally greatest. Breeding at this time also allows the mare to foal in the spring when pastures are actively growing.
- Rebreed mares 25 to 30 days after foaling, if they are in good condition with no reproductive-tract problems.
- Make sure the pregnant mares and fillies get plenty of exercise.
- Allow the mare to foal in a clean pasture, free from internal parasites, or in a large, clean pen or stall (**Figure 31-16**).
- Acclimate the mare in the facility in which she will foal several days before she is expected to foal.
- Remain out of the mare's sight when she is foaling. Mares prefer to be alone with no interruptions during the foaling process.
- Contract the services of a veterinarian if the mare shows signs of difficulty in foaling or if the position of the foal is abnormal for delivery.

Figure 31-16 Mares and foals should be kept in clean, parasite-free surroundings.
© Karel Gallas/Shutterstock.com.

Caring for Foals

Some of the approved practices for caring for a foal from birth are the following:

1. Make sure the foal is breathing. Tickling the foal's nose with a piece of straw often will stimulate it to start breathing after it is born. More drastic measures, such as artificial respiration, may be needed in serious situations.
2. Remove mucus from the nose of the foal, and dry off the foal.
3. Dip the end of the navel cord in iodine after tying it off to prevent infection from entering the foal through the open navel cord.
4. Make sure the foal nurses as soon as it can stand. Let it stand on its own. Normally, nursing occurs within one-half hour after birth.
5. Check to see the foal has a bowel movement within the first 12 hours of birth. If this does not occur naturally, the foal may need an enema.
6. Reduce the amount of milk the foal is allowed to drink if it shows signs of diarrhea,
    - Provide creep feed for the nursing foal when it reaches 10 days to 2 weeks of age.
    - Begin training the foal as soon as possible. A foal that is trained from an early age seldom needs to be broken.
    - Do not mistreat the mare or foal at any time.
    - Wean the foal at 4 to 6 months of age. The foal and the mare should not be allowed to see each other for several weeks after weaning to break the bond between mother and offspring.
    - Castrate colts that are not to be used for breeding purposes. A **colt** is a young male horse or pony. A castrated male horse is called a **gelding**. Geldings are usually much more docile and easier to handle than stallions.

General Care and Management

Approved practices for the general care and management of all horses and ponies include the following:

- Groom the horse daily or weekly at a minimum (**Figure 31-17**).
- Use only soft brushes when grooming a horse as its skin is sensitive and easily damaged by rough treatment.
- Shorten the mane and tail when needed by pulling the hairs from the underside.
- Be especially careful not to get water in the horse's ears when washing.
- Dry off the horse quickly so that it does not catch a cold in cool weather.
- Inspect and clean the horse's hooves daily and always before and after riding.
- Trim the horse's hooves every 4 to 6 weeks if the horse is not shod, or wearing shoes.
- Replace the shoes of a shod horse every 4 to 6 weeks. A **farrier** (a person who shoes horses and cares for their feet) should be engaged to perform this task.
- Shoe a horse for the first time when it is 2 years old or when it starts being worked.
- Cool down the horse thoroughly after every exercise period and after riding. Do this before the horse is allowed to drink large quantities of water.
- Do not overfeed a horse. The total daily consumption of concentrates and roughages should not total more than 2 to 2.5 percent of the horse's weight.
- Feed a horse at a regular time each day. The number of times that the horse is fed per day makes little difference, as long as it is the same every day.
- Do not abruptly change the ration. The stomach of the horse is temperamental and slowly adjusts to changes in feeding practices.
- Never give moldy feed to a horse.
- Provide unlimited access to fresh, clean water at all times.
- Maintain a strict health-care program for all horses.
- Maintain clean and sanitary facilities at all times.

A   B

Figure 31-17 (A) Grooming a horse helps it develop trust for people. It is also important to keep the horse clean. While grooming, take that opportunity to make a complete inspection of the horse. (B) Frequent inspection and care of a horse's feet are important because a lame horse has no value. It is much easier to prevent lameness than it is to cure it.

© SeventyFour/Shutterstock.com. © Jeff Banke/Shutterstock.com.

Science Connection: The Ever-Evolving Equine Influenza Virus

Scientists are seeking ways to make equine vaccinations even more effective.
© hedgehog94/Shutterstock.com

University and industry-sponsored equine research centers employ scientists with expertise in all aspects of horse health and well-being. One area of study is vaccinations to protect horses from diseases such as equine influenza virus (EIV). EIV is a highly contagious upper respiratory disease. Horses can become quite ill with EIV leading to high vet bills and long recovery times. The rate of EIV has been trending upward since 2008, but in recent years, there has been a considerable spike in cases.

Dr. Kyuyoung Lee, University of California, Davis, and other researchers are studying this trend, its cause, and the best strategies to protect horses against EIV. They have discovered three findings. (1) EIV strains are undergoing antigenic drift, or sudden and significant changes, which leads to new virus strains. The immune system doesn't recognize them, and vaccines fail. (2) The strains detected in the field differ from the ones contained in the vaccine, which also causes them to fail. (3) Vaccine choice must align with current field strains causing infection.

Researchers use genetic sequencing to understand the importance of antigenic drift and its impact on the effectiveness of vaccines. Dr. Kyuyoung focused his research on determining whether U.S. influenza vaccine failure rates result from antigenic drift or the introduction of foreign strains. He concluded that antigenic drift of EIV in the United States is the likely cause of vaccine failure indicating the necessity of developing more effective vaccines.

A wide-ranging sequencing study revealed that the EIV vaccine known as Florida '13 aligned with all the positive equine influenza samples tested. Choosing a vaccine such as this can play a big role in combating the current rising trend of EIV.

In addition to vaccine choice, the timing of vaccination is also critical. Veterinarians should give the EIV vaccine one or two months earlier in the year and ensure horses receive the necessary two doses a year.

Purchasing a Horse

When buying a horse, be sure to do the following:

1. Note blemishes and unsoundness characteristics of the horse. A **blemish** is an abnormality that does not affect the use of the horse. A condition of **unsoundness**, such as chronic lameness, is an abnormality that does affect the use of the horse.
2. Check the age of the horse by examining its teeth (**Figure 31-18**). Examining the teeth can also indicate how long a horse might be useful.
3. Determine that the horse is not blind or having other problems that may affect its value and use.
4. Insist on evidence of good health.
5. Try to determine if the horse has vices.
6. Try to determine the personality and spirit of the horse.
7. Consider the price of the horse and whether you can afford the cost of owning one.
8. Check the pedigree of the horse, if it is a purebred.

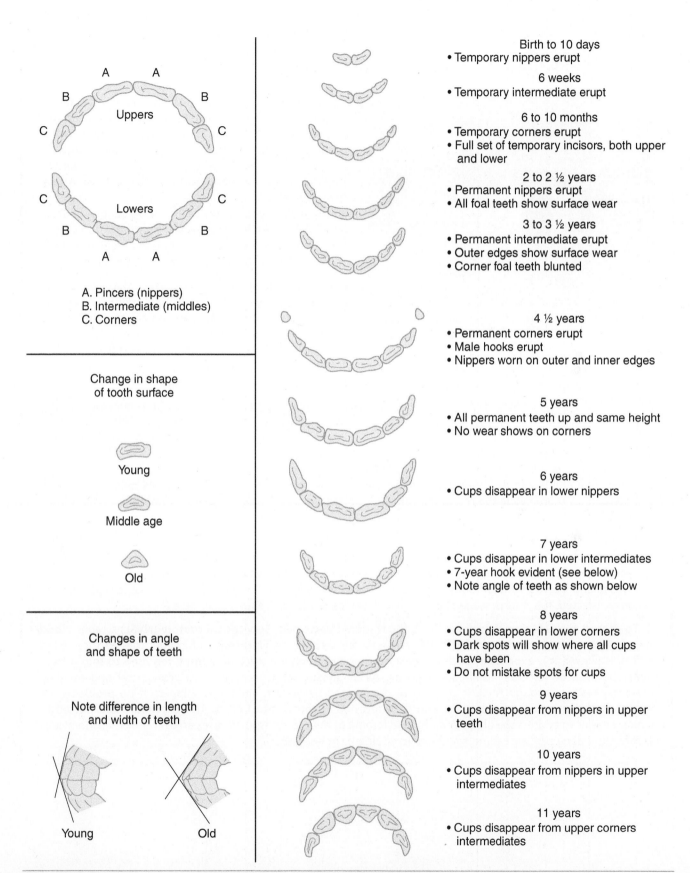

Figure 31-18 The age of a horse can be determined by the number and shape of its teeth.

Science Connection: An International Solution

Most U.S. horses are vulnerable to equine piroplasmosis, a disease which occurs regularly in several countries. Horses arriving in the United States are first screened for the disease, which has a mortality rate of up to 50 percent.
Courtesy of Barbara Lee Jensen, After Hours Farm.

U.S. veterinarians and other animal health officials are constantly on the alert for a lurking menace—equine piroplasmosis, also known as equine babesiosis. This is a malaria-like infection of horses caused by parasites. It is endemic (regularly found) in South and Central America, the Caribbean (including Puerto Rico), Africa, the Middle East, and Eastern parts of Central and Southern Europe. Horses of Brazil, Argentina, Russia, and Poland have lived consistently with the parasite, so they have developed a level of immunity, which protects them from serious consequences of the disease.

The United States, Canada, Australia, Japan, and Iceland are not considered to be endemic areas, but for this very reason such countries are at risk. Because most U.S. horses have not been exposed to the parasite, they have developed little or no resistance to it. Therefore, such horses are vulnerable, and a bout with equine piroplasmosis can be fatal.

Piroplasmosis resembles malaria in humans because, in both diseases, insects transmit the infectious agent, which then attacks and destroys red blood cells. Sick animals become feverish and lethargic and refuse to eat. Microscopic, single-celled parasites, called *Theileria equi* (formerly called *Babesia equi*) or *Babesia caballi*, enter the horse through infected blood-sucking ticks. The parasites then multiply in the bloodstream and form tiny pear-shaped or ring-shaped bodies called merozoites. The merozoites then invade the red blood cells. The horse's immune system reacts by forming antibodies in an attempt to fend off the invaders.

The United States safeguards its horse population by keeping the parasite out of the country. However, horses coming into the United States from another country may be carriers of the parasite, even though they themselves appear healthy. Several field tests have been developed to detect the presence of this parasitic disease. Each test involves taking a blood sample from a horse and testing it for specific antibodies that are produced when a horse becomes infected. It is now possible to perform a single test for both *T. equi* and *B. caballi*.

To protect the U.S. domestic horse herd, the National Veterinary Services Laboratory (NVSL) screens blood samples from equids (horses, mules, donkeys, asses, and zebras) arriving from other countries to the United States. However, it is expensive to ship a horse to an international border only to be required to ship it back home because of rejection for health reasons. As diagnostic tests become more economical and reliable, pre-entry testing through NVSL is strongly recommended before the animal is shipped. If no problems are indicated, the animal can be shipped, and then retested at the point of entry. This procedure is expected to make it easier to address the old international dilemma of how to transport horses across borders economically without the danger of spreading the dreaded equine piroplasmosis.

Chapter Review

Student Activities

1. Write the **Key Terms** and their meanings in your notebook.
2. Develop a word search using the **Key Terms**.
3. Survey students in your school regarding the number, type, and breed of horses they own.
4. Research the history and development of the horse. Write a paper, create a chart, or draw pictures outlining the physical changes that occurred in the fossil record of the modern horse.
5. Visit a tack shop and make a list of the various types of horse equipment sold there.
6. Develop a bulletin board or slideshow presentation showing as many breeds and types of horse as possible.
7. Write a report about a breed or type of horse that interests you.
8. Invite a horse breeder or owner to visit the class and talk about the care and management of horses.
9. Invite a horse veterinarian to visit the class and talk about health care and first aid procedures for horses.
10. Look at horse-breed-and-care websites or magazines to determine the uses of the various breeds of horse. Also, look for the types of job opportunities that may be available in the horse industry.
11. Demonstrate to the class the techniques of grooming a horse.

National FFA Connection — Proficiency Award In Equine Science—Placement

Teaching children proper riding skills prepares them to treat all animals with kindness.
© LanaG/Shutterstock.com

The National FFA Organization provides equine proficiency awards at local, state, and national levels. To qualify for the placement award, the student must be employed with an equine enterprise providing experiences in production, breeding, showing, marketing, and management of horses. Training of horses may also be included when horses are managed but not owned by the member. Approved training activities "such as but not only" roping, show, rodeo, racing, riding lessons, and therapeutic riding may strengthen your award application. Consult with your agriscience teacher as you develop a plan to pursue this award.

Unit 9
The Future of Food Is Now

Food science research today faces one of the most important challenges that the human race has yet encountered: How will the agricultural industry feed a growing human population even as farmland is increasingly converted to other uses? These new uses range from houses to industrial complexes. What options are available to increase future crop resources? While options appear to be limited, they must include the following:

- Focus on achieving genetic improvement and more efficient production of the major food crops.
- Re-evaluate underutilized crops. Currently, three crops—rice, wheat, and maize—supply 50 percent of the world's consumer calories, yet the pool of edible plant species numbers around 30,000.
- Investigate plant biodiversity to discover new crops.

To make the most of our resources, new foods must be developed, and better ways to package and preserve food products must be designed. Promising developments include:

- Cultured meat, derived from live animal cells.
- CellPod, a student-designed home appliance that grows food from a plant cell culture.
- CRISPR'd foods, derived through a gene-editing process that modifies the DNA of cells. Examples include naturally decaffeinated coffee beans, wheat with reduced gluten, and bananas that are resistant to fungus.
- High-pressure/low-temperature pasteurization of fruit, dairy, fish, and meat products using water pressure to destroy bacteria without damaging vitamins, flavor, or quality.
- Edipeel, an edible coating for fruits and vegetables. This water-soluble, colorless, protective coating is known to double shelf-life.
- Biodegradable packaging for food products. Made from polymers, which are molecules found in living organisms. Such packaging is designed to degrade without harm to the environment.

The need to develop a food supply that can keep pace with Earth's growing population is as urgent as it is challenging. As the above examples demonstrate, it is through diligent research, bold innovation, and a re-evaluation of the underutilized resources already available that we can—and must—rise to that challenge.

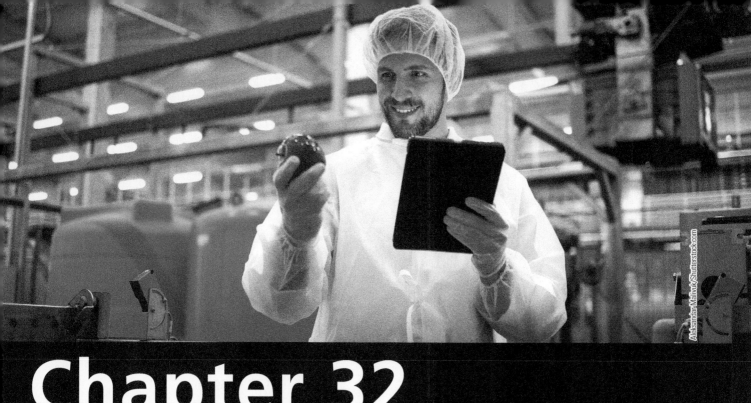

Chapter 32
The Food Industry

Objective
To explore elements, trends, and career opportunities in the food industry.

Competencies to Be Developed
After studying this chapter, you should be able to:
- explain what is meant by the term *food industry*.
- determine the importance of the food industry to the consumer.
- describe the economic scope of the food industry.
- identify government requirements and other assurances of food quality and sanitation.
- compare the major crop and animal commodity production areas in this nation and in the world.
- discuss the major food commodity groups and their predominant origins.
- explain the major operations that occur in the food industry.
- describe career opportunities in food science.
- discuss future developments predicted for the food industry.

Key Terms
food industry
retailer
wholesaler
distributor
processor
grader
packer
trucker

harvester
producer
grade
climatic condition
harvesting
maturity
underripe
overripe

microorganisms
processing
bran
endosperm
germ
edible

Generally speaking, most Americans have access to an abundant supply of food. But the Covid pandemic helped us understand what a delicate balance exists in our food processing and distribution system. For the first time in our lives, we encountered empty shelves in our grocery stores. It would be wrong to believe that in the United States there are no hungry people—about 10 percent of Americans are food insecure—but until the pandemic, the supply of food had never been in question.

Distribution of food has not always kept pace, although public and private efforts to feed children at school and to distribute food through food pantries are ongoing. For example, federal spending on the food and nutrition assistance programs of the United States Department of Agriculture (USDA) totaled $182.5 billion in 2021, 49 percent more than in 2020. Spending also increased on the Supplemental Nutrition Assistance Program (SNAP) and the Special Supplemental Nutrition Program for Women, Infants, and Children (WIC).

The **food industry** is involved in the production, processing, storage, preparation, and distribution of food for consumption. Pet and animal food as well as human food require a chain of people, places, equipment, regulations, and resources to change farm products into edible foods.

Careers are plentiful in this industry. The dollars spent on food and fiber in the United States provide jobs for approximately 16.7 percent of the country's working population (**Figure 32-1**).

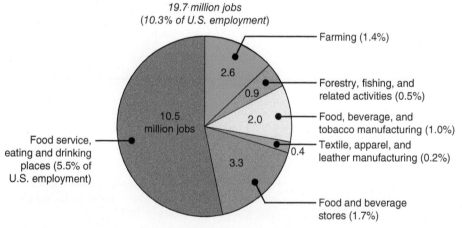

Figure 32-1 Career sectors in the food and fiber system.

Figure 32-2 The process by which food moves from producer to consumer involves many people.

Economic Scope of the Food Industry

Consider this: When you purchase groceries at the supermarket or a hamburger at the local fast-food restaurant, does most of your food dollar go directly to the farmer who raised the beef? How about the lettuce, tomato, pickle, bun, and sesame seeds that top your hamburger? Many businesses and individuals join the farmer in dividing up your food dollar. In addition, the economic chain reaction that begins with your food purchase sends signals to the retailer, wholesaler, distributor, processor, grader, packer, trucker, harvester, producer, and others to replace that food for your next purchase (**Figure 32-2**).

A **retailer** is a person or store that sells directly to the consumer. The retailer is the end of the marketing chain, whereas a **wholesaler** is a person who sells to the retailer, having purchased fresh or processed food in large quantities. A **distributor** stores the food until a request is received to transport the food to a regional market. A **processor** is anyone involved in cleaning, separating, handling, and preparing a food product before it is ready to be sold to the distributor. A **grader** is the person who inspects the food for freshness, size, and quality and determines under what criteria it will be sold and consumed. A **packer** is the person or firm responsible for putting the food into containers, such as boxes, crates, bags, or bins, for shipment to the processing plant. A **trucker** is the person responsible for transporting the product anywhere along the way from farm to consumer. A **harvester** is the person who removes the edible portions from plants in the field. Before any of this can happen, the **producer** grows the crop and determines its readiness for harvest. It seems like everyone gets a part of your food dollar (**Figure 32-3**)!

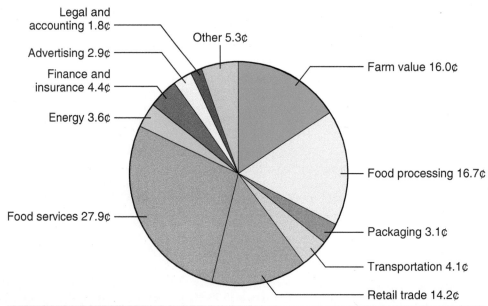

Figure 32-3 The farm share of the food dollar was reported by ERS/USDA to be 16.0 cents in 2021. Numerous other services and value-added inputs to a product account for the remaining amount of each dollar spent by consumers for food.

Where you spend your food dollar also influences who gets how much of your dollar. A meal purchased in a restaurant costs considerably more than a meal prepared from raw food products at home. This raises an interesting question: In what ways have our changes in lifestyles and the shift to families with two or more people employed outside the home influenced how and what we eat? Many more meals are eaten outside of the home than was the case a generation ago, and convenience foods for use at home are more in demand today. In the United States in 2021, the cost of food eaten away from home accounted for 44 percent of all income spent on food. In the same year, the percentage of disposable income needed for food was 10.3 percent in the United States (**Figure 32-4**).

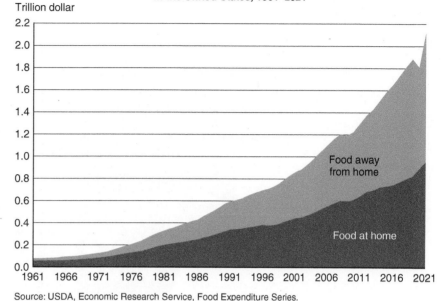

Source: USDA, Economic Research Service, Food Expenditure Series.

Figure 32-4 Proportion of disposable income used for food in the United States.

Quality Assurance

Safety Regulations

The USDA oversees food labeling and enforces regulations regarding representation on such labels (**Figure 32-5**). Additional quality-assurance programs administered by the USDA include inspection of slaughterhouses and processing plants and oversight of processing operations (**Figure 32-6**). The National Shellfish Sanitation Program, the U.S. Public Health Service, and the U.S. Food and Drug Administration work with the USDA to ensure the safety of food and food products. States, counties, and municipalities also have inspectors. They regulate local conditions to ensure sanitation and safe food handling, especially in restaurants and food-preparation areas.

Nutrition Facts

Serving Size	1 Cup (55 g/2.0 oz)
Servings per Container	13

Amount Per Serving	Cereal	Cereal with 1/2 Cup Vitamins A & D Skim Milk
Calories	170	210
Fat Calories	10	10

	% Daily Values**	
Total Fat 1.0 g*	2%	2%
Sat. Fat 0 g	0%	0%
Cholesterol 0 mg	0%	0%
Sodium 300 mg	13%	15%
Potassium 340 mg	10%	16%
Total Carbohydrate 43 g	14%	16%
Dietary Fiber 7 g	28%	28%
Sugars 17 g		
Other Carbohydrate 19 g		
Protein 4 g		
Vitamin A	15%	20%
Vitamin C	0%	2%
Calcium	2%	15%
Iron	45%	45%
Vitamin D	10%	25%
Thiamin	25%	30%
Riboflavin	25%	35%
Niacin	25%	25%
Vitamin B₆	25%	25%
Folate	25%	25%
Vitamin B₁₂	25%	35%
Phosphorus	20%	30%
Magnesium	20%	25%
Zinc	25%	25%
Copper	15%	15%

*Amount in cereal. One-half cup skim milk contributes an additional 40 calories, 65 mg sodium, 6 g total carbohydrate (6 g sugars), and 4 g protein.

**Percent Daily Values are based on a 2,000-calorie diet. Your daily values may be higher or lower depending on your calorie needs:

	Calories	2,000	2,500
Total Fat	Less than	65 g	80 g
Sat. Fat	Less than	20 g	25 g
Cholesterol	Less than	300 mg	300 mg
Sodium	Less than	2,400 mg	2,400 mg
Potassium		3,500 mg	3,500 mg
Total Carbohydrate		300 g	375 g
Dietary Fiber		25 g	

Calories per gram:
Fat 9 • Carbohydrate 4 • Protein 4

Figure 32-5 The commodity label is the consumer's best assurance of food quality and value.

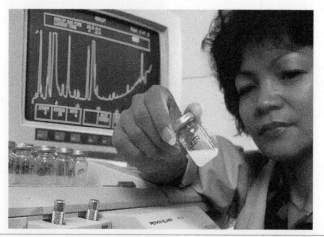

Figure 32-6 Quality-control personnel and government inspectors use USDA standards to monitor foods for cleanliness, wholesomeness, and quality.

Courtesy of USDA/ARS #K-3512-3.

Grading Standards

Americans have become accustomed to high-quality food in every state and every store. The grading system established by the USDA has provided a uniform set of trading terms known as grades. **Grades** are based on quality standards. They improve acceptability of products by the consumer.

Grade standards are established for the following food and fiber commodities: beef, pork, lamb, poultry, rabbits, goat, eggs, grain, nuts, cotton, dairy products; fruits, vegetables; rice and pulses; fish and seafood; wool and mohair. Grades indicate freshness, potential flavor, texture, and uniformity in size and weight, depending on the commodity (**Figure 32-7**).

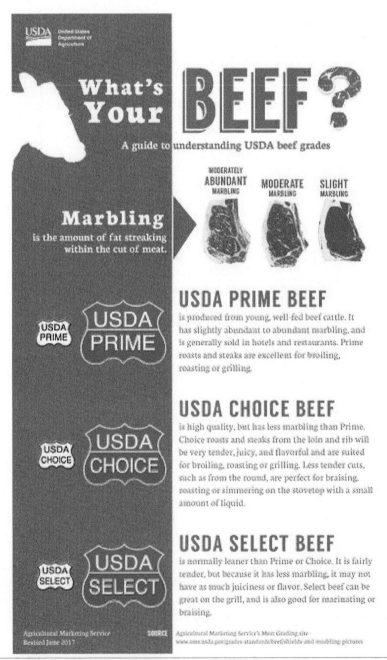

Figure 32-7 Grades of beef as established by the U.S. Department of Agriculture (USDA).
Courtesy of USDA.

Science Connection | Keeping Food Safe

Food scientists test the quality of milk and dairy products to ensure that the products are safe and wholesome.
© Parilov/Shutterstock.com.

Authorities generally agree that the food supply in the United States is among the safest in the world. Federal and state food inspectors work to ensure that our nation's food supply is safe. For example, in 2020, the Food and Drug Administration (FDA) launched a 10-year plan to improve food safety called "The New Era of Food Safety Blueprint." It provides a framework for preventing contaminated food outbreaks, improving food traceability, and perhaps the biggest challenge, fostering and strengthening food safety culture in facilities.

Avoiding contamination by common bacteria and other living organisms would be impossible in any place except a sterile environment. For this reason, from processing plants to home kitchens, food is produced in sanitary, but not sterile, environments, and any residual external biological contamination is removed by washing or peeling. As appropriate, food is then heated to sufficient temperatures to kill internal biological organisms.

Contamination by chemicals is quite a different matter. Plants grown in soils and fish grown in water containing poisonous chemicals, heavy metals, and other pollutants can result in chemical contamination of fruits, vegetables, nuts, and fish products. Similarly, animals grazing on contaminated pastures or eating contaminated feeds can have chemical contaminants in their meat and milk. Drugs used for medicines to keep animals free from diseases, parasites, and insects can contaminate meat and milk if not used according to label instructions. Unfortunately, heating does not remove or neutralize most chemical contaminants. Therefore, food inspectors and other health inspectors keep a vigilant watch with all the tools of modern science to monitor foods from source-to-table and divert unacceptable products from the food stream. In actual practice, reputable food producers, processors, and handlers protect the food supply through quality-control measures to protect their businesses and to avoid penalties by the government inspectors.

For example, rigorous scrutiny is directed toward possible residues of pesticides and medicines in meat and milk. Constant improvements in laboratory and other scientific procedures support this vigilance. A case in point is the development of high-tech probes to speed-up tests for benzimidazoles on meats. Benzimidazoles are medicines used to protect cattle, sheep, pigs, chickens, and goats from internal parasites. However, if the drug has been administered in incorrect doses or with inappropriate timing, some of it can remain in the meat.

Meat specimens are routinely inspected using high-performance liquid chromatography, a technique that allows food scientists to identify and analyze components in food. Future chemists may complement this approach with other monoclonal antibody assays, tests that allow for the detection of a vast range of foods and environmental contaminants, even in minute amounts. Such assays hold the promise of dramatically speeding up testing without compromising accuracy. Already, inexpensive monoclonal antibodies have been developed that will detect benzimidazoles in meat at concentrations as low as one part per billion—in terms of time, that is the equivalent of 1 second in 32 years!

Food Commodities

What Influences Where Foods Are Grown?

In agriculture, a commodity is any product produced for sale from plants or animals. Two key food commodities are crops and animal-based foods, such as meat, dairy, and seafood. Before exploring examples of these commodities, consider some of the factors that influence where food crops are grown.

Climatic conditions result in some foods growing better and in greater abundance in certain areas of the world. Climatic conditions refer to average temperature, number of days with a certain temperature range, length of the growing season, and amount of precipitation for a given geographic area. Food production in the United States is influenced by climate (**Figure 32-8**).

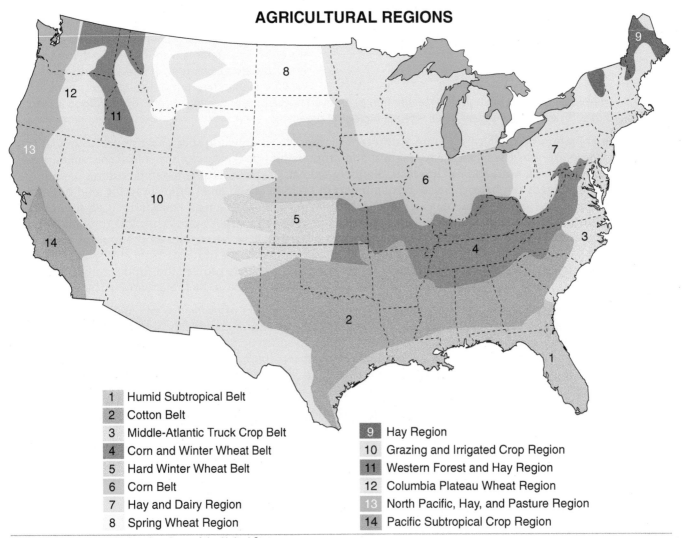

Figure 32-8 Major agricultural regions of the United States.
Courtesy of USDA.

Foods have been introduced outside the areas where they are grown naturally. The origin of the soybean, for example, can be traced back 3,000 years to China, where it is still produced and consumed. However, major growing areas for the soybean today include the United States, Brazil, and Western Europe.

Available technology is another factor that influences where food is grown. Technology refers to the equipment and scientific expertise available to cultivate, store, process, and transport the crop for consumption in a variety of forms after harvest. Modern technology has allowed producers to raise crops somewhat artificially with irrigation and in greenhouses where the conditions of temperature and moisture are controlled (**Figure 32-9**). Different varieties of food have been developed to grow under different climatic conditions, such as extreme heat or cold. Similarly, in aquaculture, seafood and fish are produced under controlled conditions (**Figure 32-10**).

Figure 32-9 Technologies such as modern irrigation systems have made it possible to produce crops in arid environments such as deserts.
Courtesy of DeVere Burton.

Figure 32-10 Controlled living environments such as fish runs make it possible to produce fish under controlled conditions.
Courtesy of DeVere Burton.

In the United States, we are accustomed to having almost every food available fresh at any time of the year; but all foods are not grown in all parts of the country. Citrus fruit, including oranges and grapefruit, require warm climates, such as those found in Texas, California, and Florida. More than 70 percent of the fresh vegetables that are consumed in the United States are grown in California, Florida, and Arizona. Every day, however, people in North Dakota and Maine enjoy the nutrients and good taste of fresh or processed citrus and vegetable products, such as orange juice. Our country is not only "America the Beautiful" but also "America the Bountiful."

Crop Commodities

Grains

Various grains have different growing requirements and are therefore produced in different parts of the United States and the world. Wheat, which originated in Asia, is grown in the cooler climates of the United States. Different varieties have been developed to accommodate different growing seasons and climatic conditions around the world. Corn is a warm-weather crop, but the many varieties and types permit its growth in every one in the United States (**Figure 32-11**). Rice, however, has special moisture requirements; therefore, its production is limited to specific areas of the country.

Figure 32-11 Corn is adapted to a wide range of growing conditions and is grown in every state.

© Zeljko Radojko/Shutterstock.com.

Oil Crops

Oil crops are sometimes thought of as the invisible food product. Soybeans, corn, cotton, flax, sunflowers, coconut, peppermint, and spearmint are all significant oil crops in the United States, which is a world leader in the production of soybeans, corn, cotton, and peanut oils (**Figure 32-12**). Soybean products are referred to in ancient Chinese literature, and the origin of the peanut may be traced to Brazil and Paraguay. Sunflowers are native to the United States and gaining importance. They are also grown in Spain, China, Russia, Bulgaria, and Mediterranean areas. Safflower oil is a relatively minor oil in terms of proportion to the total oil crop worldwide. Its origins in northern India, North Africa, and the Middle East are indicative of its drought tolerance.

Figure 32-12 Edible oils are obtained from the seeds of many different kinds of plants, making the United States a world leader in the production of vegetable oils.

Valentyn Volkov/Shutterstock.com

Sugar Crops

Sugar beets and sugarcane are the principal sugar crops in the United States. Corn is a secondary source of sugar. Sugar beets are grown in temperate areas, with most of the production in the states of Minnesota, Idaho, North Dakota, and Michigan (**Figure 32-13**). Sugarcane is grown in tropical and subtropical locations around the world. Florida, Louisiana, Texas, and Hawaii are the four largest producers of cane sugar in the United States.

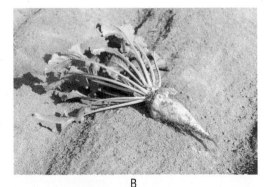

Figure 32-13 Sugar beets are grown in a temperate climate and are a significant source of processed sugar.

© iStock.com/Paul Jackson. LUGOSTOCK/Adobe Stock Photos

Citrus

Oranges, limes, lemons, and grapefruit all require warm temperatures to survive. They cannot tolerate freezing conditions. Consequently, the southern states with warmer climates, such as Florida, California, Texas, and Arizona, are the major producers of citrus. The industry provides consumers with both fresh citrus and frozen concentrate products.

Tree Fruits

The many varieties of fruit that grow on trees require specific weather conditions. Therefore, various fruits are adapted to different parts of the country. Apples and pears require cooler temperatures and do well in mountainous areas (**Figure 32-14**). Washington is particularly well adapted for these fruits. Bananas require very warm conditions and grow best in tropical areas. The United States imports most of the bananas we eat. Hawaii is the only state with significant commercial production.

Figure 32-14 Apples are popular with U.S. consumers, who ate an average of 16.18 pounds per person in 2021.

Marek Mnich/Shutterstock.com.

Vegetables and Berries

Vegetables and berries are consumed shortly after harvest, or processed by canning, drying, or freezing for future consumption. Vegetables that require cooler climates include cabbage, broccoli, potatoes, and cauliflower. Vegetables requiring warmer environments include beans, tomatoes, and sweet corn. Vegetable and berry production occurs in most regions of the country.

Animal-Based Commodities

Animals, like crops, are typically raised in locations with some regard to climatic conditions. Artificial cooling or heating of livestock is practiced in some regions, but it is expensive. Where fewer artificial conditions are introduced, the cost of production is minimized.

The type, cost, and availability of livestock feed are important factors influencing where animals are raised in the United States and around the world. Large amounts of water must be available for livestock. This can be a limiting factor for animal production in some areas.

Beef

Most beef is raised near corn, the main feed source for feedlot cattle. More than half of the corn in the United States is grown in Iowa, Nebraska, Illinois, and Minnesota. Therefore, beef is raised extensively in the Midwest. The open ranges in the western part of the United States provide other important areas where beef is raised.

Pork

Corn is also the primary food of hogs. Similar to beef, the primary area where hogs are raised is the Midwestern part of the United States. The mid-Atlantic and southern states are also important hog-production areas.

Lamb

Sheep require large amounts of grazing area. Therefore, they are raised extensively in the range states of the far West. However, as is true with beef and pork, lamb products are produced in other states.

Poultry

Poultry can be raised in a variety of settings. Typically, chickens and turkeys are raised indoors, where ventilation and temperature are carefully controlled. Most poultry are raised in the mid-Atlantic and Southern states. However, important poultry-producing areas are also found in California and other states.

Dairy Products

Wisconsin is known as the "Dairy State." Many dairy products are produced there and in other northern-tier states. California also has a large dairy industry with many cooperatives and processing plants. These provide dairy products to consumers nationwide (**Figure 32-15**). Dairy animals prefer moderate environments, so the industry is extensive in the northern part of the United States. In recent years, however, the industry has expanded significantly in Idaho, New Mexico, and other western states where high-quality forages are available.

Figure 32-15 The U.S. dairy industry provides a wide range of milk products to consumers.
JulijaDmitrijeva/Shutterstock.com.

Seafood

States that border the Atlantic and Pacific Oceans and the Gulf of Mexico are the primary suppliers of seafood in the United States. However, the science of aquaculture has expanded the production of fish products in interior states.

Workplace Safety

Accident Awareness

Accidents occur far too often in the food industry. They result in lost time on the job, but they are also responsible for injuries that make it impossible for the victim to return to work. In the worst instances, workers may die due to the severity of their injuries. In addition, noninjury accidents occur that affect consumers, such as failure to detect contaminants in food products. Losses due to injuries and mistakes include medical expenses, lost productivity, legal costs (liability), and the costs of recalling defective products from store shelves. Safety awareness training is vital for individual workers. All levels of the food industry will benefit by developing work safety plans and programs.

Fatal accidents on farms and ranches rank third behind the construction and transportation industries. In 2019, there were 410 workplace deaths in crop and animal production and the aquaculture industries, according to the National Institute for Occupational Safety & Health. In addition, approximately 4,000 injuries to youth occur on farms and ranches each year.

Some of the most common farm accidents include or occur due to:

- **Overturning tractors and machinery** – The U.S. Department of Labor has estimated that nearly half of all farm-related accidents involve a rolled tractor.
- **Falls** – Climbing trees, barns, silos, and other heights too often results in injuries.
- **Toxic chemical exposure** – Both single and long-term exposure to toxins can cause deaths and disabilities.
- **Suffocation** – Working in grain bins, silos and silage pits, manure storage structures, and other poorly ventilated areas is dangerous due to lack of oxygen or the presence of toxic gases.
- **Heat stress** – Too much sun or a hot work environment may result in heat stroke, dehydration, heart problems, and other illnesses.
- **Crushed limbs** – Injuries occur when a worker gets entangled in moving parts of machinery such as hay balers and other mechanically powered equipment.
- **Animal-related injuries** – Horses, bulls, dogs, and other livestock cause serious injuries and even death to their handlers.

Some accidents happen because workers become less attentive to their work environment. For example, accidents occur most often when workers are tired. Patterns show that accidents occur more frequently between 11 and 12 o'clock in the morning, just before lunchtime. Friday is the most accident-prone day of the week, while fewer accidents occur on Tuesday and Wednesday when workers are rested and more alert.

Making the Workplace Safer

First aid is the care given to a victim immediately following an injury or the onset of illness. First aid training is a critical need in every work setting because such incidents often occur without warning. The first aid policy is determined by the employer, but it is defined and driven by worker safety legislation and enforced by state and federal agencies.

Employers frequently focus on two related approaches for keeping employees safe from harm in the workplace. In the first approach, the work area is carefully evaluated for safety concerns. This assessment is followed by correcting hazardous conditions and targeting safety training to employees based upon how potential hazards intersect with their specific duties. All employees should be trained in all aspects of the work environment where they might reasonably be expected to be engaged.

First Aid Training

The second approach to employee safety is to provide first-aid training for the workplace.

For example, an agribusiness may provide basic first aid training to each employee and specialized training such as Emergency Medical Technician (EMT) training to enough individuals to ensure that trained personnel are located nearby when workers are present. Specialty training goes beyond basic first aid to provide expertise in trauma response. For example, such training might include the use of an on-site heart defibrillator (functions to reset heart rhythm back to normal).

Both basic first aid training and the more advanced levels of emergency response training are provided by professional instructors who are approved to offer certifications to workers successfully completing training programs. Certifications from accredited training organizations may be required by state and federal governments to protect workers. In addition, the issue of liability for employee safety is a major concern of employers. Work environment inspections and training for emergency responses are important ways to limit employer liability and increase worker safety.

No amount of safety training can take the place of people who know what to do when an injury accident occurs. Life-threatening accidents and incidents can and do happen wherever people are located. For example, safety training is not likely to prevent a heart attack or stroke from happening to someone who is at work, but an employee who is on-site with up-to-date first aid training could save a life. For this reason, many employers provide first aid training to at least some of their employees.

One of the most important responses to an incident is to call for professional help first. It is critical that medical help arrives as quickly as possible, so the person who administers first aid should ensure that someone calls for help immediately. First aid is limited care that lasts only until more qualified help arrives. Once a first aid procedure is started, this level of care should be continued until the patient can be placed in the care of medical personnel such as a paramedic, nurse, or doctor.

Basics of First Aid

Before rendering aid to a distressed person, it is important to assess the situation quickly and accurately. Do not place yourself in danger by entering an unsafe area too quickly. For example, electrical shock, hazardous materials, danger of collapse, unsafe machinery, and other hazards may claim the lives of those who fail to evaluate the cause of injury or illness to someone else.

Pause to ask questions and to observe danger signs. Avoid causing further harm to the injured person. Moving someone with a neck or spine injury may cripple them for life or even contribute to death. Of course, there are times when an injured person must be removed to a place of safety (fire danger, dangerous fumes).

Once a decision has been made to administer first aid, prioritize the order of care:

A. **Airway**—make sure the airway is not obstructed.
B. **Breathing**—determine whether breathing has stopped and administer aid if there is a breathing problem.
C. **Circulation**—identify and treat sources of bleeding and symptoms of shock.
D. **Disabling injury**—check for burns, broken bones, or similar injuries.

A person who is not breathing must be given resuscitation immediately. Current practice is to perform cardiopulmonary resuscitation (CPR). It would be a mistake to first treat the accident victim by immobilizing a broken bone before restoring the circulation of oxygen-rich blood to the brain and heart. It is critical to remember that the greatest threat to life should be treated first.

Once the injury victim is breathing and bleeding has been stopped, provide appropriate care while you wait for professional help. Make the patient as comfortable as possible, doing what you can to calm the patient and ease his or her pain. Do not leave the patient alone; keep someone with them until medical help arrives.

Personal Protective Equipment for Food Processing

People who work in food processing are exposed to a variety of potential hazards that could lead to serious injuries. Protective equipment and clothing can go a long way toward protecting those who work around moving machinery in a processing environment. Several government agencies are charged with responsibility for protecting the health and safety of workers in the United States. Their agents conduct regular compliance audits of food processors to ensure that protective devices are installed on machinery and that they function as they should. They also regulate the types of safety clothing and equipment that is used by workers.

Industrial safety clothing for most uses includes the following:
- Fiber-metal safety helmets are standard gear in hazardous areas within a processing facility. Workers in low hazardous areas frequently wear disposable hats or caps.
- Safety glasses
- Ear plugs or industrial ear protection
- Gloves
- Back braces for heavy lifting
- Safety boots and shoes
- Respiratory protective devices
- Coverall, lab coat, or wet suit
- Stress protection against workplace heat.

Operations in the Food Industry

Harvesting

Harvesting occurs when products such as seeds or tubers are gathered from the plants where they were grown or produced. This may involve taking potatoes out of the ground, picking oranges off a tree, harvesting bean pods from bean plants, or removing and threshing grain from stalks (**Figure 32-16**). It is most important that the crop be harvested in a timely and careful fashion. The plant should be harvested at the correct stage of maturity. **Maturity** means the state or quality of being fully developed or ripened. Harvesting a crop when it is of proper maturity means that it is not underripe, overripe, or spoiled. **Underripe** means that it has not reached maturity. **Overripe** means that the plant is past the optimum maturity; stalks or limbs can easily break or shatter, or fruit can drop. Spoiled means that chemical changes have taken place in the food or food product that either reduce its nutritional value or render it unfit to eat.

Figure 32-16 Harvesting is the timely removal of crops from the field. Most crops are harvested mechanically using specialized machines, such as this potato-harvesting equipment.

Courtesy of DeVere Burton.

Spoilage is usually caused by microorganisms. **Microorganisms** that contribute to food spoilage include bacteria, fungi, and nematodes. Bacteria consist of a group of single-celled plants. Fungi are plants that lack chlorophyll and obtain their nourishment from other plants, thus contributing to rot, mold, and plant diseases. Nematodes are small worms that feed on or in plants or animals. They are found in nature in moist soil, water, or decaying matter. They pierce the cells of plants and animal products, feeding on tissues, fluids, and juices.

The moisture content of a product determines how well it can undergo processing. It also dictates the types of handling procedures and storage facilities that are required. Some fruits, such as bananas and tomatoes, continue to ripen after they have been picked from the tree and vine. Other foods, such as beans and oranges, do not continue to ripen once picked, and they must be handled accordingly. Knowledge of the complete growth process and handling requirements of each commodity is essential for the producer and harvester.

Harvesting involves the use of equipment and labor. Because timing is critical, migratory labor is often hired at harvest time. Migratory labor is provided by workers who move from place to place, following the harvest as it occurs throughout the year. As crops ripen, laborers migrate to new geographic locations where the harvest season is just beginning. Usually, workers move from the southern tier of states to central and northern regions as crops mature. Many of these workers return to the same farms year after year, and they become efficient and skilled. For example, picking fruit is usually done by hand, and the crop must be harvested within a very short period. A dependable crew is needed, but only for a short season. Local laborers are not always available, nor are they as skilled as workers who pick fruit for several months each year. When dependable workers are not available, fruit and vegetable crops are at risk of being left unharvested.

Some food crops are adapted to mechanical harvest methods. Engineers and technicians continue to develop and market machines to perform many harvesting operations. Such machines make harvesting much more efficient than it was in the past (**Figure 32-17**). However, machines may not always be as gentle with the fruits and vegetables as human hands. Consequently, plant breeders have developed new crop varieties that lend themselves better to mechanical harvesting. Such varieties may not be as appealing to the human touch or taste, though. For instance, tomato varieties having skins tough enough for mechanical harvesting are harder to slice and may not be as juicy or flavorful as varieties suitable for the home garden.

Figure 32-17 Mechanical harvesters replace the backbreaking and difficult work of hand harvesting. This machine is separating cranberries from other plant materials.

Courtesy of USDA/ARS #K-4416-14.

Processing and Handling

The steps involved in turning the raw agricultural product into an attractive and consumable food are collectively known as **processing**. Processing factories, or plants function to clean, dry, weigh, refrigerate, preserve, store, and convert raw commodities into a variety of other products (**Figure 32-18**). For example, wheat is cleaned, dried, weighed, and graded for quality. It is then ground into flour. However, before this occurs, it may be separated into bran and germ. The skin or covering of a wheat kernel is known as **bran**. Inside the bran is the **endosperm**, which will become flour, and the **germ**, which is a new wheat plant inside the kernel. Wheat flour is used for breads, cereals, cakes, and pasta. Other grains have similar parts and are used to make similar products.

Processing of tomatoes results in a variety of products (**Figure 32-19**). Some people claim that the fresh tomato defines summer. The peak of the North American growing season occurs about the time tomatoes mature and the fruits ripen. To the gardener, this means picking tomatoes directly from the vine in a backyard garden and consuming them immediately. However, tomatoes are harvested year-round somewhere in the world, and the food industry delivers them to us in edible form. **Edible** means fit and safe to eat. Processing tomatoes makes it possible to preserve them for future use.

After tomatoes are cleaned and separated for size and quality, they may be canned whole; chopped, cooked, and strained for juice; or made into other products. Such products include salsa, spaghetti and hamburger sauces, paste, relish, catsup, and many other foods.

Figure 32-18 Processing is one of many intermediate steps between the producer and the consumer.
Courtesy of FFA.

Figure 32-19 Tomatoes can be eaten fresh or processed into a variety of products.
Courtesy of W. Altee Burpee Company.

Science Connection: Benefits from Blemished Tomatoes

A valuable antioxidant, lycopene, can be recovered from tomatoes with blemishes. The extracted lycopene is marketed as a health product.
© Emin Ozkan/Shutterstock.com

The vitamin and mineral industry could mean big money for tomato growers. A valuable antioxidant, lycopene, is found in tomatoes. This is the substance responsible for the red color of the tomato. Lycopene helps protect cells from harm that can be caused by the oxidation process. It has been found to reduce the risk of prostate cancer and other diseases. For many years, the substance was expensive, costing as much $2,500 a kilo in its pure form.

This changed when agriscientists at the University of Florida created an inexpensive way to extract lycopene from tomatoes. This discovery may have helped the tomato industry earn millions of dollars. Prior to this breakthrough, many tomato growers lost money on tomatoes deemed unfit for the market due only to superficial blemishes, which consumers found visually unappealing. With the advent of the new extraction technique, growers gained an opportunity to market blemished tomatoes to the health industry. The technique provided consumers with a less expensive product and earned growers a healthier profit, a definite win-win.

Transporting

Trucks, planes, boats, cars, trains, carts, and bicycles are vehicles used by the food industry in various parts of the world (**Figure 32-20**). The transporting of fresh and processed food products composes 4.1 percent of the cost of getting food from the producer to the consumer in the United States. Timing and the distance foods must travel contribute to the ultimate cost of the foods. The efficiency of transportation also influences food quality in terms of freshness and spoilage. Insulated and refrigerated trucks enable food products to move in the fresh form to most parts of the country year-round (**Figure 32-21**). This luxury is not available to most people in the world.

Approximately 70.5 percent of our perishable food is shipped by truck. Much of the less-perishable foods, such as wheat, potatoes, and beets, are shipped by rail. Air transportation allows us to enjoy perishable foods from distant regions and different countries.

Figure 32-20 Food commodities are moved around the world by nearly every mode of transportation.

Figure 32-21 Refrigerated trucks permit year-round supplies of fresh foods to every part of the United States.
© Robert Pernell/Shutterstock.com

For example, pineapples and papayas from Hawaii are enjoyed all over the world because of air transportation. Because of modern transportation, fresh seafood is enjoyed in many areas far from ocean environments.

The consumer provides the final link from farm to table. How far the food is shipped, how the food was packaged for transportation, how long the food was in transit, and how warm the food became during transport all influence ultimate food quality. From the milk truck driver taking milk from the farm to the processing center, to the trucker delivering products to the local store, transportation is a key component in the food industry. Knowledgeable and competent employees are great assets when the cash value of payloads on their trucks or rail cars is considered. A delay could be costly for both business and the consumer.

Even if food is in perfect condition when you buy it, the quality can decrease substantially before you get it home if precautions are not taken. Perishable food should be packaged correctly in the store, kept cool in the car or truck, and refrigerated or frozen on arrival home.

Marketing

Wholesalers purchase food products from packing houses, processors, fish markets, and produce terminals. They, in turn, sell to retailers and institutions such as hospitals, schools, restaurants, and retail stores (**Figure 32-22**). Grocery stores and fast-food chains are important links in the food chain before the food is purchased by the consumer.

Figure 32-22 Wholesale terminals provide the facilities for trucks and trains to bring food commodities together, allowing buyers to obtain commodities for their retail outlets.
Courtesy of FFA, Photo by Bill Stagg.

Figure 32-23 Superstores may stock 15,000 or more items.
Courtesy of FFA.

Consumers can purchase their food items from many types of retail stores. Such marketing sources meet the needs of consumers in different locations and situations. Superstores carrying 35,000 or more items, conventional supermarkets, limited-assortment and box stores, convenience stores, unconventional food stores, food cooperatives, farmers markets, roadside stands, pick-your-own businesses, and other farm outlets comprise the most common places consumers purchase their food items (**Figure 32-23**). The major differences among the various types of stores are the numbers of items stocked and the physical sizes of the facilities.

Hot Topics in Agriscience — Artificial Intelligence in Agriculture

With AI, a robotic arm is programmed to recognize and pick only mature fruit, while leaving immature fruit to ripen.
© luchschenF/Shutterstock.com.

Artificial Intelligence (AI), the simulation of human intelligence processes by machines, is expected to play an increasingly important role in future food production, processing, packaging, inventory control, shipping, storage, and even financial functions. In a system of machine learning, AI allows for the integration of computers, cameras, and optics with machines that can perform complicated, repetitive tasks.

One example is the use of sensor-based systems to sort food items by size and color using sensor-based systems. Features such as infrared sensors and cameras perform visual tasks that require recurring responses and sequences of actions that high-tech robots are programmed to perform. Just imagine the effect of AI control on a sequence of events in a milk powder processing plant:

1. Bags of milk powder arrive at the pallet-loading area on a conveyor belt.
2. The bags are detected by an electric sensor.
3. The pallet-loading machine is activated and positions an empty pallet on the loading dock.
4. A positioning camera activates a robotic arm that places each bag of powder in an exact position on the pallet until the first level is completed.
5. The pallet is lowered to allow the loading sequence to load the second level.
6. The loading sequence is repeated until the desired number of layers is properly loaded.
7. The pallet is raised to its original position.
8. The rollers beneath the pallet are activated, rolling the loaded pallet to a conveyor system.
9. The conveyer system moves the pallet to a machine to be wrapped with plastic.
10. The conveyer moves the completed pallet to the freezer platform to await transport to storage.

The sequence can be repeated continuously until it runs out of bagged powder or until the freezer platform runs out of space.

Career Opportunities in the Food Industry

Each of the areas discussed in this chapter and Chapter 33 requires people who manage, operate, and carry out the many and varied functions of the food industry. As with all careers in agriscience, those in the food industry present many challenges and rewards. Career opportunities await individuals at the local, county, state, national, and international levels. Careers in the food industry can be divided into eight, often overlapping, categories (**Figure 32-24**). Career opportunities in each area are numerous.

Some Careers in Food Science and the Food Industry

Business
Accountant
Buyer
Distributor
Financial analyst
Loan officer
Marketing specialist
Salesperson
Statistician

Communications
Advertising specialist
Broadcaster
Media specialist
TV producer/Demonstrator
Writer

Education
College professor
Extension specialist
Industry educator
Dietician
Teacher

Processing
Butcher
Efficiency expert
Engineer
Plant line worker
Plant supervisor
Refrigeration specialist
Safety expert

Quality Assurance
Food analyst
Grader
Inspector
Lab technician
Quality-control supervisor
Quarantine officer

Research and Development
Distribution analyst
Biochemist
Microbiologist
Packaging specialist
Process engineer

Retailing/Food Service
Baker
Cook/Pizza maker
Counter salesperson
Deli operator
Meat cutter
Nutritionist
Produce specialist
Restaurant owner/operator
Waiter/waitress

Transportation
Dispatcher
Trucker
Rail operator
Merchant marine

Figure 32-24 Career opportunities in food science and the food industry are many and varied.

Agri-Profile Career Area: Scientist/Inspector/Quality Controller/Buyer/Seller/Processor/Trucker/Wholesaler/Retailer

Processors convert raw farm produce to products that are convenient for consumers.
iStock.com/Leezsnow

The food industry is a large-scale system that covers both plant and animal products. Its complex network includes the producers, processors, distributors, wholesalers, retailers, fast-food establishments, restaurants that together supply the food the world's population consumes, including home kitchens. While many career opportunities in the food industry have been discussed in previous chapters, there are even more options to explore. Here are some examples.

On the land, field supervisors and coordinators direct the work of crews to harvest crops at the peak of their quality and transport them to processing or packing plants. In addition, the meat processing-and-packaging industry employs 500,00 workers in the United States. On the water, crews take fish, oysters, clams, lobsters, and other seafood from the production habitat to processing centers. The work needed to operate huge packing and/or processing machines takes place in the fields, orchards, and on boats.

Across all of these categories, quality-control personnel collect food specimens, label them, test them, and maintain records to ensure quality control on each batch of food coming out of the plant. There are also career opportunities in food science, store management, produce management, meat cutting, laboratory testing, field supervision, research, diet and nutrition, health and fitness, and promotion. Overall, given the massive scale of, and daily demands on, the food industry, there's a "feast" of rewarding employment possibilities.

The Food Industry of the Future

The food industry is ever-changing, with new developments occurring each day (**Figure 32-25**). Areas of special interest to the food researcher include new food products, new processing and preserving techniques, and new equipment for harvesting labor-intensive crops.

Aquaculture is meeting the increasing demand for fish and will continue to supplement the catches of commercial fishermen. The use of extreme heat and cold in processing has contributed to the development of items that meet the demand for convenience foods. The convenience of grocery shopping by app has begun to play a larger role in the food chain. Economic efficiency in convenience foods and convenience stores is under constant review. Meanwhile, the USDA and other agencies continue their vigilance regarding safety and nutritional standards at all steps of the food chain.

A

B

C

Figure 32-25 The future of food is now: (A) Cultured meat grown in the laboratory from live animal cells, (B) plant-based "meat," and (C) fruits treated with protective film.

© New Africa/Shutterstock.com. © nevodka/Shutterstock.com. Courtesy of USDA/ARS #K-3517-5.

Chapter Review

Student Activities

1. Write the **Key Terms** and their meanings in your notebook.
2. Keep a food-dollar diary to document where your food dollars are spent. Record the cost of meals and snacks eaten in and outside the home.
3. Do a cost comparison of meals prepared at home and similar meals consumed at fast-food places and restaurants.
4. Trace the activities that occur in transforming wheat in the field into a hamburger roll consumed in your home.
5. Draw a diagram tracing the individual food components of a deluxe hamburger back to the places where the components were produced. Label each component, process, and commodity along the way.
6. Ask your instructor to arrange a field trip to a butcher shop or supermarket to observe demonstrations on meat cutting and packaging.
7. Create a multimedia presentation or a collage illustrating the various activities of the food industry.
8. Prepare a report on some of the safety regulations that are in place in the United States to prevent illness caused by food contamination.
9. Invite a safety compliance officer from a local workplace to address the class on the topic of "Planning and implementing an effective workplace emergency response."
10. **SAE Connection:** Refer to Figure 32-24, which lists a variety of career opportunities in the food industry. Choose a career that interests you. Find out what steps you will have to take to get the job in which you are most interested. Be sure to include all training and education that will be required.

Inquiry Activity | Overview of Food System

Courtesy of W. Altee Burpee Company.

After watching the *Overview of Food System* video available on MindTap, answer the following questions.

1. What are the five components of the food system?
2. Describe how a raw product moves through the food system to reach the consumer.
3. How do consumers affect other parts of the food system?
4. What is the role of partner industries in the food system?
5. In which components of the food system is transportation involved?

Chapter 33
Food Science

Objective
To explore the nutrient requirements for human health and the processes used in food science to ensure an adequate and wholesome food supply.

Competencies to Be Developed
After studying this chapter, you should be able to:
- discuss nutritional needs of humans and the food groups that meet these needs.
- categorize foods in the U.S. Department of Agriculture *MyPlate* nutrition initiative.
- discuss food customs of major world populations.
- relate methods used in processing and preserving foods.
- list the major steps used in slaughtering meat animals.
- list the major cuts of red-meat animals.
- identify methods of processing fish.
- understand the grades and market classes of animal meats.

Key Terms

nutrient	retortable pouch	vacuum pan
fermentation	irradiation	shackle
blanching	dry-heat cooking	hide
canning	moist-heat cooking	viscera
dehydration	convection oven	carcass
freeze-drying	dehydrator	split carcass
oxidative deterioration	smoker	shroud
dehydrofrozen product	casein	age (ripen)

block beef
disassembly process
fabrication and boxing
giblet

kosher
dressing percentage
sweetbreads
tripe

tankage
collagen

"You are what you eat." Have you ever paused to consider what that common saying means? Food is the material needed by the body to sustain life. It consists of carbohydrates, fats, proteins, and supplementary substances, such as minerals and vitamins. Food is used to sustain growth, repair cells, sustain vital processes, and furnish energy to the body. This chapter explores the foods humans need to maintain health and sustain growth (**Figure 33-1**). It also examines the processes by which foods that begin as raw products arrive at our tables.

Figure 33-1 The well-being of children is largely dependent on the nutrition they receive. Proper nutrition and healthy habits help keep a body fit.
Courtesy of USDA.

Carbohydrates
Minerals
Proteins
Water
Fats
Vitamins

Figure 33-2 Nutrients are classified into six categories. Each category of nutrients supports different functions in the body.

Nutritional Needs

The body is a complex system with many nutritional demands. **Nutrients** are substances that are necessary for the functioning of an organism. More than 50 specific nutrients are required for bodily functions. Nutrition involves a combination of processes by which all body parts receive and use materials necessary for function, growth, and renewal. It also includes the release of energy, the building up of body tissues (both hard and soft), and the regulating of body processes. After food is digested, basic nutrients enter the bloodstream, which then transports them to the cells of the body. Nutrients are classified into six major groups, each supporting different functions in the body (**Figure 33-2**).

Carbohydrates

Carbohydrates serve as the main source of energy for the body. They contain four calories per gram. There are three different types of carbohydrates: sugars, starches, and fiber (**Figure 33-3**). Sugars are simple carbohydrates and are found naturally in many foods, such as fruit, milk, and peas. Refined sugar, the sweet substance added to many house-

Figure 33-3 Fruits, vegetables, and grains are healthful sources of carbohydrates.
Courtesy of Price Chopper Supermarkets.

hold foods, comes from sugar beets and sugarcane. Starch is a complex carbohydrate that is found in foods such as bread, potatoes, rice, and vegetables. Starches and sugars are converted to glucose in the body and serve as the major body fuel. Some of the fuel that is generated is stored by the body for later use. This occurs when the glucose is not fully used by the body and it is converted to fat. Fiber is also a complex carbohydrate and is found in the walls of plant cells. Humans are unable to digest fiber, yet it performs two key functions, moving food through the body and expelling waste after digestion.

Fats

Fats are another source of energy for the body. Fats contain nine calories per gram. They are a compact source of energy because they have 2.25 times the number of calories as the other two energy sources—carbohydrates and proteins.

Fats are present in differing amounts in most foods. Foods that are known to be high in fat content include cheese, meat, poultry skin, avocados, and numerous others. Some foods we are accustomed to eating require using fats as part of their preparation. Baked goods, such as cakes and cookies; salad dressing; and fried foods acquire fats through preparation (**Figure 33-4**).

Fat can be highly beneficial. For example, it supports the functionality of certain vitamins. Known as fat-soluble vitamins, these nutrients are retained in the body by becoming dissolved in fats, and they require fats to carry them to the parts of the body where they are needed. On the other hand, too much fat results in obesity and is associated with serious diseases, such as heart ailments, diabetes, and high blood pressure.

Figure 33-4 Animal products provide high-quality protein, but their skin and fatty layers are best removed to avoid excess fat in the diet.
Courtesy of USDA/ARS #K-4285-2.

Proteins

The body needs food with proteins to build and rebuild its cells. Proteins have four calories per gram. Hair, skin, teeth, and bones are all parts of your body that require protein. Proteins are in a continuous cycle of building up and breaking down. Approximately 3 to 5 percent of your body's protein is rebuilt each day. Beans, peanut butter, meats, eggs, and cheese are high-protein foods.

Vitamins

Vitamins are also essential to the functions of the body. Fat-soluble vitamins are not required in the diet each day because they can be stored in the body. These include vitamins A, D, E, and K. Nine other vitamins—vitamin C and eight B vitamins—are water soluble and must be replenished daily. The various vitamins perform specific functions in the body, and some foods are known to be rich in particular vitamins (**Figure 33-5**).

VITAMINS

A	B	C	D	E	K
Functions					
vision bones skin healing wounds	using protein, carbohydrates, and fats to keep eyes, skin, and mouth healthy brain nervous system	wound healing blood vessels bones teeth other tissues works with minerals	needed for using calcium and phosphorus bones teeth	preserve cell tissue	blood clotting
Sources					
yellow, orange, and green vegetables	whole-grain and enriched cereals, breads, meats, beans	citrus fruits, melons, berries, leafy green vegetables, broccoli, cabbage, spinach	fatty fish, liver, eggs, butter, added to most milk	vegetable oils, whole-grain cereals	leafy green vegetables, peas, cauliflower, whole grains

Figure 33-5 Vitamins have specific functions, so it is important to choose foods that supply them in adequate amounts.

Minerals

More than 20 minerals are needed by the body. The amounts needed may be small, but they are required, nonetheless. Essential minerals can be divided into two groups, major minerals, and trace minerals. Even though trace minerals are needed in smaller amounts, both groups of minerals are equally important. Some minerals are required for healthy bones, others regulate bodily functions or aid in making special materials for cells, while still others trigger important chemical reactions in the body (**Figure 33-6**).

Water

The human body is more than 50 percent water. Water carries nutrients to cells, removes waste, and maintains the body's proper temperature. Fluid foods, such as milk and juice, obviously help supply the body with water. However, foods such as meat and bread also provide water.

Food Groups That Meet Needs

Each food is different in the types of nutrients it contains and ultimately provides to the body. In order to make educated food choices, many people rely on MyPlate, the nutrition guide published by the U.S. Department of Agriculture's Center for Nutrition Policy and Promotion.

Functions of Minerals			
Bone Development	Fluid Regulation	Materials for Cells	Trigger Other Reactions
Sources of Minerals			

calcium
  milk products
magnesium
  nuts, seeds, dark-green vegetables, whole-grain products
phosphorus
  no specific food group
fluorine
  some seafood, some plants, may be added to drinking water
	sodium
	  salt
	potassium
	  bananas
	chlorine
	  salt
		iron
		  meats, liver, beans, leafy green vegetables, grains
		  works with vitamin C
		iodine
		  iodized salt
		  added to salt
			zinc
			  whole-grain breads and cereals, beans, meats, shellfish, eggs
			copper
			  fish, meats, nuts, raisins, oils, grains

Figure 33-6 Careful selection of foods helps ensure a correct balance of minerals in one's daily diet.

This helpful guide has come a long way since it was launched in 2011. Today, MyPlate free resources include an app, personalized food plans, shopping tools, the MyPlate Kitchen video site, and more. MyPlate nutrition tips can even be enabled for use on smart speakers, making it easy for users to ask—literally—for nutrition advice even as they are about to prepare food at home. All of the information provided is based on the Dietary Guidelines for Americans, 2020-2025.

In addition to nutritional guidelines, *MyPlate* also includes science-based recommendations for physical activity and the consumption of discretionary calories. These recommendations can be personalized according to physical characteristics of individual users.

Above all, the guide is easy to understand. MyPlate divides foods into five major food groups, which represent the nutritional needs of the body. The widely known MyPlate graphic, which depicts the five groups as a place setting with a plate and glass, presents an at-a-glance reminder of the relative proportions that each group should make up in a daily diet (**Figure 33-7**).

The five food groups are the following:
- Fruits: Any fruit or 100 percent fruit juice.
- Vegetables: Any vegetable or 100 percent vegetable juice.
- Grains: Any food made of wheat, rice, oats, corn, barley, or other cereal grains.
- Protein Foods: All foods made from seafood; meat, poultry, and eggs; nuts, seeds, and soy products. Note: Beans, peas, and lentils are unique in that they are part of both the Vegetable group and the Protein group.
- Dairy: Fluid milk and many foods made from milk that retain their calcium content. Calcium-fortified soymilk, lactose-free milk, and many cheeses and yogurts belong in this food group. However, milk products that have little or no calcium and a high-fat content, such as cream, cream cheese, sour cream, and butter, do not belong.

Essential oils are not one of the food groups, but essential nutrients are provided by them. Fats that are liquid at room temperature are oils, including oils from plants and fish, are part of this category.

You generally need not seek foods in the essential oils because plenty of these are used to prepare foods, and they show up in adequate amounts in the diets of most people in the United States. The names of the groups suggest some of the typical foods that they include. However, the quality of diet can be increased by selecting the most nutritious items from each group (**Figure 33-8**).

Figure 33-7 The well-known MyPlate icon presents a simple but effective reminder for consumers to eat healthfully from all five food groups. Search "MyPlate" online for a wealth of free resources.

Courtesy of USDA.

POULTRY, FISH, MEAT, AND EGGS

Lower-scoring foods tend to be high in calories, cholesterol (eggs), fat (red meat), or sodium (processed meats). The foods near the top are relatively low in fat. Most of the foods are rich in protein and iron. (All servings are 4 ounces broiled, baked, or roasted, unless noted otherwise.)

Food	NUTRITION SCORE
clams, steamed	19
turkey breast, skinless	10
tuna canned in water (3 oz.)	6
cod	1
egg white (1 large)	1
salmon, canned (3 oz.)	1
scallops, steamed	1
flounder	−1
lobster meat, boiled	−4
salmon fillet	−5
blue crab meat, steamed	−6
chicken breast, skinless	−8
turkey breast luncheon meat, 3 slices (2 oz.)	−13
tuna canned in oil (3 oz.)	−18
shrimp, steamed	−21
chicken breast with skin	−23
Canadian bacon, fried, 2 slices (2 oz.)	−24
veal cutlet, breaded, pan-fried	−25
round steak, trimmed (5 oz.)	−29
ham, luncheon meat, 2 slices (2 oz.)	−32
pork chops	−48
bacon, fried, 4½ slices (1 oz.)	−54
egg (1 large)	−59
shrimp, fried	−63
bologna, 2 slices (2 oz.)	−70
leg of lamb	−76
salami, luncheon meat (2 oz.)	−80
hamburger, lean	−81
sirloin steak (5 oz.)	−86
hamburger, regular	−92
round steak, untrimmed (5 oz.)	−97
chicken thighs, fried, home recipe (2)	−103
sausage links, 2 (2 oz.)	−112
pot roast	−114

GRAIN FOODS

Contrary to myth, starchy grain foods are not fattening. Most people would do well to eat more grain foods in place of meat. Grains, especially whole grains, are a nicely balanced, low-fat source of carbohydrate, vitamins, minerals, and protein. (All serving sizes are 1 cup cooked, unless noted otherwise.)

Food	NUTRITION SCORE
bulgur (cracked wheat)	69
wheat germ (¼ cup)	61
pearled barley	60
brown rice	45
spaghetti or macaroni	45
oatmeal	38
hominy grits	35
whole-wheat bread (2 slices)	31
hamburger or hotdog roll (1)	18
corn muffin (1)	1

FRUITS

Fruits can give you, naturally, all the sweetness you want, plus fiber, vitamins A and C, and other nutrients. Go easy on the dried fruits! Their sugars are sticky and promote tooth decay. (All servings are one medium piece, unless noted otherwise.)

Food	NUTRITION SCORE
papaya (½ medium)	74
cantaloupe (¼ medium)	67
strawberries (1 cup)	65
orange	62
prunes, uncooked (5)	51
dried apricots (5)	49
tangerine	41
watermelon (2 cups cubed)	36
apple	36
pear	36
blueberries (1 cup)	36
pink grapefruit (½)	35
pineapple, fresh (1 cup)	34
banana	32
cherries (1 cup)	32
honeydew melon (¹⁄₁₀ melon)	31
raisins (1 oz.)	28
plums (2)	27
applesauce, unsweetened (½ cup)	18
peach	17
grapes (30)	16
peaches in heavy syrup (½ cup)	2

DAIRY

While most dairy foods are rich in protein and calcium, the lower-scoring foods are high in saturated fat, cholesterol, and sodium.

Food	NUTRITION SCORE
yogurt, nonfat (1 cup)	58
milk, skim (8 oz.)	40
yogurt, plain lowfat (1 cup)	36
milk, 1% lowfat (8 oz.)	28
milk, 2% lowfat (8 oz.)	16
yogurt, fruit-flavored lowfat (1 cup)	13
chocolate milk, 2% lowfat (8 oz.)	6
cottage cheese, 1% fat (½ cup)	3
sour cream, lowfat (2 Tbsp.)	−2
ricotta cheese, part skim (1 oz.)	−3
yogurt, plain (1 cup)	−5
mozzarella cheese, part skim (1 oz.)	−5
milk, whole (8 oz.)	−7
cheddar cheese, reduced fat (1 oz.)	−7
nondairy powder coffee creamer (2 tsp.)	−12
half and half cream (2 Tbsp.)	−15
Swiss cheese (1 oz.)	−15
cottage cheese, 4% fat (½ cup)	−16
mozzarella cheese (1 oz.)	−19
sour cream (2 Tbsp.)	−26
cheddar cheese (1 oz.)	−32
American cheese (1 oz.)	−34
whipped cream (2 Tbsp.)	−59

VEGETABLES

Most vegetables are great sources of vitamins—especially A and C—and minerals. Try a new vegetable today! (All serving sizes are ½ cup cooked, unless noted otherwise.)

Food	NUTRITION SCORE
sweet potato, baked (1 medium)	184
potato, baked (1 medium)	83
spinach	76
kale	55
mixed vegetables, frozen	52
broccoli	52
winter squash (acorn, butternut), baked	44
Brussels sprouts	37
cabbage, chopped, raw (1 cup)	34
green peas	33
carrot (1)	30
okra	30
corn on the cob (1 ear)	27
tomato (1 medium)	27
green pepper (½)	26
cauliflower, raw	25
artichoke (½)	24
romaine lettuce, raw (1 cup)	24
collard greens	23
asparagus	22
celery (four 5" pieces)	19
green beans	18
turnips	16
sauerkraut	15
summer squash (zucchini)	12
green beans, canned	10
iceberg lettuce, raw (1 cup)	8
bean sprouts (¼ cup)	7
onion, chopped, raw (¼ cup)	7
eggplant	6
cucumber slices, raw	4
mushrooms, raw (¼ cup)	2
dill pickle (½ large)	−3
avocado (½ medium)	−25

LEGUMES

Beans are excellent sources of dietary fiber, protein, vitamins, and minerals. They are also very low in fat. (All serving sizes are ¾ cup cooked, unless noted otherwise.)

Food	NUTRITION SCORE
kidney beans	91
navy beans	82
black beans, black-eyed peas, or lima beans	78
lentils	74
chickpeas	68
split peas	56
tofu/bean curd (4 oz.)	33

DESSERTS

Most desserts are high in fat, sugar, and calories. Next time, try fresh fruit or nonfat frozen yogurt for a change. (Serving sizes are 1 cup, unless noted otherwise.)

Food	NUTRITION SCORE
angelfood cake (2 oz.)	1
chocolate pudding (½ cup)	−2
Jell-O (½ cup)	−7
brownie with nuts (1¾" square)	−23
sherbet	−37
vanilla ice cream	−73
cheesecake (4½ oz.)	−161

Figure 33-8 Estimated relative nutritional values of selected food items.

Eating from each of the major food groups daily will ensure a well-balanced diet and provide the essential nutrients needed for growth, development, and maintenance. At different stages in life, requirements within each group may vary to some degree, but no food group should dominate, or be eliminated from, the diet.

Food Customs of Major World Populations

What people eat from each of the major food groups varies by culture around the world. Food habits reflect what is most readily available.

For example, in hot and wet climates such as in Southeast Asia, rice provides 50 percent of calories for the population. After all, Southeast Asia produces 30 percent of the world's rice harvest. Likewise, in regions of the world where corn grows well, many food items contain corn in some form. For instance, in Mexico, which annually produces 27 million metric tons of corn, the Mexican daily diet relies heavily on corn. It is used to make such foods as tortillas, tamales, and a grilled and seasoned corn on the cob known as *elote*.

Other factors include the availability of animal-based food sources and technology. For instance, introducing dairy products in countries where dairy cows are not raised would present challenges in transportation, processing, and consumer education or marketing.

Hot Topics in Agriscience — Life Expectancy and Access to Safe Food

A newborn baby in the United States has a life expectancy of 75.1 years for boys and 80.2 years for girls.
© cath5/Shutterstock.com.

A newborn baby in the United States is entering the highest-risk period in life—infancy and early childhood. However on average, boys are likely to live 76.1 years, while girls live 81.1years. These calculations are based on the records and charts of government and life insurance companies, which insure against death on the basis of their charts. For every year that a human survives, these projected numbers improve, unless one takes up habits (such as smoking) that are known to reduce survival rates.

The average life expectancy of an individual depends on adequate medical care, clean drinking water, sanitary facilities, and food of sufficient amount and quality. Countries in which these resources are unavailable usually experience high death rates, especially among infants and small children. Other factors such as wars also influence the average number of years a person may reasonably expect to live. Countries with the greatest life expectancy are those that are wealthy enough to feed and care for their people.

Despite the pandemic, overall average life expectancy at birth has been gradually increasing over time in many countries of the world. Even, so a comparison of selected examples of life expectancy at birth illustrates the importance of providing access to good-quality food, medical care, clean water, and sanitary facilities:

Life Expectancy at Birth	
Chad: 55.17 years	United States: 79.11 years
Mozambique: 62.13 years	Germany: 81.88 years
Haiti: 64.32	Japan: 85.03 years
India: 70.42 years	Iceland: 83.52 years.

Based on data from the World Health Organization.

Methods of Processing, Preserving, and Storing Foods

Two of the most time-honored ways to preserve food for delayed use are fermenting and pickling. **Fermentation** is a chemical change that involves foaming as gas is released. Humans across many cultures have been using fermentations for thousands of years, ever since it was determined that some foods did not spoil when allowed to ferment naturally or when fermented liquids were added to the foods.

Figure 33-9 Fermentation is a process whereby bacteria convert sugars to acids or alcohol that protects the food from spoilage.
Courtesy of USDA.

Today, controlled fermentation is used to produce cheeses, wines, beers, vinegars, pickles, and sauerkraut (**Figure 33-9**). The main objective of processing and preserving is to change raw commodities into stable forms. With refrigeration and various processing techniques, we now expect that almost all foods should be available at any time during the year and in most parts of the world.

Regardless of method, slowing deterioration is the primary goal of food preservation. Tomatoes and cucumbers that will be sold raw are waxed to slow down any shriveling while the produce is on sale at the supermarket. Apples may also be treated with a decay inhibitor, such as an "antibrowning" dip extracted from natural sources. Table grapes are fumigated with sulfur dioxide to control mold. Similarly, silos in which grains are stored are purged with 60 percent carbon dioxide to control insects.

Carbon dioxide can be used to inhibit the growth of bacteria. Controlled atmosphere (CA) is the process whereby oxygen and carbon dioxide are adjusted to preserve or enhance particular foods. One example of food preservation by CA is the transporting of cut lettuce in CA storage containers to prevent the edges from turning brown. As it travels long distances from field to market, the time-sensitive lettuce is safely adjusted for slow ripening by reducing oxygen levels and increasing the concentration of carbon dioxide.

Refrigeration is an important key to many processing and preservation techniques. Refrigeration is the process of chilling or keeping cool. Low temperatures reduce or stop processes that contribute to the deterioration of products. Refrigeration retards respiration, aging, ripening, textural and color change, moisture loss and shriveling, insect activity, and spoilage from bacteria, fungi, and yeasts. Refrigeration, commonly used in developed countries, is effective, but expensive.

Figure 33-10 Food storage and preservation through ice, refrigeration, and freezing are the cornerstones of milk, meat, fish, fruit, and vegetable handling today.
Courtesy of USDA.

When crops are harvested in the field, their temperatures are between 70° and 80° F. The goal of refrigeration is to quickly reduce that temperature to near 32° F. How quickly the food is cooled depends on the type of cooling technique used. For examples, some vegetables are precooled in the field by cold air blasts, vacuum, or hydrocooling. Hydrocooling means cooling with water and is also used on dairy farms, where milk is quickly cooled in refrigerated pipelines and tanks. One cooling technique familiar to virtually everyone is the use of ice. For example, freshly harvested fish and other seafoods are packed in ice until they can be processed and sold (**Figure 33-10**).

These are just some of the many processing and preservation techniques in use. Together, they make it possible for consumers to enjoy a wide variety of foods throughout the year and around the globe.

After a product has been cooled, one logical next step is to maintain the cooling until it is frozen (**Figure 33-11**). When foods are kept at a temperature of 0° F or lower, little deterioration occurs. However, even frozen foods have storage limits. Fruits and vegetables should be consumed within 1 year after freezing. Meats should be consumed within 3 to 6 months. In addition, vegetables that are to be frozen often require blanching before freezing. **Blanching** is the scalding of food for a brief period before freezing it. This process inactivates enzymes that would otherwise cause undesirable changes when plant cells are frozen.

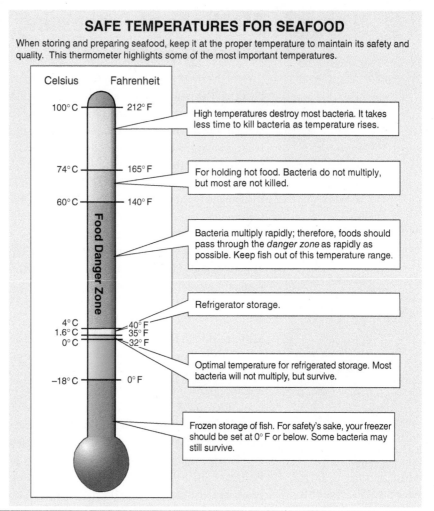

Figure 33-11 It is important to choose the appropriate heat intensity and duration for the purpose.

Yet another preservation technique has long been used in home kitchens as well as commercially. **Canning** involves putting food in airtight containers and sterilizing the food to kill all living microorganisms that could cause spoilage. Temperatures of 212° to 250° F are required to successfully can food products. In commercial practice, metal cans are coated to reduce chemical reactions between the metal can and its contents to extend the shelf life of the food. Shelf life refers to the amount of time before spoilage begins. A 2-year shelf life for canned food is considered normal. Despite its name, home canning relies on the use glass jars that are then vacuum-sealed.

Another popular way to process and preserve food is dehydration. **Dehydration** occurs as the moisture content of a food is reduced to inhibit the growth of microorganisms. Moisture can be removed by the sun, by the use of indoor tunnel or cabinet dehydrators, or by freeze-drying. **Freeze-drying** is a method of dehydration that involves the removal of moisture by rapid freezing of food at very low temperatures. When foods are dehydrated, they have a moisture content of 2 to 10 percent. Accordingly, dehydrated foods are lighter in weight and lower in volume than the same whole foods. The shelf life of dehydrated foods may be as long as 2 years. Common examples of dehydrated foods include dried soup mixes, packaged salad dressing, spices, and dried fruits.

Packaging is an important consideration in food dehydration. This is because oxidative deterioration, a loss of quality due to reaction with oxygen, can occur when air reaches dried foods. Glass or metal containers are best because they are more airtight than plastic. Even when dried foods are packaged in flexible pouches that appear to be paper or plastic, the interior of the pouch contains a barrier that contains foil.

Another processing technique is dehydrofreezing. A dehydrofrozen product is one that is processed by precooking, evaporating water, and freezing. For example, potatoes can be precooked as cubes or slices, after which the water is evaporated, reducing the weight of the product by 50 percent. The potatoes are then frozen.

Another factor to consider in the processing and handling of foods is level of humidity to be maintained during storage and transportation. Humidity is the amount of moisture in the air. For example, high-moisture products such as meats and vegetables require a humidity level of 90 to 95 percent, while dried onions require only 75 percent humidity for optimum storage.

A wide array of wrappings and packing materials are used in the processing, preservation, and transportation of foods. Cardboard boxes, wood boxes, molded pulp trays to reduce bruising, and plastic wraps in a variety of thicknesses or plies all meet different needs. Retortable pouches provide protection from light, heat, moisture, and oxygen transfer, all in one wrapping. Retortable pouches are flexible packages consisting of two layers of film (or plastic) with a layer of foil between them. Although the cost is high, the benefits include a shelf life of 1 to 2 years.

Irradiation a widely practiced food-processing procedure that uses gamma rays to kill insects, bacteria, fungi, and other organisms in food products. Gamma rays pass through food without heating or cooking it. As a result, no heat-sensitive nutrients are lost in the process (**Figure 33-12**).

The cost of the preceding processes and techniques varies widely and must be carefully considered as one of the operating costs of a processing plant. In plants where energy costs are low, more sophisticated techniques may be used without incurring excessive costs. Meanwhile, food scientists continue to research ways of preserving food more economically in relation to energy costs.

Figure 33-12 Irradiation is an approved method of preventing *Salmonella* bacteria in chicken.
Courtesy of USDA/ARS #K-3783-20.

Science Connection: Irradiation—A Food Safety Solution

Irradiation of fruits and vegetables has proven to be a safe and effective way to control the spread of insects that infect these foods.
Courtesy of USDA.

After a lengthy period of inquiry and research, the USDA and World Health Organization approved the irradiation of raw packaged poultry as a safe and effective procedure for protection against food-borne illness. Irradiation of raw food is now an accepted procedure in many countries.

Science had long sought a way to eliminate food-borne pathogens without altering or damaging the product or leaving poisonous chemicals on the food. Irradiation proved to be an effective answer. It provides the same benefits as processing food by heat, refrigeration, or freezing. It also replaces treatments with chemicals as it destroys insects, fungi, and/or bacteria that cause food to spoil or cause human disease. The treatment process involves passing food through an irradiation field, but the food itself never contacts a radioactive substance.

Notably, the approved level of irradiation does not kill all microorganisms—in other words, it does not sterilize food. Because of this, food-preserving measures must be followed when storing irradiated food and preparing it for consumption. In addition, proper cooking is still the last stage in food handling to eliminate microorganisms that could cause food poisoning or other illness to those who consume the food.

Regardless of whether or not food is irradiated, the use of water for washing and cooling it during and after processing can spread harmful microorganisms throughout the batch and leave every item with low concentrations of the microorganisms. In other words, all of the food will have low-level contamination, and the organisms will multiply and increase their hazards if conditions are right to do so. Handling raw meat that potentially has pathogens on its surface carries the threat of contamination of hands, countertops, cooking areas, and, ultimately, cooked food.

In short, irradiation of food followed by proper handling, storing, and cooking of food are all essential to ensuring the safety of the foods we eat.

Irradiated foods from the United States bear a special logo developed by the Cordex Alimentarius, an international committee on food safety. The logo is green and includes the words "Treated with Radiation" or "Treated by Irradiation."

Food Additives to Enhance Sales of Food Commodities

Processing of some foods may reduce their natural nutritional value. To compensate for that loss, vitamins and minerals are added to foods to restore their nutritional value. Commercially processed bread, noodles, and rice all have vitamins and minerals added to them before they are packaged. Likewise, vitamins A and D may be added to fluid milk before it leaves the processor.

As the name suggests, a food additive is anything that is added to a food product during processing and before it goes into a package. In addition to food additives that restore nutritional value, preservatives are also added to foods to extend their shelf life. Food additives also enhance the color or appearance of foods. Other food additives reduce the cooking time of foods such as oatmeal. Sugar is probably one of the most widely used food additives. It is found on the labels of many cereals and beverages. Food labels identify the contents of food products. The order of ingredients on a food label indicates the proportions of each, in descending order (**Figure 33-13**). The government is continually testing and evaluating the positive and negative effects of food additives. Some food additives are known to cause harm, and they are banned by the government: Among them are various artificial food colorings, cobalt salts, cinnamyl anthranilate, calamus extracts and oils, calcium cyclamate, coumarin, and safrole.

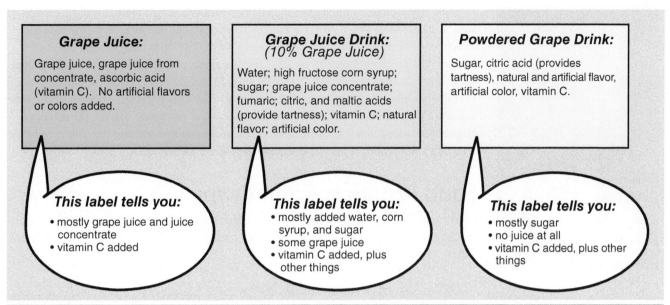

Figure 33-13 Some products may look alike, but close reading of the labels reveals the differences. Ingredients are listed in order from the highest amount to the lowest amount found in the products.

Food Preparation Techniques

Some foods can be eaten as they come in the package. Crackers and raisins are examples of foods that do not require additional processing at home to ensure safe eating. However, cooking is often required for other foods, while heating may only improve flavor in others that are either ready-to-eat or pre-cooked. Raw meat should be cooked before it is eaten to ensure safety. How food is cooked is determined by the type of food and the appliances available. Two basic methods are used to cook foods. The appliances used to accomplish these methods vary from household to household. **Dry-heat cooking** involves surrounding the food with dry air in the oven, air fryer, or under the broiler. This method is usually used for tender cuts of meat having little connective tissue and for vegetables having high-moisture content, such as potatoes. **Moist-heat cooking** involves surrounding the food with hot liquid or by steaming, braising, boiling, or stewing the food. The warm moisture breaks down the connective tissues. Moist-heat cooking is a popular method used for less-tender cuts of meat and for vegetables with low-moisture content.

The goal of preparing good food in a short time period without expending excessive energy has led to the development of several appliances that supplement the gas and electric range and oven. For example, pressure cookers concentrate moisture by sealing it in, thus reducing cooking time. Crockpots and oven bags allow slow, moist cooking without the operator continually being near the cooking process.

Conventional ovens, which feature heating elements on the top and bottom of the oven interior, are used to cook small amounts of foods by dry heat. This requires less energy than larger, more traditional ovens. Microwave ovens use electromagnetic waves to heat and cook food. They offer a more energy-efficient way to cook foods that require both dry and moist heat. Approximately 50 percent of the energy goes to the food in the microwave, whereas only 6 to 14 percent of the energy used by a conventional range actually goes to the food.

Convection ovens heat food with the forced movement of hot air. Air fryers, which surged in popularity after 2020, are countertop versions of convection ovens. Ovens that combine convection and microwave functions are also available, designed to save space as well as energy costs.

Figure 33-14 Baked goods from grains form the basic diet in many cultures.

Courtesy of Price Chopper Supermarkets.

Less common home appliances include **dehydrators**, which remove moisture from food, and **smokers**, which preserve food by keeping smoke in contact with the food for prolonged periods. Refrigeration is still required for most foods after they are dried or smoked. Until recently, home smoking of foods was not particularly popular in the United States. This changed significantly between April 2020 and February 2021, when more than 14 million outdoor home smokers and grills were sold. This is the same period when home lockdown was mandated in response to the coronavirus pandemic. Having the means to prepare foods outdoors offered families a break from confinement and the chance to safely visit with a few others, if only from across the fence or six feet apart.

Food Products From Crops

Food from plant sources helps meet body requirements for food in four of the five food groups and in essential oils. Fruits, vegetables, grains, protein, and oils all come from crops (**Figure 33-14**).

Fruits, Vegetables, and Nuts

Once they are harvested, fruits, vegetables, and nuts are nearly ready to eat. Preparation can be as simple as pick, wash, and eat for items such as leafy vegetables, berries, and fruits. These products may also be processed for storage for a few hours or many years. Food processing can be carried out at home or commercially for billions of consumers.

For some such foods, however, the journey from field to table may entail several stages, selected from a wide array of processes. The first stage may be to simply cool the product and hold it in a temperature- and humidity-controlled environment until the next stage, or it may be to process the food without first precooling. When processing occurs immediately, the product typically passes through washing equipment designed to avoid injury to the product. Washing may involve flotation, rotary, water-jet, or other cleaning procedures. Skins and hulls may be removed by hot water bath, steam, flame, or cutters. The product may then be trimmed, halved, quartered, sectioned, sliced, diced, or crushed for drying, canning, freezing, or processing into ready-to-cook or precooked products.

At every stage, nutritionists, chemists, inspectors, public health officials, and others monitor the process, including packaging. All processed foods must be safely packaged and correctly labeled. The law requires that food be safe for consumption and that each food product must conform to the weight, quality, and grade that is specified on its label.

Cereal Grains

Cereal grains make up the major diet of most of the world's people. Rice, wheat, corn, barley, oats, and other grains are consumed as whole grains, cracked or rolled grains, flour, bran, and many other products. Grains are economical to process because they can be left on the plant until nearly dry enough to prevent spoilage. If stored in a cool, dry place, properly dried grains can remain edible for many years. However, much of the world's food grain is lost every year to rotting, whether in the field or in storage, or as a result of consumption or destruction by insects, birds, and rodents.

Processing grain for human consumption generally means separating or milling the grain into its basic components—hulls, bran, flour, and germ. These components are then used to make breakfast cereals, breads, pastries, pasta, and thousands of other products that fill grocery shelves.

Oil Crops

Soybeans, cottonseed, peanuts, rape, safflower, sunflowers, flax, palm nuts, coconuts, olives, and corn are examples of crops that are rich in vegetable oils. These are used for cooking, frying, baking, and other food processes, as well as for many food products such as dressings, coffee creamers, and shortening. These oils are also used in the manufacture of paints, lacquers, plastics, and many products of industry. The seeds or nuts and other oil-rich plant parts are crushed or ground and then heated. Next, the oil is extracted by solvents and purified for food and industrial uses. The meal is dried and ground mostly for livestock feed (**Figure 33-15**).

Thanks in part to oil crops, consumers who do not eat meat products can still meet their daily requirements for food from all five food groups. People who have allergies to milk products can also meet many of their nutritional needs for protein and other nutrients from soybeans, coconuts, and other oil-crop plant sources.

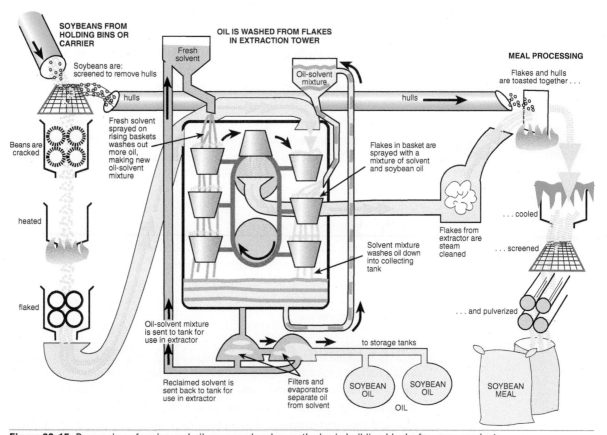

Figure 33-15 Processing of grains and oil-crop seeds releases the basic building blocks for many products.

Food Products From Animals

Meats, fish, poultry, and dairy products provide high-quality protein and are nutritious foods from animal sources (**Figure 33-16**).

Dairy Milk

On average, Americans consume 141 pounds of fluid milk per person (**Figure 33-17**). While recommendations for milk consumption are greatly influenced by calcium needs, milk actually contains some amount of all of the essential food nutrients—not only minerals, but water, fat, protein, and natural sugar in the form of lactose.

Figure 33-16 Meats, fish, poultry, eggs, and dairy products are nutritious foods from animal sources.
Africa Studio/Shutterstock.com. Goskova Tatiana/Shutterstock.com.

Figure 33-17 Milk is pasteurized to kill bacteria and homogenized to keep the milk fat evenly suspended.
© istock.com/Jason Lugo.

The nutritional completeness of milk has led to its reputation as nature's most nearly perfect food. Where it falls short in minor ways, science and the food-processing industry have intervened to improve the options for the consumer.

Fluid milk sold in the United States is pasteurized to ensure safety from disease-causing organisms. Milk is also homogenized to keep the milk fat in suspension so that it does not require stirring before each use. Whole milk is marketed with 3.5 percent milk fat. Consumers who desire less fat intake can also purchase milk with 2 percent, 1 percent, or no milk fat. The latter is called skim milk. Milk may also be purchased as fortified milk with vitamins A and D added.

Milk's Water Content

Although cow's milk is a fluid that is about 88 percent water, it also contains about 12 percent total solids. This is comparable to the solid content of many other foods. Because milk is a food specifically prepared by nature for the nourishment of the very young, it provides the water necessary for life. The water also acts as a carrier for dissolved, suspended, and emulsified components.

Milk's Protein Content

Milk provides a substantial proportion of the total protein in our food supply. **Casein** is the most abundant protein in milk. It is found only in milk and makes up about 82 percent of the total milk protein. Casein exists in suspended form and is easily coagulated by the action of acids and enzymes. Milk proteins are high-quality proteins that contain all of the essential amino acids in proper balance for good nutrition.

Milk Fat

As previously stated, fats are concentrated sources of energy. Several different fatty acids in milk give it the distinctive, pleasing flavor that complements many prepared foods. Milk also contains vitamins that are fat soluble—A, D, E, and K. Milk fat exists in a highly emulsified state, which facilitates its digestion.

Lactose and Minerals

Lactose is sometimes referred to as "milk sugar," because it is only found naturally in the milk of dairy animals. Lactose is milk's major carbohydrate and accounts for about half of the nonfat solids in milk. The relative sweetening power of lactose is about one-sixth of sucrose, the common table sugar. For consumers who have an intolerance to lactose, lactose-free milk is widely available.

Milk contains seven minerals as major constituents and many more in minor or trace amounts. Calcium and phosphorus are essential in human nutrition for building bony structures and for certain metabolic processes. Milk is the chief source of food calcium in the diets of people in the United States. It has the added advantage of containing phosphorus in the same biological ratio to calcium as occurs in the growing skeleton. It is difficult to provide the recommended daily amounts of dietary calcium without using milk or milk products because calcium is poorly distributed among other foods.

All of the vitamins known to be required by humans are found in milk. Some are fat soluble and are associated with butterfat. Those that are water soluble and are found in the nonfat portion of milk. Vitamin A and carotene are present in high concentration in milk fat. Carotene, from which vitamin A is formed in the body, gives milk fat its characteristic color. The vitamin D content of fresh milk is low. However, most commercially pasteurized milk is fortified with vitamin D to balance the product for best nutrition.

Milk is an abundant source of riboflavin (vitamin B_2) and an important source of niacin. Although the niacin content of milk is low, it is in a fully available form. Milk contains significant amounts of thiamine (vitamin B_1). Other vitamins of the vitamin B complex occurring in milk include pantothenic acid, pyridoxine (vitamin B_6), biotin, vitamin B_{12}, folic acid, and choline.

Other essential dietary components found in milk are amino acids and enzymes. Amino acids are the building blocks of proteins and can only be obtained from food. Enzymes help speed up chemical reactions in the body and are needed for digestion, liver function, and more.

Processed Milk Products

The various methods used to process milk make possible a wide variety of milk-based products (**Figure 33-18**). Three of the most widely used milk products are cream, butter, and cheese.

Cream, a component of milk, is recovered from whole milk by concentrating the fat portion of the milk, known as butterfat. This is accomplished by passing milk through a cream separator. The separated cream contains up to 40 percent butterfat. Butter, which is made from cream, contains about 80 percent fat, as does whipping cream contains about 40 percent fat, while "half-and-half" contains approximately 12 percent.

Figure 33-18 Milk is a basic ingredient of many processed food items.
Courtesy of USDA.

Figure 33-19 Cheese is manufactured from milk by coagulating the milk solids to form curd. The curd is ground and pressed into larger blocks of cheese.

Courtesy of DeVere Burton.

Cheese is made by exposing milk to certain bacterial fermentations or by treating it with enzymes. Both methods are designed to coagulate some of the proteins found in milk. Bacteria act on the cheese to impart characteristic flavor that makes each variety unique. (**Figure 33-19**).

Many other products are made by processing milk, including yogurt, butter, frozen foods, dried whole milk, cottage cheese, evaporated milk, condensed milk, ice cream, ice milk, sherbet, and nonfat dried milk.

Cottage cheese is made from skimmed milk. Condensed and evaporated milks, sold as canned products, are each produced by removing large portions of water from the whole milk through a machine called a vacuum pan. Condensed milk is further treated by adding sugar. The sugar content makes condensed milk an important ingredient in the baking and ice cream industries. Nonfat dried milk is used both as human food and animal feed. It is frequently used as an ingredient in dairy and other food products.

Meat Products

Like milk, meat products are processed in a variety of forms. Animal species vary, but the procedures for slaughtering and processing are similar. *Slaughter* in this context means to kill and process or dress animals for market purposes. There are common by-products from the processing of animals along with the meats that are most familiar to American consumers.

Beef

Preparing beef cattle for market is a complex procedure. The first step is to render the animal insensible to pain. There are several methods of accomplishing this that comply with the Humane Slaughter Act of 1958. Packers that do not comply with provisions of this act are prohibited from selling meat to the federal government. To be approved, methods must be rapid and effective. These include a single blow or gunshot, electrical current, or the use of carbon dioxide gas. Rendering insensible may be accomplished by following the ritual requirements of religious faiths.

Once the animal is insensible to pain, it is shackled, hoisted, and stuck to permit bleeding. Shackles are mechanical devices that confine the legs and prevent movement; to *hoist* is to raise into position. To *stick* is to cut a major artery in order to permit blood to drain from the body. A large artery is severed for efficient bleeding out. The head of the animal is removed during or after the bleeding-out process.

The next step in the process is the removal of the hide, or skin. The hide is cut open at the median, or midline, of the belly of the animal, and tools called hide pullers are used to remove the hide in one piece. The breast and rump bones are split at this time by sawing.

The term viscera refers to organs located in the body cavity of the animal. Organs that are removed include the heart, liver, and intestines. The kidneys are not removed at this time. Plants that are regulated by the USDA must have the carcass and viscera inspected to confirm good animal health. Carcass refers to the body meat of the animal—the part that is left after the offal has been removed. Offal consists of the non-meat material that is converted to by-products. It includes the blood, head, shanks, tail, viscera, hide, and loose fat.

Next, the carcass is split by cutting through the center of the backbone and removing the tail. The split carcass (the sides of the animal) is then washed with warm water under pressure.

A high-quality carcass should have a smooth appearance after it is cooled. To accomplish this, the hot carcass is wrapped with a large cloth called a shroud. Sides are cooled for a minimum of 24 hours before ribbing and further processing (**Figure 33-20**). The meat is kept at 34°F until it is sold and consumed.

Figure 33-20 After slaughter, the carcasses of meat animals are inspected by USDA inspectors and then cooled before being turned into block or retail cuts.

Courtesy of USDA/ARS #K-4284-12.

Fresh beef is often processed and shipped within a few days. Small processors usually allow carcasses to hang in the cooler to allow them to **age**, or **ripen**, terms that mean to leave the carcass undisturbed for a period so that minor biological changes can take place while the beef cools. One reason for this is that fresh beef is not in its most tender state immediately after slaughter. Aging helps tenderize beef and make it more flavorful.

Three methods are most commonly used to age beef—traditional aging, fast aging, and vacuum packaging. Differences in time, temperature, and technology distinguish one method from another. While the beef is aging, evaporation (loss of moisture) and discoloration are kept to a minimum. The fairly thick covering of fat on the carcass also helps the aging process.

Beef carcasses are generally disposed of in three ways: as block beef; fabricated, boxed beef; and processed meats. Traditionally, meat is shipped in exposed halves, quarters, or wholesale cuts to be cut into retail cuts in supermarkets. In this condition, it is referred to as **block beef**. It is ready for sale "over the block" or counter. This traditional method has sometimes raised concerns over sanitation, shrinkage, spoilage, and discoloration. Therefore, more packers are using the disassembly process. The **disassembly process** means that the carcass is divided into smaller cuts, vacuum sealed, boxed, moved into storage, and shipped to retailers. This process is also known as **fabrication and boxing**. Processed meats are made from scraps of meat that are not in suitable form for sale over the block. Such meats have the bones removed and are then sold as boneless cuts. They can also be canned, made into sausage, dried, or smoked.

Sheep

Sheep are slaughtered in much the same way as beef. They are first rendered insensible and bled. Next, the front feet are removed, the pelt is removed, and the hind feet and head are removed. The opening of the carcass and removal of the viscera, called *evisceration*, are similar to the procedures outlined for beef animals. In view of the small size of a lamb, the forelegs are folded at the knees and are held in place by a skewer after evisceration. Washing and cooling procedures are similar to those used for beef.

Hogs

The procedure for slaughtering hogs is a little different from that used for cattle and sheep. Hogs are rendered insensible, shackled, hoisted, and bled. The carcasses are then plunged into hot water at 150° F for about 4 minutes. This process is required to loosen the hair and dry skin.

The hair of hogs is removed by mechanical scraping. A de-hairing machine can remove the hair from about 500 hogs per hour. After the hair is removed, the hog is returned to overhead racks and processing continues.

The hog is washed and singed before the removal of the head. Singe means to burn lightly to remove any remaining hair. Next, the carcass is opened and eviscerated before being split or halved with a cleaver or electric saw. The leaf fat is removed. Leaf fat consists of layers of fat inside the body cavity. Before the carcass is washed, the kidneys and facing hams are inspected. After washing, the carcass is sent to coolers at 34° F.

Unlike the fat from beef cattle, lard is considered a product along with the meat. Lard is the fat from hogs. It is used for a variety of cooking and baking products. Lard is often combined with other animal fats and with vegetable oils, such as cottonseed, soybean, peanut, and coconut. Such mixes are extensively used for baking, cooking, frying, and other food preparations and for commercial products.

Poultry

The steps in poultry processing are similar in many ways to those required for other animals. For example, the feathers that cover a bird, like the hair that covers the hog, must be removed before evisceration.

The process begins by securing the bird on a conveyor belt and bleeding out. Next, the bird is scalded before feather removal or picking. Singeing or lightly burning the skin is required to remove the fine hairs that cover a bird under its feathers. After the feathers and hair have been removed, the bird is washed and eviscerated and the giblets cleaned. **Giblets** are the heart, liver, and gizzard of a bird. The bird is then cooled to 40° F, usually with ice (**Figure 33-21**).

Figure 33-21 The technology and design of processing plants make it possible to process, inspect, and package the millions of broilers that are consumed fresh or frozen for later use.
Courtesy of USDA.

Fish

After fish are caught or harvested, they, too, must be prepared for processing and consumption. The procedure depends on the type of fish. Evisceration usually takes place after the scales and head have been removed. Washing and cooling follow.

Fish, like other foods, are processed and consumed in a variety of ways. For example, whether tuna is to be consumed whole or processed for eating later determines whether the fish is left whole or is cut up. Some shellfish, such as crab and lobster, are kept alive until they are cooked. The heat of cooking kills them. The meat can then be removed and consumed or processed.

Kosher Slaughter

The word **kosher** means "right and proper." In order for food processing of animals to be kosher, it must be carried out in accordance with Jewish laws. This requires that animals be killed by a *shohet* (also spelled *shochet*), a person officially licensed in writing by the authority of a rabbi for this purpose. The methods, as well as the time by which the meat of the animal must be sold, are based on ritual that relates to concerns for sanitation. Meat must be consumed quickly. Neither packers nor retailers are permitted to hold kosher meat for more than 216 hours (9 days). Washing is required every 72 hours.

Major Cuts of Meat

After an animal is slaughtered, the meat is further prepared for use. Different areas of the animal are useful for different purposes. Cuts of beef, lamb, and pork, with wholesale and retail terms, are shown in **Figures 33-22–33-24**.

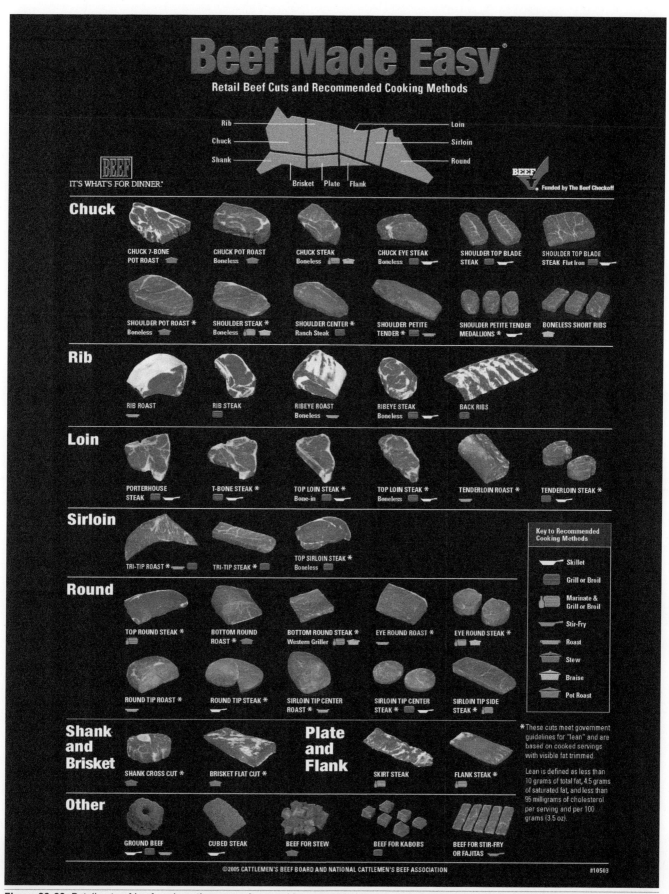

Figure 33-22 Retail cuts of beef—where they come from and how to cook them.

Chart courtesy of the Beef Checkoff Program.

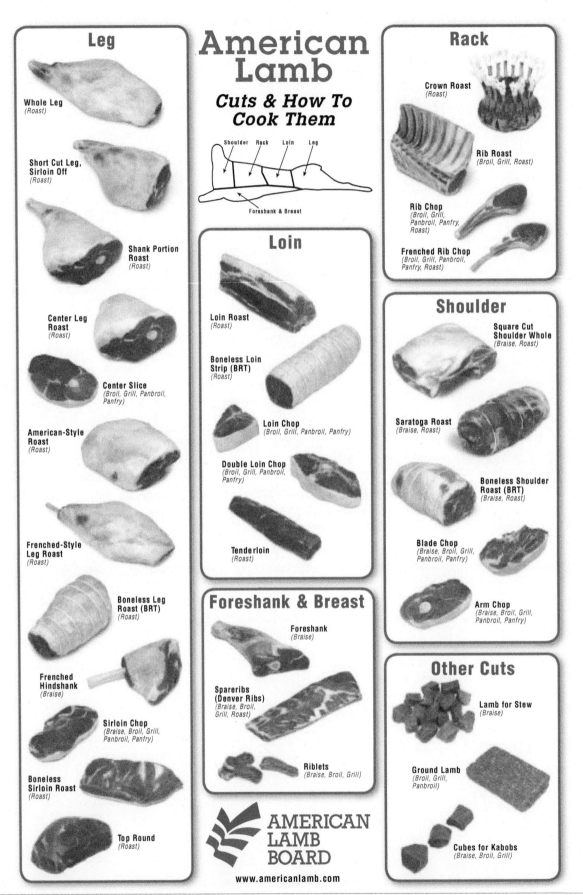

Figure 33-23 Retail cuts of lamb—where they come from and how to cook them.
Courtesy of the American Lamb Board.

Figure 33-24 Retail cuts of pork—where they come from and how to cook them.
Courtesy of the National Pork Board.

Grades and Market Classes of Animals

Cattle

Beef animals are classified as either calves or cattle. Calves are younger than 1 year, whereas cattle are older than 1 year. Calves are further divided into veal calves, feeder calves, and slaughter calves. Veal calves are younger than 3 months and are slaughtered for meat that is also called *veal*. They usually weigh less than 200 lbs.

Calves between 3 months and 1 year old that are marketed for meat are called slaughter calves. They have usually been fed at least some grain. Feeder calves are 6 months to 1 year old and are sold to people who feed them to market weight as slaughter cattle. The sex classes for feeder and slaughter calves are steers, heifers, and bulls. Steers are castrated male calves, heifers are young female calves, and bulls are unaltered male calves.

Cattle are divided into feeder cattle and slaughter cattle. Feeder cattle are further categorized into age groups of yearlings and 2 year olds and older. Yearlings are between 1 and 2 years old, and 2 year olds are 2 or more years old. These two classes of cattle can also be divided into six sex classes—steer, heifer, bull, bullock, cow, and stag. A cow is a female animal that has had a calf; a heifer is a female that has not had a calf. A steer is a male castrated before sexual maturity. A bull is a mature male; a bullock is a young intact male; and a stag is a male animal that was castrated after reaching sexual maturity.

Slaughter cattle are marketed for the purpose of being processed for meat. They are divided into the same age and sex classes as feeder cattle. They are also divided into quality and yield grades. Quality grades refer to the amount and distribution of finish (fat) on the animal. The quality grades for cattle are prime, choice, select, standard, commercial, utility, canner, and cutter (**Figure 33-25**). Yield grades are based on the amount of lean meat an animal will yield in relation to fat and bone (**Figure 33-26**). The yield grades are 1 through 5, with yield grade 1 producing the largest proportion of lean meat.

Swine

Two classes of swine are feeder pigs and slaughter hogs. In general, pigs are swine younger than 4 months old, whereas hogs are swine older than 4 months. Feeder pigs are sold to be fed to higher weights before being slaughtered. Slaughter hogs are sold for immediate slaughter.

The sex classes of swine are gilt, barrow, boar, sow, and stag. Gilts are young female swine. A barrow is a castrated male swine. Boars are unaltered males; stags are mature male swine that have been castrated after sexual maturity. Sows are mature female swine.

Swine are also graded according to quality, with the official USDA grades being U.S. No. 1 through U.S. No. 4. Animals with lean meat of an unacceptable quality are graded U.S. Utility. The highest grade of swine is U.S. No. 1 (**Figures 33-27** and **33-28**).

Sheep

Market sheep are classified according to age, use, sex, and weight. The age classes are lambs, yearlings, and sheep. Lambs are young sheep and can be divided into hothouse lambs, spring lambs, and lambs. Hothouse lambs are marketed below 3 months of age, usually for the Christmas or Easter holidays. Spring lambs are 3 to 7 months old, and lambs are 7 to 12 months old. Lambs can also be classified as feeder lambs or slaughter lambs, depending on whether they are to be fed to heavier weights or slaughtered immediately.

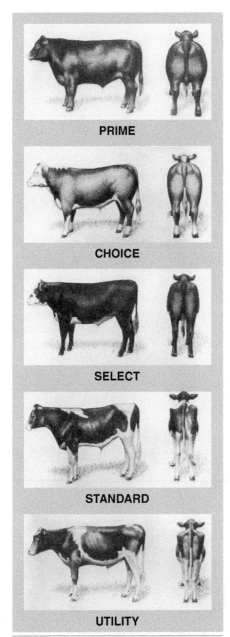

Figure 33-25 Quality grades of slaughter cattle are based on age or maturity and the amount and distribution of fat on the carcass.
Courtesy of USDA.

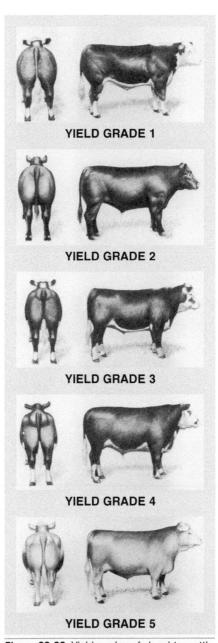

Figure 33-26 Yield grades of slaughter cattle are based on the yield of lean meat in proportion to the amount of fat and bone, from grade 1 (highest) to grade 5 (lowest).
Courtesy of USDA.

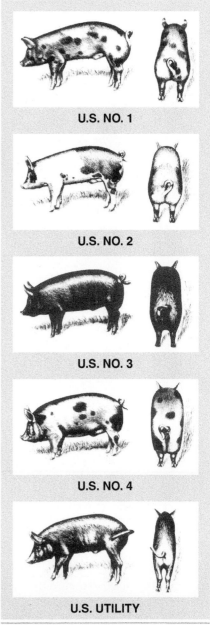

Figure 33-27 Quality grades of slaughter hogs.
Courtesy of USDA.

Yearlings are between 1 and 2 years of age, and sheep are older than 2 years. Sheep are divided into three sex classes: ram, ewe, and wether. Rams are unaltered male sheep, ewes are female sheep, and wethers are castrated male sheep.

Sheep are also graded according to yield and quality. The yield grades are 1 through 5, with 1 yielding the highest proportion of lean meat. Quality grades are prime, choice, good, utility, and cull (**Figure 33-29**).

A review of useful terms related to cattle, swine, and sheep is provided in **Figure 33-30**.

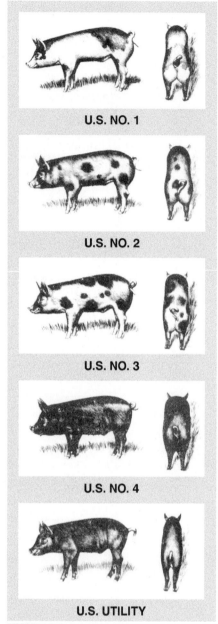

Figure 33-28 Quality grades of feeder pigs.
Courtesy of USDA.

Figure 33-29 Quality grades of slaughter lambs.

	Cattle	Swine/Hogs	Sheep
Young, immature	Calf	Pig	Lamb
Young, but maturing	Feeder or yearling	Feeder	Yearling
Castrated male	Steer or stag	Barrow or stag	Wether
Mature breeding male	Bull	Boar	Ram
Mature breeding female	Cow	Gilt or sow	Ewe
Meat of the young	Veal	Pig or pork	Lamb
Meat of the mature	Beef	Pork	Mutton

Figure 33-30 Terms given to meat animals of various sexes and maturities.

By-products—How Waste Products Are Used

Although most of the animal is consumed by humans, some parts are inedible. The **dressing percentage** is a term used to indicate the percentage or yield of hot carcass weight to the weight of the animal on foot. To arrive at the dressing percentage, the offal is removed from the live animal. The formula is the following: hot carcass weight divided by the live weight times 100. The offal may be 40 percent of the live weight of the animal.

What happens to the offal accounts for many products that are used daily (**Figure 33-31**). By-products can generally be divided into 12 categories:

1. Hides: Leather from animal hides is used to make a variety of consumer products, such as shoes, harnesses, saddles, belting, clothing, sports equipment, furniture coverings, coats, hats, and gloves (**Figure 33-32**).
2. Fats: These are used to make products such as oleomargarine, soaps, animal feeds, lubricants, leather dressing, candles, and fertilizers.
3. Variety meats: The heart, liver, brains, kidneys, tongue, cheek meat, tail, feet, **sweetbreads** (thymus and pancreatic glands), and **tripe** (pickled rumen, or stomach, of cattle and sheep) are sold over the counter as variety or fancy meats.
4. Hair: Brushes for artists are made from the fine hairs on the inside of the ears of cattle. Other hair from cattle and hogs is used for toothbrushes, paintbrushes, mattresses, upholstery for residential and commercial furniture, and air filters.
5. Horns and hoofs: These items are used as a carving medium and are fashioned into decorative knife and umbrella handles, goblets, combs, and buttons.
6. Blood: Blood is used in making animal feeds, fertilizers, and shoe polish.
7. Meat scraps and muscle tissue—After separation from the fat, meat scraps and muscle tissue are most often made into meat-meal or tankage. **Tankage** is the dried animal residue used as fertilizer and feed for animals other than cattle. The "mad cow disease" incident resulted in some recycled meat products being banned in cattle rations.

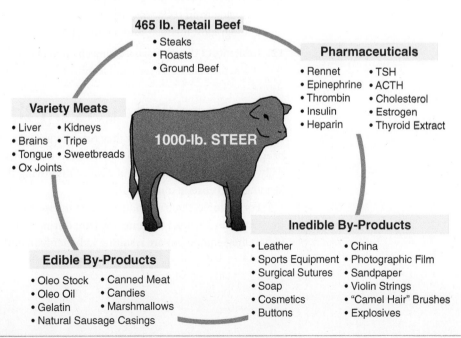

Figure 33-31 Many useful by-products come from animals.

Figure 33-32 Leather, a by-product of meat processing, is used in many products, including sporting goods, shoes, apparel, furniture, upholstery, and saddles.
© iStockphoto/susaro.

8. Bones: Bones are put to some of the same uses mentioned for horns and hoofs. In addition, bones are converted into stock feed, fertilizers, and glue.
9. Intestines and bladders: Sausage, lard, cheese, and snuff all use the intestines and bladders from cattle. Strings for musical instruments and tennis rackets are also made from these by-products.
10. Glands: The pharmaceutical industry relies heavily on animal glands for many medicinal drugs.
11. Collagen: Collagen is the chief component of the connective tissues. Glue and gelatin are made from collagen. These products, in various forms, are used in the furniture, photography, medical, and baking industries.
12. Contents of the stomach—Stomach contents of slaughtered animals are used primarily in the production of feed and fertilizer.

New Food Products on the Horizon

The foods that we eat and how they reach us are part of one of the most exciting branches of agriscience. New foods, and new versions of familiar foods, are arriving on the market daily. Research continues to find ways of improving the nutritional values of foods and keeping food costs at a minimum.

For example, the fruit industry uses an electronic, fruit-shaped beeswax sensor to log the bumps and bruises sustained by fruit during shipment. This and other improvements in handling equipment are resulting in better fruit and fewer losses for producers.

Research also influences dietary trends. At one time, eggs were thought to be problematic because of their high cholesterol content (185-212 mg on average), The belief was that would contribute to high blood cholesterol levels, which is bad for one's heart. Later research showed that eggs raise the level of HDL ("good") cholesterol and change LDL ("bad") cholesterol to a subtype that is not as associated with increased risk of heart disease. Meanwhile, researchers have also found ways to reduce the cholesterol level in eggs by changing the rations fed to laying hens.

As for the foods of the future, one thing is certain: They will reflect improvements in quality and convenience, thanks to the ongoing discoveries of agriscience research.

Agri-Profile Career Area: Food Scientist/Food Technician/Nutritionist/Dietician

Food scientists develop better ways to process, handle, package, prepare, and market food products.
Courtesy of USDA/Photo by Deb Dutcher.

Food scientists can find career opportunities in a wide variety of employers, including food production and retail companies, universities, research centers, government agencies, pharmaceutical companies, and even television and streaming media. A sampling of job titles—animal nutritionist, food microbiologist, flavor chemist, recipe developer for a syndicated cooking show—reflects the breadth of career possibilities.

Food scientists' work environments range from research laboratories and food processing facilities to the offices of businesses and nonprofit organizations. Some food scientists specialize in one area of the food industry, such as meats, fruits, vegetables, baked goods, dairy products, wine, or other beverages. Others find careers in product research and development, quality assurance, or food safety regulation.

Food production and retail companies rely on food scientists and technicians to develop new products to meet the ever-changing needs of consumers. Convenience products, such as instant- and freeze-dried coffee, dried and vacuum-packed fruits and meats, processed chicken tenders, boneless rolled meat, yogurt, ice cream, pasteurized and homogenized milk, shelf-safe milk, low-calorie and low-fat food are just some of the products that have been introduced to the market thanks to food science research and advances in technology.

To obtain a food science degree, students must develop proficiencies in such subject areas as food chemistry, biology of food, the hygiene of food handling, food processing, and food process engineering. On entering the professional field, college graduates will find that opportunities exist for food scientists in product research and development, quality assurance, food safety regulation, and university extension specialist. For example, a qualified graduate may sometimes gain experience as both an assistant professor and extension specialist who provides instruction in agricultural communication to students, local extension agents, and the community.

Advancement on the food science career path often requires master and PhD degrees. Employment opportunities are trending steadily higher in most sectors of the food industry. As of 2021, median compensation for food scientists with a bachelor's degree is $74,000. Earning potential extends into the six-figure range for scientists with graduate degrees.

Chapter Review

Student Activities

1. Write the **Key Terms** and their meanings in your notebook.
2. Keep a diary of everything you eat for a week. Note how each food represents one of the major food groups.
3. SAE Connection: With your teacher, arrange a visit to a meat department of a local grocery store to observe meat processing.
4. Compare the end result when a food has been processed in a variety of ways—for example, fresh, canned, frozen, dehydrated, and freeze-dried potatoes.
5. Make a collage of items that exist that are derived from animal by-products.
6. Using an outline of an animal, identify the major cuts of meat and where they come from on the animal.
7. Research the steps that must be taken to prepare kosher foods from food products other than meats. Write a brief summary outlining your findings.
8. Create a game that will help you remember the number of servings a person should eat daily from each food group. Use the *MyPlate* initiative as your model.

National FFA Connection — CDE Video: Meats Evaluation and Technology

Courtesy of National FFA; FFA #148

One National FFA event challenges competitors' readiness for careers in the meat animal industry.
Courtesy of USDA.

One of the most challenging competitions FFA members can compete in is the National FFA Meats Evaluation and Technology Career Development Event. The event is designed to help interested students develop employment skills geared to careers in the meat animal industry. It also helps students become knowledgeable consumers. To simulate real-world working conditions, student competitors come to the event dressed in protective clothing and prepared to work in a cold storage facility.

An introductory video to the career development event for meats evaluation and technology is available on the National FFA website. View the video to become acquainted with the skills that you will need to develop as you participate in this event. Local and state competitions will require many of these same skills. Express your interest in participating in this CDE to your local agriscience teacher.

Inquiry Activity | Food Nutrients

Courtesy of the USDA.

After watching the *Food Nutrients* video available on MindTap, answer the following questions.

1. What are the six main nutrients needed by the human body?
2. What are the three functions of nutrients? Which nutrients satisfy each function?
3. What influences the nutrient value of food?

Check Your Progress

Can you . . .

- discuss nutritional needs of humans and the food groups that meet these needs?
- categorize foods in the U.S. Department of Agriculture MyPlate nutrition initiative?
- relate methods used in processing and preserving foods?
- recall many or most of the steps used in slaughtering meat animals?
- give examples of the major cuts of red-meat animals?
- identify methods of processing fish?
- identify grades and market classes of animal meats?

If you answered "no" to any of these questions, review the chapter for those topics.

Unit 10

Planning with Purpose

Looking at the big picture, agriscience is the industry that feeds and clothes the people of our own nation and distributes the surplus food and other resources that sustain people around the world. The United States shares the results and products of research in agriscience. Our land-grant system of colleges and universities has provided a system of research in agriscience that is the envy of the world. In many countries, the greatest prize in education is to be admitted to, and graduate from, a land-grant college or university in the United States.

Looking forward, how will we sustain the accomplishments of the past? How will we acquire the resources to stay abreast of the needs and demands of the rapidly expanding world population? What business models will enable the discovery, expansion, marketing, and distribution of the "resources of life" to the hungry people of Planet Earth? Of course, the resources of the entire world must be directed toward meeting this worldwide challenge. But American agriscience capabilities in research, innovation, agribusiness, and marketing along with a healthy dose of the entrepreneur's optimism will find ways to cross political and international boundaries.

The driving forces for meeting the challenge of "abundant and enough food for all" are planning, management, and communication. In no other sector of the economy is the profit incentive of the free-enterprise system in greater evidence. In a system in which one can receive and use or accumulate the proceeds from work or innovation, the human spirit and initiative accelerates; the results are demonstrated in greater productivity. The adage that "where there's a will, there's a way" is proved correct every day as people apply their education, inventiveness, innovation, and energy in a workplace where a reasonable amount of the proceeds come back to the individual. Whether one is the owner or a stockholder receiving the profits of the organization, a salesperson or manager in line for bonuses or commissions, a pieceworker or assembly-line worker whose income is tied directly to output, or a per-hour worker with the option to seek a better-paying position, the free enterprise system rewards productivity.

This unit addresses the application of planning, communication, and management skills in agriscience. These skills transcend agriscience and culminate in the marketplace where the fruits of our labor are sold. The marketplace yields the cash to buy labor and other inputs for business and yields profits for the inventors, entrepreneurs, and other risk takers.

Chapter 34

Agribusiness Planning

Objective
To define management, determine management performance, determine how decisions are made, and describe economic principles that affect management.

Competencies to Be Developed
After studying this chapter, you should be able to:
- define management.
- describe the importance of management.
- describe the kinds of agriscience management decisions.
- list eight steps in decision making.
- describe the economic principles of supply and demand, diminishing returns, comparative advantage, and resource substitutions.
- use capital and credit wisely in business management.

Key Terms

agribusiness management
capital
credit
diminishing returns
comparative advantage
resource substitution
long-term loan
intermediate-term loan
capital investment
short-term loan
federal land bank (FLB)
Production Credit Association (PCA)
Commodity Credit Corporation (CCC)
Farmers Home Administration (FHA)
Small Business Administration (SBA)
promissory note
discount loan
add-on loan
amortized loan

Common Business Structures

Agriculture-related businesses drive huge sectors of domestic and international economies. Many of these businesses are owned by private individuals and families, and the need for business structure and management skills is evident in the industry. Different types of business structure are found across the spectrum of agricultural industry. However, most of these businesses have similar structural needs as those of business organizations that provide products and services. Some of the most common business structures are addressed in this chapter.

Sole Proprietorship

This is the most common business structure. It is simple to organize with fewer legal controls, and taxes are generally lower. Its use is restricted to an individual alone or in joint ownership with his or her spouse. The owner of this type of business is liable personally for all debts of the business.

Partnership

A partnership is a business in which two or more individuals work together and share profits or losses jointly and equally. Such a business is easily organized and is similar to a sole proprietorship with legal liability shared by the partners.

Limited Liability Company

A limited liability company (LLC) incorporates some of the advantages of a corporation in that owners are not personally responsible for legal liabilities and debts. Like sole proprietorships and partnerships, they are simple to organize.

Corporation

Corporations are companies that are controlled by a group of owners who are also shareholders. Profits are divided based on the number of shares that are owned by an individual shareholder. Owners are not subject to personal liability, and the company continues to exist after its founders are no longer involved in its operation. Personal control of a corporation is shared with the other shareholders, reducing the ability of companies to react quickly to a changing business climate.

S Corporation

Profits and losses and tax deductions of S corporations are passed along to individual shareholders instead of being absorbed by the company. Earnings are reported as individual income for tax purposes. Such companies have no more than 100 shareholders and are based in the United States.

Cooperative

A cooperative or co-op is a business structure that has been used by farmers, ranchers, and others in which the combined buying and selling power of multiple individuals or organizations is used to gain economic advantages. Agricultural service cooperatives gain volume discounts by buying fuel, seeds, fertilizers, and services for its members. Marketing cooperatives provide distribution, packaging, marketing, and transportation services for livestock and crops.

Agribusiness Management

Organizing or restructuring an agribusiness is much like building a functional home. It starts with a plan or blueprint that is thoroughly examined for flaws. A prospective agribusiness owner should engage competent advisors to identify the strengths and weaknesses of the business plan in much the same way that building inspectors examine the plans for a house. A powerful advantage is gained when flaws are identified and corrected in a business plan before capital is invested.

Agribusiness management is the human element that carries out a plan to meet goals and objectives in an agricultural business or enterprise, generally referred to as an agribusiness. Management decides the types of business or production activities in which the business will engage, such as horticulture, aquaculture, and farm supplies. Agribusinesses include farming, ranching, nursery operations, landscaping, retail stores, service enterprises, marketing businesses, lending institutions, veterinary services, consulting activities, and a variety of others.

According to most reports, businesses that fail usually do so because of poor management and communication. Mismanagement is frequently related to a lack of set objectives and goals or a failure to communicate them clearly through a well-developed action plan. Clearly, management can be considered good when maximum profits are achieved from the available resources, but for such an outcome to occur, measurably effective management and communication must be taking place at every stage of the business process (**Figure 34-1**).

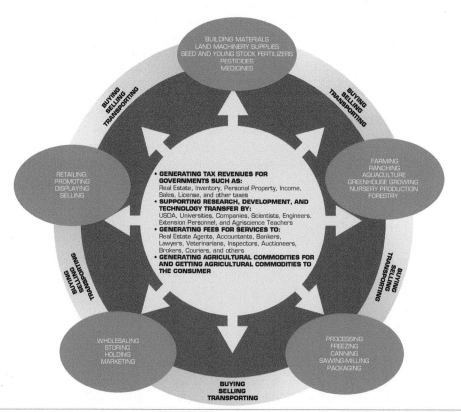

Figure 34-1 Agribusiness has many facets, and communications and management transcend the entire industry.

Science Connection: How New Tech Keeps Transforming Agribusiness

Office technology today goes way beyond desktop computers, wireless phones, and—thanks to remote access—the office itself.
© NAN728/Shutterstock.com

The personal computer or microcomputer has served modern agribusiness planning and operations well and has become indispensable to those who wholeheartedly adopted it. New integrated software upgrades in tandem with tremendous data storage capabilities have contributed to more informed decisions in all aspects of agribusiness.

The rapid advances in technology have been exponential since the early computers were introduced. Today, the trend is to place a computer within easy access of nearly all critical operations.

While a small business can still operate with just one or more computers and high-quality printers, today's agribusiness workplace is also likely to feature mobile devices like smartphones and tablets, wearable tech such as smart watches and Bluetooth earphones, and smart conference rooms set up for video conferencing.

Office software has also come a long way from the days when only one person at a time could edit the master version of a schedule, inventory report, or other key document. In the past, this led to such problems as duplication of work, discrepancies in information due to multiple versions of documents, and the likelihood that if even one person's computer crashed, an entire master document could be lost.

Today's cloud-based collaboration tools allow teams to work together on a common task—all without compromising on security. Because updates to key business documents can be made in real time, no one has to wait for the document to be emailed to them, or to log out so that they can log in and add their input. As a result, the information agribusiness employees need is always accessible and current. In agribusiness, this means that business owners can more easily track and manage inventory, payroll, accounting, tax records, and other data.

Mobile devices, apps, online conferencing, precision technology tools for collecting data, lightning-fast large-file transfer, shipping software, and other and remote access tools have also boosted productivity, enabling salespeople, buyers, executives, field technicians, plant managers, customer relationship managers, and other employees to stay connected to the business hub.

Thanks to all of this new technology, office computers have become just one workhorse among many—all working at a gallop to keep agribusiness ahead of the competition.

Influences on Agribusiness Management

Agribusiness management is influenced by the members who make up the boards of directors in corporations and cooperatives. Land, labor, and capital, being in limited supply, also influence management decisions. For instance, limited **capital**, which is money or property, may prevent the agribusiness manager from buying a new piece of equipment that would make the operation more efficient or protect the operation from excessive losses (**Figure 34-2**). Similarly, limited land will influence the agribusiness manager's selection of enterprises. For example, a cow–calf beef enterprise will need more land than a feedlot operation. Limited labor will influence the manager's selection of enterprises. For instance, if labor is limited, it will prevent the use of crop enterprises with high labor requirements, such as watermelons or tobacco. In contrast, grain production requires minimal labor.

Figure 34-2 Modern irrigation equipment permits profitable farming in areas that would otherwise be inefficient or too risky because of crop loss by drought.
© TFoxFoto/Shutterstock.com.

Estimating a Manager's Performance

The performance or ability of the agribusiness manager can be estimated in several ways. Dollar income is the measurement most often used. According to a study conducted in Ohio, the managerial ability of an individual can be determined by the manager's economic orientation, decisiveness, ability to identify alternatives, and extent of social activities or distractions from business goals (**Figure 34-3**).

Agribusiness is constantly changing, and it is expected to change even faster in the future. As a business becomes more complex, errors in management will be more costly. There will always be a need for managers who are well educated in agribusiness management, and who surround themselves with people whose skills are complementary to their own.

- Income goal (Economic orientation)
- Willingness or tendency to make decisions (Decisiveness)
- Manager's ability to recognize alternatives and opportunities
- The extent of social activities or distractions from business goals

Figure 34-3 Factors for predicting the managerial performance of agribusiness managers.

Characteristics of Decisions in Agribusiness

Although there are various types of agribusiness decisions, most decisions of managers are of an organizational or operational nature (**Figure 34-4**). Both types of decisions affect the success of the business. The following discussion covers characteristics of a decision as well as steps in decision making.

Importance

The importance of a decision may be determined by measuring the potential loss or gain that is likely to occur in consequence of the decision. For example, the selection of an agribusiness enterprise or business venture is likely to be more important than the selection of a brand of a given commodity. The manager must spend sufficient time selecting a type of business because this decision cannot be changed easily after it is implemented. In contrast, if sales of a certain brand of wire or species of plant are not successful, a different brand or species can be added or substituted at little additional cost.

Agribusiness Decisions	
Organizational	**Operational**
• Should I rent more land, or should I borrow money and purchase land? • Should the business be operated as a partnership or corporation? • What lender offers the best source of borrowed capital?	• Should cattle be sold this week or next? • Should I use high-magnesium lime or regular lime? • When should I start planting corn? Soybeans? Small grain? • When should I change the photoperiod for my poinsettias?

Figure 34-4 Different types of agribusiness decisions raise different questions.

Frequency

The frequency, or how often decisions are made, can vary greatly. Determining the cost per day to rent a truck may be used as an example. The cost of renting the truck only for 1 day, for a one-time job, might not be an important decision. However, the cost of renting the truck every few weeks as an ongoing expense makes the decision more important.

Urgency

It is important that certain decisions be made immediately, while some decisions can be delayed. The urgency of a decision depends on the cost of waiting. Two examples of decisions with different degrees of urgency might be hiring a new employee and buying a new file cabinet. If it is a busy time of year, an additional employee may be needed immediately and should be hired, since delay could result in a loss of business. However, delay in buying a file cabinet is not likely to affect the overall management and profit of the business.

Agri-Profile — Career Area: Owner/Manager/Consultant

Agribusinesses are the economic backbones of many communities.
© Richard Thornton/Shutterstock.com.

Agribusiness activities cut across the food and nonfood spectrums of agriculture. This includes the supply side, as well as the output side, of production. Career opportunities exist in every sector where goods are bought or sold.

Agribusinesses sell products such as feed, seed, pesticides, fertilizers, tools, equipment, plants, or animals. They also sell services such as animal care, crop spraying, or recreational fishing privileges. Careful planning is important before starting an agribusiness because many new businesses end in failure. Careful planning increases the chances of success.

Career opportunities in agribusiness planning include financial services such as banking, accounting services, management services, teaching, university extension service, marketing, market analysis, product specialization, product engineering, sales, ownership, and management. Agricultural economics, business management, accounting, finance, and personnel management are college programs that can lead to careers in agribusiness.

Available Choices

Some situations offer several choices. If several choices are available, the manager should delete the least-likely courses of action and focus on the more promising ones. The manager should take time to gather important information regarding the options that are available. If no choices are available, there is no decision to be made.

Steps in Decision Making

Making decisions based on facts will increase your success as a manager. The agribusiness manager must determine the best decision-making process. The following eight steps should be helpful in developing a management process (**Figure 34-5**):

1. Start the management process with a situation or established goal.
2. Gather all available facts and information for accurate analysis of the problem.
3. Analyze the available resources. Human resources are labor, time, skills, and interest. Material resources are land, equipment, and capital. Reevaluate your goal and adjust it if appropriate.
4. Determine the possible ways of accomplishing the goal or solving the problem.
5. Make an informed decision concerning the direction(s) that will be taken.
6. Follow through with a plan of action.
7. Assume responsibility for implementing the plan.
8. Evaluate the results to determine whether the goals were accomplished. If not, could a better implementation plan have been selected? Should the goals be modified?

After going through these eight steps, the manager should determine whether the process worked well. If not, where were the weak links? What changes should be made to improve the process? These and other questions will enable managers to fine-tune the decision-making process until it aligns with their particular work style, that of their staff members, and the structure of the organization.

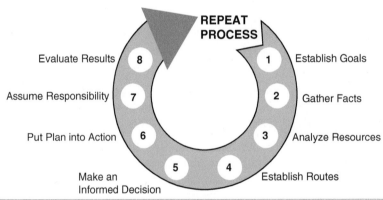

Figure 34-5 Eight steps in the decision-making process.

Fundamental Principles of Economics

Price, Supply, and Demand

Price is the amount received for an item or service. The price is determined by three price-making factors: (1) the supply of the item, (2) the demand for the item, and (3) the general price level. No one factor can be used to explain all price changes.

The general price of a commodity is influenced by the supply or availability of the item, the demand for it, the influence of wars, recessions, depressions, and many other factors. When supply and demand are in balance, a general price is established. An increase or decrease in either supply or demand is likely to influence the prevailing price.

The quantity of a product that is available to buyers at a given time is called supply. A producer of vegetables has control over some factors that influence the vegetable supply. However, one factor that the manager cannot control is the weather. If the rains do not come, the supply of vegetables will be reduced. The ability to buy items needed to produce the vegetables can also increase or reduce the supply. The ability to buy these items is also influenced by their prices. Cost and availability of credit may also influence supply. **Credit** is borrowed money. It permits the agribusiness manager to buy items needed for production or for business operations. Using credit spreads the cost of an item over time, allowing users to generate income while making payments.

Demand may be defined as the quantity of a product that buyers will purchase at a specific price at a given time. The quantity that a buyer is willing to purchase depends on the quantity available and the price. Both the desire and the ability to purchase tend to influence the extent of demand. Expanding the number of customers may also influence demand. As the size of the customer base increases, the demand for products and services can also be expected to change.

Diminishing Returns

The term **diminishing returns**, often used in economics, refers to the amount of profit generated by additional inputs. This term is often misunderstood, and a clear understanding of the law, or principle, of diminishing returns can be extremely helpful to the agribusiness manager in decision making. The principle of diminishing returns has two parts: physical returns and economic returns.

First, consider the physical aspect. Satisfaction from eating may be used to illustrate diminishing physical returns (**Figure 34-6**). When eating, each bite of food represents an additional unit of input. Each bite helps satisfy a part of the hunger. However, the added amount of satisfaction of hunger diminishes as you eat more. Each mouthful results in less satisfaction. This tendency is known as diminishing physical returns. At a certain point, the amount of hunger satisfied becomes negative with each bite taken. This means that each additional bite takes away from the satisfaction gained by previous bites.

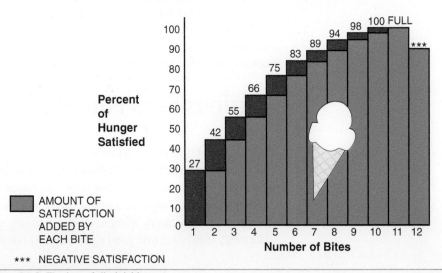

Figure 34-6 The law of diminishing returns.

Effect of Nitrogen on Corn Yields*		
Nitrogen Applied (lb.)	Yield (bu.)	Yield Increase (bu.)
0	67	—
60	103	+36
120	133	+30
180	154	+21
240	173	+19
300	169	-4

*Hypothetical and not based on research

Figure 34-7 Effect of increased nitrogen inputs on corn yields.

Now consider the economic aspect. One example of diminishing economic returns is the addition of units of nitrogen fertilizer as inputs and its effect on corn yields as outputs (**Figure 34-7**). Each additional input requires an additional cost. It may be observed that adding the first unit of nitrogen produced the greatest increase in yield. However, the rate of increase in yield diminishes as more units of input are added. Thus, the decision that the manager needs to make is how many units of nitrogen should be applied to obtain the greatest profit from growing the corn. If the number of inputs is too low or too high, the greatest potential income will not be achieved.

Comparative Advantage

The United States has nine major farming areas that have developed over a period of years. These areas have developed because of changes in demand or other factors. Within each area, different commodities are produced. Most operations in an area produce similar commodities and have similar systems of production. The reason for the similarity in operations is the comparative advantage found in following the programs that have evolved in that area.

Comparative advantage is the emphasis, in a given area, on where the most returns can be achieved. Comparative advantage may be illustrated in several ways. One example from agricultural history is the change in livestock production in New York State. At one time, most of the farms in New York raised sheep. Today, many of these farms are producing milk because of the high demand and favorable prices for milk in New York. It has become more advantageous to produce milk than to produce lambs and wool.

A similar situation exists for broiler production on farms. Fifty years ago, most farms raised enough chickens to supply their families with eggs and chicken for Sunday dinner. Today, few farms raise chickens anymore. The production of eggs and broilers has become so competitive that it is usually less expensive for farm families to buy eggs and broilers at the grocery store than it is to produce them in small numbers on their farm.

Resource Substitution

The term resource substitution refers to the use of one resource or item to replace another, when the results are the same. In other words, it is often possible to substitute a less expensive item for a more expensive one.

For example, it may be possible to substitute barley for corn in making a cheaper dairy feed. This substitution can be made without affecting total milk production. Barley is frequently less expensive than corn. Because barley and corn have slightly different feeding values for dairy cattle, barley must sell for $9.00 or less per 100 lbs. (hundredweight, or cwt) to be a better buy than corn selling for $10.00 per cwt.

The application of the principle of resource substitution is not limited to feeding livestock, but can also be used in many or most management situations. Can you think of other examples of resource substitutions?

Agribusiness Finance

Importance and Uses of Credit

The management of finances is the single most important function of the agribusiness manager. While possessing a great deal of technical knowledge and know-how is important, this alone will not make agribusiness managers successful—they must also be competent money managers. Without careful planning of finances above all other kinds of planning, some otherwise good agribusinesses can fail.

Credit is borrowed money. At one time, the manager who sought credit was considered a poor manager. Today, this is no longer true. The manager who approaches a lending institution to negotiate for credit need not do so in an apologetic manner. The lending institution must sell its commodity—money—to stay in business. However, the use of credit is a privilege that must not be abused. Once a company is labeled a poor credit risk, that reputation is difficult to overcome. Credit managed wisely, however, is a valuable tool in business.

Classifications of Credit

Credit is classified according to its period of use. Loans may be classified as long term, intermediate term, or short term (**Figure 34-8**).

Classifications of Credit		
Long-Term Loans	**Intermediate-Term Loans**	**Short-Term Loans**
Used to purchase land and buildings, and are for a period of 8 to 40 years	Extend for a period of 1 to 7 years for the purchase of breeding livestock, farm equipment, tractors, and similar items	Usually written for a period of 1 year or less to cover the cost of feeder livestock, feed, fertilizer, seed, fuel, and so forth

Figure 34-8 Classifications of agricultural credit.

Long-term loans are used to purchase land and buildings. The loan period ranges from 8 to 40 years. Interest rates for this type of loan are usually less than they are for other types. These loans are made by federal land banks, the Farmers Home Administration (FHA), insurance companies, local banks, and individuals.

Intermediate-term loans are made for periods of 1 to 7 years. Capital investments are usually made with the money from this type of loan. **Capital investment** is money spent on commodities that are kept 6 months or longer. Examples are breeding stock, tractors, store equipment, and warehouse equipment. Production credit associations, the FHA, finance companies, and local banks are sources of this type of credit.

Short-term loans are made for a period of 1 year or less. They are often referred to as production or operating loans. Managers who borrow short-term money are usually required to pledge something of value for security. Managers and businesses with good reputations may be allowed to borrow without security. Money borrowed without security is referred to as a signature loan. The borrower signs a promissory note that the loan will be paid on or before a specified date. Production credit associations and local banks are sources of short-term loans. It is not uncommon for local agribusinesses to permit their most reliable customers to make purchases on a short-term credit basis by simply signing a sales slip.

Types of Credit

Two types of credit are productive credit and consumptive credit (**Figure 34-9**). Productive credit is used to increase production or income. This is justifiable when the estimated increase in production will increase profits. This type of credit is used to purchase supplies, plants, flowers, livestock, land, equipment, storage facilities, seed, fertilizer, labor, and other materials.

Figure 34-9 Types of credit based on how the loans are used: Productive credit is used to pay for the cost of raising (A) livestock, (B) crops, and (C) equipment or to purchase livestock or products for resale. Consumptive credit is used to purchase goods or services such as (D) cars, (E) clothing, or (F) recreation that are used up with no expectation of earning a profit.

© LivingCanvas/Shutterstock.com. © Elena Elisseeva/Shutterstock.com. © Krivosheev Vitaly/Shutterstock.com. © Maksim Toome/Shutterstock.com. arka38/Shutterstock.com. © Daniel Padavona/Shutterstock.com.

Consumptive credit is used to purchase consumable items used by the individual; it does not contribute to the business income. It is relatively easy for a family to abuse the use of consumptive credit. This type of credit can also limit the amount of productive credit available to a family business.

A common form of consumer credit that is widely used is credit card purchases. Most credit card purchases paid in 30 days or fewer are interest free. However, the cost of consumer credit is high if the balances are not paid off within the first billing cycle. Federal statutes require firms charging interest on credit card purchases to publish their interest rates.

Sources of Credit

Today, many sources of credit are available, and many differences exist in lending policies. Because of this, knowledge and understanding of credit and credit practices is helpful in securing credit. Lenders differ in the interest rates they charge, the length of the loan period, and the purposes for which money is loaned.

When seeking a loan, remember that the interest rate is not the only important factor. Of substantial importance is the lender's willingness to extend a line of credit should an unexpected event occur. Another important factor is the lender's knowledge of the agribusiness.

Some agencies and institutions make only certain types of loans (**Figure 34-10**). Commercial banks are the most important source of credit. About 41 percent of the total agricultural debt is owed to banks. This percentage represents about 45 percent of the

Sources of Agricultural Credit and Typical Rates				
Source	Length of Loan	Annual Interest Rate*	Percent of Appraisal Loan Value	Purpose of Loan
Commercial Banks and Trust Companies	6–12 Months	8–13%	To 100%	Farm Production Items
	2–3 Years	8–12%	70–80%	Machinery, Equipment, and Livestock
	10–20 Years	9–11%	60–75%	Real Estate
Federal Land Banks	20–35 Years	8½% Variable	Up to 85%	Real Estate
Production Credit Association	1 Year	9% Variable	To 100%	Production Items
	3–7 Years	9%	Varies	Machinery, Equipment, and Livestock
Farmers Home Administration	40 Years	5%	To 100%	Real Estate
	7 Years	7–9%	To 100%	Machinery, Equipment, and Livestock
Insurance Companies	20–35 Years	Varies	To 75%	Real Estate
Individuals	15–30 Years	6–7%	To 75%	Real Estate
	2–5 Years	6–8%	To 80%	Machinery, Equipment, and Livestock
Equipment Manufacturers	1 Month– 5 Years	17.5%	100%	Machinery and Equipment

*Annual interest rates fluctuate based on the current prime rate. These percentages are for example only and should not be considered current.

Figure 34-10 Sources of agribusiness credit.

real-estate loans and 33 percent of other loans. Commercial banks lead in loan volume because they make all types of loans. Individuals are important sources of agribusiness credit. About 17 percent of agricultural non–real-estate mortgages are held by individuals.

Retail merchants also supply non–real-estate credit. Their credit is usually in the form of an open account. Sometimes loans are made available by a business for purchasing equipment. The interest rates on such loans are usually higher from these sources than they are from commercial banks. The popularity of this type of credit is based on its convenience and availability.

The **federal land bank (FLB)** was created by Congress in 1916 to provide long-term credit for agriculture. This network of regional cooperative banks was formed to make long-term loans to ranchers and farmers. Today, these banks finance loans for a wide range of agribusinesses as well as home purchases for rural buyers. Regulated by the Farm Credit Administration (FCA), these banks are excellent sources of credit for the purchase of real estate because of the favorable interest rates.

Production Credit Associations (PCAs) were established by an Act of Congress in 1933. The purpose was to provide favorable, short-term credit for agriculture. These funds are secured from the federal intermediate credit banks, who in turn secure their money from private lenders. Life insurance companies have been excellent sources of credit for financing long-term real-estate loans; however, in recent years, they have cut back sharply on agricultural real-estate investments. Loans are made through brokers, correspondents, and company representatives. Insurance companies hold about 3.8 percent of agricultural real-estate mortgages.

The **Commodity Credit Corporation (CCC)** is administered as an agency of the U.S. Department of Agriculture (USDA). The CCC is a government-owned corporation. Loans are made on eligible commodities, such as grain, cotton, peanuts, and tobacco. Farmers use the commodities as security for loans.

The **Farmers Home Administration (FHA)** was created by the government during the Depression. The FHA's original purpose was to assist tenant farmers in becoming landowners. The FHA provides financing to farmers who are unable to secure credit from any other sources. The advantages of borrowing money from the FHA are as follows:

- A large percentage of the total cost of the property can be borrowed.
- The repayment plan is based on the borrower's ability to repay.
- Supervision of the loan and assistance with planning are provided.

The **Small Business Administration (SBA)** also provides loans to agribusinesses. The SBA is authorized to make agricultural production loans. Interest rates tend to be close to those of commercial banks.

Cost of Credit

Interest is the greatest expense in borrowing money. However, when making credit decisions, keep in mind that borrowers are also required to pay other fees. These fees include commissions, recording fees, title certification charges, insurance, and service charges. Many formulas are used for calculating interest (**Figure 34-11**).

A simple-interest loan refers to one in which the full amount of a loan is received by the borrower and is paid back with interest after a short period. The borrower generally signs a **promissory note** agreeing to the terms of the loan. If payments are made several times throughout the duration of the loan, the interest is paid to date and is charged on the remaining balance of the principal.

In the case of a **discount loan**, interest is subtracted from the principal at the time the loan is made. For example, on a $1,000 loan, the borrower would receive only $900 of the $1,000 if the interest were 10 percent. Because the full amount of the principal is not received, the resulting true interest rate is 11 percent.

	A Comparison of the Different Methods of Calculating Interest	
	Example: A Loan of $600.00 for 12 Months at 10%. Repaid in 12 Equal Payments.	
	FORMULAS FOR CALCULATING INTEREST	INTEREST CALCULATED
SIMPLE INTEREST	Interest (I) = Principal X Rate X Time (I = P X R X T)	$600.00 X .10 X 1 = $60.00 = 10% True Interest
DISCOUNTED LOAN	Rate (R) = $\frac{\text{Interest (dollar cost)}}{\text{Principal X Time}}$ $\left(R = \frac{I}{P \times T}\right)$	$\frac{60.00}{540.00}$ X 1 = .11, or 11% True Interest
ADD-ON LOAN	Rate (R) = $\frac{2 \times \text{No. of Payments} \times \text{Interest Charged in \$}}{\text{Beginning Principal} \times \text{Years} \times (\text{No. of Payments} + 1)}$	$600.00 X .10 = $60.00 660.00 = Total Payments $\frac{2 \times 12 \times 60}{600 \times 1 \times 13}$ = .184, or 18.4% True Interest

Figure 34-11 Formulas for calculating interest.

The **add-on loan** method is used for calculating interest on consumer loans. The interest is charged for the entire amount of the principal for the entire length of time. The total principal plus total interest is divided into combined equal installments. This results in a very high rate of true interest. For example, a $1,000 loan at 10 percent interest would require repayment of $1100 to be repaid in monthly installments of $91.67 ($1,000 principal + $100 interest = $1100/12 = $91.67).

The **amortized loan** is generally used for the purchase of land, buildings, and other expensive items. Payments are made monthly and are computed so the interest owed plus the payment on the principal is equal throughout the repayment time. By this method, almost all of the payment at the beginning of the repayment period is for interest, whereas most of the payment near the end is for principal. In other words, the amount being paid on the principal increases proportionately as the amount due for interest decreases. A careful analysis of amortization schedules for a loan over various lengths of time will help the manager determine the cost of borrowing and will aid in making wise decisions on when, how much, at what rate, and for how long to borrow capital (**Figure 34-12**).

Seeking a Loan

Most agribusinesses must borrow money to expand their operations or increase income. The decision to incur debt is not easily reached. A careful examination of available alternatives must be made, and the best alternative should be chosen.

When seeking a loan, the agribusiness manager must be prepared to fully explain the benefits and risks to potential lenders. Many loan applications are not approved because the borrower does not present the details of the business. Other loan applications are not approved because the borrower does not effectively present the advantages of expanding. To be effective, the presentation must be organized, and the details of the operation should be put in writing. Lenders are concerned about the benefits and risks associated with providing a loan. Lenders are also concerned about a logical presentation of facts and the details concerning specific agreements in the loan contract.

Information that should be included in the presentation includes the following: (1) an agribusiness plan, (2) business records (income statements, expense records, net worth statement, and financial history), (3) terms of the loan, and (4) method of repayment.

A. 48 PAYMENTS OF $368.67 EACH

Pmt	Principal	Interest	Balance	Total Interest	Pmt	Principal	Interest	Balance	Total Interest
		12.000%	14,000.00				12.000%	14,000.00	
1	228.67	140.00	13,771.33	140.00	25	290.35	78.32	7,541.63	2,758.38
2	230.96	137.71	13,540.37	277.71	26	293.25	75.42	7,248.38	2,833.80
3	233.27	135.40	13,307.10	413.11	27	296.19	72.48	6,952.19	2,906.28
4	235.60	133.07	13,071.50	546.18	28	299.15	69.52	6,653.04	2,975.80
5	237.95	130.72	12,833.55	676.90	29	302.14	66.53	6,350.90	3,042.33
6	240.33	128.34	12,593.22	805.24	30	305.16	63.51	6,045.74	3,105.84
7	242.74	125.93	12,350.48	931.17	31	308.21	60.46	5,737.53	3,166.30
8	245.17	123.50	12,105.31	1,054.67	32	311.29	57.38	5,426.24	3,223.68
9	247.62	121.05	11,857.69	1,175.72	33	314.41	54.26	5,111.83	3,227.94
10	250.09	118.58	11,607.60	1,294.30	34	317.55	51.12	4,794.28	3,329.06
11	252.59	116.08	11,355.01	1,410.38	35	320.73	47.94	4,473.55	3,377.00
12	255.12	113.55	11,099.89	1,523.93	36	323.93	44.74	4,149.62	3,421.74
13	257.67	111.00	10,842.22	1,634.93	37	327.17	41.50	3,822.45	3,463.24
14	260.25	108.42	10,581.97	1,743.35	38	330.45	38.22	3,492.00	3,501.46
15	262.85	105.82	10,319.12	1,849.17	39	333.75	34.92	3,158.25	3,536.38
16	265.48	103.19	10,053.64	1,952.36	40	337.09	31.58	2,821.16	3,567.96
17	268.13	100.54	9,785.51	2,052.90	41	340.46	28.21	2,480.70	3,596.17
18	270.81	97.86	9,514.70	2,150.76	42	343.86	24.81	2,136.84	3,620.98
19	273.52	95.15	9,241.18	2,245.91	43	347.30	21.37	1,789.54	3,642.35
20	276.26	92.41	8,964.92	2,338.32	44	350.77	17.90	1,438.77	3,660.25
21	279.02	89.65	8,685.90	2,427.97	45	354.28	14.39	1,084.49	3,674.64
22	281.81	86.86	8,404.09	2,514.83	46	357.83	10.84	726.66	3,685.48
23	284.63	84.04	8,119.46	2,598.87	47	361.40	7.27	365.26	3,692.75
24	287.48	81.19	7,831.98	2,680.06	48	365.26	3.65	0.00	3,696.40

B. 24 PAYMENTS OF $659.03 EACH

Pmt	Principal	Interest	Balance	Total Interest	Pmt	Principal	Interest	Balance	Total Interest
		12.000%	14,000.00				12.000%	14,000.00	
1	519.03	140.00	13,480.97	140.00	13	584.86	74.17	6,832.53	1,399.92
2	524.22	134.81	12,956.75	274.81	14	590.70	68.33	6,241.83	1,468.25
3	529.46	129.57	12,427.29	404.38	15	596.61	62.42	5,645.22	1,530.67
4	534.76	124.27	11,892.53	528.65	16	602.58	56.45	5,042.64	1,587.12
5	540.10	118.93	11,352.43	647.58	17	608.60	50.43	4,434.04	1,637.55
6	545.51	113.52	10,806.92	761.10	18	614.69	44.34	3,819.35	1,681.89
7	550.96	108.07	10,225.96	869.17	19	620.84	38.19	3,198.51	1,720.08
8	556.47	102.56	9,699.49	971.73	20	627.04	31.99	2,571.47	1,752.07
9	562.04	96.99	9,137.45	1,068.72	21	633.32	25.71	1,938.15	1,777.78
10	567.66	91.37	8,569.79	1,160.09	22	639.65	19.38	1,298.50	1,797.16
11	573.33	85.70	7,996.46	1,245.79	23	646.04	12.99	652.46	1,810.15
12	579.07	79.96	7,417.39	1,325.75	24	652.46	6.52	0.00	1,816.67

Figure 34-12 Comparison of a loan for $14,000 at 12 percent interest paid over 4 years versus the same loan repaid in 2 years.

A borrower should investigate several sources of credit before selecting a lender. The final selection should be one that best meets the needs of the borrower.

Selecting a Lender

Factors that should be considered in selecting a lender include the following:
- Lending institution representative's knowledge of agribusiness problems and practices
- Lending institution representative's experience in handling agricultural credit of a similar nature
- Reputation of the lending institution
- Loan policies (interest rate, repayment schedule, closing costs, penalty clause, optional prepayment clause, and policy regarding failure to meet payment because of circumstances beyond borrower's control)
- Date the loan would be advanced
- Possibility of increasing the loan
- Availability of credit for other purposes

TO BORROW... OR NOT

Pros:

- **Maintain Business Ownership**

  Infuse cash without interference in the management of the business.

- **Tax Deductions**

  Principal and interest payments are classed as business expenses, and they can be deducted from the income tax for the business.

- **Lower Interest Rates**

  After tax deductions, you may be eligible for a lower interest rate.

Cons:

- **Debt Repayment**

  Business financing is risky, especially when revenues are too low to meet repayment obligations.

- **Interest Payments**

  Interest rates may vary depending on credit history and loan type. High interest rates cut into business profits.

- **Credit Rating**

  Borrowing affects your credit rating, especially when large amounts are involved. Lower credit ratings equal higher interest rates.

- **Cash Flow**

  Lack of a stable cash flow may lead to defaults or late payments on debt obligations. Either of these problems can contribute to long-term damage to your personal and business credit.

Figure 34-13 "Pros and Cons of Borrowing."

Using Credit Wisely

There is a chance that the borrower will fail to meet the financial obligations of a loan. In this case, the borrower may lose part or all that is owned. Because of this, the borrower takes a much greater risk than the lender. The borrower, however, can minimize risk by applying certain rules:

- Production loans should be used to generate new income.
- The borrower should limit the amount borrowed on new or unfamiliar business ventures.
- The borrower should keep debt as low as possible while still maintaining efficiency.
- The borrower should keep abreast of markets and trends.
- A debt-to-net-worth ratio of 1:1 or less is a good rule of thumb.
- A proper debt-to-income relationship should be maintained. The income must be greater than the principal and interest payments.
- Dependability and terms of the loan should be considered when selecting a lender.
- The borrower should have a definite repayment plan and schedule supported by evidence of an adequate cash flow.
- The borrower should be businesslike, fair, and frank.
- The borrower should have adequate insurance to reduce the lender's risk. Property, liability, and crop and life insurance provide partial protection against risk.

Management is a vital part of the business and commerce associated with supplies, services, production, processing, distribution, and selling of plant, animal, and natural-resource commodities. Careful planning and the application of sound agribusiness principles and procedures are necessary for success in agriscience careers.

Figure 34-14 The USDA Farm Service Agency provides loans to help agricultural producers start or expand a family farming operation, purchase equipment and storage structures, and meet cash flow needs. Options include direct, lower-interest loans and guaranteed loans through commercial lenders at rates set by those lenders.

Courtesy of USDA/Photo by Bob Nichols.

Agri-Profile: Grants ... Free Money?

Livestock promotion is just one of the many categories of financial assistance available to agricultural grant seekers.
Courtesy of USDA/Photo by Scott Bauer.

A grant is money that is given to a person or organization to finance a particular activity. Although this money rarely has to be repaid, a grant is not really free money. The organizations or individuals providing the funding must see a success story. They need to know that their money has been used for a good purpose. Often, the grantor asks for documentation that shows that grant money has been used wisely.

Many grants are available to individuals in agriculture, and include such categories as educational, livestock promotion, value-added, and rural rehabilitation. These grants can be of great benefit to people whose goals match those of the grantor and who are willing to work hard to obtain results. While this rightly suggests a certain competitive aspect, many grantors accept multiple applications for each available grant category.

Finding a grant for which you qualify will take time and effort. Grant information is widely available on the Internet, in government publications, and in trade magazines. Once a person finds a grant for which they qualify, an application must be completed. Fulfilling this process with attention to detail is the most important step in getting a grant. Keep in mind, too, that you are not just "filling out a form"—your writing must engage the reader. Educating yourself on grant writing will increase your chances of securing a grant. Many resources, including books, online webinars, and articles, are available that provide valuable information on how to write successful grant proposals. Local colleges and nonprofit organizations also offer classes and workshops for grant writing.

Chapter Review

Student Activities

1. Write the **Key Terms** and their meanings in your notebook.
2. Explain how the eight steps in decision making can be used in planning an FFA or other school money-raising activity.
3. Explain how the principle of resource substitution may be used in making an agricultural mechanics project.
4. Discuss with a banker the procedures used in applying for an agribusiness loan.
5. **SAE Connection:** Arrange to have a banker discuss agribusiness capital and credit in a class presentation.
6. Arrange for an accountant to talk to the class about taxes to consider when planning an agribusiness.
7. Make an outline of this unit. Phrases and words that are bold should be included, together with a brief description of the key points of the text.
8. Conduct a search for grants for which you may qualify to help you set up or expand a business. Exercise caution to avoid potential scams. For example, do not supply any of your personal data before verifying with your teacher or another adult that a potential grant source is legitimate. For example, if a website that claims to offer grants is not that of a registered nonprofit and/or a local, state, or government agency, you should probably steer clear. Discuss with your teacher how to identify reputable grant sources and what steps you would need to take to apply for a grant.

> ### Check Your Progress
>
> Can you . . .
>
> - define management?
> - briefly explain the importance of management?
> - describe kinds of agriscience management decisions?
> - identify the steps involved in decision making?
> - describe such economic principles as supply and demand, diminishing returns, comparative advantage, and resource substitutions?
> - summarize how to use capital and credit wisely in business management?
>
> If you answered "no" to any of these questions, review the chapter for those topics.

National FFA Connection: American Star in Agribusiness

Courtesy of National FFA; FFA #148

The National FFA Organization recognizes high-achieving high school graduate members.
UM-UMM/Shutterstock.com.

The highest ranking level of FFA student membership is the American FFA Degree. It is reserved for high school graduate members who show evidence of "leadership, community service, academic success, and outstanding SAE programs." Each year, the American Star in Agribusiness recipient is selected from among those FFA members who have achieved the American FFA Degree. The award recognizes the top non–production-based SAE program in agribusiness for that year. Corresponding Star agribusiness awards are made to high school members at local and state levels.

Thinking about building an agribusiness of your own? Visit the National FFA website to find engaging interviews with finalists for the American Star in Agribusiness. Representing a wide range of agribusinesses and experiences, they offer a wealth of advice and inspiration to help you get started. Just search "American Star in Agribusiness Finalists."

Chapter 35
Marketing in Agriscience

Objective
To determine the strategies and procedures for marketing agricultural commodities, products, and services to maximize profits.

Competencies to Be Developed
After studying this chapter, you should be able to:
- describe the components of a marketing strategy and a marketing plan.
- distinguish between a marketing strategy and a marketing plan
- describe various pricing strategies that maximize profits.
- distinguish between direct, intermediate, and retail markets.
- discuss advantages and disadvantages of direct, intermediate, and retail markets.
- recognize fees, commissions, and other costs of marketing.
- recognize industry trends and price cycles.
- describe the use of futures in agriscience marketing.

Key Terms

marketing	industry trend	position
product	target market	message
service	demographics	marketing plan
marketing strategy	contingency plan	budget
consumer	offering	key performance indicator
customer	value	profit
executive summary	valued-added product	promotion

supply
demand
direct marketing
intermediate marketing

middleman
retail marketing
commission
vertical integration

commodity exchange
futures market

Free enterprise is an economic system that allows individuals and businesses to make their own economic decisions with little government interference. They can make their own choices in buying products or services, selling their products, and owning a private for-profit business. Businesses can freely determine their products and prices.

Responsible businesses strive to deliver quality commodities, products, and services to customers and to provide fair salaries and benefits to employees while generating a solid profit. Marketing plays an essential role in this process. **Marketing** refers to the activities a business undertakes to promote the buying and selling of their commodities, products, or services. A commodity is raw material or an agricultural product that is produced to be bought and sold. A **product** is a tangible item put on the market to be purchased and consumed, while a **service** is an intangible item that arises from the output of one or more individuals. To reach their customers, businesses must have a marketing strategy and a marketing plan.

A number of methods are used to market agricultural commodities. The methods chosen by an individual producer are often a matter of what is available and what the producer prefers. Care should be taken to carefully choose the means of marketing. Intelligent marketing practices often make the difference between profit and loss in a competitive business.

Marketing Starts with a Strategy

Once a business has created their business strategy describing how they can make profits and achieve growth by making internal changes, they can begin drafting a marketing strategy. A **marketing strategy** is a long-term plan for achieving a company's goals by reaching consumers and turning them into customers. A **consumer** is always the person who uses a product but may or may not have purchased it. On the other hand, a **customer** always purchases a product but may or may not use it. A customer may be an individual or a business. For example, a customer may buy a crate of oranges and resell them in their convenience store. A consumer purchases one of the oranges to eat for a snack.

A marketing strategy explains the reason for marketing. Its purpose is to describe how the business's marketing goals will help them achieve their business goals. The marketing strategy should provide answers to these questions: What product will you deliver or service will you offer? Who will you deliver or offer it to? How will you deliver it? Who are your competitors? Once complete, the marketing strategy helps a business make the most out of their investment, keeps the marketing focused, and measures the sales results.

The components of a marketing strategy are the following:
- executive summary
- background
- market analysis
- target market
- situational analysis
- offering
- position and message
- selling

Executive Summary

An **executive summary** provides a brief overview of the marketing strategy and should highlight each section. Although the executive summary is presented at the beginning of a marketing strategy, it is the last component written.

Background

The background describes the business goals, marketing goals, and challenges. Goals create focus and motivation to achieve set targets. Successful businesses set goals, share them with employees, and gauge their progress regularly. Draw the business goals from the business strategy. Then, identify both short- and long-term objectives for sales, finances, and marketing. The goals should be quantified and easily measurable. For example, to sell 600 units of product A and 1,000 units of product B by the end of the growing season.

Market Analysis

Market analysis is an assessment of the marketing environment within an industry, such as agriculture. This analysis will better inform business decisions including marketing strategies. An analysis should answer the following questions (**Figure 35-1**):

- Do you want to expand, invest, reassess, create, or reposition offerings?
- How large is the consumer population?
- Who is the ideal customer?
- What external factors could shape of change the market?

The answers to these questions provide the information a business needs regarding opportunity, market sizing, target market, and impacts on the market, such as industry trends, economics, and seasons.

Important Questions for Market Analysis

- Do you want to expand, invest, reassess, create, or reposition offerings?
- How large is the consumer population?
- Who is the ideal customer?
- What external factors could shape of change the market?

Figure 35-1 Knowing who the customers are and what the market is are essential to developing an effective marketing strategy.

An **industry trend** refers to anything that changes the market in which the business operates. Some examples of industry trends are the price consumers are willing to pay for a product or service, consumer response to marketing strategies, and the types of items consumers are purchasing. Industry trends should be identified through research. Staying current with the trends in a specific industry, assists businesses in attracting new consumers while maintaining loyal customers.

Target Market

Not all consumers are alike, which is a challenge for businesses. Businesses use marketing to reach customers, but running an ad is not enough if the ones who see it don't use or prefer their product. Market segmentation helps businesses target the people who are most likely to buy the product. A market is segmented by splitting it up into groups with similar characteristics, which allows for more precisely targeted marketing and personalized content. A specific group of people to whom a business's product or service is directed is a **target market**.

One way consumers are segmented is based on **demographics**, which includes differences in gender, age, ethnic background, income, occupation, education, household size, and other relevant information. For example, a company might identify the age of the ideal consumer as 18 to 24. Other ways consumers are split up are based on interests, activities, opinions, values, and attitudes; on product usage rates, brand loyalty, and customer history; and on market size, region, population density, and climate.

Knowing who the target market is—and who is not—is essential to implementing the marketing strategy and selecting the marketing plan.

Situational Analysis

Situational analysis involves researching competitors to learn about their products, sales, and marketing. In addition, an analysis of the competitor's customer experience, positioning, pricing, and social media should be completed. A business needs to identify what makes their product different from the competitor's product as well as opportunities where they can outsell them.

A SWOT analysis should also be performed. SWOT stands for Strengths, Weaknesses, Opportunities, and Threats. A SWOT analysis is a management technique many businesses use to understand these elements for themselves and their competitors. Strengths are internal factors, such as skills and expertise, that give a business or product an advantage over the competition. Weaknesses are internal factors that create disadvantages for a business, such as outdated machinery or inefficient production methods, and give the competition an advantage. External opportunities provide space to grow a business, such as following a new trend or entering a new market. Threats are external factors that may harm a business, such as natural disasters or a new competitor. Understanding both the business's and the competitor's SWOTs is a highly valuable decision-making tool.

Contingency Plan

A **contingency plan** is pre-planning for unexpected situations that are out of the control of the business. Contingency plans are used to keep a marketing strategy on track. A benchmark is a pre-determined point of reference that can be used to determine progress toward a goal. When benchmarks are not met, the contingency plan is put into action. Contingency plans deal with ifs. For example, "if _____ happens, then the business will do _____." A business's contingency plan helps a business prepare, deal with, and recover when things do not go as planned. Just as all businesses are different, no two contingency plans are the same. However, all plans should factor in the following steps:

1. Understand the critical functions or mission of the business.
2. Recognize the resources required to support critical functions.
3. Anticipate potential problems.
4. Prioritize the potential outside risks and focus on risks in order of severity.
5. Create a contingency plan for each scenario.
6. Share, practice, adjust, and maintain the plan.

Offering

Offering are commodities, products, and services created to deliver value to customers. Offering go beyond fulfilling a consumer's needs, satisfying their wants, or both. They also include convenience, quality, and support. For example, a farmer sells boxes of produce directly to their customers. Customers can pick the boxes up for free from the farm or have them delivered for a small fee. Since customers needs are different, ones looking to save money may see free-pickup as a benefit. For those who prioritize convenience, they may see a benefit in the delivery service despite the cost. Offering can increase sales and promote customer loyalty.

In the marketing strategy, a business should include offering, what the need is, the features and benefits for different groups within the target market, and how they will deliver the features and benefits.

Value-Added Products

One way a business can stand out and get ahead is offering value-added products (**Figure 35-2**). In business, **value** is the difference between the price charged for a product or service and the benefits the customer perceives they receive. The U.S. Department of Agriculture (USDA) Rural Business Development defines a **valued-added product** as follows:

- a change in the physical state or form of the product, such as making strawberries into jam
- the production of a product in a manner that enhances its value, such as organic products
- the physical segregation of an agricultural commodity or product in a manner that results in the enhancement of the value of that commodity or product, such as grading different classes of grains to charge a higher price

Value-added products prevent produce from going to waste and create the ability to sell products off-season. These products are more appealing to the consumer, and the consumer is more willing to pay a premium over similar but undifferentiated products. These products can help a business generate higher return, break into of a new high-value market, or give an edge over the competition. They can also help create brand identity or develop brand loyalty.

Figure 35-2 Value-added products allow the producer to capture additional shares of the profits from the commodity.
Courtesy of DeVere Burton.

Science Connection — How Much Would You Pay for Improved Health?

A healthy family requires foods that furnish all of the required nutrients.
© merzzie/Shutterstock.com.

People are willing to pay more for health-promoting genetically modified (GM) food, according to a University of Purdue study. Jayson Lusk, an associate professor of agricultural economics at the University of Purdue, sent out a mail survey that asked consumers how much they would pay for golden rice. Golden rice is a GM food that has a gene from a daffodil in it. This gene triggers rice to produce a chemical that the human body can change into vitamin A.

In his questionnaire, Lusk outlined the benefit of golden rice from a consumer's standpoint. Earlier studies indicated that consumers would pay more for foods that were not GM foods. In these studies, the benefits to farmers were outlined, but the potential benefits for consumers were not mentioned. The results of Lusk's study show that consumers will pay a greater price for golden rice if they understand the benefit.

A valuable lesson can be learned from this study. As products like these become available to consumers, a shift in marketing must follow. Consumers need to be informed about the personal benefits they will get from the GM food. If they are not informed of the benefit, they will have no incentive to buy the food. Without proper marketing, GM foods tend to meet resistance. Conversely, when advertising is well executed, foods such as golden rice can be a big success.

Position and Message

As part of the marketing strategy, the business needs to first assess its **position**, or the consumers' perception of their product. They have to establish what sets their product apart from their competitor's product. A business needs to also clearly identify what they are selling, why they are selling it, to whom they are selling it, and any hows. From here, they can build their messaging. In marketing, a **message** is a statement that effectively communicates why the the product or service is unique and superior compared to similar products.

Selling

This section of the strategy details how the business plans to deliver the product or service to their customers, the price, and details about promotion. These topics are covered in depth later in the chapter.

Marketing Plan

The marketing strategy is the roadmap for the marketing plan. A **marketing plan** details what a business will do, where they will do it, when they will implement it, and how they will track success. The marketing plan puts the marketing strategy into action. Its purpose is to outline the marketing efforts in actionable steps. The components of the marketing plan are as follows:

- executive summary
- target market
- projected budget
- situational analysis
- marketing mix
- key performance indicators

Like the strategy, the plan begins with an executive summary and also includes target market and situational analysis sections. The executive summary is a brief overview of the marketing plan. The target market section follows. It provides an overview of the target market detailed in the strategy. The situational analysis section follows the budget section. Using the marketing strategy, the business can describe their SWOT analysis, marketing analysis, and possible challenges and impacts on their business. The sections that differ from the marketing strategy are discussed in detail in this section.

In addition to the traditional components listed, businesses need to consider the digital components of their plan, including their website and social media.

Figure 35-3 Creating a budget is an important component of a marketing plan.

© Andrey_Popov/Shutterstock.com

Projected Budget

A clear budget is vital for the success of a marketing plan as it provides direction when the plan is executed. A **budget** is the amount of money available for, required for, and assigned to a particular purpose, such as advertising (**Figure 35-3**). For the purposes of the marketing plan, a business needs to know how much money they have to spend on the marketing campaign and a timeline for implementation. The budget determines what types of promotions they can implement. The goal is to get the best return on investment, which is discussed later in the chapter.

The budget is created to determine if incorporating the marketing strategy will be profitable. A projected income statement summarizes a business's profit forecast, or how much the business anticipates they will earn based on known expenses. A projected income statement includes the following.

- Income, or revenue, is the amount of money produced from a given source, such as sales, government payments, and dividend earnings from investments.
- Expenses are the costs associated with conducting business, such as employee salaries, animal feed, loan payments, interest, insurance, taxes, and other production costs. It should also include the cost of associated with implementing the plan such as advertising. (See the *Expenses* section of **Figure 35-4**).
- Net income is income minus expenses. This value is the actual amount of money a business earns.

Projected Income Statement for Custom Lamb Enterprise			
	Year 1: 2021	Year 2: 2022	Year 3: 2023
SALES REVENUE			
Lamb	$46,160.00	$69,240.00	$86,550.00
TOTAL REVENUE	**$46,160.00**	**$69,240.00**	**$86,550.00**
EXPENSES			
Marketing Expenses	$2,797.78	$2,230.00	$2,430.50
Lamb Purchase	$22,400.00	$34,440.00	$49,000.00
Processing & USDA Labeling	$12,000.00	$14,400.00	$26,250.00
Packaging Materials	$2,000.00	$3,000.00	$4,375.00
Labor	-	$3,900.00	$7,800.00
Insurance	$360.00	$360.00	$360.00
Licenses and Permits	$232.00	$232.00	$232.00
Utilities	$399.00	$399.00	$420.00
Transportation	$400.00	$400.00	$400.00
Freezers	$2,400.00	-	$1,800.00
Miscellaneous	$550.00	$550.00	$550.00
TOTAL EXPENSES	**$43,538.00**	**$59,911.00**	**$67,393.75**
NET INCOME BEFORE TAXES	**$2,621.30**	**$9,329.00**	**$19,156.25**
ROI	6%	15%	28%

Figure 35-4 An example of a projected income statement for an expanding custom lamb enterprise.
Courtesy of Renee Peugh.

Marketing Mix

A marketing plan must include the four P's of marketing, also known as the marketing mix. The four P's are factors that a business has the power to control, which are product, price, place, and promotion (**Figure 35-5**).

Figure 35-5 The four components a business can control in a given market are product, price, place, and promotion.
© Dusit/Shutterstock.com.

- *Product:* The products or services a business offers its customers is also known as the offering. Factors to consider when choosing what product(s) or service(s) to sell are: the ability to fill a need, product quality, features, packaging, brand name, and value.
- *Price:* The amount of money a customer pays for a the product or service is the price. The price of a product is determined by analyzing the competition, the demand, production costs, and what consumers are willing to pay.
- *Place:* The distribution method or location where a product or service is made available to customers is the place, or channel. In agriculture, this may be direct from the farmer or from a wholesaler, broker, or retailer. Additionally, the transaction may take place in person or online.
- *Promotion:* The activities that communicate a product's or a service's benefits and features to the target market is the promotion. Promotion is often called a marketing campaign. The goal of promotion is to persuade the target market to purchase a product or a service.

The marketing plan should include the rational for these decisions. Price, place, and promotion are discussed in more detail later in the chapter.

Science Connection: Improved Quality Passed on to Consumers

Frank Scholnick and James Chen of the USDA examine the results of experimental treatments for leather.
Courtesy of USDA/ARS #K-3998-5

Agriscience research is continually introducing new products to market. For example, new colors of no-wrinkle fabrics are made with cotton, leather is tanned by less-polluting methods, and a new detergent made from animal fats is biodegradable. Cotton has always been a preferred fabric, but its tendency to wrinkle has limited its use. In the past, cotton fabric was dyed before a no-wrinkle finish was applied because the chemical bond created before the refinishing process would repel dye if applied after the process. The cotton fabric must swell to accommodate the molecules of dye, but this cannot normally occur after the fabric has been heated and treated with a no-wrinkle finish. The fabric industry needs to be able to dye fabric just before clothing is manufactured to keep up with the rapid changes in clothing fashions.

To solve the problem, scientists at the USDA Agricultural Research Service (ARS) Textile Finishing Chemistry Research Unit in New Orleans have developed techniques a no-wrinkle finish can be applied to cotton fabric before dying. By adding a variety of quaternary ammonium salts to the no-wrinkle treatment solution and a positive charge to the fabric, it will accept dye. This procedure broadens the choice of dyes that can be used, enabling the fabric industry to offer a wider range of shades with deeper, more vibrant colors.

Leather is another textile scientists seek to improve. Scientists at the ARS Eastern Regional Research Center in Philadelphia are looking for more environmentally friendly ways to decrease bacterial decay in hides and thus preserve leather. One of the most promising techniques is the use of electron beam irradiation. The technique promises to replace the salt and brine method of curing and extend the qualities of strength and elasticity for some new leather products.

Cotton-made clothing is often washed with laundry detergent in a washing machine. Scientists have long sought laundry detergents that will not pollute groundwater when they are discharged with wastewater. Tallow is beef fat and a by-product of the beef industry. Agriscientists have developed a tallow-laced soap that is environmentally friendly. The soap contains no phosphates; will not harm humans, domestic animals, or wildlife; and will usually biodegrade in 24 hours. The product is as effective as phosphate detergents and is economical, too.

Key Performance Indicators

While the budget informs how much a business has to pay the costs of a marketing campaign, **key performance indicators** (KPIs) measure the success of the campaign. KPIs help businesses make decisions on future marketing efforts. They also indicate whether the cost of the campaign was worth the investment. One type of KPI is a marketing return on invest (MROI). Calculating MROI can gauge the success of the marketing plan.

A financial return is the *profit* made from an investment. In this case, the investment is the money spent on a marketing campaign. Business owners need to see that the money they invest in a marketing campaign will bring a larger financial return. The equation for MROI is as follows:

MROI = (income generated from marketing effort − marketing investment) / marketing investment

For example, a beekeeper spent $200 on their marketing campaign. They made an additional $2,800 in sales due to product promotion. Calculating the beekeeper's MROI is as follows:

$$(\$2,800 - \$200) / \$200 = 13 \text{ or a 13:1 return ratio}$$

The interpretation of this equation is that for every dollar spent on the marketing campaign (promotion), the business earned $13. An MROI of 5 (5:1) is a good marketing ratio that shows the campaign successfully increased profits.

While the MROI is helpful when determining if the money spent on product or service promotion was beneficial, a total return on investment (ROI) provides a bigger picture. An ROI takes all the expenses resulting from the new endeavor into account and shows an overall financial return on money invested in the enterprise. The equation for ROI is as follows:

$$\text{ROI} = (\text{net profit} / \text{total investment}) \times 100$$

Using the budget from the custom lamb enterprise (see **Figure 35-4**), the net income for year three is projected to be $19,156.25, and the total expenses are projected to be $67,393.75. Calculating the custom lamb enterprise's ROI is as follows:

$$(\$19,156.25 / \$67,393.75) \times 100 = 28.42\% \text{ rounded to } 28\%$$

Promotion

Once the marketing strategy and marketing plan are completed, a business can start using the budgeted dollars for promotion. The goal of **promotion** is to communicate to the target market the necessity of the product and its price. The benefits of promotion outweigh the costs. Promotions increase brand growth, customer loyalty, market growth, and of course, increase revenue. There are seven main types of promotion, and a business may decide to use one or several in combination to reach their target market. The types of promotion strategies are as follows:

- *Direct marketing:* A business reaches out to specific customers to inform them or new products, services, or upcoming sales. E-mails, mail, phone calls, and flyers are examples. This method is fast and effective.
- *Sales promotion:* A business offers discounts as an appealing short-term incentive to increase product demand and sales, such as flash discounts, promo codes, and buy one, get one (BOGO) deals. This method is a good way to sell the remaining inventory of an old product.
- *Digital marketing:* A business uses online advertising such as social media marketing, affiliate marketing, or search engine optimization. This method reaches a wider audience.
- *Personal selling:* A representative of a business meets face-to-face, over the phone, or via video conferencing. The goal is build meaningful relationships with customers.
- *General advertising:* This method is for a wider audience and promotes general awareness of a company or product. Mass media advertising, such as magazines, billboards, radio, and television, are examples. The goal is to improve brand recognition.
- *Public relations:* Public relations helps a business maintain or enhance their image. Press releases, social media endorsements, and publicity stunts are examples.
- *Sponsorships:* A business pays a fee to a person or helps fund an event in return for publicity and advertising. Partnering with a social media influencer, celebrity, or other well-known person can can increase popularity, sales, and reputation. This method is often used on social media platforms.

In addition to these main promotion strategies, another method has emerged with the ever-growing audience for streaming video services and devices. OTT (over-the-top) advertising is similar to a TV commercial, but businesses can use them to better reach their target market. For podcasts, satellite radio, and music streaming services, there is programmatic audio advertising.

The promotion of commodities may be done in numerous ways. Product displays are common in stores and other retail outlets. Similarly, displays and sales of machinery, animals, food, flowers, ornaments, landscape designs, and other commodities and services are common at fairs, shows, open houses, trade shows, professional meetings, as well as online digital campaigns. Free samples may be offered in stores, to social media influencers, or at fairs, shows and wherever prospective customers may gather.

Hot Topics in Agriscience: Social Media Marketing

Social media influencers are paid by companies and brands to create sponsored content and post it on their social media accounts.
© PeopleImages.com-Yuri A/Shutterstock.com.

Globally, more than 4.2 billion people report using social media platforms such as Facebook, Instagram, TikTok, and Twitter. All of this adds up to an unprecedented opportunity to connect businesses with potential customers. Social media platforms expose customers to more products and services than TV, radio, magazines, and websites. Social media's digital nature helps engage and connect vast numbers of customers around the world who may have little more in common than their interest in specific products and services.

In addition to a company or brand having their own social media accounts, they pay influencers to promote their product or service through sponsored content. A social media influencer is a person who has built a reputation for their knowledge and expertise on a specific topic. Followers trust the opinions of an influencer. Thus, the influencer is able to influence the behavior of their followers including the products they buy or services they use. Some influencers have an excess of 1 million followers, a massive, captive audience to whom brands and companies can market their products and services.

Consider, too, that in just a few decades, the position of social media marketing manager has become mainstream. Today, more than 93 percent of U.S. marketers in companies larger than 100 employees use social media to engage customers with their products and services, up from 86 percent in 2013.

Research and Promotion Programs

The promotion of any one agricultural commodity from farms, ranches, commercial fisheries, nurseries, or greenhouses is expensive and difficult because there are so many small-volume producers in the United States. For example, in 2020, Iowa corn producers grew corn on 13.6 million acres, but many corn growers farm 500 acres or less. To assist farmers' in marketing and selling their products, the U.S. government has instituted research and promotion programs that collect a small fee based on the amount of the commodity each farmer produces. These programs are often called check-off programs because farmers once had to check a box to participate in them. This is no longer the case, but the nickname is still commonly used. Each commodity type pools the collected funds and uses them for research, lobbying the government, promotion, and advertising the specific commodity.

Currently, the Agricultural Marketing Service (AMS), a branch of the USDA, oversees 22 national research and promotion (R&P) boards (**Figure 35-6**). These boards empower farmers, ranchers, and agricultural businesses. Every R&P program's mission is to maintain and expand the markets for its commodity. Many states and private producer groups serve similar promotional functions as federally regulated check-offs. It is estimated that about 90 percent of all U.S. farmers contribute to some 300 federal, state, and private generic promotion programs, covering approximately 90 commodities. The spending for research and promotion by these programs is approaching $1 billion annually. In 2018, the government established the Agriculture Trade Promotion program for the purpose of promoting American commodities in foreign countries. Commodity advertising for agricultural products and services typically includes the use of the Internet, newspapers, radio, television, billboards, signs, pamphlets, transportation advertising, direct mail, breed journals, product catalogs, and magazines.

Research and Promotion Program Commodities		
Eggs	Cotton	Fluid milk
Pecans	Hass avocados	Highbush blueberries
Lamb	Mushrooms	Dairy
Beef	Honey	Mango
Christmas trees	Peanuts	Pork
Potatoes	Watermelon	Paper & packaging (food)
Popcorn	Softwood lumber	Sorghum
Soybean		

Figure 35-6 AMS oversees 22 research and promotion boards. Every research and promotion program's mission is to maintain and expand the markets for its commodity.

Commodity Pricing

Agricultural commodity prices tend to increase and decrease on a fairly well-defined cycle. The cause of general price cycling is supply and demand or government intervention. However, cycles are sometimes interrupted, accelerated, or delayed by disasters or political events. For example, the discovery of "mad cow disease" interfered with both the beef and milk price cycles.

Many factors make commodity prices unpredictable: The use of rapid transportation on a modern system of interstate highways (**Figure 35-7**); air shipping of highly perishable foods; refrigerated shipping; food-processing procedures that lock in freshness; extensive, specialized, high-technology storage facilities; and an excellent marketing system all contribute to the availability of food and fiber commodities to nearly any part of the nation on a year-round basis. These factors help maintain markets and prices that are fairly stable and predictable.

Figure 35-7 Large trucks and modern highways move agricultural commodities quickly from the farm to the processor or market.
Courtesy of the National FFA Organization.

Supply and Demand

The relationship between supply and demand determines price and whether the production and processing of a product can be profitable. **Supply** is the amount of a product available at a specific time and price. Some of the things that may determine supply include how many businesses are locally producing a specific product, how much of the product is coming into the market from other areas, and the product's history of profitability.

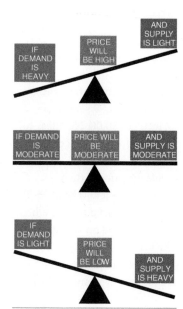

Figure 35-8 In a free-market system, the relationship between supply and demand determines price.

Similarly, demand for a product is determined by numerous factors. **Demand** is the amount of a product wanted at a specific time and price. It is often determined largely by price. The less expensive a product of a given quality is, the more demand there is for it. However, other factors influence demand for a product. The amount of money available to consumers to buy the product is a factor that is often overlooked. Competition from similar products may also reduce demand. Seasonal variations in demand also need to be considered. For example, it is easier to sell ice cream in the summer than in the winter. After deciding to sell or offer a product or service, businesses must create consumer demand. Promotional activities such as advertising, digital marketing, and providing product samples are examples of methods that create consumer demand (**Figure 35-8**).

Because of supply and demand, the final determination of prices for agricultural commodities is not always under the control of the producer or representative of the producer. Other factors are the perishable nature of unprocessed foods, the relationship between peak quality and retail prices of fruits and vegetables, and the lack of control over the production of staple food crops such as eggs, milk, and meat after the production cycle has started. Therefore, the producer may not have much control over the supply that is being collectively generated and may have to accept lower prices or loss of the product because of spoilage.

Price Cycling in Agricultural Markets

When prices are high, producers tend to increase the production of animals and animal products. At first, this action pushes the prices even higher because female animals that normally would have been marketed are retained in the breeding herd. When the results of the increased production reach the market and supply exceeds demand, the prices begin to decrease. When prices decrease to the point where production is not profitable, livestock producers reduce production by selling some of the breeding herd. Of course, this pushes prices even lower. With decreased supplies, the demand increases, and the cycle begins again.

Today, many other factors influence prices. As a result, the traditional livestock price cycles have less variation than in the past. Factors that currently influence livestock prices include:

- importation and exportation of livestock products
- development of new uses for livestock products
- increased advertising of livestock products
- world weather conditions
- general economic conditions in the world
- changes in consumer demands.

The availability of a given crop is stimulated by price, which is influenced by demand. Many grain crops have so many uses that the market for the items is complex. For instance, corn is the number one choice for most farm-animal and poultry feeds. It is frequently used in pet foods, breakfast cereals, corn oil, plastics, alcohol, and baked goods.

However, in many of these commodities, corn can be replaced by substitutes. For instance, barley and other grains are substituted if the price of corn increases above a certain level. Similarly, when the price of crude oil increases to a certain level, the price of corn becomes competitive. Gasoline refiners will then use less petroleum and more grain alcohol in the fuel. Weather and other production factors influence the size of the world grain supply and influence prices in the United States and abroad.

Pricing Strategies

The quickest and most effective way for a business to maximize profits is to get their prices right. The following pricing strategies are used to achieve this goal:

- *Psychological pricing:* a strategy designed to make the price seem lower or less significant than it is. For instance, the pricing of an item at $19.99 seems to attract many buyers that may reject the same item priced at $20.00. Needless to say, the profit from selling more items will easily offset the loss of one cent per item.
- *Dynamic Pricing:* this strategy allows companies to change the price of their product or service based on the market demand at any given moment (**Figure 35-9**). For example, many gas stations have switched to digital signage that advertises their current price for a gallon of gas. As the price of gas fluctuates based on supply and demand, gas stations are able to immediately change the price.
- *Bundle pricing:* a strategy designed for the business to sell more product by encouraging, and the customer to buy more for less money. For example, buy three for $10. This increases value perception.
- *Penetration pricing:* a strategy wherein the price of a product is set below that of competitors to entice customers to try it. Profits will be low temporarily, but they can be regained if new customers continue to buy and the price is gradually increased.
- *Price skimming:* this strategy is the opposite of penetration pricing. With this method, the price of a new product is set for unusually high profits at first, when high-income or willing customers are available. This enables the producer to recoup development expenses more quickly and gradually reduce the price as sales volume increases or other competing products enter the market.
- *Premium pricing:* a strategy for businesses that create high-quality products and market them to high-income customers. This works well for value-added products that customers consider to be of a higher value than the standard product.
- *Value-based pricing:* this strategy is similar to premium pricing. A business prices its product based on how much the customer believes the product is worth. This model is best for unique products, rather than agricultural commodities.

Figure 35-9 Automated adjustment of prices enables retailers to change prices quickly in accordance with promotional activities or changing market conditions. The price changes in their computer system, and the associated barcode for the item automatically registers the new price.

Courtesy of USDA/ARS #K-2656-2.

In general, in a nonregulated, free-enterprise system, there is a tendency to price commodities according to what the market will bear rather than what it costs to produce the items. This pricing strategy is generally regarded as fair because the system rewards risk-taking with the potential for profit. If one must be prepared to shoulder losses when offering a commodity that does not reach sales expectations, then it only seems reasonable to permit that individual or company to benefit from high-profit opportunities.

Direct Marketing

Direct marketing is selling a product directly to consumers. An increased demand for local farm-to-table products has in turn increased demand in direct markets. With the growing popularity of farm fresh products, farmers have found the need to learn more about marketing methods. Marketing campaigns, such as USDA's Know Your Farmer, Know Your Food, have been successful in drawing consumers' attention to the importance of the farmers who produce their food.

Direct marketing has two important benefits: improved economic viability to farmers and opportunities for personal connections. In addition, farmers do not need to use brokers, so they do not have to pay associated fees. Transportation costs are kept to a minimum because the buyer generally comes to the farm to make purchases. On the other hand, the farmer takes on the responsibility of marketing and selling the products to consumers rather than passing them onto a wholesaler or broker. Common direct market strategies include roadside stands, farmers' markets, and Community Supported Agriculture (CSA).

Farmers' Markets

Farmers' markets have popped up in communities all over the country. Farmers' markets give producers access to markets that would seldom be available otherwise, especially to small farms and beginning farmers. They also have better pricing opportunities than selling through a wholesaler as farmers can set their own prices. The producers also have the opportunity to educate the consumer concerning the value of good products (**Figure 35-10**).

Figure 35-10 Farmers' markets provide opportunities for producers to sell directly to customers opening their production facilities to the general public.
© Arina P Habich/Shutterstock.com

Several costs and inconveniences are associated with marketing agricultural products at farmers' markets. Farmers have to submit an application and be approved before they can sell at a farmers' market. Often, fees must be paid to cover the costs of operating farmers' markets. Competition may be greater, especially if several producers are selling the same type of product. Vehicles with heating, cooling, or both may be necessary to get products to the market as fresh as possible. Products may be subject to certain food regulations and packaging requirements to meet state and federal standards.

Community Supported Agriculture

Community supported agriculture (CSA) is a channel for farms to pre-sell shares of their products to customers by offering weekly or monthly deliveries of their products throughout the season. In addition to fruits and vegetables, value-added products such as bread, eggs, and cheese, may be included. Various CSA models are offered depending on the size, location, and products on the farm. This is an important marketing channel for small- and medium-size farms when used alongside other direct marketing channels. At the core of the CSA model, the customers, the community, support their local farm by sharing the risk each season. Customers pay the same amount for the season regardless of the harvest. The risk is worth it to their customers though as supporting a CSA ensures the farm survives over time and continues to be a source of local, healthy food. In this way, farmers grow a loyal customer base.

Cooperatives

A producer-owned marketing cooperative (co-op) is a popular production model. A group of farmers pool their production and market and distribute it through the co-op. The farmers own and control the co-op themselves. The growth of co-ops has enabled producers to negotiate prices on a large-volume basis. Some co-ops are able to afford processing plants where fresh commodities can be processed and withheld, if necessary, until better prices are offered (**Figure 35-11**). They also have the ability to arrange for transportation of products from farm to market, balance supply and demand of agricultural products, and plan advertising to increase sales.

Figure 35-11 Producer-owned marketing cooperatives give small producers more power to influence the prices they receive for their commodities.

© Disobey Art/Shutterstock.com.

Approximately 85 percent of the milk marketed in the United States in 2020 was sold through producer-owned milk-marketing co-ops. The co-ops either process the milk and sell it directly to consumers or sell it to other large processing plants.

Intermediate Marketing

Intermediate marketing refers to a producer selling a product to an individual or business in the supply chain who is not the consumer. Intermediate markets act as middlemen. A **middleman** is any intermediary between the farmer and consumer. For the purposes of this chapter, restaurants, schools, and grocers are referred to as **retail marketing**. The producer sells directly to these buyers, and the buyers then sell the product to the consumer. Wholesalers and brokers are further removed from the buyer than grocers or restaurants. Wholesalers and brokers buy from the producer, and then they resell the product to the retail market. In this chapter, wholesalers and brokers are referred to as wholesale marketing.

Wholesale Marketing

The Census Bureau's Economic Census classifies wholesalers as merchant wholesalers, manufacturing sales branches and offices (MSBOs), and brokers and agents. MSBOs are run by large grocery manufacturers or processors to market their own products.

Merchant Wholesalers

Merchant wholesalers, also referred to as third-party wholesalers, buy groceries and grocery products from processors or manufacturers, and then resell them to retailers, institutions, or other businesses.

Merchant grocery wholesalers are classified into three groups by the types of products they distribute. General-line distributors, or broadline or full-line distributors, are companies that handle a broad line of food products. Specialty distributors are companies that distribute items such as frozen foods, dairy products, meat and meat products, or fresh fruits and vegetables. They operate in niche markets, such as airlines, convenience stores, and warehouse clubs. This type of marketing makes up nearly half of grocery wholesale sales. Miscellaneous distributors sell a narrow range of dry groceries such as canned foods, coffee, bread, or soft drinks.

Brokers

Wholesale brokers are wholesale operators who buy and/or sell as representatives of others for a **commission**, or a fee for the service of selling the product. They have access to any seller in the food chain, including packing houses, processors, agribusinesses, and large-scale farms. They act as middlemen arranging the logistics of getting the product from the farm to the buyer. They often take commissions based on a percentage of the sale. Wholesale brokers typically do not own or physically handle the products.

In terminal markets, wholesale buyers are typically located in permanent stalls. Terminal markets are central markets, usually in cities, that serve as an assembly and trading place for agricultural products. Wholesale buyers purchase products from farmers, brokers, distributors, or packing houses in large quantities, and resell it to brokers, distributors, or businesses in larger or smaller quantities. The buyer and the producer negotiate a price when the producer delivers the product. Prices are usually lower than a producer can receive through direct marketing.

At auction markets, livestock is sold by public bidding on animals individually or in lots, meaning groupings (**Figure 35-12**). A live auctioneer or online auction software facilitates the sale of goods. Auctions can be in-person, virtual, or a combination of both. Virtual auctions can be live or pre-recorded and open to bidders until a pre-determined closing time. Online auctions have grown in popularity as both the buyer and the seller appreciate the convenience. A major difference between in-person auctions and a virtual one is the method of product delivery. A seller needs to be aware of any additional costs or buyer responsibility prior to bidding. Auctions represent the most common means of marketing livestock and are usually the most practical for the small business livestock producer.

As is true when selling in terminal markets, commissions are charged for selling animals through other markets. The amount of commission charged varies with the type and size of the animal. Because many auction markets are small, there may not be the same degree of buyer competition that is present at terminal markets.

Figure 35-12 Auction markets are popular for selling livestock in many communities.
Courtesy of Bill Angell.

Retail Marketing

Agricultural products flow to the retail market either directly from the producer or through an intermediary market. Products sold through retail markets then reach the consumer. The retail food industry has changed significantly over the past two decades, as warehouse clubs, drugstores, and other nontraditional food stores have increased their shares of food sales. Therefore, traditional grocery stores now face competition from nontraditional food retailers. The top four retailers in 2019 were Walmart Stores, Inc., Kroger, Albertson's, and Target. Walmart and Target are both nontraditional grocery stores.

Vertical Integration

There is another method of getting the product to consumers that is a combination of direct, wholesale, and in some cases, retail. In the poultry market, almost 99 percent of the chickens produced for meat are grown under a system called **vertical integration**.

Figure 35-13 Vertical integration in the poultry market takes the processed chicken directly to the retailer eliminating the need for live-bird markets.
Courtesy of USDA.

A vertically integrated business is one that owns and controls every step of the supply chain and production. They do not depend on suppliers or middlemen. The use of vertical integration in the production and marketing of agricultural products allows for extremely large systems of production that can be efficient. Only the number of animals that are expected to be in demand by consumers will be produced. Such production systems are completely market driven. There is less competition from other producers, and all phases of production can be controlled (**Figure 35-13**).

Futures Markets

A commodity market is a marketplace for buying, selling, and trading raw materials. These commodities are split into two categories. Hard commodities include natural resources. Soft commodities include agricultural products. A **commodity exchange** is an organization licensed to manage the process of buying and selling commodities under specific laws using a system of licensed brokers. The major U.S. commodity exchanges are ICE Futures U.S. and the CME Group, which holds four major exchanges: the Chicago Board of Trade (CBOT), the Chicago Mercantile Exchange (CME), the New York Mercantile Exchange (NYMEX), and the Commodity Exchange (COMEX), Inc. Commodity exchanges also manage futures markets. A **futures market** is a procedure conducted by commodity exchanges to provide networks and legal frameworks for sellers and buyers to work through brokers in making contracts called futures contracts, or simply futures.

Buying and Selling Futures

Futures are defined as legally binding agreements made on the trading floor of a futures exchange to buy or sell something at a future date. Futures prices are determined by competitive bidding of brokers for prospective buyers and sellers from all over the world. Anyone with appropriate assets can work through a broker and buy or sell futures on the commodity market. The futures market floats up and down by the minute while the futures exchange is "open" or in session. Digital communication permits agricultural managers, buyers, sellers, and brokers to communicate almost instantly to place their bids to buy or sell. The commodity exchange handles the legal work to complete the transactions. Many agricultural commodities, as well as other commodities, are bought and sold in this manner (**Figure 35-14**).

Figure 35-14 Commodity exchanges make possible the use of futures trading as a valuable marketing and management tool.
Image used with permission of CME Group, Inc.

Farmers, ranchers, cattle feeders, crop growers, and other agricultural owners and managers use the futures market as a tool to stabilize their businesses. The wise buying and selling of futures can protect them from losses caused by excessive price fluctuations. For instance, before deciding how much corn to grow, how much to spend on fertilizer and other inputs, or how much corn to keep for feeding cattle or hogs on the farm, the farmer should know the price per bushel for which the corn could be sold after it is grown. The other decisions are then easier to make.

Current commodity prices are at your fingertips using an app on a smartphone or going directly to a related website. By studying commodity pricing, commodity sellers can determine the futures price for commodities such as corn in a given month. If the farmer regards that price as high enough to commit a certain number of bushels for sale, then they will inform their broker to "sell" that many bushels at the rate of the futures price for the month selected. The broker will negotiate a contract that obligates the farmer to deliver that many bushels of corn on the specified date for the amount of the futures price. Now the farmer is guaranteed the price for the corn, and other decisions about the farming operation can be based on this selling price.

If the price of corn is lower in December than the $9.45 futures contract that was made, the payment on delivery will still be $9.45 per bushel, and the farmer will have successfully protected the business against a price drop. However, if the price of corn is greater than $9.45 per bushel in December, the farmer might wonder if the purchase of a futures contract was a wise decision. Here is how a futures contract might be made for 5,000 bushels of corn. Suppose the settlement price on the futures market for December is $9.45 per bushel. The value of the contract for 5,000 bushels is $9.45 × 5,000, or $47,250. The farmer is obligated to deliver the corn for $47,250 regardless of the market price of corn at the expiration of the contract, which could be higher or lower depending on the current market value.

Opening and Offsetting a Position

The purchase of a futures contract does not have to be a final transaction that plays through to the actual buying or selling of the commodity. In fact, in most cases, the products are not delivered because the participant makes another contract that offsets the first one. The initial step in the futures market is called opening a position. At a later date, before the futures contract expires, the participant can take a second position to offset the first. Such action is known as offsetting a position. When the position is offset, the participant is said to be out-of-the-market and has no obligation to deliver or take delivery. To offset the position described previously, the farmer would need to buy 5,000 bushels of corn for delivery on the same date as the first contract. This would offset the contract to sell 5,000 bushels. If the farmer were fortunate enough to find a day when the futures price was less than $9.45 and buy futures for 5,000 bushels for the specified date of delivery at that lower price, then the farmer could pocket the difference as income from the business transaction.

Export Marketing

Export marketing is the practice of selling products or services to another country. After the European Union, the United States is the world's second largest agricultural trader. China is the largest buyer of U.S. agricultural products followed by Mexico and Canada.

In 2021, soybeans, corn, and beef were the top American exports. In 2022, U.S. agricultural exports were projected to be a record $191 billion. U.S. agricultural output for many products is growing faster than domestic demand. To sustain current revenue and prices the agriculture industry has been relying on export markets, which is driving this growth.

The United States also imports agricultural commodities. This means the United States buys products from other countries to sell in the U.S. market. In 2019, China surpassed the United States and the European Union becoming the world's largest agricultural importer. However, U.S. agricultural imports have continued to grow steadily over the past two decades. In 2021, the United States imported $171 billion worth of agricultural products. The increasing demand for year-round variety in foods has driven imports of horticultural products, such as fruits and vegetables, during the offseason of U.S. production. Horticultural products accounted for more than half of U.S. agricultural imports in 2021. Sugar, coffee, cocoa, and spices were other top imports.

Both U.S. agricultural exports and imports have increased significantly over the last 25 years. Developing countries entering and engaging in the global market and the implementation of foreign and domestic policies have expanded U.S. access to foreign markets.

The Role of Global Currencies

Most commerce occurs using the currency of the country where a transaction takes place. However, international trade involves more than an exchange of commodities or products. The value of a product is also affected by the exchange rate between currencies. On any given day, the exchange rates for two currencies may fluctuate. These differences occur due to several different factors, such as different interest rates, currency supply and demand, inflation, and economic outlook.

When considering international marketing of agricultural products, a strong domestic currency may not be favorable. While travel between the two countries in question may become cheaper, domestic products may become relatively more expensive to the prospective buyer. When domestic products are not competitive in the market, entire industries may suffer economic problems despite a strong dollar. A weak currency stimulates exports, while a strong currency favors imports. Marketing agricultural products to an international market is strengthened during a time period when the dollar is weak in comparison with other currencies.

The value of a product in a domestic market is usually determined by the law of supply and demand. For example, when a surplus of butter exists, its value per unit in a domestic market can be expected to go down. The opposite is true when butter is in short supply. A shortage of a commodity drives the price up. The foreign demand for a product may exist, and the need may even become critical. When the currency exchange rate contrasts a weak foreign currency against a strong domestic currency, the actual import cost to foreign buyers increases. Price may restrict the amount of a purchase or even prevent an export sale.

Agri-Profile — Career Area: Agricultural Marketing

Soybeans are used in a variety of food products. As the top exported crop in the United States, continued marketing and promotion of soybeans is essential to the U.S. economy
Courtesy of USDA.

Agricultural marketing covers an extensive range of products, services, and occupations. The producers and breeders of vegetable and fruit crops, grains, flowers, ornamental shrubs, turfgrass, trees, pets, laboratory animals, livestock, horses, dairy animals, poultry, fish, wildlife, milk, eggs, wool, and fur all market the products of their businesses.

Marketing cooperatives, dealers, auctioneers, livestock handlers, transportation services, commodity market managers, futures traders, writers, and broadcasters, and many other agricultural professionals have some connection to marketing, whether by depending on it, developing it, or influencing and being influenced by it as part of the company culture.

The path to a marketing career often starts with postgraduate studies in plant and animal sciences, food science, agricultural economics, and of course, marketing and communications. In these careers, people use their knowledge and skills to create domestic and international marketing opportunities for U.S. soft commodities. They may work in futures trading or in a position related to the importing or exporting these commodities, or they may help ensure the quality and availability of wholesome foods and a wide range of products by working in food safety and inspection.

Chapter Review

Student Activities

1. Write the **Key Terms** and their meanings in your notebook.
2. Search several commodity exchanges marketing reports. Compare the prices received for the various classes and grades of animals listed. Compare the prices received from various markets.
3. **SAE Connection:** Interview a livestock buyer. Ask how prices are determined and where animals are bought and sold. Ask how to go about getting a job as a livestock buyer. Report your findings to the class.
4. Invite the operator of a wholesale market to class to talk about the issues involved in wholesale marketing of animals and animal products.
5. List the types of markets for agricultural commodities in your area.
6. Write a brief paper on the "best philosophy for pricing."
7. Develop an ad or other promotional plan for an agricultural commodity.
8. Compare the values of the currencies of five United States trading partners with the dollar over a period of several weeks by comparing exchange rates. Graph these values over time. Consider which of these currencies is most likely to support a strong U.S. export market.
9. Build a display for retailing an agricultural commodity.
10. Research the procedures used to buy and sell in the futures market.
11. Conduct research on one or more major international trade agreements that influence the marketing of agriscience products.
12. Write a brief marketing plan for an agricultural product you think would be useful to people in your community.

Inquiry Activity   Marketing

Courtesy of DeVere Burton.

After watching the *Marketing* video available on MindTap, discuss the following items with a classmate.

1. Work together to list the factors that should be considered when marketing agricultural products. Then, pause the video at the 00:19 mark to check that you have listed all the considerations.
2. Explain the three scenarios in supply and demand.
3. Talk about the advertisement of commodities that benefit all groups. What commodities would benefit from this type of advertising?
4. Describe the three types of marketing presented in the video.

Chapter 36
Entrepreneurship in Agriscience

Objective
To define entrepreneurship and determine considerations for planning and operating an agribusiness.

Competencies to Be Developed
After studying this chapter, you should be able to:
- define and describe entrepreneurship.
- describe the steps in planning a business venture.
- state five basic functions performed in the operation of a small business.
- select a product or service for a personal or group enterprise.
- determine the basic functions performed by small-business managers.
- analyze the outcome of a business venture.
- use small-business financial records.
- analyze the benefits of self-employment versus other types of employment.

Key Terms

entrepreneur
entrepreneurship
buying function
selling function
promoting function
distribution function
financing function
short-term plan
long-term plan
actuating
balance sheet
asset
liability
net worth
profit and loss statement
cash flow statement
inventory report

Entrepreneurship in agriscience provides extensive career possibilities at all levels of effort. It creates opportunities for the inventor, risk taker, profit seeker, owner, manager, and employee. The products and services of entrepreneurs help drive every business, including farming, factories, equipment sales, fishing, guide services, veterinarian work, teaching, insurance, real estate, finance, and numerous others. Opportunities in agriscience exist worldwide (**Figure 36-1**).

Figure 36-1 An entrepreneur must learn to recognize and take advantage of business opportunities when they occur. Agribusiness opportunities are found throughout the world, and today's young entrepreneurs will find opportunities to do business in the international arena.
© Fizkes/Shutterstock.com.

The Entrepreneur

The **entrepreneur** is the person who organizes a business or trade or improves an idea. The word comes from the French word *entreprendre*, which means "to undertake." **Entrepreneurship** is the process of planning and organizing a small-business venture. It also involves managing people and resources to create, develop, and implement solutions to problems to meet the needs of people. An inventor is the person responsible for devising something new or for making an improvement to an existing idea or product.

The entrepreneur can be the inventor as well as the small-business manager. However, in many cases, they are different people, each with distinct talents. The entrepreneur functions as the liaison between the inventor and the manager or management team. The entrepreneur brings these two groups together for the purpose of getting the invention to the individuals it will serve.

It is the entrepreneur who visualizes the venture strategy and is willing to take the risk to get the venture off the ground. Inventors or business managers are not entrepreneurs unless they organize the venture. An understanding of the differences, as well as the similarities, among these roles is important to comprehending entrepreneurship as a career option.

Entrepreneurship

Various types of business enterprises are part of the entrepreneurship system. The element of individual, partner, or corporate ownership means private control as opposed to government ownership control. The profits (or losses) from entrepreneurship go to the owners. However, owners hire managers and other employees. Therefore, entrepreneurships provide jobs for everyone (**Figures 36-2** and **36-3**).

Figure 36-2 Many businesses start with high school occupational-experience programs in agriscience.
Photo courtesy of the National FFA Organization.

Figure 36-3 Successful businesses generate job opportunities for managers and other employees.
khaleddesigner/Shutterstock.com.

A sales project conducted by a school organization such as the FFA is a type of entrepreneurship. However, it differs from a typical small-business venture in the following ways:

1. The sales project should be a valuable learning experience as well as a money-making activity.
2. The sales project is usually of a limited duration. It is completed within a short period. A small business typically is designed to operate indefinitely.
3. The sales project involves the voluntary participation of the class. The small business must hire and pay its employees.
4. The sales project usually involves little risk for financial loss to the individual. The small-business owner (entrepreneur) could lose their investment.
5. The sales project may need to be approved by the school administration, and it may need to be licensed by the government, or pay taxes.

With these distinctions in mind, you can plan, organize, and implement a sales project and gain insight into the world of entrepreneurship at the same time (**Figure 36-4**).

Figure 36-4 Group projects in school can help students develop valuable entrepreneurial skills.
© SeventyFour/Shutterstock.com.

Operating Businesses

Five basic functions are performed in the operation of both a small business and a sales project. These functions represent the basic steps in moving a product or service from the supplier to the consumer. They are the buying function, selling function, promoting function, distribution function, and financing function.

Buying Function

The **buying function** involves selecting a product or service to be marketed or sold for profit. The selection of a product or service is based on thorough marketing research, which determines consumer (customer) needs and wants. It also determines who may already be promoting the product or service.

Selling Function

The **selling function** includes studying the product or service to determine the reasons customers will want and need the product or service. This function also involves developing suitable customer approaches. Planning sales presentations, determining methods of overcoming objections, and planning for the close of a sale are all part of this function (**Figure 36-5**).

Promoting Function

The **promoting function** involves developing a plan to identify ways to make potential customers aware of the product or service to be offered. Examples of promotional activities include online, print, TV, radio, podcast, and outdoor advertising.

Distribution Function

The **distribution function** involves physically organizing and delivering the selected product or service. For products, this includes receiving, storing, and distributing the merchandise. Many of the same activities are associated with services, particularly distributing to the customer (**Figure 36-6**).

Figure 36-5 How products are displayed is important. The display of flowers at this greenhouse is visually appealing and sure to attract customers.

© Cagkan Sayin/Shutterstock.com.

Figure 36-6 Product distribution may be made to retail outlets or to other businesses, homes, farms, and many other destinations consumer sites.

© iStockphoto/Sean Locke.

Financing Function

The **financing function** includes obtaining capital for the initial inventory, recording sales, maintaining inventory, computing profit or loss, and reporting the results of the venture. Accountability to lenders is always part of the financing function, and lenders should always be informed of the progress of the business. Informed lenders are more likely to finance new ventures and work with borrowers when problems arise.

Hot Topics in Agriscience: Digital Marketing

Mobile devices and the Internet make shopping possible from anywhere.
© mrmohock/Shutterstock.com.

Digital marketing, also called online marketing or Internet marketing, allows entrepreneurs to market products directly to anyone who accesses the Internet using a computer, tablet, phone, or other devices. It allows a person to run a business from their own home, and it allows business organizations of all sizes to serve online shoppers. Digital marketing has opened up new advertising and promotion opportunities, and virtually anyone who has a product or service to sell can do so online.

Digital marketing has also introduced some new problems to the business world. Payment is made by credit card, but how do you keep credit card numbers secure from discovery by dishonest people? How do you protect online shoppers from being exploited by cybercrooks and others who develop lists of names by tracking the websites shoppers visit? What privacy rights should be protected by Internet providers? What new laws are needed to protect online consumers? What risks and benefits will blockchain technology and cryptocurrency add to the mix? These are only a few of many issues that are developing alongside this rapidly evolving marketing opportunity.

Agri-Profile: Career Area: Manager/Assistant Manager/Management Trainee

The entrepreneur is constantly seeking ways to provide better goods and services to the clientele.
© M_Agency/Shutterstock.com.

Agribusiness management is seen by many as the ideal career. It offers people the advantage of continuing in the work of a family business, in which they gain experience as they grow up. Such experience can also give career seekers the confidence to promote their management skills to potential employers.

Many high school and college students are well established in agribusinesses before they finish their educations. Some popular agribusinesses include lawn services, logging, lumber businesses, greenhouse or nursery operations, machinery repair, agricultural supplies, home and garden centers, florist shops, retail flower sales, livestock sales, farming, and ranching.

Preparation for careers in agribusiness management includes both formal education and on-the-job training. Effective programs should include classroom, laboratory research, supervised agriscience experience, and leadership development. High school agribusiness and agriscience programs provide excellent training for business, but they do not take the place of higher education in business management. Advanced agribusiness programs at technical schools, colleges, or universities can help career seekers gain a competitive edge in economics, finance, and management.

Selecting a Product or Service

The first step in the process of establishing a new business or developing a fundraising project is the selection of a product or service (buying function). This involves analyzing potential products/services to determine the level of demand (number of customers). The analysis of potential products or services should provide answers to these and other pertinent questions:

- Who are the potential purchasers of the proposed product or service? Customers can often be viewed in groups that have similar interests.
- On what basis does each group make decisions on the product or services? Typical responses to this question include price, quality, and continuing service after the sale.
- What are the competing products or services? How do they compare with the proposed product or service relative to price, quality, and service?
- What is the total estimated demand for the product or service?
- How much profit can be expected on the sale of this product or service? (Multiply the number of units that are expected to sell times the markup per unit and subtract any expenses incurred in the purchase, sale, and delivery of the product or service.)

Organization and Management

Put simply, the role of an entrepreneurial manager in a business venture is to make the right things happen so the organization's goals can be achieved. To accomplish this, managers work mainly with data and people. Managing involves getting all the parts of the business—including personnel, marketing strategies, finances, and records—to function together toward the same goal.

No two managers fulfill the same jobs in the same way. Each job is shaped by the type of venture and the personality of the individual manager. However, managers perform many of the same functions. These functions include

- planning the work,
- organizing people and resources for work,
- actuating work, and
- controlling and evaluating work.

Planning

When managers make plans, they set objectives or goals. They establish or implement policy for the business, and they recommend strategies to achieve the goals. Managers must also plan for both the short term and the long term. **Short-term plans** are accomplished in several days or weeks. **Long-term plans** are accomplished over several months or years. Plans must be constantly reviewed and updated, since no matter how thorough the plans, they alone do not guarantee success.

Organizing

After a plan is developed, the work must be organized. The entrepreneurial manager arranges for the necessary equipment, supplies, or other resources and identifies or hires the people needed to carry out the plan, and thus bring the product or service to market. Staff development may also be required.

Actuating

Actuating simply means putting the plan into action. The manager informs employees of the plan and sets expectations. All of the persons involved must understand their roles. Actuation also includes motivating people to work efficiently and effectively together, ideally by creating a respectful workplace culture that welcomes new ideas and rewards hard work.

Controlling and Evaluating

Managers must carefully control implementation of plans after the work begins. The quality and quantity of all results must be evaluated. If the results are satisfactory, work can continue. If problems arise, changes must be made and alternate plans may need to be developed. Managers must be capable of making adjustments in such areas as personnel, equipment, policies, or procedures whenever necessary (**Figure 36-7**).

In a very small-business, all management functions may be carried out by one person. As a grows and its objectives become more complex, additional managers are usually required. When new managers are hired, the organizational lines of responsibility and the written job descriptions should already be in place, to clearly convey the functions each manager is to perform.

Figure 36-7 The manager must be familiar with all the functions and able to work with all the employees of the business.

© Andrey_Popov/Shutterstock.com.

Hot Topics in Agriscience | Cooperation for the Good of All

A farmers' cooperative pools the business efforts of many farmers, giving them the advantage of volume sales and purchases that one person acting alone could not achieve.

© ruzanna/Shutterstock.com.

A cooperative, or co-op, is a type of business organization that is created, owned, and controlled by its members. There are three main types of co-ops: market, purchasing, and service based. Marketing co-ops are common in agriculture. Such a co-op receives products from its members and then resells those products for the best possible price. One producer alone may not be able to produce enough crops or other agricultural commodities to obtain a large contract. But when many producers band together, they have more leverage, or power, to compete for large-order contracts.

Purchasing co-ops can make large-quantity purchases for members. By purchasing in bulk, members can get the best possible prices for products they need. Items such as farm equipment, seed, fertilizers, dairy supplies, and much more can be purchased at reduced prices for members. Service co-ops provide low-cost service items to their members. For example, health insurance can be purchased at a much lower cost when a large number of people are becoming insured as opposed to a single family. This makes it possible to spread the cost of losses across many participants instead of just a few individuals. As the size of the insured pool increases, the cost per individual goes down. Co-ops are valuable tools to the farm and ranch families that elect to participate in them.

Small-Business Financial Records

Financial records are invaluable tools to the entrepreneur and the manager. Good financial management allows the manager to maintain control of the business venture and to increase profits or reduce losses.

Financial records can reveal which products and services are selling and which are not. These records can also measure the degree of success achieved, or the value delivered, by each person on the sales force. Financial records should indicate the amount of inventory on hand and what quantity of a product or service has been sold. How much profit is made and the total value of the venture can be computed at any time using appropriate financial records.

The small-business owner or entrepreneur relies on detailed financial records. For example, businesses regularly generate and maintain financial records for such items as buildings, fixtures and equipment, credit, debt, and return on investments. However, such records are not limited to businesses alone. Balance sheets, profit and loss statements, along with inventory and sales reports may also prove useful for such endeavors as a group educational project. Some typical financial records are discussed in the following sections (**Figure 36-8**).

Figure 36-8 Financial records are fundamental management tools.
© kurhan/Shutterstock.com.

Balance Sheet

A **balance sheet** reflects the state of the business at a specific point in time. It shows the assets, liabilities, and owner's investment on a particular date. The equation for the balance sheet follows: Assets = Net worth − Liabilities. **Assets** include everything the venture owns, including cash on hand, equipment, and inventory. **Liabilities** include both current and long-term debts. **Net worth** is the owner's investment in the business, including any profits the business yields. A sample balance sheet is shown in **Figure 36-9**.

Profit and Loss Statement

The **profit and loss statement** projects costs and other expenses against sales and revenue over time. A profit and loss statement features five key components (**Figure 36-10**):

1. Total sales
2. Cost of goods sold
3. Gross profit
4. Expenses
5. Net profit (or loss)

Balance Sheet

Date: _____

ASSETS

Current: Cash _____
Merchandise _____
Accounts Receivable _____

Fixed: Land _____
Building _____
Machinery _____
Equipment _____

Other:
1. _____
2. _____
3. _____

TOTAL ASSETS: _____

LIABILITIES

Current: Notes Payable _____
Accounts Payable _____

Noncurrent:
Debts more than one-year maturity _____

TOTAL: _____

NET WORTH: _____

TOTAL LIABILITIES AND NET WORTH: _____

Figure 36-9 A balance sheet is a statement of a business owner's assets, liabilities, and net worth.

Profit and Loss Statement

For period ending _____

TOTAL SALES $75,000

COST OF GOODS SOLD
Beginning Inventory 55,000
(Plus) Purchases 10,000
(Less) Ending Inventory................ 15,000

TOTAL COST OF GOODS SOLD 50,000
(Beginning Inventory plus Purchases minus Ending Inventory)................ 25,000

GROSS PROFIT.............................. 25,000
(Total Sales minus Total Cost of Goods Sold)

EXPENSES
Salaries 18,000
Payroll Taxes............................. 2,500
Rent.. 4,500
Advertising............................... 1,000

TOTAL EXPENSES........................... 26,000

NET PROFIT (LOSS) (before taxes) ..... (1,000)

(Gross Profit minus Total Expenses)

Figure 36-10 Sample profit and loss statement.

Cash Flow Statement

A **cash flow statement** describes the availability of funds for the purposes of running a business (**Figure 36-11**). When a positive cash flow exists, the income that is received by the business exceeds the expenses, and there is cash on hand. A negative cash flow indicates that the expenses that are generated by the business exceed the income at a given time. Nearly every business deals with periods of negative cash flow, and it is at these times that an operating loan may be needed to tide the business over until income is received. The cash flow statement allows a business owner to estimate the need for credit and to establish when the credit will be needed.

For example, a beef ranch operates with a negative cash flow for most of the year. Once the calves are ready for market, they are sold together with cows that are not being retained in the breeding herd. Most of the income is received within a few days or weeks. In contrast, a dairy farm has a fairly constant cash flow throughout the year from the sale of milk and excess animals. The cash flow statement helps the business manager and the loan officer determine how much operating cash is needed, when it is needed, and when it will be repaid.

Item No.	Description	Beginning Inventory	Items Sold	Ending Inventory	Total Cost Beg./Inv.	Total Cost End/Inv.	Total Sales	Total Profit
000	Tick Tack Tote	45	45	0	33.75	.00	45.00	11.25
001	Brown Tray	32	32	0	25.60	.00	48.00	22.40
002	Fish Tray	20	20	0	15.00	.00	30.00	15.00
003	Lion Tray	6	6	0	4.80	.00	9.00	4.20
004	Guitar Tray	8	8	0	6.40	.00	12.00	5.60
005	Asst. Animal Tray	34	34	0	25.50	.00	51.00	25.50
006	Elephant Tray	10	10	0	5.00	.00	15.00	10.00
007	Screwdriver Set	22	22	0	11.00	.00	33.00	22.00
008	Brush & Shoehorn	36	36	0	32.40	.00	81.00	48.60
009	Fully Auto. Umbrella	9	9	0	20.70	.00	31.50	10.80
010	Auto. Umbrella	8	8	0	16.00	.00	24.00	8.00
011	Dad's No. 1 Keychain	200	190	10	90.00	4.50	190.00	104.50
012	Grandpa Keychain	100	79	21	45.00	13.95	79.00	47.95
013	Dad's Pad	48	30	18	28.80	10.80	45.00	27.00
014	Mini Screwdriver Set	75	75	0	60.00	.00	150.00	90.00
015	Dad's Plaque	72	72	0	28.80	.00	108.00	79.20
016	Tic Tac Toe	36	36	0	21.60	.00	36.00	14.40
017	Dad's Pen	1	1	0	.80	.00	1.50	.70

Figure 36-11 Sample cash flow statement.

Inventory Report

The **inventory report** includes such information as how many units of each product are on hand, how many have been sold, the cost of each item, total sales, and profit. The inventory report can be used to guide decisions, such as to reduce the price of slow-selling items, reorder other items, and to determine which items are yielding the best profits.

Every producer and/or seller of products depends on up-to-date inventory reports. Advances in technology have greatly improved such reporting from the old "pen and paper" era when keeping inventory could meant shutting down a business for one or more days to carry out manual counting. Such *periodic inventory* has been increasingly replaced by *perpetual inventory*. For example, at supermarkets today, the use of computerized cash registers at checkout stands ensures that the inventory report is updated each time a sale is made. A computerized perpetual inventory system can also generate new orders for products whenever inventory for a particular item falls below a predetermined level.

Forms of Employment

The decision to either work for someone else or to open a business of one's own is a challenging one. A helpful way to make that decision is to consider the advantages and disadvantages of both options.

Being an Employee

The advantages of being an employee, working for someone else, center on security. Salaried employees have no personal financial risk or responsibility to the company for which they work. Employees generally work regular hours. If they work additional hours, they may be paid overtime for those hours. In addition, employees are often guaranteed paid time off for vacation and sick leave, and other substantial benefits, such as health insurance, tuition reimbursement, and a company retirement or profit-sharing plan. An employee can count on general stability, with a fairly accurate idea of what the income will be from year to year. In addition, the employee may move up the career ladder within the company, gaining new opportunities and greater compensation.

However, there are also some disadvantages to working for someone else. The company has little or no long-term financial responsibility to the employee, such as if a recession occurs and the company decides to reduce the number of personnel. Raises or other increases in compensation aren't always based on merit or individual negotiation. Some companies set a fixed salary scale, so an employee could remain at a particular salary level until the right combination of years or experience is met. This is also true of promotions that could be based on years accumulated rather than on merit. In addition, some employees experience "burn-out": as the work pattern becomes routine, or worse, overwhelming. Employees wishing to make a change might find themselves waiting—perhaps for years—for an in-company position to open up, or else having to discreetly seek a position with another company. Management controls many or most of these options, not the employee.

Being Self-Employed

Figure 36-12 Ownership of a small business encourages innovation and independence.
© mangostock/Adobe Stock Photos

One of the major reasons cited for opening a business is that it gives owners control over their own destiny. The owner has the opportunity to set personal goals and recruit a team to help carry out those goals. Success in meeting those goals can yields a sense of achievement and independence along with financial rewards (**Figure 36-12**). Unlike a salaried employee, an owner's ability to earn money is not restricted to a pre-set level.

The disadvantages of owning a business may not be as obvious By taking control of a business, an owner also takes on the demands and consequences of that business. For example, the necessary capital outlay may jeopardize family savings or even one's family home. The number of hours required to run the business mean a definite commitment from the owner as well as from any family members involved. Should the company have difficulty, responsibility for both the management and the financial problems rests with the owner.

Contributions of Small Businesses

Entrepreneurs are considered the cornerstone of the U.S. enterprise system. According to the United States Small Business Administration, small businesses account for two out of every jobs added to the economy.

The role of entrepreneurs in the United States is highlighted by the following:

- Most businesses in the United States (99 percent) are classified as small by the Small Business Administration.
- New businesses are formed at a rapid rate (about 600,000 per year).
- Small businesses generate almost half (43.5 percent) of the U.S. gross national product (GNP).
- Small businesses employ 41.7 percent of all U.S. workers.

- Small businesses created approximately 64 percent of the new jobs in our economy in recent years.
- Small businesses have historically produced up to 13 times as many patents per employee (products, services, techniques) as large businesses (**Figure 36-13**). However, this trend has declined in recent years.

The U.S. economy fluctuates up and down in response to many influences. However, as the economy continues toward an emphasis on services, the role and importance of small business and entrepreneurship is expected to increase. This is likely to occur because small businesses are especially dominant in the service sector of the economy.

New businesses can be the center of innovation because they are not generally tied to existing ways of doing things. They have a sense of energy, urgency, and vitality that comes from the entrepreneurial spirit.

The contribution of small businesses to the U.S. private-enterprise system is important, and it is likely to remain so in the future.

Figure 36-13 The small businesses that dot the landscape of the United States are highly productive and responsive to the needs of their customers.
© LightField Studios/Shutterstock.com.

Chapter Review

Student Activities

1. Write the **Key Terms** and their meanings in your notebook.
2. Develop a collage on the bulletin board for classroom display illustrating entrepreneurship opportunities in agriscience.
3. Make a table showing the advantages and disadvantages of being (1) an employee and (2) an entrepreneur. The following format is suggested:

Being an Employee	Being Self-employed
Advantages	Advantages
1.	1.
2.	2.
etc.	etc.
Disadvantages	Disadvantages
1.	1.
2.	2.
etc.	etc.

4. Organize a cooperative business within your class or FFA for buying or marketing a product of interest to the group.
5. Become familiar with the record book and record-keeping system used by members of your local FFA chapter or 4-H club for keeping records on individual projects.

6. **SAE Connection:** Supervised Agricultural Experience (SAE) is often a student's first exposure to entrepreneurship. The SAE component of agriscience education creates the expectation that students will combine classroom instruction with actual hands-on work that reinforces what was learned. Often, this results in ownership of an agricultural enterprise. The concept is based on true principles and "doing" is still one of the best ways to learn. List three to five types of SAE entrepreneurship projects that interest you. (Use the Appendix to find examples.) Then select one such project and conduct research to find out more about the steps involved to turn it into a profitable business. As part of your online research, challenge yourself to find real-life examples of students who have developed such an entrepreneurship project. Summarize your findings and present them to the class.
7. Become an entrepreneur—own and operate your own business venture.
8. Get a summer or after-school job with a company similar to one you may be considering for your own future career. While working for that company, learn as much about running a business as you can.
9. Create an individual business plan using the information you have learned from this chapter. Discuss with your parents and trusted advisors what steps you will need to take to make your business plan a reality.
10. Initiate an Internet search to discover how technology is applied to agriculture, food, and natural resource systems. Start your search using one or more of the following sets of search words: a. agriculture internet of things, b. smart foods, c. smart environment. Prepare a short written summary of your findings.
11. Join other FFA members across America who share the collaborative record-keeping software known as "Agricultural Education Online Recordkeeping System, AET." Encourage your teacher and classmates to collaborate in tracking all of your FFA educational and financial activities. The software summarizes these experiences into standard FFA award applications and reports.
12. Plan and implement a senior project, agriscience project, or similar activity to validate and explain the uses and benefits of computer-based equipment in agriculture, food, and natural resources.

National FFA Connection: Entrepreneurship and the FFA

Courtesy of National FFA; FFA #148

National FFA programs provide opportunities for students to practice entrepreneurial skills and principles that were learned in the classroom.
Kaspars Grinvalds/Shutterstock.com

FFA has always promoted learning by doing, and this mindset carries over into life skills and business skills. Entrepreneurship is all about developing an idea and taking it to the world. One category of the FFA proficiency awards recognizes students who establish ownership enterprises. Through a Supervised Agriculture Experience, the FFA member "plans, implements, operates, and assumes financial risk to produce products or provide services."

Other FFA events offer student members to hone their employment skills. For example, participants in the National FFA Employment Skills Leadership Development Event (LDE) submit cover letters and resumés, practice interviewing for jobs, and even negotiating offers.

Inquiry Activity: Entrepreneurship

© iStockphoto/Sean Locke.

After watching the *Entrepreneurship* video available on MindTap, answer the following questions:

1. What is the first consideration when opening a business?
2. Is following your passion the best basis for developing a business? Why or why not?
3. What are one or two questions to ask when determining demand for the business?
4. Complete these sentences: An organized business is one in which _____ . For example, _____.
5. What is even more important than possessing a great deal of knowledge in your chosen product or service? Give examples of how to successfully manage this aspect of your business.

Appendix A Reference Tables

The information in this appendix includes a variety of useful conversions, conversion factors, measurement standards, and common measures.

Table A-1

Conversion Tables for Common Weights and Measures	
Nonmetric Unit	Metric Equivalent
1 pound	454 grams
2.2 pounds	1 kilogram
1 quart	1 liter
15.43 grains	1 gram
2,205 pounds	1 metric ton
1 inch	2.54 centimeters
0.39 inch	10 millimeters or 1 centimeter
39.37 inches	1 meter
1 acre	406 hectares

Table A-2

Weight Conversions	
Measurements	Equivalent Amount
8 tablespoons	1/2 cup
3 teaspoons	1 tablespoon
1 pint	2 cups
2 pints	1 quart
4 quarts	1 gallon or 8 pounds of water
2,000 pounds	1 ton
16 ounces	1 pound
27 cubic feet	1 cubic yard
1 peck	8 quarts
1 bushel	4 pecks

Other Conversions	
1%	0.01
	10,000 ppm
1 Megacalorie (M-cal)	1,000 calories
1 calorie (big calorie)	1,000 calories (small calorie)
1 M-cal	1 therm

Table A-3

Common Measures and Approximate Equivalents	
1 liquid teaspoon	5 milliliters (ml)
3 liquid teaspoons	1 liquid tablespoon = 15 ml
2 liquid tablespoons	1 liquid ounce = 30 ml
8 liquid ounces	1 liquid cup = 0.24 liter
2 liquid cups	1 liquid pint = 0.47 liter
2 liquid pints	1 liquid quart = 0.9463 liter
4 liquid quarts	1 liquid gallon (U.S.) = 3.7854 liters

Table A-4

Fahrenheit to Centigrade Temperature Conversions					
°F	°C	°F	°C	°F	°C
100	37.8	77	25.0	54	12.2
99	37.2	76	24.4	53	11.7
98	36.7	75	23.9	52	11.1
97	36.1	74	23.3	51	10.6
96	35.6	73	22.8	50	10.0
95	35.0	72	22.2	49	9.4
94	34.4	71	21.7	48	8.9
93	33.9	70	21.1	47	8.3
92	33.3	69	20.6	46	7.8
91	32.8	68	20.0	45	7.2
90	32.2	67	19.4	44	6.7
89	31.7	66	18.9	43	6.1
88	31.1	65	18.3	42	5.6
87	30.6	64	17.8	41	5.0
86	30.0	63	17.2	40	4.4
85	29.4	62	16.7	39	3.9
84	28.9	61	16.1	38	3.3
83	28.3	60	15.6	37	2.8
82	27.8	59	15.0	36	2.2
81	27.2	58	14.4	35	1.7
80	26.7	57	13.9	34	1.1
79	26.1	56	13.3	33	0.6
78	25.6	55	12.8	32	0.0

Formulas used: °C = (°F − 32) × 5/9 *or* °F = (°C × 9/5) + 32

Table A-5

Conversion Factors for English and Metric Measurements					
To Convert English	**To the Metric Multiply by**	**To Convert Metric**	**Multiply by**	**To get English**
acres	0.4047	hectares	2.47	Acres
acres	4047	m²	0.000247	Acres
BTU	1055	joules	0.000948	BTU
BTU	0.0002928	kwh	3415.301	BTU
BTU/hr	0.2931	watts	3.411805	BTU/hr
bu	0.03524	m³	28.37684	bu
bu	35.24	L	0.028377	bu
ft³	0.02832	m³	35.31073	ft³
ft³	28.32	L	0.035311	ft³
in³	16.39	cm³	0.061013	in³
in³	1.639×10^{-5}	m³	61012.81	in³
in³	0.01639	L	61.01281	in³
yd³	0.7646	m³	1.307873	yd³
yd³	764.6	L	0.001308	yd³
ft	30.48	cm	0.032808	ft
ft	0.3048	m	3.28084	ft
ft/min	0.508	cm/sec	1.968504	ft/min
ft/sec	30.48	cm/sec	0.032808	ft/sec
gal	3785	cm³	0.000264	gal
gal	0.003785	m³	264.2008	gal
gal	3.785	L	0.264201	gal
gal/min	0.06308	L/sec	15.85289	gal/min
in	2.54	cm	0.393701	in
in	0.0254	m	39.37008	in
mi	1.609	km	0.621504	mi
mph	26.82	m/min	0.037286	mph
oz	28.349	gm	0.035275	oz
fl oz	0.02947	L	33.93281	fl oz
liq pt	0.4732	L	2.113271	liq pt
lb	453.59	gm	0.002205	lb
qt	0.9463	L	1.056747	qt
ft²	0.0929	m²	10.76426	ft²
yd²	0.8361	m²	1.196029	yd²
tons	0.9078	tonnes	1.101564	tons
yd	0.0009144	km	1093.613	yd
yd	0.9144	m	1.093613	yd

Table A-6

More Conversion Factors for Metric and English Units
Length
1 mile = 1.609 kilometers; 1 kilometer = 0.621 mile
1 yard = 0.914 meter; 1 meter = 1.094 yards
1 inch = 2.54 centimeters; 1 centimeter = 0.394 inch
Area
1 square mile = 2.59 square kilometers; 1 square kilometer = 0.386 square mile
1 acre = 0.00405 square kilometer; 1 square kilometer = 247 acres
1 acre = 0.405 hectare; 1 hectare = 2.471 acres
Volume
1 acre/inch = 102.8 cubic meters; 1 cubic meter = 0.00973 acre/inch
1 quart = 0.946 liter; 1 liter = 1.057 quarts
1 bushel = 0.352 hectoliter; 1 hectoliter = 2.838 bushels
Weight
1 pound = 0.454 kilogram; 1 kilogram = 2.205 pounds
1 pound = 0.00454 quintal; 1 quintal = 220.5 pounds
1 ton = 0.9072 metric ton; 1 metric ton = 1.102 tons
Yield or Rate
1 pound/acre = 1.121 kilograms/acre; 1 kilogram/acre = 0.892 pound/acre
1 ton/acre = 2.242 tons/hectare; 1 ton/hectare = 0.446 ton/acre
1 bushel/acre = 1.121 quintals/hectare; 1 quintal/hectare = 0.892 bushel/acre
1 bushel/acre = (60#) = 0.6726 quintal/hectare; 1 quintal/hectare = 1.487 bushel/acre (60#)
1 bushel/acre = (56#) = 0.6278 quintal/acre; 1 quintal/acre = 1.597 bushels/acre (56#)
Temperature
To convert Fahrenheit (F) to Celsius (C): $0.555 \times (F - 32)$
To convert Celsius (C) to Fahrenheit (F): $1.8 \times (C + 32)$

Table A-7

A Comparison of the Different Methods of Calculating Interest	
Simple Interest	Interest (I) = Principal X Rate X Time (I = P X R X T)
Discounted Loan	Rate (R) = $\dfrac{\text{Interest (dollar cost)}}{\text{Principal X Time}}$ $\left(R = \dfrac{I}{P \times T}\right)$
Add-On Loan	Rate (R) = $\dfrac{2 \times \text{No. of Payments} \times \text{Interest Charged in \$}}{\text{Beginning Principal} \times \text{Years} \times (\text{No. of Payments} + 1)}$

Appendix B Supervised Agricultural Experience

Supervised Agricultural Experience (SAE) is an integral part of a total program of agricultural education. It is where you plan, propose, conduct, document, and evaluate your programmatic experiential learning activities. It is where you apply and test what you have learned in your classes and experiences from your participation in the National FFA Organization (FFA) events. An SAE will help you become well-versed in recordkeeping for your own portfolio. It will also aid in your personal leadership development and in the strategic planning process for teams, groups, organizations, and programs.

Key Terms

FFA Degree program
Portfolio
Supervised Agricultural Experience (SAE)
Ownership/Entrepreneurship SAE
Placement/Internship
Research SAE
Foundational SAE
School-Based Enterprise SAE
Service-Learning SAE
Improvement Project
Proficiency awards

Learning Objectives

- Define Supervised Agricultural Experience (SAE).
- Differentiate between the Foundational and Immersion SAE.
- Identify the five components of the Foundational SAE.
- Describe the various types of Immersion SAE projects.
- Identify the essential parts of the SAE plan.
- Identify what is needed to successfully conduct and document an SAE project.
- Discuss how each of the SAE projects are evaluated.
- Describe options for enhancing your SAE.
- Identify the benefits of SAE.

SAE Overview

Supervised Agricultural Experience (SAE) is a program of experiential learning activities conducted outside of the regular agricultural education class time. The student-led, instructor-supervised, work-based learning experience is an integral part of a total program of agricultural education

including SAE, FFA, and classroom and laboratory activities. It is designed to help students develop and apply the knowledge and skills learned in an agricultural education classroom.[1] SAE is based upon Agriculture, Food, and Natural Resources (AFNR); technical standards; and career ready practices aligned to your Career Plan of study.

SAE allows you to apply what you have learned in your courses while doing things you love and exploring potential careers. In your SAE, you will practice skills that will lead to success in your career and life in general. SAE is where you plan, propose, conduct, document, and evaluate experiential learning activities. As the *SAE for All* model suggests, students start with a Foundational SAE and then progress to working on one or more Immersion SAEs.

The *SAE for All* model illustrates the path you will take in the SAE program, from when you first enroll in agricultural education, to the point at which you choose your Immersion SAE, to when you complete your high school education. Ultimately, the goal of the *SAE for All* program is to prepare you to be ready for the next steps after high school. The *SAE for All* pathway can be viewed at https://SAEforall.org/sae-for-all-program/?wizard.

SAE for All will expose you to a comprehensive set of real-life situations that can be used in any career path you choose. It will also aid in applying your SAE to personal leadership development and in the strategic planning process for teams, groups, organizations, and programs in the world of work. Visit https://saeforall.org to discover tools and resources to support learning.

Foundational SAE—Get It Started

The career-related Foundational SAE is where an SAE begins. The Foundational SAE results in the development of a plan to begin an Immersion SAE. A Foundational SAE is conducted by all students in the agricultural education program, whether they are on a four-year sequence or enrolled for only one or two semesters. Agricultural education teachers assist students with the plan and working through concepts and activities to achieve the following five components of a Foundational SAE:

- Career exploration and planning
- Employability skills for college and career readiness
- Financial management and planning
- Workplace safety
- Agricultural literacy[2]

Immersive SAE—Next Steps

SAE projects can be completed in a variety of categories, always beginning with the Foundational SAE, and then transitioning to usually one of the following based on career plans and interests:

- Ownership/Entrepreneurship
- Placement/Internship
- Research: Experimental, Analysis, or Invention
- School-Based Enterprise
- Service Learning

Ownership/Entrepreneurship—Work for Self

In the **Ownership/Entrepreneurship SAE**, you plan, implement, operate, and assume all or some of the financial risk in a productive or service activity or agriculture, food, or natural resources-related business.[3] If you have this type of SAE, you own the materials and other inputs, and you keep financial records to determine return on investment. The following are some examples of SAE experiential learning activities that can make up an SAE program:

- Grow grapes.
- Raise bees.
- Grow an acre of corn.

Placement/Internship—Work for Someone Else

If you work for someone else on a farm or ranch, in an agricultural business, or in a verified nonprofit organization providing a "learning-by-doing" environment, your SAE qualifies for the **Placement/Internship SAE** type. This type of experience may be paid or nonpaid.[4] Not everyone reading this can have an Ownership/Entrepreneurship SAE, but almost everyone can seek out a paid or unpaid "working for someone else" experience for the purpose of learning and applying your agricultural knowledge and skills. The following are some examples of Placement/Internship SAEs:

- Work in a flower shop.
- Work on Saturdays at a local stable.
- Work at a grocery store.

Research: Experimental, Analysis, or Invention—Solve a Problem

The agriculture industry is becoming a highly scientific field full of problems to be solved by you and your classmates. Problems range from world hunger to climate change and even social ideals and political policies related to the agriculture industry. A **Research SAE** helps you prepare for a long and productive career solving these important issues for our society. The Research SAE involves a program of extensive activities where you plan and conduct experiments or other forms of scientific evaluation using the scientific process. There are three types of Research SAEs: Experimental, Analytical, and Invention.[5]

Experimental Research SAE

According to the National Council for Agricultural Education, the Experimental Research SAE involves an extensive activity where the student plans and conducts a major agricultural experiment using the scientific process. The scientific process is a way of answering scientific questions through observations, making assumptions, experimenting, and drawing conclusions based on your analysis. The purpose of the experiment is to provide students firsthand experience in verifying, learning, or demonstrating scientific principles in agriculture, discovering new knowledge, and using the scientific process. In an Experimental SAE, there is a hypothesis and a control group, and variables are manipulated.[6]

Analytical Research SAE

In an Analytical Research SAE, you choose a real-world agriculture, food, or natural resource-related problem that is not amenable to experimentation and design a plan to investigate and analyze the problem. You will gather and evaluate data from a variety of sources and then produce some type of finished product. The product may include a marketing display or marketing plan for a commodity, product, or service; a series of newspaper articles; a land-use plan for a farm; a detailed landscape design for a community facility; an advertising campaign for an agribusiness, and so on. A student-led Analytical SAE is flexible enough so that it could be used in any type of agricultural class, provides valuable experience, and contributes to the development of critical thinking skills.[7]

Invention Research SAE

In the Invention Research SAE, you identify a need in an agriculture, food, or natural resource-related industry and perform research and analysis to solve a problem or increase efficiency by developing or adapting a new product or service to the industry. You plan, document, and develop your innovation through the iterative processes of design, prototyping, and testing with the goal of creating a marketable product or service.

The following are examples of research SAEs:
- Compare the effect of various plant food on plant growth.
- Demonstrate the impact of different levels of soil acidity on plant growth.
- Determine if different rations of feed improve the growth rate of pigs.

School-Based Enterprise SAE—Manage It at School

The **School-Based Enterprise SAE** is an entrepreneurial operation that you manage. Your operation, however, takes place in a school setting that provides not only facilities, but also goods and services that meet the needs of an identified market.[8] To give you the most educational value, this type of SAE should replicate the real world of work as much as possible. This type of SAE is usually cooperative in nature and management decisions are made by you in cooperation with your teacher. Activities in this type of SAE "may include, but are not limited to: cooperative livestock raising, school gardens and land labs, production greenhouses, school based agricultural research, agricultural equipment fabrication, equipment maintenance services, or a school store."[9]

Service-Learning SAE—Plan a Service Project

Service learning is one of the highest forms of leadership. Our goal should always be to leave things better than we found them. The goal of the **Service-Learning SAE** is to make the community in which you live better. This SAE type combines community service activities with structured reflection.

As part of this type of SAE, you can become involved in the development of a needs assessment, planning goals and objectives, creating budgets, implementing the activity, and promoting and evaluating a chosen project. The project could be in support of a school, community organization, religious institution, or non-profit organization. You would be responsible for raising necessary funds for the project (if funds are needed). The project must stand by itself and not be part of an ongoing chapter project or community fundraiser. The project must be somewhat challenging and require your awesome leadership.[10] The following are examples of Service-Learning SAEs:
- Test water wells in community for contamination.
- Design a web page for your FFA chapter or for a local organization.
- Design and install a landscape plan for a local business.

Plan Your SAE Program

Planning is an important part of any team, business, project, or program. Your SAE program isn't any different. To properly plan, you will need to understand the steps in planning, as well as the guidelines for planning the SAE.

Steps In Planning Your SAE

Step One	Identify your career interests in agriculture. Your SAE program, experiential learning activities, and the career you choose someday needs to be something about which you can get excited. If you like the SAE or your future career choice, you are more likely to stick with it and be successful.
Step Two	Review the job responsibilities of career interest areas you may have. You might like the idea of a particular project or career, but if you don't like specific tasks involved, you might need to choose another project or program. For example, if you like the idea of a Placement SAE as a Veterinarian Technical Assistant, but you don't want to clean out cages, it may not be the best fit for you.
Step Three	Identify SAE programs of interest by interviewing friends who have an SAE or by viewing suggestions found in various resources.
Step Four	Develop a timeline for your SAE program. In other words, which projects will happen first, second, third, and so on? What is the completion date for each activity? Electronic portfolios, recordkeeping systems, or simple calendar systems can be used for this step.
Step Five	Building on step four, design a long-range plan for the SAE program. Remember, projects or activities happen in a shorter time span, but an SAE program, which is necessary to be competitive for FFA proficiency and Star awards, happens over a longer frame and grows in scope and diversity.
Step Six	Develop the first-year (annual) plan, starting now. After you know the long-range plan and the timeline for the different experiential learning activities/projects, you are ready to put the first-year plan together. Again, think about completion dates and specific strategies for reaching those goals.
Step Seven	Replan on a regular basis. Part of good planning is reviewing your activities in light of your plan and then adjusting your plan. This should happen fairly often, but not so often that you are always planning instead of completing tasks.[11]

Guidelines to Planning SAE

The steps outlined previously should get you moving in the right direction, but the following guidelines will help ensure a successful SAE:

- Plan for year-round experiences.
- Include ownership and/or placement projects either at home, at school, or in the community.
- Identify a number of improvement projects and supplementary skills for the year.
- Develop a budget.
- Plan ownership projects with some form of profit.
- Explore different locations to gain desired experiences.
- Discuss the SAE program with parents or guardians.
- Plan the scope of the program to earn enough profit to qualify for advanced FFA degrees.
- Provide for a variety of activities and experiences.
- Increase the SAE scope annually.
- Choose SAE program experiences that relate to career interest areas.[12]

Parts of an Annual SAE Plan

A yearlong SAE program plan will keep you on track and help you make decisions about your SAE. The annual plan consists of a calendar, description of projects, budget, improvement projects, and supplementary skills. For entrepreneurship/ownership projects, prepare a description of the size/scope of your enterprise, the location of your projects, the nature of the business or enterprise, partners involved, methods of marketing, facilities needed, and months involved in the project. For placement projects, detail the location, beginning and ending dates, and pay schedule. It is also important to include a budget and to keep up with income and receipts as the year progresses. You will also want to include a description of improvement projects you wish to complete and supplemental skills you would like to develop. An **improvement project** is an activity that improves the appearance, convenience, efficiency, safety, or value of a home, farm, ranch, agribusiness, or other agriculture facility. Your annual plan should include specific activities of the improvement project as well as hours of labor committed and an estimate of costs.[13]

Propose A SAE Program

Following your SAE plan, it is a good practice to write a statement justifying your plans with a proposal. This proposal should explain why you chose the SAE you did as well as discuss why your plan is laid out the way it is. Your proposal should reiterate the importance of completion dates on your timeline. The SAE proposal should also detail specific agricultural or leadership concepts that you hope to achieve because of your SAE program.

Conduct Your SAE Program

The first step to conducting a quality SAE program is getting started. All of the planning and decision-making can take time, but it is worth it. Once you've started, you are going to have a great time, but you have some work ahead of you as well. Specifically, you will need to become skilled at documenting your SAE program as well as evaluating it.

Document Your SAE Program

Documenting your SAE is important because it gives you an opportunity to see the planning come to life. Documenting your SAE makes a permanent record from which you can learn and improve your project. It shows the problems encountered and solved, which applies to future activities and provides us with confidence to continue in our SAE.

Calendar

Using a calendar can help with more than planning. You can use a calendar, such as the one on your smartphone, to document hours invested, skills learned, and even tasks performed.

Journal or Portfolio

An alternative to using a simple calendar system is to document your SAE with a journal or portfolio, paper or electronic. A paper system could be kept using a notebook or ledger, but an electronic system is much easier to use. Many states have an online journal or portfolio that allows you to keep up with important documentation. You will want to check with your teacher on the preferred method of documentation for your school.

A **journal** is a written record of experiences and observations. It may also be a ledger in which transactions have been recorded as they occurred.[14] A **portfolio**, usually kept online, houses all of your work (skills, projects, activities, service, finances, and so on) compiled over a period of time. It is also advisable to utilize your portfolio for recordkeeping as well.

Apply Proper Recordkeeping Skills

Recordkeeping is the process of keeping a journal or portfolio of everything you have done. As stated previously, you will need to make notes in your SAE whenever you do or learn something new, and you need to document time and money spent on your activities, projects, and program. The skill of recordkeeping will serve you well in your career.[15]

At its core, recordkeeping is quite simple. For every project, you will need to capture the following for all activities or items: the date, the name or description, the hours you spent, and expenses or income that resulted. Be sure to add up the hours you have invested in all activities. Time is money, and each one of your experiences represents new knowledge and skills you've developed. You will also add up expenses and income to determine your profit and loss (P&L). You can use state-endorsed, pre-programmed spreadsheets or websites to determine your P&L and to track hours invested or keep records on paper. It is up to your personal preference, available resources, and the wishes of your advisor or business partner.

Learning to keep records for and through your SAE has many benefits. Keeping accurate records can help you determine if you made or lost money. It can also help ensure that you are fairly compensated with each transaction. Your knowledge and experience in recordkeeping will help you determine which parts of the business are doing well and which parts are not. Becoming proficient at recordkeeping will also help you make management decisions, document your net worth for loans, prepare tax returns, and plan for future events. Being a good recordkeeper for your SAE will also help document your activities for FFA recognitions and degree purposes. Good records can also protect you legally and help you plan a budget for the following year.[16]

Evaluate Your SAE Program

When your teacher or adviser comes to visit you and your SAE, they are there to answer any questions you may have. The teacher is also there to help you evaluate your SAE. The evaluation looks different for different types of SAEs. Lesson 8 from the SAE Handbook resource available online (http://harvest.cals.ncsu.edu/site/WebFile/IIB8.pdf) breaks down specific components of SAE evaluation.

Entrepreneurship/Ownership

The following are evaluation components for an Entrepreneurship/Ownership SAE:
- Accuracy of records
- Neatness of records
- Dates of records
- Net income (total income minus total expenses)
- Are good management practices being used?
- Efficiency factors (yield per acre, number of offspring raised, etc.)

- Improvements made since the last observation
- Cleanliness of facilities
- Customer satisfaction
- What skills were learned

Placement/Internship

The following are evaluation components for a Placement/Internship SAE:

- Satisfaction of the employer
- Number of hours worked
- Accuracy of records
- Neatness of records
- Currency of records
- Does student report to work on time?
- Ability to get along with the other workers
- Ability to get along with customers
- Skill in performing the expected tasks
- Attitude of the student

Research

The following are evaluation components for a Research SAE:

- Was the scientific process used?
- How well was the job done?
- What was learned?
- What was the reaction of the beneficiaries of the service-learning project?
- Were records kept on the activity, and what shape are they in?

Exploratory

The following are evaluation components for an Exploratory SAE:

- How many different activities were conducted?
- How many hours were involved?
- What was learned?
- Accuracy of records
- Neatness of records
- Dates of records

School-Based Enterprise

The following are evaluation components for a School-Based Enterprise SAE:

- How many hours were worked?
- What was the quality of the work done?
- What was the scope or size of the activity?
- Accuracy of records
- Neatness of records
- Dates of records

Service Learning

The following are evaluation components for a Service-Learning SAE:

- How many hours were involved?
- How well was the job done?
- What was learned?
- What was the reaction of the beneficiaries of the service-learning project?
- Were records kept on the activity, and what shape are they in?[17]

Enhancing Your SAE With Other Opportunities

SAE in not only financial recordkeeping, but also a record of skills, knowledge, credentials, certifications, experiences, career planning, reflection, and leadership development.[18] A solid SAE will provide you with the opportunity to receive FFA degrees and compete for different proficiency awards.

The **FFA degree program** is FFA's primary recognition program. SAE helps students apply for each degree, and every degree has certain SAE requirements. The following are the five degrees of FFA membership: Discovery, Greenhand, Chapter, State, and American. These five degree areas recognize you for your overall participation in FFA. Everything you do in FFA, when combined, helps you move toward a degree.[19]

Proficiency awards are awards that recognize students' excellence in SAE.[20] Developing your SAE into a proficiency award is a time-consuming but rewarding task. You should apply for the proficiency award in the specific area and career pathway for which you are strongest and have the most experience. For instance, if you have worked for a livestock producer for three years and only raised goats for one year, it would be best to apply in the placement area for Animal Systems rather than Entrepreneurship for the same system. The career pathways for your SAE and proficiency awards are as follows:

- **Agribusiness Systems**: The study of business principles—including management, marketing, and finance—and their application to enterprises in agriculture, food, and natural resources
- **Agricultural Education:** The study of agricultural education and extension, focusing on the education of youth and the community in specific areas of agriculture.

- **Animal Systems:** The study of animal systems, including life processes, health, nutrition, genetics, management, and processing, through the study of small animals, aquaculture, livestock, dairy, horses or poultry.
- **Biotechnology Systems:** The study of data and techniques of applied science for the solution of problems concerning living organisms.
- **Environmental Service Systems:** The study of systems, instruments, and technology used in waste management and their influence on the environment.
- **Food Products and Processing Systems:** The study of product development, quality assurance, food safety, production, sales and service, regulation and compliance, and food service within the food science industry.
- **Natural Resource Systems:** The study of the management of soil, water, wildlife, forest, and air as natural resources.
- **Plant Systems:** The study of plant life cycles, classifications, functions, and practices through the study of crops, turfgrass, trees and shrubs, or ornamental plants.
- **Power, Structural, and Technical Systems:** The study of agricultural equipment, power systems, alternative fuel sources, and precision technology, as well as woodworking, metalworking, welding, and project planning for agricultural structures.[21]

Agribusiness Systems Project Examples

If you are interested in pursuing the Agribusiness Systems Career Pathway, there are many project ideas that you can explore as you select your Immersion SAE. The following project examples can serve as a starting point as you consider what you are passionate about and what brings you a sense of fulfillment. Good luck!

Placement/Internship Projects

- Work as an agricultural news consultant for local radio or newspapers.
- Work in food distribution or the restaurant industry.
- Work with a local florist.
- Job shadow agribusiness professionals, visit agribusinesses to interview personnel, take educational tours, etc.

Ownership/Entrepreneurship Projects

- Provide a hand-weeding crew for local peanut/vegetable farmers.
- Start a farm-sitting business for vacationing farmers.
- Form a cooperative with other students and share in the profits of a greenhouse crop.
- Start a recycling business (collecting newspapers and plastics to sell to recycling plants).
- Purchase and resell aerial photographs from the tax office to local landowners.
- Provide a custom barbecue service for your community.
- Provide a co-op program for an agribusiness.
- Write news articles on agriculture or FFA for local newspapers.
- Start a company to promote agricultural businesses (sell custom caps and t-shirts with farm or ag business names or logos).
- Provide a farm sign business (manufacture, sell, install, and maintain).
- Start an agricultural photography service and photograph animals, show animals, equipment, barns, families, or children with animals.
- Contract with the local Chamber of Commerce to conduct county tours for prospective businesses.

Research Projects

- Conduct a study of input costs over time for a given agricultural business.
- Conduct a study of farmer buying habits over a certain amount of time.
- Conduct a study to determine challenges farmers face when adapting to new technology.
- Examine the impact of fuel costs for agricultural trucking agencies.
- Survey local producers about the variables that make the most difference in profit and loss on their operation.

School-Based Enterprise Projects

- Start a local farm produce newspaper and sell ads to farmers.
- Write "how to" pamphlets to sell at local garden supply stores (e.g., How to Grow Tomatoes, etc.).
- Develop a weekly podcast where local farmers and agribusinesses can come to your school and be interviewed about their crops and products.

- Plan, grow, market, and sell plants through the school greenhouse.

Service-Learning Projects

- Provide a small-engine maintenance and repair service.
- Provide systematic maintenance and service on outdoor power equipment at customers' homes or at school-provided facilities.
- Offer a custom-parts or custom-supplies delivery business to farms in your county.
- Remove pesticide jugs monthly from farms and transport to landfill.
- Start a kerosene route for homeowners (probably little demand in the summertime).

Other Project examples

Agribusiness Systems is just one career pathway within the Agriculture, Food, or Natural Resources cluster. The other pathways are Agricultural Education; Animal Systems; Biotechnology Systems; Environmental Service Systems; Food Products and Processing Systems; Plant Systems; and Power, Structural, and Technical Systems. Project examples for each of these career pathways are available on MindTap.

Summary

Supervised Agricultural Experience (SAE) is a program of experiential learning activities conducted outside of the regular agricultural education class time. There are different types of SAEs and related opportunities and rewards programs, so every student in agricultural education and FFA can learn and develop from the authentic experience. Recordkeeping and documentation of the valuable experiences are very important and will serve you and your future career well.

1. *Supervised Agricultural Experience, SAE for All: Student Guide.* The National Council for Agricultural Education, 2017, p. 2. ffa.app.box.com/s/m572qkpzzbdb9ew04haovv434k33pb9o
2. Ibid., p. 6.
3. "Supervised Agricultural Experience." The National Council for Agricultural Education, www.ffa.org/about/supervised-agricultural-experiences Accessed February 18, 2016.
4. Ibid.
5. Ibid.
6. *Philosophy and Guiding Principles for Execution of the Supervised Agricultural Experience Component of the Total School Based Agricultural Education Program.* National Council for Agricultural Education, March 31, 2015, pp. 2–3. ffa.app.box.com/s/i8ntesw8zsajaxxdnj5cle6zaf0a6za3 Accessed February 18, 2016.
7. Ibid, p. 3.
8. Ibid.
9. Ibid.
10. Ibid, pp. 3–4.
11. "Getting Started in an SAE Program." *Agricultural Core Curriculum, Lesson 612a.* California Agricultural Education, www.calaged.org/sites/default/files/resources/64833_Supervised%20Agriculture%20Experiences.zip Accessed February 20, 2016.
12. Ibid, p. 3.
13. Ibid, p. 4.
14. National FFA Organization, ed. "What Is a Journal and How Do We Prepare One?" *SAE Handbook, Lesson RK.3*, p. 2. harvest.cals.ncsu.edu/site/WebFile/lp3.pdf Accessed February 22, 2016.
15. National FFA Organization, ed. "Record Keeping." *Life Knowledge, Lesson MS.69*, p. 3. harvest.cals.ncsu.edu/site/WebFile/MS69.PDF, Accessed February 22, 2016.
16. National FFA Organization, ed. "Why Do We Keep Records?" *SAE Handbook, Lesson 7*, p. 4. harvest.cals.ncsu.edu/site/WebFile/IIB7.pdf Accessed February 23, 2016.
17. National FFA Organization, ed. "How Will My SAE Be Evaluated?" *SAE Handbook, Lesson 8*, p. 3. harvest.cals.ncsu.edu/site/WebFile/IIB8.pdf Accessed February 23, 2016.
18. *Philosophy and Guiding Principles*, p. 4.
19. National FFA Organization, ed. "Proficiency Awards and SAE." *Life Knowledge, Lesson MS.70*, p. 3. harvest.cals.ncsu.edu/site/WebFile/MS70.PDF Accessed February 23, 2016.
20. Ibid, p. 3.
21. National FFA Organization, ed. *National FFA Agricultural Proficiency Awards: A Special Project of the National FFA Foundation*, 2016, p. 11. ffa.app.box.com/v/Library/file/289984566159 Accessed February 23, 2016.

Appendix C: Ensuring Safety in SAE Lab Work

Guidelines for Safety in the Laboratory

The following rules must be observed to ensure your safety and the safety of others in the agricultural education classroom and laboratory.

1. Your concern for safety should begin even before the first laboratory or shop activity. Always read and think about safety before starting an activity.
2. Perform laboratory work only when your teacher is present. Unauthorized or unsupervised laboratory experimenting is not allowed.
3. Know the location and use of all safety equipment in your laboratory. These should include the eye wash station, first-aid kit, fire extinguisher, and safety shower (if available).
4. Wear safety glasses or goggles at all times in the wood and metal laboratory. Wear protective clothing along with ear-and-eye protection accessories at all times as directed by the instructor. Tie back loose hair and secure loose clothing.
5. Check chemical labels twice to make sure you have the correct substance. Pay attention to the hazard classification shown on the label.
6. Avoid unnecessary movement and talk in the laboratory.
7. Gum, food, or drinks should not be brought into the laboratory.
8. Never sniff chemicals, and do not place your nose near the opening of a chemical container.
9. Any laboratory accident, however small, should be reported immediately to your teacher.
10. In case of a chemical spill on your skin or clothing, rinse the affected area with plenty of water. If the eyes are affected, water-washing must begin immediately and continue for 10 to 15 minutes or until professional assistance is obtained.
11. When discarding used chemicals or hazardous materials, carefully follow the instructions provided.
12. Know where the Material Safety Data Sheet forms are located and learn how to use them.
13. No horseplay, running, and so on are allowed at any time in the laboratory/shop or classroom.
14. Do not use equipment unless you have been instructed on its use and safety precautions and have been given permission to do so.
15. Keep all tools stored in their correct locations while not in use.
16. Make sure all safety devices, guards, and other items of protective equipment are in place and working at all times.
17. Make sure compressed gas cylinders are kept chained to a wall or on a secure rack with caps tightly on.
18. Before making adjustments on any piece of equipment, make sure it is turned off and unplugged.
19. Keep your work area clean and clear of hazards.
20. If water has been spilled on the floor, clean it up immediately. If a hazardous chemical has been spilled, notify your teacher but do not attempt to clean it up until you are told to do so and advised of the proper procedures.
21. Do not put hands and fingers closer than 2 inches to a blade while using saws, drills, and other similar equipment.
22. Know where all tools, chemicals, and equipment are located in the shop.
23. Always use the proper tool for the job.
24. Do not stand closer than 6 feet to power equipment being operated by someone else.

25. Recognize the color coding for shops, and observe all regulations regarding color coding:

Safety RED	=	DANGER
Safety ORANGE	=	WARNING
Safety YELLOW	=	CAUTION
Safety BLUE	=	INFORMATION
Safety GREEN	=	SAFETY
Safety WHITE	=	TRAFFIC MARKINGS

Portions of Appendix B were adapted from the Georgia Agricultural Education Curriculum Guide, Atlanta, GA.

26. Lift heavy objects safely using proper lifting techniques.
27. Never throw anything in the classroom or laboratory.
28. Take responsibility for the safety of yourself and others. Follow tool and equipment safety manuals. Obey rules and guidelines as well as the instructions of your teacher. Use good judgment for safety's sake. If in doubt, ask!

Scientists use protective equipment for their personal safety and to avoid contaminating samples.
Courtesy of USDA.

Appendix D Building Birdhouses for SAE

Construction Project: Flicker or Woodpecker Houses

Plans

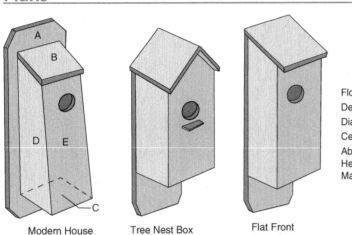

Modern House   Tree Nest Box   Flat Front Nest Box

DIMENSIONS

Floor of House:	7" · 7 ½"
Depth of House:	7 ½"
Diameter of Hole:	2½"
Center of Hole	14½" (15¼" Outside)
Above Floor:	18' to 20'
Height Above Ground:	¾"
Material	

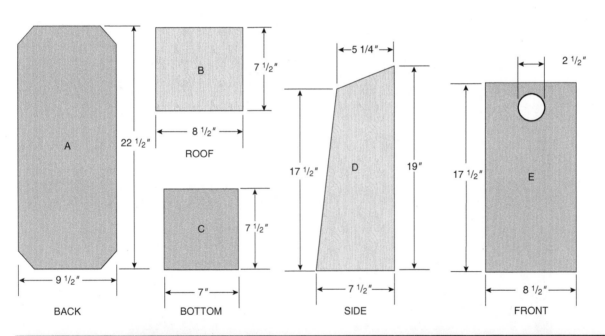

BACK   BOTTOM   SIDE   FRONT

Bill of Materials (for Modern House)			
Materials Needed	Quantity	Dimensions	Description or Use
Lumber (rough and rustic)	1	3/4" × 8–1/2" × 17–1/2"	Front (A)
"	2	3/4" × 7–1/2" × 19"	Sides (B)
"	1	3/4" × 7" × 7–1/2"	Bottom (C)
"	1	3/4" × 7–1/2" × 8–1/2"	Roof (D)
"	1	3/4" × 9–1/2" × 22–1/2"	Back (E)
Box nails or finishing nails		4d or 6d	Attach

The flicker will nest readily in a well-placed birdhouse of the proper dimensions. Roughen the interior of the nest box to assist the young woodpeckers to reach the entrance hole. Cover the bottom with sawdust so that the mother bird can shape the nest for eggs and young birds. Material may be one-half-inch instead of three-fourths-inch thick, in which case the width of the front and roof will be 8 inches.

Construction Procedure

1. Select pine, redwood, or other lumber that will withstand weather without paint. Rough lumber is generally preferred and will provide a rustic appearance. Planed lumber should be used if the house is to be painted for decorative purposes, but this is not recommended for most species except Martin houses.
2. Cut all parts from a 3/4"×9–1/2"×8' board.
3. Cut the sides to the proper angle.
4. Bevel the top edge of the front to conform to the angle of the sides.
5. Bevel the upper edge of the roof to conform to the back and sides.
6. Nail the front, sides, and bottom together.
7. Nail the back to the sides and bottom. (Center the assembly on the back for a good appearance. Leave 1 inch at the bottom and 2.5 inches at the top.)
8. Attach the roof with two hinges or nail it lightly so that the roof is easily removed for periodic cleaning of the nest area.

Construction Project: Nesting and Den Boxes

Plans

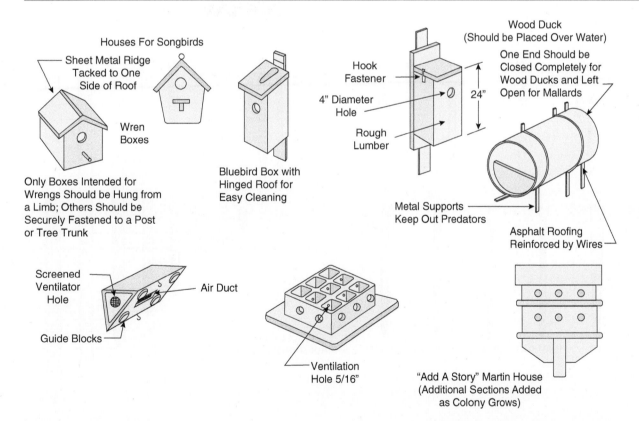

Note: Bills of materials are not provided with these nesting and den boxes because of the variety of types pictured. Nesting and den boxes should be planned from the information provided. An appropriate bill of materials may then be developed for the plan.

Characteristics of Common Woods

Species	Hardness	Known For	Some Major Uses
Birch	Hard	Surface veneer for panels	Cabinets and doors
Cedar, red	Medium	Pleasant odor	Furniture, chests, and birdhouses
Cherry	Hard	Red grain	Fine furniture
Cypress	Medium	Rot resistance	Structural material in wet places, birdhouses
Fir and Hemlock	Soft	Light, straight, strong	Construction framing, siding, sheathing
Locust, black	Hard	Rot resistance	Fence posts, birdhouses
Maple	Hard	Light grain	Floors, bowling alleys, durable furniture
Mahogany	Medium	Reddish color	Fine furniture
Oak	Hard	Toughness, strength	Floors, barrels, wagon bodies, feeders, farm buildings
Pine, yellow	Medium	Wear resistance, tough	Floors, stairs, trim
Pine, white	Soft	Easy to work, straight	Shelving, siding, trim
Redwood	Soft	Excellent rot resistance	Yard posts, fences, birdhouses
Walnut, black	Hard	Brown grain	Fine furniture
Willow, black	Soft	Brown grain, easy to work, walnut look	Furniture

Types of Nails

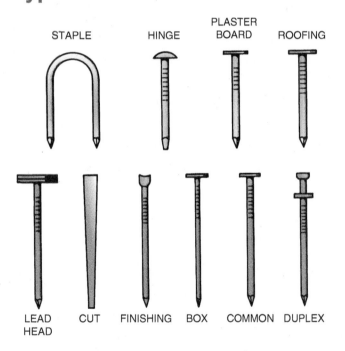

STAPLE, HINGE, PLASTER BOARD, ROOFING, LEAD HEAD, CUT, FINISHING, BOX, COMMON, DUPLEX

Length of Nails*

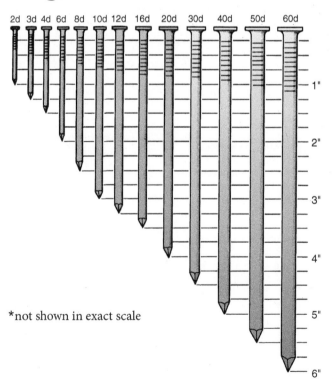

2d 3d 4d 6d 8d 10d 12d 16d 20d 30d 40d 50d 60d

*not shown in exact scale

Gauges and Types of Wood Screws*

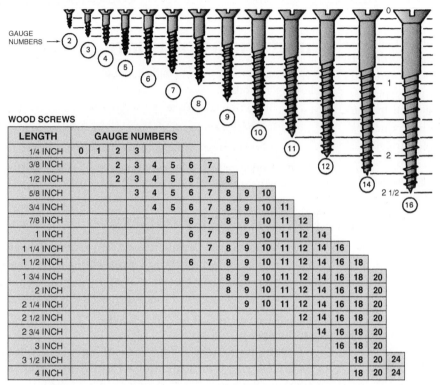

WOOD SCREWS

LENGTH	GAUGE NUMBERS														
1/4 INCH	0	1	2	3											
3/8 INCH			2	3	4	5	6	7							
1/2 INCH			2	3	4	5	6	7	8						
5/8 INCH				3	4	5	6	7	8	9	10				
3/4 INCH					4	5	6	7	8	9	10	11			
7/8 INCH							6	7	8	9	10	11	12		
1 INCH							6	7	8	9	10	11	12	14	
1 1/4 INCH								7	8	9	10	11	12	14	16
1 1/2 INCH							6	7	8	9	10	11	12	14	16	18
1 3/4 INCH									8	9	10	11	12	14	16	18	20
2 INCH									8	9	10	11	12	14	16	18	20
2 1/4 INCH										9	10	11	12	14	16	18	20
2 1/2 INCH													12	14	16	18	20
2 3/4 INCH														14	16	18	20
3 INCH															16	18	20
3 1/2 INCH																18	20	24
4 INCH																18	20	24

When You Buy Screws Specify (1) Length, (2) Gauge Number, (3) Type of Head–Flat, Round, or Oval, (4) Material–Steel, Brass, Bronze, etc., (5) Finish–Bright, Steel Blued, Cadmium, Nickel, or Chromium Plated.

*not shown in exact scale

Appendix E Developing a Personal Budget

One of the most important parts of becoming an adult is creating a budget to handle your personal finances. A budget is a plan of action that includes projections of income and expenses for all or part of business or personal expenses. The process in this appendix is designed to help you understand a personal budget. Many of the same principles can be used when developing your supervised agricultural experience (SAE).

There are several reasons for developing and using a budget, including that it

- helps you plan for the lifespan of assets
- is an excellent device for organizing
- is useful to obtain credit
- allows experimenting with different outcomes
- identifies costs and income
- helps refine and organize
- is a good management tool for anyone with limited experience

Limitations of Budgets

People sometimes hesitate to budget, or fail to budget, for the following reasons:

- Budgeting takes time.
- It is necessary to search for accurate information.
- It is sometimes difficult to predict future actions.
- Managers tend to underestimate costs and overestimate income.

The Budget Process

Budgets do not need to be complicated. The process can be as simple as determining, listing, and summarizing all of your anticipated expenses and income. Obviously, the goal should be for your income to exceed your expenses. If this is not the case, you need to change the level of either your income or your expenses.

A simple worksheet follows that may help you organize a budget. List your estimated costs for the items that apply to you or your business.

Developing a personal budget leads to lifelong rewards.
© Joke_Phatrapong/Shutterstock.com

MONTHLY EXPENSES

Home:	Rent/Own	_____
	Maintenance	_____
Utilities:	Phone/Internet	_____
	Electricity	_____
	Cable	_____
	Gas	_____
Insurance:	Home	_____
	Vehicle	_____
	Life	_____
	Health	_____
	Disability	_____
Vehicle:	Payment	_____
	Gas	_____
	Oil/Tires/Maintenance	_____

Food: _____
Clothing: _____
Savings: _____
Entertainment/Recreation: _____
Emergencies: _____
Cleaning/Toiletry/Beauty: _____
Gifts/Contributions: _____
Medical Care: _____
Installment Payments: Credit Cards _____
                              Loans _____
                              Retail shopping _____

MONTHLY INCOME

Salary: _____
Side jobs: _____
Gifts: _____
Sales: _____
Other Income: _____
Total Income: _____
BALANCE: *(income minus expenses)*: _____

Portions of Appendix A were adapted from the Georgia Agricultural Education Curriculum Guide, Atlanta, GA.

References

"Alfalfa Weevil." Alfalfa Weevil | Integrated Crop Management, 2022. https://crops.extension.iastate.edu/encyclopedia/alfalfa-weevil.

"Alfalfa Weevil - [Fact Sheet]." Penn State College of Agricultural Sciences. https://agresearch.montana.edu/wtarc/producerinfo/entomology-insect-ecology/AlfalfaWeevil/PennStateFactSheet.pdf."Why Animal Research?" Animal Research at Stanford, 2021. https://med.stanford.edu/animalresearch/why-animal-research.html.

"Aquaculture and Fisheries Biotechnology." University of Maryland Baltimore County Department of Marine Biotechnology (UMBC), 2020. https://marinebiotechnology.umbc.edu/aquaculture-and-fisheries-biotechnology/.

ASKUSDA, January 15, 2022. https://ask.usda.gov/s/article/How-much-meat-and-poultry-do-Americans-eat-yearly.

Benjamin Pinguet "The Role of Drone Technology in Sustainable Agriculture." Global Ag Tech Initiative, May 25, 2021. https://www.globalagtechinitiative.com/in-field-technologies/drones-uavs/the-role-of-drone-technology-in-sustainable-agriculture/.

California Agricultural Experiment Station Circular, 347. Intergovernmental Panel on Climate Change, 1996.

"Climate Change 2021: The Physical Science Basis." Working Group I Contribution to the IPCC Sixth Assessment Report. Cambridge University Press, April 2021. https://www.ipcc.ch/report/ar6/wg1/.

Dessler, Andrew. *Introduction to Modern Climate Change.* Cambridge: Cambridge University Press, 2022.

"Euonymus Scale: A Constant Threat for Ornamental Plantings." Penn State Extension, April 25, 2021. https://extension.psu.edu/euonymus-scale-a-constant-threat-for-ornamental-plantings.

"Farm Share of U.S. Food Dollar Rose One Cent in 2020, Largest Increase in Nearly a Decade, as Food-at-Home Spending Increased." USDA ERS - Chart Detail, March 21, 2022. https://www.ers.usda.gov/data-products/chart-gallery/gallery/chart-detail/?chartId=103547.

"Field Identification of Citrus Blight." Food and Agriculture Organization of the United Nations, 2021. https://www.fao.org/state-of-food-agriculture/en/.

Freese, Betsy. "10 Predictions for the Future of Gene Editing in Livestock." Successful Farming. Successful Farming, November 29, 2018. https://www.agriculture.com/livestock/10-predictions-for-the-future-of-gene-editing-in-livestock.

Futch, Stephen H., Kenneth S. Derrick, and Ronald H. Briansky. "Field identification of citrus blight." University of Florida, October 26, 2020. https://edis.ifas.ufl.edu/publication/HS241.

Green, Jessica. "Understanding the Genetics behind Rabbit Coat Colors: Part 1 - Introduction." OSU Extension Service. Oregon State University Extension Service, June 10, 2022. https://extension.oregonstate.edu/animals-livestock/poultry-rabbits/understanding-genetics-behind-rabbit-coat-colors-part-1.

Gross, Kevin C., Chien Yi Wang, and Mikal Saltveit. "The Commercial Storage of Fruits, Vegetables, and Florist and Nursery Stocks," Revised February 2016. https://www.ars.usda.gov/ARSUserFiles/oc/np/CommercialStorage/CommercialStorage.pdf.

Haas, Michael J. "New Gene Could Help Improve Tomato Flavor and Shelf Life." Cornell Chronicle, October 19, 2021. https://news.cornell.edu/stories/2021/10/new-gene-could-help-improve-tomato-flavor-and-shelf-life.

Hansen, Maureen. "Dairying in the Next Half-Century" Dairy Herd, January 19, 2022. https://www.dairyherd.com/news/business/dairying-next-half-century.

Hershey, David R. "Solution Culture Hydroponics: History & Inexpensive Equipment." University of California Press. University of California Press, February 1, 1994. https://online.ucpress.edu/abt/article/56/2/111/15156/Solution-Culture-Hydroponics-History-amp.

"History of FCA." History of FCA | Farm Credit Administration. https://www.fca.gov/about/history-of-fca.

"Horses Returning to the U.S." USDA APHIS. https://www.aphis.usda.gov/aphis/ourfocus/importexport/animal-import-and-export/equine/returning-to-the-us.

"Iacuc Committees." Foundation for Biomedical Research, July 14, 2020. https://fbresearch.org/animal-care/animal-testing-regulations/iacuc/.

"Impact of Technology on Agriculture." National Geographic Society. https://education.nationalgeographic.org/resource/impact-technology-agriculture.

"Imported Fire Ants." USDA APHIS. https://www.aphis.usda.gov/aphis/ourfocus/planthealth/plant-pest-and-disease-programs/pests-and-diseases/imported-fire-ants/CT_Imported_Fire_Ants.

"'Larger than Usual': This Year's Ozone Layer Hole Bigger Than Antarctica." *The Guardian*, September 15, 2021. https://www.theguardian.com/environment/2021/sep/16/larger-than-usual-ozone-layer-hole-bigger-than-antarctica.

"Latest Issue: The State of Food and Agriculture (Sofa) 2021." FAO. https://www.fao.org/publications/sofa/sofa-2021/en/.

"List of Pests of Significant Public Health Importance." Environmental Protection Agency, December 21, 2021. https://www.epa.gov/insect-repellents/list-pests-significant-public-health-importance.

Morgan, Lynette. "Background and History of Hydroponics and Protected Cultivation." *Hydroponics and protected cultivation: A practical guide*, 2021, 1–10. https://doi.org/10.1079/9781789244830.0001.

"NASA Ozone Watch: Latest Status of Ozone." National Aeronautics and Space Administration (NASA), October 18, 2018. https://ozonewatch.gsfc.nasa.gov/meteorology/annual_data.html.

"National Bioengineered Food Disclosure Standard." Federal Register, December 21, 2018. https://www.federalregister.gov/documents/2018/12/21/2018-27283/national-bioengineered-food-disclosure-standard.

"New USDA Report Finds Fewer People Faced Hunger in 2019." Feeding America, September 10, 2020. https://www.feedingamerica.org/about-us/press-room/new-usda-report-finds-fewer-people-faced-hunger-2019.

"Pocket K No. 18: Ethics and Agricultural Biotechnology." International Service for the Acquisition of Agri-biotech Applications (ISAAA), 2022. https://www.isaaa.org/resources/publications/pocketk/18/default.asp.

Sass, Jennifer. "Busy as a Bee: Pollinators Put Food on the Table." NRDC, June 3, 2019. https://www.nrdc.org/policy-library/busy-bee-pollinators-put-food-table.

"The Sources and Solutions: Agriculture." Environmental Protection Agency, November 4, 2021. https://www.epa.gov/nutrientpollution/sources-and-solutions-agriculture.

Sustainable Management of Food . Environmental Protection Agency, March 8, 2022. https://www.epa.gov/sustainable-management-food.

"Understanding Epds and Genomic Testing in Beef Cattle." Penn State Extension, May 5, 2020. https://extension.psu.edu/understanding-epds-and-genomic-testing-in-beef-cattle.

"United in Science 2022: A Multi-Organization High-Level Compilation of ..." WMO, UNEP, GCP, UK Met Office, IPCC, UNDRR, September 13, 2022. https://www.undrr.org/publication/united-science-2022-multi-organization-high-level-compilation-latest-climate-science.

"U.S. Dairy Consumption Beats Expectations in 2020 and Continues to Surge Upward despite Disruption Caused by Pandemic." IDFA, October 5, 2021. https://www.idfa.org/news/us-dairy-consumption-beats-expectations-in-2020-and-continues-to-surge-upward-despitedisruption-caused-by-pandemic.

"USDA Animal Care." United States Department of Agriculture (USDA), September 9, 2022. https://www.aphis.usda.gov/aphis/ourfocus/animalwelfare/usda-animal-care-overview.

"USDA Map of Africanized Honey Bee Spread Updated." United States Department of Agriculture (USDA), February 2017. https://www.ars.usda.gov/news-events/news/research-news/2007/usda-map-of-africanized-honey-bee-spread-updated/.

"What Work Requires of Schools. A Scans Report for America 2000." U.S. Department of Labor, May 31, 2017. https://eric.ed.gov/?id=ED332054.

"Wool Profile." Agricultural Marketing Resource Center, November 2021. https://www.agmrc.org/commodities-products/livestock/lamb/wool-profile.

"World Malaria Report 2021." World Health Organization, December 6, 2021. https://www.who.int/teams/global-malaria-programme/reports/world-malaria-report-2021.

Zhou, Xue, Won Suk Lee, Yiannis Ampatzidis, Yang Chen, Natalia Peres, and Clyde Fraisse. "Strawberry Maturity Classification from UAV and Near-Ground Imaging Using Deep Learning." Smart Agricultural Technology. Elsevier, September 16, 2021. https://www.sciencedirect.com/science/article/pii/S2772375521000010.

Glossary / Glosario

A

A horizon layer near the soil surface consisting of mineral and organic matter.

abiotic nonliving disease.

abortion loss of a fetus before it is viable.

accent distinctive feature or quality.

accent color attention-getting color.

acid pH of less than 7.0.

acidity sourness or the tendency toward a low pH, less than 7.0.

active ingredient a component that achieves one or more purposes of the mixture.

actuating putting a plan into action.

acute toxicity a measurement of the immediate effects of a single exposure to a chemical.

adaptation a process that occurs as heritable traits favoring the survival of an organism are passed from one generation to the next.

add-on loan the method used for calculating interest on consumer loans. The loan is repaid in installments of equal payments. Interest is added on at the beginning of the payments.

adenine (A) a base in genes designated by the letter *A*.

adjourn a motion used to close a meeting.

adventitious root root other than the primary root or a branch of a primary root.

aeration the mixing of air into water or soil to improve the oxygen supply of plants and other organisms.

horizonte A capa cerca de la superficie que consta de materia mineral y orgánica.

abiótica enfermedad causada por factores no vivientes.

aborto pérdida de un feto antes de que sea viable.

acento rasgo o calidad distintiva.

color acentuado color llamativo.

ácido pH inferior a 7.0.

ácidez sabor ácido o la tendencia a tener un pH bajo, inferior a 7.0.

ingrediente activo componente que alcanza uno o más de los objetivos de la mezcla.

accionar poner un plan en acción.

toxidad aguda una medida de los efectos inmediatos de una sola exposición a una sustancia química.

adaptación un proceso que ocurre a medida que las características que favorecen la supervivencia de un organismo se pasan de una generación a la siguiente.

préstamo adicional el método que se usa para calcular el interés sobre los préstamos al consumidor. El préstamo se paga a plazos en cuotas iguales. El interés se agrega al comienzo de los pagos.

adenino (A) una base en los genes designado con la letra «A».

aplazar una moción que se usa para levantar una reunión.

raíz adventicia una raíz diferente de la raíz fundamental o una rama de la raíz fundamental.

airear mezclar el aire con agua o tierra para mejorar el suministro de oxígeno a las plantas y otros organismos.

aeroponics the growing of plants in which their roots hang in the air and are misted regularly with a nutrient solution.

agar a plant nutrient medium in which plant tissues are placed during the tissue-culture process.

age (ripen) to leave undisturbed for a period.

agribusiness commercial firms that have developed with or stem from agriculture.

arborist a specialist in the cultivation and care of trees and shrubs.

agribusiness management the human element that carries out a plan to meet goals and objectives in an agriscience business.

agricultural business, employment, or trade in agriculture, agribusiness, or renewable natural resources.

agricultural economics management of agricultural resources, including farms and agribusinesses.

agricultural education teaching and program management in agriculture.

agricultural engineering application of engineering principles in agricultural settings.

agricultural literacy education in or understanding about agriscience.

agricultural mechanics design, operation, maintenance, service, selling, and use of power units, machinery, equipment, structures, and utilities in agriscience.

agricultural processing, products, and distribution industry that hauls, grades, processes, packages, and markets commodities from production sources.

agricultural supplies and services businesses that sell supplies and agencies that provide services for people in agriscience.

agriculture activities concerned with the production of plants and animals, and the related supplies, services, mechanics, products, processing, and marketing.

agriscience the application of scientific principles and new technologies to agriculture.

agriscience professions professional jobs dealing with agriscience situations.

cultivo aeropónico el cultivo de plantas en el cual las raíces de tales se cuelgan al aire y se rocían con regularidad con una solución nutritiva.

agar medio de nutrición de plantas donde se colocan los tejidos de las plantas durante el proceso de cultivo de tejidos.

madurar dejar sin tocar por un período.

agroindustria las empresas comerciales que se han desarrollado con o se derivan de la agricultura.

arborista especialista en el cultivo y cuidado de árboles y arbustos.

administración de agroindustria el elemento humano que lleva a cabo un programa para cumplir con las metas y los objetivos de un negocio de agriciencia.

agrícola negocio, empleo o industria de la agricultura, la agroindustria o los recursos naturales renovables.

economía agrícola administración de recursos agrícolas, incluyendo granjas y agroindustrias.

educación agrícola enseñanza y administración de programa en la agricultura.

ingeniería agrícola aplicación de los principios de ingeniería dentro del ambiente agrícola.

educación agrocientífica el estudio o la comprensión de la agrociencia.

mecánica agrícola diseño, funcionamiento, mantenimiento, servicio, venta y uso de aparatos mecánicos, maquinaria, equipos, estructuras y servicios públicos en la agriciencia.

procesamiento, productos y distribución agrícolas industria que transporta, clasifica, empaca y comercializa productos básicos de las fuentes de producción.

provisiones y servicios agrícolas los negocios que venden provisiones y las agencias que prestan servicios para las personas en la industria de la agrociencia.

agricultura actividades que tratan de la producción de plantas y de animales, y de los suministros, servicios, maquinaria, productos, procesamientos y mercadeo relacionados con estas.

agriciencia la aplicación de principios científicos y nuevas tecnologías a la agricultura.

profesiones de agrociencia puestos profesionales que se ocupan de los asuntos de agrociencia.

agriscience research project an original agricultural research project consisting of identifying a problem, a review of scientific literature, conducting a research project, and reporting the results.

agronomy science and economics of managing land and field crops.

air colorless, odorless, and tasteless mixture of gases.

air layering plant propagation by girdling a plant stem, wrapping with sphagnum peat, and protecting with plastic.

alkaline pH of more than 7.0.

alkalinity sweetness, or the tendency toward a high pH greater than 7.0.

alluvial deposit soil transported by streams.

amend a type of motion used to add to, subtract from, or strike out words in a main motion.

ammonia/nitrite/nitrate the chemical components generated during biological breakdown of animal wastes.

amortized loan loan repaid in equal installments of principal and interest.

amphibian(s) organisms that complete part of their life cycle in water and part on land.

anatomy the structure and arrangement of the various parts of the body of an animal or plant.

angiosperm a plant with its seeds enclosed in a pod or seed case.

Angora wool from wool-producing rabbits or goats.

animal science animal growth, care, and management.

animal science technology use of modern principles and practices in animal growth and management.

anion ion that is negatively charged.

annual a plant with a life cycle that is completed in one growing season.

annual ring ring in a cross section of a tree root, trunk, or limb representing 1 year's growth.

annual weed a weed that completes its life cycle within 1 year.

anther portion of the male part that contains the pollen.

antibiotic substance used to help prevent or control infections and diseases of animals.

apiary area where beehives are kept.

proyecto de investigación de agrociencia un proyecto de investigación agrícola original que consiste en identificar un problema, analizar literatura científica, llevar a cabo un proyecto de investigación y reportar los resultados.

agronomía ciencia y economía de la administración de tierras y cultivos del campo.

aire mezcla de gases incoloras, inodoras e insípidas.

acodo al aire propagación de las plantas mediante la realización de una incisión anular en el tallo, la envoltura del tallo con turba de esfagno y su protección con plástico.

alcalino que tiene valor pH superior a 7.0.

alcalinidad sabor dulce o la tendencia a un alto nivel de pH superior a 7.0.

depósito aluvial depósito de tierras transportadas por arroyos.

enmendar un tipo de moción que se usa para agregar, substraer o quitar palabras de una moción principal.

amoníaco/nitrito/nitrato los componentes químicos que se producen durante la descomposición biológica de los desechos animales.

préstamo amortizado préstamo que se paga en cuotas iguales del capital principal e el interés.

anfibios organismos que cumplen parte del ciclo de vida en el agua y otra parte en la tierra.

anatomía la estructura y composición de las diversas partes del cuerpo de un animal o una planta.

angiosperma planta cuyas semillas están envueltas por una vaina o una cápsula.

Angora lana de los conejos o cabras que dan la lana.

ciencia de animales la cría, el cuidado y la administración de animales.

tecnología de las ciencias de animales uso de principios y métodos modernos en la cría y la administración de animales.

anión ion con carga negativa.

anual una planta cuyo ciclo de vida se completa en una temporada de crecimiento.

anillo de desarrollo anual anillo formado en un corte transversal de raíz, tronco o rama de un árbol, que representa el crecimiento de un año.

mala hierba anual una mala hierba que completa su ciclo de vida en un año.

antera porción de la parte macho que contiene el polen.

antibiótico sustancia que se usa para evitar o controlar infecciones y enfermedades de los animales.

apiario sitio en donde se tienen las colmenas.

apiculture beekeeping.	**apicultura** arte de criar abejas.
aquaculturist trained professional involved in the production of aquatic plants and animals.	**acuiculturista** profesional capacitado que se dedica a la producción de plantas y animales acuáticos.
aquaculture raising of finfish, shellfish, and other aquatic animals under controlled conditions. Also, the management of the aquatic environment for production of plants and animals.	**acuicultura** la crianza de peces, mariscos, crustáceos y otros animales acuáticos en circunstancias controladas. Adicionalmente, la administración del ambiente acuático para la producción de plantas y animales.
aquifer water-bearing rock formation.	**acuífero** formación rocosa que contiene agua.
arachnid (arthropod) a living creature such as a spider or mite that is distinguished from insects by having eight legs.	**arácnido** criatura viva tal como las arañas o los ácaros que se distingue de los insectos por tener ocho patas.
arboriculture care and management of trees for ornamental purposes.	**arboricultura** cuidado y administración de árboles para fines ornamentales.
arid an area deficient in rainfall; dry.	**árido** un área que carece de precipitación; seca.
asbestos heat- and friction-resistant material.	**asbesto** materia que es resistente al calor y la fricción.
aseptic sterile or free of microorganisms.	**aséptica** estéril o libre de microorganismos.
asexual reproduction propagation using a part or parts of one parent plant.	**reproducción asexual** reproducción que utiliza una o más partes de una planta parental.
assets anything the business owns.	**activos** total de lo que posee una empresa.
asymmetrical not equal on both sides of center.	**asimétrico** que ambos lados no son iguales relativos al centro.
auction market market where products are sold by public bidding.	**mercado de subasta** mercado donde se venden los bienes mediante subastas públicas.
axil the upper angle between the leaf or flower stem and the stalk of the plant.	**axila** el ángulo superior entre el tallo de la hoja o la flor y el tallo de la planta.

B

B horizon soil below the A horizon or topsoil and generally referred to as subsoil.	**horizonte B** la tierra debajo del horizonte A o de la capa superficial del suelo, generalmente conocido con el nombre de subsuelo.
bacteria one-celled, microscopic organisms.	**bacteria** microorganismos unicelulares.
balance state of quality and calm between items.	**equilibrio** estado de bienestar y tranquilidad entre los ítems.
balance sheet a statement of the assets and liabilities of a business on a specific date.	**hoja de balance** un extracto de cuenta de los bienes y deudas de un negocio en una fecha específica.
balled and burlapped (B & B) plants that have been dug in the field, wrapped in burlap, and laced with a heavy twine.	**empacado y harpillerado** plantas que han sido arrancadas del campo, envueltas en arpillera y atadas con cordel grueso.
balling gun a device used to place a pill in an animal's throat.	**pistola para bolos** aparato que se usa para administrar medicamentos a un animal.
bare-root plant that is dug for transplanting with little soil remaining on the roots.	**planta de raíces expuestas** plantas que son desenterradas para el transplante con solo un poco de tierra que queda en las raíces.
bases genetic material that connects strands of DNA.	**bases** material genético que conecta fibras de ADN.
basic color background color.	**color básico** color de fondo.
bedrock the area below horizon C consisting of large soil particles.	**roca de fondo** el área debajo del horizonte C que consiste en partículas de tierra más grandes.

BelRus superior baking potato bred to grow well in the Northeast.	**BelRus** patata de hornear de calidad superior producida para cultivarse bien en el noreste de los EE.UU.
biennial a plant that takes two growing seasons or 2 years from seed to complete its life cycle.	**bianual** una planta que toma dos temporadas de crecimiento o dos años desde semilla para completar su ciclo de vida.
biennial weed a weed that will live for 2 years.	**mala hierba bienal** una mala hierba que vive por dos años.
binomial having two names.	**binomio** que tiene dos nombres.
bio life or living.	**bio** vida o viviente.
biochemistry chemistry as it applies to living matter.	**bioquímica** química que se aplica a la materia viviente.
biological control pest control that uses natural control agents.	**control biológico** control de los insectos o animales nocivos que utiliza agentes de control natural.
biology basic science of the plant and animal kingdoms.	**biología** la ciencia básica de los reinos vegetal y animal.
biotechnology use of cells or components of cells to produce products or processes.	**biotecnología** uso de células o componentes de células para producir productos o procesos.
biotic disease disease caused by living organisms.	**enfermedad biótica** enfermedad causada por organismos vivientes.
blade the upper portion of the grass leaf.	**brizna** la parte superior de la hoja de pasto.
blanching the brief scalding of food before freezing.	**escaldar** el sumergir los alimentos por un tiempo breve en agua hirviendo antes de congelarlos.
blemish in horses, any abnormality that does not affect the use of the horse.	**defecto** en los caballos, cualquier abnormalidad que no afecte el uso del caballo.
block beef meat sold over the counter to consumers.	**carne descuartizada** carne de res que se vende al por menor a los consumidores.
boar male animal of the swine family.	**verraco** el macho de la familia de los porcinos.
board foot a unit of measurement for lumber that equals $1 \times 12 \times 12$ inches.	**pie tabla** unidad para medir madera que equivale a $1 \times 11 \times 12$ pulgadas.
border planting a planting that is used to separate some part of the landscape from another. It might also be used as a fence or a windbreak.	**cantero de plantas** una sección de plantas que se usa para separar alguna parte del paisaje de la otra. Se puede usar como cerca o también protección contra el viento.
bovine somatotropin (BST) hormone that stimulates increased milk production in cows.	**somatotropina bovina** una hormona que estimula un aumento de secreción láctea en las vacas.
brackish water waters influenced by tide and river flow with intermediate salinity of 3 to 22 percent.	**agua salobre** aguas afectadas por la marea y el flujo de río, que tienen salinidad del 3 al 22 por ciento.
bran skin or covering of a wheat kernel.	**salvado** cascarilla de un grano de trigo.
broiler young chicken grown for meat.	**pollo tomatero** gallina jóven que se cría para comer.
buck male animal of the goat, deer, or rabbit family.	**cabrón** macho de la cabra. **conejo** macho de la coneja. **gamo** ciervo macho.
bud grafting the union of a small piece of plant tissue containing a bud and a plant rootstock.	**injerto de brote** la unión de un pequeño pedazo de tejido vegetal que contiene un brote y una rama de una planta enraízada.
buffer a substance in a solution that tends to stabilize the pH.	**estabilizador** substancia en una solución que tiende a estabilizar el pH.
bulb short underground stem surrounded by many overlapping, fleshy leaves.	**bulbo** tallo, subterráneo corto rodeado por muchas hojas carnosas y traslapadas.

bull male animal of the cattle family.	**toro** macho del ganado vacuno.
business meeting a gathering of people working together to make decisions.	**reunión** asamblea de personas que trabajan juntas para tomar decisiones.
buying function selection of a product or service to be marketed or sold for a profit.	**función adquisitiva** selección de un producto o servicio que se comercializará o venderá para sacarle por ganancia.

C

C horizon soil below the B horizon; it is important for storing and releasing water to the upper layers of the soil.	**horizonte C** la tierra debajo del horizonte B; es importante para guardar y soltar agua a las capas superiores de la tierra.
calf young member of the cattle family.	**becerro** un jóven del ganado vacuno.
calyx group of sepals of a flower.	**cáliz** grupo de los sépalos de la flor.
cambium growth layer in a tree root, trunk, or limb.	**cambium** capa de crecimiento de raíz, tronco o rama del árbol.
cane cutting(s) stems are cane-like and cuttings are cut into sections that have one or two eyes, or nodes.	**esqueje a la caña** los tallos son parecidos a la caña y se cortan en secciones que tienen uno o dos nudos.
cane mature wood in grapevines and some other fruits that has produced fruit and lost the leaves. The next year's new shoots and fruit grow from the cane.	**caña** madera madura en viñedos u otras frutas que han producido y perdido sus hojas. Los retoños del próximo año crecen de las cañas.
canning storing food in airtight containers.	**enlatado** guardar los comestibles en recipientes herméticos.
cannula blunt needle.	**cánula** aguja sin punta afilada.
canopy the top of the plant that has the framework and leaves.	**copa o bóveda de ramas** la parte superior de la planta que tiene la estructura y las hojas.
capability class soil classification indicating the most intensive, but safe, land use.	**clase de capacidad** clasificación de suelos indicando el uso más intensivo pero seguro de la tierra.
capillary water water held by soil particles and available for plant use.	**agua capilar** agua sujetada por partículas de tierra y disponible para el uso de las plantas.
capital money or property.	**capital** dinero o propiedad.
capital investment money spent on commodities that are kept 6 months or longer.	**inversión de capital** dinero que se ha gastado en productos básicos que se guardan por seis meses o más.
carbohydrate starches and sugars that provide energy in the diet.	**carbohidrato** almidones y azúcares que proporcionan energía en la dieta.
carbon monoxide colorless, odorless, and highly poisonous gas; carbon dioxide and water combine to make plant food and release oxygen.	**monóxido de carbono** gas incoloro, inodoro y sumamente tóxico; el bióxido de carbono y el agua se combinan para hacer alimento vegetal y soltar oxígeno.
carcass body of meat after the animal has been eviscerated.	**carne en canal** el cuerpo de carne después de haber sido destripado el animal.
carcinogen a chemical capable of producing a tumor.	**carcinógeno** una sustancia química capaz de producir un tumor.
career a person's occupation or profession.	**carrera** la profesión u ocupación de una persona.
career plan plan developed by students and supporting adults whereby intermediate goals and strategies are developed that lead to a specific career.	**plan de carrera** un plan desarrollado por estudiantes y adultos de apoyo mediante el cual se desarrollan metas y estrategias intermedias que conducen a una carrera específica.

career exploration learning about occupations and jobs as part of the process of choosing one's life work.

carrying capacity the number of animals that a pasture will provide feed for.

casein predominant protein in milk.

cash crop a crop grown for cash sale.

cash flow statement a financial document that describes the availability of operating funds for a business at different points in time.

cation ion that is positively charged.

cell a unit of protoplasmic material with a nucleus and cell walls.

cell membrane a thin tissue surrounding the contents of a cell through which nutrients pass in and waste materials pass out.

cellulose woody fiber parts that make up plant cell walls.

central nervous system the brain and the spinal cord.

cereal crop grasses grown for their edible seeds.

cheese milk that is exposed to bacterial fermentation.

chemical control the use of pesticides for pest control.

chemistry science dealing with the characteristics of elements or simple substances.

chevon meat from goats.

chick newborn chicken or pheasant.

chlorofluorocarbon (CFC) any of a group of compounds consisting of chlorine, fluorine, carbon, and hydrogen used as aerosol propellants and refrigeration gas.

chlorophyll green pigment in leaves.

chloroplast membrane-bound body inside a cell containing chlorophyll pigment; necessary for photosynthesis.

chlorosis yellowing of the leaf.

chromosome the rod-like carrier for genes.

chronic toxicity a measurement of the effect of a chemical over a long period and under lower exposure doses.

circulatory system the system that provides food and oxygen to the cells of the body and filters waste materials from the body.

exploración de carrera aprender sobre ocupaciones y trabajos como parte del proceso de elegir el trabajo de la vida de un individuo.

capacidad de carga la cantidad de animales que pueden alimentarse en un pastizal.

caseína proteina prevalente de la leche.

cultivo comercial una cosecha cultivada para vender por pago al contado.

estado de dinero efectivo en caja documento financiero que describe los fondos disponibles a un negocio en ciertos momentos temporales.

catión ion de carga positiva.

célula unidad de materia protoplásmica con núcleo y paredes.

membrana celular un tejido delgado que rodea el contenido de una célula, a través del cual entran los nutrientes y salen los materiales de desecho.

celulosa partes leñosas de fibra que constituyen las paredes de célula vegetal.

sistema nervioso central el cerebro y la médula espinal.

cultivo de cereal hierbas que son cultivadas por sus semillas comestibles.

queso leche que se expone a la fermentación bacterial.

control químico el uso de pesticidas para control at los insectos y las plagas nocivas.

química la ciencia que trata de las características de los elementos o sustancias sencillas.

carne de chivo carne de ganado caprino.

polluelo gallina o faisán recién nacido.

cloroflurocarbonos cualquiera de un grupo de compuestos que se consisten en cloro, flúor, carbono e hidrógeno usados como propulsores de aerosol y de gas de refrigeración.

clorofila pigmento verde de las hojas.

cloroplasto dentro de la célula, cuerpo envuelto por una membrana que contiene la clorofila; necesario para la fotosíntesis.

clorosis amarilleo de la hoja.

cromosoma cuerpo en forma de bastoncillo que lleva los genes.

toxidad crónica medición del efecto de una sustancia química durante un período largo de tiempo y a dosis de exposición más bajas.

aparato circulatorio el sistema que proporciona los alimentos y oxígeno a las células del cuerpo, y filtra los residuos del cuerpo.

English	Español
citizenship functioning and interacting in a society in a positive way.	**ciudadanía** funcionamiento e interacción en una sociedad en una forma positiva.
clay smallest of soil particles; less than 0.002 mm.	**arcilla** las partículas más pequeñas del suelo; menos de 0,002 mm.
clear cut removal of all marketable trees from an area.	**corte raso** el cortar y eliminar todos los árboles comerciales de un área.
climate the weather conditions of a specific region.	**clima** las condiciones atmosféricas de una región específica.
climatic conditions temperature, temperature range, and precipitation.	**condiciones climáticas** temperatura, variedad de temperatura y precipitación.
clod a lump or mass of soil.	**terrón** masa de tierra.
clone exact duplicate.	**clon** una duplicación exacta.
cockerel young male chicken or pheasant.	**gallo jóven** gallo jóven. **faisán jóven** faisán jóven.
cold frame a bottomless wood box with a sloping glass in which early-season plants are protected from freezing temperatures.	**bastidor frío** caja de madera sin fondo con un vidrio inclinado en el cual se protegen de las temperaturas congeladas a las plantas de temporada temprana.
collagen chief component of connective tissue.	**colágeno** componente principal del tejido conjuntivo.
collar light green or white banded area on the outside of a leaf blade.	**collar** banda verde pálida o blanca en la parte exterior de la hoja.
colluvial deposit soil deposited by gravity.	**depósito coluvial** suelo depositado por la gravedad.
color breed breed of horses based on color.	**cría selectiva basada en el color** cría selectiva de caballos basada en su color.
colostrum first milk produced by mammals; high in antibodies.	**calostro** primera leche producida por mamíferos; tiene niveles elevados de anticuerpos.
colt young male horse or pony.	**potro** caballo o poni macho jóven.
combine machine that is used to cut and thresh seed crops such as grain.	**segadora trilladora** máquina que siega y trilla el grano en el campo.
commensalism one type of wildlife living in, on, or with another without either harming or helping it.	**comensalismo** una clase de fauna que vive en, encima de, o con, otra clase sin dañarla ni tampoco ayudarla.
commission fee for selling a product.	**comisión** honorario por la venta de un producto.
Commodity Credit Corporation (CCC) institution that lends money for production of farm commodities.	**corporación de crédito para productos básicos** institución que presta dinero para la producción de mercancías agrícolas.
commodity exchange organization licensed to manage the buying and selling of commodities.	**bolsa de comercio** organización autorizada para administrar la compra y venta de productos básicos.
comparative advantage the ability to produce a commodity with greater returns than those of a competitor because of a favorable condition such as climate.	**ventaja comparativa** la habilidad de producir un producto básico con mayor ganancia que la del competidor debido a condiciones favorables tales como el clima.
competition two types of wildlife eating the same source of food.	**competición** dos clases de fauna que consumen la misma fuente de alimentación.
compost mixture of partially decayed organic matter.	**composte** mezcla de material orgánico parcialmente descompuesto.

compound a chemical substance that is composed of more than one element.	**compuesto** una sustancia química compuesta por más de un elemento.
compound leaf two or more leaves arising from the same part of the stem.	**hoja digitada** dos o más hojas que brotan del mismo tallo.
concentrate feed high in total digestible nutrients and low in fiber.	**concentrado** alimento con alto nivel de nutrientes digeribles totales y bajo en fibra.
condominium apartment building or unit in which the apartments are individually owned.	**condominio** edificio de apartamentos o unidad en la cual los apartamentos tiene propietarios individuales.
conifer evergreen tree that has needle-like leaves and produces cones.	**conífera** árboles de hoja perene con forma de aguja que produce piñas.
conservation tillage techniques of soil preparation, planting, and cultivation that disturb the soil the least and leave the maximum amount of plant residue on the surface.	**labranza de conservación** técnicas de preparación del terreno, siembra y cultivo que menos remueven la tierra y dejan la cantidad máxima de residuos vegetales en la superficie.
consumer demographics categories of information about preferences of consumers or potential consumers.	**datos demográficos del consumidor** categorías de información sobre preferencias de los consumidores o clientes potenciales.
consumer person who uses a product.	**consumidor** persona que usa un producto.
contact herbicide a herbicide that will not move or translocate within the plant.	**herbicida de contacto** un herbicida que no se mueve ni se desplaza dentro de la planta.
contagious diseases that can be spread by contact.	**contagiosas** enfermedades que se pueden transmitir por contacto.
containerized plant plant that is grown in a pot or other type of container and shipped in the container.	**cultivadas en recipiente** plantas que crecen en macetas u otros tipos de recipientes y son transportadas en el recipiente.
contaminate to add material that will change the purity or usefulness of a substance.	**contaminar** añadir material que cambiará la pureza o utilidad de una sustancia.
continuous-flow system the nutrient solution flows constantly over the plant roots.	**sistemas de flujo continuo** la solución nutritiva fluye continuamente sobre las raíces de las plantas.
contour level line around a hill in which all points along the line are the same elevation.	**contorno** línea de nivel alrededor de una colina en la cual todos los puntos a lo largo de la línea tienen la misma elevación.
contour farming operations such as plowing, discing, planting, cultivating, and harvesting across the slope and on the level.	**cultivo del contorno** operaciones tales como el arado, el enterramiento, sembrar, cultivar, y la recolección de la pendiente y el nivel.
convection oven an oven that heats food by circulating hot air.	**horno de convección** un horno que calienta la comida mediante la circulación de aire caliente.
conventional tillage land is plowed, turning over all crop residues.	**labranza clásica** se ara la tierra, volteando todos los residuos de la cosecha.
Cooperative Extension System an educational agency of the U.S. Department of Agriculture and an arm of land-grant state universities.	**System Cooperativo de Extensión** una entidad educativa del Departamento de Agricultura de los Estados Unidos, y un ramo de las universidades estatales de concesión de terrenos.
cooperative group of producers who join together to market a commodity and/or to purchase supplies.	**cooperativa** grupo de productores que se reunen para comercializar un producto básico y/o para comprar abastecimientos.

corm short, flattened underground stem surrounded by scaly leaves.	**bulbo** tallo subterráneo, corto y aplanado que está envuelto por hojas escamosas.
corn picker machine that removes ears of corn from stalks.	**recolectora de maíz** máquina que recoge las espigas de maíz de los tallos.
cotton gin machine that removes cotton seed from cotton fiber.	**desmotadora** máquina que saca la semilla a la fibra del algodón.
courage willingness to proceed under difficult conditions.	**valor** la voluntad de seguir adelante a pesar de las circunstancias difíciles.
cow female of the cattle family that has given birth.	**vaca** hembra de la familia vacuna que ha parido.
credit money borrowed.	**crédito** dinero prestado.
crop rotation planting of different crops in a given field every year or every several years.	**rotación de cultivos** siembra de cultivos diferentes en un campo determinado cada año o cada dos o tres años.
crop science use of modern principles in growing and managing crops.	**ciencia de cultivos** el uso de principios modernos en el cultivo y la administración de los cultivos.
crown an unelongated stem of major meristematic tissue of turfgrass.	**corona** un tallo no alargado de tejido meristemático mayor del césped.
crustacean aquatic organism with an exoskeleton that molts during growth.	**crustáceo** organismo acuático con un dermatoesqueleto que se muda a medida que el organizmo va creciendo.
cultivar a group of plants with a particular species that has been cultivated and is distinguished by one or more characteristics; through sexual or asexual propagation, it will keep these characteristics.	**variedad obtenida por selección** un grupo de plantas de una especie particular que ha sido cultivada y que se distingue por una o más características, las características se mantendrán con la propagación sexual o asexual.
cultivation the act of preparing and working soil.	**cultivo** la acción de preparar y labrar la tierra.
cultural control pest control that adapts farming practices to better control pests.	**control del cultivo** control de las plagas o los insectos nocivos que adapta los métodos agrícolas para controlar las mejor.
cuticle topmost layer of the leaf; waxy protective covering of the leaf.	**cutícula** la superficie de la hoja; capa protectora cerosa de la hoja.
cutting vegetative part removed from the parent plant and managed so it will regenerate itself.	**tala** la parte vegetativa que se quita de la planta progenitora para que se regenere.
cytosine (C) a base in genes designated by the letter C.	**citosino** una base en los genes denominada con la letra «C».

D

deciduous plants that lose their leaves every year.	**caducifolio** plantas cuyas hojas se caen cada año.
decomposer an organism that is capable of breaking down dead plant and animal matter into soil components.	**descompositor** un organismo que es capaz de descomponer las plantas y animales muertos y convertirlos en componentes del suelo.
deficiency disease condition resulting from improper levels or balances of nutrients.	**enfermedade por carencia** enfermedad resultante de niveles bajos o desequilibrios de los nutrientes.
degree program an educational line of study that leads to the completion of a degree such as an Associate, Bachelor, Master, or Doctorate.	**programa educativo** un plan de estudios que tiene de objetivo obtener un título universitario como un grado de asociado, licenciatura, maestría o doctorado.

dehydration removing moisture with heat.	**deshidratación** extraer la humedad usando calor.
dehydrator a device for drying food.	**deshidratador** un aparato para secar alimentos.
dehydrofrozen product removing moisture after partial cooking, and then freezing.	**producto deshidrocongelaado** extraer la humedad después de cocinar parcialmente y luego congelar.
demand the amount of a product wanted at a specific time and price.	**demanda** cantidad de producto pedido en un tiempo específico y por un precio específico.
dicotyledon (dicot) plant with two seed leaves.	**dicotiledónea** plantas con dos hojas de semilla.
digestive system system that provides food for the body of the animal and for all of its systems.	**aparato digestivo** sistema que proporciona los alimentos al cuerpo del animal y para todos sus sistemas.
diminishing returns point at which each additional unit of input decreases the output or returns.	**rendimientos decrecientes** el punto en donde cada unidad adicional de insumo disminuye el rendimiento o los ingresos.
direct sale the selling of crops or animals directly to a processor by the producer.	**venta directa** la venta de animales o cultivos directamente a los procesadores por el productor.
disassembly process dividing the carcass into smaller cuts.	**carnear** tomar la carne en canal y descuartizarla.
discount loan a loan in which interest is subtracted from the principal at the time the loan is made.	**préstamo con descuento** un préstamo en que se substrae el interés del principal en el momento en que se realiza el préstamo.
disease triangle the term applied to the relationship of the host, pathogen, and the environment in disease development.	**triángulo de enfermedad** aplícase a la relación entre el portador, el agente patógeno y el medio ambiente en el desarrollo de la enfermedad.
disinfectant material that destroys infective agents such as bacteria and viruses.	**desinfectante** material que mata a los agentes infecciosos como bacterias y virus.
distribution function physical organization and delivery of product or service.	**función de distribución** la organización física y la entrega de un producto o servicio.
distributor person or business storing food for transport to regional markets.	**distribuidor** persona o empresa que almacena alimentos para transportarlos a mercados regionales.
DNA (deoxyribonucleic acid) coded genetic material in a cell.	**ácido desoxirribonucléico (ADN)** la materia genética codificada de una célula.
dock remove or shorten tails of certain animals.	**cortar la cola** quitar o acortar las colas de ciertos animales.
doe female of the goat, deer, or rabbit family.	**cabra** hembra del cabrón. **coneja** hembra del conejo.
domestic a tame animal for use by humans.	**domestico** un animal amansado que usan los seres humanos.
dominant gene that expresses itself to the exclusion of other genes.	**dominante** el gen que se manifiesta a exclusión de otros genes.
donkey member of the horse family with long ears and a short erect mane.	**burro** miembro de la familia de los caballos con orejas grandes y crines erectas.
dormant resting stage, no active growth.	**latente** estado de descanso; no hay crecimiento activo.
double-eye cutting used when the plant has leaves that are opposite.	**esqueje de yema doble** se emplea cuando la planta tiene hojas a simétricas.
draft animals used for work.	**tiro** animales que se usan para cargar.
draft horse type of horse bred for work.	**caballo de tiro** tipo de caballo criado para trabajar.
drake male duck.	**pato** el macho de los patos.
drench a process of administering drugs orally to animals.	**administración oral** el proceso de administrar medicamentos oralmente a los animales.

dressing percentage the proportion of the live weight of an animal to the weight of the carcass before cooling.

drip irrigation the use of small tubes to deliver irrigation water to the roots of crop plants in a uniform manner.

drip line the edge of the tree where the branches stop.

drone male bee.

drupe a stone fruit.

dry-heat cooking surrounding food with dry air while cooking.

dry matter material left after all water is removed from a feed material.

duckling young duck.

dwarf tree that has rootstock that limits growth to 10 feet or fewer.

E

economic threshold level the level of pest damage to justify the cost of a control measure.

edible fit to eat or consume by mouth.

element a uniform substance that cannot be further decomposed by ordinary means.

embryo a fertilized egg.

endocrine or hormone system a group of ductless glands that release hormones into the body.

endoplasmic reticulum a cell structure that stores proteins and facilitates their movement to other parts of the cell as needed.

endosperm interior of a wheat kernel that will become wheat flour.

enterprise commercial business to generate profits.

enthusiasm energy to do a job and inspiration to encourage others.

entomology science of insect life.

entomophagous insects that feed on other insects.

entrepreneur person organizing a business, trade, or entertainment.

entrepreneurship process of planning and organizing a business.

entreprenuership SAE enterprises that are supervised by teachers and conducted by students as owners or managers of businesses based on agriscience.

porcentaje de rendimiento del canal la proporción del peso vivo del animal en relación al peso del canal antes de refrigerarse.

irrigación de goteo el uso de pequeños tubos que llevan agua de irrigación de manera uniforme a las raíces de plantas de cultivo.

línea de goteo el límite del árbol donde terminan las ramas.

zángano macho de la abeja.

drupa fruta que contiene un carozo.

cocinar al calor seco rodear la comida con aire seco para cocinarla.

materia seca materia que queda después de quitar toda el agua del material de alimento.

patito cría del pato.

enano árbol que tiene un rizoma que limita el crecimiento a 10 pies o menos.

precio de umbral el nivel de daño causado por plagas o insectos nocivos que justifica el costo de una medida de control.

comestible apropiado para comerse o tomar por boca.

elemento una sustancia uniforme que no puede descomponerse por medios ordinarios.

embrion óvulo fecundado.

sistema endocrino u hormonal conjunto de glándulas de secreción interna las cuales secretan hormonas en el cuerpo.

retículo endoplásmico una estructura celular que almacena las proteínas y facilita su circulación a otras partes de la célula como sea necesario.

endosperma parte interior de un grano de trigo que llegará a ser harina de trigo.

empresa establecimiento comercial que produce ganancias.

entusiasmo energía para hacer una obra y inspiración para animar a otras personas.

entomología ciencia que tiene que ver con la vida de los insectos.

entomófagos insectos que se alimentan con otros insectos.

empresario persona que organiza un negocio, comercio o espectáculo.

capacidad empresarial el proceso de planear y organizar un negocio.

SAE empresarial empresas supervisadas por maestros y dirigidas por estudiantes como dueños o gerentes de negocios que se basan en la agrociencia.

environment all the conditions, circumstances, and influences surrounding and affecting an organism.

epidermis surface layer on the lower and upper sides of the leaf.

equine having characteristics of a horse.

equitation the art of riding on horseback.

eradicant fungicide a fungicide applied after disease infection has occurred.

eradication complete control or removal of a pest from a given area.

erosion wearing away.

estrogen hormone that regulates the heat period.

estrus heat period or time when female animal is receptive to the male animal.

estuary ecological system including bays, streams, and tidal areas influenced by brackish water.

evergreen plants that do not lose their leaves on a yearly basis.

ewe female animal of the sheep family.

exoskeleton the external body wall of an insect.

exploratory SAE a program in which the teacher conducts activities that allow students to become involved in learning a variety of subjects about agriscience and careers related to agriculture.

extemporaneous speaking a form of public speaking wherein some preparation is made, but the speech is not written or memorized.

extravaginal growth turfgrass growth in which growth originates from an axillary bud on the crown.

medio ambiente todas las circunstancias, condiciones e influencias que rodean y afectan al organismo.

epidermis capa de la cara superior e inferior de la hoja.

equino que tiene características de un caballo.

equitación el arte de montar caballo.

fungicida erradicador un fungicida aplicado después de que ha ocurrido la infección.

erradicación control completo o extirpación de una plaga o insecto dañino de un área determinada.

erosión desgaste o deterioro.

estrógeno hormona que regula el período en que el animal está en celo.

celo período o tiempo fecundo cuando la hembra está receptiva al macho.

estuario sistema ecológico que incluye bahías, arroyos y tierras bajas del litoral que son afectados por agua salobre.

planta perenne plantas que no se deshojan cada año.

oveja hembra del carnero.

dermatoesqueleto caparazón o esqueleto externo de un insecto.

SAE exploratoria un programa en el cual el maestro conduce actividades que permiten a los estudiantes participar en el aprendizaje de una variedad de materias sobre la agrociencia y las carreras relacionadas a la agricultura.

discurso improvisado una forma de discurso público en donde se hace una cierta preparación, pero el discurso no se pone por escrito ni se memoriza.

crecimiento extravaginal crecimiento de césped que brota de un botón axilar en la corona.

F

fabrication and boxing vacuum sealing of meat in boxes before shipment.

famine widespread starvation.

Farmers Home Administration (FmHA) government agency that assists farmers to become landowners.

farrier a person who shoes horses.

farrow giving birth in the swine family.

fat nutrients that have 2.25 times as much energy as carbohydrates.

fabricación y empaque envasar al vacío las cajas de carne antes de transportarlas.

hambruna escasez generalizada de alimentos.

Departamento de Hogares de Agricultores organismo de gobierno que ayuda a los granjeros a volverse propietarios de tierra.

herrador de caballos persona que clava y ajusta las herraduras a un caballo.

parir a cerdos parir a cerdos.

grasa alimentos nutritivos que tienen 2.25 veces más energía que los carbohidratos.

federal land bank lending institution that provides long-term credit for agriculture.

feed additive a nonnutritive substance added to feed to improve growth, increase feed efficiency, or maintain health.

feedstuff any edible material used for animal feed.

fermentation a chemical change that results in gas release.

fertilizer material that supplies nutrients for plants.

fertilizer grade percentages of primary nutrients in fertilizer.

fetus embryo from the time of attachment to the uterine wall until birth.

FFA a national intracurricular organization for students enrolled in agriscience programs in their schools.

fibrous root one of the two major root systems, consisting of many fine, hair-like roots.

field crop a class of crop that includes grains, oil crops, and specialty crops.

filament structure that supports the anther.

filly young female horse or pony.

financing function obtaining capital for a business.

fine-textured (clay) soil soil that usually forms very hard lumps or clods when dry; plastic when wet.

floriculture production and distribution of cut flowers, potted plants, greenery, and flowering herbaceous plants.

flower reproductive part of the plant.

flowering bud a terminal bud on a plant that produces flowers.

fluke very small, flat worm that is a parasite.

foal newborn horse or pony. Also, to give birth in horses.

food additive anything added to food before packaging.

food chain interdependence of plants or animals on each other for food.

food industry production, processing, storage, preparation, and distribution of food.

banco federal de tierras institución prestamista que provee crédito a largo plazo para la agricultura.

aditivos para alimentos una sustancia no nutritiva añadida al alimento para mejorar el crecimiento, aumentar la eficacia del alimento o para mantener la salud.

forraje cualquier material comestible que sirve para el alimento.

fermentación un cambio químico que resulta en el desprendimiento de gases.

abono material que proporciona nutrientes para las plantas.

grado del abono proporción de nutrientes en el abono.

feto embrión desde el momento en el que se pega a la pared uterina hasta el nacimiento.

AFF una organización dentro del programa para estudiantes matriculados en los programas de agriciencia.

raíz fibrosa uno de dos sistemas principales del raíz que consiste en muchas raíces finas.

cultivo de campo una clase de cultivo que incluye granos, cultivos de aceite y cultivos de especialización.

filamento estructura que sostiene la antera.

potra hembra jóven del caballo o poni.

función de financiar obtener capital para un negocio.

terreno de textura fina (arcilla) usualmente forma terrones duros cuando se seca; es plástico cuando está mojado.

floricultura producción y distribución de flores cortadas, de plantas en maceta, ramas y hojas verdes y plantas herbáceas florecientes.

flor parte reproductora de la planta.

brote de flor un botón terminal en una planta que produce flores.

trematodos gusanos de cuerpo plano muy pequeños que son parásitos.

potro caballo o poni recién nacido.

suplemento aditivo cualquiera cosa añadida al alimento antes de envasarlo.

cadena trófica dependencia recíproca de plantas o animales para alimentarse.

industrias alimenticias producción, procesado, almacenaje, preparación y distribución de alimentos.

foot-candle amount of light found 1 foot from a standard candle or candela.	**candela por pie cuadrado** cantidad de luz que se encuentra de un pie de una candela o vela estándar.
forage crop plants grown for their vegetative growth and fed to animals.	**forraje** plantas cultivadas por su vegetación y dadas de comer a los animales.
Foliage leafy portion of a plant.	**Follaje** parte frondosa de una planta.
forest large group of trees and shrubs.	**bosque** gran conjunto de árboles y arbustos.
forest land land at least 10 percent stocked by forest trees.	**tierra forestal** tierra poblada con por lo menos 10 por ciento de árboles forestales.
forestry industry that grows, manages, and harvests trees for lumber, poles, posts, panels, paper, and many other commodities.	**silvicultura** industria que cultiva, administra, y cosecha los árboles para maderos, varas, palos, tableros, papel y muchos otros productos.
formal equal in size and number.	**formal** igual en cantidad y tamaño.
formulation the physical properties of the pesticide and its inert ingredients.	**formulación** las propiedades físicas del pesticida y sus ingredientes inertes.
foundational SAE the beginning part of an SAE where the student develops a detailed plan for conduction of a Supervised Agricultural Experience.	**SAE fundamental** el inicio de una SAE donde el estudiante desarrolla un plan detallado para dirigir una experiencia agrícola supervisada.
foundational SAE a supervised agricultural experience in which all students learn to plan and implement career experiences as they progress through three levels: awareness, intermediate, and advanced.	**SAE fundamental** una experiencia agrícola supervisada en la que todos los estudiantes aprenden a planificar e implementar experiencias profesionales a medida que avanzan a través de tres niveles: conocimiento, intermedio y avanzado.
4-H network of youth clubs directed by Cooperative Extension System personnel to enhance personal development and provide skill development in many areas.	**4-H** una red de asociaciones para los jóvenes que es dirigida por el personal del Servicio Cooperativo de Extensión para fomentar el desarrollo personal y enseñar técnicas en muchas áreas.
free water water that drains out of soil after it has been wetted.	**agua libre** agua que se drena del suelo después de mojar el suelo.
freeze-drying removing moisture with cold.	**deshidratar por congelación** extraerle a algo la humedad por congelarla.
fresh water water that flows from the land to oceans and contains little or no salt.	**agua dulce** agua que corre desde la tierra hasta los mares y que tiene poca o ninguna sal.
fructose simple fruit sugar.	**fructosa** azúcar simple de frutas.
fruit mature ovary; seed.	**fruto** óvulo maduro, semilla.
fry small, newly hatched fish.	**alevines** peces recién salidos del huevo.
fungal endophyte microscopic plant growing within a plant.	**endofito fungoso** planta microscópica que crece dentro de una planta.
fungi members of a major group of lower plant life that lack chlorophyll and obtain nourishment from either live or decaying organic matter.	**hongos** miembros de un mayor grupo de plantas inferiores que no producen clorofila y obtienen nutrición de organismos vivos o materias orgánicas en descomposición.
fungicide a material used to destroy fungi or protect plants against their attack.	**fungicida** una sustancia que se usa para destruir hongos o proteger a las plantas de su ataque.
furrow groove made in the soil.	**surco** hendidura hecha en la tierra.
futures market legal frameworks for sellers and buyers to buy and sell futures contracts.	**mercado de futuros** estructura legal para vende-dores y compradores para comprar y vender contratos a futuro.

G

gait way of moving.

galactose simple milk sugar.

gamete a reproductive cell.

gander male goose.

gavel a small wooden hammer-like tool that is used by the presiding officer to direct a meeting.

gelding castrated male animal of the horse family.

gene a unit of hereditary material located on a chromosome.

gene mapping finding and recording the locations of genes in a cell.

generation the collective offspring of common parents.

gene splicing the process of removing and inserting genes in cells.

genetic control pest control using resistant varieties of crops.

genetic engineering movement of genes from one cell to another.

genetics the biology of heredity.

genotype the genetic makeup of an individual or group of organisms.

genus (plural, *genera*) a closely related and definable group of animals or plants comprising one or more species.

geography a social science related to the interactions of people and the different features of the earth.

germ new wheat plant inside the kernel.

germinate a seed sprouting or starting to grow.

gestation length of pregnancy.

giblets heart, liver, and gizzard of poultry.

gills organ of aquatic animal that absorbs oxygen from the water.

gilt female of the swine family that has not given birth.

ginning the process of removing the seeds from cotton.

Girl Scouts and Boy Scouts youth organizations that provide opportunities for leadership development and skill development.

glacial deposit soil deposited by ice.

glucose simple sugar and the building blocks for other nutrients.

goose female of the goose family.

gosling young goose.

paso manera de moverse.

galactosa azúcar simple de leche.

gameto una célula reproductiva.

ganso macho de la gansa.

martillo un pequeño martillo de madera usado por un funcionario que dirige una reunión.

caballo castrado caballo al cual se le han extirpado los testículos.

gene unidad de materia hereditaria ubicada en un cromosoma.

trazar un mapa de genes localización y registro de las posiciones de genes en una célula.

generación la descendencia colectiva de padres comunes.

empalme de genes el proceso de extirpar y meter genes en las células.

control genético control de plagas o insectos dañinos usando variedades resistentes de cultivos.

ingeniería genética traslado de genes de una célula a otra.

genética la biología de la herencia.

genotipo el material genético de un individuo o grupo de organismos.

género un grupo definible y estrechamente relacionado de plantas o animales que consta de una o más especies.

geografía una ciencia social que trata de las interrelaciones de la gente con las características distintas de la tierra.

germen progenie de trigo dentro del grano.

germinar una semilla que brota o empieza a crecer.

gestación tiempo que dura el embarazo.

menudillos corazón, higadillo y molleja de las aves.

agalla órgano de animal acuático que absorbe el oxígeno del agua.

cerda jóven nulípara hembra de la familia porcina que no ha parido.

desmotar el proceso de sacar la semilla al algodón.

Niñas Exploradoras y Niños Exploradores asociaciones para los jóvenes que dan oportunidades de desarrollo de capacidades de liderazgo y capacidades generales.

depósitos glaciales tierras depositadas por hielo.

glucosa azúcar simple y molécula fundamental de otros nutrientes.

gansa hembra del ganso.

ansarino ganso jóven.

grade quality standard.	**grado** estándar de calidad.
grader person who inspects food for freshness, size, and quality.	**clasificador** persona que inspecciona los alimentos para evaluar su frescura, tamaño y calidad.
gradient a measurable change in an amount over time or distance.	**índice** un cambio mensurable de una cantidad en el tiempo o la distancia.
grafting joining two plant parts together so that they will grow as one.	**injertar** juntar dos partes de plantas para que crezcan como una sola planta.
graft union the location where a scion and rootstock meet when parts of two plants are grafted together.	**unión de injerto** sitio donde una vara y un tallo de dos plantas diferentes se juntan.
grass waterway strip of grass growing in the low area of a field where water can gather and cause erosion.	**estepa de gramíneas** franja de hierba que crece en la parte baja del campo donde el agua puede acumularse y causar erosión.
gravitational water water that drains out of soil after it has been wetted.	**agua de gravitación** agua que se drena del suelo después de mojarlo.
greenhouse effect a buildup of heat at the Earth's surface caused by energy from sunlight becoming trapped in the atmosphere.	**efecto invernadero** concentración de calor en la superficie de la tierra debido a la energía de la luz del sol que queda atrapada en la atmósfera.
green manure (crop) crop grown to be plowed under to provide organic matter to the soil.	**plantas para abono verde** plantas cultivadas para ser enterradas con el objetivo de proporcionar materia orgánica al suelo.
Green Revolution process in which many countries became self-sufficient in food production.	**revolución verde** proceso en el que muchos países logran a ser autosuficientes en la producción de alimentos.
group planting trees or shrubs planted together so as to point out some special feature or to provide privacy or a small garden area.	**siembra por agrupación** árboles o arbustos plantados juntos para señalar alguna característica especial o para proveer intimidad o un jardín pequeño.
guanine (G) a base in genes designated by the letter G.	**guanino (G)** una base en los genes denominada con la letra «G».
guard cells cells that surround the stoma.	**células guardias** células que rodean el estoma.

H

habitat area or type of environment in which an organism or biological population normally lives.	**hábitat** área o tipo de medio ambiente en que normalmente vive un organismo o una población biológica.
hand unit of measurement for horses equal to 4 inches.	**palmo** unidad de medir a los caballos que es igual a cuatro pulgadas.
hardness wood's resistance to compression.	**dureza de la madera** la resistencia de la madera a la compresión.
hardwood wood from deciduous trees.	**madera dura** madera de los árboles caducos.
hardiness the ability of a plant to survive and grow in a given environment.	**resistente** la capacidad de una planta de sobrevivir y crecer en un ambiente determinado.
harvester person responsible for taking products from plants in the field.	**cosechador** persona responsable de recoger el fruto del campo.
harvesting taking a product from the plant where it was grown or produced.	**cosechar** recoger un producto de la planta donde fue cultivado o producido.
hay forages that have been cut and dried to a low level of moisture, used for animal feed.	**heno** forrajes que han sido cortados y secados a un nivel bajo de humedad, usado para alimentar animales.

haylage silage made from forages dried to 40 to 55 percent moisture.

heartwood inactive core of a tree trunk or limb.

heel cutting a shield-shaped cut made about halfway through the wood around the leaf and the axial bud.

heel in place plant roots in a trench that is deep enough to cover the roots with moist soil to protect them until they can be permanently planted.

heifer female animal of the cattle family that has not given birth.

heritability the capacity to be passed down from parent to offspring.

hen adult female chicken, duck, turkey, and pheasant.

herb plant kept for aroma, medicine, or seasoning.

herbaceous a plant that has a stem that does not turn woody; it is more or less soft and succulent, and often lacks winter hardiness.

herbicide a substance for killing weeds.

heredity transmission of characteristics from parent to offspring.

heterozygous pairs of genes that are different.

hide skin of an animal.

high technology use of electronics and ultra-modern equipment to perform tasks and control machinery and processes.

hinny cross between a stallion and a jennet.

home gardening production of vegetables for use by a single family.

homozygous pairs of genes that are alike.

horizon layer.

hormone chemical that regulates activities of the body.

horse member of the horse family 14.2 or more hands tall.

horticulture the science of producing, processing, and marketing fruits, vegetables, and ornamental plants.

host animal a species of animal in or on which diseases or parasites can live.

hotbed a cold frame that has a source of artificial heat.

hybrid plant or animal offspring from crossing two different species or varieties.

hierba presecada y ensilada ensilaje hecho de forrajes secados a 40 o 55 por ciento de humedad.

duramen núcleo inactivo de un tronco o rama de un árbol.

escudete un corte en forma de escudo hecho a medio camino a través de la madera alrededor de la hoja y el botón axilar.

taconear colocar y tapar con tierra húmeda las raíces de plantas en una zanja profunda para protegerlas hasta que puedan plantarse permanentemente.

vaquilla hembra de ganado vacuno que no ha parido.

capacidad hereditaria la capacidad de ser transmitido del progenitor a la progenie.

gallina hembra adulta del gallo.

hierba planta guardada para fines aromáticos, medicinales o como condimento.

herbáceo una planta que tiene un tallo que no se vuelve leñoso; es más o menos blando y carnoso, tampoco es resistente al invierno.

herbicida una sustancia empleada para matar las malas hierbas.

herencia la transmisión de rasgos de los progenitores a la progenie.

hetercigótico pares de genes que son diferentes.

cuero piel de los animales.

alta tecnología el uso de electrónica y equipo ultramoderno para desempeñar funciones y controlar maquinaria y procesos.

burdégano hijo de caballo y burra.

jardinería en el hogar producción de vegetales para el uso de una sola familia.

homocigótico pares de genes que son semejantes.

horizonte capa.

hormona sustancia química que regula las actividades del cuerpo.

caballo miembro de la familia de los caballos que miden 14.2 o más palmos de alto.

horticultura la ciencia de producir, procesar y comercializar frutas, verduras y plantas ornamentales.

animal huésped una especie de animal en o sobre el cual viven parásitos o enfermedades.

superficie caliente un bastidor frío que tiene un surtidor de calor artificial.

híbrido progenie vegetal o animal que procede del cruzar dos distintas especies o razas.

hybrid vigor offspring of greater strength and potential resulting when two different breeds or varieties are crossed.	**vigor híbrido** progenie de más fuerza y potencial que resulta de cruzar dos distintas razas o variedades.
hydrocarbon organic compound containing hydrogen and carbon.	**hidrocarburo** compuesto orgánico que contiene hidrógeno y carbono.
hydrocooling removal of heat by immersing in cold water.	**enfriamiento con agua** quitar el calor por inmersión en agua fría.
hydroponics the practice of growing plants without soil.	**cultivo hidropónico** el proceso de cultivar plantas sin tierra.
hygroscopic water water that is held too tightly by soil particles for plant roots to absorb.	**agua higroscópica** agua sujetada con demasiado rigor por las partículas de tierra para que las raíces de la planta la absorban.
hyphae the thread-like vegetative structure of fungi.	**hifas** la estructura vegetal filiforme de l os hongos y mohos.

I

imbibition the absorption of water into the cell causing it to swell.	**embebición** la absorción de agua en la célula, lo cual hace que se hinche.
immune not affected by.	**inmune** que no es afectado.
immersion SAE a student SAE that expands and builds on the agricultural literacy component of the foundational SAE.	**SAE de inmersión:** un SAE para estudiantes que se expande y se basa en el componente de alfabetización agrícola del SAE fundamental.
implant a substance placed under the skin and slowly absorbed to improve growth of animals.	**implantar** una sustancia comercializads bajo la piel y lentamente absorbida para mejorar el crecimiento de los animales.
improvement activity project that improves beauty, convenience, safety, value, or efficiency learned outside the regularly scheduled classroom or laboratory classes.	**actividad de mejoramiento** proyecto que mejora la belleza, conveniencia, seguridad, valor o eficacia y que se aprende fuera de las clases o laboratorios programados normalmente.
improvement by selection picking the best parents for the next generation.	**mejoramiento por selección** seleccionar a los mejores padres para la generación siguiente.
inbreeding mating of animals that are related.	**procrear en consanguinidad** apareamiento de animales emparentados.
incomplete dominance neither gene expresses itself to the exclusion of the other.	**dominancia incompleta** ninguno de los genes se manifiesta a exclusión al otro.
induction a specific set of conditions must occur to cause flower development.	**inducción** deben darse circunstancias específicas para causar el desarrollo de la flor.
inflorescence the flowers and ultimately the seed area of the plant.	**florescencia** son las flores y en última instancia el área de las semillas de la planta.
informal in landscaping, planting areas that contain elements that are not equal in number, size, or texture.	**informal** cuando se refiere al paisajismo, se refiere a las áreas que contienen elementos que no son iguales en cantidad, tamaño ni textura.
infusion the process for treating udder problems through the teat canal.	**infusión** el proceso usado para tratar problemas de la ubre por el canal de la teta.
inner bark contains the cambium from which the tree grows and the bark is renewed; transports food within a tree.	**corteza interna** contiene el cambium del que crece el árbol y la corteza se renueva; transporta alimentos dentro del árbol.
inorganic compound a compound that does not contain carbon.	**compuesto inorgánico** un compuesto que no contiene carbono.

insecticide a material used to kill insects or protect against their attack.

insemination the placement of semen in the female reproductive tract.

instar the state of the insect during the period between molts.

institutional advertising advertising designed to create a favorable image of the firm or institution.

integrated pest management (IPM) pest-control program based on multiple control practices.

integrity capable of personally upholding a high moral standard.

intermediate-term loan loan with payment periods that range from 1 to 7 years.

internode area between two nodes.

internship a career experience that places a student in a business for a specific period to learn technical skills through actual work experiences.

intradermal between layers of skin.

intramuscular in a muscle.

intraperitoneal in the abdominal cavity.

intraruminal in the rumen.

intravaginal growth a type of growth in which shoots develop within the lower leaf sheath at a crown's axillary bud.

intravenous in a vein.

inventory report units received and sold, unit cost, total sales, and profit.

involuntary muscle muscle that operates in the body without control by the will of the animal.

ion atom or a group of atoms that has an electrical charge.

irradiation treating food with gamma rays.

irrigation addition of water to plants to supplement that provided by rain or snow.

J

jack male donkey or mule.

jennet female donkey or mule.

journal an accurate record of events and observations, such as weight gain in livestock, personal accomplishments, data collection, etc.

jungle fowl wild ancestor of the chicken.

---

insecticida un material que se usa para matar insectos o proteger contra su ataque.

inseminación la introducción de semen en las vías genitales de la hembra.

estados larvarios la fase del insecto durante el período entre las mudas.

publicidad institucional publicidad creada para establecer un concepto favorable de la empresa o institución.

control integrado de plagas y animales nocivos programa de control de los insectos y plagas nocivos basado en múltiples métodos de control.

integridad capaz de sostener personalmente un alto nivel moral.

préstamo de plazo intermedio préstamo que tienen plazos de pago que oscilan entre uno y siete años.

entrenudos área entre dos nudos.

internado una experiencia que coloca al estudiante en un negocio por un plazo específico de tiempo para aprender habilidades técnicas a través de experiencias reales en el trabajo.

intradérmico entre las capas de la piel.

intramuscular en los músculos.

intraperitoneal en la cavidad abdominal.

intraruminal en el rumen.

crecimiento intravaginal un tipo de renuevo que desarrolla brotes en el interior de la vaina de la hoja inferior, en la corona del botón axilar.

intravenoso en la vena.

inventario unidades recibidas y vendidas, coste por unidad, ventas totales y ganancia.

músculo involuntario músculo que funciona en el cuerpo sin control de la voluntad del animal.

iones átomo o grupo de átomos que tienen una carga eléctrica.

irradiación tratamiento de comestibles con rayos gamma.

irrigación agregado de agua a los cultivos para suplementar la proporcionada por la lluvia o la nieve.

burro macho de la burra o mula.

burra hembra del burro o mula.

diario un registro preciso de los acontecimientos y las observaciones, como el aumento de peso del ganado, los logros personales, la recolección de datos, etc.

ave de la selva antepasado salvaje de la gallina.

K

Katahdin popular potato variety of the 1930s.

kid young goat.

knowledge familiarity, awareness, and understanding.

kosher prepared in accordance with Jewish dietary laws.

L

lactation period period when mammals are producing milk.

lactose compound milk sugar.

lacustrine deposit soil deposited by lakes.

lamb member of the sheep family younger than 1 year; also, meat from young sheep.

land degradation the decline of soil quality or reduction in productivity.

larvae mobile organisms that become fixed and grow into nonmobile adults.

laser an intense, narrow beam of light that is used to measure nutrient deficiencies in plants from an orbiting satellite.

lay on the table a motion used to stop discussion on a motion until the next meeting. The way to table a motion is to say, "I move to table the motion."

layer chicken developed for the purpose of laying eggs.

LC_{50} lethal concentration of a pesticide in the air required to kill 50 percent of the test population.

LD_{50} lethal dose of a pesticide required to kill 50 percent of a test population.

leadership the capacity or ability to lead.

leaf blade the wide portion of a leaf in which photosynthesis occurs.

leaf cutting a cutting made from a leaf without a petiole.

leaf petiole cutting a plant consisting of a leaf and petiole on which a rooting compound is applied and the cutting is placed in a soil medium to develop into a new plant.

leaf section cutting a section of a plant leaf containing a vein that is placed in a plant growth medium to generate a new plant.

Katahdin variedad popular de patata de la década de 1930.

chivo cabrío jóven.

conocimiento familiaridad, conciencia y entendimiento.

kosher preparado según las leyes dietéticas judaicas.

período de lactancia período en que los mamíferos producen leche.

lactosa compuesto de azúcar de leche.

depósito lacustre el suelo depositado por los lagos.

cordero miembro de la familia de las ovejas que tiene menos de un año de edad; también la carne de la oveja jóven.

deterioro del suelo el empeoramiento de la calidad del suelo o la baja productividad de la tierra.

larvas organismos movibles que se vuelven fijos y se desarrollan y convierten en adultos no móviles.

laser un haz de luz intenso, estrecho que se utiliza para medir deficiencias en nutrientes en plantas desde un satélite en órbita.

colocar sobre la mesa una moción usada para detener el debate sobre una moción hasta la próxima reunión. La manera de presentar esta moción es decir, "Propongo que se coloque el asunto sobre la mesa".

gallina ponedora una gallina criada con objetivo de poner huevos.

CM_{50} concentración mortal de un pesticida en el aire necesaria para matar el 50 por ciento de una población prueba.

DM_{50} dosis mortal de un pesticida necesaria para matar el 50 por ciento de una población prueba.

dirección la capacidad o habilidad de guiar.

limbo la porción de una hoja donde ocurre la fotosíntesis.

corte de hoja un corte hecho de una hoja sin pecíolo.

esqueje de pecícolo de hojas una planta que consiste de una hoja y un pecíolo en la cual se aplica un compuesto para enrraizar, y el esqueje se coloca en un medio de tierra para desarrollar.

esqueje de sección de hoja una sección de una hoja de la planta que contiene un nervio, que se coloca en un medio de crecimiento para generar una nueva planta.

legume plant in which certain nitrogen-fixing bacteria use nitrogen gas from the air and convert it to nitrates that the plant can use as food.

lenticels pores in the stem that allow the passage of gases in and out of the plant.

liabilities current and long-term debts.

light horse type of horse developed for riding.

light intensity brightness of light.

ligule a membranous or hairy structure located on the inside of the leaf.

lime material that reduces the acid content of soil and supplies nutrients such as calcium and magnesium to improve plant growth.

linen fabric produced from fibers in flax plants.

linseed oil oil produced from flax seed.

lipid fat droplets inside a cell.

listen to concentrate on hearing and understanding what others are communicating to you.

loam a granular soil containing a good balance of sand, silt, and clay.

loess deposit soil deposited by wind.

long-term loans loans with payment periods that range from 8 to 40 years.

long-term plans plans accomplished over months or years.

loss-leader pricing commodity offered for sale at prices less than the cost level.

loyalty reliable support for an individual, group, or cause.

lumber boards cut from trees.

M

macronutrient element used in relatively large quantities.

main motion a basic motion used to present a proposal for the first time. The way to state it is to get recognized, and then say, "I move that …"

malting process of preparing grain for the production of beer and alcohol.

maltose compound milk sugar.

mammal milk-producing animal.

mange crusty skin condition caused by mites.

legumbre planta en la cual ciertas bacterias fijadora del nitrógeno utilizan el gas de nitrógeno del aire y lo convierten a nitratos que la planta puede usar para alimentarse.

lenticelas poros en el tallo o el pedúnculo que permiten que los gases entren y salgan de la planta.

pasivos deudas actuales y a largo plazo.

caballo de silla tipo de caballo criado para montar.

intensidad de la luz luminosidad de la luz.

lígula está ubicada en la vaina foliar interior y es una estructura membranosa y pilosa.

cal material que reduce el contenido de ácido del suelo y proporciona minerales tales como calcio y magnesio para mejorar el crecimiento de las plantas.

lino tejido producido de las fibras de las plantas lináceas.

aceite de linaza aceite producido de la semilla de lino.

lípidos gotitas de grasa dentro de una célula.

escuchar oír con atención y comprender lo que dicen los demás.

suelo arcilloso un suelo granular que contiene una mezcla balanceada de arena, légamo y arcilla.

depósito loésico tierra depositada por el viento.

préstamos a largo plazo préstamos cuyos plazos de pago oscilan entre 8 a 40 años.

planes a largo plazo planes realizados por meses o años.

precio de lanzamiento producto básico que se ofrece a la venta a un precio menor al nivel de costo.

lealtad apoyo seguro para un individuo, un grupo o una causa.

maderos trozos cortados de los árboles.

macronutriente elemento usado en cantidades relativamente grandes.

moción principal la moción básica para presentar una propuesta por primera vez. La manera de declararla es ser reconocido y luego decir, «Yo propongo que …

malteado proceso de preparar el grano para la producción de cerveza y alcohol.

maltosa azúcar de la leche compuesto.

mamífero animal que produce la leche.

sarna enfermedad de la piel que es causada por los acáridos.

mare adult female horse or pony.	**yegua** hembra adulta del caballo o poni.
margin edge of the leaf.	**reborde** el borde de una hoja.
market gardening production of a wide variety of vegetables that are generally sold in local roadside markets.	**jardinería comercial** producción de una amplia variedad de vegetales que generalmente se venden en mercados locales ubicados al costado de los caminos.
mathematics a discipline that deals with quantity (numbers), space, structure, and their relationships.	**matemáticas** la ciencia que trata de la relación entre las cantidades númericas, espaciales y estructurales.
maturity the state or quality of being fully grown.	**madurez** estado o cualidad de estar completamente desarrollado.
mechanical pest control pest control that affects the pest's environment or the pest itself.	**control mecánico de plagas o insectos nocivos** control de plagas o insectos nocivos que afecta el ambiente de la plaga o insecto nocivo o la plaga o insecto en sí.
medium surrounding environment in which something functions and thrives.	**medio ambiente** los alrededores en los cuales funciona y crece alguna cosa.
meiosis cell division that results in the formation of gametes.	**meiosis** división celular que resulta en la formación de los gametos.
meristematic tissue plant tissue responsible for plant growth.	**tejido meristemático** tejido vegetal que causa el crecimiento de la planta.
mesophyll tissue of the leaf where photosynthesis occurs.	**mesofilo** tejido de la hoja en donde ocurre la fotosíntesis.
metamorphosis the change in growth stages of an insect.	**metamorfosis** el cambio en las etapas de crecimiento del insecto.
microbe living organism that requires the aid of a microscope to be seen.	**microbio** organismo viviente que requiere la ayuda del microscopio para verse.
micronutrient element used in very small quantities.	**micronutriente** elemento usado en cantidades muy pequeñas.
microorganism tiny plant or animal that may contribute to food spoilage.	**microorganismo** microbio animal o vegetal que puede contribuir a la putrefacción de los comestibles.
middlemen people who handle an agricultural product between the farm and the consumer.	**intermediarios** personas que manejan un producto agrícola entre la granja y el consumidor.
milking machine machine that milks cows and goats.	**máquina de ordeñar** aparato o máquina que extrae la leche de la ubre de las vacas y cabras.
mineral element essential for normal body functions.	**mineral** elemento esencial para las funciones normales del cuerpo.
minimum tillage soil is worked only enough so that seed will germinate.	**labranza mínima** la tierra es labrada justo lo suficiente para que germinen las semillas.
minutes the official written record of a business meeting.	**actas** el registro oficial escrito de una reunión de negocios.
mitochondrion plays a role in converting animal food to usable energy.	**mitocondria** juega un papel importante en la conversión de los alimentos de origen animal energía aprovechable.
mitosis simple cell division for growth.	**mitosis** división celular simple que resulta en el crecimiento.
mohair hair from Angora goats used to make a shiny, heavy, woolly fabric.	**mohair** pelo de la cabra de Angora usado para hacer un tejido lanoso, grueso y con brillo.
moist-heat cooking surrounding the food with liquid while cooking.	**cocinar con calor húmedo** rodear la comida con líquido mientras se cocina.
moldboard plow plow with a curved bottom that will turn prairie soils.	**arado de vertedera** arado con orejera capaz de arar el suelo de llanura.

molt softening and cracking of the exoskeleton so that crustaceans may escape and grow a larger exoskeleton.

monoclonal antibody natural substance in blood that fights diseases and infections.

monocotyledon (monocot) plant with one seed leaf.

monogastric animal with a single-compartment stomach.

mosaic a virus disease of plants in which a leaf shows a symptom of light and dark-green mottling of the foliage.

motion a proposal for group action that is presented in a meeting to be acted on by an organization.

mulch material placed on soil to break the fall of raindrops (preventing erosion), prevent weeds from growing, or improve the appearance of the area.

mule cross between a jack and a mare.

muscular system the lean meat of the animal.

mutation change in genes.

mutton meat from mature sheep.

mutualism two types of wildlife living together for the mutual benefit of both.

mycelium a collection of fungal hyphae.

MyPlate MyPlate illustrates the five food groups that are the building blocks for a healthy diet using a familiar image—a place setting for a meal. USDA's Choose My Plate campaign features selected messages to focus consumers on key behaviors such as: balancing calories by reducing portion sizes; increasing consumption of healthy foods, such as fruits, vegetables, and whole grains; and identifying foods to reduce, including high sodium foods and sugary drinks.

mudar proceso de ablandamiento y agrietamiento del dermatoesquéleto para que los crustáceos puedan escaparse y desarrollar un dermatoesquéleto más grande.

anticuerpo monoclonal substancia natural en la sangre que lucha contra enfermedades e infecciones.

monocotiledóneo planta con solamente una hoja de semilla.

monogástrico animal con estómago de un solo compartimento.

mosaico una enfermedad viral de las plantas en donde la hoja manifiesta un síntoma de vetas manchas de verde claro y oscuro en el follaje.

moción una propuesta para acción de un grupo que es presentada en una junta para ser puesta en acción por la organización.

pajote material puesto encima del suelo para evitar que la caída de las gotas de lluvia causen la erosión, para prevenir el crecimiento de malas hierbas o para mejorar el aspecto del área.

mula el hijo de un burro macho y una yegua.

sistema muscular la carne magra del animal.

mutación cambios en los genes.

cordero carne de cordero mayor.

mutualismo dos clases de fauna que viven juntas para el beneficio mutuo de las dos.

micelio un grupo de hifa fungosa.

Mi plato Mi plato utiliza una imagen conocida para mostrar los cinco grupos de alimentos que son la base fundamental de una dieta saludable: el entorno de una comida. La campaña "Choose My Plate" [Elige Mi Plato] del Departamento de Agricultura de Estados Unidos presenta mensajes seleccionados para que los consumidores se enfoquen en comportamientos clave, tal como equilibrar las calorías mediante la reducción del tamaño de las porciones; aumentar el consumo de alimentos saludables, como frutas, verduras y granos integrales e identificar los alimentos que se deben reducir, incluyendo los alimentos ricos en sodio y las bebidas azucaradas.

N

nematode a type of tiny roundworm that causes damage to specific kinds of plants.

net worth the value of an owner's assets minus liabilities (debts).

neutral neither acidic nor alkaline.

nitrate a form of nitrogen used by plants.

nematodo un pequeño tipo de ascáride que causa daños a un tipo específico de plantas.

valor neto el valor de los activos de un propietario menos los pasivos.

neutro ni ácido ni alcalino.

nitrato un tipo de nitrógeno usado por plantas.

nitrogen fixation conversion of nitrogen gas to nitrate by bacteria.	**fijación del nitrógeno** la conversión de gas de nitrógeno en nitrato realizada por bacterias.
nitrous oxide a compound containing nitrogen and oxygen; they make up about 5 percent of the pollutants in automobile exhaust.	**óxido nitroso** un compuesto que contiene nitrógeno y oxígeno; constituye aproximadamente el 5 por ciento de los agentes contaminantes en el gas de escape de los automóviles.
node portion of the stem that is swollen or slightly enlarged that gives rise to buds.	**nudo** la parte del tallo o pedúnculo que está hinchada o levemente agrandada que produce los botones.
nomenclature a systematic method of naming plants or animals.	**nomenclatura** un método sistemático de dar nombres a las plantas o los animales.
noncontagious diseases that cannot be spread to other animals.	**no contagiosas** enfermedades que se transmiten a otros animales.
no-till seed is planted directly into the residue of the previous crop, without exposing the soil surface.	**sin labranza** dícese de cuando se siembra la semilla directamente en el residuo de la cosecha previa, sin exponer la superficie del suelo.
noxious weed plant that is highly damaging to an environment and that is controlled under authority of state law.	**hierba tóxica** planta que es sumamente dañina al medio ambiente que es controlada bajo la autoridad de la ley estatal.
nuclear membrane a protective membrane that forms a barrier surrounding the cell nucleus.	**membrana nuclear** una membrana protectora que forma una barrera en torno al núcleo de la célula.
nucleic acid original name for deoxyribonucleic acid.	**ácido nucleico** nombre original del ácido desoxirribonucléico.
nucleus a cell structure that contains pairs of chromosomes on which genes are located at specific locations.	**núcleo** una estructura celular que contiene pares de cromosomas en las cuales los genes están ubicados en lugares específicos.
nurse crop a crop used to protect another crop until it can get established.	**cultivo protector** un cultivo que se usa para proteger a otro cultivo hasta que este pueda establecerse.
nursery a place where young trees, shrubs, and other plants are grown.	**vivero** un lugar donde se crían árboles, arbustos y otras plantas jóvenes.
nutrient substance necessary for the functioning of an organism.	**nutriente** sustancia necesaria para el funcionamiento del organismo.
nutrition the process whereby all body parts receive materials needed for function, growth, and renewal.	**nutrición** el proceso en que todas las partes del cuerpo reciben materiales necesarios para función, crecimiento y renovación.

O

O horizon the soil layer on the surface that is composed of organic matter and a small amount of mineral matter.	**horizonte O** la capa de tierra que está en la superficie compuesta de substancias orgánicas y una pequeña cantidad de substancias minerales.
offsetting a position taking a second position on the futures market to offset a first.	**compensar una posición** tomar una posición secundaria en el mercado de futuros para compensar la primera.
oilseed crop crop produced for the oil content of their seeds.	**cosecha de semilla oleaginosa** cultivo producido por el contenido de aceite de sus semillas.
olericulturist one who studies the cultivation of vegetables.	**oleicultor** uno que estudia el cultivo de verduras.
olericulture the cultivation of vegetables.	**oleicultura** el cultivo de verduras.

online video merchandising the use of online video recordings of livestock that are offered for sale at distant locations, thus enabling a potential buyer to bid for the livestock online or via telephone.	**comercialización por vídeo en línea** el uso de las grabaciones de ganado en línea que se ponen a la venta en lugares lejanos, permitiendo así que un comprador potencial haga una oferta por el ganado en línea o por teléfono.
on-the-job training experience obtained while working in an actual job setting.	**entrenamiento práctico** experiencia obtenida mientras se trabaja en un empleo verdadero.
opening a position initial step in the futures market.	**una posición inicial** paso inicial en el mercado de futuros.
order of business the items and sequence of activities conducted at a meeting.	**orden del día** los ítems y la secuencia de actividades conducidos en una reunión.
organic compound a compound that contains carbon.	**compuesto orgánico** un compuesto que contiene carbono.
ornamental a plant grown for its appearance and beauty.	**ornamentales** planta que se cultiva por su apariencia y belleza.
osmosis the flow of a fluid through a semipermeable membrane separating two solutions, which permits the passage of the solvent but not the dissolved substance. The liquid will flow from a weaker to a stronger solution, thus tending to equalize concentrations.	**osmosis** el flujo de un líquido a través de una membrana semipermeable que separa dos soluciones, que permite el paso del solvente pero no de la sustancia disuelta. El líquido fluirá de una solución más débil a una más fuerte, así tendiendo a igualar concentraciones.
out-of-the-market when a futures position is offset by a second position.	**fuera del mercado** cuando una posición de futuros es compensada por una segunda posición.
ova female reproductive cells or eggs from the ovary of an animal.	**óvulo** células reproductivas femeninas o huevos del ovario de un animal.
ovary female organ that produces eggs or female sex cells; also, that portion of the flower that contains the ovules or seeds.	**ovario** órgano de la hembra que produce los óvulos o células sexuales femeninas; además aquella parte de la flor que contiene los óvulos o semillas.
overripe beyond maturity.	**demasiado maduro** más allá de madurez.
overseeding seeding a second crop into one that is already growing.	**sobresembrar** sembrar un cultivo encima de otro que ya está creciendo.
ovulation process of releasing mature eggs from the ovary.	**ovulación** el proceso de liberar óvulos maduros del ovario.
ovule unfertilized seed.	**óvulo (vegetal)** semilla ya no fecundada.
ownership/entrepreneurship SAE a student program through which a student owns an enterprise such as livestock, crops, etc.	**la propiedad/el espíritu empresarial SAE** un programa estudiantil a través del cual un estudiante es dueño de una empresa como ganado, cosechas, etc.
ownership/entrepreneurship SAE supervised activities associated with ownership and management of a for-profit agribusiness or enterprise.	**propiedad/emprendimiento SAE** actividades supervisadas asociadas con la propiedad y administración de una agroindustria o empresa con fines de lucro.
ownership/internship SAE supervised activities of students as owners or entrepreneurs of their own for-profit business.	**propiedad/pasantía SAE** actividades supervisadas de los estudiantes como propietarios o empresarios de su propio negocio con fines de lucro.
oxidative deterioration decay resulting from exposure to air.	**deterioro oxidativo** el deterioro que resulta al exponerse al aire.
ozone compound that exists in limited quantities about 15 miles above the Earth's surface.	**ozono** compuesto que existe en cantidades limitadas aproximadamente quince millas encima de la superficie de la tierra.

P

packer person or firm responsible for preparing commodities for shipment.

parasite organism that lives in or on another organism with no benefit to the host.

parasitism one type of wildlife living and feeding on another without killing it.

parent material the horizon of unconsolidated material from which soil develops.

parliamentary procedure a system of guidelines or rules for making group decisions in business meetings.

particulate a small particle that is suspended in the air.

parturition the birthing process.

pasture forages that are harvested by the livestock itself.

pathogen organism that produces disease.

peat moss a type of organic matter made from sphagnum moss.

penetration pricing a strategy in which price is set less than that of competitors.

percolation movement of water through the soil.

perennial a plant that lives from year to year.

perennial weed a weed that lives for more than 2 years.

perfect flower flower containing all of the parts: stamen, pistil, petals, and sepals.

peripheral nervous system system that controls the functions of the body tissues, including the organs.

perlite natural volcanic glass material having water-holding capabilities.

pesticide chemical used to control pests.

pesticide resistance the ability of a pest to tolerate a lethal level of a pesticide.

pest resurgence ability of a pest population to recover.

petal brightly colored, sometimes fragrant portion of the flower.

petiole the slender leaf stock that supports the blade, attaching it to the stem of a plant.

pH measurement of acidity or alkalinity from 1 to 14.

empaquetador la persona o la empresa responsable de preparar productos básicos para envío.

parásito organismo que vive en o encima de otro organismo sin beneficio al huésped.

parasitismo una clase de fauna que vive en y se alimenta de otra sin matarla.

material matriz el horizonte de material no consolidado del que se forma el suelo.

procedimiento parlamentario un sistema de principios o normas para tomar decisiones de grupo en reuniones de negocio.

partícula una pequeña partícula que queda suspendida en el aire.

alumbramiento el proceso de dar a luz.

pasto forrajes que se cosechan por el ganado mismo.

patógeno organismos que producen las enfermedades.

turba un tipo de materia orgánica formada de esfagno.

precio de penetración una estrategia donde se fija el precio más bajo que el de los competidores.

filtración paso del agua a través de la tierra.

perene una planta que vive de año a año.

mala hierba perenne una mala hierba que vive más de dos años.

flor perfecta flor que consta de todas las partes: estambre, pistilo, pétalos y sépalos.

sistema nervioso periférico sistema que controla las funciones de los tejidos de cuerpo, incluyendo los órganos.

perlita material vítreo volcánico natural que tiene capacidades de contener el agua.

pesticida sustancia química empleada para combatir plagas e insectos nocivos.

resistencia al pesticida la capacidad de plagas o insectos nocivos para tolerar un nivel mortal de pesticida.

resurgimiento de plaga o insecto nocivo la capacidad de una población de plagas o insectos nocivos de recuperarse.

pétalo parte de la flor de color vivo y a veces fragrante.

pecíolo el esbelto rabo de la hoja que sostiene la brizna, uniéndola al tallo de la planta.

valor pH medida de la acidez o alcalinidad desde 1 a 14.

phenotype physical appearance of an individual.	**fenotipo** características externas, o aspecto de un individuo.
pheromone a chemical secreted by an organism to cause a specific reaction by another organism of the same species.	**feromona** una sustancia química secretada por un organismo para causar una reacción específica por otro organismo de la misma especie.
phloem cells forming conductive tissues that carry manufactured food to areas of the plant where it is stored or used.	**floema** las células que forman los tejidos conductores que transportan los productos alimenticios que se han elaborado a las zonas de la planta en las que se almacenan o se utilizan.
photodecomposition chemical breakdown caused by exposure to light.	**fotodescomposición** la descomposición química causada por la exposición a la luz.
photosynthesis process in which chlorophyll in green plants enables those plants to use light to manufacture sugar from carbon dioxide and water.	**fotosíntesis** proceso en la cual la clorofila de las plantas verdes las capacita para usar la luz para fabricar azúcar del dióxido de carbono y agua.
phototropism a process by which a plant leaf is capable of adjusting its angle of exposure to the sun.	**fototropismo** un proceso por el cual la hoja de una planta es capaz de ajustar su ángulo de exposición al sol.
physiology study of the functions and vital processes of living creatures and their organs.	**fisiología** el estudio de las funciones y los procesos vitales de criaturas vivientes y sus órganos.
pistil female part of the flower consisting of stigma, style, ovary, and ovule.	**pistilo** parte femenina de las flores que consta de estigma, estilo, ovario y óvulo.
placement/Internship SAE a student experience where the student works with a business, farm, or other place in order to learn skills	**colocación/Práctica SAE** un estudiante experiencia donde el estudiante trabaja con una empresa, granja u otro lugar para aprender habilidades
placement/internship SAE The student works for a business, enterprise, or agency related to agriculture, forestry or natural resources.	**colocación/pasantía SAE** El estudiante trabaja para un negocio, empresa o agencia relacionada con la agricultura, la silvicultura o los recursos naturales.
placement supervised agricultural experience career experiences that place students with employers who are conducting agricultural business such as farming ranching, greenhouses, and others.	**colocación supervisada de experiencia agrocientífica** experiencias de carrera que colocan a estudiantes con empleadores que conducen negocios agriculturales tales como agricultura, ganadería, viveros y otros.
plan to think through.	**planear** pensar muy bien en cómo se va a hacer algo.
plant hardiness zone map a map developed by the U.S. Department of Agriculture, dividing the country into zones based on average winter temperatures.	**mapa de zonas de resistencia de las plantas** mapa desarrollado por el Departamento de Agricultura de los EE.UU., que divide al país en zonas basándose en las temperaturas medias de invierno.
plant nutrition provision of elements to plants.	**nutrición vegetal** el suministro de elementos a las plantas.
plantscaping the design and arrangement of plants and structures in an indoor area.	**paisajismo de interiores** el diseño y el arreglo de plantas y de estructuras en un área de interior.
plugging establishment of turf by using small pieces of existing turfgrass.	**cubrir de trasplantes de tepe** establecer el césped usando pequeños pedazos de césped ya existente.
plywood construction material made of thin layers of wood glued together.	**madera contrapechada** material de construcción hecho de hojas delgadas de madera adheridas.
point of order a procedure used to object to some item in or about a meeting that is not being done properly. The procedure to use is to say, "point of order." The presiding officer should then recognize the member by saying, "State your point."	**cuestión de orden** un procedimiento que sirve para oponerse a algún ítem de o sobre la reunión que no está haciéndose adecuadamente. El procedimiento es decir, «¡Cuestión de orden!» Entonces el funcionario que preside reconoce al miembro y dice «Exprese su cuestión.»

pollen small male sperm or grains that are necessary for fertilization in the flower.

pollination transfer of pollen from anther to stigma.

polluted containing harmful chemicals or organisms.

pome fleshy fruits with embedded core and seeds.

pomologist a fruit grower or scientist.

pony member of the horse family less than 14.2 hands tall.

porcine somatotropin (PST) hormone that increases meat production in swine.

pore spaces between soil particles through which plant roots penetrate, and in which air, water, and nutrients are stored.

pork meat from swine.

portfolio a collection of papers and other material that represents a body of work over a period of time.

postemergence herbicide a herbicide applied after the weed or crop is present.

potable drinkable—that is, free of harmful chemicals and organisms.

poult young turkey.

ppm parts per million.

ppt parts per thousand.

precipitate a chemical action that results in the dropout of solids in a solution.

pre-cooling rapid removal of heat before storage or shipment.

predation a way of life where one type of wildlife eats another type.

predator an animal that feeds on a smaller or weaker animal.

preemergence herbicide a herbicide applied before weed or crop germination.

presiding officer a president, vice president, or chairperson who is designated to lead a business meeting.

prestige pricing pricing to buyers with special desires for quality, fashion, or image.

prey animal eaten by another animal.

primary nutrients in agriculture, nitrogen, phosphorus, and potash.

processing turning raw agricultural products into consumable food.

polen esperma pequeño o polvillo macho que es necesario para la fecundación de la flor.

polinización transporte del polen del antero hasta el estigma.

contaminado que contiene sustancias químicas u organismos dañinos.

pomo frutos carnosos con carozo y semillas.

pomólogo científico o cultivador de frutos.

poni miembro de la familia de los caballos que mide menos de 14.2 palmos de alto.

somatotropino porcino una hormona que aumenta la producción de carne en los cerdos.

poro espacio entre partículas de tierra en los que penetran las raíces y donde se almacenan el aire, el agua y los nutrientes.

puerco carne del cerdo.

portafolio una colección de documentos y otro material que representa un cuerpo de trabajo durante un periodo de tiempo.

herbicida postemergente un herbicida aplicado después de que se presenta la mala hierba o el cultivo.

potable que puede beberse, o sea que está libre de organismos y sustancias químicas dañinas.

pavipollo pavo jóven.

partes por millón partes por millón.

partes por mil partes por mil.

precipitado una reacción química que resulta en el depósito de sedimento en una solución.

prerefrigeración eliminación rápida del calor antes de almacenamiento o transporte.

depredación una manera de vivir en que un tipo de fauna se alimenta de otro tipo.

predador un animal que se alimenta con otro animal más pequeño o más débil.

herbicida pre-emergente un herbicida aplicado antes de la germinación de malas hierbas o cultivos.

funcionario el presidente, vicepresidente o presidente interino designado para dirigir una sesión de negocio.

precio de prestigio valoración para los compradores que tienen deseos especiales de calidad, moda o imagen.

presa un animal que otro animal se ha comido.

nutrientes principales en la agricultura, nitrógeno, fósforo y potasa.

procesado convertir los productos agrícolas en bruto en alimentos comestibles.

processor person or business cleaning, separating, handling, and preparing a product before it is sold to a distributor.

producer person who grows a crop.

product advertising advertising that focuses on the product itself.

production agriculture farming and ranching.

Production Credit Association(s) (PCA) lending institution that provides short-term credit for agriculture.

production or productive enterprise project conducted for wages or profit.

profession occupation requiring an education, especially law, medicine, teaching, or the ministry.

proficiency awards awards that are given to students based on superior Supervised Agricultural Experiences.

profit proceeds from a business transaction.

profit and loss statement projection of costs and expenses against sales and revenue over time.

progeny offspring.

progesterone hormone that prevents estrus during pregnancy and causes development of the mammary system.

project a series of activities related to a single objective or enterprise.

promissory note note signed by borrower agreeing to the terms of the loan.

promoting function plan to make potential customers aware of the product or service.

propagation process of increasing the numbers of a species or perpetuating a species.

protectant fungicide a fungicide applied before disease infection.

protein nutrient made up of amino acids and essential for maintenance, growth, and reproduction.

protoplasm all of the contents within the walls of a living cell.

protozoa microscopic, one-celled animals that are parasites of animals.

pruning the removal of dead, broken, unwanted, diseased, or insect-infested wood.

---

procesador persona o empresa que limpia, separa, maneja y prepara un producto antes de que se venda al distribuidor.

productor persona que cultiva el cultivo.

publicidad de producto publicidad que se concentra en el producto mismo.

agricultura de producción cultivo y ganadería.

Asociación(es) de Crédito para Productores institución de préstamo que provee crédito a corto plazo para la agricultura.

producción o empresa productiva proyecto conducido por salarios o lucro.

profesión vocación que exige una enseñanza superior, especialmente el derecho, la medicina, la enseñanza o el ministerio.

premios de competencia los premios que se conceden a los estudiantes con experiencias agrícolas supervisadas superiores.

beneficio ganancias de una transacción comercial.

estado de ganancias y pérdidas proyección de los costos y gastos comparados con las ventas e ingresos por un tiempo.

progenie descendencia.

progesterona hormona que impide el estro durante el embarazo y causa el desarrollo del sistema mamario.

proyecto una serie de actividades relativas a un solo objetivo o empresa.

pagaré obligación escrita firmada por el prestatario en la cual acepta las condiciones y los plazos del préstamo.

función de promocionar plan para hacer que los clientes potenciales se den cuenta del producto o servicio.

propagación el proceso de aumentar la población de una especie o de perpetuar una especie.

fungicida preventivo un fungicida aplicado antes de la infección.

proteína nutriente que consta de aminoácidos y que es esencial para mantener la vida, el crecimiento y la reproducción.

protoplasma todos los contenidos dentro de las paredes de una célula viva.

protozoarios microorganismos unicelulares que son parásitos de otros animales.

podar eliminar la madera muerta, trozada, no deseada, enferma o plagada por los insectos.

psychology the science of human and animal behavior.	**psicología** la ciencia del comportamiento humano y animal.
psychological pricing a strategy designed to make a price seem lower or less significant.	**valoración psicológica** una estrategia creada para hacer que un precio parezca más bajo o menos significativo.
pullet young female chicken.	**polla** hembra jóven del pollo.
pulpwood wood used for making fiber for paper and other products.	**madera para pasta de papel** madera usada para hacer fibras para papel y otros productos.

Q

quality grade grade based on amount and distribution of finish on an animal.	**nivel de calidad** nivel basado en la cantidad y distribución del acabado de engorde de un animal.
quarantine isolation of pest-infested material.	**cuarentena** aislamiento de material plagado.
queen only fertile, egg-laying female bee in the hive.	**reina** la única abeja hembra del panal que es fecunda y pone huevos.

R

radioactive material material that is emitting radiation.	**material radioactivo** material que emite la radiación.
radon colorless, radioactive gas formed by disintegration of radium.	**radón** gas radioactivo incoloro, formado por la desintegración del rádium.
ram male member of the sheep family.	**carnero** macho de la oveja.
ration the amount of feed fed in one day.	**ración** la porción de alimento que se da de comer en un día.
real-world experience an activity conducted in the daily routine of our society.	**experiencia de la vida diaria** una actividad realizada dentro de la rutina diaria de nuestra sociedad.
reaper machine that cuts grain.	**segador** máquina que corta cereales.
recessive a gene that remains hidden and expresses itself only in the absence of a dominant gene.	**recesivo** un gen que queda oculto y se expresa solo en ausencia de un gen dominante.
recombinant DNA technology gene splicing.	**tecnología de recombinación del ADN** empalme de genes.
recordkeeping the process of keeping a journal or portfolio of everything you have done.	**mantenimiento de datos** el proceso de escribir en un diario o armar una carpeta de todo lo que se ha hecho.
recordkeeping for every project, students capture and record the following data: date, name or description, hours spent, and expenses or income that resulted.	**mantenimiento de registros:** para cada proyecto, los estudiantes capturan y registran los siguientes datos: fecha, nombre o descripción, horas dedicadas y gastos o ingresos resultantes.
recuperative potential the ability of a plant to recover after being damaged.	**potencial de recuperación** la capacidad de la planta para reestablecerse después de ser dañada.
refer a motion used to refer some other motion to a committee or person for finding more information and/or taking action on the motion on behalf of the members. The way to state a referral is to say, "I move to refer this motion to . . ."	**referir** una moción usada para referir otra moción a una comisión para averiguar más información y/o para tomar acción sobre la moción en nombre de los socios. La manera de proponer una referencia es decir: decir, «Yo propongo referir esta moción a . . .»
renewable natural resources resources provided by nature that can replace themselves.	**recursos naturales renovables** recursos suministrados por la naturaleza que pueden recuperarse solos.
reproduction the generation of a new plant or animal.	**reproducción** la creación de una planta o animal nuevo.

Research SAE a student experience where the student conducts and reports on a research experiment or topic.

research SAE students are involved in the investigation of materials, processes, and information, which they use to establish new knowledge or to validate previous research.

residual soil parent material formed in place.

resource substitution the use of a resource or item for another when the results are the same.

respiration a process in which energy and carbon dioxide are released due to digestion or the breakdown of plant tissues during periods of darkness.

respiratory system system that provides oxygen to the blood of the animal.

resumé a summary of a person's education, technical skills, and career experiences.

retailer person or store that sells directly to the consumer.

retail marketing the selling of a product directly to consumers.

retortable pouch package with foil between two layers of plastic.

rhizome horizontal, underground stems.

ribosome a cell structure that is responsible for synthesizing proteins.

rooster adult male chicken or pheasant.

rootbound roots restricted by a container.

root cap the outermost part of a root that protects the tender tip of a growing root as it penetrates the soil.

root crop crop grown for the thick, fleshy storage root that it produces.

root cutting section of a root cut used for propagation purposes.

root hairs small microscopic roots that arise from the cells located on the surface of a plant root.

rooting hormone chemical used to stimulate root formation on a cutting.

Investigación SAE una experiencia estudiantil en la cual el estudiante realiza y hace un informe sobre una investigación o un tema.

investigación SAE: los estudiantes participan en la investigación de materiales, procesos e información, que utilizan para establecer nuevos conocimientos o para validar investigaciones anteriores.

suelo residual materiales parentales del suelo formados en el lugar.

sustitución de recursos el uso de un recurso o cuerpo en vez de otro cuando los resultados son semejantes.

respiración un proceso que libera energía y bióxido de carbono gracias a la digestión o la descomposición de tejidos vegetales durante períodos de oscuridad.

sistema respiratorio sistema que proporciona oxígeno a la sangre del animal.

currículum vitae un resumen de la educación de una persona, sus habilidades técnicas y experiencias de carrera.

minorista persona o tienda que vende directamente al consumidor.

comercialización minorista la venta de un producto directamente al consumidor.

embalaje de retortas paquete que tiene una hoja de aluminio entre dos hojas de plástico.

rizoma tallo subterráneo, horizontal.

ribosoma una estructura celular que es responsable de la síntesis de las proteínas.

gallo adulto macho del pollo.
faisán adulto macho de los faisanes.

raíces restringidas por el recipiente raíces restringidas por un recipiente.

capucha de raíz la parte más externa de una raíz que protege la tierna punta de una raíz en crecimiento mientras penetra la tierra.

cosecha de vegetales de raíces cultivo realizado para obtener el tubérculo grueso carnoso que produce.

estaca de raíz cuando se corte parte de la raíz y se usa para fines de propagación.

peluza de raíz pequeñas raíces microscópicas que surgen de las células localizadas en la superficie de la raíz de una planta.

hormona radicular sustancia química usada para estimular la formación de raíces en un esqueje.

root pruning systematic cutting of the roots by hand or machine to encourage the roots to develop within the root ball range.	**poda de raíces** el podar sistemáticamente las raíces a mano o a máquina para alentar a que las raíces se desarrollen dentro del área de la raíz.
rootstock root system and stem of a plant on which another plant is grafted.	**portainjerto** sistema radicular y tallo de una planta sobre la cual se injerta otra planta.
roughage grass, hay, or silage and other feeds high in fiber and low in TDN.	**forraje** hierba, heno o ensilaje y otros alimentos con alto contenido de fibra y bajo contenido de NDT.
roundworm slender worm that is tapered on both ends.	**ascáride** lombriz delgado que tiene aspecto afilado en sus dos extremos.
rumen one of the compartments of the stomach in cattle, deer, and sheep that is responsible for the breakdown of cellulose in the feed that is consumed.	**rumen** uno de los compartimentos del estómago de ganado, venados y ovejas responsable de descomponer la celulosa que contiene el alimento consumido.
ruminant animal that has a stomach with four digestive compartments.	**rumiante** animal que tiene un estómago con cuatro compartimentos.
russet a baking potato variety best adapted for growth in sandy vocanic soils.	**russet** una variedad de la patata para el horno adaptada para el crecimiento en suelos volcánicos arenosos.

S

SAE agreement This document defines the scope of the experience, responsibilities, and roles, and identifies safety issues and is planned and signed by the student and all participating adults.	**Acuerdo SAE:** este documento define el alcance de la experiencia, las responsabilidades y los roles, e identifica los problemas de seguridad y está planificado y firmado por el estudiante y todos los adultos participantes.
salmonid any of the family of soft-finned fish such as salmon or trout that have the last vertebrae upturned.	**salmónicos** cualquiera de los integrantes de la familia de peces de aletas blandas, tales como el salmón y la trucha, que tienen la última vértebra doblada hacia arriba.
saltwater marsh the ecological system of plants and animals influenced by tidal waters with salinity of 15 to 34 parts per million.	**salina** sistemas ecológicos de plantas y animales influidos por mareas con salinidad de 15 a 34 partes por mil.
sand largest soil particles; 1 to 0.05 mm.	**arena** las partículas de tierra más grandes; de 1 a 0.05 mm.
sapwood transports water and minerals upward in tree roots, trunks, and stems.	**sámago** traslada agua y minerales hacia arriba en las raíces, los troncos y tallos de árbol.
scale size of items.	**escala** tamaño de un cuerpo.
scarify to soak, keep moist, or mechanically scrape a seed coat to aid germination.	**escarificar** empapar, mantener húmedo o raspar a máquina la capa de una semilla para alentar la germinación.
school-based enterprise SAE a Supervised Agricultural Experience that is conducted on the school campus.	**actividad académica SAE** una experiencia agrícola supervisada que se lleva a cabo en la escuela.
school-based enterprise SAE similar to an Entrepreneurship SAE, but this operation is based at the school or FFA chapter and may involve a group of students working cooperatively.	**SAE empresarial basado en la escuela:** similar a un SAE de Emprendimiento, pero este la operación se basa en la escuela o el capítulo de la FFA y puede involucrar a un grupo de estudiantestrabajando cooperativamente.
scientific method a procedure for investigating problems of a scientific nature.	**método científico** un procedimiento para investigar problemas de naturaleza científica.

scientific process a step-by-step process that a researcher uses to ensure that the research and outcome is correctly completed.	**proceso científico** un proceso desarrollado paso a paso que un investigador utiliza para garantizar que la investigación y sus resultados se han completado correctamente.
scion the top of the stem and leaves of a plant that is to be propagated.	**codo o varilla de injerto** la punta del tallo y hojas de una planta que se va a propagar.
scurvy vitamin C deficiency disease.	**escorbuto** una enfermedad causada por la deficiencia de vitamina C.
seasonal pertaining to a particular season of the year.	**estacional** que pertenece a una estación determinada del año.
secondary host a plant or animal that carries a disease or parasite during part of the life cycle.	**huésped secundario** una planta o un animal que porta una enfermedad o parásito durante parte del ciclo de la vida.
secondary nutrient nutrients required by plants in moderate amounts; they include calcium, magnesium, and sulfur.	**nutrientes secundarios** los nutrientes requeridos por las plantas en cantidades moderadas; incluyen el calcio, el magnesio y azufre.
seed blend combination of different cultivars of the same species.	**mezcla de semilla** combinación de cultivos diferentes de la misma especie.
seed culm stem that supports the inflorescence of the plant.	**caña de semilla** tallo que permite la florescencia de la planta.
seed legume crop crop that is nitrogen-fixing and produces edible seeds.	**legumbre** cultivo que es fijador del nitrógeno y produce semillas comestibles.
seedling young plant grown from seed.	**plántula** planta joven cultivada de semilla.
seed mixture combination of two or more species.	**mezcla de semillas** combinación de dos o más especies.
selective breeding mating adults who have characteristics desired in the offspring.	**crianza selectiva** selección de padres para apareamiento que tienen los rasgos que se desean para la descendencia.
selflessness placing the desires and welfare of others above oneself.	**abnegación** colocar los deseos y el bienestar de los demás por encima de los de uno mismo.
selling function market research, sales plans, and sales closures.	**función de venta** las investigaciones del mercado, las ventas, los planes y cierres de venta.
semen the sperm cells and accompanying fluids.	**semen** las células de los espermatazoides y los líquidos que los acompañan.
semiarid an area partially deficient in rainfall; dry.	**semiárido** un área que parcialmente carece de precipitación; que está seco.
semidwarf tree that has rootstock that limits growth to 15 feet or fewer.	**semienano** árbol que tiene un rizoma que limita su crecimiento a 15 pies o menos.
seminal root root formed at the time of seed germination to anchor the seed in soil.	**raíz seminal** raíz formada en el momento de la germinación de la semilla para fijar la semilla en el suelo.
semipermeable membrane membrane that permits a solution to move through it.	**membrana semipermeable** membrana que permite que la penetre una solución.
sepal small, green, leaf-like structure found at the base of the flower.	**sépalo** cuerpo verde pequeño y de aspecto de hoja que se encuentran en la base de la flor.
service learning SAE a Supervised Agricultural Experience where the student does a significant amount of community service.	**servicio de aprendizaje SAE** una experiencia agrícola supervisada donde el estudiante realiza una cantidad significativa de servicio a la comunidad.

service-learning SAE students are involved in solving or improving community and agricultural issues or problems.	**aprendizaje-servicio SAE** los estudiantes están involucrados en resolver o mejorar asuntos o problemas comunitarios y agrícolas.
sequence related or continuous series.	**sucesión** series continuas o relacionadas.
sewage system receives and treats human waste.	**sistema de cloacas** recibe y trata los desechos humanos.
sex-linked genes carried on chromosomes that determine sex.	**(herencia) ligada al sexo** genes que se transmiten por los cromosomas que determinan el sexo.
sexual reproduction union of an egg and sperm to produce a seed or fertilized egg.	**reproducción sexual** unión de un óvulo y un esperma para producir una semilla u óvulo fecundado.
shackles mechanical devices that restrict movement.	**grilletes** dispositivos mecánicos que restringen el movimiento.
sheath lower portion of the turfgrass leaf.	**vaina foliar** parte a la base de la hoja de césped.
sheet erosion removal of soil from broad areas of the land.	**erosión laminar** eliminación del suelo de áreas amplias del terreno.
shellfish any aquatic animal having a shell or shell-like exoskeleton.	**mariscos** cualquier animal acuático que tiene cáscara o dermatoesqueleto parecido a un cascarón.
short-term loan loan with a payment period of 1 year or fewer.	**préstamo a corto plazo** préstamo con plazos de pago de un año o menos.
short-term plan plan accomplished in days or weeks.	**plan a corto plazo** plan realizado dentro de días o semanas.
shrinkage changes in dimensions of wood as it reacts to changes in humidity and temperature.	**contracción** disminución de volumen de la madera cuando reacciona a los cambios de humedad y temperatura.
shrouded wrapped tightly with a cloth.	**amortajado** envuelto ajustadamente con un paño.
shrub a woody perennial plant that normally produces many stems or shoots from the base and does not reach more than 15 feet in height.	**arbusto** plantas perennes leñosas que normalmente producen muchas ramas o brotes desde la base, y no alcanzan a más que 15 pies de alto.
signal word required words on the label that denotes the relative toxicity of the product.	**palabra señalizadora** palabras que se deben incluir en la etiqueta para indicar la toxicidad relativa del producto.
silage feed resulting from the storage and fermentation of green crops in the absence of air.	**ensilaje** alimento formado por la colocación en silos y fermentación de cultivos verdes en ausencia de aire.
silo airtight storage facility for silage.	**silo** almacenamiento estanco para el ensilaje.
silt intermediate soil particles; 0.05 to 0.002 mm.	**légamo** partículas de tierra intermedias; 0.05 a 0.002 mm.
silviculture the scientific management of forests.	**silvicultura** la administración científica de los bosques.
simple layering stem is bent to the ground, held in place, and covered with soil.	**acodo simple** el tallo es doblado hasta tocar la tierra, sujetado y cubierto con tierra.
simple leaf a single leaf arising from a plant stem.	**hoja simple** una hoja singular que nace de un tallo de planta.
single-eye cutting cutting made with the node of a plant with alternate leaves.	**injertos con un solo nudo** injerto realizado con la yema de una planta que tiene hojas alternadas.
skeletal system bones joined together by cartilage and ligaments.	**sistema esqueletal** huesos unidos por cartílago y ligamentos.

skimming setting the price of a new product for unusually high profits at first when affluent and willing customers are available.

Small Business Administration (SBA) institution that provides loans to agribusinesses.

smoker device used to add smoke flavor and taste to food.

sociology focuses on human issues and problems.

sodding removal of a portion of the soil and turfgrass plant for vegetative establishment purposes.

softwood wood from conifers.

soil top layer of the Earth's surface suitable for the growth of plant life.

soil amendment additive that improves the soil.

soil profile a cross-sectional view of soil.

soil science study of the properties and management of soil to grow plants.

soil structure the way soil granules bind together in clumps; includes the pore arrangement in the aggregates.

sow a mature female pig that has given birth.

spawn egg-laying process of fish.

species the basic unit in the classification system whose members have similar structure, common ancestors, and maintain their characteristics; subgroup of genus.

specimen plant plant that is used as a single plant to highlight it or some other special feature of the landscape.

sperm male reproductive units.

sphagnum pale and ashy mosses used to condition soil.

split carcass halves of an animal after it is killed.

split-vein cutting cutting made by slitting a large leaf on the veins before placing in a rooting medium.

sprigging planting a section of a rhizome or stolon (sprig) in the soil.

sprinkler device that sprays water on crops; its effect is much like rain.

fijación de precios de nivel elevado fijar el precio de un producto nuevo para que se saquen ganancias extraordinariamente altas al principio, cuando hay clientes adinerados y dispuestos.

Departamento de Pequeñas Empresas hace empréstitos a las agroindustrias.

secador al humo dispositivo que se usa para añadir el sabor ahumado a los comestibles.

sociología ciencia que se enfoca en los asuntos y problemas de los humanos.

cubrir con césped excavación de una parte del suelo y césped para fines de establecimiento vegetal.

madera blanda la madera de las coníferas.

tierra, suelo capa de la superficie de la tierra que es propicia para el crecimiento de la vegetación.

enmienda sustancia que rectifica los suelos.

pérfil del suelo una vista del suelo por corte transversal.

ciencia de estudio de los suelos el estudio de las propiedades y administración de los suelos para cultivar plantas.

estructura de los suelos la manera en la que los gránulos de la tierra se unen para formar terrones; incluye el arreglo de los agregados.

cerda un cerdo hembra adulto que ha dado a luz.

poner huevos en los peces, el proceso de poner huevos.

especie la unidad básica del sistema de clasificación cuyos miembros tienen estructura semejante, antepasados comunes y mantienen sus características; subgrupo de género.

planta maestra planta que se usa como planta singular para mostrarla o para mostrar otro rasgo especial del paisaje.

esperma unidades reproductivas del macho.

esfagno musgo pálido y ceniciento usado para acondicionar el suelo.

carne en canal las mitades del animal después de la matanza.

esqueje de nervio partido un injerto realizado cortando una hoja grande por sus nervios antes de colocarla en un medio para echar raíces.

propagación por estacas de raíz plantar una parte de un rizoma o estolón (ramito) en el suelo.

aspersor dispositivo que rocía agua al aire para regar los cultivos; tiene casi el mismo efecto que la lluvia.

sprinkler irrigation the use of pipes to deliver water under pressure to a sprinkler head for the purpose of watering plants that are remote from a water source.

squab young pigeon that is butchered for meat before it is old enough to fly.

stake driving a wooden or metal rod into the ground near the plant and tying the plant to the rod.

stallion adult male horse or pony.

stamen male part of the flower that contains the pollen, anther, and filament.

starch major energy source in livestock feeds.

starter solution diluted mixture of single or complete fertilizers used when plants are transplanted.

statistics a branch of mathematics that is used to collect large amounts of data, organize it in number form, and analyze it.

steer castrated male member of the cattle family.

stem section cutting a section of plant stem containing a node that is placed in a plant medium to generate a new plant.

stem tip cutting cutting taken from the end of a stem or branch, normally including the terminal bud.

stigma part of the pistil that receives the pollen.

stimulant crop crop that stimulates the senses of users.

stolon a stem that grows above ground.

stolonizing type of vegetative establishment in which sprigs are broadcast onto the soil surface.

stoma small openings, usually on the lower side of the leaf, that control movement of gases.

strip cropping alternating strips of row crops with strips of close-growing crops.

style enlarged terminal part of the pistil.

subcutaneous under the skin.

subsoil a soil layer that corresponds to the B horizon that is composed almost entirely of mineral with generally large chunky soil structure.

succulent in horticulture, thick, fleshy leaves or stems that store moisture.

sucrose compound cane sugar.

riego por aspersión el uso de cañerías para entregar el agua bajo presión a una cabeza de aspersión con el fin de regar plantas que están lejos de una fuente de agua.

pichón aves jóvenes que son matadas por su carne antes de ser suficientemente maduras para volar.

estaca clavar una estaca o varilla de metal en el suelo cerca de la planta y amarrar la planta a la varilla o estaca.

semental adulto macho del caballo o poni.

estambre órgano sexual masculino de la flor que consiste en el polen, antero y filamento.

almidón fuente principal de energía en los alimentos.

abono de material inicial una mezcla diluida de abonos individuales o completos que se usan cuando se trasplantan las plantas.

estadística una rama de las matemáticas que se utiliza para recolectar, organizar y analizar enorme cantidades de datos.

bueyezuelo macho castrado de la familia vacuna.

corte de sección del tallo una sección del tallo de la planta que contiene un nudo que es colocado en un medio de cultivo para generar una nueva planta.

esqueje de punta de tallo un esqueje sacado del extremo de un tallo o ramo, normalmente incluye el botón terminal.

estigma parte del pistilo que recibe el polen.

cosecha estimulante cosecha que estimula los sentidos de los usuarios.

estolón un tallo que crece encima de la tierra.

sembrar estolones al voleo tipo de establecimiento vegetal en que los ramitos son sembrados al voleo en la superficie del suelo.

estomas aberturas pequeñas, usualmente en la vaina exterior de la hoja, que controlan el movimiento de gases.

cultivo en fajas fajas alternas de cultivos en hilera con fajas de cultivos que crecen muy juntos.

estilo parte terminal agrandada del pistilo.

subcutáneo debajo de la piel.

subsuelo una capa del suelo que corresponde al horizonte B que está compuesto casi completamente por minerales y grandes trozos de tierra.

planta suculenta en la horticultura: hojas gruesas y carnosas que almacenan agua.

sucrosa azúcar de caña compuesto.

sulfur pale-yellow element occurring widely in nature.	**azufre** elemento amarillo pálido muy común en la naturaleza.
Supervised Agricultural Experience SAE activities of the student outside the agricultural class or laboratory done to develop agricultural skills.	**experiencia agrícola supervisada** (SAE, por sus siglas en inglés) actividades que realiza el estudiante fuera de la clase de agricultura o del laboratorio para desarrollar habilidades en agricultura.
supplement a nutrient added to supply a deficiency in plants and animals.	**suplemento** un nutriente agregado para suplir una deficiencia en plantas y animales.
supply the quantity of a product that is available to buyers at a given time.	**oferta** la cantidad de un producto que está disponible a los consumidores en un momento dado.
surface irrigation water flows over the soil surface to the crop.	**riego de superficie** el agua fluye sobre la superficie del suelo al cultivo.
sweetbreads thymus and pancreatic glands of animals.	**mollejas** glándulas de timo y páncreas de los animales.
symbol a warning illustration such as a skull and crossbones located on a chemical container that warns of chemical toxicity to humans and animals.	**símbolo** una ilustración de advertencia como una calavera y huesos localizados en un recipiente de químicos que advierte de toxicidad química para animales y humanos.
symmetrical in landscaping, a planting that is equal in number, size, and texture on both sides.	**simétrico** en el paisajismo, colocación de plantas igual cantidad, tamaño y textura a ambos lados de un eje central.
syringing a light application of water to a turfgrass.	**aplicación con mango de riego** una aplicación ligera de agua al césped.
systemic herbicide a herbicide that is absorbed by the roots of a plant and translocated throughout the plant.	**herbicida sistemático** un herbicida que es absorbido por las raíces de la planta y transportado por la planta.

T

tack equipment or gear for horses.	**arreo** guarniciones o aparejo para caballos.
tact skill of encouraging others in positive ways.	**tacto** característica o facilidad de animar a los demás de manera positiva.
TAN measurement of total ammonial nitrogen.	**NAT** medida de nitrógeno amoniacal total.
tankage dried animal residue after slaughter.	**fertilizante orgánico** los desechos secos del animal después de la matanza.
taproot large main root of the system; usually has little or no branch roots.	**raíz eje** gran raíz primaria del sistema radicular; usualmente tiene pocas o ninguna raíz secundaria.
targeted pest identified pest that, if introduced, poses a major economic threat.	**plagas objetivo** plaga o insecto nocivo que, si se introduce, representa una amenaza económica mayor.
taxonomy systematic classification of plants and animals.	**taxonomía** clasificación sistemática de plantas y animales.
T-budding a grafting method in which a bud is placed on a rootstock using a vertical cut in the stem with a horizontal cut across the top forming a slit into which the bud is placed.	**retoños en T** un método de injerto en el cual un codo es unido a un tallo usando un corte vertical en el tallo con un corte horizontal a lo largo de la parte superior formando una ranura dentro de la cual se coloca el injerto.
TDN total digestible nutrients; the measure of digestibility of feed.	**NDT** nutrientes digeribles totales; medición de la digestibilidad del alimento.
technology application of science to an industrial or commercial objective; also, the equipment and expertise to cultivate, harvest, store, process, and transport crops for consumption.	**tecnología** la aplicación de la ciencia a un objetivo industrial o comercial; también, el equipo y pericia para cultivar, cosechar, almacenar, procesar y transportar las cosechas para su consumo.

terminal growing at the end of a branch or stem.	**terminal** dícese de algo que está creciendo en el extremo de una rama o un tallo.
terminal bud bud at the end of a twig or branch.	**botón terminal** yema que está en el extremo de una rama o ramita.
terminal market a stockyard that acts as a place to hold animals until they are sold to another party.	**mercado terminal** un corral que sirve para guardar a los animales hasta que se vendan a otro individuo.
terrace soil or wall structure built across the slope to capture water and move it safely to areas where it will not cause erosion.	**bancal** estructura de tierra o pared construida a lo largo del declive para coger el agua y transportarla seguramente a las áreas en donde no causará erosión.
terrestrial land organism.	**terrícola** organismo que vive en la tierra.
testosterone male sex hormone.	**testosterona** hormona sexual masculina.
tetraethyl lead colorless, poisonous, and oily liquid.	**plomo tetraetílico** líquido incoloro, tóxico y aceitoso.
texture visual or surface quality.	**textura** cualidad visual o de la superficie de algo.
thatch building of organic matter on the soil and around turfgrass plants.	**fertilización del suelo** la fomentación de materia orgánica en el suelo y alrededor de las plantas de césped.
thermal requirement classification of plants according to growing season required.	**requisito termal** clasificación de las plantas según la temporada de cultivo necesaria.
thymine (T) a base in genes designated by the letter *T*.	**timino** una base en los genes nominada por la letra «T».
tidewater water that flows up the mouth of a river as the ocean tide rises and comes in.	**agua de marea** agua que fluye a la boca de un río mientras la marea sube y entra.
tiller a young plant shoot that grows from the crown of a plant.	**retoño** brote joven de una planta que emerge del botón axilar de la corona de una planta.
timberland forest land capable of producing more than 20 cf3 of industrial wood per year.	**bosque productor de madera** tierra forestal capaz de producir más de 20 pies cúbicos de maderos industriales por año.
tip layering tip of a shoot placed in media and covered.	**acodo de punta** la punta un brote que se mete en un medio de cultivo y se tapa.
tissue culture plant reproduction using small, actively growing plant parts under sterile conditions and medium.	**cultivo de tejidos** reproducción vegetal usando partes pequeñas y de crecimiento activo de la planta bajo condiciones y medios estériles.
tofu food made by boiling and crushing soybeans and letting them coagulate into curds.	**tofu** comestible elaborado hirviendo y triturando la soja y dejando que se coagule para formar requesón.
tom male turkey.	**pavo** macho de la pava.
topsoil desirable proportions of plant nutrients, chemicals, and living organisms located near the surface that support good plant growth.	**capa superficial del suelo** porciones apropiadas de nutrientes vegetales, sustancias químicas y organismos vivos ubicados cerca de la superficie que permiten un buen crecimiento vegetal.
townhouse one of a row of houses connected by common side walls.	**vivienda condominio** una de una hilera de casas conectadas por paredes laterales comunes.
toxicity a measurement of how poisonous a chemical is.	**toxicidad** una medida de la toxadad de una sustancia química.
toxin(s) a poisonous substance, causing injury to animals or plants.	**toxina** una sustancia capaz de producir efectos tóxicos, resultando en daño a animales o plantas.
tractor source of power for belt-driven machines, as well as for pulling.	**locomotora de tracción** fuente de potencia para máquinas propulsadas por correa, y también se usa para jalar.

transpiration process by which a plant loses water vapor.

transplant plant grown from seed in a special container.

tree a woody plant that produces a main trunk and has a more or less distinct and elevated head (a height of 15 feet or more).

tripe the pickled rumen of cattle and sheep.

truck cropping large-scale production of a few selected vegetable crops that are shipped to wholesale markets.

trucker person transporting commodities from farm to consumer.

tuber specialized food-storage stem that grows underground.

turbidity measure of suspended solids.

turfgrass grass that is mowed frequently for a short and even appearance.

turgor swollen or still condition as a result of being filled with liquid.

U

underripe as applied to vegetation, fruits or seeds, any that is not mature.

unsoundness in horses, any abnormality that affects the use of the horse.

urinary system system that removes waste materials from the blood.

V

vaccine a preparation introduced into the body to prevent a disease by stimulating antibody production against it.

vacuum pan a device used to remove water from milk.

variegated a leaf having streaks, marks, or patches of color.

veal young calves or the meat from young calves.

vector a living organism that transmits a disease.

vegetable any herbaceous plant whose fruit, seeds, roots, tubers, bulbs, stems, leaves, or flower parts are used as food.

vegetative bud a terminal plant bud that produces stem and leaf growth.

vegetative propagation in plants other than by seed.

transpiración proceso por el cual la planta pierde vapor del agua.

trasplante planta cultivada de semilla en un recipiente especial.

árbol planta leñosa que produce un tronco principal y que tiene una copa más o menos diferenciada y elevada (de 15 pies o más de alto).

callos el rumen adobado de ganado y corderos.

producción agrícola a granel producción de gran escala de pocos vegetales seleccionados de cultivos que son enviados a mercados de abasto.

camionero persona que transporta productos básicos de la finca al consumidor.

tubérculos radiculares raíz especializada de almacenamiento de alimentación que crece subterráneamente.

turbiedad medida de sólidos suspendidos.

césped hierbas que se siegan frecuentemente para obtener un aspecto corto y plano.

turgencia condición hinchada o rígida resultante de llenarse de líquido.

U

insuficientemente madura según aplica a la vegetación, cualquiera que no esté madura.

defectuoso en los caballos, cualquier abnormalidad que afecta el uso del caballo.

sistema urinario conjunto de órganos que elimina los desechos de la sangre.

V

vacuna una preparación que se introduce en el cuerpo para prevenir una enfermedad estimulando la producción de anticuerpos contra ella.

tacho de vacío un dispositivo usado para quitar el agua de la leche.

jaspeado una hoja que tiene rayas, marcas o manchas de color.

ternera terneros jóvenes o el carne de ellos.

vector un organismo viviente que transmite una enfermedad.

vegetal cualquier planta herbácea cuyos frutos, semillas, raíces, bulbos, tallos, hojas o partes de flor, son usados como comestibles.

botón vegetativo el botón de una planta terminal que produce crecimiento de tallo y hoja.

propagación vegetativa reproducción de las plantas mediante un método distinto al de la semilla.

veneer thin sheet of wood used in paneling and furniture.	**chapa** hoja delgada de madera que se usa en revestimiento de madera y en mueblería.
vermiculite mineral matter used for starting plant seeds and cuttings.	**vermiculita** materia mineral usada para empezar semillas e injertos de plantas.
vertebrate an animal with a backbone.	**vertebrados** animales que tienen columnas vertebrales.
vertical integration occurs when several steps in the production, marketing, and processing of animals are joined together.	**integración vertical** ocurre cuando varios pasos en la producción, mercadeo y procesamiento de animales son unidos.
veterinarian a person who practices medicine with animals.	**veterinario** una persona que practica la medicina con animales
vice bad habit in horses.	**vicio** mala costumbre de un caballo.
virus pathogenic entity consisting of nucleic acid and a protein sheath.	**virus** seres patógenos que consisten de ácido nucléico y una vaina de proteína.
viscera internal organs of an animal, including heart, liver, and intestines.	**víscera** órganos internos del animal incluyendo el corazón, higado e intestinos.
vitamin complex chemical essential for normal body functions.	**vitamina** sustancias químicas esenciales para funciones normales del cuerpo.
voluntary muscle muscle that can be controlled by animals to do things such as walk and eat food.	**músculo voluntario** músculo que puede ser controlado por los animales para hacer cosas como caminar y comer alimentos.

W

warp the tendency of wood to bend permanently because of moisture change.	**alabearse** la tendencia de la madera de torcerse permanentemente debido al cambio de la humedad.
water clear, colorless, tasteless, and nearly odorless liquid.	**agua** líquido claro, incoloro, insípido y casi inodoro.
water cycle movement of water to surface, to atmosphere, to surface.	**ciclo del agua** movimiento del agua desde la superficie, hasta la atmósfera y regreso a la superficie.
waterfowl ducks and geese.	**aves acuáticas** patos y gansos.
water resources all aspects of water conservation and management.	**recursos de agua** todos los aspectos de la conservación y administración del agua.
water table level below which soil is saturated or filled with water.	**nivel hidrostático** nivel debajo del cual el suelo está saturado de agua.
watershed a large land area from which water flows or from which it is absorbed from rain or melting snow, and from which water drains as it emerges to the surface through springs.	**cuenca** una área grande de tierra de donde fluye el agua o donde es absorbida de la lluvia o nieve derretida, y de donde el agua se drena a medida que emerge a la superficie por manantiales.
wetland a lowland area often associated with ponds or creeks that is saturated with freshwater.	**tierras mojadas** una región de tierra baja que muchas veces es asociada con charcos o arroyos, que está saturada con agua dulce.
wholesale marketing the marketing of a product through a middleman.	**comercialización al por mayor** el mercadeo de un producto por un intermediario.
wholesaler person who sells to the retailer.	**mayorista** persona que vende al minorista.
wildlife animals that are adapted to live in a natural environment without the help of humans.	**fauna** animales que se han adaptado a vivir en un ambiente natural sin la ayuda de los seres humanos.

woodlot small, privately owned forest.

wool modified hair obtained from sheep and some other animals. It is a fiber with good insulating qualities that is used to make cloth.

worker female bee that does the work in the hive.

X

X-Gal compound that causes marked bacteria to turn blue.

xylem vessels of the vascular bundle that transport the water and nutrients within plants from roots to leaves.

Y

yield grade grade based on the amount of lean meat in relation to the amount of fat and bone in cattle and sheep.

Z

zygote fertilized egg.

bosque privado bosque pequeño que tiene propietario particular.

lana pelo modificado obtenido de borregos y otros animales. Es una fibra con muy buenas cualidades de aislamiento que se usa para hacer telas.

obrera abeja hembra que hace el trabajo de la colmena.

X-Gal un compuesto que causa que se vuelva azul la bacteria designada.

xilema los buques de los haces vasculares que transportan el agua y nutrientes en las plantas de raíces en las hojas.

proporción carne/grasa niveles basados en la cantidad de carne magra en proporción con la cantidad de grasa y hueso en el ganado vacuno y las ovejas.

cigoto óvulo fecundado.

Index

Page numbers in *italics* refer to figures.

A

Abiotic (nonliving) diseases, 266
Acetanilides, 279
Acidity ranges, *332*
Active ingredients, 180
Adaptation, 244
Additives, 686
Add-on loans, 721
Adenine, 52
Adenosine triphosphate (ATP), 594
Advertising, 737–739
Aeroponics, 185, *185*
African swine fever (Asf), 566
African violets, 356, 471
Age, horses, *649*
Age profiles, 34
Aggregates, 155, 172
Aggregate cultures, 184, *185*
Agribusiness. **See** Business
Agricultural literacy, 86
Agriculture
   -agribusiness, 7–8
   careers, 65–66
   definition, 7
   distribution, 67–68
   economics, 16
   engineering, 10
   environmental impact, 26
   global aspects, 12
   jobs in, 8
   mechanization, 36–38
   precision, 40
   processing, 67–68
   production, 7–8, 66–67
   publications, 14
   services, 71–72
   supervised experience, 16
   supplies, 71–72
   worker shortage, 17
Agriscience, 16–17
   aspects, 10, 14–16
   breakthroughs in, 42–45
   definition, 7–11, 65
   everyday influence, 11–14
   facets, *710*
   food supply, 6
   future advances, 45
   impact of, 7
   opportunities, 16, 755
   project, 15
Agronomists, 438, 456
Agronomy, 14

Agropyron cristatum, 501
Agrostis palustris, 501
A horizon, 171
AI. **See** Artificial insemination
Air
   consumption, 135
   currents, 125
   definition, 125
   layering, 350
   plants and, 328
   pollution, 125–129, 135
   seed germination, 341–342
Air quality
   control, 129
   improving, 136–137
   maintaining, 136–137
   threats to, 125–126
Alfalfa
   bacteria, 449
   characterization, 437
   high-protein, 460
Algae, 144, 173
Alkalinity
   definition, 175
   excessive, correcting, 179
   ranges, 330
Alluvial deposits, 165
Aluminum sulfate, 179
American beech, *205*
American Star Awards, 726
American sycamore, *205*
Amino acids, 546
Ammonia, 246
Amortized loans, 722
Amphibians, 242
Angora, 598
Animals. **See also** Wildlife; **specific species**
   anatomy, 535–553
   aquatic, 244, 247
   breeding, 40–42
   byproducts, 701–702
   circulatory system, 540–541
   companion, 604–607
   competition, 223
   digestive system, 542
   domestication, 49
   endocrine system, 542
   energy content, 547
   environment and, 28–29
   foods from, 689–697
   gene changes, 576–577
   genetic manipulation, 54

   manager, 538
   manure, 180
   market trends, 745–747
   medical treatment, basic, 562
   microscopic, 56
   muscular system, 539–540
   nervous system, 541–542
   physiology, 541–551
   predators, 221
   prey, 221
   respiration, 135
   respiratory system, 541
   restraining, 565
   science, 10, 16
   selective breeding, 50, 572
   skeletal system, 538–539
   tallow, 43
   technician, 606
   transgenic, 55
   urinary system, 542
   vertebrate, 221
   waste, 622
   woodlot grazing, 211
Anions, 330
Annual wormwood, 444
Anther, 315
Antibiotics, 549
Aphids, 273, *374*
Apiarys, 601
Apiculture. **See** Honeybees
Apples
   cross-section, 316
   fruiting spur, *412*
   harvesting, *417*
   storage, 419
   tree pruning, 415
Applied science, 14
Aquaculture
   biology, 244–245
   caged, 249
   competition in, 244
   crop selection, 247–248
   definition, 10, 241
   food chain, 244
   hatcheries, 251
   production, 245–247, 253
   research, 251, 252
   resource management, 250–251
   salinity gradient, 242
   trends, 253
Aquaculturists, 241, 247
Aquifer, 156
Arboriculture, 211

Artemisinin, 444
Artificial insemination
    advantages, 571, 582
    definition, 571
    disadvantages, 583
    process, 571
Artificial nesting, 232
Asbestos, 130
Aseptic technique
    description, 356–357
    hood for, 357
    materials preparation, 357–358
    practice activities, 356–357
    procedure, 358–359
Asexual propagation
    division, 350–351
    grafting, 351–353
    layering, 349
    leaf cuttings, 346
    root cuttings, 348–349
    stem cuttings, 344–346
    tissue cultures, 356
Ash, *205*, 206
Asian lady beetle, 527
Aspen, 203, 204
Assets, 761
Auctioneers, 743
Auction markets, 743
Avian influenza, 535
Axil, 309

B

Babesia equii, 650
Babesiosis, equine, 650
Bacillus popilliae, 273
Bacteria. **See also** specific species
    alfalfa roots, 449
    characterization, 268
    soil, 175
    in waste management, 56–57
Bakewell, Robert, 572
Balance sheets, 761
Baldwin, Anna, 37
Balled plants, 525–526
Balsam fir, 201, *202*
Banks, 717
Bare-rooted plants, 524–525
Barley, 426–428
Beech, *205*, 206
Beef
    aging, 692
    block, *692*
    boxing, 693
    fabrication, 693
    retail cuts, *692*
    ripening, 692
Beef cattle
    associated careers, 620
    breeds, 617–619
    carcasses, 692
    commodities, 663
    disassembly process, 693
    hides, 701
    history, 616
    origins, 616
    processing, 692–693
    production practices, 619–620
    products, 692–693, 701
    by products, 616, 701–702
    reproductive system, 578–579
    slaughter, 692–693
    species, 616
    types, 617–619
Beekeepers, 601–602, 606
Bees. **See** Honeybees
Beetles, 280
Beets, sugar, 45
Bell, Marion, 289
Beneficial Insects Introduction Research Laboratory, 374
Bermuda grass, 453–454, 500
B horizon, 172
Bighorn sheep, 227
Binomial systems, 319
Biochemistry, 14
Biodegradable detergents, 734
Bio-diesel, 43–44, 433
    vs. feed grains, 427
Biodiversity, 212
Biological attractants, 42
Biological control, 273
Biological pest control, 272
Biology, 14
Biotechnology. **See also** Genetic engineering
    benefits, 39
    definition, 10
    designer foods, 440
    environmental cleanup, 236
    ethics in, 58–59
    history, 49
    improvements from, 49–51
    pest control application, 273
    pesticides, 30, 289
    reproduction. **See** Artificial insemination
    safety, 58
    waste management, 56–57
Biotic diseases, 267
Birch, 203, *204*, 522
Birds, artificial nesting, 231
Bird's-foot trefoil, 450
Bitternut hickory, *205*
Black cherry, *205*
Black walnut, *205*, 207
Black willow, *205*
Bladder products, 702
Blanching, 683
Block beef, *692*
Blood products, 701
Blueberries, 42
Blue rose, 485
Bone marrow, 539
Bones, 702
Border plantings, 519
Botany, 319
Bottomland Hardwoods Forest, 199
Bovine somatotropin
    consumer resistance to, 55
    production, 55
Boxing, beef, 693
Boy Scouts of America, 104
Brackish water, 243
Bracts, 315
Bradford pear, 520
Bran, 669
Breeding
    AI, 572
    dairy cattle, 612
    genetic engineering and, 578
    horses, 645
    methods, 582–583
    natural selection, 576–577
    papers, 580
    planned selection, 575–576
    selective, 572
    systems of, 580–582
Broadcast planters, 439
Brokers, 438
Bromegrass, 451
Bromus tectorum, 263
BST. **See** Bovine somatotropin
Buchloe dactyloides, 502
Buds, grafting, 352
Buffalo grass, 502
Bulbs, 308, 335
Burlapped plants, 525–526
Business. **See also** Entrepreneurship
    careers, 713, 758
    decisions, 712–714
    economics, 714–716
    employment options, 763–765
    finance, 716–725, 758
    management, 710–712
    operations, 757
    owning, 764
    records, 761–763
    small, contributions, 765
    software, 711
    structures, 709
Byproducts, 701–702

C

Cacti, 476
Calcium, 330
Calcium nitrate, 330
Calgene, 42
Cane cuttings, 345
Canning, 684

Canola, 431
Capillary water, 147
Carbamates, 283
Carbohydrates
  characterization, 546–547
  function, 540
  nutritional role, 677–678
  sources, 677–678
Carbon dioxide
  in aquaculture, 244
  characterization, 133
  global warming and, 133–134
  measuring, 324
  in photosynthesis, 135, *135*
  shortage, 324
Carbon monoxide, 128
Carcasses, 692
Carcinogens, 297
Career(s)
  AFNR, 85
  agriculture, 67–68
  agriscience, 10, 16, 75–78
  agronomist, 456
  animal technician, 606
  aquaculturists, 247
  beekeepers, 600
  biology, 319
  botany, 319
  broker, 438
  business, 713, 758
  cattle industry, 616
  education, 74
  elevator manager, 438
  entomology, 267
  environmental management, 33
  exploration, 84
  FFA events, 108–109
  foliage plant-associated, 474
  food industry, 655–656, 663, 672
  forage crops, 456
  forestry, 69–70, 207
  fruit/nut production, 407
  gardening, 379
  genetics, 50–51, 571
  golf course-associated, 152, 493, 513
  grain handler, 438
  home garden, 383–384
  horse-associated, 638
  horticulture, 68–69
  hydroponics, 184
  landscaping, 518
  marketing, 751
  mechanics, 71–72
  natural resources, 70–71
  nutrition, 703
  nutrition-associated, 538
  opportunities, 17, 65–66
  path, 87
  pest control, 281
  planning for, 75–78, 86–87
  plant breeding, 345

  plant pathology, 267
  plant physiologists, 328
  plantscaping, 486–489
  professions, 73–74
  requirements, 65
  SAE, 86–87
  skills, 76
  soil science, 74
  taxonomy, 219
  teaching, 75
  truck cropping, 390
  turfgrass-associated, 495
  vacancies, 65
  veterinarians, 538, 556
  wildlife biologist, 230
Career Exploration and Planning
  component, 85
Carson, Rachel, 30, 268
Cash flow statement, 762
Catfish, *234*
Cations, 330
Cats, *605*
Cattle. **See** Beef cattle; Dairy cattle
Causal agents, 266–267
CCC. **See** Commodity Credit
  Corporation
Cedar, *202*, 203
Cells
  definition, 573
  division, 306, 574
  maturation, 308
  structure, 573, *574*
Cellulose, 544
CELSS. **See** Controlled ecological
  life support system
Central Broad-leaved, 199
Cereal crops, 437
CFCs. **See** Chlorofluorocarbons
Chain saws, *134*
Cheatgrass, 263
Check-off program, 738
Cheese, 692
Chemicals
  impact, 30
  spills, 150
Chemistry, description, 14
Cherry, *205*, *206*
Chickens. **See** Poultry
Chilocorus kuwande, 527
China
  forests, 31
  populations, 34–35
  soil erosion, 152
Chinch bugs, 265
Chlorinated hydrocarbons, 282
Chlorofluorocarbons
  characterization, 129
  ozone hole and, 129
Chlorophyll, *136*, 324
Chloroplasts, 314, 324
Chlorosis, 333

C horizon, 172
Chromatids, 342
Chromosomes, 572, 573
Circulatory system, 540–541
Classroom instruction, 83
Clay soils, 172
Clear cutting, 212
Climate
  change, 131–133
  crops and, 369
  definition, 27
  soil and, 162
Clones
  African violets, 356
  aseptic technique, 356–357
  definition, 54
  embryo, 579
  Mule, 57
  potato, 354
Closebreeding, 582
Clovers, 450
Cockroaches, 29
Cold frames, 376–377
Collagen, 702
Colluvial deposits, 165
Colorado potato beetles, 54, 280
Colorado River Basin, *13*
Colors, plantscaping, 486
Columbus Center for Marine
  Sciences, 252
Columbus, Christopher, 611, 622
Commensalism, 223
Commodities
  exchanges, 748
  fruits, 663
  futures market, 748–749
  grains, 662
  meat, 663–664
  oil, 662
  pricing, 739–740
  promotion of, 738–739
  safety, 659
  vegetables, 663
Commodity Credit Corporation, 720
Communication
  leadership and, 103
  specialists, 751
Community gardens, 365
Community, influence, 25
Comparative advantage, 716
Composts
  building, 162
  definition, 12
  description, 162
  materials, 162
Comprehensive agriscience program, *83*
Computers, 74
Condominium, 25
Conservation
  tillage, 155
  water, 146

Consumers
　demographics, 733
　direct sales, 744
　power of, 731
Container growth, 526
Contamination, 28
Continuous-flow systems, 185, *185*
Controlled ecological life support system (CELSS), 379
Contour practice, 155
Convection ovens, 687
Cooking methods, 687
Cooperative extension system, 372
Cooperatives, 709
　description, 760
　managers, 751
　pricing, 739–740
Corms, 308, 335
Corn
　characterization, 424
　classifications, 424
　crop, 12
　cultivation, 424
　high-lysine, 546
　nutrient deficiencies, *181*
　origins, 424
　picker, 38
　taxonomy, 319
Cornus florida, 520
Corolla, 316
Corporations, 709
Cortez, Hernán, 626
Cotton
　boll, *259*
　characterization, 434
　importance, 434
　rainbow, 734
Cotton gin, 36
Cotton seeds, 434
Crawfish, 245, 247
Cream, 691
Credit. **See also** Loans
　classification, 717
　cost of, 721–722
　obtaining, 717
　rates, *719*
　risk, minimizing, 722
　sources, 719–721
　types, 717–719
　use of, 717, 722
Creed speaking, 111
Creeping bentgrass, 501
Crested wheatgrass, 501
Crockpots, 687
Crops. **See also** Foods
　aquaculture, 247–248
　climate and, 660
　commodities, 662–663
　cool-season, 437
　edible, 669
　field. **See** Field crops
　foods from, 688–689
　frost susceptibility, 411
　handling, 669
　harvest temperatures, 683
　high-value, *185*
　major regions, *660*
　marketing, 671
　market trends, 745–747
　maturity, 667
　nurse, 455
　processing, 669
　rotation, 155, 184
　science, 10
　sod, 184
　soil pH range, 175–176
　spoilage, 668
　transporting, 670–671
　truck, 383, 387, 390
　wildlife and, 228
Crossbreeding, 580
Crowns, 496
Crustaceans, 245
Cucurbitaceae, 384
Culm, 496
Cultures, 184
Currency, global, 748
Cuscuta campestris, 306
Cuticles, plant, 314
Cuttings
　cane, 345
　leaf, 346
　roots, 348–349
　stem, 344
Cynodon dactylon, 502
Cytosine, 52

D

Dairy cattle
　breeding, 612
　breeds, 617–619
　cars risk to, 631
　culling, 614
　dry period, 614
　economics, 612
　evaluation and management, 633
　history, 611
　origin, 611
　production practices, 612
　transponders for, 631
　types, 612
Dairy products. **See** Milk
Dairy Replacement Heifer Project, 91
Dallis grass, 455
Dams, 582
DDT, 30, 282
Dealers, 751
Decomposers, 175
Decorative plants. **See** Ornamentals
Deere, John, 36
Deer management, *230*
Degradation, land, 154
Dehydration, 684
Dehydrofrozen, 685
Demand
　determining, 731
　principle, 714–715
Demographics, 733
Deserts, 146
Diameter limit cutting, 213
Dicotyledons, 384
Diet. **See** Nutrients
Dieticians, 703
Digestive system, 542
Diminishing returns, 715–716
Dinitroanilines, 280
Dipping, 565
Direct sales, 742
Disassembly process, 693
Discount pricing, 740
Diseases. **See also** Health
　abiotic, 266
　animal, spread of, 29
　biotic, 237
　classification, 559–562
　common symptoms, 560–561
　contagious, 559
　control, 267
　deficiency, 538
　insect, spread of, 29–30
　noncontagious, 560
　parasites-induced, 561–562
　plant, 266–267
　poultry, 535
　resistance to, 55
　trees, 531
　triangle, 266
　whirling, 251
Dissolved oxygen, 246
Distribution, 67–68, 757
Division, 350–351
DNA
　bases, 52
　components, *52*
　definition, 52
　fingerprinting, 59
　recombinant, 54
　shape, 53
　strands, 52–53
　traits, 52
Dodder, 306
Dogs, 603
Dogwood, flowering, 520
Domestication
　honeybees, 599–600
　poultry, 589
　rabbits, 595
Donkeys, 635, 640
Dosage, lethal, 287

Douglas fir, 201
Drainage, 482
Drea, John J., 527
Drenching, 564
Drill planters, 439
Drip irrigation, 397, 441
Drugs
    administering, 562
    forms of, 564
    withdrawal, 563
Dry-heat cooking, 687
Ducks, 231
Dust bowl, 146
Dust, radioactive, 128–129

E

Earth
    circumference, 125
    crust, 145
    sun proximity to, 130
Earthworms, 391
Ecology, definition, 224
Economics
    beef production, 619–620
    dairy production, 612
    definition, 16
    food industry, 655–656
    goat production, 630–631
    honey production, 600–601
    horse breeding, 645
    poultry production, 594
    principles, 714–716
    rabbit production, 595
    sheep production, 628
    small business, 765
    swine production, 612
    threshold level, 270
Edible crops, 669
Eggs
    characterization, 574
    preparation, 703
    production, 583
Electrical conductivity meter, 189
Electricity, rural, 36
Electron microscopes, 310
Elevator manager, 438
Embryos
    cloning, 583
Employability skills, 86, *86*
Employment options, 763–765
Endocrine system, 542
Endosperm, 669
English ivy, 306
Entomologists, 267
Entomology, 14
Entomophagous insects, 265
Entrepreneurship. **See also** Business
    buying/selling, 757
    definition, 755
    management, 759–760
    operating, 757
    organization, 759–760
    product selection, 759
    profits/losses, 761
    SAE, 88–91, 93
    sales projects, 756
    service selection, 759
    types, 756
Environment
    agricultural impact, 26
    animals and, 28–29
    chemicals and, 30
    climate and, 27–28
    community, 25
    definition, *11*
    family, 25
    food supply, 25
    health and, 557
    homes, 22–25
    humans and, 28–29
    influencing factors, 28–31
    insects and, 29–30
    manipulation, 270
    neighborhood, 25
    sharing, 31
    success story, 126
    topography and, 27
Environmental management, 33
Environmental Protection Agency, 137
Epidermis
    plant, *310*, 314
    roots, 326
Equipment
    agricultural, 72
    computers, 74
    garden, 133, 376
    harvesting, 36
    planters, 444
    riding, 642–644
    spades, 526
    sterilizing, 354–355
    tillage, *443*, 509
    turfgrass, 508
Eradication, 272
Erosion
    controlling, 148, 150
    gully, 154
    prevention, 154
    reducing, 154–156
    sheet, 154
    types, 154
Erwinia amylovora, 602
Escherichia coli, 56, 560
Esters, 435
Euonymus scale, 527
Euphorbia pulcherrima, 473
Evisceration, 693
Export markets, 750

F

Fabrication, beef, 693
Fabric industry, 734
Family, impact, 25
Farmers, 739
Farmers Home Administration (FHA), 720
Farmers' markets, 401, 742
Farms. **See also specific crops**
    contour practice, 155
    crop rotation, 155
    migratory labor, 688
    retailing at, 741–743
    slash and burn, 152
    strip cropping, 155
    turf, 68, *69*
    wildlife, 228–229
Farriers, 638
Fats
    animal, 701
    bio-diesel from, *44*
    characterization, 548
    nutritional role, 678
    reducing, 44–45
    sources, 549
Federal land banks, 717
Feed. **See also** Nutrition
    additives, 549, 562
    classification, 551
    composition, 549–550
    concentrates, 551
    definition, 538
    formulator, 538
    roughage, 543, 551
Fermentation
    definition, 683
    function, 683
    process, 463–464
Ferns, 474
Fertilization
    definition, 329
    fruit/nut crops, 412
    ornamentals, 482
    rotation, 184
    trees, 528–529
    turfgrass, 505–506
    vegetable gardens, 390
Fertilizers
    applications, 182
    complete, 179
    by crop rotation, 184
    definition, 155
    function, 180
    grades, 180
    holes, 528–529, *529*
    ingredients, 179–180
    inorganic, 180–182
    manure, 182–183
    nutrient content, 181, *506*
    organic, 180
    plant nutrients, *182*

superphosphate, 181
unmixed, 181
Fescues, 453, 499
Festuca arundinacea, 500
Fetus, definition, 580
FFA
   aim, 106
   career development, 108–109
   colors, 106
   creed, 107
   degree requirements, 108
   description, 105–107
   emblem, 106
   member profile, 110
   motto, 107
   National FFA Organization, 83, 87
   proficiency awards, 92
   programs, 83
   purpose, 106
   salute, 107
   speech competitions, 109
   student handbook, 119
FHA. **See** Farmers Home Association
Field crops
   classification, 437
   grains, 423–430
   harvesting, 444–445
   irrigation, 441–442
   medical applications, 444
   oilseed, 444
   planting, 440–441
   seedbed preparation, 438–439
   seeds, 436
   specialty, 433–435
   spoilage, 443
   storage, 444–445
Finance. **See also** Credit
   balance sheet, 761
   business, 716–725, 758
   cash flow statement, 762
   profit/loss statement, 761
   records, 761
Fire ant control, 44
Fire blight, 602
Fires, 213–214
First aid, 665–666
Fish
   adaptation, 244
   artificial stocks, 234
   in caged culture, 249
   freshwater, 243–244
   fry, 243
   genes, 252
   growth, control, 252
   migrating, 247
   oxygen needs, 246
   -plants, relationship, 244
   processing, 694
   saltwater, 243–244
   shellfish, 245, 252, 253

sport, 234
supplies, 664
temperatures, *684*
trout, 246, 247
Fisheries
   catfish, 234
   crop selection, 247–248
   natural, 241
   trout, 247
Flax, 433
Floriculture, 471
Flowering dogwood, 520
Flowers
   beauty, 315
   buds, 309
   characterization, 315–317
   exotic, 44
   growing, 471
   indoor, 471–473, 484
   induction, 497
   ovary, 316
   parts, *315*
   perfect, 316
   removal, 344
   structure, 315–316
   varieties, selecting, 372
Flukes, 561
Foliage
   definition, 471
   light and, 478
   types, 474–477
Food chains
   aquatic, 244
   description, 143
   models, *143*
   origins, 143
   plants role, 303
Food industry
   careers, 655–656, 672, 673
   dairy products, 689–692
   economics, 655–656
   function, 655–656
   future developments, 674
   grading, 658
   handling, 669
   harvesting, 667–668
   inspecting, 658
   marketing, 671
   meat products, 692–694
   new products, 702–703
   operations within, 661–662
   processing, 669
   quality assurance, 657–658
   sanitation, 657
   transportation, 670–671
Foods. **See also** Crops
   additives, 686
   animal-based, 689–697
   borne pathogens, 686
   convenience, 656

cooking methods, 687
crop-based, 688–689
designer, 43, 440
function, 594
genetically modified, 43, 739
global aspects, 662
groups, 679–680, *681*
irradiation, 685
labels, 686
life expectancy and, 682
microorganisms in, 49
nutritional values, 42, 679–680, *681*
organic, 11, 297
plant storage, 335–336
poisoning, 59
preparation, 687
preservation, 377, 683, 684
processing, 669, 683
safety, 659
scientist, 703
storing, 683
supply, 6, 25
surpluses, 11
technician, 703
varieties, 661
Food science degrees, 679–680, *681*
Foot-candles, 480
Forage crops
   associated careers, 456
   corn as, 455
   fermentation, 463–464
   function, 437
   harvesting, 459–464
   importance, 459
   Kenaf, 465
   manager, 456
   maturity, 460
   pest control, 459
   planting, 457–459
   rangeland, 455
   seedbeds, 457
   selection, 456–457
   sheep, *626*
   storage, 459–464
   types, 449–464
   uses, 464–465
Forestry, 69–70, 207
Forests. **See also** Trees; Woodlots
   acreage, 236
   biodiversity, 212
   characterization, 195
   China's, 31
   composition, 195–196
   fires, 213–214
   natural resources and, 200
   North American regions, 196–200
   products, 195
   resources, 518–519
   water cycle and, 200
   wildlife, 200, 226, 229–230

Formulations, 285–286
Foundational SAE
    components, *85*, 85–86
    definition, 85
Freemartin, 580
Free water, 147
Freeze-drying, 684
Frost, 411
Fruit production
    careers, 407
    fertilization, 414–415
    harvest, 416–418
    planning, 411–412
    planting, 413–414
    soil factors, 411
    soil/site preparation, 412
    tree pruning, 415
Fruits. **See also** specific crops
    categories, 407
    characterization, 308–310
    climate factors, 410
    commodities, 663
    common, 408
    crop pollination, 411
    fertility, 411
    foods from, 688
    growth patterns, 412
    small-bush, 418
    trees, 414–416
    true, 316
    vine, 409
    washing, 688
Fry, 243
Fuchsias, 472
Fungi
    benefits, 394
    characterization, 267–268
    soil, 175
Fungicides
    characterization, 283
    chemical structure, 284
    classification system, 284
    eradicant, 283
    protectant, 283
Furrows, 367
Future Farmers of America (FFA). **See** FFA
Futures
    buying/selling, 748–749
    definition, 748
    description, 748
    positions, 750

G

Galls, 531
Gardeners, 379
Gardenias, 472
Gardens. **See** Home gardens
Gasoline, *127*, 150
Geese, Canadian, *226*

Generation, 50
Genes
    definition, 50, 572
    expression, 572–573
    fish, 252
    function, 572–573
    heterozygous, 575
    homozygous, 575
    horse color, 641
    lean beef, 619
    location, 575
    mapping, 54
    splicing, 54
    traits, 53
    traits and, 575
Genetic engineering. **See also** Biotechnology
    benefits, 39
    in breeding, 578
    definition, *10*
    foods, 43, 739
    pest resistant plants, 280
    process, 56
Genetics
    animal improvement, 576–577
    careers in, 51, 585
    code, 51–54
    definition, 51
    Mendelian, 50–51
    principles, 572–578
Genus, 317
Geography, 16
Geraniums, 472
Germ, 669
Germination
    air and, 342–343
    chamber, *190*
    hydroponics, 190
    light and, 343
    temperature, 343
    water and, 342
Gills, 245
Glacial deposits, 165
Glands, 702
Glidden, Joseph, 36
Global currency, 748
Global markets, 747–750
Global warming
    cause, 131–133
    debate over, 131–133
Goats
    breeds, 629–630
    economics, 629
    history, 629
    origins, 629
    production practices, 628
    types, 629
Golden rice, 440
Golf courses. **See also** Turfgrass
    maintenance supervisor, 513
    playability level, 493

Graders, 656, 751
Grades, 581, 658
    beef, 658
    cattle, 698
    cows, 617
    fertilizers, 180
    quality, 699, *699–700*
    sheep, 626
    swine, 698
Gradient effect, 247
Grading up, 581
Grafting, 351–353
Grain handler, 438
Grains
    by-products, 551
    commodities, 662
    drying, 445
    foods from, 688
    types, 423–433
Grants, 721
Grape vines, *417*
Grasscycling, 507
Grasses
    identifying, 498
    planting, 457–459
    turf. **See** Turfgrass
    types, 451–455
    waterways, 155
Gravitational water, 147
Grazing, 214, 455
Greenhouse effect, 131
Greenhouses, 378
Green Revolution, 41
Grooming, 483–484
Groundskeepers, 495
Groundwater, 156
    pollution, 156
    safeguarding, 157
    types, 147
Group plantings, 520
Growth
    habit, 408
    intravaginal, 497
    patterns, 412
    tree, 207–209
    types, 521, *521522*
Guanine, 52
Guard cells, 314
Guide dogs, 603
Gully erosion, 154
Guying, 528
Gypsum, 179

H

Hair products, 701
Hand mating, 582
Hardness, wood, 209
Hardwood forests, 203
Hartwig, Edgar E., *432*

Harvesters, 462
Harvesting, food, 667–668
Harvests
    aspects, 667–668
    definition, 667
    field crops, 444–445
    forage, 459–464
    fruit/nut crops, 416
    hay, 459–462
    machines, *445*
    market gardens, 401
    pastures, 459
    silage, 462–463
    temperatures, 683
    woodlots, 214
Hatcheries, 250
Hawaiian Forest, 200
Hay
    cubes, 462
    definition, 449
    drying, 459
    harvesting, 459
    managers, 459
    maturity levels, *459*
    peanut, 451
    raking, 460
    storage, 461–462
4-H Clubs, 103
Health. **See also** Diseases
    environmental factors, 557–559
    housing and, 558
    isolation and, 558–559
    maintaining, 555, 563
    manure handling, 558
    monitoring, 556
    pasture rotation, 559
    paying for, 739
    pest control and, 293–297, 558
    sanitation and, 557
    signs of, 555–556
Hedera helix, 306
Heel cuttings, 346
Heeling in, 524
Heliothis, 289
Hemlock, 199, 200
Herbicides
    characterization, 279–281
    chemical families, 279
    classification systems, 279
    contact, 279
    nonselective, 279
    postemergence, 279
    preemergence, 279
    selective, 279
    systemic, 279
Heritability, 577
Hickory, *205*, 206
Hides, 692, 701
Hives, 600
Hoagland solution, 188, *189*

Hogs. **See** Swine
Holt, Benjamin, 37
Homarus sp., 245
Home gardens
    career opportunities, 383–384
    cold frames, 393
    common crops, 369
    community, 365
    cultivating, 372
    definition, 383
    enterprises, 383
    family's needs, 365
    greenhouses, 378
    harvesting, 375–376
    hotbeds, 378
    layouts, *367*
    locating, 366
    pest management, 374
    plan, 366
    planting, 367
    raised beds, 367
    size, 365
    soil preparation, 366–367
    tools, *368*
    watering, 373
    weeding, 373
Homes
    middle/upper class, 25
    modest incomes, 22–25
    poor, 22
Honeybees
    Africanized, 257
    biocontrol role, 602
    domestication, 600
    economics, 600–601
    history, 599–600
    pollination by, 411
    production practices, 601–603
Hoof products, 701
Hormones, 542
Horn products, 701
Horseback riding
    attire, 642–643, *642*
    basic gaits, 644
    English equitation, 642–643
    saddles, 642, *644*
    safety rules, 645
    Western horsemanship, 643–644
Horses
    age, *649*
    anatomy, 640
    associated careers, 638
    breeding, 645
    care, 646
    color of, 641
    economics, 636
    grooming, 647
    history, 635
    origins, 635
    restraining, 565

    types, 638–641
    wild, 636
Horticulture
    careers, 68–69
    description, 14
Host, secondary, 561
Hotbeds, 378
Hothouse tomatoes, 185
Housing, 558
Humane Slaughter Act of 1958, 692
Human Genome Project, 573
Humans
    environment and, 28–29
    intelligence, 5
    life expectancy, 682
    life spans, 34–35
    population, 34–35
    wildlife and, 221, 224–225
Humidity, 685
Hunger, 33
Hunting, 123
Hutch, 599
Hybrids, 581
Hydrocarbons, 126
Hydrocooling, 403
Hydroponics, 161
    aeration, 188–189
    careers in, 191
    classroom, 187–188
    future of, 190
    germination, 190
    laboratory, 187–188
    maintenance, 189
    nutrient solutions, 188
    oxygen, 186
    plants, 189–190
    requirements, 184–187
    seeds for, 190
    solution culture, 185
    sunlight, 187
    systems, 184
    temperature, 187
    water, 186
Hygroscopic water, 147
Hyphae, 267

I

Ice-minus, 54
Ictalurus spp., 247
Imbibition, 342
Immersion SAE, 87
Impatiens, 44, 473
Improvement activity, 92
Inbreeding, 581–582
Indicator species, 150
Industrial technology, 9
Inflorescence, 497
Influenza, avian, 535
Infusions, 564

Injections, 564
Inorganic fertilizers, 180–182
Insecticides
    botanical, 282
    characterization, 282–283
    classification systems, *282*
    production, 285
    synthetic, 282–283
Insects
    adaptation, 263
    anatomy, 263–264
    beneficial, 265, 374
    Colorado potato beetle, 54
    definition, 259
    development, 264–266
    economic impact, 263
    entomophagous, 265
    environment and, 29–30
    feeding damage, 263
    fire ants, 44
    male sterilization, 274
Instars, 264
Insulin, 55–56, *56*
Integrated pest management. **See also** Pesticides
    aspects, 10–11
    biological control, 272–273
    chemical control, 274
    cultural control, 273
    definition, 10–11
    ecosystem manipulation, 270
    environmental impact, 294
    history, 268
    key pests, 269
    male sterilization, 274
    mechanical control, 273
    monitoring, 271
    pest resistance, 272
    physical control, 273
    principles, 269
    regulatory control, 271–272
    threshold levels, 270
Intelligence, gift of, 5
Intergovernmental Panel on Climate Change, 131
Internet marketing, 758
Internode, 309
Internships, 88–89, 90, 91, 93, 255, 513
Interviews, 99, 110
Intestine products, 692
Intravaginal growth, 497
Inventors, 36–38
Inventory reports, 763
In vitro mating, 583
Involuntary muscles, 540
Ions, 330, *332*
IPCC. **See** Intergovernmental Panel on Climate Change
IPM. **See** Integrated pest management
Irradiation, 685

Irrigation
    definition, 146
    field crops, 441–442
    history, 441–442
    turfgrass, 507–508
    types, 397
    vegetable production, 396–397

J

Jefferson, Thomas, 36
Jockeys, 638
Jog, 644

K

Kenaf, 465
Kentucky bluegrass
    characterization, 453
    cultivars, 499
    fertility requirements, *506*
    uses, 499–500
Kosher slaughter, 694
Kreb cycle, 594

L

Labor, migratory, 668
Lacustrine deposits, 165
Lakes, wildlife, 234–236
Lambs. **See also** Sheep
    production practices, 626
    retail cuts, 696
Land. **See also** Soils
    cultivation, 423
    degradation, 154
    demand for, 141
    erosion, 151–157
    function, 146–148
    ownership, 123
    as reservoir, 147
    as resource, 141, *141*
    -water relationship, 146
Land capability maps
    classes, 168–169
    description, 168
    subclasses, 169
    units, 169
    use of, 170–171
Landscape architects, 518
Landscape maintenance technician, 495
Landscaping. **See** Plantscaping
Lawn mowers, *134*
Lawns. **See** Turfgrass
Layering, 349–353
Lead, 127–128, 229
Leadership
    communication and, 103
    definition, 101
    development opportunities, 103–111

    importance, 101–102
    at meetings, 115–119
    speaking skills, 111
    traits, 102–103
Leaves
    characterization, 310
    compound, 313
    cuttings, 346
    function, 314
    internal structure, 314
    margins, *311*
    mold, 162
    parts, 313
    scorch, 334
    shape/form, 312–313
    simple, 313
    types, 313
Legumes
    examples, 183
    nitrogen fixation, 183
    seeds, planting, *458*
    types, 449–451
Lenticels, 309
Lespedeza, 451
Lesquerella, 444
Lethal dose, 287
Life expectancy, 682
Life spans, 34–35
Light
    for foliage, 485
    in hydroponic systems, 187
    intensity, 324
    measuring, 480
    for ornamentals, 480–481
    seed germination, 343
Lime, 155
Limestone, standard ground, 179
Liming, 392
Limited liability company (LLC), 709
Linebreeding, 582
Linseed oil, 433
Liquids, 564
Listeria, 59
Living conditions. **See** Environment
Loamy soils, 172
Loans
    add-on, 721
    amortized, 722
    comparison chart, *721*
    long-term, 717, *717*
    production, 717
    seeking, 722
    short-term, 717, *717*
    signature, 717
    simple-interest, 721
Loblolly pine, *202*
Lobsters, 245
Loess deposits, 165
Lolium perenne, 500–501
Long-term loans, 717

Lope, 644
Loss-leader pricing, 740
Lumber
   definition, 195
   measure formula, 195–196
   seasoning, 216–217
Lycopene, 670
Lymph glands, 540
Lysine, 546

M

Macronutrients, 329
Magnesium, 334
Manage, definition, 101
Management, 710
   definition, 710
   influence, 711
   woodlot, 211–216
Manure, 180–183, 558
   and methane, 148
Maple, 203, *204*
Mapping, 54
Maps. **See** Land capability maps
Marginal burn, 334
Marketing
   advertising, 738–739
   approaches to, 731–732
   careers, 751
   commissions, 745
   cycles, 745–747
   fees, 745
   global, 747–750
   Internet, 758
   promotions, 738–739, 758
      retail, *740*
   strategies, 741
   techniques, 744–745
   trends, 745–747
   vegetables, 401
   wholesale, 671, 743–745
Markets, 743
   auction, 743–744
   careers, 73
   companies, 38
   export, 750
   gardening, 383, 388
   inventors, 38
   roadside, 742
   technologies, 36
   terminal, 743
   undeveloped countries, 39
Marshes, 242–243, *242*
Mathematics, 16
Mating. **See** Breeding
Maturity, crop, 667
McCormick, Cyrus, 36
Meat. **See** specific types
Meat cutters, 751
Mechanization, 11, 35, 38–40, 74, 462

Media. **See** Plant-growing media
Medical research, *598*, 607
Meetings
   conducting, 117–119
   officers, 116
   procedures, 115, 117
   requirements, 115
Meiosis, 574
Membranes, semipermeable, 327
Mendel, Gregor Johann, 50–51, 572
Merozoites, 650
Mesophyll, 314
Methane
   global warming and, 134
   from manure, 148
   origins, 134
   use, 134
Microbes, 56
Micronutrients, 329
Microorganisms, 49, 668
Microwave ovens, 687
Migratory labor, 668
Milk
   characterization, 612
   goats, 43
   nutrients, 690–691
   processed products, 691–692
   production, 535
   products, 691–692
   secreting glands, 44
   shelf life, 685
   water content, 690
Minerals
   characterization, 547
   function, *679*
   in hydroponic systems, 186–187
   matter, 171–172
   in milk, 690–691
   nutritional role, 679
   sources, 547
Mississippi River Delta, 151
Mistletoe, 306
Mix well, 179
Moist-heat cooking, 687
Moisture, soil, 391
Molt, 245
Monoclonal antibodies, 43
Monocotyledons, 310, 384
Monogastric system, 544
Montana Range Days program, 235
Moss, peat, 162
Motions, 119
Mowers, 460
Mulch, *154*
Mulching
   application, 528
   characterization, 528
   vegetable production, 397–398
Mules, 640
Muscular system, 539–540

Mutualism, 222
Mycelium, 267

N

National Council for Agricultural
   Education (NCAE), 84
National FFA Organization, 83
Natural resources
   careers, 70–71
   forests and, 200
   renewable, 8
Natural service, 582
Neighborhoods, influence, 25
Nematodes, 175, 267, 268, 668
Nervous system, 541–542
Netting, 510
Net worth, 761
Newcastle disease, 535
Nitidulid, 527
Nitrates, 183, 250
Nitrites, 250
Nitrogen
   fixation, 183
   function, 333
   ions, 330
   nonprotein, 548
   total ammonial nitrogen, 246
Nitrous oxides
   characterization, 127–128
   emissions, 134
   global warming and, 134
Nomenclature, 523
North American Forests, 196–200
North Appalachian Experimental
   Watershed, 391
Northern Coniferous Forest, 198
Northern Hardwoods Forest, 196, 198
No-till planters, 457
Nurse crops, 455
Nut production
   careers, 407
   disease control, 416
   fertilization, 414–415
   harvest, 416–418
   pest control, 416
   planning, 409–412
   planting, 411–412
   soil/site preparation, 412
   tree pruning, 415–416
Nutricultures, 185
Nutrients
   classes, 545–548
   deficiencies, *181*, 330, *550*
   essential, 329
   milk, 690–691
   plant. **See** Plant nutrients
   soil, 327
   solutions, 188
   sources, 548–549

Nutrients, (*Continued*)
  stock solution, 188
  supplemental, 227
Nutrition
  basic needs, 677–680, *681*
  carbohydrates and, 677–678
  careers in, 538, 703
  definition, 537–538
  fats and, 678
  food groups, 679–680, *681*
  health and, 537–538
  minerals and, 679
  proteins and, 678
  values, 42, *681*
  vitamins and, 679
  water and, 679
Nutritionists, 703
Nuts
  characterization, 316–317
  climate factors, 410
  foods from, 688
  growth patterns, 412
  types, 409

O

Oaks, 203, *205*, 206
Oats, 423
Oil meals, 551
Oils. **See also** Field crops
  commodities, 662
  cosmetic, 444
  crops, 437
  foods from, 688
  as fuel, 433
Oilseed crops, 431
Orchard grass, 451
Orchards, 413
Organic farming, 399
Organic fertilizers, 180
Organic foods, 10, 297
Organic matter, 155, 172, 392
Organisms
  fingerprinting, 59
  in soil, 164, *165*, 173
  water/soil benefits, 148
Organophosphate, 283
Ornamentals
  aspects, 471
  definition, 14
  design. **See** Plantscaping
  drainage, 482
  as economic indicator, 469
  fertilizing, 482
  flowering, 471–473, 485
  foliage, 474–477, 485
  grooming, 483–484
  growing, 480–481
  light needs, 478, 480–481
  pest control, 469
  phototropism, 481

propagation, 484
selecting, 477–478
temperature needs, 478, 481
uses, 479–480
water needs, 482
Osmosis, 243, 327
Outcrossing, 582
Ova. **See** Eggs
Ovens, 687
Overripe, 667
Ovulation, 579
Oxidative deterioration, 685
Oxygen
  in aquaculture, 235
  consumption, 135
  dissolved, 247
  in hydroponic systems, 186
  measuring, 246
Ozone, 129, 131

P

Pacific Coast Forest, 198
Packers, 692, 694
Palisade cells, 314
Paper birch, 203, *204*
Parasites
  characterization, 272–273
  classification, 561
  control of, 561
  external, 565
  horse-associated, 645
  internal, 561–562, 586
Parasitism, 222
Parent material, 165
Parliamentary procedures, 115, 117
Particles, soil, *147*
Particulates, 129–130
Partnership, 709
Parturition, 579
Pastures
  crops, 449–451
  harvesting, 459
  mating, 581
  renovating, 457–459
  rotation, 559
PCA. **See** Production Credit Associations
PCBs. **See** Polychlorinated biphenyl
Peanut hay, 451
Peanuts, 431
Pears, Bradford, 520
Pears, diseased, 268
Peat moss, 162, 366
Pedigree, 584
Pelts, 597
Penetration pricing, 740
Percolation, 166
Perennial ryegrass, 500–501
Perlite, 163
Permeable soils, 168
Pesticides

breaking down of, 283, *284*
career areas, 281
classification, 285
common names, 286
controversy, 277
directions, 287
disposal, 290
environmental impact, 30, 294
EPA registration numbers, 288
exposure, limiting, 291
formulations, 285–286
fungicides, 283
general-use, 285
in groundwater, 157
health concerns, 294–297
herbicides, 279–281
history of, 278
hygiene, 292
ingredients, 286
inorganic, 278
insecticides, 282–283
labels, 284–290
lethal dose, 287
limitation notice, 290
manufactures' address, 288
misuse, 285, 290
movement of, 294
net contents, 287
organic, 278
precautionary statements, 288
protective gear, 292
reentry information, 290
resistance to, 280
restricted-use, 285
risk management, 290–291
safer, 289
safety, 297
signal words, 287–288
spray, 130
storage, 290, 293
symbols, 287–288
trade names, 285
vapor drift, 294
water quality and, 149
Pest management
  biological, 443
  chemical, 444
  cultural, 443
  field crops, 442–444
  forage, 459
  fruit production, 416
  genetic, 443–444
  good practices, 259
  health and, 558
  integrated. **See** Integrated pest management
  mechanical, 443
  nut production, 416
  ornamentals, 487
  trees, 527
  vegetable production, 398

Pests
  definition, 130, 260
  garden, 374
  insects, 263
  key, 269
  management. **See** Integrated pest management
  population equilibrium, 270
  resistance, 272, 359
  resurgence, 274
  weeds, 262–263
  woodlots, 214
Petals, 316
Pet care, 604–607
Petiole tests, 178
pH
  definition, 178–179
  nutrition and, *332*
  scale, 175, *332*
  soil, 330
  tests, 178–179
Phenoxys, 280
Pheromone, 272
Phloem, 306, 308
Phosphorus, 329, 330
Photodecomposition, 280
Photosynthesis
  factors, 324
    improving, 136–137
    process, 136, *136*, 324
Phototropism, 481
Physiology, animal, 538–539
Physiology, plant, 456
Pigs. **See** Swine
Pills, 563
Pine trees. **See specific species**
Placement/internship, 88–89, 90, 91, 93, 255, 513
Plan, definition, 101
Planters, 440–441, 457
Plant-growing media
  amending, 175–178
  decomposed, 162
  hydroponic, 184–186
  liming, 176
Plant hardness zone map, *519*
Plant nutrients
  absorption, *183*
  deficiencies, 181, 333
  essential, 329
  excess, 333
  function, 11, 333
Plant reproduction
  asexual propagation, 344–349
  definition, 339
  methods, *339*
  process, 342
  sexual propagation, 339–343
Plants. **See also** Trees
  air and, 328
  annuals, 369

  aquatic, 241, 242, 244
  balled, 525–526
  bare-rooted, 524–525
  beets, 45
  biennials, 369
  border, 519
  bracts, 315
  breeding, 345
  burlapped, 525–526
  cell parts, 323–324, *323*
  coarse-textured, 487
  composition, 303
  container-grown, 526
  cuttings, 344–346
  defoliators, 263
  diseases, 266–267
  diversity, 314
  exotic flowers, 44
  explorers, 44
  fertilization, 329
  -fish, relationship, 244
  flowers, 303, 315–317
  food storage, 335–336
  fruits, 316–317
  gas exchange, 326
  genetic manipulation, 54
  germination, 190
  group, 523
  growth, *154*
  growth requirements, 186
  hardiness, 520
  hydroponic education, 190
  indoor. **See** Ornamentals
  inflorescence, 497
  leaves, 310–314
  microscopic, 56
  necessities, 303
  new varieties, 340
  nomenclature, 523
  nuts, 316–317
  pathologists, 267
  perennials, 369
  perfecting, 359
  performance, improving, 39–40, 45
  photosynthesis, 135, 137, 324
  physiology, 323, 329
  potatoes, 41
  residue, 155
  respiration, *135*, 325
  rice, 45
  root systems, 303, 304–306
  science careers, 3
  seasonal, 365
  selective breeding, 43, 50
  shape, *523*
  soils and, 327–328, 330
  soybeans, 41, *41*
  specimen, 477, 519
  stems, 308–310
  taxonomy, 317–319

  tomato, 42
  transpiration, 326
  use, 359
  vegetables, 316–317
  water absorption, 148, 328
  Wisconsin Fast, 190
Plantscaping
  accent, 487
  balance/scale, 488
  careers, 489
  colors in, 486
  definition, 486
  design aspects, 486–489
  design process, 488
  form, 486
  maintenance, 488
  sequence, 487
  texture, 487
  trees in, 517
Plowing, 391
Plugging, 512
Poa pratensis
  cultivars, 499–500
  fertility requirements, 511
  uses, 499–500
Poinsettia, 469. **See also** *Euphorbia pulcherrima*
Poisoning, lead, 229
Poison ivy, 306
Pollen, 315
Pollination, 411, *601*
Pollution
  air, 125–137
  ground water, 141
  particulate, 129–130
  water, 148–149, 226
  wetland, 231–232
Polychlorinated biphenyl, 56
Polygastric system, 543–544
Ponderosa pine, *202*, 203
Ponds
  constructed, 249
  roll overs, 249
  wildlife, 234–236
Ponies, 640
Poplar, 203, *204*
Populations
  age profiles, 34
  China, 34
  dynamics, 227
  growth rate, 32
  life spans, 34
  patterns, changes, 34–35
  wildlife, 221
  worlds, 21
Porcine somatotropin, 55
Pork. **See also** Swine
  byproducts, 663
  retail cuts, 697
Potassium, 334
Potassium nitrate, 330

Potatoes, 41, 354
Poultry
  classes, 589
  commodities, 664
  definition, 589
  digestive system, 545
  diseases, 538
  domestication, 589
  economics, 591
  history, 589
  processing, 694
  production practices, 594
  small breeds, 41
  types, 591–593
  uses, 591–593
Powders, 564
Prawns, 245
Precipitate, 330
Precipitation, 146
Precision agriculture, 40
Pre-cooling, 403
Predation, 223
Predators, 223
Preservation, 683–685
Presiding officer, 116, *116*
Prestige pricing, 740
Prey, 221
Pricing, 714–715, 739–740
Procambarus spp., 245
Processed meats, 692–693
Processing
  agriculture, 67
  beef, 692–693
  careers in, 66
  crops, 669
  dairy products, 690–692
  food, 667, 669, 683–685
Processors, 673
Production agriculture enterprise, 89
Production Credit Associations, 720
Production loans, 717
Profit maximization, 741
Progeny, 50
Progesterone, 579
Project, 87
Promissory note, 721
Promotions, 738–739, 764
Propagation, 484
Proteins
  in alfalfa, 460
  characterization, 548
  nutritional role, 678
  sources, 548
Protozoa, 175, 561
Pruning
  definition, 530
  purpose, 415
  requirements, 530
  root, 525
  techniques, 530

  timing, 530
  woodlot, 211
PST. **See** Porcine somatotropin
Public lands, 123
Public speaking. **See** Speeches
Pulse, taking, 556
Purebreeding, 580
Purify, definition, 148
Pyrus calleryana, 520

Q

Quaking aspen, *204*
Quarantine Act of 1912, 271
Quarantines, 271
Quincy, Edmund, 36

R

Rabbits
  breeds, 597, *597*
  domestication, 595
  economics, 595
  fur, 597
  history, 595
  production practices, 599
  in research, 607
  types, 597–599
  uses, 597–599
Radioactive dust, 128–129
Radon, 128
Ranchers, 620
Rangeland forage, 455
Ranger manager, 456
Rape. **See** Canola
Reaper, 36
Recombinant DNA, 440
Recordkeeping, 96
Records, financial, 761–763
Red clover, 450
Red Fescue, 499
Red-spotted ladybug, *527*
Reed canary grass, 453
Refrigeration, 683
Registration papers, 580, 584, *585*
Render, 622
Rendering insensible, 692
Renewable resources, 9, *9*
Repotting, 484
Reproduction. **See also** Breeding
  female system, 579–580
  male system, 578–579
  plant, 339
  problems, 580
Research
  assistant opportunities, 93
  medical, *598*, 607
  rabbits, 607
Research SAE, 91

Residual soil, 165
Resistance
  pesticides, 274
  pests, 272, 359
Resources
  land, 146–148
  management, 250–251
  renewable, 9, *9*
  substitution, 716
Respiration
  in aquaculture, 244
  definition, 135
  plant, 325
  rates, 555
  seed, 342–343
Respiratory system, 541
Resumés, 763
Retailer, 656
Retail marketing, 741–743
Rhizomes, 308, 335, 496
Rhus radicans, 306
Rice
  characterization, 429
  golden, 440
  hybrids, 45
  importance, 429
Riparian areas, 233
Risk management, 290–291
Roadside markets, 742
Rocky Mountain Forest, *198*
Roll overs, 249
Root crops, 437
Roots
  adventitious, 306
  caps, 306
  cells, 308
  cuttings, 344
  description, 304
  elongation, 308
  epidermis, 327
  fibrous, 304–306
  food storage, 335
  hairs, 308, 327
  pruning, 525
  soil space, 304
  structure, 307
  systems, 304–306
  taproot, 304
  tissues, 306
  turfgrass, 495
  types, 304
Rootstocks, 408, *408*
Rotary mowers, 460
Rotenone, 282, *283*
Roughage, 543
Roundworms, 561
Row crop planters, 439
Rubber production, 42–43
Ruminants, 543–544
Rye, 429

S

SAE for All, 84
SAE Programs. **See** Supervised agricultural experience programs
Safety
    biotechnology, 58
    food, 657
    pesticides, 291
    riding, 645
    workplace, 664–665
Safflower, 432–433
Sales, direct, 744
Salinity gradient, 242–243
Salmon, 243, 244
Salmonids, 247
Salvage harvesting, 213
Sandy soils, 172
Sanitation, 557, 657
Scarify, 342
School-Based Enterprise, 92–93
Scorch, 334
S corporations, 709. **See also** Corporations
Seafood. **See** Fish
Seasons, 386
Seaweeds, 243
Secchi disc, 246, *246*
Secretaries, 116
Seedbeds, 438–439, 457
Seeding rates, *510*
Seedlings, 394–395, *525*
Seeds
    cotton, 434
    crops, 457
    culm, 496
    food storage, 336
    forage, 457
    germination, 341–342
    growing from, 393
    labels, 510
    legume crops, 437
    legumes, planting, 450
    planters, 439
    planting, 393
    post-harvest, 436
    propagation, 339–343
    respiration, 342–343
    scarify, 342
Seed-tree cutting, 213
Selective breeding, 50, 572
Self-employment, 763–765
Semen, 582
Semipermeable membrane, 327
Sepals, 316
Septic systems, 150
Service-Learning SAE, 94, *94*
Sewage disposal, 23
Sexual propagation, 339–343
Shackles, 692
Sheep. **See also** Lambs
    breeds, 626–627
    byproducts, 701
    commodities, 663
    economics, 626
    forage, *626*
    history, 626
    Merino, 572
    origins, 626
    slaughter, 693
    types, 626–627
Sheep Experiment Station, 535
Sheet erosion, 154
Shelf life, 685
Shellfish, 245, 246, 252, 253
Shelterwood cutting, 213
Shepherds, 626
Shoot systems, 496
Short-term loans, 717
Shrimp, 245
Shrinkage, wood, 209
Shrubs. **See** Trees
Signature loans, 717
Silage
    acids from, 456
    definition, 449
    description, *49*
    fermentation, 463–464
    harvesting, 450
    spoilage, 462–463
    storage, 462–463
Silent Spring, 30, 268
Silos, 462–463
Silviculture, 207
Simple-interest loans, 721
Simple layering, 349
Sires, 581
Site-specific farming, 40
Sitka spruce, *202*, 203
Skeletal system, 538–539
Skills, 106
    career, 76
    communication, 103, 111
    employability, 84, 86, 86
    leadership, 102, 108, 120
Skimming, 740
Slash and burn farming, 152
Slaughter
    beef, 692–693
    hogs, 693
    kosher, 694
    poultry, 694
Slope, 166
Small Business Administration (SBA), 720
Sociology, 16
Sod crops, 184
Sodding, 511
Softwoods, 200–201, *202*
Soils. **See also** Land; Plant-growing media
    acidity, 330
    aggregates, 172
    alkalinity, 175, 179, 330
    amendments, 509
    characteristic, 178
    chemistry, 178–184
    classification, 168–171
    clay, 171
    climate, 164
    composition, 164–168
    conditioning, 366
    conservation, 154–156
    decomposer, 175
    definition, 162
    deposits, 165
    drainage, 166
    elements, 330
    erosion, 150, 151–152
    fertilizing, 179–180
    function, 327–328
    gypsum, 179
    horizons, 165, 171
    loamy, 172
    location, *170*
    mechanical analysis, 172
    modification, 509
    moisture, 391
    nutrients in, 327
    organic matter, 155, 173
    organisms in, 171, 175
    origins, 164–168
    parent material, 164
    particles, *147*
    particle size, 172
    percolation, 166
    permeable, 168
    pH, 175, 178–179
    pores, 147
    profile, 165, *166*, 171–172
    residual, 165
    sampling process, 175–178
    sandy, 172
    saturation, 147
    science, 10, 74
    sponge qualities, *146*
    structure, 192
    sulfur, 179
    tests, 180
    texture, 172
    time factors, 167–168
    topography, 166–167
    watering, 150
    as water reservoir, 184
    weathering, 168
    wet, 167
Soil science career, 74
Sole proprietorship, 709
Solution cultures, 190
Sorghum, 429
Southern forests, 199
Southern pine, 203
Southern Weed Science Laboratory, 444

Soybeans
    characterization, 431
    pest-damage index, 270
    secrets of, 41
    uses, 431–432
Spades, 526
Spawning, 247
Specialty crops, 433–435
Species, definition, 317
Species, indicator, 150
Specimen plants, 477, 519
Speeches
    competitions, 108
    effective, 111
    evaluating, *114*
    extemporaneous, 111
    outlines, 112, *112*
    planning, 112
    prepared talks, 111
    types, 111
Sperm, 571
Spermatogenesis, *575*
Sphagnum, 162
Split carcass, 692
Split-vein cuttings, 348
Spoilage, 683
Sprigging, 512
Sprinkler irrigation, *397*, 441
Staking, 528
Stamen, 315, 316
Standard ground limestone, 179
Starch, 677
Starter solutions, 182
Statistics, 16
St. Augustine grass, 501, 502–503
Stems
    cross section view, 335
    cuttings, 344
    definition, 308
    food storage, 335
    herbaceous, 309
    key parts, 310
    parts, 310
    turfgrass, 496–497
    types of, 309
Stenotaphrum secundatum, 501, 502–503
Sterility, 580
Sterilization, 274, 354–355
Stiffness, wood, 210
Stimulant crops, 437
Stolonizing, 512
Stolons, 496
Stoma, 314
Stomachs, four-chamber, 544
Storage
    cool, 376
    field crops, 444–445
    food, 335–336, 670–671
    forage, 459–464
    fruit crops, 419
    hay, 459–462
    nut crops, 419
    silage, 462–463
    vegetable crops, 376, 401–403
    warm, 376
Stream-side forestation, 233
Streams, management, 226
Strength, wood, 210
Strip cropping, 155
Subirrigation, 397
Succulents, 473, 476
Sugar
    as additive, 686
    beets, 45
    cane, 434
    characterization, 434
    crops, 437, 662
    function, 677
    maple, 203, *204*
Sulfur
    characterization, 126
    function, 334
    as insecticide, 282
    soil, 179
Sunflowers, 433
Sunlight effects, 130
Superior oils, 282, 283
Superstores, *671*
Supervised agricultural experience programs, 83
    agreement, 89, *90*
    agricultural work opportunities, *84*
    agriscience program, 83
    classroom instruction and, 83
    comprehensive agriscience program, *83*
    conduct on, 95–97
    document, 96
    effective, 83
    evaluate an, 97
    Experimental Research, 92
    Foundational, 83, 85–87
    immersion, 87–88
    meaningful, 95
    National FFA Organization, 83
    ownership/entrepreneurship, 87, 89–91
    placement/internship, 88–89
    plan, *88*, 95
    propose an, 95
    purpose, 95
    research, 91–92
    school-based enterprise, 92–93
    service learning, 94
Supply, definition, 731
Supply, principle, 714–715
Surface irrigation, 397, 441
Swathers, 460
Sweet clovers, 450
Sweetgum, 203, *204*
Swine. **See also** Pork
    breeds, 624
    death, cause of, 560
    economics, 622–623
    history, 622
    origins, 622
    production practices, 625
    types, 624
    viruses, 535
Sycamore, *205*, 206
Symptoms, 264
Syringe, 557

T

Tall fescue, 476, 499, *500*
Tallow, 43
TAN. **See** Total ammonial nitrogen
Taxonomy
    binomial system, 319
    definition, 317
    Latin use in, 319
    plant, 319
T-budding, 352, *353*
Technology
    definition, 9
    high, 9
    industrial, 9
Temperatures
    aquaculture, 246
    average annual, 27
    global patterns, 131–133
    harvest, 683
    in hydroponic systems, 187
    lakes/ponds, 235
    for ornamentals, 476, 478
    photosynthesis and, 324
    seafood, 684
    seed germination, 343
    taking, 556
    vegetable storage, 403
Terminal bud, 309
Terminal markets, 743
Testes, 579
Testosterone, 579
Tetraethyl lead gasoline, 127
Thymine, 52
Tidewater, 143
Tillage
    equipment, *509*
    forage, 457
    minimum, 438
    turfgrass, 508–510
    types, 438
Time factors, 167–168, 525
Timothy, 452–453, *452*
Tip layering, 349–350
Tissue cultures
    advantages, 353
    aseptic technique, 356–357
    description, 353
    laboratory, 356

preparing, 355
roots, 306
sterile media for, 353-354
storage, 355
Tobacco, 434-435
Tomatoes
   genetically engineered, 42
   processing, 669-670
   rotten, 670
Topography, 27, 166-167
Total ammonial nitrogen, 246
Toughness, wood, 210
Toxins, 247
Traits
   definition, 52
   genes and, 575-576
   selecting, 573
Transgenic animals, 55
Transpiration, 326
Transplants, 394-395
Transponders, 631
Transporting, 670-671
Trees. **See also** Lumber; Wood
   biocontrol, 527
   characterization, 195
   classification, 195
   color changes, 324
   cutting, 213
   disease control, 416
   diseases, 531
   dwarf, 408, *408*
   fertilizing, 528-529
   fruit, 408
   gall infestation, 531
   geographic location, 520
   growth, 207-209
   growth habit, 408
   growth types, 520
   guying, 528
   hardwood, 203
   indoor, 471-473
   in landscapes, 517
   major parts, *208*
   mulching, 528
   names, 523-524
   nuts, 410
   obtaining, 524-526
   pest control, 416
   pests, 527
   planting, 519
   production, 415
   products, 195
   pruning, 211, 415-416, 530
   rings, *208*
   rootstocks, 408
   semidwarf, 408, 428
   shape, *523*
   shrubs *vs.*, 518
   site location, 520

size, 195, 518, *522*
softwood, 196, 200-201, *202*
spades, 526
staking, 528
standard, 408
uses, *519*
water absorption, 208
Triazines, 280
Trickle irrigation, 397
Tropical Forest, 198, 200
Trout
   farmed, 248
   oxygen needs, 247
   whirling disease, 251
Truck cropping. **See also** Vegetable
   production
   careers in, 383-384
   definition, 383
   site selection, 387
True clovers, 450
Tuber crops, 437
Tubers, 308
Turbidity, 246, 247
Turf farming, *69*
Turfgrass
   adaptation zones, 499
   associated careers, 495
   cool-season, 495, 497, 499
   establishment, 508-513
   fertilization, 505-506
   fertilizers, 506
   irrigation, 507-508
   mixing, 506
   mowing, 503
   mowing heights, 505
   netting, 510
   planting, 510
   plugging, 512
   poor quality, *511*
   root system, 495-498
   seeding, 510-511
   selection, 508
   shoot system, 496
   site preparation, 508-510
   sodding, 511
   species, 499-502
   sprigging, 512
   stolonizing, 512
   tillers, 497
   warm-season, 495, 501
Turkeys. **See** Poultry

U

Underripe, 667
United States Department of Agriculture
   food safety oversight, 657
   organic food standards, 297

shortage study, 17
Universal solvent, 144
Urinary system, 542
USDA. **See** United States Department of
   Agriculture

V

Vaccinations, 566
Vapor drift, 294
Variety, 319
Vascular bundles, 310, 314
Vectors, 259
Vegetable production
   cultivation, 396
   harvesting, 399-400
   hydroponics, 399
   marketing, 401
   organic, 399
   planting, 392
   site size, 388
   soil preparation, 390
   storage, 401-403
Vegetables. **See also specific crops**
   characterization, 316-317
   classification, 384
   commodities, 663
   cool season, 386
   edible parts, 384
   foods from, 680
   growing season, 386
   harvesting, 375-376
   hothouse, 185
   ideal temperatures, 376
   planting guide, *370*
   seeds, 431-433
   storage, 376
   warm season, 386
   washing, 569
Vegetative bud, 309
Vegetative spreading, 497
Ventilation systems, *128*
Vertebrates, 221
Vertical integration, 745, *745*
Veterinarians, 563, 567, 585
Veterinary skills, 566
Video merchandising, 744
Vines, 476
Viruses, 268
Viscera, 692
Vitamins
   characterization, 547-548
   function, *679*
   in milk, 690
   nutritional role, *677*
   sources, 547-548
Volatilization, 280
Voluntary muscles, 540

W

Warp, wood, 209
Waste disposal
    animal, 622
    biotechnology and, 56
    sewage, 23, *23*
Water. **See also** Irrigation
    animal component, 538
    brackish, 243
    capillary, *147*
    conserving, 148–150
    control projects, 13
    cultures, 190
    drainage, 166
    drinking, 24
    free, 147
    fresh, 143
    gravitational, 147
    ground. **See** Groundwater
    hardness, 246
    hydroponic, 184–186
    hygroscopic, *147*
    imbibitions, 342
    importance, 148
    -land relationship, 146
    management, 228
    in milk, 679
    natural flow, 244
    nutritional role, *677*
    ocean, 144
    for ornamental, 479
    oxygen content, 244
    pesticides in, 157
    pollution, 148–149, 226
    potable, 142
    purity levels, 144, 148
    quality, 148–150
    quality, testing, 246
    reservoir, soil as, 184
    runoff, 150
    salt, 143
    soil leaching, 167
    tree absorption, 207–209
    turbidity in, 246, 247
    as universal solvent, 144
Water cycles
    description, 144–145
    forests and, 200
    function, 242
Watershed, 146
Waterways, 150, 155
Weeding, 373
Weeds. **See also** specific species
    annual, 260
    biennial, 261
    definition, 260
    nonselective control, 508
    noxious, 262–263
    perennial, 261–262
    vegetable gardens, 388
Weight, wood, 209
Wetlands
    constructed, 156
    pollution, 226
    wildlife, 226–227
Wheat
    characterization, 426
    processing, 669
Whirling disease, 251
White ash, *205*
Whiteflies, 487
White oak, *204*
White pine, *202*, 203
Whitney, Eli, 36
Wholesale marketing, 743–745
Wholesalers, 656, 673
Wildlife
    biologists, 227
    characteristics of, 221–222
    commensalism, 223
    declines, *221*
    dynamics, 227
    on farms, 228–229
    forests, 229–230
    future, 236–237
    humans and, 221, 224–225
    lakes practices, 234–236
    mutualism, 222
    parasitism, 222
    ponds practices, 234–236
    predation, 223
    relationships, 222–223
    supplemental feeding, 227
    wetlands, 231–232
Wildlife management
    classifications, 226–227
    farm practices, 228–229
    forest practices, 229–230
    stream practices, 232–234
    wetland practices, 231–232
Wisconsin Fast Plants, 190
Wood
    bending strength, 210
    decay resistance, 210
    ease of working, 209
    hardness, 209
    nail holding, 209
    paint holding, 209
    properties, 209–210
    shrinkage, 209
    stiffness, 210
    surface, 210
    toughness, 210
    warp, 209
Woodlots
    fire, 213–214
    growing, 211
    harvest cutting, 211
    harvesting, 214
    pests, 214
    protection, 213
    restocking, 211
Wool, 598, *598*, *617*, 626–628
Workplace safety, 664–667
Worms. **See** Nematodes
Wormwood, 444
Wrappings, 685, 692

X

Xylem, *308*
Xylem cells, 308

Y

Yeast, 175
Yellow poplar, *204*

Z

Zoysia japonica, 502, *521*